D0142074

The Industrial Electronics Handbook

SECOND EDITION

INTELLIGENT SYSTEMS

The Industrial Electronics Handbook
SECOND EDITION

FUNDAMENTALS OF INDUSTRIAL ELECTRONICS

POWER ELECTRONICS AND MOTOR DRIVES

CONTROL AND MECHATRONICS

INDUSTRIAL COMMUNICATION SYSTEMS

INTELLIGENT SYSTEMS

The Electrical Engineering Handbook Series

Series Editor
Richard C. Dorf
University of California, Davis

Titles Included in the Series

The Avionics Handbook, Second Edition, Cary R. Spitzer
The Biomedical Engineering Handbook, Third Edition, Joseph D. Bronzino
The Circuits and Filters Handbook, Third Edition, Wai-Kai Chen
The Communications Handbook, Second Edition, Jerry Gibson
The Computer Engineering Handbook, Vojin G. Oklobdzija
The Control Handbook, Second Edition, William S. Levine
CRC Handbook of Engineering Tables, Richard C. Dorf
Digital Avionics Handbook, Second Edition, Cary R. Spitzer
The Digital Signal Processing Handbook, Vijay K. Madisetti and Douglas Williams
The Electric Power Engineering Handbook, Second Edition, Leonard L. Grigsby
The Electrical Engineering Handbook, Third Edition, Richard C. Dorf
The Electronics Handbook, Second Edition, Jerry C. Whitaker
The Engineering Handbook, Third Edition, Richard C. Dorf
The Handbook of Ad Hoc Wireless Networks, Mohammad Ilyas
The Handbook of Formulas and Tables for Signal Processing, Alexander D. Poularikas
Handbook of Nanoscience, Engineering, and Technology, Second Edition, William A. Goddard, III, Donald W. Brenner, Sergey E. Lyshevski, and Gerald J. Iafrate
The Handbook of Optical Communication Networks, Mohammad Ilyas and Hussein T. Mouftah
The Industrial Electronics Handbook, Second Edition, Bogdan M. Wilamowski and J. David Irwin
The Measurement, Instrumentation, and Sensors Handbook, John G. Webster
The Mechanical Systems Design Handbook, Osita D.I. Nwokah and Yidirim Hurmuzlu
The Mechatronics Handbook, Second Edition, Robert H. Bishop
The Mobile Communications Handbook, Second Edition, Jerry D. Gibson
The Ocean Engineering Handbook, Ferial El-Hawary
The RF and Microwave Handbook, Second Edition, Mike Golio
The Technology Management Handbook, Richard C. Dorf
Transforms and Applications Handbook, Third Edition, Alexander D. Poularikas
The VLSI Handbook, Second Edition, Wai-Kai Chen

The Industrial Electronics Handbook

SECOND EDITION

INTELLIGENT SYSTEMS

Edited by

Bogdan M. Wilamowski

J. David Irwin

CRC Press is an imprint of the
Taylor & Francis Group, an **informa** business

CRC Press
Taylor & Francis Group
6000 Broken Sound Parkway NW, Suite 300
Boca Raton, FL 33487-2742

Printed in the United States of America on acid-free paper
10 9 8 7 6 5 4 3 2 1

International Standard Book Number: 978-1-4398-0283-0 (Hardback)

Library of Congress Cataloging-in-Publication Data

Intelligent systems / editors, Bogdan M. Wilamowski and J. David Irwin.
p. cm.
"A CRC title."
Includes bibliographical references and index.
ISBN 978-1-4398-0283-0 (alk. paper)
1. Intelligent control systems. 2. Neural networks (Computer science) I. Wilamowski, Bogdan M. II. Irwin, J. David, 1939- III. Title.

TJ217.5.I54477 2010
006.3'2--dc22 2010020581

Visit the Taylor & Francis Web site at
http://www.taylorandfrancis.com

and the CRC Press Web site at
http://www.crcpress.com

Contents

PART V Applications

Preface

The field of industrial electronics covers a plethora of problems that must be solved in industrial practice. Electronic systems control many processes that begin with the control of relatively simple devices like electric motors, through more complicated devices such as robots, to the control of entire fabrication processes. An industrial electronics engineer deals with many physical phenomena as well as the sensors that are used to measure them. Thus, the knowledge required by this type of engineer is not only traditional electronics but also specialized electronics, for example, that required for high-power applications. The importance of electronic circuits extends well beyond their use as a final product in that they are also important building blocks in large systems, and thus the industrial electronics engineer must also possess knowledge of the areas of control and mechatronics. Since most fabrication processes are relatively complex, there is an inherent requirement for the use of communication systems that not only link the various elements of the industrial process but are also tailor-made for the specific industrial environment. Finally, the efficient control and supervision of factories require the application of intelligent systems in a hierarchical structure to address the needs of all components employed in the production process. This need is accomplished through the use of intelligent systems such as neural networks, fuzzy systems, and evolutionary methods. The Industrial Electronics Handbook addresses all these issues and does so in five books outlined as follows:

1. *Fundamentals of Industrial Electronics*
2. *Power Electronics and Motor Drives*
3. *Control and Mechatronics*
4. *Industrial Communication Systems*
5. *Intelligent Systems*

The editors have gone to great lengths to ensure that this handbook is as current and up to date as possible. Thus, this book closely follows the current research and trends in applications that can be found in *IEEE Transactions on Industrial Electronics*. This journal is not only one of the largest engineering publications of its type in the world, but also one of the most respected. In all technical categories in which this journal is evaluated, it is ranked either number 1 or number 2 in the world. As a result, we believe that this handbook, which is written by the world's leading researchers in the field, presents the global trends in the ubiquitous area commonly known as industrial electronics.

An interesting phenomenon that has accompanied the progression of our civilization is the systematic replacement of humans by machines. As far back as 200 years ago, human labor was replaced first by steam machines and later by electrical machines. Then approximately 20 years ago, clerical and secretarial jobs were largely replaced by personal computers. Technology has now reached the point where intelligent systems are replacing human intelligence in decision-making processes as well as aiding in the solution of very complex problems. In many cases, intelligent systems are already outperforming human activities. The field of computational intelligence has taken several directions. Artificial neural networks are not only capable of learning how to classify patterns, for example, images or sequences of events, but they can also effectively model complex nonlinear systems. Their ability to classify sequences of events is probably more popular in industrial applications where there is an inherent need to model nonlinear system behavior—as an example, measuring the system parameters that are easily obtainable and using a neural network to evaluate parameters that are difficult to measure but essential for system control. Fuzzy systems have a similar application. Their main advantage is their simplicity and ease of implementation. Various aspects of neural networks and fuzzy systems are covered in Parts II and III. Part IV is devoted to system optimization, where several new techniques including evolutionary methods, swarm, and ant colony optimizations are covered. Part V is devoted to several applications that deal with methods of computational intelligence.

For MATLAB® and Simulink® product information, please contact

The MathWorks, Inc.
3 Apple Hill Drive
Natick, MA, 01760-2098 USA
Tel: 508-647-7000
Fax: 508-647-7001
E-mail: info@mathworks.com
Web: www.mathworks.com

Acknowledgments

The editors wish to express their heartfelt thanks to their wives Barbara Wilamowski and Edie Irwin for their help and support during the execution of this project.

Editorial Board

Editors

Bogdan M. Wilamowski received his MS in computer engineering in 1966, his PhD in neural computing in 1970, and Dr. habil. in integrated circuit design in 1977. He received the title of full professor from the president of Poland in 1987. He was the director of the Institute of Electronics (1979–1981) and the chair of the solid state electronics department (1987–1989) at the Technical University of Gdansk, Poland. He was a professor at the University of Wyoming, Laramie, from 1989 to 2000. From 2000 to 2003, he served as an associate director at the Microelectronics Research and Telecommunication Institute, University of Idaho, Moscow, and as a professor in the electrical and computer engineering department and in the computer science department at the same university. Currently, he is the director of ANMSTC—Alabama Nano/Micro Science and Technology Center, Auburn, and an alumna professor in the electrical and computer engineering department at Auburn University, Alabama. Dr. Wilamowski was with the Communication Institute at Tohoku University, Japan (1968–1970), and spent one year at the Semiconductor Research Institute, Sendai, Japan, as a JSPS fellow (1975–1976). He was also a visiting scholar at Auburn University (1981–1982 and 1995–1996) and a visiting professor at the University of Arizona, Tucson (1982–1984). He is the author of 4 textbooks, more than 300 refereed publications, and has 27 patents. He was the principal professor for about 130 graduate students. His main areas of interest include semiconductor devices and sensors, mixed signal and analog signal processing, and computational intelligence.

Dr. Wilamowski was the vice president of the IEEE Computational Intelligence Society (2000–2004) and the president of the IEEE Industrial Electronics Society (2004–2005). He served as an associate editor of *IEEE Transactions on Neural Networks*, *IEEE Transactions on Education*, *IEEE Transactions on Industrial Electronics*, the *Journal of Intelligent and Fuzzy Systems*, the *Journal of Computing*, and the *International Journal of Circuit Systems and IES Newsletter*. He is currently serving as the editor in chief of *IEEE Transactions on Industrial Electronics*.

Professor Wilamowski is an IEEE fellow and an honorary member of the Hungarian Academy of Science. In 2008, he was awarded the Commander Cross of the Order of Merit of the Republic of Poland for outstanding service in the proliferation of international scientific collaborations and for achievements in the areas of microelectronics and computer science by the president of Poland.

J. David Irwin received his BEE from Auburn University, Alabama, in 1961, and his MS and PhD from the University of Tennessee, Knoxville, in 1962 and 1967, respectively.

In 1967, he joined Bell Telephone Laboratories, Inc., Holmdel, New Jersey, as a member of the technical staff and was made a supervisor in 1968. He then joined Auburn University in 1969 as an assistant professor of electrical engineering. He was made an associate professor in 1972, associate professor and head of department in 1973, and professor and head in 1976. He served as head of the Department of Electrical and Computer Engineering from 1973 to 2009. In 1993, he was named Earle C. Williams Eminent Scholar and Head. From 1982 to 1984, he was also head of the Department of Computer Science and Engineering. He is currently the Earle C. Williams Eminent Scholar in Electrical and Computer Engineering at Auburn.

Dr. Irwin has served the Institute of Electrical and Electronic Engineers, Inc. (IEEE) Computer Society as a member of the Education Committee and as education editor of *Computer*. He has served as chairman of the Southeastern Association of Electrical Engineering Department Heads and the National Association of Electrical Engineering Department Heads and is past president of both the IEEE Industrial Electronics Society and the IEEE Education Society. He is a life member of the IEEE Industrial Electronics Society AdCom and has served as a member of the Oceanic Engineering Society AdCom. He served for two years as editor of *IEEE Transactions on Industrial Electronics*. He has served on the Executive Committee of the Southeastern Center for Electrical Engineering Education, Inc., and was president of the organization in 1983–1984. He has served as an IEEE Adhoc Visitor for ABET Accreditation teams. He has also served as a member of the IEEE Educational Activities Board, and was the accreditation coordinator for IEEE in 1989. He has served as a member of numerous IEEE committees, including the Lamme Medal Award Committee, the Fellow Committee, the Nominations and Appointments Committee, and the Admission and Advancement Committee. He has served as a member of the board of directors of IEEE Press. He has also served as a member of the Secretary of the Army's Advisory Panel for ROTC Affairs, as a nominations chairman for the National Electrical Engineering Department Heads Association, and as a member of the IEEE Education Society's McGraw-Hill/Jacob Millman Award Committee. He has also served as chair of the IEEE Undergraduate and Graduate Teaching Award Committee. He is a member of the board of governors and past president of Eta Kappa Nu, the ECE Honor Society. He has been and continues to be involved in the management of several international conferences sponsored by the IEEE Industrial Electronics Society, and served as general cochair for IECON'05.

Dr. Irwin is the author and coauthor of numerous publications, papers, patent applications, and presentations, including *Basic Engineering Circuit Analysis*, 9th edition, published by John Wiley & Sons, which is one among his 16 textbooks. His textbooks, which span a wide spectrum of engineering subjects, have been published by Macmillan Publishing Company, Prentice Hall Book Company, John Wiley & Sons Book Company, and IEEE Press. He is also the editor in chief of a large handbook published by CRC Press, and is the series editor for Industrial Electronics Handbook for CRC Press.

Dr. Irwin is a fellow of the American Association for the Advancement of Science, the American Society for Engineering Education, and the Institute of Electrical and Electronic Engineers. He received an IEEE Centennial Medal in 1984, and was awarded the Bliss Medal by the Society of American Military Engineers in 1985. He received the IEEE Industrial Electronics Society's Anthony J. Hornfeck Outstanding Service Award in 1986, and was named IEEE Region III (U.S. Southeastern Region) Outstanding Engineering Educator in 1989. In 1991, he received a Meritorious Service Citation from the IEEE Educational Activities Board, the 1991 Eugene Mittelmann Achievement Award from the IEEE Industrial Electronics Society, and the 1991 Achievement Award from the IEEE Education Society. In 1992, he was named a Distinguished Auburn Engineer. In 1993, he received the IEEE Education Society's McGraw-Hill/Jacob Millman Award, and in 1998 he was the recipient of the

IEEE Undergraduate Teaching Award. In 2000, he received an IEEE Third Millennium Medal and the IEEE Richard M. Emberson Award. In 2001, he received the American Society for Engineering Education's (ASEE) ECE Distinguished Educator Award. Dr. Irwin was made an honorary professor, Institute for Semiconductors, Chinese Academy of Science, Beijing, China, in 2004. In 2005, he received the IEEE Education Society's Meritorious Service Award, and in 2006, he received the IEEE Educational Activities Board Vice President's Recognition Award. He received the Diplome of Honor from the University of Patras, Greece, in 2007, and in 2008 he was awarded the IEEE IES Technical Committee on Factory Automation's Lifetime Achievement Award. In 2010, he was awarded the electrical and computer engineering department head's Robert M. Janowiak Outstanding Leadership and Service Award. In addition, he is a member of the following honor societies: Sigma Xi, Phi Kappa Phi, Tau Beta Pi, Eta Kappa Nu, Pi Mu Epsilon, and Omicron Delta Kappa.

Contributors

Sabeur Abid
Ecole Superieure Sciences et Techniques Tunis
University of Tunis
Tunis, Tunisia

Filipe Alvelos
Algoritmi Research Center
and
Department of Production and Systems
University of Minho
Braga, Portugal

Christian Blum
ALBCOM Research Group
Universitat Politècnica de Catalunya
Barcelona, Spain

Oleg Boulanov
Department of Electrical and Computer Engineering
University of Calgary
Calgary, Alberta, Canada

Tak Ming Chan
Algoritmi Research Center
University of Minho
Braga, Portugal

Mo-Yuen Chow
Department of Electrical and Computer Engineering
North Carolina State University
Raleigh, North Carolina

Kun Tao Chung
Department of Electrical and Computer Engineering
Auburn University
Auburn, Alabama

Carlos A. Coello Coello
Departamento de Computación
Centro de Investigación y de Estudios Avanzados del Instituto Politécnico Nacional
Mexico City, Mexico

Nicholas Cotton
Panama City Division
Naval Surface Warfare Centre
Panama City, Florida

Mehmet Önder Efe
Department of Electrical and Electronics Engineering
Bahçeşehir University
İstanbul, Turkey

Åge J. Eide
Department of Computing Science
Ostfold University College
Halden, Norway

Farhat Fnaiech
Ecole Superieure Sciences et Techniques Tunis
University of Tunis
Tunis, Tunisia

Nader Fnaiech
Ecole Superieure Sciences et Techniques Tunis
University of Tunis
Tunis, Tunisia

Hani Hagras
The Computational Intelligence Centre
University of Essex
Essex, United Kingdom

Barrie W. Jervis
Department of Electrical Engineering
Sheffield Hallam University
Sheffield, United Kingdom

Józef Korbicz
Institute of Control and Computation Engineering
University of Zielona Góra
Zielona Góra, Poland

Sam Kwong
Department of Computer Science
City University of Hong Kong
Kowloon, Hong Kong

Thomas Lindblad
Physics Department
Royal Institute of Technology
Stockholm, Sweden

Manuel López-Ibáñez
IRIDIA
Université Libre de Bruxelles
Brussels, Belgium

Kim Fung Man
Department of Electronic Engineering
City University of Hong Kong
Kowloon, Hong Kong

Milos Manic
Department of Computer Science
University of Idaho, Idaho Falls
Idaho Falls, Idaho

Michael Margaliot
School of Electrical Engineering
Tel Aviv University
Tel Aviv, Israel

Marcin Mrugalski
Institute of Control and Computation Engineering
University of Zielona Góra
Zielona Góra, Poland

Andrzej Obuchowicz
Institute of Control and Computation Engineering
University of Zielona Góra
Zielona Góra, Poland

Teresa Orlowska-Kowalska
Institute of Electrical Machines, Drives and Measurements
Wroclaw University of Technology
Wroclaw, Poland

Guy Paillet
General Vision Inc.
Petaluma, California

Witold Pedrycz
Department of Electrical and Computer Engineering
University of Alberta
Edmonton, Alberta, Canada

and

System Research Institute
Polish Academy of Sciences
Warsaw, Poland

Ioannis Pitas
Department of Informatics
Aristotle University of Thessaloniki
Thessaloniki, Greece

Valeri Rozin
School of Electrical Engineering
Tel Aviv University
Tel Aviv, Israel

Vlad P. Shmerko
Electrical and Computer Engineering Department
University of Calgary
Calgary, Alberta, Canada

Elsa Silva
Algoritmi Research Center
University of Minho
Braga, Portugal

Adam Slowik
Department of Electronics and Computer Science
Koszalin University of Technology
Koszalin, Poland

Adrian Stoica
Jet Propulsion Laboratory
Pasadena, California

Krzysztof Szabat
Institute of Electrical Machines, Drives and Measurements
Wroclaw University of Technology
Wroclaw, Poland

Ryszard Tadeusiewicz
Automatic Control
AGH University of Science and Technology
Krakow, Poland

Kit Sang Tang
Department of Electronic Engineering
City University of Hong Kong
Kowloon, Hong Kong

Anastasios Tefas
Department of Informatics
Aristotle University of Thessaloniki
Thessaloniki, Greece

J.M. Valério de Carvalho
Algoritmi Research Center
and
Department of Production and Systems
University of Minho
Braga, Portugal

Juyang Weng
Department of Computer Science and Engineering
Michigan State University
East Lansing, Michigan

Paul J. Werbos
Electrical, Communications and Cyber Systems Division
National Science Foundation
Arlington, Virginia

Bogdan M. Wilamowski
Department of Electrical and Computer Engineering
Auburn University
Auburn, Alabama

Tiantian Xie
Department of Electrical and Computer Engineering
Auburn University
Auburn, Alabama

Ronald R. Yager
Iona College
New Rochelle, New York

Svetlana N. Yanushkevich
Department of Electrical and Computer Engineering
University of Calgary
Calgary, Alberta, Canada

Gary Yen
School of Electrical and Computer Engineering
Oklahoma State University
Stillwater, Oklahoma

Hao Yu
Department of Electrical and Computer Engineering
Auburn University
Auburn, Alabama

I

Introductions

1

Introduction to Intelligent Systems

Ryszard Tadeusiewicz
AGH University of Science and Technology

1.1 Introduction

Numerous intelligent systems, described and discussed in the subsequent chapters, are based on different approaches to machine intelligence problems. The authors of these chapters show the necessity of using various methods for building intelligent systems. Almost every particular problem needs an individual solution; thus, we can study many different intelligent systems reported in the literature. This chapter is a kind of introduction to particular systems and different approaches presented in the book. The role of this chapter is to provide the reader with a bird's eye view of the area of intelligent systems. Before we explain what intelligent systems are and why it is worth to study and use them, it is necessary to comment on one problem connected with the terminology.

The problems of equipping artificial systems with intelligent abilities are, in fact, unique. We always want to achieve a general goal, which is a better operation of the intelligent system than one, which can be accomplished by a system without intelligent components. There are many ways to accomplish this goal and, therefore, we have many kinds of artificial intelligent (AI) systems. In general, it should be stressed that there are two distinctive groups of researches working in these areas: the AI community and the computational intelligence community. The goal of both groups is the same: the need for artificial systems powered by intelligence. However, different methods are employed to achieve this goal by different communities.

AI [LS04] researchers focus on the imitation of human thinking methods, discovered by psychology, sometimes philosophy and so-called cognitive sciences. The main achievements of AI are traditionally rule-based systems in which computers follow known methods of human thinking and try to achieve similar results as human. Mentioned below, and described in detail in a separate chapter, expert systems are good examples of this AI approach.

Computational intelligence [PM98] researchers focus on the modeling of natural systems, which can be considered as intelligent ones. The human brain is definitely the source of intelligence; therefore, this area of research focuses first on neural networks, very simplified but efficient in practical application models of small parts of the biological brain. There are also other natural processes, which can be used (when appropriately modeled) as a source of ideas for successful AI systems. We mention, e.g., swarm intelligence models, evolutionary computations, and fuzzy systems.

The differentiation between AI and computational intelligence (also known as soft computing [CM05]) is important for researchers and should be obeyed in scientific papers for its proper classification. However, from the point of view of applications in intelligent systems, it can be disregarded. Therefore, in the following sections, we will simply use only one name (artificial intelligence) comprising both artificial intelligence and computational intelligence methods. For more precise differentiations and for tracing bridges between both approaches mentioned, the reader is referred to the book [RT08].

The term "artificial intelligence" (AI) is used in a way similar to terms such as mechanics or electronics but the area of research and applications that belong to AI are not as precisely defined as the other areas of computer science. The most popular definitions of AI are always related to the human mind and its emerging property: natural intelligence. At times it was fashionable to discuss the general definition of AI as follows: Is AI at all possible or not? Almost everybody knows Turing's answer to this question [T48], known as "Turing test," where a human judge must recognize if his unknown-to-him partner in discussion is an intelligent (human) person or not. Many also know Searle's response to the question, his "Chinese room" model [S80]. For more information about these contradictions, the reader is referred to the literature listed at the end of this chapter (a small bibliography of AI), as well as to a more comprehensive discussion of this problem in thousands of web pages on the Internet. From our point of view, it is sufficient to conclude that the discussion between supporters of "strong AI" and their opponents is still open—with all holding their opinions.

For the readers of this volume, the results of these discussions are not that important since regardless of the results of the philosophical roll outs—"intelligent" systems were built in the past, are used contemporarily, and will be constructed in the future. It is because intelligent systems are very useful for all, irrespective of their belief in "strong AI" or not. Therefore in this chapter, we do not try to answer the fundamental question about the existence of the mind in the machine. We just present some useful methods and try to explain how and when they can be used. This detailed knowledge will be presented in the next few sections and chapters. At the beginning, let us consider neural networks.

1.2 Historical Perspective

This chapter is not meant to be a history of AI because the users are interested in exploiting mature systems, algorithms, or technologies, regardless of long and difficult ways of systematic development of particular methods as well as serendipities that were important milestones in AI's development. Nevertheless, it is good to know that the oldest systems, solving many problems by means of AI methods, were neural networks. This very clever and user-friendly technology is based on the modeling of small parts of real neural system (e.g., small pieces of the brain cortex) that are able to solve practical problems by means of learning. The neural network theory, architecture, learning, and methods of application will be discussed in detail in other chapters; therefore, here we only provide a general outline.

One was mentioned above: neural networks are the oldest AI technology and it is still the leading technology if one counts number of practical applications. When the first computers were still large and clumsy, the neurocomputing theorists, Warren Sturgis McCulloch and Walter Pitts, published "A logical calculus of the ideas immanent in nervous activity" [MP43], thus laying foundations for the field of artificial neural networks. This paper is considered as the one that started the entire AI area. Many books written earlier and quoted sometimes as heralds of AI were only theoretical speculations. In contrast, the paper just quoted was the first constructive proposition on how to build AI on the basis of mimicking brain structures. It was a fascinating and breakthrough idea in the area of computer science.

During many years of development, neural networks became the first working AI systems (Perceptron by Frank Rosenblatt, 1957), which was underestimated and it lost "steam" because of the (in)famous book by Marvin Minski [MP72], but returned triumphantly as an efficient tool for practical problem solving with David Rumelhart's discovery of backpropagation learning method [RM86]. Since the mid-1980s the power and importance of neural networks permanently increased, reaching now a definitely leading position in all AI applications. However, its position is somehow weakened because of the increase in popularity and importance of other methods belonging to the so-called soft computing. But if one has a problem and needs to solve it fast and efficiently—one can still choose neural networks as a tool that is easy to use, with lots of good software available.

The above comments are the reason we pointed out neural networks technology in the title of this chapter with the descriptive qualification "first." From the historical point of view, neural network was the first AI tool. From the practical viewpoint, it should be used as the first tool, when practical problems need to be solved. It is great chance that the neural network tool you use turns out good enough and you do not need any more. Let me give you advice, taken from long years of experience in solving hundreds of problems with neural networks applications. There are several types of neural networks elaborated on and discovered by hundreds of researchers. But the most simple and yet successful tool in most problems is the network called MLP (multilayer perceptron [H98]). If one knows exactly the categories and their exemplars, one may use this network with a learning rule such as the conjugate gradient method. If, on the other hand, one does not know what one is expecting to find in the data because no prior knowledge about the data exists, one may use another popular type of neural network, namely, the SOM (self-organizing map), also known as Kohonen network [K95], which can learn without the teacher. If one has an optimization problem and needs to find the best solution in a complex situation, one can use the recursive network, known as the Hopfield network [H82]. Experts and practitioners can of course use also other types of neural networks, described in hundreds of books and papers but it will be a kind of intellectual adventure, like off-road expedition. Our advice is like signposts pointing toward highways; highways are boring but lead straight to the destination. If one must solve a practical problem, often there is no time for adventures.

1.3 Human Knowledge Inside the Machine—Expert Systems

Neural networks discussed in the previous section, of which detailed descriptions can be found in the following chapter, are very useful and are effective tools for building intelligent systems but they have one troublesome limitation. There is a huge gap between the knowledge encoded in the neural network structure during the learning process, and easy for human understanding knowledge presented in any intelligible form (mainly based on symbolic forms and natural language statements). It is very difficult to use knowledge that is captured by the neural network during its learning process, although sometimes this knowledge is the most valuable part of the whole system (e.g., in forecasting systems, where neural networks sometimes are—after learning—a very successful tool, but nobody knows how and why).

The above-mentioned gap is also present when going in the opposite way, e.g., when we need to add man's knowledge to the AI system. Sometimes (and, in fact, very often) we need to have in an automatic intelligent system some part of this knowledge embedded, which can be obtained from the human expert. We need to insert this knowledge into an automatic intelligent system because it is often easier and cheaper to use a computer program instead of constantly asking humans for expert opinion or advice.

Such design with computer-based shell and human knowledge inside it is known as an expert system [GR89]. Such a system can answer the questions not only searching inside internal knowledge representation but can also use methods of automatic reasoning for automatic deriving of conclusions needed by the user. The expert system can be very helpful for many purposes, combining the knowledge elements extracted from both sources of information: explicit elements of human expert wisdom collected in the form of the knowledge base in computer memory, and elements of user knowledge hidden in the system and emerged by means of automatic reasoning methods after questioning the system.

The main difference between the expert systems and neural networks is based on the source and form of knowledge, which is used in these two AI tools for practical problem solving. In neural networks, the knowledge is hidden and has no readable form but can be collected automatically on the base of examples forming the learning data set. Results given by neural networks can be true and very useful but never comprehensible to users, and therefore must be treated with caution. On the other hand, in the expert system, everything is transparent and intelligible (most of such systems can provide explanations of how and why the particular answer was derived) but the knowledge used by the system must be collected by humans (experts themselves or knowledge engineers who interview experts), properly formed (knowledge representation is a serious problem), and input into the system's knowledge base. Moreover, the methods of automatic reasoning and inference rules must be constructed by the system designer and must be explicit to be built into the system's structure. It is always difficult to do so and sometimes it is the source of limitations during the system's development and exploitation.

1.4 Various Approaches to Intelligent Systems

There are various approaches to intelligent systems but fundamental difference is located in the following distinction: the methods under consideration can be described as symbolic versus holistic ones.

In general, the domain of AI (very wide and presented in this chapter only as a small piece) can be divided or classified using many criteria. One of the most important divisions of the whole area can be based on the difference between the symbolic and holistic (pure numerical) approach. This discriminates all AI methods but can be shown and discussed on the basis of only two technologies presented here—neural networks and expert systems. Neural networks are technology definitely dedicated toward quantitative (numerical) calculations. Signals on input, output, and, most importantly, every element inside the neural network, are in the form of numbers even if their interpretation is of a qualitative type. It means that we must convert qualitative information into quantitative representation in the network. This problem is out of the scope of this chapter; therefore, we only mention a popular way of such a conversion, called "one of *N*." The merit of this type of data representation is based on spreading one qualitative input to *N* neurons in the input layer, where *N* is a number of distinguishable quantitative values, which can be observed in a considered data element. For example, if a qualitative value under consideration is "country of origin" and if there are four possible countries (let us say the United States, Poland, Russia, Germany) we must use for representation of this data four neurons with all signals equaling zero, except one input, corresponding to the selected value in input data, where the signal is equal 1. In this representation, 0,1,0,0 means Poland, etc. The identical method is used for the representation of output signals in neural networks performing a classification task. Output from such a network is in theory singular, because we expect only one answer: label of the class to which a classified object belongs given the input of the network at this moment. But because the label of the class is not a quantitative value—we must use in the output layer of the network as many neurons as there are classes—and the classification process will be assessed as successful when an output neuron attributed to the proper class label will produce a signal much stronger than other output neurons.

Returning to the general categorization of AI methods: qualitative versus quantitative we point out expert systems as a typical tool for the processing of qualitative (symbolic) data. The source of power in every expert system is its knowledge base, which is constructed from elements of knowledge obtained from human experts. Such elements of knowledge are different from the merit point of view because the expert system can be designed for solving different problems. Also, the internal representation of the human knowledge in a particular computer system can be different, but always in its symbolic form (sometimes even linguistic [natural language sentences]).

The methods of symbolic manipulations were always very attractive for AI researchers because the introspective view of human thinking process is usually registered in a symbolic form (so-called internal speech). Thus, in our awareness, almost every active cognitive process is based on symbol manipulations. Also, from the psychological point of view, the nature of activity of the human mind is defined as analytical-synthetical. What is especially emphasized is the connection between thinking and speaking (language), as the development of either of these abilities is believed to be impossible to exist in separation one from another.

Therefore "founding fathers" of AI in their early works massively used symbolic manipulations as tools for AI problem solving. The well-known example of this stream of works was the system named GPS (General Problem Solver) created in 1957 by Herbert Simon and Allen Newell [NS59]. It was a famous example, but we stress that a lot of AI systems based on symbolic manipulations and applying diverse approaches have been described in the literature. They were dedicated to automatic proving of mathematical theorems, playing a variety of games, solving well-formalized problems (e.g., Towers of Hanoi problem), planning of robot activities in artificial environments ("blocks world"), and many others. Also, early computer languages designed for AI purposes (e.g., LISP) were symbolic.

The differentiation between symbolic manipulations (as in expert systems) and holistic evaluation based on numerical data (like in neural networks) is observable in AI technology. It must be taken into account by every person who strives for the enhancement of designed or used electronic systems powering them by AI supplements.

We note one more surprising circumstance of the above discussed contradiction. Our introspection suggests a kind of internal symbolic process, which is accompanied with every metal process inside the human brain. At the same time, neural networks that are models of the human brain are not able to use symbolic manipulation at all!

AI methods and tools are used for many purposes but one of the most important areas where AI algorithms are used with good results is for problems connected with pattern recognition. The need of data classification is very popular because if we can classify the data, we can also better understand the information hidden in the data streams and thus can pursue knowledge extraction from the information.

In fact, to be used in AI automatic classification methods, we must take into account two types of problems and two groups of methods used for problem solution.

1.5 Pattern Recognition and Classifications

The first one is a classical pattern recognition problem with many typical methods used for its solving. At the start of all such methods, we have a collection of data and—as a presumption—a set of precisely defined classes. We need a method (formal algorithm or simulated device like neural network) for automatic decision making as to which class a particular data point belongs. The problem under consideration is important from a practical point of view because such classification-based model of data mining is one of the most effective tools for discovering the order and internal structure hidden in the data. This problem is also interesting from the scientific point of view and often difficult to solve because in most pattern recognition tasks, we do not have any prior knowledge about classification rules. The relationship between data elements and the classes to which these data should be classified is given only in the form of collection of properly classified examples. Therefore, all pattern recognition problems are examples of inductive reasoning tasks and need some machine learning approach that is both interesting and difficult [TK09].

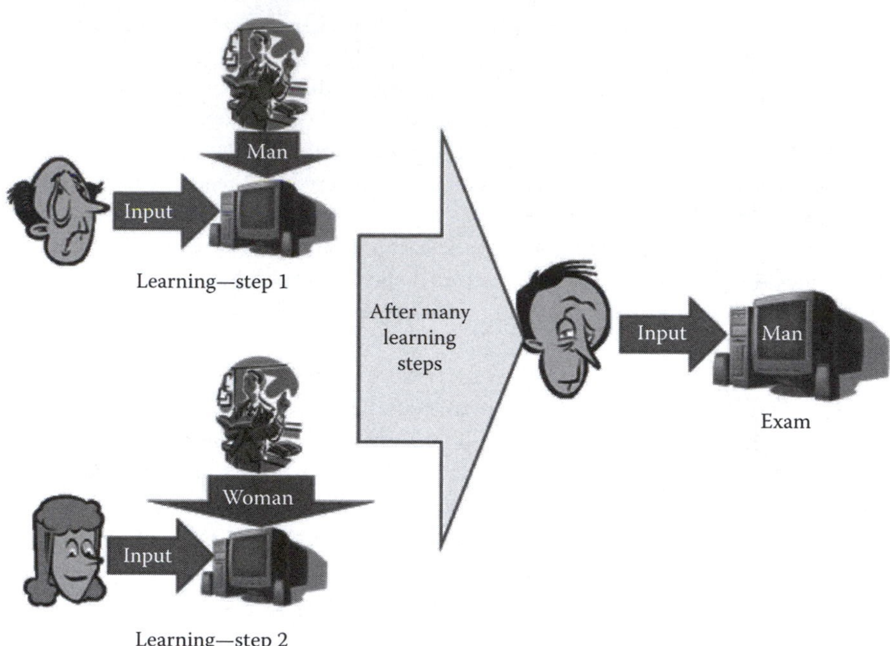

FIGURE 1.1 Pattern recognition problem with supervised learning used.

Machine learning methods can be divided into two general parts. The first part is based on supervised learning while the second part is related to unsupervised learning, also called self-learning or learning without teacher.

An example of supervised learning is presented in Figure 1.1. The learning system (represented by computer with learning algorithm inside) receives information about some object (e.g., man's face). The information about the object is introduced through the system input when the teacher guiding the supervised learning process prompts proper name of the class, to which this object should be numbered among. The proper name of the class is "Man" and this information is memorized in the system. Next another object is shown to the system, and for every object, teacher gives additional information, to which class this object belongs. After many learning steps, system is ready for exam and then a new object (never seen before) is presented. Using the knowledge completed during the learning process, the system can recognize unknown objects (e.g., a man).

In real situations, special database (named learning set) is used instead of human teacher. In such database, we have examples of input data as well as proper output information (results of correct recognition). Nevertheless, the general scheme of supervised learning, shown in Figure 1.2, is fulfilled also in this situation.

Methods used in AI for pattern recognition vary from simple ones, based on naïve geometrical intuitions used to split data description space (or data features space) into parts belonging to different classes (e.g., *k*-nearest neighbor algorithm), through methods in which the computer must approximate borders between regions of data description space belonging to particular classes (e.g., discriminant function methods or SVM algorithms), up to syntactic methods based on structure or linguistics, used for description of classified data [DH01].

A second group of problems considered in AI and related to the data classification tasks is cluster analysis [AB84]. The characteristics of these problems are symmetrical (or dual) to the above-mentioned pattern recognition problems. Whereas in pattern recognition we have predefined classes and need a method for establishing membership for every particular data point into one of such classes, in cluster analysis, we only have the data and we must discover how many groups are in the data. There are many

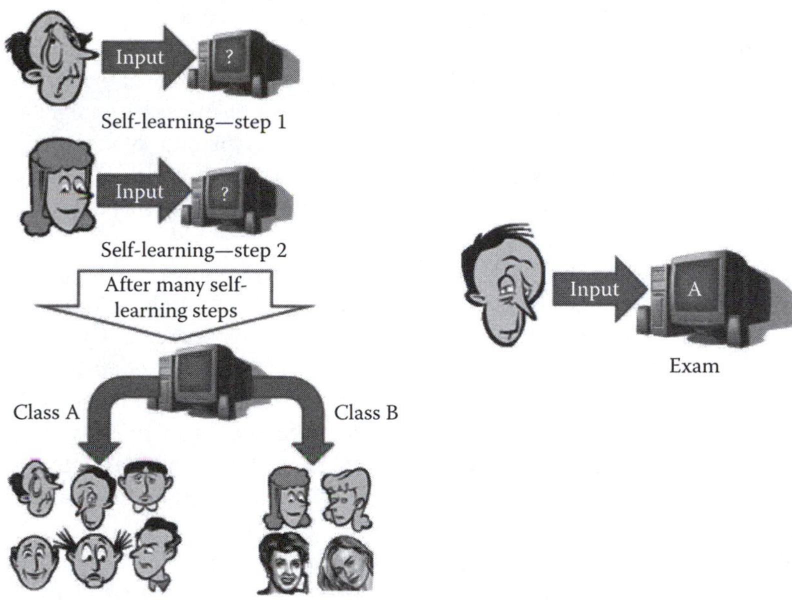

FIGURE 1.2 Classification problem with unsupervised learning used.

interesting approaches to solving clustering problems, and this problem can be thought of as the first step in building automatic systems capable of knowledge discovery, not only learning [CP07].

Let us discuss unsupervised learning scheme used for automatic solving of classification problems. During self-learning, the learned algorithm also receives information about features of the objects under consideration, but in this case, this input information is not enriched by accompanying information given by the teacher—because teacher is absent. Nevertheless, self-learning algorithm can perform classification of the objects using only similarity criteria and next can recognize new objects as belonging to particular self-defined classes.

1.6 Fuzzy Sets and Fuzzy Logic

One of the differences between the human mind and the computer relates to the nature of information/knowledge representation. Computers must have information in precise form, such as numbers, symbols, words, or even graphs; however, in each case, it must be an exact number, or a precisely selected symbol, or a properly expressed word or graph plotted in a precisely specified form, color, and dimension. Computers cannot accept a concept such as "integer number around 3," or "symbol that looks like a letter," etc. In contrast, human minds perform very effective thinking processes that take into account imprecise qualitative data (e.g., linguistic terms) but can come up with a good solution, which can be expressed sharply and precisely.

There are many examples showing the difference between mental categories (e.g., "young woman") and precisely computed values (e.g., age of particular people). Definitely, the relation between mathematical evaluation of age and "youngness" as a category cannot be expressed in a precise form. We cannot precisely answer the question, at which second of a girl's life she transforms to a woman, or at which exact hour her old age begins.

In every situation, when we need to implement in an intelligent system a part of human common sense, there is a contradiction between human fuzzy/soft thinking and the electronic system's sharp definition of data elements and use of precise algorithms. As is well known, the solution is to use fuzzy set and fuzzy logic methods [Z65]. Fuzzy set (e.g., the one we used above, "young woman") consists of the elements that, for sure (according to human experts), belongs to this set (e.g., 18-year-old graduate of high school), and the elements that absolutely do not belong to this fuzzy set (e.g., 80-year-old grandma), but take into account the elements that belong to this set only partially. All elements that have degrees of membership different from zero belong to this particular fuzzy set. Some of them have membership function with values of 1—they belong to the set unconditionally. Elements with membership function have values of 0—they are outside of the set. For all other elements, the value of membership function is a real number between 0 and 1. The shape of membership function is defined by human experts (or sometimes from available data) but for practical computations, the preferred shapes are either triangular or trapezoidal.

Fuzzy logic formulas can be dually expressed by if ... then ... else ... statements but they are expressed by means of fuzzy formulas. It is worth mentioning that fuzzy logic came into being as an extension of Lukasiewicz's multimodal logic [L20]. Details of this approach are described in other chapters.

It is worth mentioning here a gap between rather simple and easy-to-understand key ideas used in fuzzy data representation as well as simple fuzzy logic reasoning methods and rather complex practical problems solved in AI by means of fuzzy systems. It can be compared to walking in high mountains—first we go through a nice flowering meadow but after a while the walk transforms into extreme climbing.

Not all AI researchers like fuzzy methods. A well-known AI expert commented that this approach can be seen as "fuzzy theory about fuzzy sets." But in fact, the advantages of using fuzzy methods are evident. Not only the knowledge-based systems (i.e., expert systems) broadly use fuzzy logic and fuzzy representation of linguistic terms, but the fuzzy approach is very popular in economic data assessment, in medical diagnosis, and in automatic control systems. Moreover, their popularity increases because in many situations they are irreplaceable.

1.7 Genetic Algorithms and Evolutionary Computing

Figure 1.3 shows an example how the property of face image can be categorized. The face can be wide or narrow, can have large or small mouth, and eyes can be close or far. Once these categories are selected, each image of a face can be considered as a point in the three-dimensional space, as shown in Figure 1.4. Of course, often in the object we can distinguish more than just three properties and this would be a point

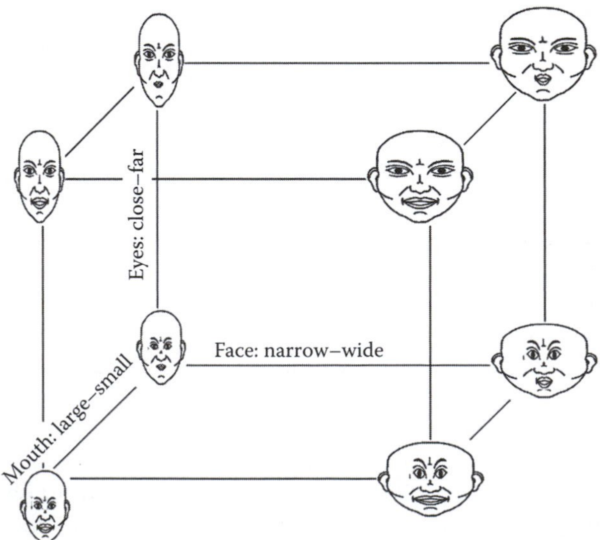

FIGURE 1.3 Example features that can be used for categorization and recognition of faces.

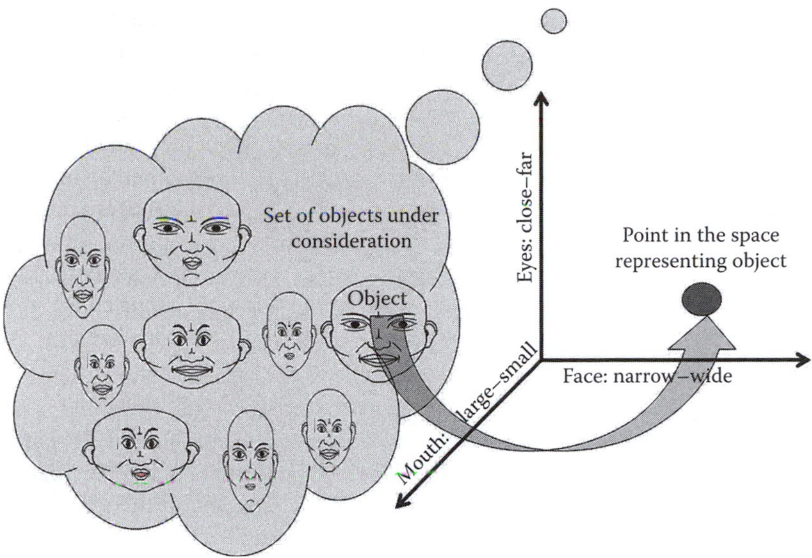

FIGURE 1.4 Relation between image of face and point in three-dimensional space.

in multidimensional space. Fuzzy systems may handle well the classifications when dimensionaliy they are limited to three. For problems with larger dimensions, neural networks have a significant advantage.

While describing neural networks, which are popular AI technology, we stressed their biological origin as a crude model of a part of a brain. Thanks to this fact, artificial neural networks exhibit brain-like behavior: they can learn and self-organize, generalize, and be used as predictors/classifiers, arrange information on the base of auto- and hetero-associative criteria, perform holistic and intuitive analysis of complex situations, are robust, etc. On the basis of the neural network example, we show the effectiveness of translating the biological knowledge into technological applications. Neural networks are obviously not a unique example of such biology-to-technology transmission of ideas. Another very well-known example is evolutionary computation [M96].

1.8 Evolutionary Computations and Other Biologically Inspired Methods for Problem Solving

The biological theory of evolution in many details (especially connected with the origin of humans) is still the area of hot discussions but no one questions the existence of evolution as a method of natural species optimization. In technology, we also seek for ever-better optimization methods. Existing optimization algorithms can be divided (freely speaking) into two subgroups. The first subgroup is formed by methods based on goal-oriented search (like fastest decrease/increase principle); an example is the gradient descent algorithm. The second group is based on random search methods; an example is the Monte Carlo method.

Both approaches to optimization suffer serious disadvantages. Methods based on goal-oriented search are fast and efficient in simple cases, but the solution may be wrong because of local minima (or maxima) of the criteria function. It is because the search process in all such methods is driven by local features of the criterion function, which is not optimal in the global sense. There is no method, which can be based on local properties of the optimization functionality and at the same time can effectively find the global optimum. On the other hand, the methods that use random searches can find proper solutions (optimal globally), but require long computational times. It is because the probability of a global optimum hit is very low and is increased only by means of performing a large number of attempts.

AI methods based on evolutionary computations combine random searches (because of using crossover and mutation) with goal-oriented searches (maximization of the fitness function, which is a functional to be optimized). Moreover, the search is performed simultaneously on many parallel paths because of several individuals (represented by chromosomes) belonging to every simulated population. The main idea of the evolutionary computing is based on defining all parameters and finding their optimal values. We generate (randomly) initial value for chromosomes (individuals belonging to the initial population) and then artificial evolution starts. Details of evolutionary computing are given in other chapters. It is worth to remember that evolutionary computing is a more general term than, e.g., genetic algorithms. Every user of genetic algorithms is doing evolutionary computing [K92].

In the title of this section, we mentioned that there exist other biologically inspired methods for problem solving. We note below just two that are very popular.

The first one is the ant colony optimization method that is used for solving many optimization problems, and is based on the ant's behavior. Like the neural network, it is a very simplified model of a part of the human brain, while genetic algorithms work on the basis of evolution, the ant's calculations use simplified model of the social dependences between ants in an ant colony. Every particular ant is a primitive organism and its behavior is also primitive and predictable. But the total ant population is able to perform very complicated tasks like the building of the complex three-dimensional structure of the anthill or finding the most efficient way for transportation of food from the source to the colony. The most efficient way can sometimes be equivalent to the shortest path; it takes into account the structure of the ground surface for minimizing the total effort necessary for food collection. Intelligence of the ant colony is its emerging feature. The source of very clever behavior observed sometimes for the whole ant population is located in rather simple rules controlling behavior of each particular ant and also in the simple rules governing relations and "communication" between ants. Both elements (e.g., mechanisms of single ant activity control as well as communication schemes functioning between ants) are easily modeled in the computer. Complex and purposeful behavior of the entire ant population can then be converted into an intelligent solution of a particular problem by the computer [CD91].

The second (too previously discussed) bio-inspired computational technique used in AI is an artificial immune systems methodology. The natural immune system is the strongest anti-intruder system that defends living organisms against bacteria, viruses, and other alien elements, which try to penetrate the organism. Natural immune systems can learn and must have memory, which is necessary for performing the above-mentioned activities. Artificial immune systems are models of this biological system that are able to perform similar activities on computer data, programs, and communication processes [CT02].

1.9 Intelligent Agents

Over many years of development of AI algorithms dedicated to solving particular problems, there was a big demand (in terms of computer calculation power and in memory). Therefore, programs with adjective "intelligent" were hosted on big computers and could not be moved from one computer to the other. An example is Deep Blue—a chess-playing computer developed by IBM—which, on May 11, 1997, won the chess world championship against Garry Kasparov.

In contemporary applications, AI, even the most successful, located in one particular place is not enough for practical problem solving. The future is distributed AI, ubiquitous intelligence, which can be realized by means of intelligent agents.

Agent technology is now very popular in many computer applications, because it is much easier to achieve good performance collaboratively, with limited costs by using many small but smart programs (agents) that perform some information gathering or processing task in a distributed computer environment working in the background. Typically, a particular agent is given a very small and well-defined task. Intelligent cooperation between agents can lead to high performance and high quality of the resulting services for the end users. The most important advantage of such an AI implementation is connected with the fact that the intelligence is distributed across the whole system and is located in these places (e.g., Web sites or network nodes) when necessary [A02].

AI methods used on the base of agent technology are a bit similar to the ant colony methods described above. But an intelligent agent can be designed on the base of neural networks technology, can use elements taken from expert systems, can engage pattern recognition methods as well as clustering algorithms. Almost every earlier mentioned element of AI can be used in the intelligent agent technology as a realization framework.

The best-known applications of distributed AI implemented as a collection of cooperating but independent agents are in the area of knowledge gathering for Internet search machines. The second area of intelligent agent applications is related to spam detection and computer virus elimination tasks. Intelligent agent technology is on the rise and possibly will be the dominating form of AI in the future.

1.10 Other AI Systems of the Future: Hybrid Solutions

In the previous sections, we tried to describe some "islands" from the "AI archipelago." Such islands, like neural networks, fuzzy sets, or genetic algorithms are different in many aspects: their theoretical background, technology used, data representation, methods of problem solving, and so on. However, many AI methods are complementary, not competitive. Therefore many modern solutions are based on the combination of these approaches and use hybrid structures, combining the best elements taken from more than one group of methods for establishing the best solution. In fact, AI elements can be combined in any arrangement because they are flexible. The very popular hybrid combinations are listed below:

- Neuro-fuzzy systems, which are based on fuzzy systems intuitive methodology combined with neural networks power of learning
- Expert systems powered by fuzzy logic methods for conclusion derivations
- Genetic algorithms used for the selection of the best neural network structure

Hybridization can be extended to other combinations of AI elements that when put together work more effectively than when used separately. Known are hybrid constructions combining neural networks with other methods used for data classification and pattern recognition. Sometimes, expert systems are combined not only with fuzzy logic but also with neural networks, which can collect knowledge during its learning process and then put it (after proper transformation) as an additional element in the knowledge base–powered expert system. Artificial immune systems can cooperate with cluster analysis methods for proper classification of complex data [CA08].

Nobody can foretell how AI will develop in the future. Perhaps AI and computational intelligence will go toward automatic understanding technologies, developed by the author and described in [TO08]? This chapter was meant to provide a general overview of AI and electronic engineering, and enriched with this information the reader can hopefully be better suited to find proper tools for specific applications.

References

[A02] Alonso E, AI and agents: State of the art, *AI Magazine*, 23(3), 25–30, 2002.

[AB84] Aldenderfer MS and Blashfield RK, *Cluster Analysis*, Sage, Newbury Park, CA, 1984.

[CA08] Corchado E, Abraham A, and Pedrycz W (eds), *Proceedings of the Third International Workshop on Hybrid Artificial Intelligence Systems (HAIS 2008)*, Burgos, Spain, Lecture Notes in Computer Science, Vol. 5271, Springer, Berlin, Germany, 2008.

[CD91] Colorni A, Dorigo M, and Maniezzo V, Distributed optimization by Ant Colonies, *Actes de la première conférence européenne sur la vie artificielle*, Paris, France, Elsevier Publishing, Amsterdam, the Netherlands, 1991, pp. 134–142.

[CM05] Cios KJ, Mamitsuka H, Nagashima T, and Tadeusiewicz R, Computational intelligence in solving bioinformatics problems [Editorial paper], *Artificial Intelligence in Medicine*, 35, 1–8, 2005.

[CP07] Cios KJ, Pedrycz W, Swiniarski R, and Kurgan L, *Data Mining: A Knowledge Discovery Approach*, Springer, Heidelberg, Germany, 2007.

[CT02] de Castro LN and Timmis J, *Artificial Immune Systems: A New Computational Intelligence Approach*, Springer, Heidelberg, Germany, 2002, pp. 57–58.

[DH01] Duda RO, Hart PE, and Stork DG, *Pattern Classification* (2nd edn), Wiley, New York, 2001.

[GR89] Giarrantano J and Riley G, *Expert Systems—Principles and Programming*, PWS-KENT Publishing Company, Boston, MA, 1989.

[H82] Hopfield JJ, Neural networks and physical systems with emergent collective computational abilities, *Proceedings of the National Academy of Sciences of the USA*, 79(8), 2554–2558, April 1982.

[H98] Haykin S, *Neural Networks: A Comprehensive Foundation* (2nd edn), Prentice Hall, Upper Saddle River, NJ, 1998.

[K92] Koza J, *Genetic Programming: On the Programming of Computers by Means of Natural Selection*, MIT Press, Cambridge, MA, 1992.

[K95] Kohonen T, *Self-Organizing Maps*, Series in Information Sciences, Vol. 30, Springer, Heidelberg, Germany, 1995.

[L20] Lukasiewicz J, O logice trojwartosciowej (in Polish), *Ruch filozoficzny*, 5, 170–171, 1920. English translation: On three-valued logic, in Borkowski L (ed), *Selected Works by Jan Łukasiewicz*, North-Holland, Amsterdam, the Netherlands, 1970, pp. 87–88.

[LS04] Luger G and Stubblefield W, *Artificial Intelligence: Structures and Strategies for Complex Problem Solving* (5th edn), The Benjamin/Cummings Publishing Company, Inc., Redwood City, CA, 2004.

[M96] Mitchell M, *An Introduction to Genetic Algorithms*, MIT Press, Cambridge, MA, 1996.

[MP43] McCulloch WS and Pitts W, A logical calculus of the ideas immanent in nervous activity, *Bulletin of Mathematical Biophysics*, 5, 115–133, 1943.

[MP72] Minsky M and Papert S, *Perceptrons: An Introduction to Computational Geometry*, The MIT Press, Cambridge, MA, 1972 (2nd edn with corrections, first edn 1969).

[NS59] Newell A, Shaw JC, and Simon HA, Report on a general problem-solving program, *Proceedings of the International Conference on Information Processing*, Paris, France, 1959, pp. 256–264.

[PM98] Poole D, Mackworth A, and Goebel R, *Computational Intelligence: A Logical Approach*, Oxford University Press, New York, 1998.

[RM86] Rumelhart DE, McClelland JL, and the PDP Research Group, *Parallel Distributed Processing: Explorations in the Microstructure of Cognition*, Vol. 1: *Foundations*, MIT Press, Cambridge, MA, 1986.

[RT08] Rutkowski L, Tadeusiewicz R, Zadeh L, and Zurada J (eds), *Artificial Intelligence and Soft Computing—ICAISC 2008*, Lecture Notes in Artificial Intelligence, Vol. 5097, Springer-Verlag, Berlin, Germany, 2008.

[S80] Searle J, Minds, brains and programs, *Behavioral and Brain Sciences*, 3(3), 417–457, 1980.

[T48] Turing A, Machine intelligence, in Copeland BJ (ed), *The Essential Turing: The Ideas That Gave Birth to the Computer Age*, Oxford University Press, Oxford, U.K., 1948.

[TK09] Theodoridis S and Koutroumbas K, *Pattern Recognition* (4th edn), Elsevier, Amsterdam, the Netherlands, 2009.

[TO08] Tadeusiewicz R, Ogiela L, and Ogiela MR, The automatic understanding approach to systems analysis and design, *International Journal of Information Management*, 28(1), 38–48, 2008.

[Z65] Zadeh LA, Fuzzy sets, *Information and Control*, 8(3), 338–353, 1965.

2

From Backpropagation to Neurocontrol*

Paul J. Werbos
National Science Foundation

This chapter provides an overview of the most powerful practical tools developed so far, and under development, in the areas which the Engineering Directorate of National Science Foundation (NSF) has called "cognitive optimization and prediction" [NSF07]. For engineering purposes, "cognitive optimization" refers to optimal decision and control under conditions of great complexity with use of parallel distributed computing; however, the chapter will also discuss how these tools compare with older tools for neurocontrol, which have also been refined and used in many applications [MSW90]. "Cognitive prediction" refers to prediction, classification, filtering, or state estimation under similar conditions.

The chapter will begin with a condensed overview of key tools. These tools can be used separately, but they have been designed to work together, to allow an integrated solution to a very wide range of possible tasks. Just as the brain itself has evolved to be able to "learn to do anything," these tools are part of a unified approach to replicate that ability, and to help us understand the brain itself in more functional terms as a useful working system [PW09]. Many of the details and equations are available on the web, as you can see in the references.

The chapter will then discuss the historical background and the larger directions of the field in more narrative terms.

2.1 Listing of Key Types of Tools Available

2.1.1 Backpropagation

The original form of backpropagation [PW74,PW05] is a *general closed-form* method for calculating the derivatives of some outcome of interest with respect to all of the inputs and parameters to any differentiable complex system. Thus if your system has N inputs, you get this information for a cost N times less than traditional differentiation, with an accuracy far greater than methods like perturbing the inputs. Any real-time sensor fusion or control system which requires the

* This chapter does not represent the views of NSF; however, as work performed by a government employee on government time, it may be copied freely subject to proper acknowledgment.

use of derivatives can be made much faster and more accurate by using this method. The larger that N is, the more important it is to use backpropagation. It is easier to apply backpropagation to standardized subroutines like artificial neural networks (ANN) or matrix multipliers than to custom models, because standardized "dual" subroutines can be programmed to do the calculations. Backpropagation works on input–output mappings, on dynamical systems, and on recurrent as well as feedforward systems.

2.1.2 Efficient Universal Approximation of Nonlinear Functions

Any general-purpose method for nonlinear control or prediction or pattern recognition must include some ability to approximate unknown nonlinear functions. Traditional engineers have often used Taylor series or look-up tables (e.g., "gain scheduling") or radial basis functions for this purpose; however, the number of weights or table entries increases exponentially as the number of input variables grows. Methods like that can do well if you have only one to three input variables, or if you have a lot of input variables whose actual values never leave a certain hyperplane, or a small set of cluster points. Beyond that, the growth in the number of parameters increases computational cost, and also increases error in estimating those parameters from data or experience.

By contrast, Andrew Barron of Yale has proven [Barron93] that the well-known multilayer perceptron (MLP) neural network can maintain accuracy with more inputs, with complexity rising only as a polynomial function of the number of inputs, if the function to be approximated is smooth. For nonsmooth functions, the simultaneous recurrent network (SRN) offers a more universal Turing-like extension of the same capabilities [PW92a,CV09,PW92b]. (Note that the SRN is not at all the same as the "simple recurrent network" later discussed by some psychologists.)

In actuality, even the MLP and SRN start to have difficulty when the number of true independent inputs grows larger than 50 or so, as in applications like full-scale streaming of raw video data, or assessment of the state of an entire power grid starting from raw data. In order to explain and replicate the ability of the mammal brain to perform such tasks, a more powerful but complex family of network designs has been proposed [PW98a], starting from the cellular SRN (CSRN) and the Object Net [PW04,IKW08,PW09]. Reasonably fast learning has been demonstrated for CSRNs in performing computational tasks, like learning to navigate arbitrary mazes from sight and like evaluating the connectivity of an image [IKW08,YC99]. MLPs simply cannot perform these tasks. Simple feedforward implementations of Object Nets have generated improvements in Wide-Area Control for electric power [QVH07] and in playing chess. Using a feedforward Object Net as a "critic" or "position evaluator," Fogel's team [Fogel04] developed the world's first computer system, which *learned* to play master-class chess without having been told specific rules of play by human experts.

All of these network designs can be trained to minimize square error in predicting a set of desired outputs $\mathbf{Y}(t)$ from a set of inputs $\mathbf{X}(t)$, over a database or stream of examples at different times t. Alternatively, they can be trained to minimize a different measure of error, such as square error plus a penalty function, or a logistic probability measure if the desired output is a binary variable. Efficient training generally requires a combination of backpropagation to get the gradient, plus any of a wide variety of common methods for using the gradient. It is also possible to do some scaling as part of the backwards propagation of information itself [PW74], but I am not aware of any work which has followed up effectively on that possibility as yet. When the number of weights is small enough, training by evolutionary computing methods like particle swarm optimization [dVMHH08,CV09] works well enough and may be easier to implement with software available today.

These designs can also be trained using some vector of (gradient) feedback, F_Y(t), to the output of the network, in situations where desired outputs are not known. This often happens in control applications. In situations where desired outputs and desired gradients are both known, the networks can be trained to minimize error in both. (See Gradient Assisted Learning [PW92b].) This can be the most efficient way to fit a neural network to approximate a large, expensive modeling code [PW05].

2.1.3 More Powerful and General Decision and Control

The most powerful and general new methods are adaptive, approximate dynamic programming (ADP) and neural model-predictive control (NMPC). The theorems guaranteeing stability for these methods require much weaker conditions than the theorems for traditional or neural adaptive control; in practical terms, that means that they are much less likely to blow up if your plant does not meet your assumptions exactly [BDJ08,HLADP,Suykens97,PW98b]. More important, they are optimizing methods, which allow you to directly maximize whatever measure of performance you care about, deterministic or stochastic, whether it be profit (minus cost), or probability of survival in a challenging environment. In several very tough real-world applications, from low-cost manufacturing of carbon-carbon parts [WS90], to missile interception [HB98,HBO02,DBD06] to turbogenerator control [VHW03] to automotive engine control [SKJD09,Prokhorov08], they have demonstrated substantial improvements over the best previous systems, which were based on many years of expensive hand-crafted effort. Reinforcement learning methods in the ADP family have been used to train anthropomorphic robots to perform dexterous tasks, like playing ice hockey or performing tennis shots, far beyond the capacity of traditional human-programmed robots [Schaal06].

NMPC is basically just the standard control method called model predictive control or receding horizon control, using neural networks to represent the model of the plant and/or the controller. In the earliest work on NMPC, we used the term "backpropagation through time (BTT) of utility" [MSW90]. NMPC is relatively easy to implement. It may be viewed as a simple upgrade of nonlinear adaptive control, in which the backpropagation derivative calculations are extended over time in order to improve stability and performance across time. NMPC assumes that the model of the plant is correct and exact, but in many applications it turns out to be robust with respect to that assumption. The strong stability results [Suykens97] follow from known results in robust control for the stability of nonlinear MPC. The Prokhorov controller for the Prius hybrid car is based on NMPC.

In practical terms, many engineers believe that they need to use adaptive control or learning in order to cope with common changes in the world, such as changes in friction or mass in the engines or vehicles they are controlling. In actuality, such changes can be addressed much better and faster by inserting time-lagged recurrence into the controller (or into the model of the plant, if the controller gets to input the outputs of the recurrent neurons in the model). This makes it possible to "learn offline to be adaptive online" [PW99]. This is the basis for extensive successful work by Ford in "multistreaming" [Ford96,Ford97,Ford02]. The work by Ford in this area under Lee Feldkamp and Ken Marko was extremely diverse, as a simple web search will demonstrate.

ADP is the more general and brain-like approach [HLADP]. It is easier to implement ADP when all the components are neural networks or linear systems, because of the need to use backpropagation to calculate many derivatives or "sensitivity coefficients"; however, I have provided pseudocode for many ADP methods in an abstract way, which allows you to plug in any model of the plant or controller—a neural network, a fixed model, an elastic fuzzy logic module [PW93], or whatever you prefer [PW92c,PW05].

Workers in robust control have discovered that they cannot derive the most robust controller, in the general nonlinear case, without "solving a Hamilton–Jacobi–Bellman" equation. This cannot be done exactly in the general case. ADP can be seen as a family of numerical methods, which provides the best available approximation to solving that problem. In pure robust control, the user trains the controller to minimize a cost function which represents the risk of instability and nothing else. But in practical situations, the user can pick a cost function or utility function which is a sum of such instability terms plus the performance terms which he or she cares about. In communication applications, this may simply mean maximizing profit, with a "quality of service payment" term included, to account for the need to minimize downtime.

Some ADP methods assume a model of the plant to be controlled (which may itself be a neural network trained concurrently); others do not. Those which do not may be compared to simple trial-and-error approaches, which become slower and slower as the number of variables increases. Researchers who have only studied model-free reinforcement learning for discrete variables sometimes say that reinforcement learning is too slow to give us anything like brain-like intelligence; however, model-based ADP designs for continuous variables have done much better on larger problems. Balakrishnan has reported that the best model-based methods are relatively insensitive to the accuracy of the model, even more so than NMPC is.

One would expect the brain itself to use some kind of hybrid of model-free and model-based methods. It needs to use the understanding of cause-and-effect embedded in a model, but it also needs to be fairly robust with respect to the limits of that understanding. I am not aware of such optimal hybrids in the literature today.

2.1.4 Time-Lagged Recurrent Networks

Time-lagged recurrent networks (TLRNs) are useful for prediction, system identification, plant modeling, filtering, and state estimation. MLPs, SRNs, and other static neural networks provide a way to approximate any nonlinear function as $\underline{\mathbf{Y}} = \underline{\mathbf{f}}(\underline{\mathbf{X}}, \mathbf{W})$, where $\mathbf{W}$ is a set of parameters or weights. A TLRN is any network of that kind, augmented by allowing the network to input the results of its own calculations from previous time periods.

There are many equivalent ways to represent this idea mathematically. Perhaps, the most useful is the oldest [PW87a,b]. In this representation, the TLRN is a combination of two functions, $\underline{\mathbf{f}}$ and $\underline{\mathbf{g}}$, used to calculate:

$$\underline{\mathbf{Y}}(t) = \underline{\mathbf{f}}(\underline{\mathbf{Y}}(t-1), \underline{\mathbf{R}}(t-1), \underline{\mathbf{X}}(t-1), W) \tag{2.1}$$

$$\underline{\mathbf{R}}(t) = \underline{\mathbf{g}}(\underline{\mathbf{Y}}(t-1), \underline{\mathbf{R}}(t-1), \underline{\mathbf{X}}(t-1), W) \tag{2.2}$$

People often say that the vector $\underline{\mathbf{R}}$ is a collection of variables "inside the network," which the network remembers from one time to the next, as these equations suggest. However, if this TLRN is trained to predict a motor to be controlled, then it may be important to send the vector $\underline{\mathbf{R}}$ to the controller as well. If the network is well-trained, then the combination of $\underline{\mathbf{Y}}$ and $\underline{\mathbf{R}}$ together represents the *state vector* of the motor being observed. More precisely, in the stochastic case, it is a compact optimized representation of the "belief state" of what we know about the state of the motor. *Access to the full belief state is often essential to good performance in real-world applications. Neural network control of any kind can usually be improved considerably by including it.*

Feldkamp and Prokhorov have done a three-way comparison of TLRNs, extended Kalman filters (EKF) and particle filters in estimating the true state vectors of a partially observed automotive system [Ford03]. They found that TLRNs performed about the same as particle filters, but far better than EKF. TLRNs were much less expensive to run than particle filters complex enough to match their performance. (The vector $\underline{\mathbf{R}}$ represents the full belief state, because the full belief state is needed in order to minimize the error in the updates; the network is trained to minimize that error.)

Ironically, the Ford group used EKF *training* to train their TLRN. In other words, they used backpropagation to calculate the derivatives of square error with respect to the weights, and then inserted those derivatives into a kind of EKF system to adapt the weights. This is also the only viable approach now available on conventional computers (other than brute force evolutionary computing) to train cellular SRNs [IKW08].

TLRNs have also been very successful, under a variety of names, in many other time-series prediction applications. Mo-Yuen Chow has reported excellent results in diagnostics of motors [MChow93] and of their components [MChow00]. Years ago, in a performance test funded by American Airlines, BehavHeuristics found that ordinary neural networks would sometimes do better than standard univariate time-series models like ARMA(p,q) [BJ70], but sometimes would do worse; however, TLRNs could do better consistently, because Equations 2.1 and 2.2 are a universal way to approximate what statisticians call multivariate NARMAX models, a more general and more powerful family of models. Likewise, in the forecasting competition at the *International Joint Conference on Neural Networks 2007 (IJCNN07)*, hard-working teams of statistician students performed much better than hard-working teams of neural network students, but a researcher from Ford outperformed them all, with relatively little effort, by using their standard in-house package for training TLRNs.

At *IJCNN07*, there was also a special meeting of the *Alternative Energy Task Force of the IEEE Computational Intelligence Society*. At that meeting, engineers from the auto industry and electric power sector all agreed that the one thing they need most from universities is the training of students who are fully competent in the use of TLRNs. (ADP was the next most important.)

For a student textbook building up to the use of TLRNs with accompanying software, see [PEL00].

In practical applications today, TLRNs are usually trained to minimize square error in prediction. However, in applications in the chemical industry, it has been found that "pure robust training" commonly cuts prediction errors by a factor of three. More research is needed to develop an optimal hybrid between pure robust training and ordinary least squares [PW98b].

The TLRN and other neural networks provide a kind of global prediction model $\underline{\mathbf{f}}$. But in some pattern classification applications, it is often useful to make predictions based on what was seen in the closest past example; this is called precedent-based or memory-based forecasting. Most "kernel methods" in use today are a variation of that approach. For full brain-like performance in real-time learning, it is essential to *combine* memory-based capabilities with global generalization, and to use both in adapting both; in other words "generalize but remember." I have discussed this approach in general terms [PW92a]; however, in working implementations, the closest work done so far is the work by Atkeson on memory-based learning [AS95] and the part of the work by Principe which applies information theoretic learning (related to kernel methods) to the residuals of a global model [PJX00,EP06]. Clustering and associative memory can play an important role in the memory of such hybrids.

2.1.5 Massively Parallel Chips Like Cellular Neural Network Chips

When NSF set up a research program in neuroengineering in 1988, we *defined* an ANN as any general-purpose design (algorithm/architecture) which can take full advantage of massively parallel computing hardware. We reached out to researchers from all branches of engineering and computer science willing to face up squarely to this challenge.

As a result, all of these tools were designed to be compatible with a new generation of computer chips, so that they can provide real-time applications much faster and cheaper than traditional algorithms of the same level of complexity. (Neural network approximation also allows models and controllers of reduced complexity.) For example, a group at Oak Ridge learned about "backpropagation" and renamed it the "second adjoint method" [PW05]. Engineers like Robert Newcomb then built some chips, which included "adjoint circuits" to calculate derivatives through local calculations on-board a specialty chip. Chua's group [YC99] has shown in detail how the calculations of backpropagation through time map into a kind of cellular neural network (CNN) chip, allowing thousands of times acceleration in performing the same calculation.

From 1988 to about 2000, there were few practical applications which took real advantage of this capability. At one time, the Jet Propulsion Laboratory announced a major agreement between Mosaix LLC and Ford to use a new neural network chip, suitable for implementing Ford's large TLRN diagnostic and control systems; however, as the standard processors on-board cars grew faster and more powerful, there was less and less need to add anything extra. Throughout this period, Moore's law made it hard to justify the use of new chips.

In recent years, the situation has changed. The speed of processor chips has stalled, at least for now. New progress now mainly depends on being able to use more and more processors on a single chip, and on getting more and more general functionality out of systems with more and more processors. That is exactly the challenge which ANN research has focused on for decades now. Any engineering task which can be formulated as a task in prediction or control can now take full advantage of these new chips, by use of ANN designs.

CNN now provide the best practical access to these kinds of capabilities. CNNs have been produced with thousands of processors per chip. With new memristor technology and new ideas in nanoelectronics and nanophotonics, it now seems certain that we can raise this to millions of processors per chip or more. A new center (CLION) was created in 2009 at the FedEx Institute of Technology, in Memphis, under Robert Kozma and myself, which plans to streamline and improve the pipeline from tasks in optimization and prediction to CNN implementations based on ANN tools.

2.2 Historical Background and Larger Context

The neural network field has many important historical roots going back to people like Von Neumann, Hebb, Grossberg, Widrow, and many others. This section will focus on those aspects most important to the engineer interested in applying such tools today.

For many decades, neural network researchers have worked to "build a brain," as the Riken Institute of Japan has put it. How can we build integrated intelligent systems, which capture the brain's ability to learn to "do anything," through some kind of universal learning ability?

Figure 2.1 reminds us of some important realities that specialized researchers often forget, as they "miss the forest for the trees."

The brain, as a whole system, is an information-processing system. As an information-processing system, its entire function as a whole system is to calculate its outputs. Its outputs are actions—actions like moving muscles or glandular secretions. (Biologists sometimes call this "squeezing or squirting.") The brain has many important subsystems to perform tasks like pattern recognition, prediction and memory, among others; however, these are all internal subsystems, which can be fully understood based on what they contribute to the overall function of the entire system. Leaving aside the more specialized preprocessors and the sources of primary reinforcement, the larger challenge we face is very focused: how can we build a general-purpose intelligent controller, which has all the flexibility and learning abilities of this one, based on parallel distributed hardware? That includes the development of the required subsystems—but they are just part of the larger challenge here.

In the 1960s, researchers like Marvin Minsky proposed that we could build a universal intelligent controller by developing general-purpose reinforcement learning systems (RLS), as illustrated in Figure 2.2.

We may think of an RLS as a kind of black box controller. You hook it up to all the available sensors ($\underline{\mathbf{X}}$) and actuators ($\underline{\mathbf{u}}$), and you also hook it up to some kind of performance monitoring system which gives real-time feedback (U(t)) on how well it is doing. The system then learns to maximize performance over time. In order to get the results you really want from this system, you have to decide on what you really want the system to accomplish; that means that you must translate your goals into a kind of metric performance or "cardinal utility function" U [JVN53]. Experts in business decision making have developed very extensive guidelines and training to help users to translate what they want into a utility function; see [Raiffa68] for an introduction to that large literature.

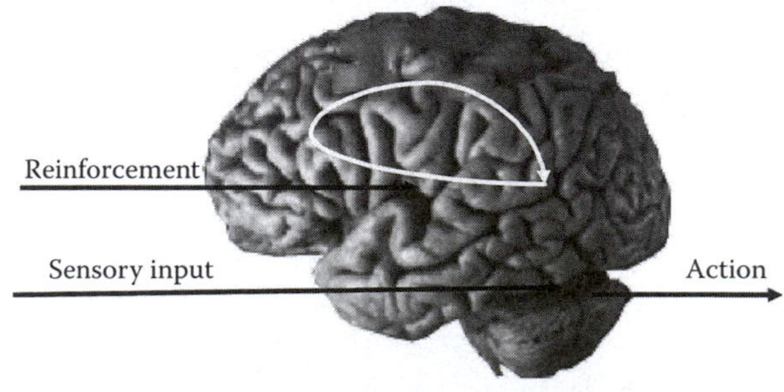

FIGURE 2.1 Brain as a whole system is an intelligent controller. (Adapted from NIH.)

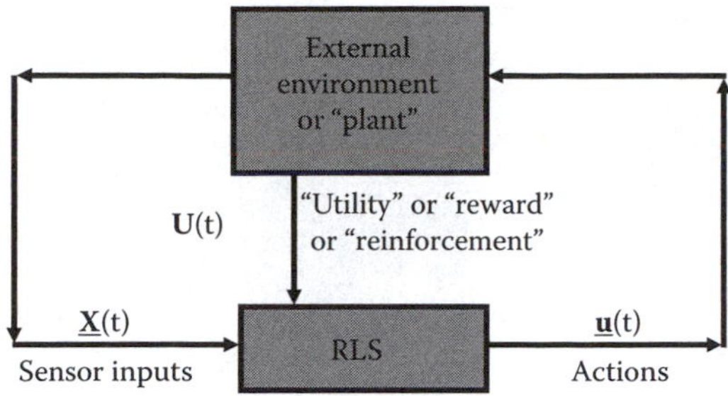

FIGURE 2.2 Reinforcement learning systems.

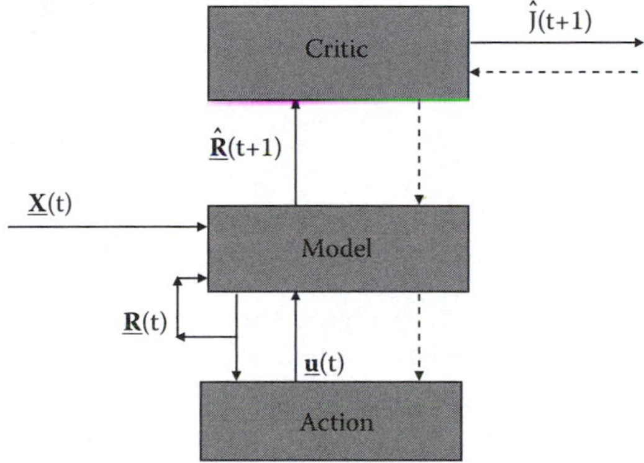

FIGURE 2.3 First general-purpose ADP system. (From 1971–72 Harvard thesis proposal.)

The earlier work on RLS was a great disappointment to researchers like Minsky. Trial-and-error methods developed on the basis of intuition were unable to manage even a few input variables well in simulation. In order to solve this problem, I went back to mathematical foundations, and developed the first reinforcement learning system based on adaptive, ADP, illustrated in Figure 2.3.

The idea was to train all three component networks in parallel, based on backpropagation feedback. There were actually three different streams of derivatives being computed here—derivatives of J(t+1) with respect to weights in the Action network or controller; derivatives of prediction error, in the Model network; and derivatives of a measure of error in satisfying the Bellman equation of dynamic programming, to train the critic. The dashed lines here show the flow of backpropagation used to train the Action network. Equations and pseudocode for the entire design, and more sophisticated relatives, may be found in [PW92a,b,c]. More recent work in these directions is reviewed in [HLADP], and in many recent papers in neural network conferences.

There is a strong overlap between reinforcement learning and ADP, but they are not the same. ADP does not include reinforcement learning methods which fail to approximate Bellman's equation or some other condition for optimal decision making *across time*, with foresight, allowing for the possibility of random disturbance. ADP assumes that we (may) know the utility function $U(\underline{\mathbf{X}})$ itself (or even a recurrent utility function), instead of just a current reward signal; with systems like the brain, performance is improved enormously by exploiting our knowledge that U is based on a variety of variables, which we can learn about directly.

All of the ADP designs in [PW92c] are examples of what I now call vector intelligence. I call them "vector intelligence" because the input vector $\underline{\mathbf{X}}$, the action vector $\underline{\mathbf{u}}$ and the recurrent state information $\underline{\mathbf{R}}$ are all treated like vectors. They are treated as collections of independent variables. Also, the upper part of the brain was assumed to be designed around a fixed common sampling time, about 100–200 ms. There was good reason to hope that the complexities of higher intelligent could be the emergent, learned

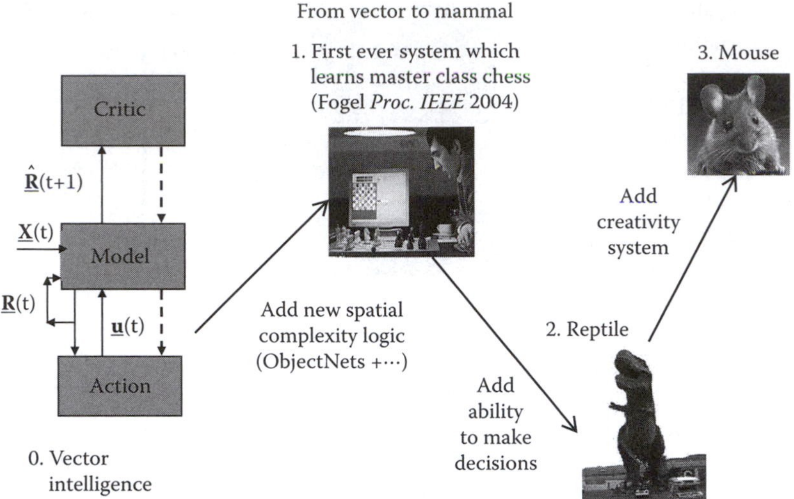

FIGURE 2.4 Levels of intelligence from vector to mammal.

result of such a simple underlying learning system [PW87a,PW09]. All of these systems are truly intelligent systems, in that they should always converge to the optimal strategy of behavior, given enough learning time and enough computing capacity. But how quickly? It is possible to add new features, still consistent with general-purpose intelligence, which make it easier for the system to learn to cope with spatial complexity, complexity across time (multiple time intervals), and to escape local minima. This leads to a new view illustrated in Figure 2.4.

In 1998 [PW98], I developed mathematical approaches to move us forward all the way from vector intelligence to mammal-level intelligence. However, as a practical matter in engineering research, we will probably have to master the first of these steps much more completely before we are ready to make more serious progress in the next two steps.

See [PW09] for more details on this larger picture, and for thoughts about levels of intelligence beyond the basic mammal brain.

References

[AS95] CG Atkeson and S Schaal, Memory-based neural networks for robot learning, *Neurocomputing*, 9(3), 243–269, December 1995.

[Barron93] AR Barron, Universal approximation bounds for superpositions of a sigmoidal function, *IEEE Transactions on Information Theory*, 39(3), 930–945, 1993.

[BDJ08] SN Balakrishnan, J Ding, and FL Lewis, Issues on stability of ADP feedback controllers for dynamical systems, *IEEE Transactions on Systems, Man, and Cybernetics*, 38(4), 913–917, August 2008. See also very extensive material posted at http://arri.uta.edu/acs/, on the web site of the Automation and Robotics Research Institute of the University of Texas at Arlington.

[BJ70] GEP Box and GM Jenkins, *Time-Series Analysis: Forecasting and Control*, Holden-Day, San Francisco, CA, 1970.

[CV09] R Cleaver and GK Venayagamoorthy, Learning functions nonlinear with MLPs and SRNs trained with DEPSO, *Proceedings of the International Joint Conference on Neural Networks*, Atlanta, GA, June 14–19, 2009.

[DBD06] V Durbha, SN Balakrishnan, and W Dyer, Target interception with cost-based observer, AIAA 2006-6218, *Presented at AIAA Guidance, Navigation, and Control Conference and Exhibit*, Keystone, Colorado, August 21–24, 2006.

[dVMHH08] Y del Valle, GK Venayagamoorthy, S Mohagheghi, JC Hernandez, and RG Harley, Particle swarm optimization: Basic concepts, variants and applications in power systems, *IEEE Transactions on Evolutionary Computation*, 12(2), 171–195, April 2008.

[EP06] D Erdogmus and J Principe, From linear adaptive filtering to non-linear signal processing, *IEEE SP Magazine*, 23, 14–33, 2006.

[Fogel04] DB Fogel, TJ Hays, SL Han, and J Quon, A self-learning evolutionary chess program, *Proceedings of the IEEE*, 92(12), 1947–1954, December 2004.

[Ford02] DV Prokhorov, LA Feldkamp, and IYu Tyukin, Adaptive behavior with fixed weights in RNN: An overview, *Proceedings of the International Joint Conference on Neural Networks (IJCNN), WCCI'02*, Honolulu, HI, May 2002. Posted at http://home.comcast.net/~dvp/fwadap1.pdf and http://pdl.brain.riken.go.jp/media/pdfs/Tyukin__first_-Prokhorov__WCCI2002.pdf

[Ford03] LA Feldkamp and DV Prokhorov, Recurrent neural networks for state estimation, *Proceedings of the Workshop on Adaptive and Learning Systems*, Yale University, New Haven, CT (ed. KS Narendra), 2003. Posted with permission at http://www.werbos.com/FeldkampProkhorov2003.pdf

[Ford96] GV Puskorius, LA Feldkamp, and LI Davis, Dynamic neural network methods applied to on-vehicle idle speed control, *Proceedings of the International Conference on Neural Networks: Plenary, Panel and Special Sessions*, Washington DC, 1996.

[Ford97] GV Puskorius and LA Feldkamp, Multi-stream extended Kalman filter training for static and dynamic neural networks, *Proceedings of the IEEE Conference on Systems, Man and Cybernetics*, Orlando, FL, Vol. 3, pp. 2006–2011.

[HB98] D Han and SN Balakrishnan, Adaptive critic based neural networks for agile missile control, *1998 AIAA Guidance Navigation, and Control Conference and Exhibit*, Boston, MA, August 10–12, 1998, pp. 1803–1812.

[HBO02] D Han, SN Balakrishnan, and EJ Ohlmeyer, Optimal midcourse guidance law with neural networks, *Proceedings of the IFAC 15th Triennial World Congress*, Barcelona, Spain, 2002.

[HLADP] J Si, AG Barto, WB Powell, and D Wunsch (eds), The general introduction and overview, Chapter 1, *Handbook of Learning and Approximate Dynamic Programming* (IEEE Press Series on Computational Intelligence), Wiley-IEEE Press, New York, 2004. Posted at www.werbos.com/Werbos_Chapter1.pdf

[IKW08] R Ilin, R Kozma, and PJ Werbos, Beyond feedforward models trained by backpropagation: A practical training tool for a more efficient universal approximator, *IEEE Transactions on Neural Networks*, 19(6), 929–937, June 2008.

[JVN53] J Von Neumann and O Morgenstern, *The Theory of Games and Economic Behavior*, Princeton University Press, Princeton, NJ, 1953.

[MChow00] B Li, M-Y Chow, Y Tipsuwan, and JC Hung, Neural-network-based motor rolling bearing fault diagnosis, *IEEE Transactions on Industrial Electronics*, 47(5), 1060–1069, October 2000.

[MChow93] MY Chow, RN Sharpe, and JC Hung, On the application and design of artificial neural networks for motor fault detection—Part II, *IEEE Transactions on Industrial Electronics*, 40(2), 181–196, April 1993.

[MSW90] WT Miller, R Sutton, and P Werbos (eds), *Neural Networks for Control*, MIT Press, Cambridge, MA, 1990, now in paper.

[NSF07] National Science Foundation, Cognitive optimization and prediction: From neural systems to neurotechnology, *Emerging Frontiers in Research Initiation 2008 (EFRI-2008)*, NSF document NSF 07-579. www.nsf.gov/pubs/2007/nsf07579/nsf07579.pdf

[PEL00] JC Principe, NR Euliano, and W Curt Lefebvre, *Neural and Adaptive Systems: Fundamentals through Simulations*, Wiley, New York, 2000.

[PJX00] JC Principe, JW Fisher, and D Xu, Information theoretic learning, in *Unsupervised Adaptive Filtering* (ed. S Haykin), Wiley, New York, 2000, pp. 265–319.

[Prokhorov08] D Prokhorov and Prius HEV neurocontrol and diagnostics, *Neural Networks*, 21, 458–465, 2008.

[PW04] P Werbos, *Object Nets*, U.S. Patent 6,708,160 B1, granted March 16, 2004. As with all U.S. patents, it is easily found on Google scholar.

[PW05] P Werbos, Backwards differentiation in AD and neural nets: Past links and new opportunities, *Automatic Differentiation: Applications, Theory and Implementations* (eds. H Martin Bucker, G Corliss, P Hovland, U Naumann, and B Norris), Springer, New York, 2005. Posted at www.werbos.com/AD2004.pdf

[PW09] P Werbos, Intelligence in the brain: A theory of how it works and how to build it, *Neural Networks*, 22(3), 200–212, April 2009. Related material is posted at www.werbos.com/Mind.htm

[PW74] P Werbos, Beyond regression: New tools for prediction and analysis in the behavioral sciences, PhD thesis, Committee on Applied Mathematics, Harvard University, Cambridge, MA, 1974. Reprinted in its entirety in P Werbos, *The Roots of Backpropagation: From Ordered Derivatives to Neural Networks and Political Forecasting*, Wiley, New York, 1994.

[PW87a] P Werbos, Building and understanding adaptive systems: A statistical/numerical approach to factory automation and brain research, *IEEE Transactions on Systems, Man and Cybernetics*, 17(1), 7–20, January/February 1987.

[PW87b] P Werbos, Learning how the world works: specifications for predictive networks in robots and brains, *Proceedings of the Systems, Man and Cybernetics* (*SMC87*), New York, 1987.

[PW92a] P Werbos, Neurocontrol and supervised learning: An overview and evaluation, *Handbook of Intelligent Control* (eds. D White and D Sofge), Van Nostrand, New York, 1992. Posted at http://www.werbos.com/HICChapter3.pdf

[PW92b] P Werbos, Neural networks, system identification and control in the chemical process industries, *Handbook of Intelligent Control* (eds. D White and D Sofge), Van Nostrand, New York, 1992. Posted at http://www.werbos.com/HIC_Chapter10.pdf

[PW92c] P Werbos, Approximate dynamic programming for real-time control and neural modeling, *Handbook of Intelligent Control* (eds. D White and D Sofge), Van Nostrand, New York, 1992. Posted at http://www.werbos.com/HICChapter13.pdf

[PW93] P Werbos, Elastic fuzzy logic: A better fit to neurocontrol and true intelligence *Journal of Intelligent and Fuzzy Systems*, 1, 365–377, 1993. Reprinted and updated in *Intelligent Control* (ed. M Gupta), IEEE Press, New York, 1995.

[PW98a] P Werbos, A brain-like design to learn optimal decision strategies in complex environments, *Dealing with Complexity: A Neural Networks Approach* (ed. M Karny, K Warwick, and V Kurkova), Springer, London, U.K., 1998. Also in S Amari and N Kasabov, *Brain-Like Computing and Intelligent Information Systems*, Springer, Singapore, 1998.

[PW98b] P Werbos, Stable adaptive control using new critic designs. Posted as adap-org 9810001 at arXiv.org (nlin archives), 1998.

[PW99] P Werbos, Neurocontrollers, *Encyclopedia of Electrical and Electronics Engineering* (ed. J Webster), Wiley, New York, 1999. Posted at http://www.werbos.com/Neural/Neurocontrollers_1999.htm

[QVH07] W Qiao, G Venayagamoorthy, and R Harley, DHP-based wide-area coordinating control of a power system with a large wind farm and multiple FACTS devices, *Proceedings of the International Joint Conference on Neural Networks* (*IJCNN07*), Orlando, FL, 2007.

[Raiffa68] H Raiffa, *Decision Analysis*, Addison-Wesley, Reading, MA, 1968. See also H Raiffa and R Schlaifer, *Applied Statistical Decision Theory*, Wiley-Interscience, New York; New Edition, May 15, 2000.

[Schaal06] H Kimura, K Tsuchiya, A Ishiguro, and H Witte (eds), Dynamic movement primitives: A framework for motor control in humans and humanoid robotics, *Adaptive Motion of Animals and Machines*, Springer, Tokyo, Japan, 2006.

[SKJD09] P Shih, BC Kaul, S Jagannathan, and JA Drallmeier, Reinforcement-learning-based output-feedback control of nonstrict nonlinear discrete-time systems with application to engine emission control, *IEEE Transactions on Systems, Man, and Cybernetics—Part B: Cybernetics*, 39(5), 1162–1179. Numerous related papers by Sarangapani are available on the web.

[Suykens97] JA Suykens, B DeMoor, and J Vandewalle, NLq theory: A neural control framework with global asymptotic stability criteria, *Neural Networks*, 10(4), 615–637, 1997.

[VHW03] GK Venayagamoorthy, RG Harley, and DC Wunsch, Implementation of adaptive critic based neurocontrollers for turbogenerators in a multimachine power system, *IEEE Transactions on Neural Networks*, 14(5), 1047–1064, September 2003.

[WS90] D White and D Sofge, Neural network based control for composite manufacturing, *Intelligent Processing of Materials: Winter Annual Meeting of the American Society of Mechanical Engineers*, Dallas, TX, vol. 21, 1990, pp. 89–97. See also chapters 5 and 6 in D White and D Sofge (eds), *Handbook of Intelligent Control*, Van Nostrand, New York, 1992.

[YC99] T Yang and LO Chua, Implementing back-propagation-through-time learning algorithm using cellular neural networks, *International Journal of Bifurcation and Chaos*, 9(9), 1041–1074, June 1999.

3

Neural Network–Based Control

Mehmet Önder Efe
Bahçeşehir University

3.1 Background of Neurocontrol

The mysterious nature of human brain with billions of neurons and enormously complicated biological structure has been a motivation for many disciplines that seem radically different from each other at a first glance, for example, engineering and medicine. Despite this difference, both engineering and medical sciences have a common base when the brain research is the matter of discussion. From a microscopic point of view, determining the building blocks as well as the functionality of those elements is one critical issue, and from a macroscopic viewpoint, discovering the functionality of groups of such elements is another one. The research in both scales has resulted in many useful models and algorithms, which are used frequently today. The framework of artificial neural networks has established an elegant bridge between problems, which display uncertainties, impreciseness with noise and modeling mismatches, and the solutions requiring precision, robustness, adaptability, and data-centeredness. The discipline of control engineering with tools offered by artificial neural networks have stipulated a synergy the outcomes of which is distributed over an enormously wide range.

A closer look at the historical developments in the neural networks research dates back to 1943. The first neuron model by Warren McCulloch and Walter Pitts was postulated and the model is assumed to fire under certain circumstances, [MP43]. Philosophically, the analytical models used today are the variants of this first model. The book entitled *The Organization of Behaviour* by Donald Hebb in 1949 was another milestone mentioning the synaptic modification for the first time [H49]. In 1956, Albert Uttley reported the classification of simple sets containing binary patterns, [U56], while in 1958 the community was introduced to the perceptron by Frank Rosenblatt. In 1962, Rosenblatt postulated several learning algorithms for the perceptron model capable of distinguishing binary classes, [R59], and another milestone came in 1960: Least mean squares for the adaptive linear element (ADALINE) by Widrow and Hoff [WH60]. Many works were reported after this and in 1982, John J. Hopfield proposed a neural model that is capable of storing limited information and retrieving it correctly with partially true initial state [H82]. The next breakthrough resulting in the resurgence of neural networks research is the discovery of error backpropagation technique [RHW86]. Although gradient descent was a known technique of numerical analysis, its application was formulated for feedforward neural networks by David E. Rumelhart, Geoffrey E. Hinton, and Ronald J. Williams in 1986. A radically different viewpoint for activation scheme, the radial basis functions, was proposed by D.S. Broomhead and D. Lowe, in 1988. This approach opened a new horizon particularly in applications requiring clustering of raw data. As the models and alternatives enriched, it became important to prove the universal approximation properties associated with each model. Three works published in 1989, by Ken-Ichi Funahashi, Kurt Hornik, and George Cybenko, proved that the multilayer feedforward networks are universal approximators performing the superpositions of sigmoidal functions to approximate a given map with finite precision [HSV89,F89,C89].

The history of neural network–based control covers mainly the research reported since the discovery of error backpropagation in 1982. Paul J. Werbos reported the use of neural networks with backpropagation utility in dynamic system inversion, and these have become the first results drawing the interest of control community to neural network–based applications. The work of Kawato et al. and the book by Antsaklis et al. are the accelerating works in the area as they describe the building blocks of neural network–based control [KFS87,APW91]. The pioneering work of Narendra and Parthasarathy has been an inspiration for many researchers studying neural network–based control, or neurocontrol [NP91]. Four system types, the clear definition of the role of neural network in a feedback control system, and the given examples of Narendra and Parthasarathy have been used as benchmarking for many researchers claiming novel methods. Since 1990 to date, a significant increase in the number of neural network papers has been observed. According to Science Direct and IEEE databases, a list showing the number of published items containing the words *neural* and *control* is given in Table 3.1, where the growing interest to neural control can be seen clearly.

In [PF94], decoupled extended Kalman filter algorithm was implemented for the training of recurrent neural networks. The justification of the proposed scheme was achieved on a cart-pole system, a bioreactor control problem, and on an idle speed control of an engine. Polycarpou reports a stable, adaptive neurocontrol scheme for a class of nonlinear systems, and demonstrates the stability using Lyapunov theorems [M96]. In 1996, Narendra considers neural network–based control of systems having different types of uncertainties, for example, the mathematical details embodying the plant dynamics are not known, or their structures are known but parameters are unavailable [N96]. Robotics has been a major implementation area for neural network controllers. In [LJY97], rigid manipulator dynamics is studied with an augmented tuning law to ensure stability and tracking performance. Removal of certainty equivalence and the removal of persistent excitation conditions are important contributions of the cited work. Wai uses the neural controller as an auxiliary tool for improving the tracking performance of a two-axis robot containing gravitational effects [W03]. Another field of research benefiting from the possibilities offered by neural networks framework is chemical process engineering. In [H99,EAK99], a categorization of schemes under titles predictive control, inverse model–based control, and adaptive control methods are presented. Calise et al. present an adaptive output feedback control approach utilizing a neural network [CHI01] while [WH01] focuses on enhancing the qualities of output regulation in nonlinear systems. Padhi et al. make use of the neural network–based control in distributed parameter systems [PBR01]. Use of neural network–based adaptive controllers in which the role of neural network is to provide nonlinear functions is a common approach reported several times in the literature [AB01,GY01,GW02,GW04,HGL05,PKM09]. Selmic and Lewis report the compensation of backlash

TABLE 3.1 Number of Papers Containing the Keywords Neural and Control between 1990 and 2008

Year	Items in Science Direct	Items in IEEE
2008	988	1355
2007	820	1154
2006	732	1254
2005	763	843
2004	644	869
2003	685	737
2002	559	820
2001	565	670
2000	564	703
1999	456	749
1998	491	667
1997	516	737
1996	489	722
1995	380	723
1994	326	731
1993	290	639
1992	234	409
1991	187	408
1990	176	270

with neural network–based dynamic inversion exploiting Hebbian tuning [SL01], Li presents a radial basis function neural network–based controller acting on a fighter aircraft [LSS01]. Chen and Narendra present a comparative study demonstrating the usefulness of a neural controller assisted by a linear control term [CN01]. A switching logic is designed and it is shown that neural network–based control scheme outperforms the pure linear and pure nonlinear versions of the feedback control law. Another work considering the multiple models activated via switching relaxes the condition of global boundedness of high-order nonlinear terms and improves the neurocontrol approach of Chen and Narendra [FC07]. An application of differential neural network–based control to nonlinear stochastic systems is discussed by Poznyak and Ljung [PL01]; nonlinear output regulation via recurrent neural networks is elaborated by [ZW01], and estimation of a Lyapunov function is performed for stable adaptive neurocontrol in [R01]. Predictive control approach has been another field of research that used the tools of neural networks framework. In [WW01], a particular neural structure is proposed, and this scheme is used to model predictive control. Yu and Gomm consider the multivariable model predictive control strategy on a chemical reactor model [YG03]. Applications of neurocontrol techniques in discrete-event automata is presented in [PS01], in reinforcement learning is reported in [SW01], and those in large-scale traffic network management is handled [CSC06]. Unmanned systems are another field that benefit from the neural network–based approaches. Due to the varying operating conditions and difficulty of constructing necessary process models, numerical data-based approaches become preferable as in the study reported in [K06], where a neural controller helps a proportional plus derivative controller and handles the time-dependent variations as its structure enables adaptation. This makes it sure that a certain set of performance criteria are met simultaneously. In [HCL07], a learning algorithm of proportional-integral-derivative type is proposed and a tracking control example is given in second-order chaotic system. Prokhorov suggests methods to train recurrent neural models for real-time applications [P07], and Papadimitropoulos et al. implement a fault-detection scheme based on an online approximation via neural networks [PRP07]. Two of the successful real-time results on spark ignition engines are reported in [CCBC07,VSKJ07], where neural network models are used as internal model controller in [CCBC07] and as observer and controller in [VSKJ07]. In [YQLR07], tabu search is adapted for neural network training to overcome the problem of convergence to local minimum. Alanis et al. [ASL07] propose high-order neural networks used with backstepping control technique. Discrete time output regulation using neural networks is considered in [LH07], and use of neurocontrollers in robust Markov games is studied by [SG07]. Recently, radial basis function neural networks were employed for the control of a nonholonomic mobile robot in [BFC09] and fine-tuning issues in large-scale systems have been addressed in [KK09]. Another recent work by Ferrari reports an application of dynamic neural network–forced implicit model following on a tailfin-controlled missile problem [F09]. Direct adaptive optimal control via neural network tools is considered in [VL09], model predictive control for a steel pickling process is studied in [KTHD09], and issues in the sampled data adaptive control are elaborated in [P09].

In brief, in the 1980s, the first steps and models were introduced while the 1990s were the years of stipulating the full diversity of learning schemes, architectures, and variants. According to the cited volume of research, outcomes of this century are more focused on applications and integration to other modules of feedback systems. The approaches seem to incorporate the full power of computing facilities as well as small-size and versatile data acquisition hardware. Since the introduction of the McCulloch–Pitts neuron model, it can be claimed that parallel to the technological innovations in computing hardware, progress in neural network–based control systems seem to spread over wider fields of application. In what follows, the learning algorithms and architectural possibilities are discussed. The methods of neural network–based control are presented and an application example is given.

3.2 Learning Algorithms

A critically important component of neural network research is the way in which the parameters are tuned to meet a predefined set of performance criteria. Despite the presence of a number of learning algorithms, two of them have become standard methods in engineering applications, and are elaborated in the sequel.

3.2.1 Error Backpropagation

Consider the feedforward neural network structure shown in Figure 3.1. The structure is called feedforward as the flow of information has a one-way nature. In order to describe the parameter modification rule, a variable sub- and superscripting convention needs to be adopted. In Figure 3.2, the layers in a given network are labeled, and the layer number is contained as a superscript. The synaptic weight in between the node i in layer $k + 1$ and node j in layer k is denoted by w_{ij}^{k}. Let an input vector and output vector at time t_0 be defined as $I_0 = (u_1(t_0)u_2(t_0)\ldots u_m(t_0))$ and $O_0 = (y_1(t_0)y_2(t_0)\ldots y_n(t_0))$, respectively. An input–output pair, shortly a *pair* or sample, is defined as $S_0 = \{I_0, O_0\}$. Consider there are P pairs in a given data set, which we call training set. When the training set is given, one must interpret it as follows: When I_0 is presented to a neural network of appropriate dimensions, its response must be an

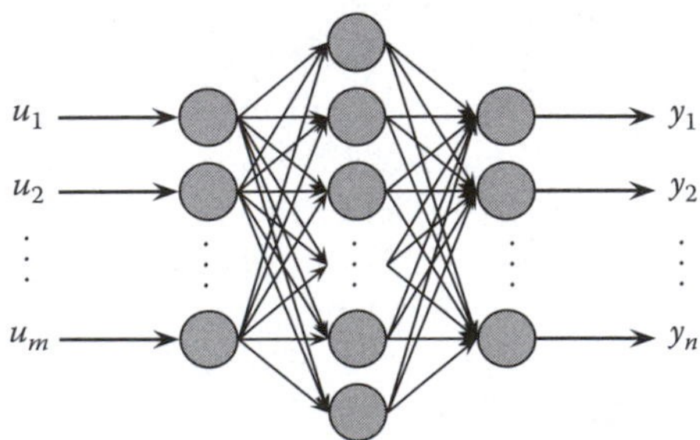

FIGURE 3.1 Structure of a feedforward neural network.

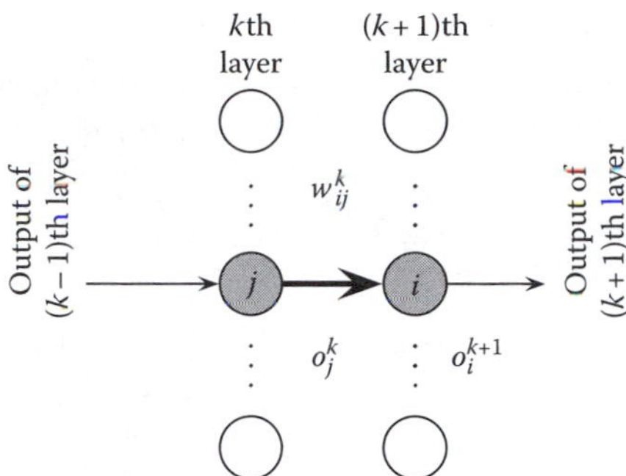

FIGURE 3.2 Sub- and superscripting of the variables in a feedforward neural network.

approximate of O_0. Based on this, the response of a neural network to a set of input vectors can be evaluated and the total cost over the given set can be defined as follows:

$$J(w) = \frac{1}{2P}\sum_{p=1}^{P}\sum_{i=1}^{n}\left(d_i^p - y_i^p(I_p, w)\right)^2 \tag{3.1}$$

where d_i^p is the target output corresponding to the ith output of the neural network that responds to pth pattern. The cost in (3.1) is also called the mean squared error (MSE) measure. Similarly, y_i^p is the response of the neural network to I_p. In (3.1), the generic symbol w stands for the set of all adjustable parameters, that is, the synaptic strengths, or weights.

Let the output of a neuron in the network be denoted by o_i^{k+1}. This neuron can belong to the output layer or a hidden layer of the network. The dependence of o_i^{k+1} to the adjustable parameters is as given below. Define $S_i^{k+1} := \sum_{j=1}^{n_k} w_{ij}^k o_j^k$ as the net sum determining the activation level of the neuron:

$$o_i^{k+1} = f\left(S_i^{k+1}\right) \tag{3.2}$$

where

$f(\cdot)$ is the neuronal activation function

n_k is the number of neurons in the kth layer

Gradient descent prescribes the following parameter update rule:

$$w_{ij}^k(t+1) = w_{ij}^k(t) - \eta\frac{\partial J(w)}{\partial w_{ij}^k(t)} \tag{3.3}$$

where

η is the learning rate chosen to satisfy $0 < \eta < 1$

the index t emphasizes the iterative nature of the scheme

Defining $\Delta w_{ij}^k(t) = w_{ij}^k(t+1) - w_{ij}^k(t)$, one could reformulate the above law as

$$\Delta w_{ij}^k(t) = -\eta\frac{\partial J(w)}{\partial w_{ij}^k(t)} \tag{3.4}$$

which is known also as the MIT rule, steepest descent, or the gradient descent, all referring to the above modification scheme.

3.2.1.1 Adaptation Law for the Output Layer Weights

Let $(k+1)$th layer be the output layer. This means $y_i = o_i^{k+1}$, and the target value for this output is specified explicitly by d_i. For the pth pair, define the output error at the ith output as

$$e_i^p = d_i^p - y_i^p(I_p, w) \tag{3.5}$$

After evaluating the partial derivative in (3.4), the updating of the output layer weights is performed according to the formula given in (3.6), where the pattern index is dropped for simplicity.

$$\Delta w_{ij}^k = \eta \delta_i^{k+1} o_j^k \tag{3.6}$$

where δ_i^{k+1} for the output layer is defined as follows:

$$\delta_i^{k+1} = e_i f'(S_i^{k+1}) \tag{3.7}$$

where $f'(S_i^{k+1}) = \partial f / \partial S_i^{k+1}$. One could check the above rule from Figure 3.2 to see the dependencies among the variables involved.

3.2.1.2 Adaptation Law for the Hidden Layer Weights

Though not as straightforward to show as that for the output layer, the weights in the hidden layer need an extra step as there are many paths through which the output errors can be backpropagated. In other words, according to Figures 3.1 and 3.2, it is clear that a small perturbation in w_{ij}^k will change the entire set of network outputs causing a change in the value of J. Now we consider $(k+1)$th layer as the hidden layer (see Figure 3.3). The general expression given by (3.6) is still valid and after appropriate manipulations, we have

$$\delta_i^{k+1} = \left(-\sum_{h=1}^{n_{k+2}} \delta_h^{k+2} w_{hi}^{k+1} \right) f'(S_i^{k+1}) \tag{3.8}$$

It is straightforward to see that the approach presented is still applicable if there are more than one hidden layers. In such cases, output errors are backpropagated until the necessary values shown in (3.8) are obtained.

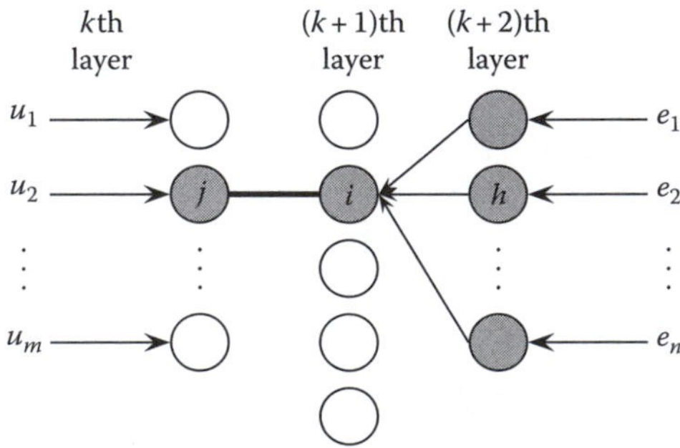

FIGURE 3.3 Influence of the output errors on a specific hidden layer weight.

Few modifications to the update law in (3.6) were proposed to speed up the convergence. Error back-propagation scheme starts fast, yet as time passes, it gradually slows down particularly in the regions where the gradient is small in magnitude. Adding a momentum term or modifying the learning rate are the standard approaches whose improving effect were shown empirically [H94].

3.2.2 Second-Order Methods

Error backpropagation was used successfully in many applications; however, its very slow convergence speed has been the major drawback highlighted many times. Being the major source of motivation, increasing the convergence speed has been achieved for feedforward neural networks in 1994 [HM94]. Hagan and Menhaj applied the Marquardt algorithm to update the synaptic weights of a feedforward neural network structure. The essence of the algorithm is as follows.

Consider a neural network having n outputs, and N adjustable parameters denoted by the vector $w = (w_1\ w_2\ \dots\ w_N)$. Each entry of the parameter vector corresponds to a unique parameter in an ordered fashion. If there are P pairs over which the interpolation is to be performed, the cost function qualifying the performance is as given in (3.1) and the update law is given in (3.9).

$$w_{t+1} = w_t - \left(\nabla_w^2 J(w_t)\right)^{-1} \nabla_w J(w_t) \tag{3.9}$$

where t stands for the discrete time index. Here, $\nabla_w^2 J(w_t) = 2H(w_t)^{\mathrm{T}} H(w_t) + g(H(w_t))$ with $g(H(w_t))$ being a small residual, and $\nabla_w J(w_t) = 2H(w_t)^{\mathrm{T}} E(w_t)$ with E and H being the error vector and the Jacobian, as given in (3.10) and (3.11), respectively. The error vector contains the errors computed by (3.5) for every training pair, and the Jacobian contains the partial derivatives of each component of E with respect to every parameter in w.

$$E = \begin{pmatrix} e_1^1 & \cdots & e_n^1 & e_1^2 & \cdots & e_n^2 & \cdots & e_1^P & \cdots & e_n^P \end{pmatrix}^{\mathrm{T}} \tag{3.10}$$

$$H(w) = \begin{pmatrix} \frac{\partial e_1^1(w)}{\partial \omega_1} & \frac{\partial e_1^1(w)}{\partial \omega_2} & \cdots & \frac{\partial e_1^1(w)}{\partial \omega_N} \\ \frac{\partial e_2^1(w)}{\partial \omega_1} & \frac{\partial e_2^1(w)}{\partial \omega_2} & \cdots & \frac{\partial e_2^1(w)}{\partial \omega_N} \\ \vdots & \vdots & & \vdots \\ \frac{\partial e_n^1(w)}{\partial \omega_1} & \frac{\partial e_n^1(w)}{\partial \omega_2} & \cdots & \frac{\partial e_n^1(w)}{\partial \omega_N} \\ \vdots & \vdots & & \vdots \\ \frac{\partial e_1^P(w)}{\partial \omega_1} & \frac{\partial e_1^P(w)}{\partial \omega_2} & \cdots & \frac{\partial e_1^P(w)}{\partial \omega_N} \\ \frac{\partial e_2^P(w)}{\partial \omega_1} & \frac{\partial e_2^P(w)}{\partial \omega_2} & \cdots & \frac{\partial e_2^P(w)}{\partial \omega_N} \\ \vdots & \vdots & & \vdots \\ \frac{\partial e_n^P(w)}{\partial \omega_1} & \frac{\partial e_n^P(w)}{\partial \omega_2} & \cdots & \frac{\partial e_n^P(w)}{\partial \omega_N} \end{pmatrix} \tag{3.11}$$

Based on these definitions, the well-known Gauss–Newton algorithm can be given as

$$w_{t+1} = w_t - \left(H(w_t)^{\mathrm{T}} H(w_t)\right)^{-1} H(w_t)^{\mathrm{T}} E(w_t) \tag{3.12}$$

and the Levenberg–Marquardt update can be constructed as

$$w_{t+1} = w_t - \left(\mu I + H(w_t)^{\mathrm{T}} H(w_t)\right)^{-1} H(w_t)^{\mathrm{T}} E(w_t) \tag{3.13}$$

where

$\mu > 0$ is a user-defined scalar design parameter for improving the rank deficiency problem of the matrix $H(w_k)^{\mathrm{T}}H(w_k)$

I is an identity matrix of dimensions $N \times N$

It is important to note that for small μ, (3.13) approximates to the standard Gauss–Newton method (see (3.12)), and for large μ, the tuning law becomes the standard error backpropagation algorithm with a step size $\eta \approx 1/\mu$. Therefore, Levenberg–Marquardt method establishes a good balance between error backpropagation and Gauss–Newton strategies and inherits the prominent features of both algorithms in eliminating the rank deficiency problem with improved convergence. Despite such remarkably good properties, the algorithm in (3.13) requires the inversion of a matrix of dimensions $nP \times N$ indicating high computational intensity. Other variants of second-order methods differ in some nuances yet in essence, they implement the same philosophy. Conjugate gradient method with Polak–Ribiere and Fletcher–Reeves formulas are some variants used in the literature [H94]. Nevertheless, the problem of getting trapped to local minima continues to exist, and the design of novel adaptation laws is an active research topic within the realm of neural networks.

3.2.3 Other Alternatives

Typical complaints in the application domain have been to observe very slow convergence, convergence to suboptimal solutions rendering the network incapable of performing the desired task, oversensitivity to suddenly changing inputs due to the gradient computation, and the like. Persisting nature of such difficulties has led the researchers to develop alternative tuning laws alleviating some of these drawbacks or introducing some positive qualities. Methods inspired from variable structure control are one such class showing that the error can be cast into a phase space and guided toward the origin while displaying the robustness properties of the underlying technique [YEK02,PMB98]. Another is based on derivative-free adaptation utilizing the genetic algorithms [SG00]. Such algorithms refine the weight vector based on the maximization of a fitness function. In spite of their computational burden, methods that do not utilize the gradient information are robust against the sudden changes in the inputs. Aside from these, unsupervised learning methods constitute another alternative used in the literature. Among a number of alternatives, reinforcement learning is one remarkable approach employing a reward and penalty scheme to achieve a particular goal. The process of reward and penalty is an evaluative feedback that characterizes how the weights of a neural structure should be modified [SW01,MD05,N97].

3.3 Architectural Varieties

Another source of diversity in neural network applications is the architecture. We will present the alternatives under two categories: first is the type of connectivity, second is the scheme adopted for neuronal activation. In both cases, numerous alternatives are available as summarized below.

3.3.1 Structure

Several types of neural network models are used frequently in the literature. Different structural configurations are distinguished in terms of the data flow properties of a network, that is, feedforward models and recurrent models. In Figure 3.1, a typical feedforward neural network is shown. Three simple versions of recurrent connectivity are illustrated in Figure 3.4. The network in Figure 3.4a has recurrent neurons in the hidden layer, Figure 3.4b feeds back the network output and established recurrence externally, and Figure 3.4c contains fully recurrent hidden layer neurons.

Clearly, one can set models having multiple outputs as well as feedback connections in between different layers but due to the space limit, we omit those cases and simply consider the function of a neuron having feedback from other sources of information as in Equation 3.2 but now we have the net sum as follows:

$$S_i^{k+1} = \sum_{j=1}^{n_k} w_{ij}^k o_j^k + \sum_{i=1}^{R} \rho_i \xi_i \tag{3.14}$$

where

the second sum is over all feedback connections

ρ_i denotes the weight determining the contribution of the output ξ_i from a neuron in a layer

Clearly, the proper application of error backpropagation or Levenberg–Marquardt algorithm is dependent upon the proper handling of the network description under feedback paths [H94].

3.3.2 Neuronal Activation Scheme

The efforts toward obtaining the best performing artificial neural model has also focused on the neuronal activation scheme. The map built by a neural model is strictly dependent upon the activation function. Smooth activation schemes produce smooth hypersurfaces while sharp ones like the sign function produce hypersurfaces having very steep regions. This subsection focuses on the dependence of performance on the type of neuronal activation function. In the past, this was considered to some extent in [KA02,HN94,WZD97,CF92,E08]. Efe considers eight different data sets, eight different activation functions with networks having 14 different neuron numbers in the hidden layer. Under these conditions, the research conducted gives a clear idea about when an activation function is advisable. The goal of such a research is to figure out what type of activation function performs well if the size of a network is small [E08]. Some of the neuronal activation functions mentioned frequently in the literature are

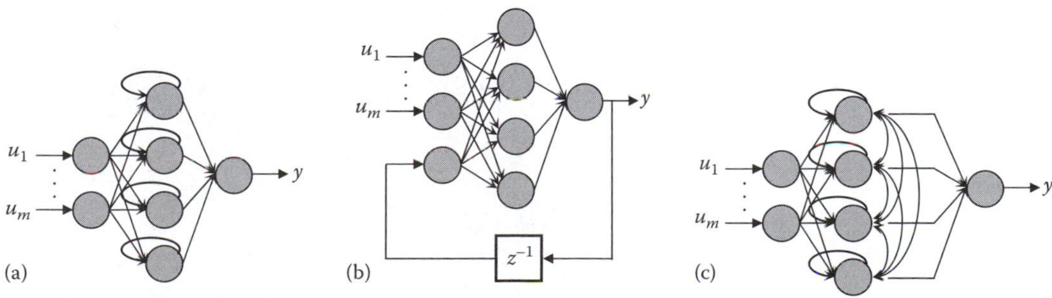

FIGURE 3.4 Three alternatives for feedback-type neural networks having single output.

TABLE 3.2 Neuronal Activation Functions

Name	Activation Function	Adjustables
Hyperbolic tangent	$f(x) = \tanh(x)$	—
Damped polynomial	$f(x) = (ax^2 + bx + c)\exp(-\lambda x^2)$	a, b, c, λ
Multilevel	$f(x) = \frac{1}{2M+1}\sum_{k=-M}^{M} \tanh(x - \lambda k)$	M, λ
Trigonometric	$f(x) = a\sin(px) + b\cos(qx)$	a, b, p, q
Arctangent	$f(x) = \text{atan}(x)$	—
Sinc	$f(x) = \begin{cases} \frac{\sin(\pi x)}{\pi x} & x \neq 0 \\ 1 & x = 0 \end{cases}$	—
Logarithmic	$f(x) = \begin{cases} \ln(1+x) & x \geq 0 \\ -\ln(1-x) & x < 0 \end{cases}$	—

tabulated in Table 3.2, which shows the diversity of options in setting up a neural network, and choosing the best activation scheme is still an active research involving the problem in hand as well.

3.4 Neural Networks for Identification and Control

3.4.1 Generating the Training Data

A critically important component in developing a map from one domain to another is dependent upon the descriptive nature of the entity that lies between these domains and that describes the map implicitly. This entity is the numerical data, or the raw data to be used to build the desired mapping. To be more explicit, for a two-input single-output mapping, it may be difficult to see the global picture based on the few samples shown in Figure 3.5a, yet it is slightly more visible from Figure 3.5b and it is more or less a plane according to the picture in Figure 3.5c. Indeed, the sample points in all three points were generated using $y = 3u_1 + 4u_2$. This simple experiment shows us that in order to describe the general picture, the data must be distributed well around the questioned domain as well as it must be dense enough to deduce the general behavior.

In the applications reporting solutions to synthetic problems, the designer is free to generate as much data as he or she needs; however, in real-time problems, collecting data to train a neural network may be a tedious task, or even sometimes a costly one. Furthermore, for problems that have more than two inputs, utilizing the graphical approaches may have very limited usefulness. This discussion amounts to saying that if there are reasonably large number of training data describing a given process, a neural network–based model is a good alternative to realized the desired map.

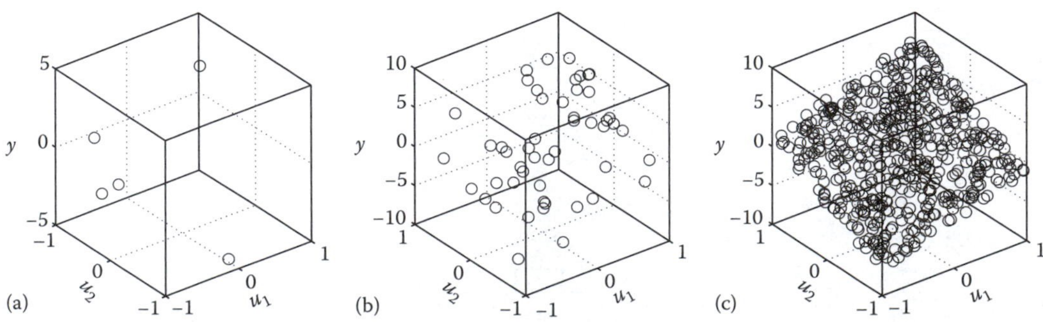

FIGURE 3.5 Changing implication of some amount of data.

3.4.2 Determining the Necessary Inputs: Information Sufficiency

Previous discussion can be extended to the concept of information sufficiency. For the plane given above, a neural model would need u_1 and u_2 as its inputs. Two questions can be raised:

- Would the neural model be able to give acceptable results if it was presented only one of the inputs, for example, u_1 or u_2?
- Would the neural model perform better if it was presented another input containing some linear combination of u_1 and u_2, say for example, $u_3 = \alpha_1 u_1 + \alpha_2 u_2$, $(\alpha_1, \alpha_2 \neq 0)$?

The answer to both questions is obviously no, however, in real-life problems, there may be a few hundred variables having different amounts of effect on an observed output, and some—though involved in the process—may have negligibly small effect on the variables under investigation. Such cases must be clearly analyzed, and this has been a branch of neural network–based control research studied in the past [LP07].

3.4.3 Generalization or Memorization

Noise is an inevitable component of real-time data. Consider a map described implicitly by the data shown in Figure 3.6. The circles in the figure indicate the data points given to extract the shown target curve. The data given in the first case, the left subplot, does not contain noise and a neural network that memorizes, or overfits, the given data points produce a network that well approximates the function. Memorization in this context can formally be defined as $J = 0$, and this can be a goal if it is known that there is no noise in the data. If the designer knows that the data is noisy as in the case shown in the middle subplot, a neural network that overfits the data will produce a curve, which is dissimilar from the target one. Pursuing memorization for the data shown in the right subplot will produce a neural model that performs even worse. This discussion shows that memorization, or overfitting, for real life and possibly noisy data is likely to produce poorly performing neural models; nevertheless, noise is the major actor determining the minimum possible value of J for a given set of initial conditions and network architecture.

In the noisy cases of Figure 3.6, a neural network will start from an arbitrary curve. As training progresses, the curve realized by the neural model will approach the target curve. After a particular instant, the neural model will produce a curve that passes through the data points but is not similar to the target curve. This particular instant is determined by utilizing two sets during the training phase. One is used for synaptic modification while the other is used solely for checking the cost function. If the cost functions computed over both sets decrease, the network is said to generalize the given map. If the cost for training set continues to decrease while that for the test set starts increasing, the neural network is said to start overfitting the given data, and this instant is the best instant to stop the training.

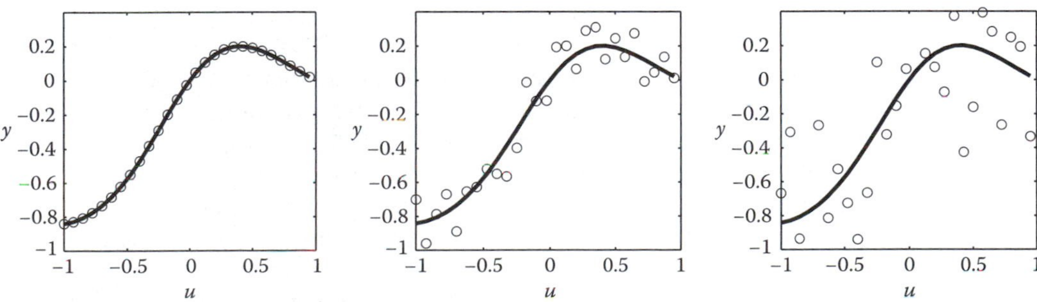

FIGURE 3.6 Effect of noise variance on the general perception.

3.4.4 Online or Offline Synthesis

A neural network–based controller can be adaptive or nonadaptive depending on the needs of the problem in hand and the choice of the user. Nonadaptive neural network controller is trained prior to installation and its parameters are frozen. This approach is called offline synthesis (or training). The other alternative considers a neurocontroller that may or may not be pretrained but its weights are refined during the operation. This scheme requires the extraction of an error measure to set up the update laws. The latter approach is more popular as it handles the time variation in the dynamical properties of a system and maintains the performance by altering the controller parameters appropriately. This scheme is called online learning. A major difference between these methods is the number of training data available at any instant of time. In the offline synthesis, the entire set of data is available, and there is no time constraint to accomplish the training, whereas in the online training, data comes at every sampling instant and its processing for training must be completed within one sampling period, and this typically imposes hardware constraints into the design.

3.5 Neurocontrol Architectures

Consider the synthetic process governed by the difference equation $x_{t+1} = f(x_t, u_t)$, where x_t is the value of state at discrete time t and u_t is the external input. Assume the functional details embodying the process dynamics are not available to follow conventional design techniques, and assume it is possible to run the system with arbitrary inputs and arbitrary initial state values. The problem is to design a system that observes the state of the process and outputs u_t, by which the system state follows a given reference r_t. A neural network–based solution to this problem prescribes the following steps.

- Initialize the state x_t to a randomly selected value satisfying $x_t \in \mathcal{X} \in \Re$, where $\mathcal{X}$ is the interval we are interested in.
- Choose a stimulus satisfying $u_t \in \mathcal{U} \in \Re$, where $\mathcal{U}$ stands for the interval containing likely control inputs.
- Measure/evaluate the value of x_{t+1}.
- Form an input vector $I_t = (x_{t+1}\ x_t)$ and an output $O_t = u_t$ and repeat these four steps P times to obtain a training data set.
- Perform training to minimize the cost in Equation 3.1 and stop training when a stopping criterion is met.

A neural network realizing the map $u_t = \text{NN}(x_{t+1}, x_t)$ would answer the question "What would be the value of the control signal if a transition from x_{t+1} to x_t is desired?" Clearly, having obtained a properly trained neural controller, one would set $x_{t+1} = r_{t+1}$ and the neurocontroller would drive the system state to the reference signal as it had been trained to do so.

If $x_t \in \Re^n$, then we have $r_t \in \Re^n$ and naturally $I_t \in \Re^{2n}$. Similarly, multiple inputs in a dynamic system would require a neurocontroller to have the same number of inputs as the plant contains and the same approach would be valid. A practical difficulty in training the controller utilizing the direct synthesis approach is that the designer decides on the excitation first, that is, the set $\mathcal{U}$, however, it can be practically difficult to foresee the interval to which a control signal in real operation belongs. This scheme can be called direct inversion. In the rest of this section, common structures of neurocontrol are summarized.

In Figure 3.7, identification of a dynamic system is shown. The trainer adjusts the neural network identifier (shown as NN-I) parameters in such a way that the cost given in (3.1) is minimized over the given set of training patterns. The identification scheme depicted in Figure 3.7, can also be utilized to identify an existing controller acting in a feedback control system. This approach corresponds to the mimicking of a conventional controller [N97].

In Figure 3.8, indirect learning architecture is shown. An input signal u is applied to the plant, it responds to the signal, which is denoted by y. A neural network controller, (NN-C in the figure) shown

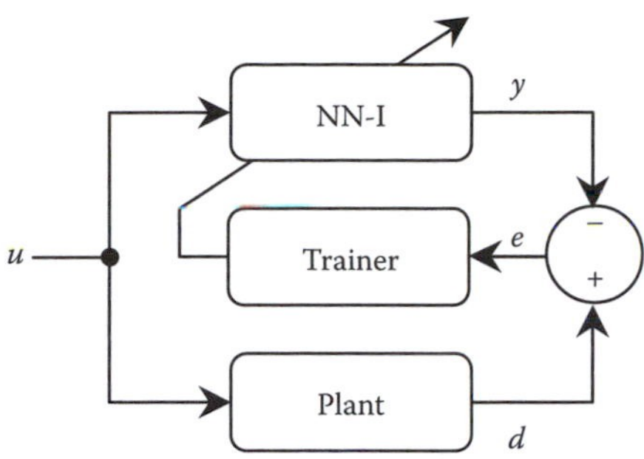

FIGURE 3.7 NN-based identifier training architecture.

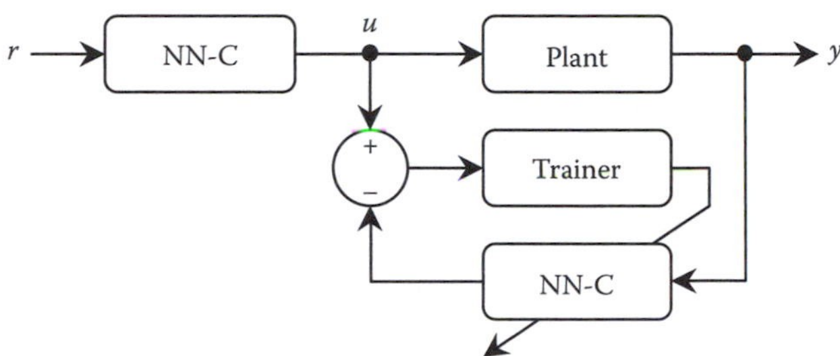

FIGURE 3.8 Indirect learning architecture.

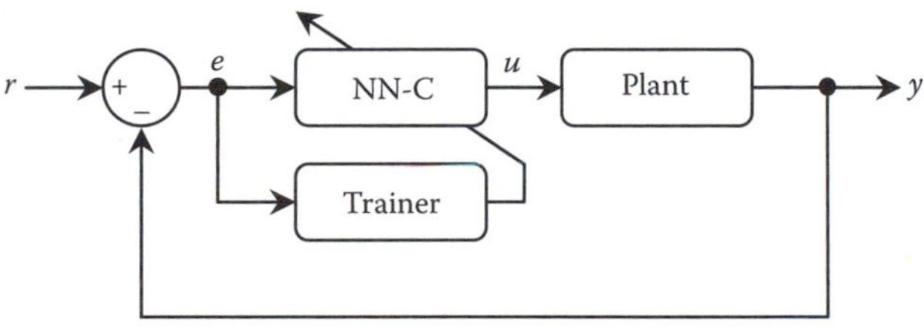

FIGURE 3.9 Closed-loop direct inverse control.

at the bottom, receives the response of the plant and tries to reconstruct the input u based on this observation. In the same time, the modified neural network is copied in front of the plant, which inverts the plant dynamics as time passes [PSY88,N97].

In Figure 3.9, closed-loop direct inverse control is depicted. The neural network acts as the feedback controller, and it receives the tracking error as the input. A trainer modifies the neural network parameters to force the plant response to what the command signal r prescribes. The tracking error is used directly as a measure of the error caused by the controller, and the scheme is called direct inversion in closed loop.

Open-loop version of the inversion method shown in Figure 3.9 is illustrated in Figure 3.10, which is also called the specialized learning architecture in the literature [PSY88]. According to the shown connectivity, the neural inverter receives the command signal (r) and outputs a control signal (u), in response to which the plant produces a response denoted by y and the error defined by $r - y$ is used to tune the parameters of the neural controller.

As highlighted by Narendra and Parthasarathy, indirect adaptive control using neural networks is done as shown in Figure 3.11. An identifier develops and refines the forward model of the plant, and a controller receives the equivalent value of the output error after passing through the neural model utilizing backpropagation technique. Such a method extracts a better error measure to penalize the

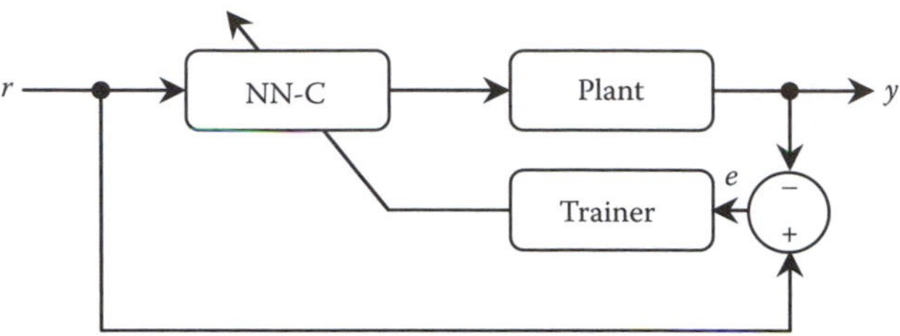

FIGURE 3.10 Specialized learning architecture.

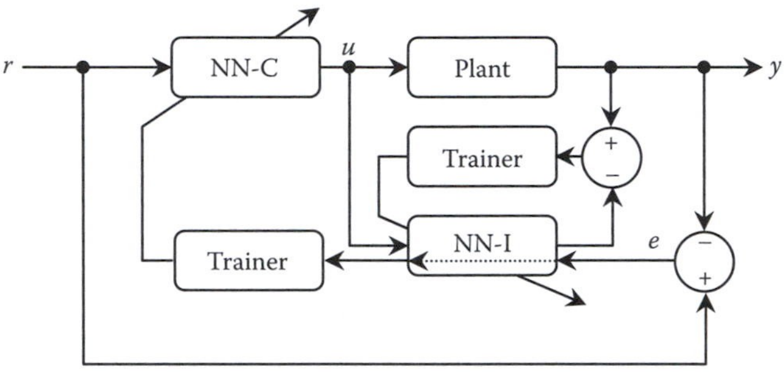

FIGURE 3.11 Indirect adaptive control scheme.

controller parameters. Since the controller tuning needs an extra propagation of information via the neural identifier, the scheme is called indirect adaptive control [NP91], or forward and inverse modeling [N97].

In many applications of neurocontrol, the plant, when discretized has the form $y_{t+1} = f(y_t, y_{t-1}, \ldots, y_{t-L_1}) + g(y_t, y_{t-1}, \ldots, y_{t-L_2})u_t$. Such models can enjoy the feedback linearization technique and when some set of consecutively collected numerical observations are available, one can proceed to develop the functions $f(\cdot)$ and $g(\cdot)$ separately, as shown in Figure 3.12. A classical control scheme supplied by the estimates of $f(\cdot)$ and $g(\cdot)$ can drive the system output to a given command signal adaptively.

Feedback error learning architecture proposed by [KFS87] is given in Figure 3.13, where a conventional controller is placed to stabilize the plant. It is assumed that the stabilizing effect provided

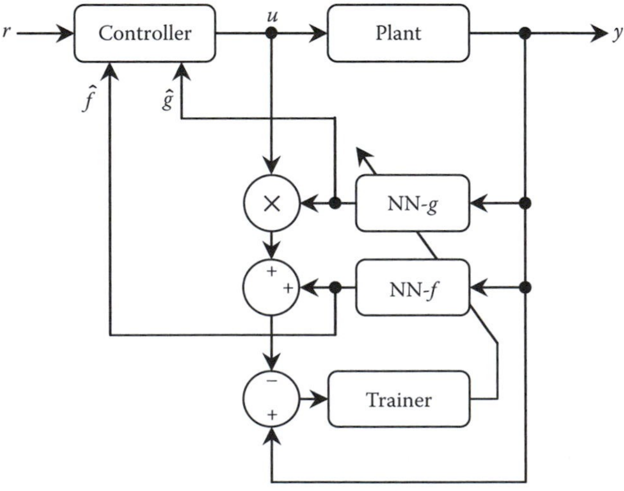

FIGURE 3.12 Feedback linearization via neural networks.

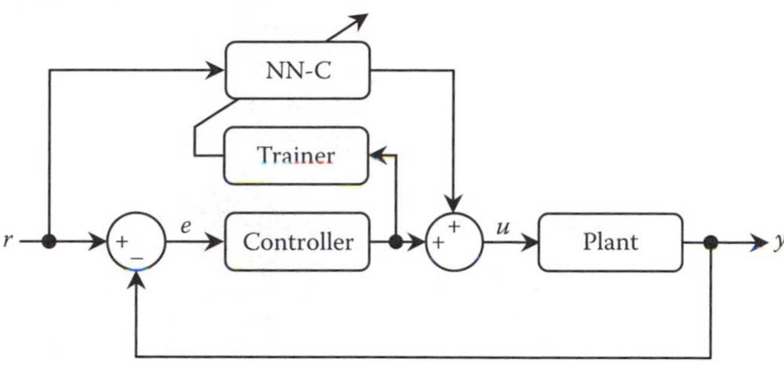

FIGURE 3.13 Feedback error learning architecture.

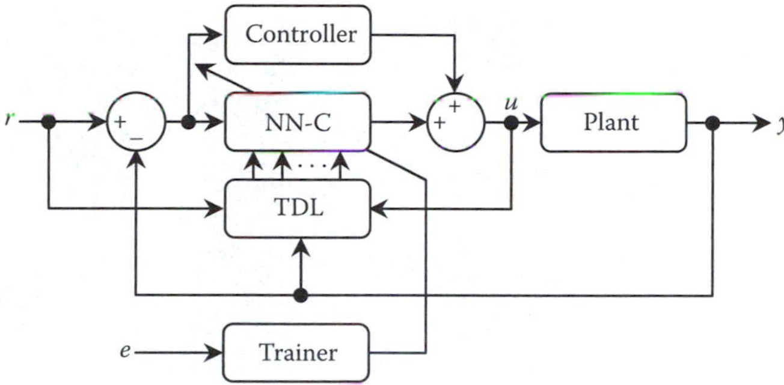

FIGURE 3.14 A typical neural network–based control architecture.

by the conventional controller may not be perfect and the neural controller provides a corrective control by acting in the feedforward path.

The typical diagrammatic view of neurocontrol is as illustrated in Figure 3.14, where there may be two controllers operated simultaneously and tapped delay line (TDL) block provides some history of relevant variables. A conventional controller—if used—maintains the stability of the closed loop and a neurocontroller complements it. Depending on the design and expectations, one of these controllers play the primary role while the other carries the auxiliary role. In every application example, a trainer should be provided an error measure quantifying the distance between the current output and the target value of it.

3.6 Application Examples

3.6.1 Propulsion Actuation Model for a Brushless DC Motor–Propeller Pair

The dynamical model of an unmanned aerial vehicle (UAV) like the one shown in Figure 3.15 could be obtained using the laws of physics. Principally, a control signal to be applied to the motors must be converted to pulse width modulation (pwm) signals, then electronic speed controllers properly drive the brushless motors, and a thrust value is obtained from each motor–propeller pair. The numerical value of the thrust is dependent upon the type of the propeller, and the angular speed of the rotor in radians is $f_i = b\Omega_i^2$ with f_i is the thrust at ith motor, b is a constant-valued thrust coefficient, and Ω_i is the angular speed in rad/s. If the control inputs (thrusts) needed to observe a desired motion were immediately available, then it would be easier to proceed to the closed-loop control system design without worrying about the effects of the actuation periphery, which introduces some constraints shaping the transient and steady-state behavior of the propulsion. Indeed, the real-time picture is much more complicated as the vehicle is an electrically powered one and battery voltage is reducing. Such a change in the battery

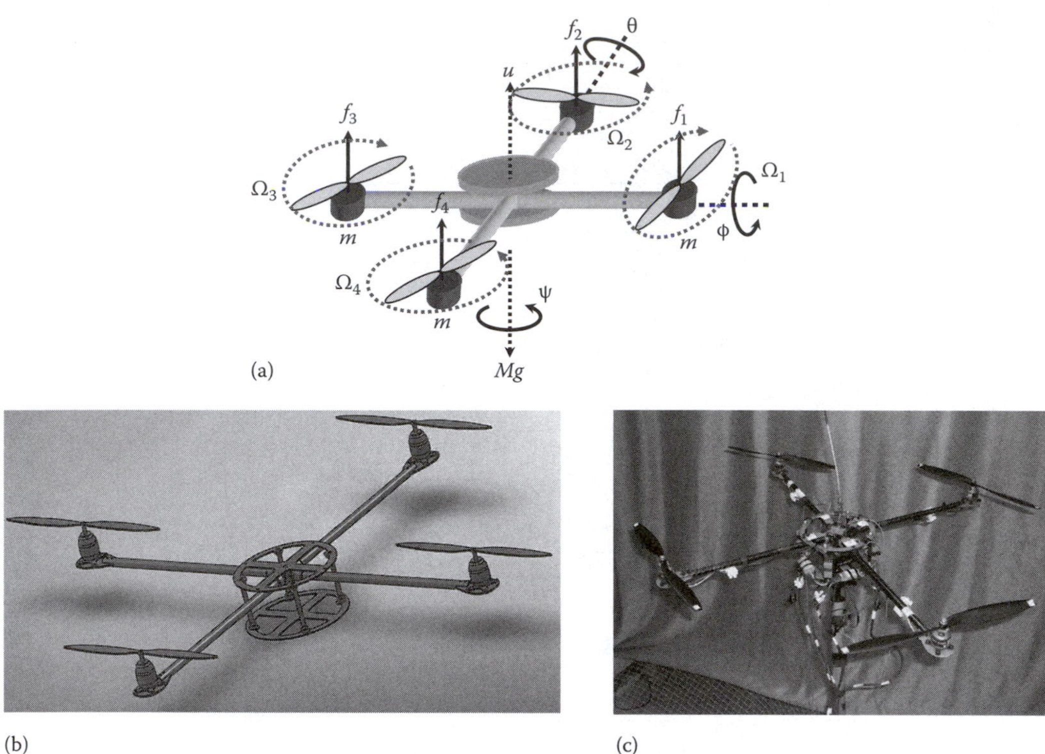

FIGURE 3.15 (a) Schematic view and variable definitions of a quadrotor-type UAV, (b) CAD drawing, and (c) real implementation.

voltage causes different lift forces at different battery voltage levels although the applied pwm level is constant, as seen in Figure 3.16. Same pwm profile is applied repetitively and as the battery voltage reduces, the angular speed at a constant pwm level decreases thereby causing a decrease in the generated thrust. Furthermore, the relation with different pwm levels is not linear, that is, same amount of change in the input causes different amounts of change at different levels, and this shows that the process to be modeled is a nonlinear one.

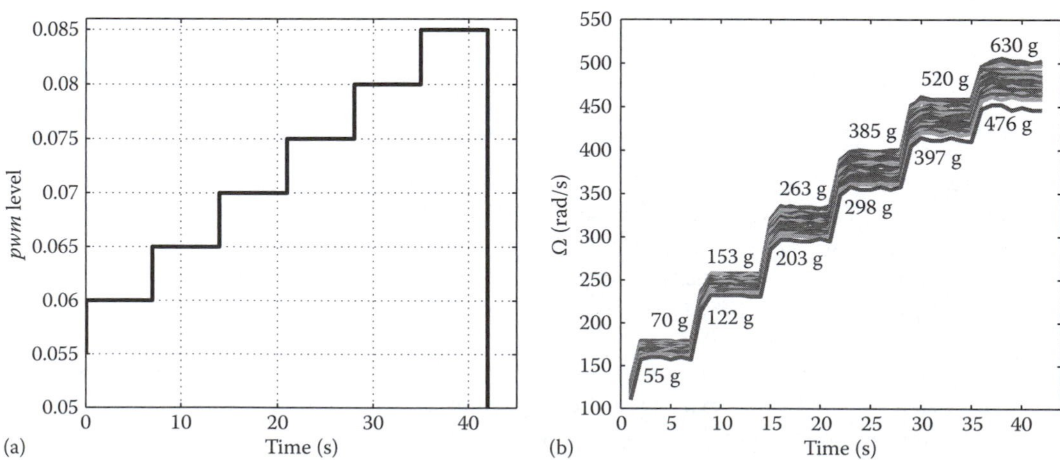

FIGURE 3.16 (a) Applied pwm profile and (b) decrease in the angular speed as the battery voltage decreases.

According to Figure 3.16, comparing the fully charged condition of the battery and the condition at the last experiment displays 15 g of difference for the lowest level, 154 g at the highest level, which is obviously an uncertainty that has to be incorporated into the dynamic model and a feedback controller appropriately. Use of neural networks is a practical alternative to resolve the problem induced by battery conditions. Denoting $V_b(t)$ as the battery voltage, a neural network model performing the map $y_{pwm} = \mathrm{NN}(\Omega_c, V_b)$ is the module installed to the output of a controller generating the necessary angular speeds. Here Ω_c is the angular speed prescribed by the controller. Another neural network that implements $y_\Omega = \mathrm{NN}(V_b, pwm, \sigma_1(pwm))$ is the module installed to the inputs of the dynamic model of the UAV. The function $\sigma_1(\cdot)$ is a low-pass filter incorporating the effect of transient in the thrust value. The dynamic model contains f_is that are computed using Ω_is.

The reason why we would like to step down from thrusts to the pwm level and step up from pwm level to forces is the fact that brushless DC motors are driven at the pwm level and one has to separate the dynamic model of the UAV and the controller by drawing a line exactly at the point of signal exchange occurring at the pwm level. Use of neural networks facilitates this in the presence of voltage loss in the batteries.

In Figure 3.17, the diagram describing the role of aforementioned offline trained neural models are shown. In Figure 3.18, the results obtained with real-time data are shown. A chirp-like pwm profile was generated and some noise added to obtain a pwm signal to be applied. When this signal is applied as an input to any motor, the variation in the battery voltage is measured and filtered to guide

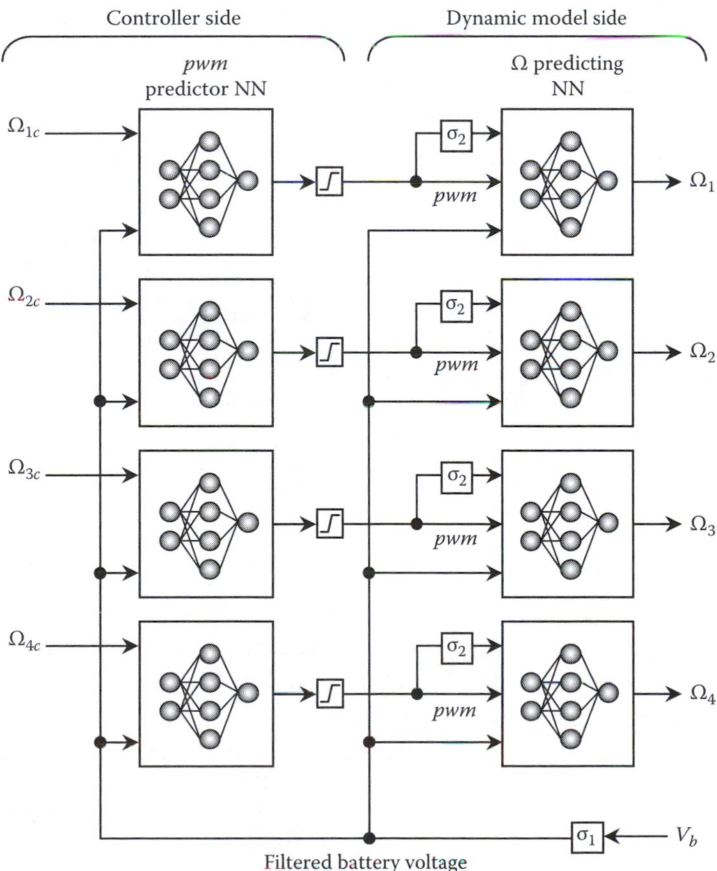

FIGURE 3.17 Installing the neural network components for handshaking at pwm level.

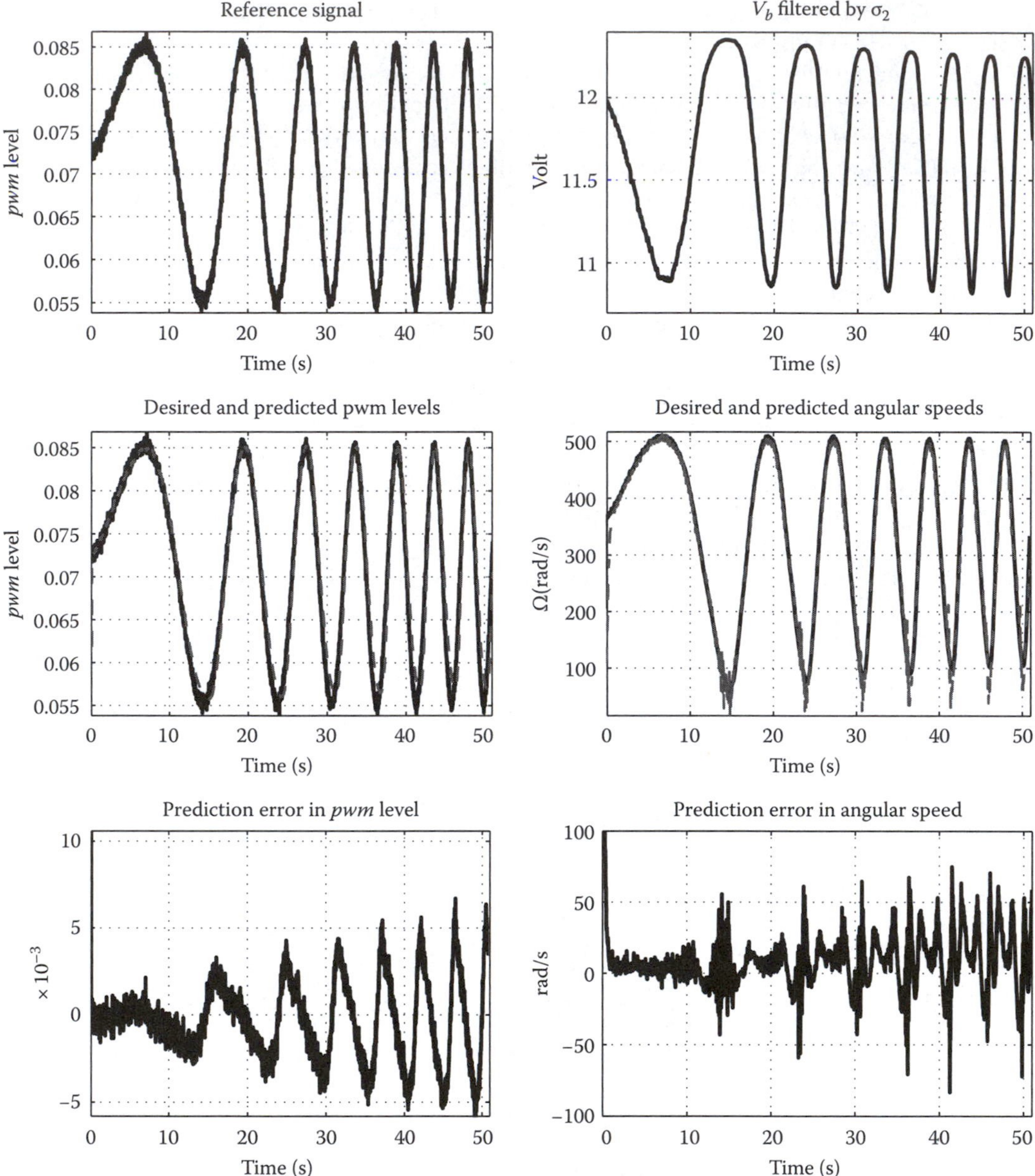

FIGURE 3.18 Performance of the two neural network models.

the neural models, as shown in the top right subplot. After that, the corresponding angular speed is computed experimentally. In the middle left subplot, the reconstructed pwm signal and the applied signal are shown together whereas the middle right subplot depicts the performance for the angular speed predicting neural model. Both subplots suggest a useful reconstruction of the signal asked from the neural networks that were trained by using Levenberg–Marquardt algorithm. In both models, the neural networks have single hidden layer with hyperbolic tangent-type neuronal nonlinearity and linear output neurons. The pwm predicting model has 12 hidden neurons with $J = 90.8285 \times 10^{-4}$ as the final cost, angular speed predicting neural model has 10 hidden neurons with $J = 3.9208 \times 10^{-4}$ as the final cost value.

Bottom subplots of Figure 3.18 illustrate the difference between the desired and predicted values. As the local frequency of the target output increases, the neural models start performing poorer yet the performance is good when the signals change slowly. This is an expected result that is in good compliance with the typical real-time signals obtained from the UAV in Figure 3.15.

3.6.2 Neural Network–Aided Control of a Quadrotor-Type UAV

The dynamic model of the quadrotor system shown in Figure 3.15 can be derived by following the Lagrange formalism. The model governing the dynamics of the system is given in (3.15) through (3.20), where the first three ODEs describe the motion in Cartesian space, the last three define the dynamics determining the attitude of the vehicle. The parameters of the dynamic model are given in Table 3.3.

$$\ddot{x} = (\cos\phi\sin\theta\cos\psi + \sin\phi\sin\psi)\frac{1}{M}U_1 \tag{3.15}$$

$$\ddot{y} = (\cos\phi\sin\theta\sin\psi - \sin\phi\cos\psi)\frac{1}{M}U_1 \tag{3.16}$$

$$\ddot{z} = -g + \cos\phi\cos\theta\frac{1}{M}U_1 \tag{3.17}$$

$$\ddot{\phi} = \dot{\theta}\dot{\psi}\left(\frac{I_{yy} - I_{zz}}{I_{xx}}\right) + \frac{j_r}{I_{xx}}\dot{\theta}\omega + \frac{L}{I_{xx}}U_2 \tag{3.18}$$

$$\ddot{\theta} = \dot{\phi}\dot{\psi}\left(\frac{I_{zz} - I_{xx}}{I_{yy}}\right) - \frac{j_r}{I_{yy}}\dot{\phi}\omega + \frac{L}{I_{yy}}U_3 \tag{3.19}$$

$$\ddot{\psi} = \dot{\theta}\dot{\phi}\left(\frac{I_{xx} - I_{yy}}{I_{zz}}\right) + \frac{1}{I_{zz}}U_4 \tag{3.20}$$

where

$$\omega = \Omega_1 - \Omega_2 + \Omega_3 - \Omega_4 \tag{3.21}$$

$$U_1 = b\Omega_1^2 + b\Omega_2^2 + b\Omega_3^2 + b\Omega_4^2 = f_1 + f_2 + f_3 + f_4 \tag{3.22}$$

TABLE 3.3 Physical Parameters of the Quadrotor UAV

L	Half distance between two motors on the same axis	0.3 m
M	Mass of the vehicle	0.8 kg
g	Gravitational acceleration constant	9.81 m/s^2
I_{xx}	Moment of inertia around x-axis	15.67e−3
I_{yy}	Moment of inertia around y-axis	15.67e−3
I_{zz}	Moment of inertia around z-axis	28.346e−3
b	Thrust coefficient	192.3208e−7 N s^2
d	Drag coefficient	4.003e−7 N m s^2
j_r	Propeller inertia coefficient	6.01e−5

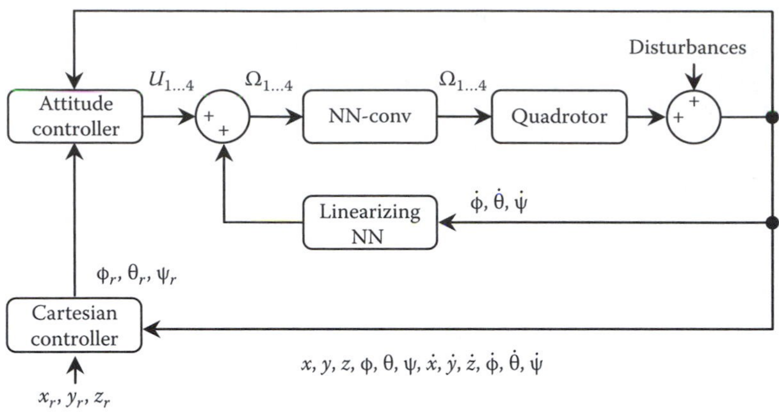

FIGURE 3.19 Block diagram of the overall control system.

$$U_2 = b\Omega_4^2 - b\Omega_2^2 = f_4 - f_2 \tag{3.23}$$

$$U_3 = b\Omega_3^2 - b\Omega_1^2 = f_3 - f_1 \tag{3.24}$$

$$U_4 = d(\Omega_1^2 - \Omega_2^2 + \Omega_3^2 - \Omega_4^2) \tag{3.25}$$

The control problem here is to drive the UAV toward a predefined trajectory in the 3D space by generating an appropriate sequence of Euler angles, which need to be controlled as well. The block diagram of the scheme is shown in Figure 3.19, where the attitude control is performed via feedback linearization using a neural network, and outer loops are established based on classical control theory.

The role of the linearizing neural network is to observe the rates of the roll ($\dot{\phi}$), pitch ($\dot{\theta}$), yaw ($\dot{\psi}$) angles, and the ω parameter, and to provide a prediction for the function given as follows:

$$\text{NN} \approx F = (\dot{\phi}, \dot{\theta}, \dot{\psi}, \omega) = \begin{pmatrix} \dot{\theta}\dot{\psi}\left(\dfrac{I_{yy} - I_{zz}}{I_{xx}}\right) + \dfrac{j_r}{I_{xx}}\dot{\theta}\omega \\ \dot{\phi}\dot{\psi}\left(\dfrac{I_{zz} - I_{xx}}{I_{yy}}\right) - \dfrac{j_r}{I_{yy}}\dot{\phi}\omega \\ \dot{\theta}\dot{\phi}\left(\dfrac{I_{xx} - I_{yy}}{I_{zz}}\right) \end{pmatrix} \tag{3.26}$$

The observations are noisy and based on the dynamic model, a total of 2000 pairs of training data and 200 validation data are generated to train the neural network. Levenberg–Marquardt training scheme is used to update the network parameters and the training was stopped after 10,000 iterations, the final MSE cost is $J = 3.3265\text{e}{-7}$, which was found acceptable. In order to incorporate the effects of noise, the training data were corrupted 5% of the measurement magnitude to maintain a good level of generality. The neural network realizing the vector function given in (3.26) has two hidden layers employing hyperbolic tangent-type neuronal activation functions, the output layer is chosen to be a linear one. The first hidden layer has 12, and the second hidden layer has 6 neurons. With the help of the neural network, the following attitude controller has been realized:

$$U_2 = \frac{I_{xx}}{L}\left(K_\phi e_\phi + 2\sqrt{K_\phi}\,\dot{e}_\phi + \ddot{\phi}_r - \text{NN}_1\right) \tag{3.27}$$

$$U_2 = \frac{I_{xx}}{L}\left(K_\theta e_\theta + 2\sqrt{K_\theta}\,\dot{e}_\theta + \ddot{\theta}_r - \mathrm{NN}_2\right) \tag{3.28}$$

$$U_4 = I_{zz}\left(K_\psi e_\psi + 2\sqrt{K_\psi}\,\dot{e}_\psi + \ddot{\psi}_r - \mathrm{NN}_3\right) \tag{3.29}$$

where

$e_\phi = \phi_r - \phi$
$e_\theta = \theta_r - \theta$
$e_\psi = \psi_r - \psi$
a variable with subscript r denotes a reference signal for the relevant variable

Since the motion in Cartesian space is realized by appropriately driving the Euler angles to their desired values, the Cartesian controller produces the necessary Euler angles for a prespecified motion in 3D space. The controller introducing this ability is given in (3.30) through (3.33). The law in (3.30) maintains the desired altitude, and upon writing it into the dynamical equations in (3.15) through (3.17) with small-angle approximation, we get the desired Euler angle values as given in (3.31) through (3.32). Specifically, turns around z-axis are not desired, and we impose $\psi_r = 0$ as given in (3.33).

$$U_1 = \frac{M(F_z + g)}{\cos\phi\cos\theta} \tag{3.30}$$

$$\phi_r = -\arctan\left(\frac{F_y}{F_z + g}\right) \tag{3.31}$$

$$\theta_r = \arctan\left(\frac{F_x}{F_z + g}\right) \tag{3.32}$$

$$\psi_r = 0 \tag{3.33}$$

where

$F_z = -4\dot{z}_r - 4(z - z_r)$
$F_y = -\dot{y}_r - (y - y_r)$
$F_x = -\dot{x}_r - (x - x_r)$
$K_\phi, K_\theta = 4, K_\psi = 9$

The results for the feedback control are obtained via simulations, as shown in Figures 3.20 and 3.21. In the upper row of Figure 3.20, the trajectory followed in the 3D space is shown first. The desired path is followed at an admissible level of accuracy under the variation of battery voltage shown in the middle subplot. The practice of brushless motor–driven actuation scheme modulates the entire system, and a very noisy battery voltage is measured. The filtered battery voltage is adequate for predicting the necessary pwm level discussed also in the previous subsection. The bottom row of Figure 3.20 shows the errors in Cartesian space. Since the primary goal of the design is to maintain a desired altitude, performance in z-direction is comparably good from the others. The errors in x- and y-directions are also due to the imperfections introduced by the small-angle approximation. Nevertheless, comparing with the trajectories in the top left subplot, the magnitudes of the shown errors are acceptable too. In Figure 3.21, the attitude of the vehicle and the errors in the Euler angles are shown. Despite the large initial errors, the controller is able to drive the vehicle attitude to its desired values quickly, and the prescribed desired attitude angles are followed with a good precision.

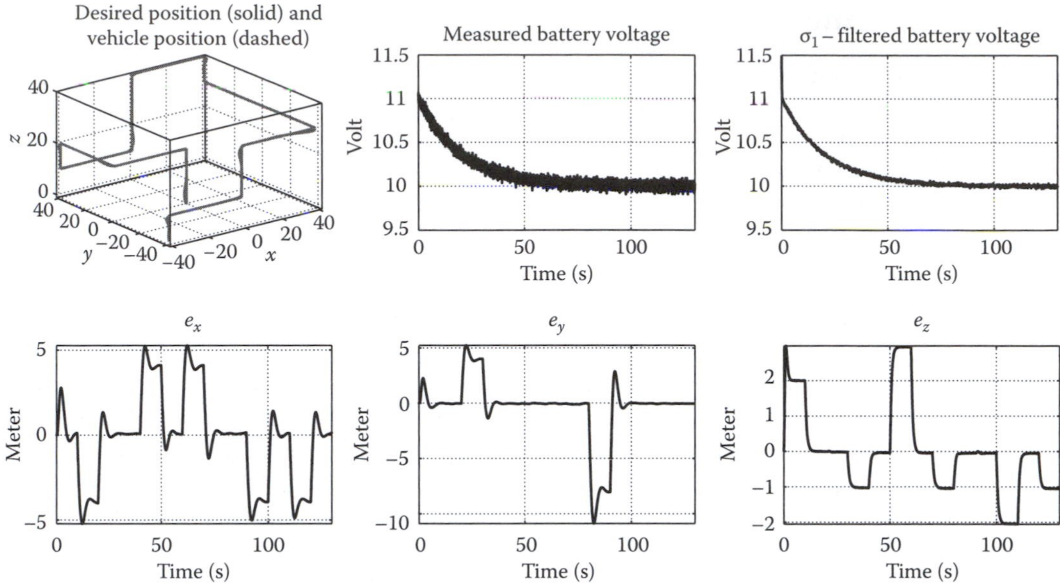

FIGURE 3.20 The trajectory followed in the Cartesian space.

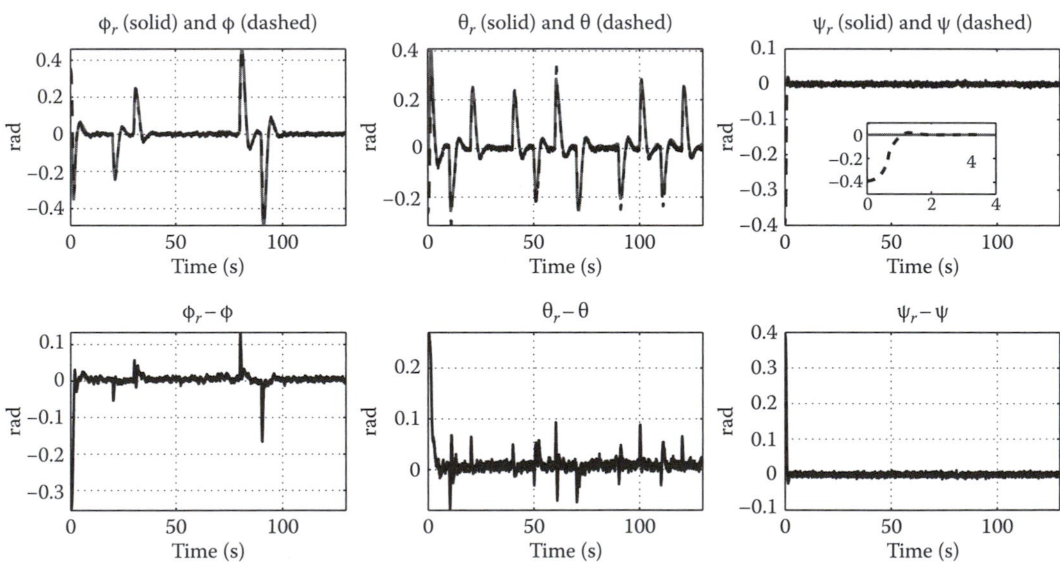

FIGURE 3.21 The attitude of the vehicle and the errors in the Euler angles.

Overall, the neural network structures in such a multivariable nonlinear feedback control example provide handshaking in between the actuation model and the dynamic model of the plant for alleviating the difficulties caused by the variations in the battery voltage; second, they provide linearization of the attitude dynamics to guide the vehicle correctly in the 3D space. In both positions, the neural networks function well enough as the application presented here is a good test bed for real-time data collection and numerical data-based approaches like neural network tools.

3.7 Concluding Remarks

Ever-increasing interest to neural networks in control systems led to the manufacturing of software tools that are of common use, hardware tools coping with computational intensity and a number of algorithms offering various types of training possibilities. Most applications of neural networks are involved with several number of variables changing in time and strong interdependencies among the variables. Such cases require a careful analysis of raw data as well as analytical tools to perform necessary manipulations. The main motivation of using neural networks in such applications is to solve the problem of constructing a function whose analytical form is missing but a set of data about it is available. With various types of architectural connectivity, activation schemes, and learning mechanisms, artificial neural networks are very powerful tools in every branch of engineering, and their importance in the discipline of control engineering is increasing parallel to the increase in the complexity being dealt with. In the future, the neural network models will continue to exist in control systems that request some degrees of intelligence to predict necessary maps.

Acknowledgments

This work is supported by TÜBİTAK, grant no. 107E137. The author gratefully acknowledges the facilities of the Unmanned Aerial Vehicles Laboratory of TOBB University of Economics and Technology.

References

[AB01] G. Arslan and T. Başar, Disturbance attenuating controller design for strict-feedback systems with structurally unknown dynamics, *Automatica*, 37, 1175–1188, 2001.

[APW91] P.J. Antsaklis, K.M. Passino, and S.J. Wang, An introduction to autonomous control systems, *IEEE Control Systems Magazine*, 11(4), 5–13, June 1991.

[ASL07] A.Y. Alanis, E.N. Sanchez, and A.G. Loukianov, Discrete-time adaptive backstepping nonlinear control via high-order neural networks, *IEEE Transactions on Neural Networks*, 18(4), 1185–1195, 2007.

[BFC09] M.K. Bugeja, S.G. Fabri, and L. Camilleri, Dual adaptive dynamic control of mobile robots using neural networks, *IEEE Transactions on Systems, Man, and Cybernetics—Part B: Cybernetics*, 39(1), 129–141, 2009.

[C89] G. Cybenko, Approximation by superpositions of a sigmoidal function, *Mathematics of Control, Signals, and Systems*, 2, 303–314, 1989.

[CCBC07] G. Colin, Y. Chamaillard, G. Bloch, and G. Corde, Neural control of fast nonlinear systems—Application to a turbocharged SI engine with VCT, *IEEE Transactions on Neural Networks*, 18(4), 1101–1114, 2007.

[CF92] C.-C. Chiang and H.-C. Fu, A variant of second-order multilayer perceptron and its application to function approximations, *International Joint Conference on Neural Networks (IJCNN'92)*, New York, vol. 3, pp. 887–892, June 7–11, 1992.

[CHI01] A.J. Calise, N. Hovakimyan, and M. Idan, Adaptive output feedback control of nonlinear systems using neural networks, *Automatica*, 37, 1201–1211, 2001.

[CN01] L. Chen and K.S. Narendra, Nonlinear adaptive control using neural networks and multiple models, *Automatica*, 37, 1245–1255, 2001.

[CSC06] M.C. Choy, D. Srinivasan, and R.L. Cheu, Neural networks for continuous online learning and control, *IEEE Transactions on Neural Networks*, 17(6), 1511–1532, 2006.

[E08] M.Ö. Efe, Novel neuronal activation functions for feedforward neural networks, *Neural Processing Letters*, 28(2), 63–79, 2008.

[EAK99] M.Ö. Efe, E. Abadoglu, and O. Kaynak, A novel analysis and design of a neural network assisted nonlinear controller for a bioreactor, *International Journal of Robust and Nonlinear Control*, 9(11), 799–815, 1999.

[F09] S. Ferrari, Multiobjective algebraic synthesis of neural control systems by implicit model following, *IEEE Transactions on Neural Networks*, 20(3), 406–419, 2009.

[F89] K. Funahashi, On the approximate realization of continuous mappings by neural networks, *Neural Networks*, 2, 183–192, 1989.

[FC07] Y. Fu and T. Chai, Nonlinear multivariable adaptive control using multiple models and neural networks, *Automatica*, 43, 1101–1110, 2007.

[GW02] S.S. Ge and J. Wang, Robust adaptive neural control for a class of perturbed strict feedback nonlinear systems, *IEEE Transactions on Neural Networks*, 13(6), 1409–1419, 2002.

[GW04] S.S. Ge and C. Wang, Adaptive neural control of uncertain MIMO nonlinear systems, *IEEE Transactions on Neural Networks*, 15(3), 674–692, 2004.

[GY01] J.Q. Gong and B. Yao, Neural network adaptive robust control of nonlinear systems in semi-strict feedback form, *Automatica*, 37, 1149–1160, 2001.

[H49] D.O. Hebb, *The Organization of Behaviour*, John Wiley & Sons, New York, 1949.

[H82] J.J. Hopfield, Neural networks and physical systems with emergent collective computational abilities, *Proceedings of the National Academy of Sciences USA*, 79, 2554–2558, 1982.

[H94] S. Haykin, *Neural Networks: A Comprehensive Foundation*, Prentice Hall PTR, Upper Saddle River, NJ, 1998.

[H99] M.A. Hussain, Review of the applications of neural networks in chemical process control—Simulation and online implementation, *Artificial Intelligence in Engineering*, 13, 55–68, 1999.

[HCL07] C.-F. Hsu, G.-M. Chen, and T.-T. Lee, Robust intelligent tracking control with PID-type learning algorithm, *Neurocomputing*, 71, 234–243, 2007.

[HGL05] F. Hong, S.S. Ge, and T.H. Lee, Practical adaptive neural control of nonlinear systems with unknown time delays, *IEEE Transactions on Systems, Man, and Cybernetics—Part B: Cybernetics*, 35(4), 849–864, 2005.

[HM94] M.T. Hagan and M.B. Menhaj, Training feedforward networks with the Marquardt algorithm, *IEEE Transactions on Neural Networks*, 5(6), 989–993, November 1994.

[HN94] K. Hara and K. Nakayamma, Comparison of activation functions in multilayer neural network for pattern classification, *IEEE World Congress on Computational Intelligence*, Orlando, FL, vol. 5, pp. 2997–3002, June 27–July 2, 1994.

[HSV89] K. Hornik, M. Stinchcombe, and H. White, Multilayer feedforward networks are universal approximators, *Neural Networks*, 2, 359–366, 1989.

[K06] V.S. Kodogiannis, Neuro-control of unmanned underwater vehicles, *International Journal of Systems Science*, 37(3),149–162, 2006.

[KA02] J. Kamruzzaman and S.M. Aziz, A note on activation function in multilayer feedforward learning, *Proceedings of the 2002 International Joint Conference on Neural Networks* (*IJCNN'02*), Honolulu, HI, vol. 1, pp. 519–523, May 12–17, 2002.

[KFS87] M. Kawato, K. Furukawa, and R. Suzuki, A hierarchical neural-network model for control and learning of voluntary movement, *Biological Cybernetics*, 57, 169–185, 1987.

[KK09] E.B. Kosmatopoulos and A. Kouvelas, Large scale nonlinear control system fine-tuning through learning, *IEEE Transactions on Neural Networks*, 20(6), 1009–1023, 2009.

[KTHD09] P. Kittisupakorn, P. Thitiyasoo, M.A. Hussain, and W. Daosud, Neural network based model predictive control for a steel pickling process, *Journal of Process Control*, 19, 579–590, 2009.

[LH07] W. Lan and J. Huang, Neural-network-based approximate output regulation of discrete-time nonlinear systems, *IEEE Transactions on Neural Networks*, 18(4), 1196–1208, 2007.

[LJY97] F.L. Lewis, S. Jagannathan, and A. Yeşildirek, Neural network control of robot arms and nonlinear systems, in O. Omidvar and D.L. Elliott, eds., *Neural Systems for Control*, pp. 161–211, Academic Press, San Diego, CA, 1997.

[LP07] K. Li and J.-X. Peng, Neural input selection—A fast model-based approach, *Neurocomputing*, 70, 762–769, 2007.

[LSS01] Y. Li, N. Sundararajan, and P. Saratchandran, Neuro-controller design for nonlinear fighter aircraft maneuver using fully tuned RBF networks, *Automatica*, 37, 1293–1301, 2001.

[M96] M.M. Polycarpou, Stable adaptive neural control scheme for nonlinear systems, *IEEE Transactions on Automatic Control*, 41(3), 447–451, 1996.

[MD05] J. Morimoto and K. Doya, Robust reinforcement learning, *Neural Computation*, 17(2), 335–359, 2005.

[MP43] W.S. McCulloch and W. Pitts, A logical calculus of the ideas immanent in nerveous activity, *Bulletin of Mathematical Biophysics*, 5, 115–133, 1943.

[N96] K.S. Narendra, Neural networks for control: Theory and practice, *Proceedings of the IEEE*, 84(10), 1385–1406, 1996.

[N97] G.W. Ng, *Application of Neural Networks to Adaptive Control of Nonlinear Systems*, Research Studies Press, Somerset, U.K., 1997.

[NP91] K.S. Narendra and K. Parthasarathy, Identification and control of dynamical systems using neural networks, *IEEE Transactions on Neural Networks*, 1, 4–27, 1990.

[P07] D. Prokhorov, Training recurrent neurocontrollers for real-time applications, *IEEE Transactions on Neural Networks*, 18(4), 1003–1015, 2007.

[P09] H.E. Psillakis, Sampled-data adaptive NN tracking control of uncertain nonlinear systems, *IEEE Transactions on Neural Networks*, 20(2), 336–355, 2009.

[PBR01] R. Padhi, S.N. Balakrishnan, and T. Randolph, Adaptive-critic based optimal neuro control synthesis for distributed parameter systems, *Automatica*, 37, 1223–1234, 2001.

[PF94] G.V. Puskorius and L.A. Feldkamp, Neurocontrol of nonlinear dynamical systems with Kalman filter trained recurrent networks, *IEEE Transactions on Neural Networks*, 5(2), 279–297, 1994.

[PKM09] J.-H. Park, S.-H. Kim, and C.-J. Moon, Adaptive neural control for strict-feedback nonlinear systems without backstepping, *IEEE Transactions on Neural Networks*, 20(7), 1204–1209, 2009.

[PL01] A.S. Poznyak and L. Ljung, On-line identification and adaptive trajectory tracking for nonlinear stochastic continuous time systems using differential neural networks, *Automatica*, 37, 1257–1268, 2001.

[PMB98] G.G. Parma, B.R. Menezes, and A.P. Braga, Sliding mode algorithm for training multilayer artificial neural networks, *Electronics Letters*, 34(1), 97–98, 1998.

[PRP07] A. Papadimitropoulos, G.A. Rovithakis, and T. Parisini, Fault detection in mechanical systems with friction phenomena: An online neural approximation approach, *IEEE Transactions on Neural Networks*, 18(4), 1067–1082, 2007.

[PS01] T. Parisini and S. Sacone, Stable hybrid control based on discrete-event automata and receding-horizon neural regulators, *Automatica*, 37, 1279–1292, 2001.

[PSY88] D. Psaltis, A. Sideris, and A.A. Yamamura, A multilayered neural network controller, *IEEE Control Systems Magazine*, pp. 17–21, April 1988.

[R01] G.A. Rovithakis, Stable adaptive neuro-control design via Lyapunov function derivative estimation, *Automatica*, 37, 1213–1221, 2001.

[R59] F. Rosenblatt, The perceptron: A probabilistic model for information storage and organization in the brain, *Psychological Review*, 65, 386–408, 1959.

[RHW86] D.E. Rumelhart, G.E. Hinton, and R.J. Williams, Learning internal representations by error propagation, in D.E. Rumelhart and J.L. McClelland, eds., *Parallel Distributed Processing: Explorations in the Microstructure of Cognition*, vol. 1, pp. 318–362, MIT Press, Cambridge, MA, 1986.

[SG00] R.S. Sexton and J.N.D. Gupta, Comparative evaluation of genetic algorithm and backpropagation for training neural networks, *Information Sciences*, 129(1–4), 45–59, 2000.

[SG07] R. Sharma and M. Gopal, A robust Markov game controller for nonlinear systems, *Applied Soft Computing*, 7, 818–827, 2007.

[SL01] R.R. Selmic and F.L. Lewis, Neural net backlash compensation with Hebbian tuning using dynamic inversion, *Automatica*, 37, 1269–1277, 2001.

[SW01] J. Si and Y.-T. Wang, On-line learning control by association and reinforcement, *IEEE Transactions on Neural Networks*, 12(2), 264–276, 2001.

[U56] A.M. Uttley, A theory of the mechanism of learning based on the computation of conditional probabilities, *Proceedings of the First International Conference on Cybernetics*, Namur, Belgium, Gauthier-Villars, Paris, France, 1956.

[VL09] D. Vrabie and F. Lewis, Neural network approach to continuous-time direct adaptive optimal control for partially unknown nonlinear systems, *Neural Networks*, 22, 237–246, 2009.

[VSKJ07] J.B. Vance, A. Singh, B.C. Kaul, S. Jagannathan, and J.A. Drallmeier, Neural network controller development and implementation for spark ignition engines with high EGR levels, *IEEE Transactions on Neural Networks*, 18(4), 1083–1100, 2007.

[W03] R.-J. Wai, Tracking control based on neural network strategy for robot manipulator, *Neurocomputing*, 51, 425–445, 2003.

[WH01] J. Wang and J. Huang, Neural network enhanced output regulation in nonlinear systems, *Automatica*, 37, 1189–1200, 2001.

[WH60] B. Widrow and M.E. Hoff Jr, Adaptive switching circuits, IRE WESCON Convention Record, pp. 96–104, 1960.

[WW01] L.-X. Wang and F. Wan, Structured neural networks for constrained model predictive control, *Automatica*, 37, 1235–1243, 2001.

[WZD97] Y. Wu, M. Zhao and X. Ding, Beyond weights adaptation: A new neuron model with trainable activation function and its supervised learning, *International Conference on Neural Networks*, Houston, TX, vol. 2, pp. 1152–1157, June 9–12, 1997.

[YEK02] X. Yu, M.Ö. Efe, and O. Kaynak, A general backpropagation algorithm for feedforward neural networks learning, *IEEE Transactions on Neural Networks*, 13(1), 251–254, January 2002.

[YG03] D.L. Yu and J.B. Gomm, Implementation of neural network predictive control to a multivariable chemical reactor, *Control Engineering Practice*, 11, 1315–1323, 2003.

[YQLR07] J. Ye, J. Qiao, M.-A. Li, and X. Ruan, A tabu based neural network learning algorithm, *Neurocomputing*, 70, 875–882, 2007.

[ZW01] Y. Zhang and J. Wang, Recurrent neural networks for nonlinear output regulation, *Automatica*, 37, 1161–1173, 2001.

4

Fuzzy Logic–Based Control Section

Mo-Yuen Chow
North Carolina State University

4.1 Introduction to Intelligent Control

For the purposes of system control, much valuable knowledge and many techniques, such as feedback control, transfer functions (frequency or discrete-time domain), state-space time-domain, optimal control, adaptive control, robust control, gain scheduling, model-reference adaptive control, etc., have been investigated and developed during the past few decades. Different important concepts such as root locus, Bode plot, phase margin, gain margin, eigenvalues, eigenvectors, pole placement, etc., have been imported from different areas to, or developed in, the control field.

However, most of these control techniques rely on system mathematical models in their design process. Control designers spend more time in obtaining an accurate system model (through techniques such as system identification, parameter estimation, component-wise modeling, etc.) than in the design of the corresponding control law. Furthermore, many control techniques, such as transfer function approach, in general, require the system to be linear and time invariant; otherwise, linearization techniques at different operating points are required in order to arrive at an acceptable control law/gain. With the use of system mathematical models, especially a linear-time invariant model, one can certainly enhance the theoretical support of the developed control techniques. However, this requirement creates another fundamental problem: How accurately does the mathematical model represent the system dynamics? In many cases, the mathematical model is only an approximated rather than an exact model of the system dynamics being investigated. This approximation may lead to a reasonable, but not necessarily good, control law for the system of interest.

For example, proportional integral (PI) control, which is simple, well known, and well suited for the control of linear time-invariant systems, has been used extensively for industrial motor control. The design process to obtain the PI gains is highly tied to the mathematical model of the motor. Engineers usually first design a PI control based on a reasonable accurate mathematical model of the motor, then use the root locus/Bode Plot technique to obtain suitable gains for the controller to achieve desirable motor performance. Then they need to tune the control gain on-line at the beginning of the use of the motor controllers in order to give acceptable motor performance for the real world. The requirement of gain tuning is mostly due to the unavoidable modeling error embedded in the mathematical models used in the design process. The motor controller may further require gain adjustments during on-line operations in order to compensate for the change in system parameters due to factors such as system degradation, change of operating conditions, etc. Adaptive control has been studied to address the changes in system parameters and have achieved a certain level of success. Gain scheduling has been studied and used in control loop [1,2] so that the motor can give satisfactory performance over a large operating range. The requirement of mathematical models imposes artificial mathematical constraints on the control design freedom. Along with the unavoidable modeling error, the resulting control laws in many cases give an overconservative motor performance.

There are also other different control techniques, such as set-point control, sliding mode control, fuzzy control, neural control, that rely less on mathematical model of the system but more on the designer's knowledge about the actual system. Especially, intelligent control has been attracting significant attention in the last few years. Different articles and experts' opinions have been reported in different technical articles. A control system, which incorporates human qualities, such as heuristic knowledge and the ability to learn, can be considered to possess a certain degree of intelligence. Such intelligent control system has an advantage over the purely analytical methods because, besides incorporating human knowledge, it is less dependent on the overall mathematical model. In fact, human beings routinely perform very complicated tasks without the aid of any mathematical representations. A simple knowledge base and the ability to learn by training seem to guide humans through even the most difficult problems. Although conventional control techniques are considered to have intelligence in a low level, we want to further develop the control algorithms from the low-level to a high-level intelligent control, through the incorporation of heuristic knowledge and learning ability via the *fuzzy* and *neural network* technologies, among others.

4.1.1 Fuzzy Control

Fuzzy control is considered an intelligent control technique and has been shown to yield promising results for many applications that are difficult to be handled by conventional techniques [3–5]. Implementations of fuzzy control in areas such as water quality control [6], automatic train operation systems [7], traffic control [13], among others, have indicated that fuzzy logic is a powerful tool in the control of mathematically ill-defined systems, which are controlled satisfactorily by human operators without the knowledge of the underlying mathematical model of the system. While conventional control methods are based on the quantitative analysis of the mathematical model of a system, fuzzy controllers focus on a linguistic description of the control action, which can be drawn, for example, from the behavior of a human operator. This can be viewed as a shift from the conventional precise mathematical control to human-like decision making, which drastically changes the approach to automate control actions.

Fuzzy logic can easily implement human experiences and preferences via *membership functions* and *fuzzy rules*, from a qualitative description to a quantitative description that is suitable for microprocessor implementation of the automation process. Fuzzy membership functions can have different shapes depending on the designer's preference and/or experience. The fuzzy rules, which describe the control strategy in a human-like fashion, are written as antecedent-consequent pairs of `IF-THEN` statements and stored in a table. Basically, there are four modes of derivation of fuzzy control rules: (1) expert experience and control engineering knowledge, (2) behavior of human operators, (3) derivation based on the fuzzy model of a process, and (4) derivation based on learning. These do not have to be mutually exclusive. In later sections, we will discuss more about *membership functions* and *fuzzy rules*.

Due to the use of *linguistic variables* and *fuzzy rules*, the fuzzy controller can be made understandable to a nonexpert operator. Moreover, the description of the control strategy could be derived by examining the behavior of a conventional controller. The fuzzy characteristics make it particularly attractive for control applications because only a linguistic description of the appropriate control strategy is needed in order to obtain the actual numerical control values. Thus, fuzzy logic can be used as a general methodology to incorporate knowledge, heuristics or theory into a controller.

In addition, fuzzy logic has the freedom to completely define the control surface without the use of complex mathematical analysis, as discussed in later sections. On the other hand, the amount of effort involved in producing an acceptable rule base and in fine-tuning the fuzzy controller is directly proportional to the number of quantization levels used, and the designer is left to choose the best tradeoff between (1) being able to create a large number of features on the control surface and (2) not having to spend much time in the fine-tuning process. The general shape of these features depends on the heuristic rule base and the configuration of the membership functions. Being able to quantize the domain of the control surface using linguistic variables allows the designer to depart from the mathematical constraints (e.g., hyperplane constraints in PI control) and achieve a control surface, which has more features and contours.

In 1965, L.A. Zadeh laid the foundations of *fuzzy set theory* [8], which is a generalization of conventional set theory, as a method of dealing with the imprecision of the real physical world. Bellman and Zadeh write "Much of the decision-making in the real world takes place in an environment in which the goals, the constraints and the consequences of possible actions are not known precisely" [9]. This "imprecision" or fuzziness is the core of fuzzy logic. Fuzzy control is the technology that applies fuzzy logic to solve control problems.

This section is written to provide readers with an introduction of the use of fuzzy logic to solve control problems; it also intends to provide information for further exploration on related topics. In this section, we will briefly overview some fundamental fuzzy logic concepts and operations and then apply the fuzzy logic technique for a dc motor control system to demonstrate the fuzzy control design procedure. The advantages of using fuzzy control is more substantial when applied to nonlinear and ill-defined systems. The dc motor control system is used as an illustration here because most readers with some control background should be familiar with this popular control example, so that they can benefit more from the fuzzy control design process explained here. If the readers are interested in getting more details about fuzzy logic and fuzzy control, please refer to the bibliography section at the end of this chapter.

4.2 Brief Description of Fuzzy Logic

4.2.1 Crisp Set

The basic principle of the conventional set theory is that an element is either a member or not a member of a set. A set that is defined in this way is called a *crisp* set, since its boundary is well defined. Consider the set, W, of motor speed operating range between 0 and 175 rad/s. The proper motor speed operating range would be written as

$$W = \{w \in W \mid 0\ \text{rad/s} \le w \le 175\ \text{rad/s}\}. \tag{4.1}$$

The set W could be expressed by its membership function $W(w)$, which indicates whether the motor is within its operating range:

$$W(w) = \begin{cases} 1; & 0\ \text{rad/s} \le w \le 150\ \text{rad/s} \\ 0; & \text{otherwise} \end{cases}. \tag{4.2}$$

A graphical representation of $W(w)$ is shown in Figure 4.1.

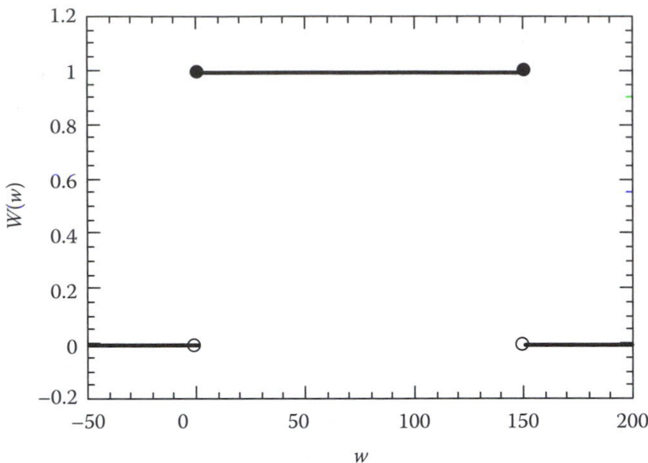

FIGURE 4.1 Crisp membership function for proper speed operating condition defined in Equation 4.1.

4.2.2 Fuzzy Set

However, in the real world, sets are not always so crisp. Human beings routinely use concepts that are approximate and imprecise to describe most problems and events. The human natural language largely reflects these approximations. For example, the meaning of the word "fast" to describe the motor speed depends on the person who uses it and the context he or she uses it in. Hence, various speeds may belong to the set "fast" to various degrees, and the boundary of the set operating range is not very precise. For example, if we wish to consider 150 rad/s as fast, then 160 rad/s is certainty fast. However, do we still say 147 rad/s is fast too? How about 145 rad/s? Conventional set theory techniques have difficulties in dealing with this type of linguistic or qualitative problems.

Fuzzy sets are very useful in representing *linguistic variables*, which are quantities described by natural or artificial language [10] and whose values are linguistic (qualitative) and not numeric (quantitative). A linguistic variable can be considered either as a variable whose value is a fuzzy number (after *fuzzification*) or as a variable whose values are defined in linguistic terms. Examples of linguistic variables are fast, slow, tall, short, young, old, very tall, very short, etc. More specifically, the basic idea underlying the fuzzy set theory is that an element is a member of a set to a certain degree, which is called the *membership grade* (*value*) of the element in the set. Let U be a collection of elements denoted by $\{u\}$, which could be discrete or continuous. U is called the *universe of discourse* and u represents the generic element of U. A *fuzzy set A* in a universe of discourse U is then characterized by a *membership function* $A(.)$ that maps U onto a real number in the interval $[A_{min}, A_{max}]$. If $A_{min} = 0$ and $A_{max} = 1$, the membership function is called a *normalized* membership function and A: $U \in [0,1]$.

For example, a membership value $A(u) = 0.8$ suggests that u is a member of A to a degree of 0.8, on a scale where zero is no membership and one is complete membership. One can then see that crisp set theory is just a special case of fuzzy set theory. A fuzzy set A in U can be represented as a set of ordered pairs of an element u and its membership value in A: $A = \{(u, A(u)) | u \in U\}$. The element u is sometimes called the support, while $A(u)$ is the corresponding membership function of u of the fuzzy set A. When U is continuous, the common notation used to represent the fuzzy set A is

$$A = \int_U \frac{A(u)}{u}, \tag{4.3}$$

and when U is discrete, the fuzzy set A is normally represented as

$$A = \sum_{i=1}^{n} \frac{A(u_i)}{u_i} = \frac{A(u_1)}{u_1} + \frac{A(u_2)}{u_2} + \cdots + \frac{A(u_n)}{u_n}, \tag{4.4}$$

where n is the number of supports in the set. It is to be noted that, in fuzzy logic notation, the summation represents union, not addition. Also, the membership values are "tied" to their actual values by the divide sign, but they are not actually being divided.

4.3 Qualitative (Linguistic) to Quantitative Description

In this section, we use *height* as an example to explain the concept of fuzzy logic. In later sections, we will apply the same concept to the motor speed control. The word "tall" may refer to different heights depending on the person who uses it and the context in which he or she uses. Hence, various heights may belong to the set "tall" to various degrees, and the boundary of the set tall is not very precise. Let us consider the linguistic variable "tall" and assign values in the set 0 to 1. A person 7 ft tall may be considered "tall" with a value of 1. Certainly, any one over 7 ft tall would also be considered tall with a membership of 1. A person 6 ft tall may be considered "tall" with a value of 0.5, while a person 5 ft tall may be considered tall with a membership of 0. As an example, the membership function μ_{TALL}, which maps the values between 4 and 8 ft into the fuzzy set "tall," is described continuously by Equation 4.5 and shown in Figure 4.2:

$$\text{TALL} = \mu_{\text{TALL}}(\text{height}) = \int_{[4,8]} \frac{1/\left(1+e^{-5(\text{height}-6)}\right)}{\text{height}}, \tag{4.5}$$

Membership functions are very often represented by the discrete fuzzy membership notation when membership values can be obtained for different supports based on collected data. For example, a five-player basketball team may have a fuzzy set "TALL" defined by

$$\text{TALL} = \mu_{\text{TALL}}(\text{height}) = \frac{0.5}{6} + \frac{0.68}{6.2} + \frac{0.82}{6.4} + \frac{0.92}{6.6} + \frac{0.98}{6.8} + \frac{1.0}{7}, \tag{4.6}$$

where the heights are listed in feet. This membership function is shown in Figure 4.3.

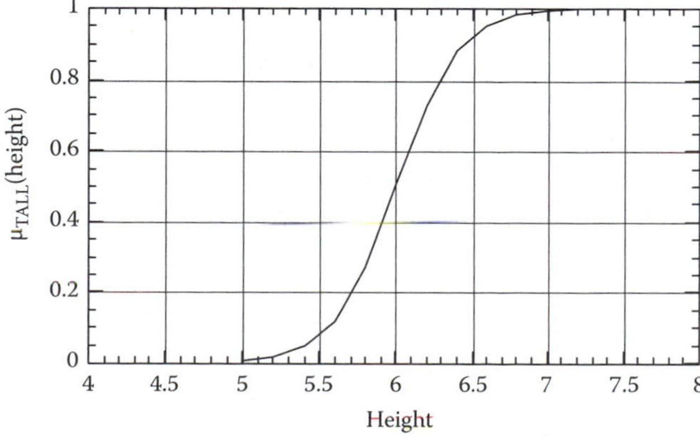

FIGURE 4.2 Continuous membership function plot for linguistic variable "TALL" defined in Equation 4.5.

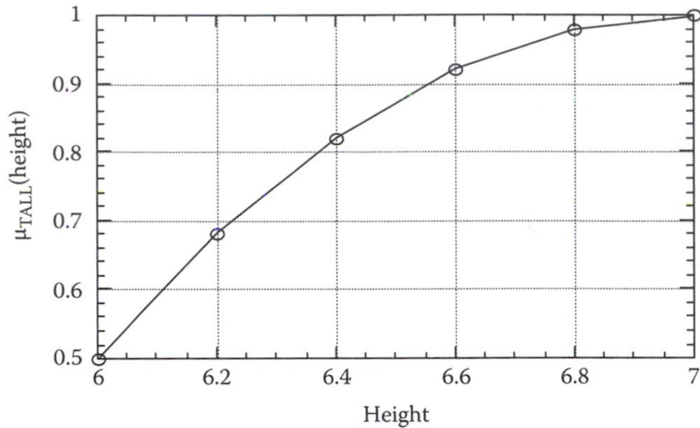

FIGURE 4.3 Discrete membership function plot for linguistic variable "TALL" defined in Equation 4.6.

In the discrete fuzzy set notation, linear interpolation between two closest known supports and corresponding membership values is often used to compute the membership value for the u that is not listed. For example, if a new player joins the team and his height is 6.1 ft, then his TALL membership value based on the membership function defined in (4.6) is

$$\mu_{\text{TALL}}(6.1) = 0.5 + \frac{6.1 - 6(0.68 - 0.5)}{6.2 - 6} = 0.59. \tag{4.7}$$

Note that the membership function defined in Equation 4.5 is normalized between 0 and 1 while the one defined in Equation 4.6 is not, because its minimum membership value is 0.5 rather than 0.

It should be noted that the choice of a membership function relies on the actual situation and is very much based on heuristic and educated judgment. In addition, the imprecision of fuzzy set theory is different from the imprecision dealt with by probability theory. The fundamental difference is that probability theory deals with randomness of future events due to the possibility that a particular event may or may not occur, whereas fuzzy set theory deals with imprecision in current or past events due to the vagueness of a concept (the membership or nonmembership of an object in a set with imprecise boundaries) [11].

4.4 Fuzzy Operations

The combination of membership functions requires some form of set operations. Three basic operations of conventional set theory are intersection, union, and complement. The fuzzy logic counterparts to these operations are similar to those of conventional set theory. Fuzzy set operations such as union, intersection, and complement are defined in terms of the membership functions. Let A and B be two fuzzy sets with membership functions μ_A and μ_B, respectively, defined for all $u \in U$. The TALL membership function has been defined in Equation 4.6, and a FAST membership function is defined as (Figure 4.4)

$$\text{FAST} = \mu_{\text{FAST}}(\text{height}) = \frac{0.5}{6} + \frac{0.8}{6.2} + \frac{0.9}{6.4} + \frac{1}{6} + \frac{0.75}{6.8} + \frac{0.6}{7}. \tag{4.8}$$

These two membership functions will be used in the next sections as examples for fuzzy operation discussions.

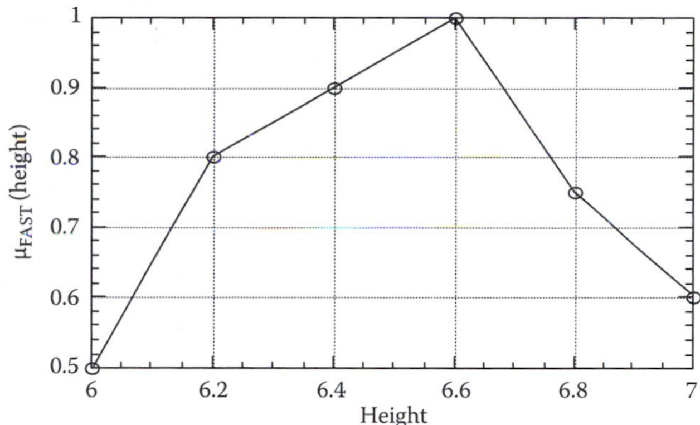

FIGURE 4.4 Discrete membership function plot for linguistic variable "FAST" defined in Equation 4.8.

4.4.1 Union

The membership function $\mu_{A\cup B}$ of the union $A \cup B$ is pointwise defined for all $u \in U$ by

$$\mu_{A\cup B}(u) = \max\{\mu_A(u), \mu_B(u)\} = \underset{u}{\vee}\{\mu_A(u), \mu_B(u)\}, \tag{4.9}$$

where $\vee$ symbolizes the max operator. As an example, a fuzzy set could be composed of basketball players who are "tall *or* fast." The membership function for this fuzzy set could be expressed in the union notation as

$$\begin{aligned}\mu_{\text{TALL}\cup\text{FAST}}(\text{height}) &= \max\left\{\mu_{\text{TALL}}(\text{height}), \mu_{\text{FAST}}(\text{height})\right\}\\ &= \frac{(0.5 \vee 0.5)}{6} + \frac{(0.68 \vee 0.8)}{6.2} + \frac{(0.82 \vee 0.9)}{6.4} + \frac{(0.92 \vee 1)}{6.6} + \frac{(0.98 \vee 0.98)}{6.8} + \frac{(0.6 \vee 1)}{7}\\ &= \frac{0.5}{6} + \frac{0.8}{6.2} + \frac{0.9}{6.4} + \frac{1}{6.6} + \frac{0.98}{6.8} + \frac{1}{7}.\end{aligned} \tag{4.10}$$

The corresponding membership function is shown in Figure 4.5.

4.4.2 Intersection

The membership function $\mu_{A\cap B}$ of the intersection $A \cap B$ is pointwise defined for all $u \in U$ by

$$\mu_{A\cap B}(u) = \min\{\mu_A(u), \mu_B(u)\} = \underset{u}{\wedge}\{\mu_A(u), \mu_B(u)\}, \tag{4.11}$$

where $\wedge$ symbolizes the min operator. As an example, a fuzzy set could be composed of basketball players who are "TALL *and* FAST." The membership function for this fuzzy set could be expressed in the intersection notation as

$$\begin{aligned}\mu_{\text{TALL}\cap\text{FAST}}(\text{height}) &= \min\left\{\mu_{\text{TALL}}(\text{height}), \mu_{\text{FAST}}(\text{height})\right\}\\ &= \frac{(0.5 \wedge 0.5)}{6} + \frac{(0.68 \wedge 0.8)}{6.2} + \frac{(0.82 \wedge 0.9)}{6.4} + \frac{(0.92 \wedge 1)}{6.6} + \frac{(0.98 \wedge 0.98)}{6.8} + \frac{(0.6 \wedge 1)}{7}\\ &= \frac{0.5}{6} + \frac{0.68}{6.2} + \frac{0.82}{6.4} + \frac{0.92}{6.6} + \frac{0.75}{6.8} + \frac{0.6}{7}.\end{aligned} \tag{4.12}$$

The corresponding membership function is shown in Figure 4.6.

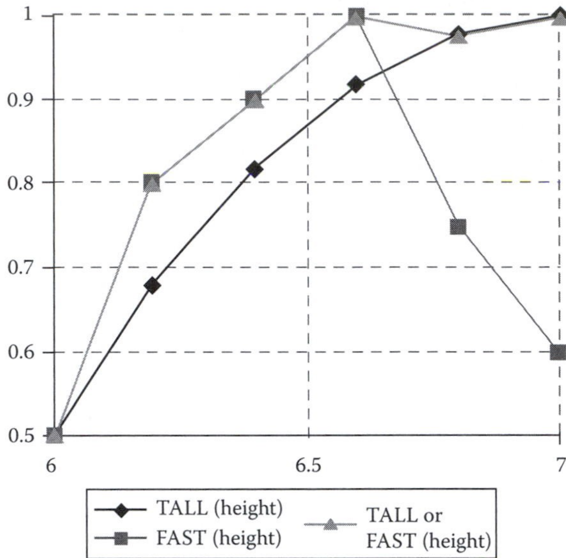

FIGURE 4.5 Membership function plot of $\mu_{\text{TALL}\cup\text{FAST}}$(height) defined in (4.10).

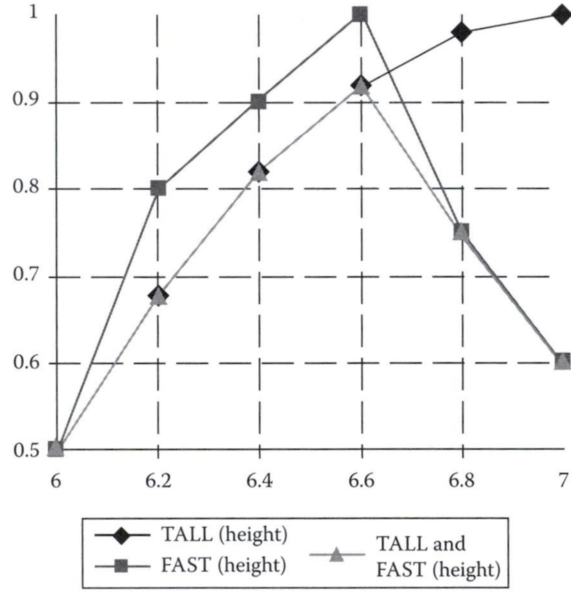

FIGURE 4.6 Membership function plot of $\mu_{\text{TALL}\cap\text{FAST}}$(height) defined in Equation 4.12.

4.4.3 Complement

Also, the membership function $\mu_{\bar{A}}$ of complement of the fuzzy set A is pointwise defined for all $u \in U$ by

$$\mu_{\bar{A}}(u) = 1 - \mu_A(u). \quad (4.13)$$

The fuzzy set "short" could be considered to be the complement of the fuzzy set TALL expressed as

$$\mu_{\text{SHORT}}(u) = \mu_{\overline{\text{TALL}}}(u) = 1 - \mu_{\text{TALL}}(u). \quad (4.14)$$

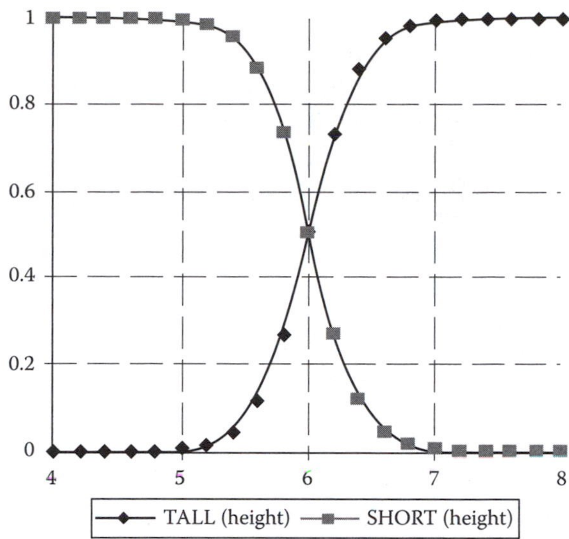

FIGURE 4.7 Membership functions μ_{SHORT}(height) and μ_{TALL}(height) defined in (4.5) and (4.14).

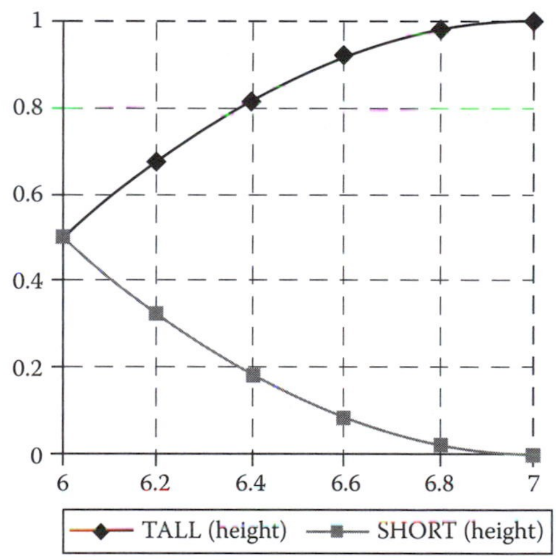

FIGURE 4.8 Membership functions μ_{SHORT}(height) and μ_{TALL}(height) defined in (4.6) and (4.14).

The SHORT fuzzy set correspondent to the TALL fuzzy set defined in Equation 4.5 is shown in Figure 4.7. The SHORT fuzzy set correspondent to the TALL fuzzy set defined in Equation 4.6 is shown in Figure 4.8.

Other operations exist for fuzzy sets. Different modifications of the union, intersection, and complement have also been proposed and used. For example, unions and intersections of various strengths can be achieved by using the Yager class (among others) to perform the fuzzy union and intersection operations. But the three operations described above represent the most popular ones.

4.5 Fuzzy Rules, Inference

4.5.1 Fuzzy Relation/Composition/Conditional Statement

Another concept of fuzzy logic is the fuzzy relation. Fuzzy logic techniques can be used to translate natural language into heuristic responses and also to combine fuzzy membership functions to formulate fuzzy rules. A *fuzzy relation* R from a set X to a set Y is defined as a fuzzy subset of the Cartesian product $X \times Y$, which is the collection of ordered pairs (x, y), where $x \in X$, and $y \in Y$. In the same way that a membership function defines an element's membership in a given set, the fuzzy relation is a fuzzy set composed of all combinations of participating sets, which determines the degree of association between two or more elements of distinct sets. It is characterized by a bivariate membership function $\mu_R(x, y)$, written as

$$R_Y(x) = \int_{X \times Y} \frac{\mu_R(x, y)}{(x, y)} \text{ in continuous membership function notation,} \tag{4.15}$$

and

$$R_Y(x) = \sum_{i=1}^{n} \frac{\mu_R(x_i, y_i)}{(x_i, y_i)} \text{ in discrete membership function notation.} \tag{4.16}$$

Let us consider a fuzzy rule:

If a player is much taller than 6.4 feet, his scoring average per game should be high. (4.17)

Let the fuzzy sets be

$$X = \text{much taller than } 6.4\,\text{ft (with } [6\,\text{ft}, 7\,\text{ft}] \text{ as the universe of discourse)}$$

$$= \frac{0}{6} + \frac{0}{6.2} + \frac{0.2}{6.4} + \frac{0.5}{6.6} + \frac{0.8}{6.8} + \frac{1}{7}, \tag{4.18}$$

and

$$Y = \text{scoring-average per game (with } [0, 20] \text{ as the universe of discourse)}$$

$$= \frac{0}{0} + \frac{0.3}{5} + \frac{0.8}{10} + \frac{0.9}{15} + \frac{1}{20} \tag{4.19}$$

If we use the min operator to form the R (other operators such as "product" is another popular choice [12]), then the consequent relational matrix between X and Y is

$$R_Y(x) = \begin{array}{c|ccccc} X \backslash Y & 0 & 5 & 10 & 15 & 20 \\ \hline 6 & 0 \wedge 0 & 0 \wedge 0.3 & 0 \wedge 0.8 & 0 \wedge 0.9 & 0 \wedge 1 \\ 6.2 & 0 \wedge 0 & 0 \wedge 0.3 & 0 \wedge 0.8 & 0 \wedge 0.9 & 0 \wedge 1 \\ 6.4 & 0.2 \wedge 0 & 0.2 \wedge 0.3 & 0.2 \wedge 0.8 & 0.2 \wedge 0.9 & 0.2 \wedge 1 \\ 6.6 & 0.5 \wedge 0 & 0.5 \wedge 0.3 & 0.5 \wedge 0.8 & 0.5 \wedge 0.9 & 0.5 \wedge 1 \\ 6.8 & 0.8 \wedge 0 & 0.8 \wedge 0.3 & 0.8 \wedge 0.8 & 0.8 \wedge 0.9 & 0.8 \wedge 1 \\ 7 & 1 \wedge 0 & 1 \wedge 0.3 & 1 \wedge 0.8 & 1 \wedge 0.9 & 1 \wedge 1 \end{array}$$

$$= \begin{array}{c|ccccc} X \backslash Y & 0 & 5 & 10 & 15 & 20 \\ \hline 6 & 0 & 0 & 0 & 0 & 0 \\ 6.2 & 0 & 0 & 0 & 0 & 0 \\ 6.4 & 0 & 0.2 & 0.2 & 0.2 & 0.2 \\ 6.6 & 0 & 0.3 & 0.5 & 0.5 & 0.5 \\ 6.8 & 0 & 0.3 & 0.8 & 0.8 & 0.8 \\ 7 & 0 & 0.3 & 0.8 & 0.9 & 1 \end{array} \tag{4.20}$$

Note that in (4.20), min $R_Y(x) = 0$, and max $R_Y(x) = 1$. Therefore, a player with height 6.8 ft and a scoring average of 10 per game should has a fuzzy relation 0.8 in a scale between 0 and 1.

4.5.2 Compositional Rule of Inference

Compositional rule of inference, which may be regarded as an extension of the familiar rule of modus ponens in classical propositional logic, is another important concept. Specifically, if R is a fuzzy relation from X to Y, and x is a fuzzy subset of X, then the fuzzy subset y of Y that is induced by x is given by the composition of R and x, in the sense of Equation 4.12. Also, if R is a relation from X to Y and S is a relation from Y to Z, then the *composition* of R and S is a fuzzy relation denoted by $R \circ S$, defined by

$$R \circ S \equiv \int_{X \times Z} \frac{\max\limits_{y}\left(\min\left(\mu_R(x,y), \mu_S(y,z)\right)\right)}{(x,z)} \tag{4.21}$$

Equation 4.21 is called the max–min composition of R and S.

For example, if a player is about 6.8 ft with membership function

$$x = \frac{0}{6} + \frac{0}{6.2} + \frac{0}{6.4} + \frac{0.2}{6.6} + \frac{1}{6.8} + \frac{0.2}{7}, \tag{4.22}$$

and we are interested in his corresponding membership function of scoring average per game, then

$$\begin{aligned}
\mu_Y = \max_{y_i}\{\min(x \circ R_Y(x))\} &= \max_{y_i}\left\{\begin{bmatrix} 0 & 0 & 0 & 0.2 & 1 & 0.2 \end{bmatrix} \wedge \begin{bmatrix} 0 & 0 & 0 & 0 & 0 \\ 0 & 0 & 0 & 0 & 0 \\ 0 & 0.2 & 0.2 & 0.2 & 0.2 \\ 0 & 0.3 & 0.5 & 0.5 & 0.5 \\ 0 & 0.3 & 0.8 & 0.8 & 0.8 \\ 0 & 0.3 & 0.8 & 0.9 & 1 \end{bmatrix}\right\} \\
&= \max_{y_i}\left\{\begin{bmatrix} 0\wedge 0 & 0\wedge 0 & 0\wedge 0 & 0\wedge 0 & 0\wedge 0 \\ 0\wedge 0 & 0\wedge 0 & 0\wedge 0 & 0\wedge 0 & 0\wedge 0 \\ 0\wedge 0 & 0\wedge 0.2 & 0\wedge 0.2 & 0\wedge 0.2 & 0\wedge 0.2 \\ 0.2\wedge 0 & 0.2\wedge 0.3 & 0.2\wedge 0.5 & 0.2\wedge 0.5 & 0.2\wedge 0.5 \\ 1\wedge 0 & 1\wedge 0.3 & 1\wedge 0.8 & 1\wedge 0.8 & 1\wedge 0.8 \\ 0.2\wedge 0 & 0.2\wedge 0.3 & 0.2\wedge 0.8 & 0.2\wedge 0.9 & 0.2\wedge 1 \end{bmatrix}\right\} \\
&= \max_{y_i}\left\{\begin{bmatrix} 0 & 0 & 0 & 0 & 0 \\ 0 & 0 & 0 & 0 & 0 \\ 0 & 0 & 0 & 0 & 0 \\ 0 & 0.2 & 0.2 & 0.2 & 0.2 \\ 0 & 0.3 & 0.8 & 0.8 & 0.8 \\ 0 & 0.2 & 0.2 & 0.2 & 0.2 \end{bmatrix}\right\} = \begin{bmatrix} 0\vee 0\vee 0\vee 0\vee 0 \\ 0\vee 0\vee 0\vee 0\vee 0 \\ 0\vee 0\vee 0\vee 0\vee 0 \\ 0\vee 0.2\vee 0.2\vee 0.2\vee 0.2 \\ 0\vee 0.3\vee 0.8\vee 0.8\vee 0.8 \\ 0\vee 0.2\vee 0.2\vee 0.2\vee 0.2 \end{bmatrix} = \begin{bmatrix} 0 \\ 0 \\ 0 \\ 0.2 \\ 0.8 \\ 0.2 \end{bmatrix}
\end{aligned} \tag{4.23}$$

4.5.3 Defuzzification

After a fuzzy rule base has been formed (such as the one shown in Equation 4.17) along with the participating membership functions (such as the ones shown in Equations 4.22 and 4.23), a defuzzification strategy needs to be implemented. The purpose of defuzzification is to convert the results of the rules and membership functions into usable values, whether it be a specific result or a control input. There are a few techniques for the defuzzification process. One popular technique is the "center-of-gravity" method, which is capable of considering the influences of many different effects simultaneously. The general form of the center of gravity method is

$$f_j = \frac{\sum_{i=1}^{N} \mu_i \times c_i}{\sum_{i=1}^{N} \mu_i}, \tag{4.24}$$

where

N is the number of rules under consideration
μ_i and c_i are the membership and the control action associated with rule i, respectively
f_j represents the jth control output

Consider the rule base

1. if X = tall and Y = fast then $C = 1$.
2. if X = short and Y = slow then $C = 0$. (4.25)

The "center-of-gravity" method applied to this rule base with $N = 2$ results in

$$f_i = \frac{\min(\mu_{\text{TALL}}(X), \mu_{\text{FAST}}(Y)) \times 1 + \min(\mu_{\text{SHORT}}(X), \mu_{\text{SLOW}}(Y)) \times 0}{\min(\mu_{\text{TALL}}(X), \mu_{\text{FAST}}(Y)) + \min(\mu_{\text{SHORT}}(X), \mu_{\text{SLOW}}(Y))}, \tag{4.26}$$

where the "min" operator has been used in association with the "and" operation. We will discuss more about the details of the defuzzification process in later sections.

4.6 Fuzzy Control

The concepts outlined above represent the basic foundation upon which fuzzy control is built. In fact, the potential of fuzzy logic in control systems was shown very early by Mamdani and his colleagues [13]. Since then, fuzzy logic has been used successfully in a variety of control applications. Since the heuristic knowledge about how to control a given plant is often in the form of linguistic rules provided by a human expert or operator, fuzzy logic provides an effective means of translating that knowledge into an actual control signal. These rules are usually of the form

IF (a set of conditions is satisfied) THEN (the adequate control action is taken),

where the conditions to be satisfied are the *antecedents* and the control actions are the *consequent* of the fuzzy control rules, both of which are associated with fuzzy concepts (linguistic variables). Several linguistic variables may be involved in the antecedents or the consequents of a fuzzy rule, depending on how many variables are involved in the control problem. For example, let x and y represent two important state variables of a process, and let w and z be the two control variables for the process. In this case, fuzzy control rules have the form

Rule$_1$: if x is A_1 and y is B_1 then w is C_1 and z is D_1
Rule$_2$: if x is A_2 and y is B_2 then w is C_2 and z is D_2
....
....
Rule$_n$: if x is A_n and y is B_n then w is C_n and z is D_n

where A_i, B_i, C_i, and D_i are the linguistic values (fuzzy sets) of x, y, w, and z, in the universes of discourse X, Y, W, and Z, respectively, with $i = 1,2,\ldots,n$. Using the concepts of fuzzy conditional statements and the compositional rule of inference, w and z (fuzzy subsets of W and Z, respectively) can be inferred from each fuzzy control rule.

Typically, though, in control problems, the values of the state variables and of the control signals are represented by real numbers, not fuzzy sets. Therefore, to convert real information into fuzzy sets, and vice versa, it is necessary to convert fuzzy sets into real numbers. These two conversion processes are generally called *fuzzification* and *defuzzification*, respectively. Specifically, a fuzzification operator has the effect of transforming crisp data into fuzzy sets. Symbolically

$$x = \text{fuzzifier}(x_o), \tag{4.27}$$

where
x_o is a crisp value of a state variable
x is a fuzzy set

Alternatively, a defuzzification operator transforms the outputs of the inference process (fuzzy sets) into a crisp value for the control action. That is

$$z_o = \text{defuzzifier}(z), \tag{4.28}$$

where
z_o is a crisp value
z is a fuzzy membership value

Referring back to the height example used previously, if a basketball player is about 6.8 ft, then after fuzzification, his height membership value is shown in Equation 4.22. Various fuzzification and defuzzification techniques are described in the references. Figure 4.9 is a block diagram representation of the mechanism of the fuzzy control described above.

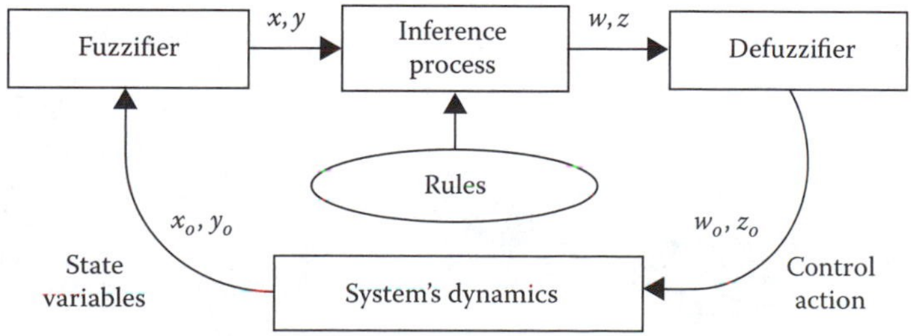

FIGURE 4.9 Block diagram representation of the concept of fuzzy logic control.

4.7 Fuzzy Control Design

4.7.1 DC Motor Dynamics—Assume a Linear Time-Invariant System

Due to the popularity of dc motors for control applications, a dc motor velocity control will be used to illustrate the fuzzy control design approach. Readers are assumed to have a basic background about dc motor operations; otherwise, please refer to [14,15]. The fuzzy control will be applied to an actual dc motor system to demonstrate the effectiveness of the control techniques. We will briefly describe the actual control system setup below.

The fuzzy controller is implemented on a 486 PC using the LabVIEW graphical programming package. The complete system setup is shown in Figure 4.10 and the actual motor control system is given in Figure 4.11. The rotation of the motor shaft generates a tachometer voltage, which is then scaled by interfacing electronic circuitry. A National Instruments data-acquisition board receives the data via an Analog Devices isolation backplane. After a control value is computed, an output current is generated by the data acquisition board. The current signal passes through the backplane and is then converted to a voltage signal and scaled by the interfacing circuitry before being applied to the armature of the motor. Load disturbances are generated by subjecting a disc on the motor shaft to a magnetic field.

For illustration purposes, the control objective concentrates on achieving zero steady-state error and smooth, fast response to step inputs. These are popular desired motor performance characteristics for many industrial applications. The parameters and their numerical values of the dc servomotor used for our simulation studies are listed in Table 4.1, obtained by conventional system identification techniques.

The parameters R_a and L_a are the resistance and the inductance of the motor armature circuit, respectively; J and f are the moment of inertia and the viscous-friction coefficient of the motor and load (referred to the motor shaft), respectively; K is the constant relating the armature current to the motor torque, and K_b is the constant relating the motor speed to the dc motor's back emf. The dc motor has an input operating range of [−15,15] volts.

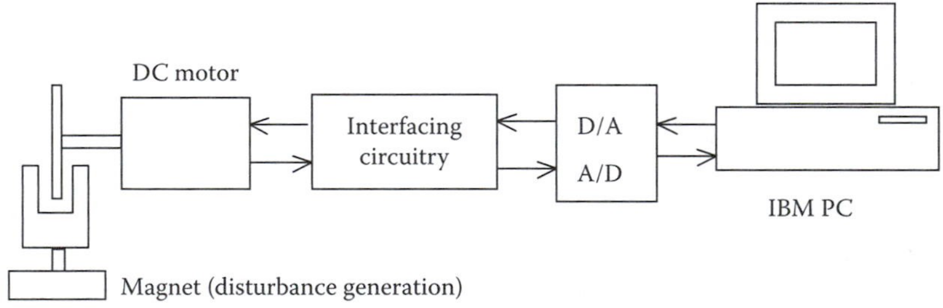

FIGURE 4.10 Schematic diagram of the experimental dc motor system.

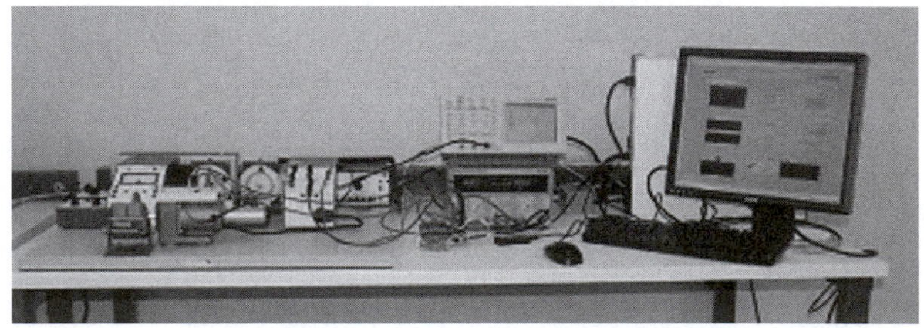

FIGURE 4.11 Motor control system setup.

4.7.2 Fuzzy Control

4.7.2.1 Initial Fuzzy Rules and Membership Function Design

In order to design a fuzzy controller, we will first need to determine the inputs, outputs, universe of discourse, membership functions, and fuzzy rules. In this section, the input variables of the fuzzy controller are the errors, ($E = e(k)$), which is the difference between the dc motor speed and the reference speed, and the change in error ($CE = e(k) - e(k-1)$), where k is the time index. The output variable of the controller is the change in the control effort ($CU = u(k) - u(k-1)$).

TABLE 4.1 DC Motor Parameters

R_a	4.67 W
L_a	170e–3 H
J	42.6e–6 kg m^2
f	47.3e–6 N m/rad/s
K	14.7e–3 N m/A
K_b	14.7e–3 V s/rad

The determination of the universe of discourse of the velocity error change and the control effort change is based on experience and knowledge about the dc motor. For example, the simulation results of the dc motor performance for different control laws based on estimated motor parameters is very helpful for the design of the fuzzy controller. The simulation results will give a rough idea about the response of the system, even though it does not give the exact system performance. Since the open-loop simulations of the system result in a possible velocity range of –500 to 500 rad/s, the minimum and maximum possible values that the error can assume are –1000 and 1000 rad/s, respectively. Hence, the universe of discourse (operating range) of the velocity error spans between –1000 and 1000 rad/s. Based on these requirements, the maximum value of error change is then set to 5.5 rad/s. Also, the maximum value for the control effort change is determined to be 1.5 V. The universes of discourse of the fuzzy variables is then partitioned into seven quantization levels (fuzzy sets), each being described by a linguistic statement such as "big," "small," etc., as listed in Table 4.2. The number of partition levels chosen is a trade-off between the resolution of the quantization and the complexity of the design problem, and is often dependent on the designer's preference.

TABLE 4.2 Fuzzy Set Definitions

PB	Positive big
PM	Positive medium
PS	Positive small
ZE	Zero
NS	Negative small
NM	Negative medium
NB	Negative big

A fuzzy membership function requires assigning a real number in the interval [0,1] to every element in the universe of discourse. This number indicates the degree to which the element belongs to a fuzzy set, such as *big* or *small*. Fuzzy membership functions can have different shapes depending on the designer's preference and/or experience. Triangular and trapezoidal shapes are popular because of simple computations and the capture of the designers' fuzzy numbers sense. Again, the choice of membership functions is a subjective matter, but prior experience can provide some useful guidelines. For example, if the measurable data is disturbed by noise, then the membership functions should be sufficiently wide to reduce noise sensitivity [23]. Some researchers and engineers also suggest that adjacent fuzzy-set values should overlap approximately 25%, and fine-tuning can be achieved by altering this overlap percentage.

Figure 4.12 shows the initial membership functions, which assign a real number in the interval [0,1] to every element in the universe of discourse used for the motor control problem. This number indicates the degree to which the element belongs to a fuzzy set, such as big or small, used in the fuzzy velocity controller.

Notice that there is a rule for every possible combination of E and CE that may arise. Since E and CE both are partitioned into 7 fuzzy sets, the fuzzy rule base table thus has a total of 49 entries, each corresponding to a different combination of input fuzzy set values. These rules have the form

$$\text{rule } i\text{: if } E = A_{E,i} \text{ and } CE = A_{CE,i} \text{ then } CU = C_i, \tag{4.29}$$

where $A_{E,i}$ and $A_{CE,i}$ are the fuzzy set values of the antecedent part of rule i for E and CE, respectively. Likewise, C_i is the fuzzy set value of the consequent part of rule i for CU.

The fuzzification process, which is the transformation of crisp inputs to fuzzy set outputs, is accomplished by using the popular correlation-product inference method. By the same token, these fuzzy set outputs were defuzzified with a centroid computation to generate an exact numerical output as stated in

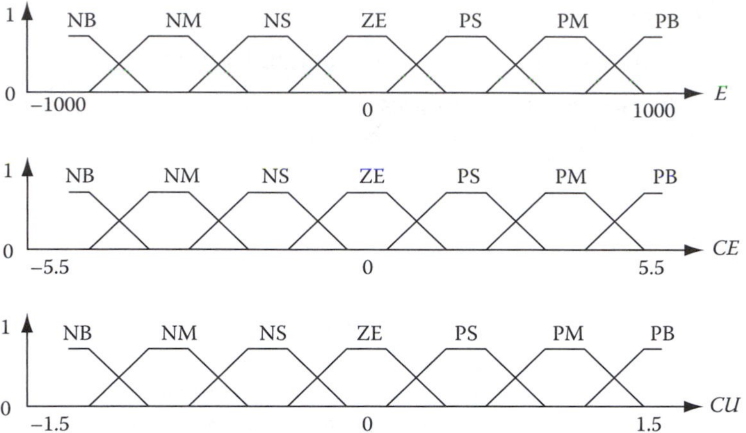

FIGURE 4.12 Initial membership functions used in the fuzzy velocity controller.

Equation 4.24. With this method, the motor's current operating point numerical values E^0 and CE^0 are required. The inference method can be described mathematically as

$$l_i = \min\{\mu_{A_{E,i}}(E^0), \mu_{A_{CE,i}}(CE^0)\}, \quad (4.30)$$

which gives the influencing factor of rule i on the decision-making process, and

$$I_i = \int \mu_{C_i}(CU)dCU, \quad (4.31)$$

gives the area bounded by the membership function $\mu_{C_i}(CU)$; thus l_iI_i gives the area bounded by the membership function $\mu_{C_i}(CU)$ scaled by l_i computed in Equation 4.30.

The centroid of the area bounded by $\mu_{C_i}(CU)$ is computed as

$$c_i = \frac{\int CU\mu_{C_i}(CU)dCU}{\int \mu_{C_i}(CU)dCU}, \quad (4.32)$$

thus $l_iI_i \in c_i$ gives the control value contributed by rule i. The control value CU^0, which combine the control efforts from all N rules, is then computed as

$$CU^0 = \frac{\sum_{i=1}^{N} l_iI_i \times c_i}{\sum_{i=1}^{N} l_iI_i} \quad (4.33)$$

In Equations 4.30 through 4.33, the subscript i indicates the ith rule of a set of N rules.

For illustration purposes, assume the motor currently has the current operating point E^0 and CE^0 and assume only two rules are used (thus $N = 2$)

if E is PS and CE *is* ZE then CU = PS,

and

if E is ZE and CE is PS then CU = ZE. (4.34)

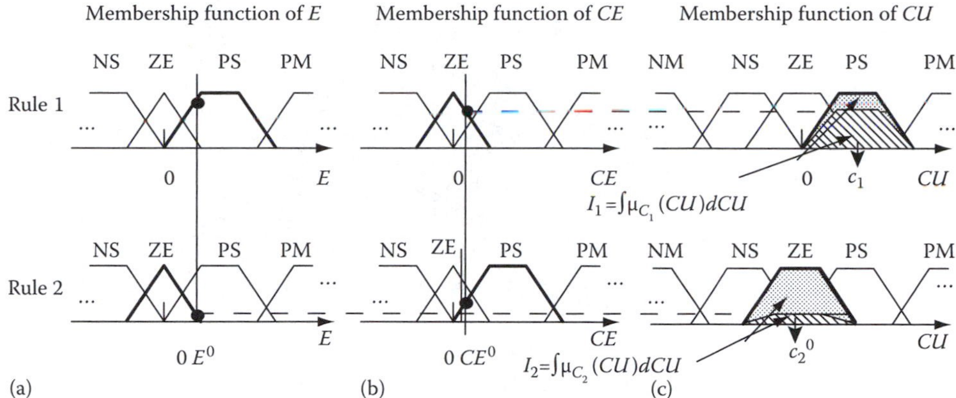

FIGURE 4.13 Graphical representation of the correlation-product inference method.

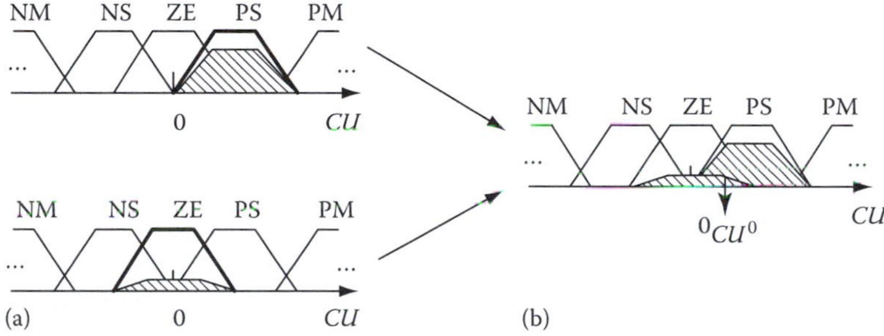

FIGURE 4.14 Graphical representation of the center-of-gravity defuzzification method.

The correlation-product inference method described in (4.30) and (4.31) and the area bounded by the inferred membership function $\mu_{C_i}(CU)$ are conceptually depicted in Figure 4.13.

By looking at rule 1, the membership value of E for PS, $\mu_{PS}(E^0)$ is larger than the membership value of CE for ZE, $\mu_{ZE}(CE^0)$, therefore

$$l_1 = \mu_{ZE}(CE^0)$$

I_1 is the area bounded by the membership function PS on CU (the hatched and the shaded areas) in Figure 4.13c and $l_1 I_1$ is only the hatched area. c_1 is computed to give the centroid of I_1. The same arguments also apply to rule 2. The defuzzification process is graphically depicted in Figure 4.14.

The scaled control membership functions (the hatched areas) from different rules (Figure 4.14a) are combined together from all rules, which form the hatched area shown in Figure 4.14b. The centroid value of the combined hatched area is then computed, to give the final crisp control value.

4.7.2.2 PI Controller

If only experience and control engineering knowledge are used to derive the fuzzy rules, the designers will probably be overwhelmed by the degrees of freedoms (number of rules) of the design, and many of the rule table entries may be left empty due to insufficient *detailed* knowledge to be extracted from

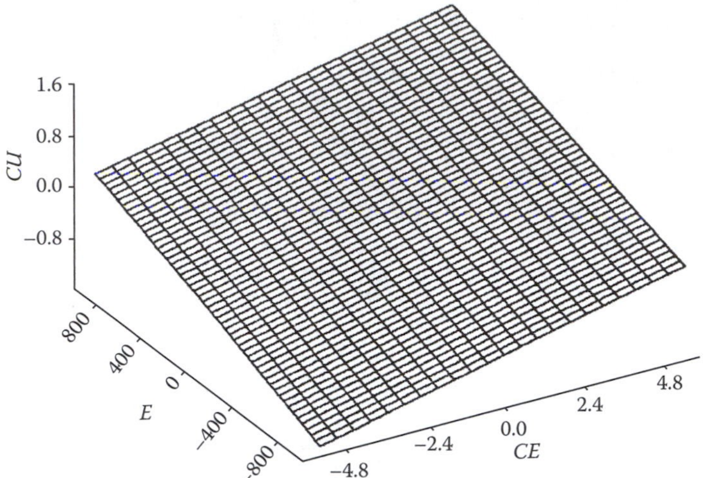

FIGURE 4.15 PI control surface.

the expert. To make the design process more effective, it will be helpful to have a structured way for the designer to follow in order to eventually develop a proper fuzzy controller.

In this section, we will illustrate how to take advantage of conventional control knowledge to arrive at a fuzzy control design more effectively [16]. The PI controller is one of the most popular conventional controllers and will be used in this section as the technique to incorporate a priori knowledge, which will eventually lead to the fuzzy controller.

The velocity transfer functions of the dc motor control can be easily obtained from many text books [17]. The velocity transfer function can be derived as

$$G_p(s) = \frac{\omega(s)}{e_a(s)} = \frac{K}{JL_a s^2 + (fL_a + JR_a)s + (fR_a + KK_b)}. \tag{4.35}$$

The general equation for a PI controller is

$$u(k) = u(k-1) + \left(K_p + \frac{K_i T_s}{2}\right)e(k) + \left(\frac{K_i T_s}{2} - K_p\right)e(k-1), \tag{4.36}$$

where K_p and K_i can be determined by the root-locus method [17]. For velocity control $K_i = 0.264$ and $K_p = 0.12$ are chosen to yield desirable response characteristics, which give an adequate tradeoff between the speed of the response and the percentage of overshoot. The PI control surface over the universes of discourse of error (E) and error change (CE) is shown in Figure 4.15.

4.7.2.3 Borrowing PI Knowledge

The PI control surface is then taken as a starting point for the fuzzy control surface. More specifically, a "fuzzification" of the PI control surface yielded the first fuzzy control surface to be further tuned. This starting point corresponds to the upper left surface of Figure 4.16 (the membership functions are identical in shape and size, and symmetric about ZE).

The initial fuzzy rule base table (Table 4.3) was specified by "borrowing" values from the PI control surface. Since the controller uses the information (E, CE) in order to produce a control signal (CU), the control action can be completely defined by a three-dimensional control surface. Any small modification to a controller will appear as a change in its control surface.

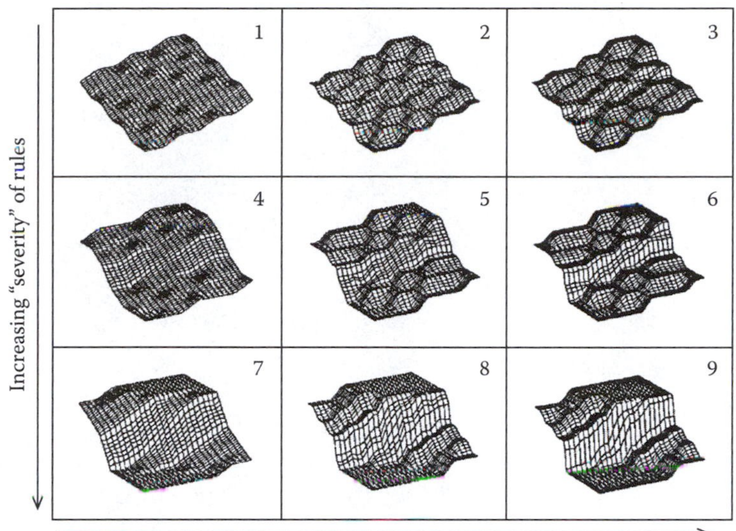

FIGURE 4.16 Effects of fine-tuning on the fuzzy control surface.

TABLE 4.3 Initial Fuzzy Rule Base Table

	PB	ZE	PS	PM	PB	PB	PB	PB
	PM	NS	ZE	PS	PM	PB	PB	PB
	PS	NM	NS	ZE	PS	PM	PB	PB
CE	ZE	NB	NM	NS	ZE	PS	PM	PB
	NS	NB	NB	NM	NS	ZE	PS	PM
	NM	NB	NB	NB	NM	NS	ZE	PS
	NB	NB	NB	NB	NB	NM	NS	ZE
		NB	NM	NS	ZE	PS	PM	PB
					E			

For example, the rule of the first row and third column of Table 4.3 (highlighted) corresponds to the statement

$$\text{if } CE = \text{PB and } E = \text{NS then } CU = \text{PM},$$

which indicates that if the error is large and is gradually decreasing, then the controller should produce a positive medium-compensating signal.

The initial fuzzy controller obtained will give a performance similar to the designed PI controller. The controller performance can be improved by fine-tuning the fuzzy controller while control signals are being applied to the actual motor. In order to fine-tune the fuzzy controller, two parameters can be adjusted: the membership functions and the fuzzy rules. Both the shape of membership functions and the severity of fuzzy rules can affect the motor performance. In general, making the membership functions "narrow" near the ZE region and "wider" far from the ZE region can improve the controller's resolution in the proximity of the desired response when the system output is close to the reference values, thus improving the tracking performance. Also, performance can be improved by changing the "severity" of the rules, which amounts to modifying their consequent part. Figure 4.16

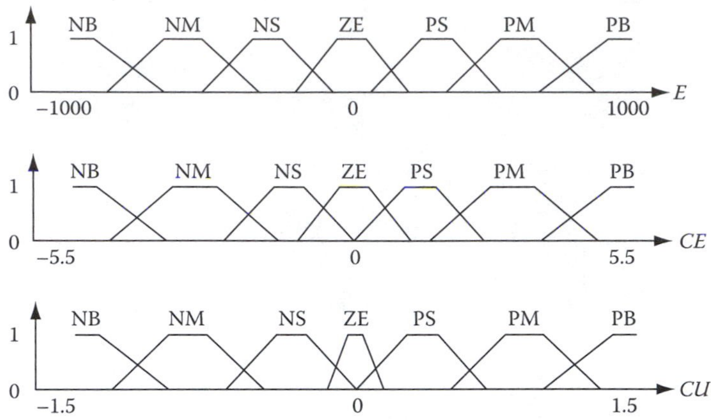

FIGURE 4.17 Intermediate membership functions used in the fuzzy velocity controller.

shows the changes in the fuzzy control surface brought upon by varying the membership functions and the fuzzy rules. The fine-tuning process begins with Figure 4.16-1. The initial control surface is similar to the PI control, which was used as a starting guideline for the fuzzy controller. The changes in control surfaces from the left-hand side to right-hand size signify the change of the shape of membership functions. The changes in control surfaces from top to bottom signify the change in rules.

In order to demonstrate the effect of fine-tuning the membership functions, we show a set of intermediate membership functions (relative to the initial one shown in Figure 4.12) in Figure 4.17 and the final membership functions in Figure 4.18. Figures 4.17 and 4.18 show that some fuzzy sets, such as the ZE in *CU*, are getting narrower, which allows *finer* control in the proximity of the desired response, while the wider fuzzy sets, such as the PB in *E*, permit *coarse* but *fast* control far from the desired response.

We also show an intermediate rule table and the final rule table during the fine-tuning process. The rule in the first row and third column (highlighted cell) corresponds to

$$\text{if } CE = \text{PB and } E = \text{NS, then } CU = \text{To be determined.}$$

In the initial rule table (Table 4.3), *CU* = PM. However, during the fine-tuning process, we found that the rule should have different action in order to give *better* performance. In Table 4.4, the *CU* becomes ZE and the final fuzzy rule table, *CU* = NS.

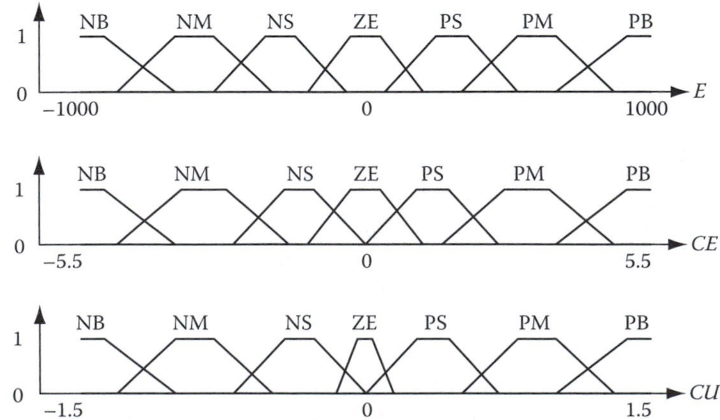

FIGURE 4.18 Final membership functions used in the fuzzy velocity controller.

TABLE 4.4 Intermediate Fuzzy Rule Base Table

	PB	NS	NS	ZE	PB	PB	PB	PB
	PM	NM	NS	ZE	PB	PB	PB	PB
	PS	NM	NS	NS	PS	PM	PB	PB
CE	ZE	NB	NM	NS	ZE	PS	PM	PB
	NS	NB	NB	NM	NS	PS	PS	PM
	NM	NB	NB	NB	NB	ZE	PS	PM
	NB	NB	NB	NB	NB	ZE	PS	PS
		NB	NM	NS	ZE	PS	PM	PB
					E			

TABLE 4.5 Final Fuzzy Rule Base Table

	PB	NM	NS	NS	PB	PB	PB	PB
	PM	NM	NM	NS	PB	PB	PB	PB
	PS	NB	NM	NM	PS	PB	PB	PB
CE	ZE	NB	NB	NM	ZE	PM	PB	PB
	NS	NB	NB	NB	NS	PM	PM	PB
	NM	NB	NB	NB	NB	PS	PM	PM
	NB	NB	NB	NB	NB	PS	PS	PM
		NB	NM	NS	ZE	PS	PM	PB
					E			

From Figure 4.16-1 and -9 it can be seen that gradually increasing the "fineness" of the membership functions and the "severity" of the rules can bring the fuzzy controller to its best performance level. The fine-tuning process is not difficult at all. The fuzzy control designer can easily get a *feel* of how to perform the correct fine-tuning after a few trials. Figure 4.16-9 exhibits the fuzzy control surface, which yielded the best results. The membership functions and fuzzy rules, which generated them, are the ones of Figure 4.18 and Table 4.5, respectively.

The performance of the controllers for the dc motor velocity control is shown in Figure 4.19 for two different references. The fuzzy controller exhibits better performance than the PI controller because of shorter rise time and settling time. The fuzzy controller was thoroughly fine-tuned to yield the best performance.

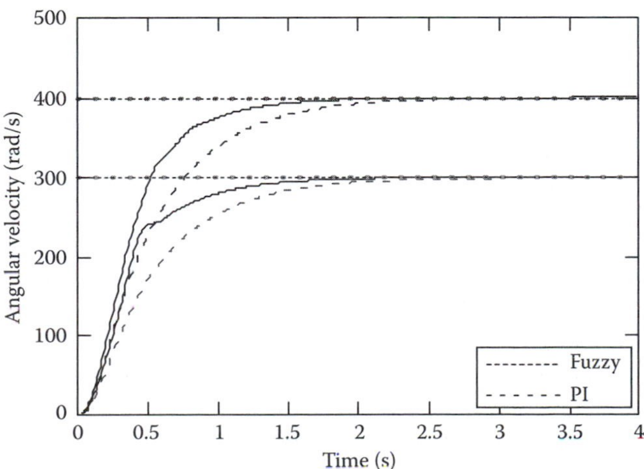

FIGURE 4.19 Velocity control. Two references are shown.

4.8 Conclusion and Future Direction

This chapter outlines the design procedures used in the design of fuzzy controllers for the velocity control of a dc motor. A comparison of these controllers with a classical PI controller was discussed in terms of the characteristics of the respective control surfaces. The PI control was shown to behave as a special case of the fuzzy controller, and for this reason it is more constrained and less flexible than the fuzzy controller. A methodology that exploits the advantages of each technique in order to achieve a successful design is seen as the most sensible approach to follow. The drawback of the fuzzy controller is that it cannot adapt. Recently, different researchers, including the author of this paper, are investigating adaptive fuzzy controllers, many of them result in implementing the fuzzy controller in a neural network structure for adaptation, which is convenient and has fast computation. In fact, a very interesting area of exploration is the possibility of combining the advantages of fuzzy logic with those of artificial neural networks (ANN). The possibility of using the well-known learning capabilities of an ANN coupled with the ability of a fuzzy logic system to translate heuristic knowledge and fuzzy concepts into real numerical values may represent a very powerful way of coming closer to intelligent, adaptive control systems.

References

1. J. T. Teeter, M.-Y. Chow, and J. J. Brickley Jr., A novel fuzzy friction compensation approach to improve the performance of a DC motor control system, *IEEE Transactions on Industrial Electronics*, 43(1), 113–120, 1996.
2. Y. Tipsuwan and M.-Y. Chow, Gain scheduler middleware: A methodology to enable existing controllers for networked control and teleoperation, *IEEE Transactions on Industrial Electronics*, 51(6), 1228–1237, 2004.
3. K. M. Passino and S. Yurkovich, *Fuzzy Control: Theory and Applications*, Addison-Wesley, Menlo Park, CA, 1998.
4. K. Michels, F. Klawonn, R. Kruse, and A. Nürnberger, *Fuzzy Control: Fundamentals, Stability and Design of Fuzzy Controllers*, Springer, Berlin, Germany, 2006.
5. J. Jantzen, *Foundations of Fuzzy Control*, Wiley, West Sussex, U.K., 2007.
6. S. A. R. Sofianita Mutalib, M. Yusoff, and A. Mohamed, Fuzzy water dispersal controller using Sugeno approach, in *Computational Science and Its Applications—ICCSA 2007*, Kuala Lumpur, Malaysia, 2007, pp. 576–588.
7. Z. Wang, Y.-h. Wang, and L.-m. Jia, Study on simulation of high-speed automatic train operation based on MATLAB, *International Journal of Heavy Vehicle Systems*, 12(4), 269–281, 2005.
8. L. A. Zadeh, Fuzzy sets, *Information and Control*, 8, 338–353, 1965.
9. R. E. Bellman and L. A. Zadeh, Decision-making in a fuzzy environment, *Management Science*, 17(4), 141–164, December 1970.
10. L. A. Zadeh, The concept of a linguistic variable and its application to approximate reasoning, Parts 1 and 2, *Information Sciences*, 8, 199–249, 301–357, 1975.
11. L. A. Zadeh, Fuzzy sets as a basis for a theory of possibility, *Fuzzy Sets and Systems*, 1, 3–28, 1978.
12. T. J. Ross, *Fuzzy Logic with Engineering Applications*, McGraw-Hill, New York, 1995.
13. C. P. Pappis and E. H. Mamdani, A fuzzy logic controller for a traffic junction, *IEEE Transactions on Systems, Man, and Cybernetics*, 7(10), 707–717, 1977.
14. W. L. Brogan, *Modern Control Theory*, 3 edn, Prentice Hall, Englewood Cliffs, NJ, 1991.
15. M.-Y. Chow and Y. Tipsuwan, Gain adaptation of networked DC motor controllers based on QoS variations, *IEEE Transactions on Industrial Electronics*, 50(5), 936–943, 2003.
16. M.-Y. Chow, A. Menozzi, and F. H. Holcomb, On the comparison of the performance of emerging and conventional control techniques for DC motor speed control subject to load disturbances, in *Proceedings of IECON'92*, San Diego, CA, November 9–12, 1992.
17. J. J. D'Azzo and C. H. Houpis, *Linear Control System Analysis and Design*, 3 edn, McGraw-Hill, New York, 1988.

II

Neural Networks

5

Understanding Neural Networks

Bogdan M. Wilamowski
Auburn University

5.1 Introduction

The fascination of artificial neural networks started in the middle of the previous century. First artificial neurons were proposed by McCulloch and Pitts [MP43] and they showed the power of the threshold logic. Later Hebb [H49] introduced his *learning rules*. A decade later, Rosenblatt [R58] introduced the perceptron concept. In the early 1960s, Widrow and Holf [WH60] developed intelligent systems such as ADALINE and MADALINE. Nilsson [N65] in his book, *Learning Machines*, summarized many developments of that time. The publication of the Mynsky and Paper [MP69] book, with some discouraging results, stopped for sometime the fascination with artificial neural networks, and achievements in the mathematical foundation of the backpropagation algorithm by Werbos [W74] went unnoticed. The current rapid growth in the area of neural networks started with the work of Hopfield's [H82] recurrent network, Kohonen's [K90] unsupervised training algorithms, and a description of the backpropagation algorithm by Rumelhart et al. [RHW86]. Neural networks are now used to solve many engineering, medical, and business problems [WK00,WB01,B07,CCBC07,KTP07,KT07,MFP07,FP08,JM08,W09]. Descriptions of neural network technology can be found in many textbooks [W89,Z92,H99,W96].

5.2 The Neuron

A biological neuron is a complicated structure, which receives trains of pulses on hundreds of *excitatory* and *inhibitory* inputs. Those incoming pulses are summed with different weights (averaged) during the time period [WPJ96]. If the summed value is higher than a threshold, then the neuron itself is generating a pulse, which is sent to neighboring neurons. Because incoming pulses are summed with time, the neuron generates a pulse train with a higher frequency for higher positive excitation. In other words, if the value of the summed weighted inputs is higher, the neuron generates pulses more frequently. At the same time, each neuron is characterized by the nonexcitability for a certain time after the firing pulse. This so-called *refractory period* can be more accurately described as a phenomenon, where after excitation, the threshold value increases to a very high value and then decreases gradually with a certain time constant. The refractory period sets soft upper limits on the frequency of the output pulse train. In the biological neuron, information is sent in the form of frequency-modulated pulse trains.

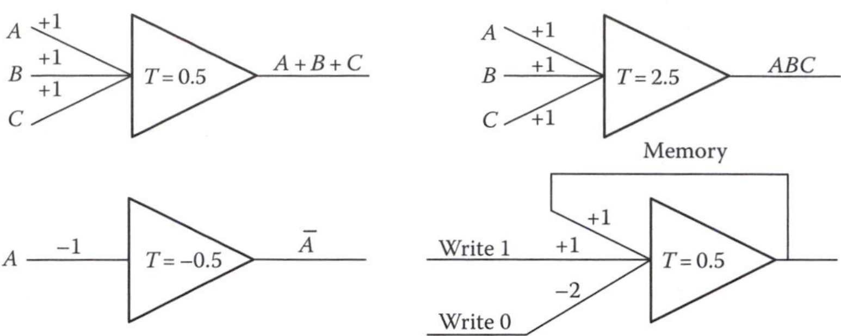

FIGURE 5.1 Examples of logical operations using McCulloch–Pitts neurons.

The description of neuron action leads to a very complex neuron model, which is not practical. McCulloch and Pitts [MP43] show that even with a very simple neuron model, it is possible to build logic and memory circuits. Examples of McCulloch–Pitts' neurons realizing OR, AND, NOT, and MEMORY operations are shown in Figure 5.1.

Furthermore, these simple neurons with thresholds are usually more powerful than typical logic gates used in computers (Figure 5.1). Note that the structure of OR and AND gates can be identical. With the same structure, other logic functions can be realized, as shown in Figure 5.2.

The McCulloch–Pitts neuron model (Figure 5.3a) assumes that incoming and outgoing signals may have only binary values 0 and 1. If incoming signals summed through positive or negative weights have a value equal or larger than threshold, then the neuron output is set to 1. Otherwise, it is set to 0.

$$out = \begin{cases} 1 & \text{if } net \geq T \\ 0 & \text{if } net < T \end{cases} \tag{5.1}$$

where

T is the threshold

net value is the weighted sum of all incoming signals (Figure 5.3)

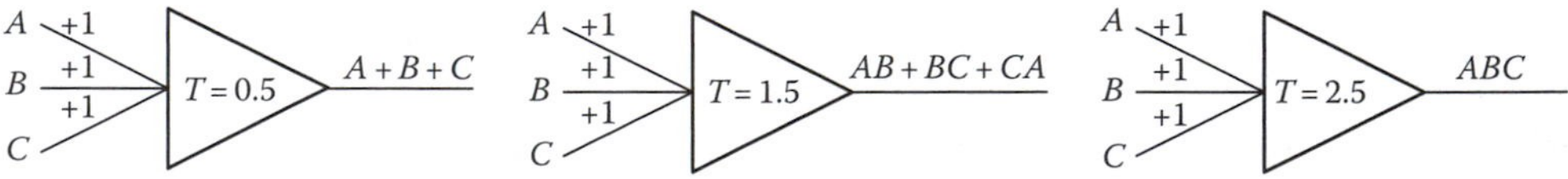

FIGURE 5.2 The same neuron structure and the same weights, but a threshold change results in different logical functions.

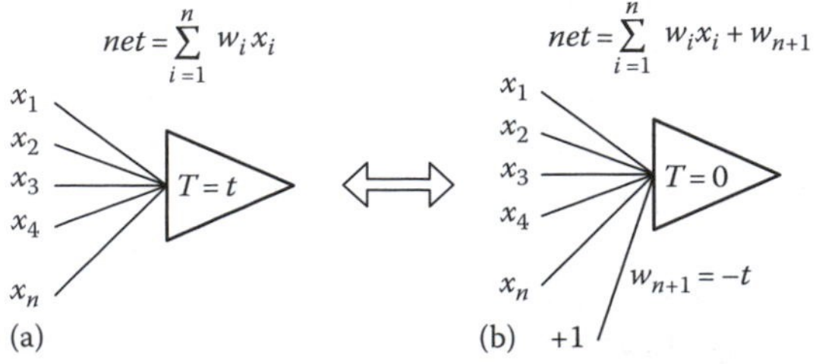

FIGURE 5.3 Threshold implementation with an additional weight and constant input with +1 value: (a) neuron with threshold T and (b) modified neuron with threshold $T = 0$ and additional weight $w_{n+1} = -t$.

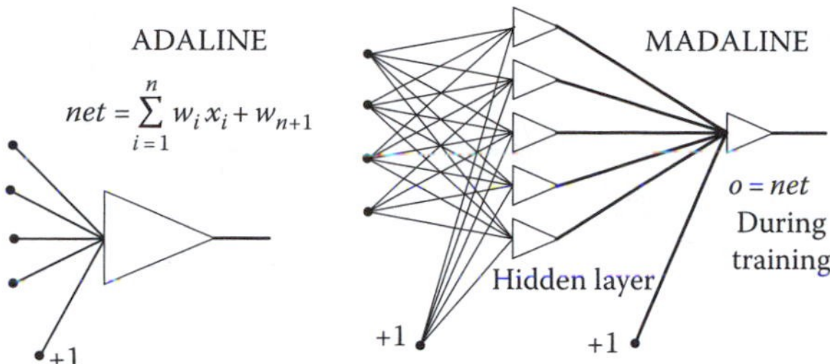

FIGURE 5.4 ADALINE and MADALINE perceptron architectures.

The perceptron model has a similar structure (Figure 5.3b). Its input signals, the weights, and the thresholds could have any positive or negative values. Usually, instead of using variable threshold, one additional constant input with a negative or positive weight can be added to each neuron, as Figure 5.3 shows. Single-layer perceptrons are successfully used to solve many pattern classification problems. Most known perceptron architectures are ADALINE and MADALINE [WH60] shown in Figure 5.4.

Perceptrons using hard threshold activation functions for unipolar neurons are given by

$$o = f_{uni}(net) = \frac{\text{sgn}(net)+1}{2} = \begin{cases} 1 & \text{if } net \geq 0 \\ 0 & \text{if } net < 0 \end{cases} \tag{5.2}$$

and for bipolar neurons

$$o = f_{bip}(net) = \text{sgn}(net) = \begin{cases} 1 & \text{if } net \geq 0 \\ -1 & \text{if } net < 0 \end{cases} \tag{5.3}$$

For these types of neurons, most of the known training algorithms are able to adjust weights only in single-layer networks. Multilayer neural networks (as shown in Figure 5.8) usually use soft activation functions, either unipolar

$$o = f_{uni}(net) = \frac{1}{1+\exp(-\lambda net)} \tag{5.4}$$

or bipolar

$$o = f_{bip}(net) = \tanh(0.5\lambda net) = \frac{2}{1+\exp(-\lambda net)} - 1 \tag{5.5}$$

These soft activation functions allow for the gradient-based training of multilayer networks. Soft activation functions make neural network transparent for training [WT93]. In other words, changes in weight values always produce changes on the network outputs. This would not be possible when hard activation

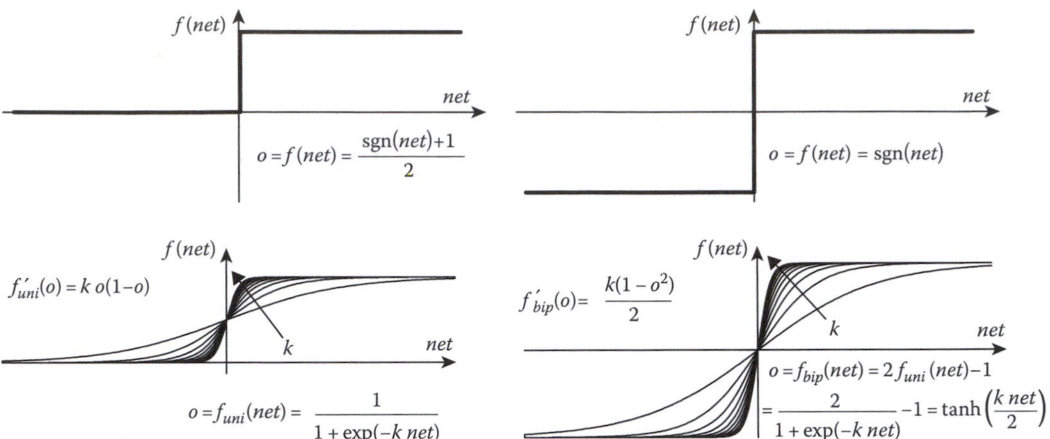

FIGURE 5.5 Typical activation functions: hard in upper row and soft in the lower row.

functions are used. Typical activation functions are shown in Figure 5.5. Note, that even neuron models with continuous activation functions are far from an actual biological neuron, which operates with frequency-modulated pulse trains [WJPM96].

A single neuron is capable of separating input patterns into two categories, and this separation is linear. For example, for the patterns shown in Figure 5.6, the separation line is crossing x_1 and x_2 axis at points x_{10} and x_{20}. This separation can be achieved with a neuron having the following weights: $w_1 = 1/x_{10}$; $w_2 = 1/x_{20}$ and $w_3 = -1$. In general, for n dimensions, the weights are

$$w_i = \frac{1}{x_{i0}} \quad \text{for } i = 1,\ldots,n; \quad w_{n+1} = -1$$

One neuron can divide only linearly separated patterns. To select just one region in n-dimensional input space, more than $n + 1$ neurons should be used.

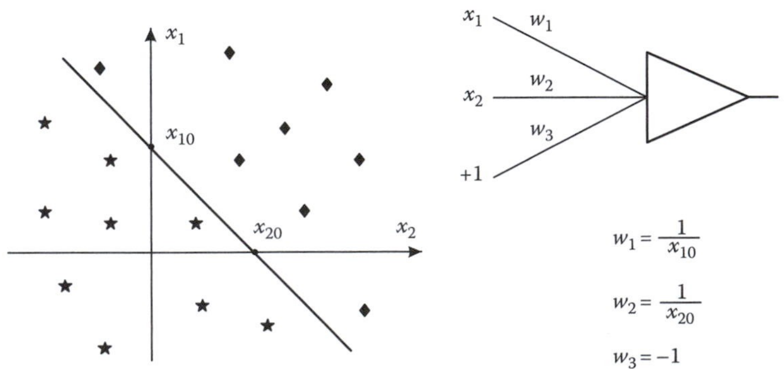

FIGURE 5.6 Illustration of the property of linear separation of patterns in the two-dimensional space by a single neuron.

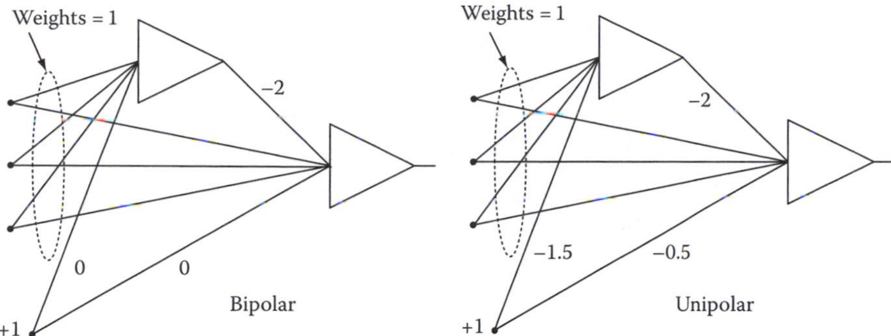

FIGURE 5.7 Neural networks for parity-3 problem.

5.3 Should We Use Neurons with Bipolar or Unipolar Activation Functions?

Neural network users often face a dilemma if they have to use unipolar or bipolar neurons (see Figure 5.5). The short answer is that it does not matter. Both types of networks work the same way and it is very easy to transform bipolar neural network into unipolar neural network and vice versa. Moreover, there is no need to change most of weights but only the biasing weight has to be changed. In order to change from bipolar networks to unipolar networks, only biasing weights must be modified using the formula

$$w_{bias}^{uni} = 0.5\left(w_{bias}^{bip} - \sum_{i=1}^{N} w_i^{bip}\right) \tag{5.6}$$

While, in order to change from unipolar networks to bipolar networks

$$w_{bias}^{bip} = 2w_{bias}^{uni} + \sum_{i=1}^{N} w_i^{uni} \tag{5.7}$$

Figure 5.7 shows the neural network for parity-3 problem, which can be transformed both ways: from bipolar to unipolar and from unipolar to bipolar. Notice that only biasing weights are different. Obviously input signals in bipolar network should be in the range from −1 to +1, while for unipolar network they should be in the range from 0 to +1.

5.4 Feedforward Neural Networks

Feedforward neural networks allow only unidirectional signal flow. Furthermore, most feedforward neural networks are organized in layers and this architecture is often known as MLP (multilayer perceptron). An example of the three-layer feedforward neural network is shown in Figure 5.8. This network consists of four input nodes, two *hidden layers*, and an output layer.

If the number of neurons in the input (hidden) layer is not limited, then all classification problems can be solved using a multilayer network. An example of such neural network, separating patterns from the rectangular area on Figure 5.9 is shown in Figure 5.10.

When the hard threshold activation function is replaced by soft activation function (with a gain of 10), then each neuron in the hidden layer will perform a different task as it is shown in Figure 5.11 and the

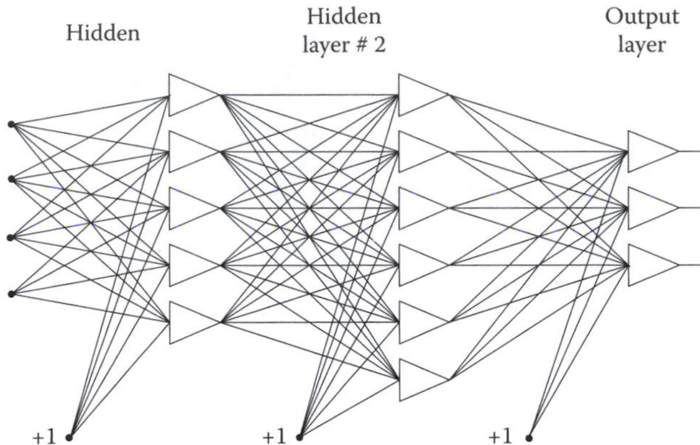

FIGURE 5.8 An example of the three-layer feedforward neural network, which is sometimes known also as MLP.

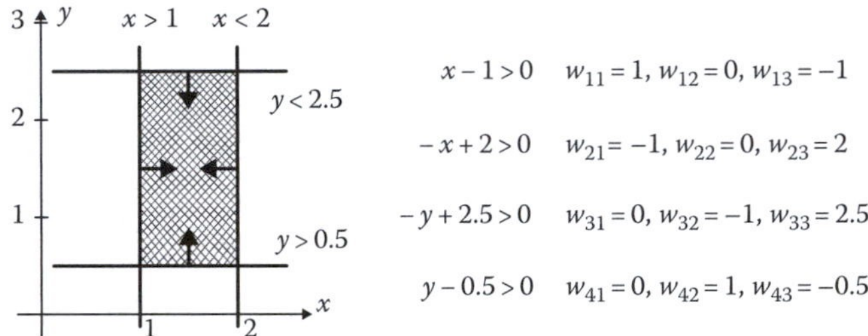

FIGURE 5.9 Rectangular area can be separated by four neurons.

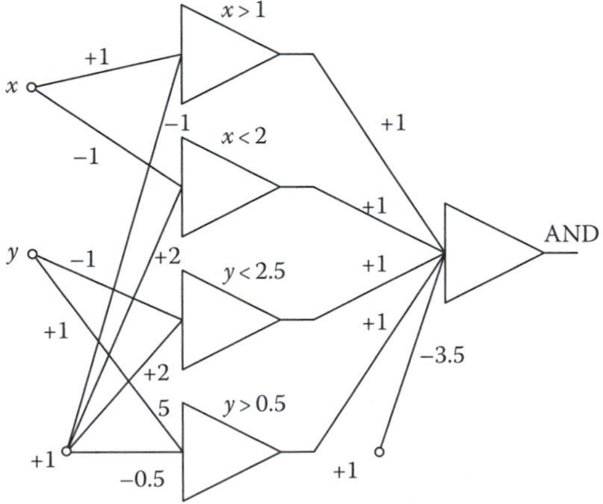

FIGURE 5.10 Neural network architecture that can separate patterns in the rectangular area of Figure 5.7.

response of the output neuron is shown in Figure 5.12. One can notice that the shape of the output surface depends on the gains of activation functions. For example, if this gain is set to be 30, then activation function looks almost as hard activation function and the neural network work as a classifier (Figure 5.13a). If the neural network gain is set to a smaller value, for example, equal 5, then the neural network performs a nonlinear mapping, as shown in Figure 5.13b. Even though this is a relatively simple example, it is essential for understanding neural networks.

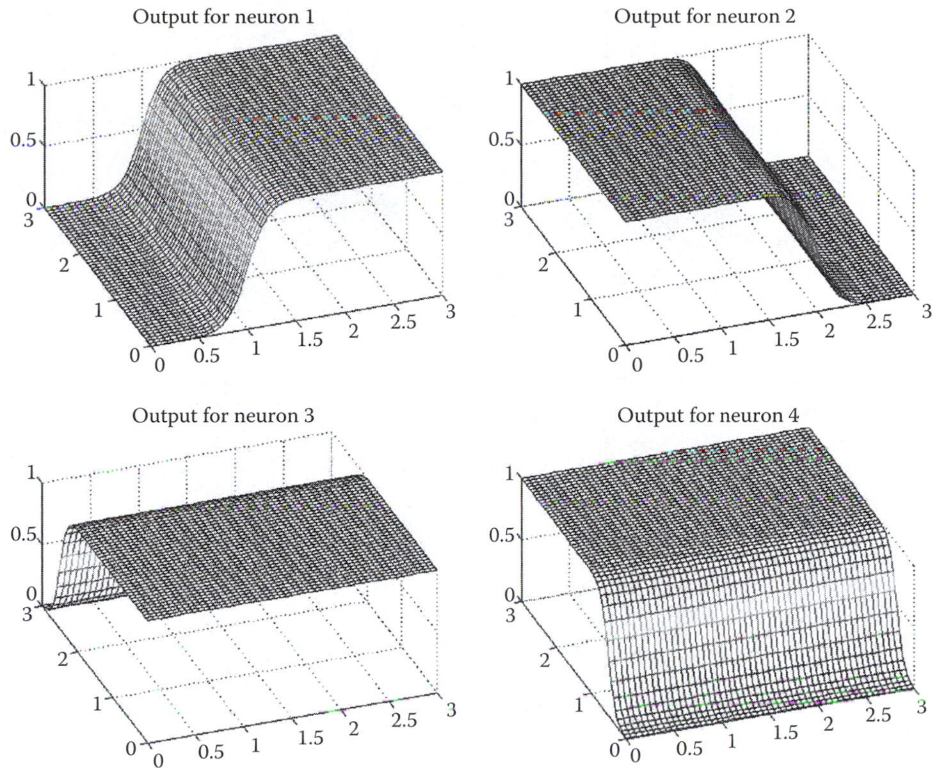

FIGURE 5.11 Responses of four hidden neurons of the network from Figure 5.10.

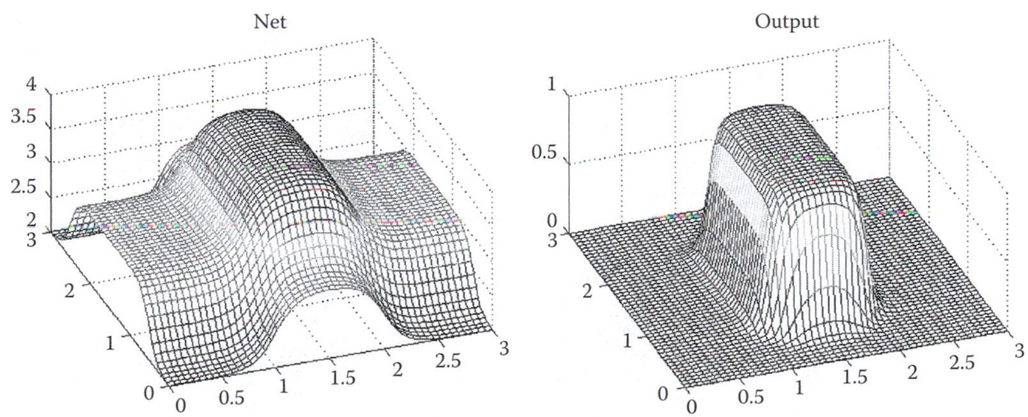

FIGURE 5.12 The net and output values of the output neuron of the network from Figure 5.10.

Let us now use the same neural network architecture as shown in Figure 5.10, but let us change weights for hidden neurons so their neuron lines are located as it is shown in Figure 5.14. This network can separate patterns in pentagonal shape as shown in Figure 5.15a or perform a complex nonlinear mapping as shown in Figure 5.15b depending on the neuron gains. In this simple example of network from Figure 5.10, it is very educational because it lets neural network user understand how neural network operates and may help to select a proper neural network architecture for problems of different complexities. Commonly used trial-and-error methods may not be successful unless the user has some understanding of neural network operation.

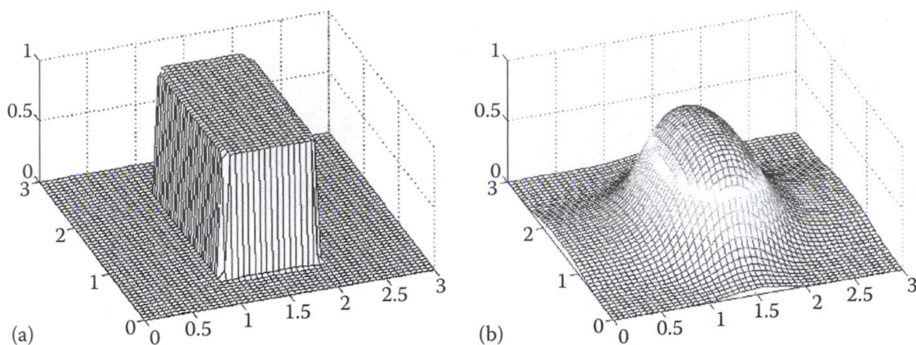

FIGURE 5.13 Response on the neural network of Figure 5.10 with different values of neurons gain: (a) gain = 30 and network works as classifier and (b) gain = 5 and network perform nonlinear mapping.

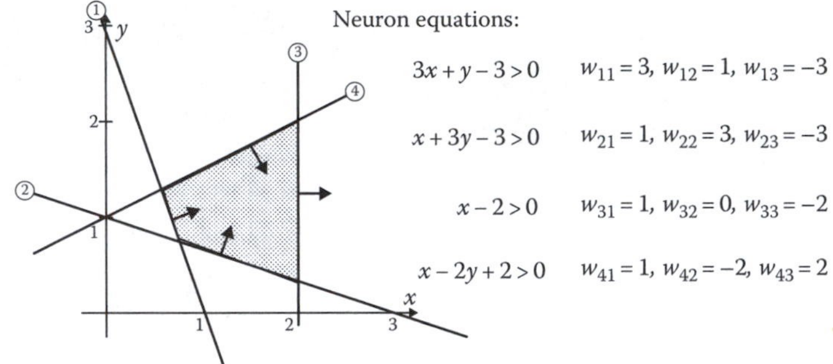

FIGURE 5.14 Two-dimensional input space with four separation lines representing four neurons.

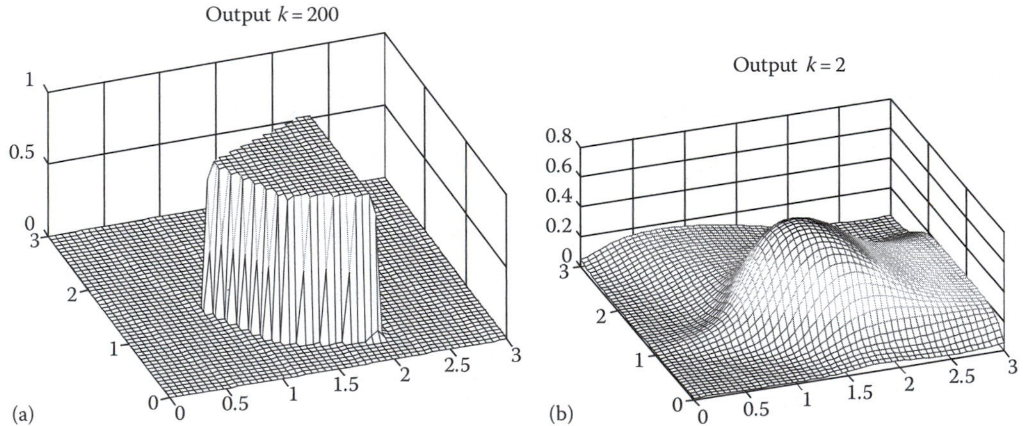

FIGURE 5.15 Response on the neural network of Figure 5.9 with weights define in Figure 5.13 for different values of neurons gain: (a) gain = 200 and network works as classifier and (b) gain = 2 and network perform nonlinear mapping.

The linear separation property of neurons makes some problems especially difficult for neural networks, such as exclusive OR, parity computation for several bits, or to separate patterns on two neighboring spirals. Also, the most commonly used feedforward neural network may have difficulties to separate clusters in multidimensional space. For example, in order to separate cluster in two-dimensional space, we have used four neurons (rectangle), but it is also possible to separate cluster with three neurons (triangle). In three dimensions we may need at least four planes (neurons) to separate space

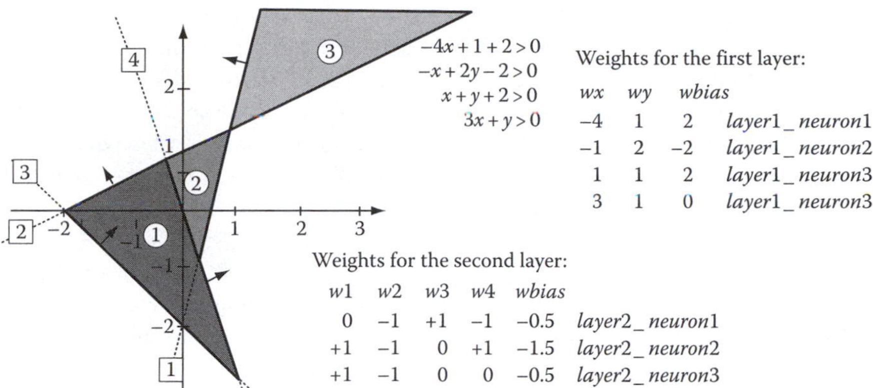

FIGURE 5.16 Problem with the separation of three clusters.

with tetrahedron. In n-dimensional space, in order to separate a cluster of patterns, there are at least $n + 1$ neurons required. However, if neural network with several hidden layers are used, then the number of neurons needed may not be that excessive. Also, a neuron in the first hidden layer may be used for separation of multiple clusters. Let us analyze another example where we would like to design neural network with multiple outputs to separate three clusters and each network output must produce +1 only for a given cluster. Figure 5.16 shows three clusters to be separated, corresponding equations for four neurons and weights for resulted neural network, as shown in Figure 5.17.

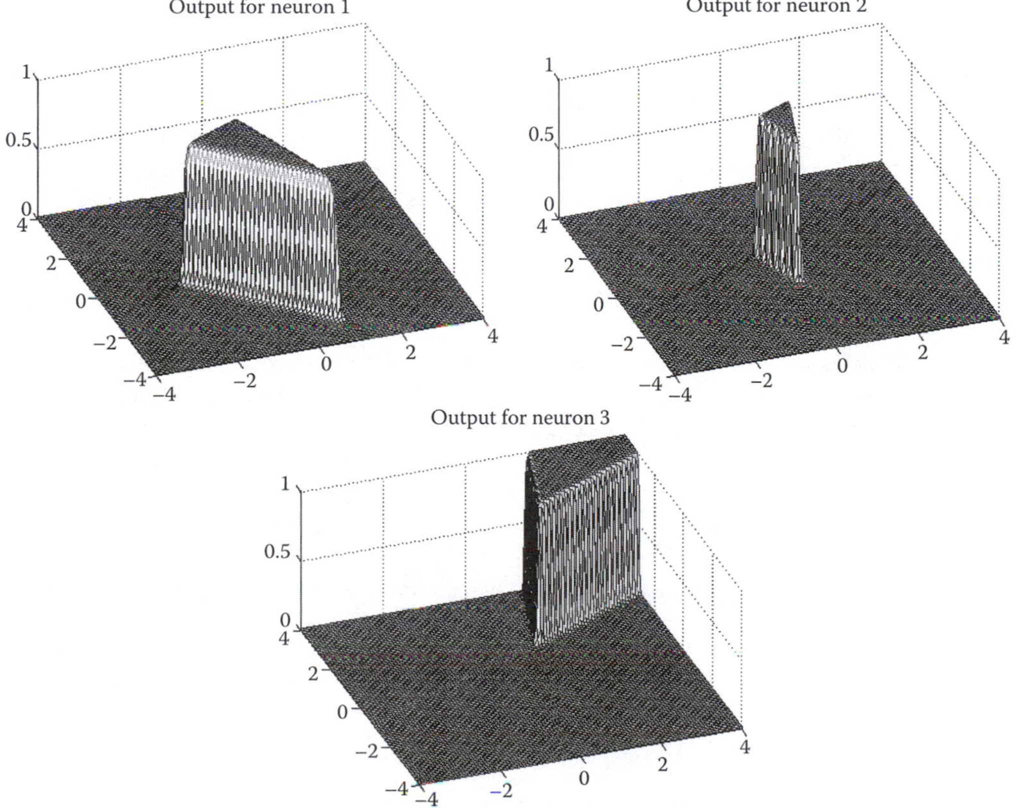

FIGURE 5.17 Neural network performing cluster separation and resulted output surfaces for all three clusters.

The example with three clusters shows that often there is no need to have several neurons in the hidden layer dedicated for a specific cluster. These hidden neurons may perform multiple functions and they can contribute to several clusters instead of just one. It is, of course, possible to develop separate neural networks for every cluster, but it is much more efficient to have one neural network with multiple outputs as shown in Figures 5.16 and 5.17. This is one advantage of neural networks over fuzzy systems, which can be developed only for one output at a time [WJK99,MW01]. Another advantage of neural network is that the number of inputs can be very large so they can process signals in multidimensional space, while fuzzy systems can handle usually two or three inputs only [WB99].

The most commonly used neural networks have the MLP architecture, as shown in Figure 5.8. For such a layer-by-layer network, it is relatively easy to develop the learning software, but these networks are significantly less powerful than networks where connections across layers are allowed. Unfortunately, only very limited number of software were developed to train other than MLP networks [WJ96,W02]. As a result, most researchers use MLP architectures, which are far from optimal. Much better results can be obtained with BMLP (bridged MLP) architecture or with FCC (fully connected cascade) architecture [WHM03]. Also, most researchers are using simple EBP (error backpropagation) learning algorithm, which is not only much slower than more advanced algorithms such as LM (Levenberg–Marquardt) [HM94] or NBN (neuron by neuron) [WCKD08,HW09,WH10], but also EBP algorithm often is not able to train close-to-optimal neural networks [W09].

References

[B07] B.K. Bose, Neural network applications in power electronics and motor drives—An introduction and perspective. *IEEE Trans. Ind. Electron.* 54(1):14–33, February 2007.

[CCBC07] G. Colin, Y. Chamaillard, G. Bloch, and G. Corde, Neural control of fast nonlinear systems—Application to a turbocharged SI engine with VCT. *IEEE Trans. Neural Netw.* 18(4):1101–1114, April 2007.

[FP08] J.A. Farrell and M.M. Polycarpou, Adaptive approximation based control: Unifying neural, fuzzy and traditional adaptive approximation approaches. *IEEE Trans. Neural Netw.* 19(4):731–732, April 2008.

[H49] D.O. Hebb, *The Organization of Behavior, a Neuropsychological Theory*. Wiley, New York, 1949.

[H82] J.J. Hopfield, Neural networks and physical systems with emergent collective computation abilities. *Proc. Natl. Acad. Sci.* 79:2554–2558, 1982.

[H99] S. Haykin, *Neural Networks—A Comprehensive Foundation*. Prentice Hall, Upper Saddle River, NJ, 1999.

[HM94] M.T. Hagan and M. Menhaj, Training feedforward networks with the Marquardt algorithm. *IEEE Trans. Neural Netw.* 5(6):989–993, 1994.

[HW09] H. Yu and B.M. Wilamowski, C++ implementation of neural networks trainer. *13th International Conference on Intelligent Engineering Systems, INES-09*, Barbados, April 16–18, 2009.

[JM08] M. Jafarzadegan and H. Mirzaei, A new ensemble based classifier using feature transformation for hand recognition. *2008 Conference on Human System Interactions*, Krakow, Poland, May 2008, pp. 749–754.

[K90] T. Kohonen, The self-organized map. *Proc. IEEE* 78(9):1464–1480, 1990.

[KT07] S. Khomfoi and L.M. Tolbert, Fault diagnostic system for a multilevel inverter using a neural network. *IEEE Trans. Power Electron.* 22(3):1062–1069, May 2007.

[KTP07] M. Kyperountas, A. Tefas, and I. Pitas, Weighted piecewise LDA for solving the small sample size problem in face verification. *IEEE Trans. Neural Netw.* 18(2):506–519, February 2007.

[MFP07] J.F. Martins, V. Ferno Pires, and A.J. Pires, Unsupervised neural-network-based algorithm for an on-line diagnosis of three-phase induction motor stator fault. *IEEE Trans. Ind. Electron.* 54(1):259–264, February 2007.

[MP43] W.S. McCulloch and W.H. Pitts, A logical calculus of the ideas imminent in nervous activity. *Bull. Math. Biophy.* 5:115–133, 1943.

[MP69] M. Minsky and S. Papert, *Perceptrons*. MIT Press, Cambridge, MA, 1969.

[MW01] M. McKenna and B.M. Wilamowski, Implementing a fuzzy system on a field programmable gate array. *International Joint Conference on Neural Networks* (*IJCNN'01*), Washington, DC, July 15–19, 2001, pp. 189–194.

[N65] N.J. Nilsson, *Learning Machines: Foundations of Trainable Pattern Classifiers*. McGraw Hill Book Co., New York, 1965.

[R58] F. Rosenblatt, The perceptron: A probabilistic model for information storage and organization in the brain. *Psych. Rev*. 65:386–408, 1958.

[RHW86] D.E. Rumelhart, G.E. Hinton, and R.J. Wiliams, Learning representations by back-propagating errors. *Nature* 323:533–536, 1986.

[W02] B.M. Wilamowski, Neural networks and fuzzy systems, Chapter 32. *Mechatronics Handbook*, ed. R.R. Bishop. CRC Press, Boca Raton, FL, 2002, pp. 33-1–32-26.

[W09] B. M. Wilamowski, Neural network architectures and learning algorithms. *IEEE Ind. Electron. Mag.* 3(4):56–63.

[W74] P. Werbos, Beyond regression: New tools for prediction and analysis in behavioral sciences. PhD dissertation, Harvard University, Cambridge, MA, 1974.

[W89] P.D. Wasserman, *Neural Computing Theory and Practice*. Van Nostrand Reinhold, New York, 1989.

[W96] B.M. Wilamowski, Neural networks and fuzzy systems, Chapters 124.1–124.8. *The Electronic Handbook*. CRC Press, Boca Raton, FL, 1996, pp. 1893–1914.

[WB01] B.M. Wilamowski and J. Binfet, Microprocessor Implementation of Fuzzy Systems and Neural Networks, *International Joint Conference on Neural Networks (IJCNN'01),* Washington DC, July 15–19, 2001, pp. 234–239.

[WB99] B.M. Wilamowski and J. Binfet, Do fuzzy controllers have advantages over neural controllers in microprocessor implementation. *Proceedings of the 2nd International Conference on Recent Advances in Mechatronics - ICRAM'99*, Istanbul, Turkey, May 24–26, 1999, pp. 342–347.

[WCKD08] B.M. Wilamowski, N.J. Cotton, O. Kaynak, and G. Dundar, Computing gradient vector and Jacobian matrix in arbitrarily connected neural networks. *IEEE Trans. Ind. Electron.* 55(10):3784–3790, October 2008.

[WH10] B.M. Wilamowski and H. Yu, Improved computation for Levenberg Marquardt training. *IEEE Trans. Neural Netw*. 21:930–937, 2010.

[WH60] B. Widrow and M.E. Hoff, Adaptive switching circuits. *1960 IRE Western Electric Show and Convention Record, Part 4*, New York (August 23), pp. 96–104, 1960.

[WHM03] B. Wilamowski, D. Hunter, and A. Malinowski, Solving parity-N problems with feedforward neural network. *Proceedings of the IJCNN'03 International Joint Conference on Neural Networks*, Portland, OR, July 20–23, 2003, pp. 2546–2551.

[WJ96] B.M. Wilamowski and R.C. Jaeger, Implementation of RBF type networks by MLP networks. *IEEE International Conference on Neural Networks*, Washington, DC, June 3–6, 1996, pp. 1670–1675.

[WJK99] B.M. Wilamowski, R.C. Jaeger, and M.O. Kaynak, Neuro-Fuzzy Architecture for CMOS Implementation, *IEEE Trans. Ind. Electron*., 46 (6), 1132–1136, December 1999.

[WJPM96] B.M. Wilamowski, R.C. Jaeger, M.L. Padgett, and L.J. Myers, CMOS implementation of a pulse-coupled neuron cell. *IEEE International Conference on Neural Networks*, Washington, DC, June 3–6, 1996, pp. 986–990.

[WK00] B.M. Wilamowski and O. Kaynak, Oil well diagnosis by sensing terminal characteristics of the induction motor. *IEEE Trans. Ind. Electron*. 47(5):1100–1107, October 2000.

[WPJ96] B.M. Wilamowski, M.L. Padgett, and R.C. Jaeger, Pulse-coupled neurons for image filtering. *World Congress of Neural Networks*, San Diego, CA, September 15–20, 1996, pp. 851–854.

[WT93] B.M. Wilamowski and L. Torvik, Modification of gradient computation in the back-propagation algorithm. Presented at *ANNIE'93—Artificial Neural Networks in Engineering*, St. Louis, MO, November 14–17, 1993.

[Z92] J. Zurada, *Introduction to Artificial Neural Systems*, West Publishing Co., St. Paul, MN, 1992.

6

Neural Network Architectures

Bogdan M. Wilamowski
Auburn University

6.1 Introduction

Different neural network architectures are widely described in the literature [W89,Z95,W96,WJK99, H99,WB01,W07]. The feedforward neural networks allow only for one directional signal flow. Furthermore, most of the feedforward neural networks are organized in layers. An example of the three layer feedforward neural network is shown in Figure 6.1. This network consists of three input nodes: two *hidden layers* and an output layer. Typical activation functions are shown in Figure 6.2. These continuous activation functions allow for the gradient-based training of multilayer networks. Usually it is difficult to predict required size of neural networks. Often it is done by trial and error method. Another approach would be to start with much larger than required neural network and to reduce its size by applying one of pruning algorithms [FF02,FFN01,FFJC09].

6.2 Special Easy-to-Train Neural Network Architectures

Training of multilayer neural networks is difficult. It is much easier to train a single neuron or a single layer of neurons. Therefore, several concepts of neural network architectures were developed where only one neuron can be trained at a time. There are also neural network architectures where training is not needed [HN87,W02]. This chapter reviews various easy-to-train architectures. Also, it will be shown that abilities to recognize patterns strongly depend on the used architectures.

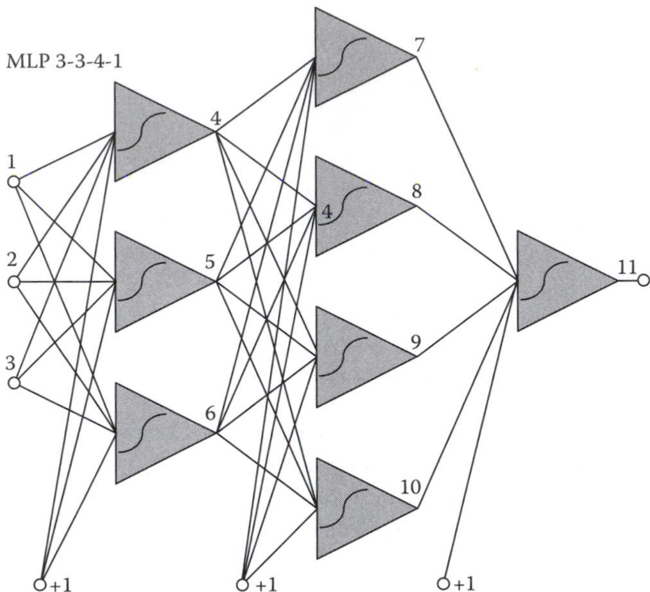

FIGURE 6.1 MLP type architecture 3-3-4-1 (without connections across layers).

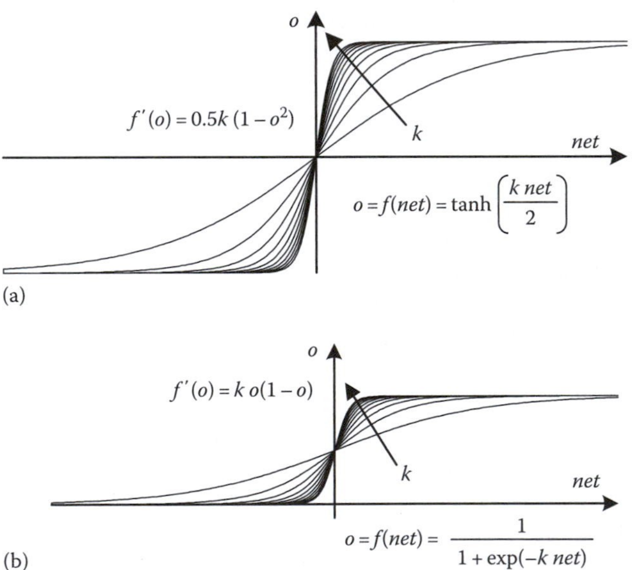

FIGURE 6.2 Typical activation functions: (a) bipolar and (b) unipolar.

6.2.1 Polynomial Networks

Using nonlinear terms with initially determined functions, the actual number of inputs supplied to the one layer neural network is increased. In the simplest case, nonlinear elements are higher order polynomial terms of input patterns.

The learning procedure for one layer is easy and fast. Figure 6.3 shows an *XOR* problem solved using functional link networks. Figure 6.4 shows a single trainable layer neural network with nonlinear polynomial terms. The learning procedure for one layer is easy and fast.

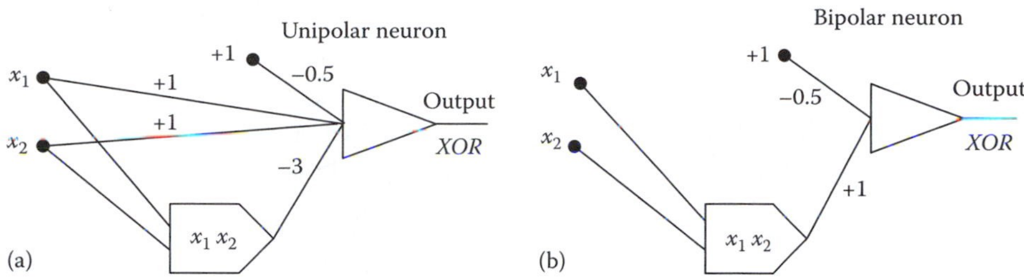

FIGURE 6.3 Polynomial networks for solution of the *XOR* problem: (a) using unipolar signals and (b) using bipolar signals.

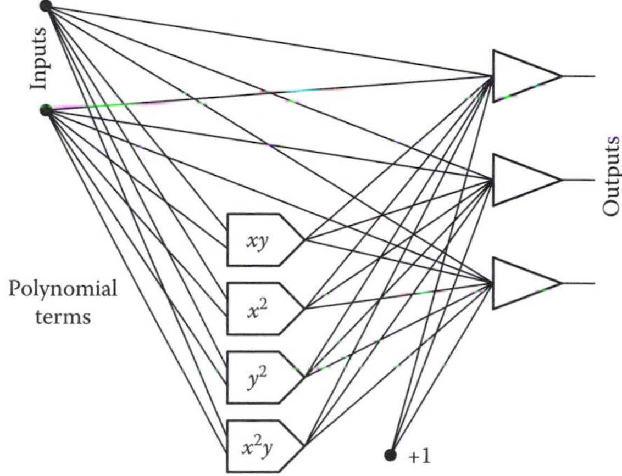

FIGURE 6.4 One layer neural network with nonlinear polynomial terms.

Note that when the polynomial networks have their limitations, they cannot handle networks with many inputs because the number of polynomial terms may grow exponentially.

6.2.2 Functional Link Networks

One-layer neural networks are relatively easy to train, but these networks can solve only linearly separated problems. One possible solution for nonlinear problems was elaborated by Pao [P89] using the functional link network shown in Figure 6.5. Note that the functional link network can be treated as a one-layer network, where additional input data are generated off-line using nonlinear transformations.

Note that, when the functional link approach is used, this difficult problem becomes a trivial one. The problem with the functional link network is that proper selection of nonlinear elements is not an easy task. However, in many practical cases it is not difficult to predict what kind of transformation of input data may linearize the problem, so the functional link approach can be used.

6.2.3 Sarajedini and Hecht-Nielsen Network

Figure 6.6 shows a neural network which can calculate the Euclidean distance between two vectors **x** and **w**. In this powerful network, one may set weights to the desired point **w** in a multidimensional space and the network will calculate the Euclidean distance for any new pattern on the input. The difficult task is the calculate $||\mathbf{x}||^2$, but it can be done off-line for all incoming patterns. A sample output for a two-dimensional case is shown in Figure 6.7.

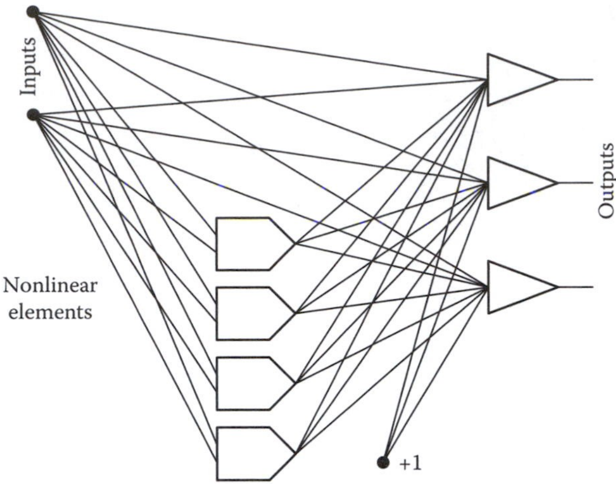

FIGURE 6.5 One layer neural network with arbitrary nonlinear terms.

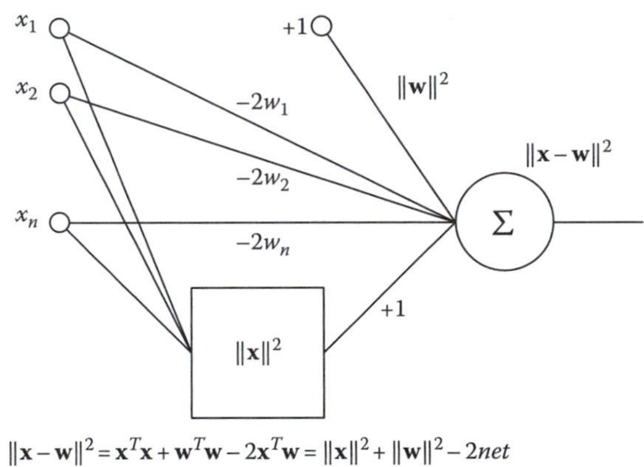

FIGURE 6.6 Sarajedini and Hecht-Nielsen neural network.

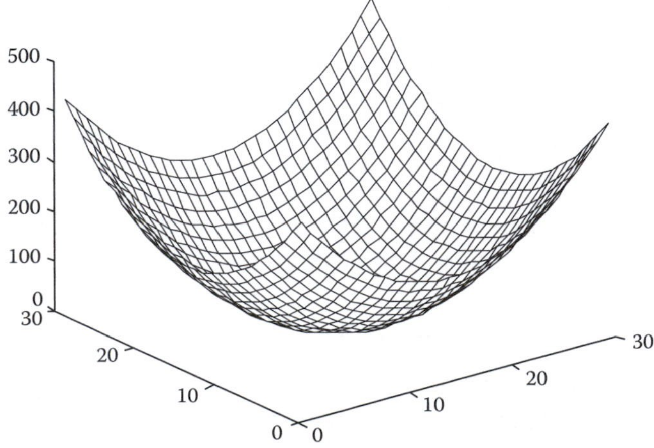

FIGURE 6.7 Output of the Sarajedini and Hecht-Nielsen network is proportional to the square of Euclidean distance.

6.2.4 Feedforward Version of the Counterpropagation Network

The counterpropagation network was originally proposed by Hecht-Nilsen [HN87]. In this chapter, a modified feedforward version as described by Zurada [Z92] is discussed. This network, which is shown in Figure 6.8, requires the number of hidden neurons to be equal to the number of input patterns, or, more exactly, to the number of input clusters.

When binary input patterns are considered, then the input weights must be exactly equal to the input patterns. In this case,

$$net = \mathbf{x}^t\mathbf{w} = \left(n - 2HD(\mathbf{x},\mathbf{w})\right) \tag{6.1}$$

where

n is the number of inputs
$\mathbf{w}$ are weights
$\mathbf{x}$ is the input vector
$HD(\mathbf{x},\mathbf{w})$ is the Hamming distance between input pattern and weights

In order that a neuron in the input layer is reacting just for the stored pattern, the threshold value for this neuron should be

$$w_{n+1} = -(n-1) \tag{6.2}$$

If it is required that the neuron must react also for similar patterns, then the threshold should be set to $w_{n+1} = -(n - (1 + HD))$, where HD is the Hamming distance defining the range of similarity. Since for a given input pattern, only one neuron in the first layer may have the value of one and the remaining neurons have zero values, the weights in the output layer are equal to the required output pattern.

The network, with unipolar activation functions in the first layer, works as a look-up table. When the linear activation function (or no activation function at all) is used in the second layer, then the network also can be considered as an analog memory (Figure 6.9) [W03,WJ96].

The counterpropagation network is very easy to design. The number of neurons in the hidden layer should be equal to the number of patterns (clusters). The weights in the input layer should be equal to the input patterns and, the weights in the output layer should be equal to the output patterns. This simple network can be used for rapid prototyping. The counterpropagation network usually has more hidden neurons than required.

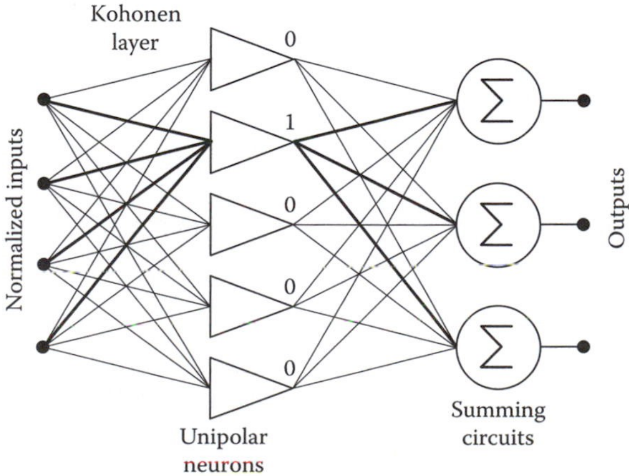

FIGURE 6.8 Counterpropagation network.

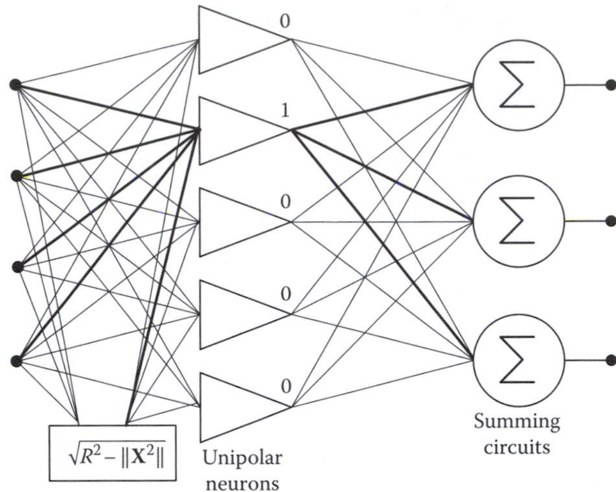

FIGURE 6.9 Counterpropagation network used as analog memory with analog address.

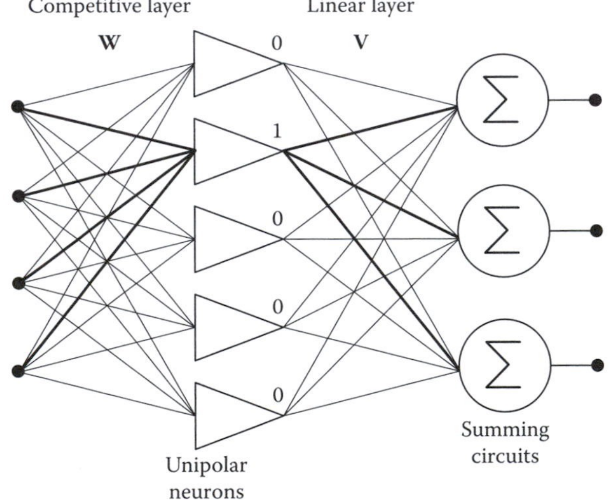

FIGURE 6.10 Learning vector quantization.

6.2.5 Learning Vector Quantization

At learning vector quantization (LVQ) network (Figure 6.10), the first layer detects subclasses. The second layer combines subclasses into a single class. First layer computes Euclidean distances between input pattern and stored patterns. Winning "neuron" is with the minimum distance.

6.2.6 WTA Architecture

The winner-take-all (WTA) network was proposed by Kohonen [K88]. This is basically a one-layer network used in the unsupervised training algorithm to extract a statistical property of the input data. At the first step, all input data is normalized so that the length of each input vector is the same, and usually equal to unity. The activation functions of neurons are unipolar and continuous. The learning process starts with a weight initialization to small random values.

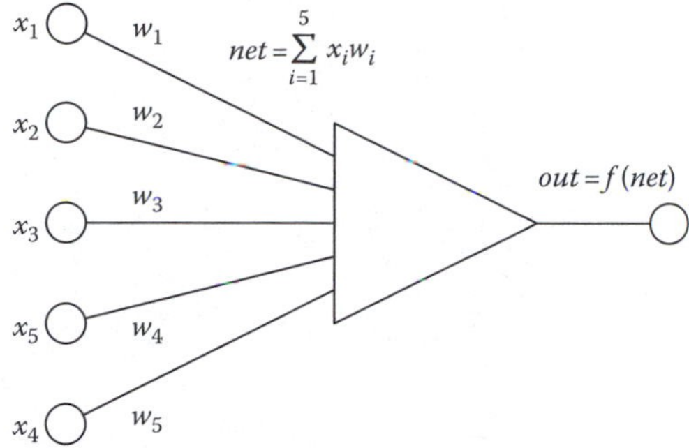

FIGURE 6.11 Neuron as the Hamming distance classifier.

Let us consider a neuron shown in Figure 6.11. If inputs are binaries, for example $\mathbf{X} = [1, -1, 1, -1, -1]$, then the maximum value of *net*

$$net = \sum_{i=1}^{5} x_i w_i = \mathbf{X}\mathbf{W}^T \tag{6.3}$$

is when weights are identical to the input pattern $\mathbf{W} = [1, -1, 1, -1, -1]$. The Euclidean distance between weight vector **W** and input vector **X** is

$$\|\mathbf{W} - \mathbf{X}\| = \sqrt{(w_1 - x_1)^2 + (w_2 - x_2)^2 + \cdots + (w_n - x_n)^2} \tag{6.4}$$

$$\|\mathbf{W} - \mathbf{X}\| = \sqrt{\sum_{i=1}^{n} (w_i - x_i)^2} \tag{6.5}$$

$$\|\mathbf{W} - \mathbf{X}\| = \sqrt{\mathbf{W}\mathbf{W}^T - 2\mathbf{W}\mathbf{X}^T + \mathbf{X}\mathbf{X}^T} \tag{6.6}$$

When the lengths of both the weight and input vectors are normalized to value of 1

$$\|\mathbf{X}\| = 1 \quad \text{and} \quad \|\mathbf{W}\| = 1 \tag{6.7}$$

then the equation simplifies to

$$\|\mathbf{W} - \mathbf{X}\| = \sqrt{2 - 2\mathbf{W}\mathbf{X}^T} \tag{6.8}$$

Please notice that the maximum value of net value $net = 1$ is when **W** and **X** are identical.

Kohonen WTA networks have some problems:

1. Important information about length of the vector is lost during the normalization process
2. Clustering depends on
   a. Order of patterns applied
   b. Number of initial neurons
   c. Initial weights

6.2.7 Cascade Correlation Architecture

The cascade correlation architecture (Figure 6.12) was proposed by Fahlman and Lebiere [FL90]. The process of network building starts with a one-layer neural network and hidden neurons are added as needed.

In each training step, the new hidden neuron is added and its weights are adjusted to maximize the magnitude of the correlation between the new hidden neuron output and the residual error signal on

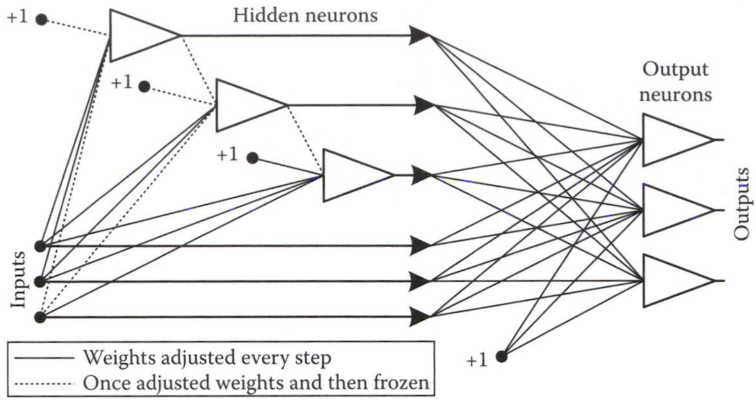

FIGURE 6.12 Cascade correlation architecture.

the network output that we are trying to eliminate. The correlation parameter S defined in the following equation must be maximized:

$$S = \sum_{o=1}^{O} \left| \sum_{p=1}^{P} (V_p - \overline{V})(E_{po} - \overline{E_o}) \right| \tag{6.9}$$

where
- O is the number of network outputs
- P is the number of training patterns
- V_p is output on the new hidden neuron
- E_{po} is the error on the network output

By finding the gradient, $\Delta S/\Delta w_i$, the weight adjustment for the new neuron can be found as

$$\Delta w_i = \sum_{o=1}^{O} \sum_{p=1}^{P} \sigma_o (E_{po} - \overline{E_o}) f_{p'} x_{ip} \tag{6.10}$$

The output neurons are trained using the delta (backpropagation) algorithm. Each hidden neuron is trained just once and then its weights are frozen. The network learning and building process is completed when satisfactory results are obtained.

6.2.8 Radial Basis Function Networks

The structure of the radial basis function (RBF) network is shown in Figure 6.13. This type of network usually has only one hidden layer with special "neurons." Each of these "neurons" responds only to the inputs signals close to the stored pattern.

The output signal h_i of the ith hidden "neuron" is computed using the formula:

$$h_i = \exp\left(-\frac{\left\| \mathbf{x} - \mathbf{s}_i \right\|^2}{2\sigma^2} \right) \tag{6.11}$$

Note that the behavior of this "neuron" significantly differs from the biological neuron. In this "neuron," excitation is not a function of the weighted sum of the input signals. Instead, the distance between the input and stored pattern is computed. If this distance is zero, then the "neuron" responds

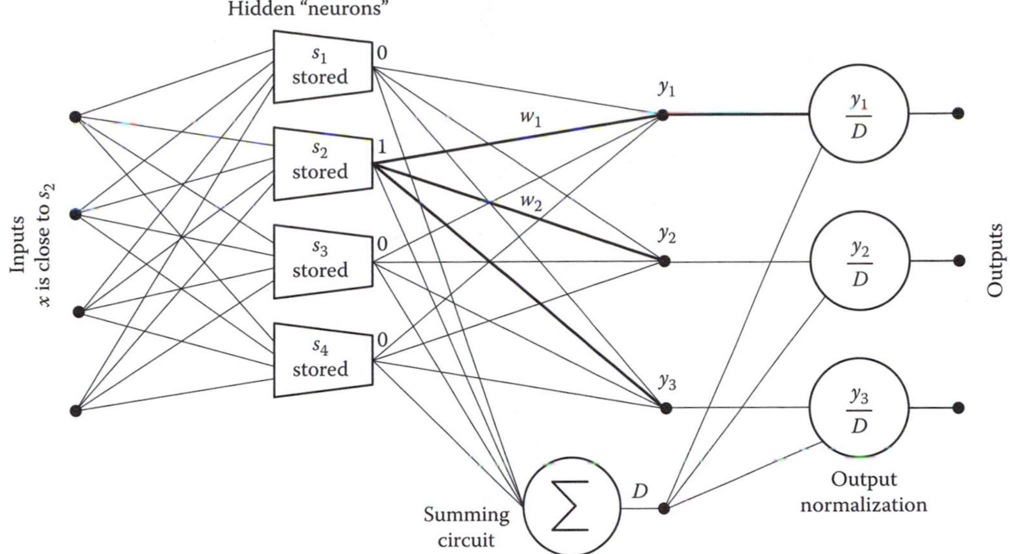

FIGURE 6.13 Radial basis function networks.

with a maximum output magnitude equal to one. This "neuron" is capable of recognizing certain patterns and generating output signals being functions of a similarity.

6.2.9 Implementation of RBF Networks with Sigmoidal Neurons

The network shown in Figure 6.14 has similar property (and power) like RBF networks, but it uses only traditional neurons with sigmoidal activation functions [WJ96]. By augmenting the input space to another dimension the traditional neural network will perform as a RBF network. Please notice that this additional transformation can be made by another neural network. As it is shown in Figure 6.15, 2 first neurons are creating an additional dimension and then simple 8 neurons in one layer feedforward network can solve the two spiral problem. Without this transformation, about 35 neurons are required to solve the same problem with neural network with one hidden layer.

6.2.10 Networks for Solution of Parity-*N* Problems

The most common test benches for neural networks are parity-*N* problems, which are considered to be the most difficult benchmark for neural network training. The simplest parity-2 problem is also known

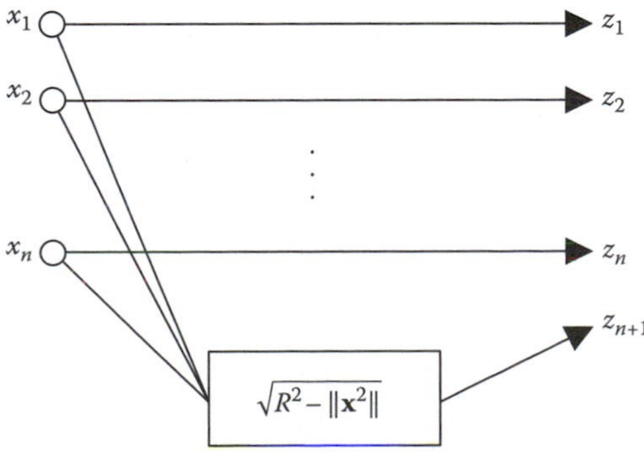

FIGURE 6.14 Transformation, which is required to give a traditional neuron the RBF properties.

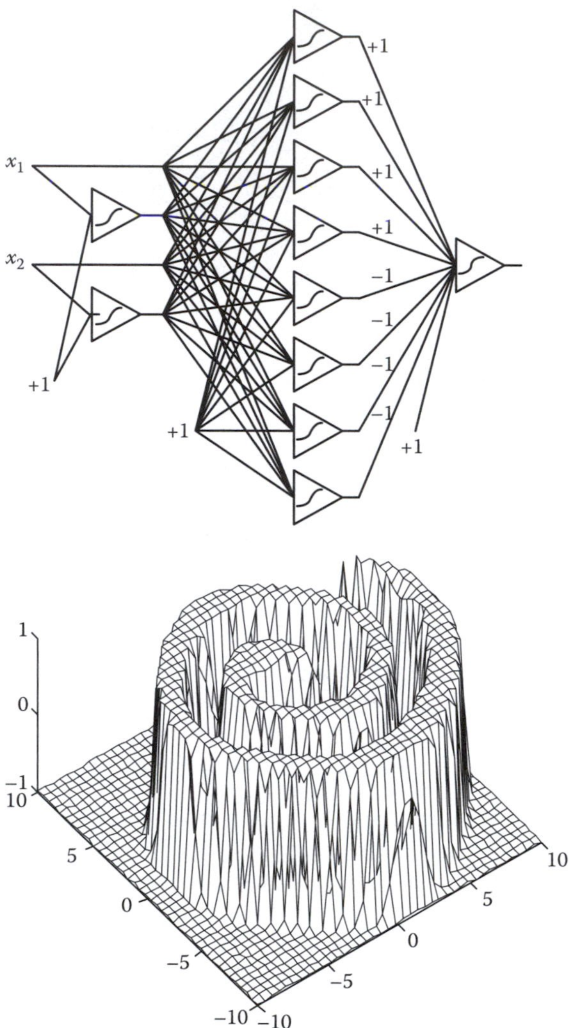

FIGURE 6.15 Solution of two spiral problem using transformation from Figure 6.14 implemented on two additional neurons.

as the *XOR* problem. The larger the *N*, the more difficult it is to solve. Even though parity-*N* problems are very complicated, it is possible to analytically design neural networks to solve them [WHM03,W09]. Let us design neural networks for the parity-7 problem using different neural network architectures with unipolar neurons.

Figure 6.16 shows the multilayer perceptron (MLP) architecture with one hidden layer. In order to properly classify patterns in parity-*N* problems, the location of zeros and ones in the input patterns are not relevant, but it is important how many ones are in the patterns. Therefore, one may assume identical weights equal +1 connected to all inputs. Depending on the number of ones in the pattern, the net values of neurons in the hidden layer are calculated as a sum of inputs times weights. The results may vary from 0 to 7 and will be equal to the number of ones in an input pattern. In order to separate these eight possible cases, we need seven neurons in the hidden layer with thresholds equal to 0.5, 1.5, 2.5, 3.5, 4.5, 5.5, and 6.5. Let us assign positive (+1) and negative (−1) weights to outputs of consecutive neurons starting with +1. One may notice that the net value of the output neuron will be zero for patterns with an odd number of ones and will be one with an even number of ones.

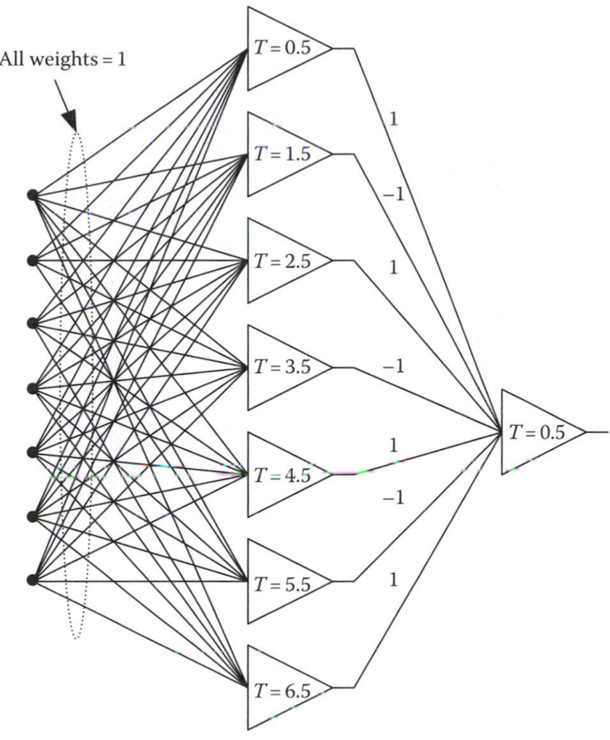

Number of ones in a pattern			0	1	2	3	4	5	6	7
net 1 through *net* 7 (from inputs only)			0	1	2	3	4	5	6	7
out1	$T = 0.5$	$w_1 = 1$	0	1	1	1	1	1	1	1
out2	$T = 1.5$	$w_2 = -1$	0	0	1	1	1	1	1	1
out3	$T = 2.5$	$w_3 = 1$	0	0	0	1	1	1	1	1
out4	$T = 3.5$	$w_4 = -1$	0	0	0	0	1	1	1	1
out5	$T = 4.5$	$w_5 = 1$	0	0	0	0	0	1	1	1
out6	$T = 5.5$	$w_6 = -1$	0	0	0	0	0	0	1	1
out7	$T = 6.5$	$w_7 = 1$	0	0	0	0	0	0	0	1
$net\ 8 = net\ 1 + \sum_{i=1}^{7} w_i * out_i$			0	1	0	1	0	1	0	1
out8 (of output neuron) $T = 0.5$			0	1	0	1	0	1	0	1

FIGURE 6.16 MLP architecture for parity-7 problem. The computation process of the network is shown in the table.

The threshold of +0.5 of the last neuron will just reinforce the same values on the output. The signal flow for this network is shown in the table of Figure 6.16.

In summary, for the case of a MLP neural network the number of neurons in the hidden layer is equal to $N = 7$ and total number of neurons is 8. For other parity-N problems and MLP architecture:

$$\text{Number of neurons} = N + 1 \tag{6.12}$$

Figure 6.17 shows a solution with bridged multilayer perceptron (BMLP) with connections across layers. With this approach the neural network can be significantly simplified. Only three neurons are needed in the hidden layer with thresholds equal to 1.5, 3.5, and 5.5. In this case, all weights associated with outputs of hidden neurons must be equal to −2 while all remaining weights in the network are equal

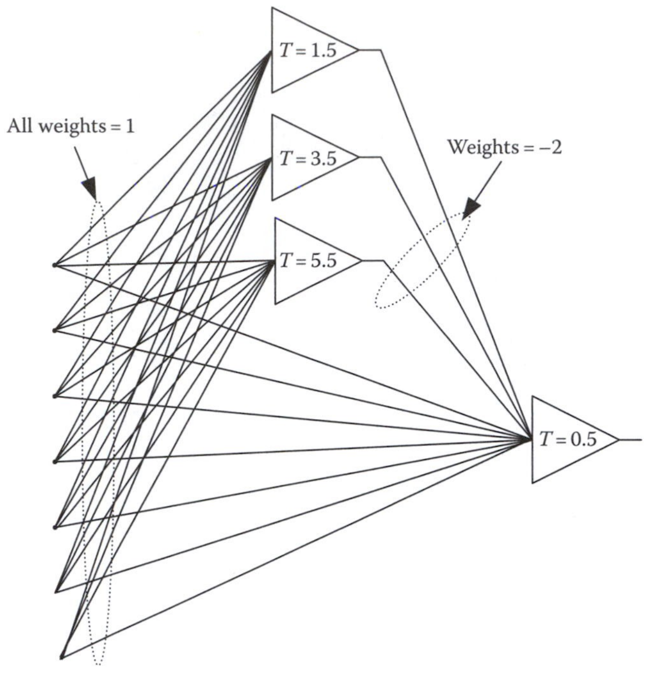

Number of ones in a pattern	0	1	2	3	4	5	6	7
*net*1 (from inputs only)	0	1	2	3	4	5	6	7
out1 $T = 1.5$	0	0	1	1	1	1	1	1
out2 $T = 3.5$	0	0	0	0	1	1	1	1
out3 $T = 5.5$	0	0	0	0	0	0	1	1
*net*1 = *ne*l − 2*(out1 + out2 + out3)	0	1	0	1	0	1	0	1
out4 (of output neuron) $T = 0.5$	0	1	0	1	0	1	0	1

FIGURE 6.17 BMLP architecture with one hidden layer for parity-7 problem. The computation process of the network is shown in the table.

to +1. Signal flow in this BMLP network is shown in the table in Figure 6.17. With bridged connections across layers the number of hidden neurons was reduced to $(N - 1)/2 = 3$ and the total number of neurons is 4. For other parity-N problems and BMLP architecture:

$$\text{Number of neurons} = \begin{cases} \dfrac{N-1}{2}+1 & \text{for odd parity} \\ \dfrac{N}{2}+1 & \text{for even parity} \end{cases} \tag{6.13}$$

Figure 6.18 shows a solution for the fully connected cascade (FCC) architecture for the same parity-7 problem. In this case, only three neurons are needed with thresholds 3.5, 1.5, and 0.5. The first neuron with threshold 3.5 is inactive (out = 0) if the number of ones in an input pattern is less than 4. If the number of ones in an input pattern is 4 or more then the first neuron becomes active and with −4 weights attached to its output it subtracts −4 from the nets of neurons 2 and 3. Instead of [0 1 2 3 4 5 6 7] these neurons will see [0 1 2 3 0 1 2 3]. The second neuron with a threshold of 1.5 and the −2 weight associated

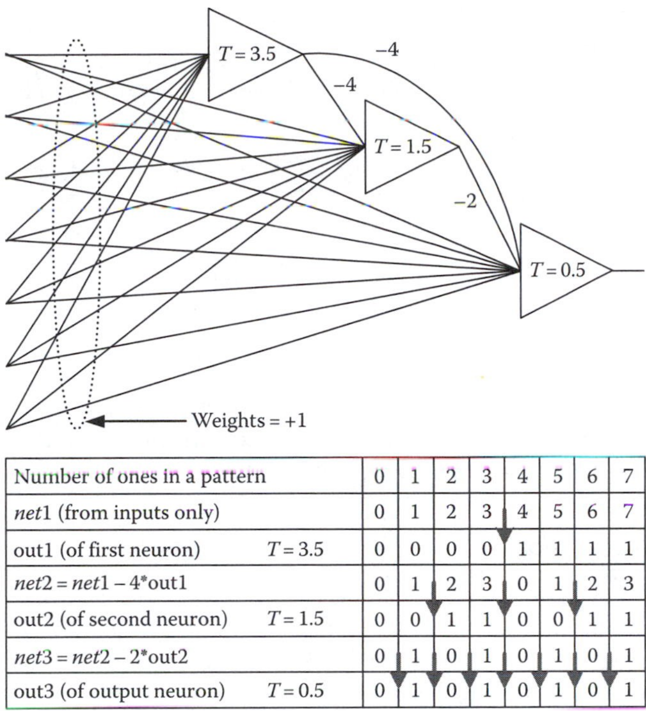

Number of ones in a pattern		0	1	2	3	4	5	6	7
*net*1 (from inputs only)		0	1	2	3	4	5	6	7
out1 (of first neuron)	$T = 3.5$	0	0	0	0	1	1	1	1
*net*2 = *net*1 − 4*out1		0	1	2	3	0	1	2	3
out2 (of second neuron)	$T = 1.5$	0	0	1	1	0	0	1	1
*net*3 = *net*2 − 2*out2		0	1	0	1	0	1	0	1
out3 (of output neuron)	$T = 0.5$	0	1	0	1	0	1	0	1

FIGURE 6.18 FCC architecture for parity-7 problem. The computation process of the network is shown in the table.

with its output works in such a way that the last neuron will see [0 1 0 1 0 1 0 1] instead of [0 1 2 3 0 1 2 3]. For other parity-N problems and FCC architecture:

$$\text{Number of neurons} = \lceil \log 2(N+1) \rceil \tag{6.14}$$

6.2.11 Pulse-Coded Neural Networks

Commonly used artificial neurons behave very differently than biological neurons. In biological neurons, information is sent in a form of pulse trains [WJPM96]. As a result, additional phenomena such as pulse synchronization play an important role and the pulse coded neural networks are much more powerful than traditional artificial neurons. They then can be used very efficiently for example for image filtration [WPJ96]. However, their hardware implementation is much more difficult [OW99,WJK99].

6.3 Comparison of Neural Network Topologies

With the design process, as described in Section 6.2 it is possible to design neural networks to arbitrarily large parity problems using MLP, BMLP, and FCC architectures. Table 6.1 shows comparisons of minimum number of neurons required for these three architectures and various parity-N problems.

As one can see from Table 6.1 and Figures 6.16 through 6.18, the MLP architectures are the least efficient parity-N application. For small parity problems, BMLP and FCC architectures give similar results. For larger parity problems, the FCC architecture has a significant advantage, and this is mostly due to more layers used. With more layers one can also expect better results in BMLP, too. These more powerful neural network architectures require more advanced software to train them [WCKD07,WCKD08,WH10]. Most of the neural network software available in the market may train only MLP networks [DB04,HW09].

TABLE 6.1 Minimum Number of Neurons Required for Various Parity-*N* Problems

	Parity-8	Parity-16	Parity-32	Parity-64
# inputs	8	16	32	64
# patterns	256	65536	4.294e+9	1.845e+19
MLP (one hidden layer)	9	17	33	65
BMLP (one hidden layer)	5	9	17	33
FCC	4	5	6	7

6.4 Recurrent Neural Networks

In contrast to feedforward neural networks, recurrent networks neuron outputs could be connected with their inputs. Thus, signals in the network can continuously be circulated. Until now, only a limited number of recurrent neural networks were described.

6.4.1 Hopfield Network

The single layer recurrent network was analyzed by Hopfield [H82]. This network shown in Figure 6.17 has unipolar hard threshold neurons with outputs equal to 0 or 1. Weights are given by a symmetrical square matrix **W** with zero elements ($w_{ij} = 0$ *for* $i=j$) on the main diagonal. The stability of the system is usually analyzed by means of the *energy function*

$$E = -\frac{1}{2}\sum_{i=1}^{N}\sum_{j=1}^{N} w_{ij}v_i v_j \tag{6.15}$$

It was proved that during signal circulation the energy E of the network decreases and system converges to the stable points. This is especially true when values of system outputs are updated in the asynchronous mode. This means that at the given cycle, only one random output can be changed to the required value. Hopfield also proved that those stable points to which the system converges can be programmed by adjusting the weights using a modified Hebbian [H49] rule

$$\Delta w_{ij} = \Delta w_{ji} = (2v_i - 1)(2v_j - 1) \tag{6.16}$$

Such memory has limited storage capacity. Based on experiments, Hopfield estimated that the maximum number of stored patterns is $0.15N$, where N is the number of neurons.

6.4.2 Autoassociative Memory

Hopfield [H82] extended the concept of his network to autoassociative memories. In the same network structure as shown in Figure 6.19, the bipolar neurons were used with outputs equal to −1 of +1. In this network pattern, $\mathbf{s}_m$ are stored into the weight matrix **W** using autocorrelation algorithm

$$\mathbf{W} = \sum_{m=1}^{M} \mathbf{s}_m \mathbf{s}_m^T - M\mathbf{I} \tag{6.17}$$

where

M is the number of stored pattern

I is the unity matrix

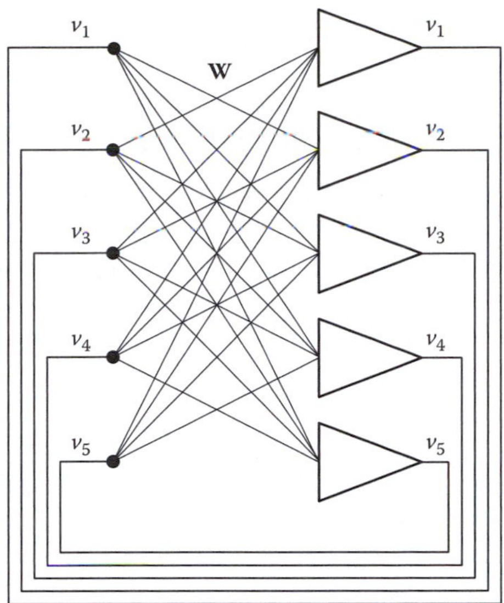

FIGURE 6.19 Autoassociative memory.

Note that **W** is the square symmetrical matrix with elements on the main diagonal equal to zero (w_{ji} *for* $i=j$). Using a modified formula, new patterns can be added or subtracted from memory. When such memory is exposed to a binary bipolar pattern by enforcing the initial network states, then after signal circulation the network will converge to the closest (most similar) stored pattern or to its complement.

This stable point will be at the closest minimum of the energy function

$$E(v) = -\frac{1}{2}\mathbf{v}^T\mathbf{W}\mathbf{v} \tag{6.18}$$

Like the Hopfield network, the autoassociative memory has limited storage capacity, which is estimated to be about $M_{max}=0.15N$. When the number of stored patterns is large and close to the memory capacity,

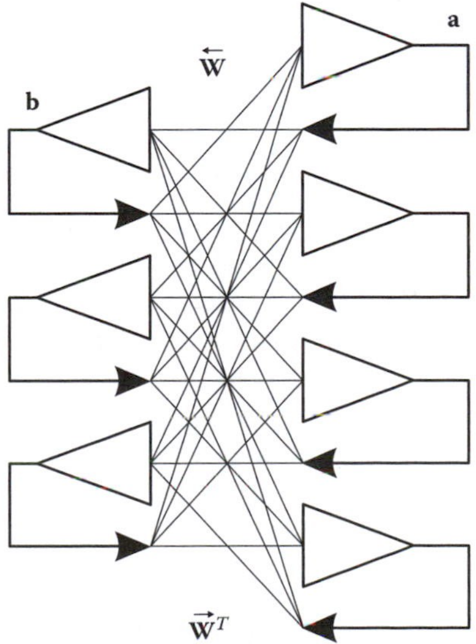

FIGURE 6.20 Bidirectional autoassociative memory.

the network has a tendency to converge to spurious states which were not stored. These spurious states are additional minima of the energy function.

6.4.3 BAM—Bidirectional Autoassociative Memories

The concept of the autoassociative memory was extended to bidirectional associative memories BAM by Kosko [K87]. This memory shown in Figure 6.20 is able to associate pairs of the patterns **a** and **b**.

This is the two layer network with the output of the second layer connected directly to the input of the first layer. The weight matrix of the second layer is $\mathbf{W}^T$ and it is $\mathbf{W}$ for the first layer. The rectangular weight matrix $\mathbf{W}$ is obtained as the sum of the cross correlation matrixes

$$\mathbf{W} = \sum_{m=1}^{M} \mathbf{a}_m \mathbf{b}_m \tag{6.19}$$

where

M is the number of stored pairs

$\mathbf{a}_m$ and $\mathbf{b}_m$ are the stored vector pairs

The BAM concept can be extended for association of three or more vectors.

References

[DB04] H. Demuth and M. Beale, *Neural Network Toolbox, User's Guide*, Version 4. The MathWorks, Inc., Natick, MA, revised for version 4.0.4 edition, October 2004. http://www.mathworks.com

[FF02] N. Fnaiech, F. Fnaiech, and M. Cheriet, A new feed-forward neural network pruning algorithm: Iterative-SSM pruning, *IEEE International Conference on Systems, Man and Cybernetics*, Hammamet, Tunisia, October 6–9, 2002.

[FFJC09] N. Fnaiech, F. Fnaiech, B. W. Jervis, and M. Cheriet, The combined statistical stepwise and iterative neural network pruning algorithm, *Intelligent Automation and Soft Computing*, 15(4), 573–589, 2009.

[FFN01] F. Fnaiech, N. Fnaiech, and M. Najim, A new feed-forward neural network hidden layer's neurons pruning algorithm, *IEEE International Conference on Acoustic Speech and Signal Processing (ICASSP'2001)*, Salt Lake City, UT, 2001.

[FL90] S. E. Fahlman and C. Lebiere, The cascade-correlation learning architecture. In D. S. Touretzky (ed.) *Advances in Neural Information Processing Systems 2*. Morgan Kaufmann, San Mateo, CA, 1990, pp. 524–532.

[H49] D. O. Hebb, *The Organization of Behavior, a Neuropsychological Theory*. Wiley, New York, 1949.

[H82] J. J. Hopfield, Neural networks and physical systems with emergent collective computation abilities. *Proceedings of the National Academy of Science* 79:2554–2558, 1982.

[H99] S. Haykin, *Neural Networks—A Comprehensive Foundation*. Prentice Hall, Upper Saddle River, NJ, 1999.

[HN87] R. Hecht-Nielsen, Counterpropagation networks. *Applied Optics* 26(23):4979–4984, 1987.

[HW09] H. Yu and B. M. Wilamowski, C++ implementation of neural networks trainer. *13th International Conference on Intelligent Engineering Systems (INES-09)*, Barbados, April 16–18, 2009.

[K87] B. Kosko, Adaptive bidirectional associative memories. *App. Opt.* 26:4947–4959, 1987.

[K88] T. Kohonen, The neural phonetic typerater. *IEEE Computer* 27(3):11–22, 1988.

[OW99] Y. Ota and B. M. Wilamowski, Analog implementation of pulse-coupled neural networks. *IEEE Transaction on Neural Networks* 10(3):539–544, May 1999.

[P89] Y. H. Pao, *Adaptive Pattern Recognition and Neural Networks*. Addison-Wesley Publishing Co., Reading, MA, 1989.

[W02] B. M. Wilamowski, Neural networks and fuzzy systems, Chap 32. In Robert R. Bishop (ed.) *Mechatronics Handbook*. CRC Press, Boca Raton, FL, 2002, pp. 33-1–32-26.

[W03] B. M. Wilamowski, Neural network architectures and learning (tutorial). *International Conference on Industrial Technology* (*ICIT'03*), Maribor, Slovenia, December 10–12, 2003.

[W07] B. M. Wilamowski, Neural networks and fuzzy systems for nonlinear applications. *11th International Conference on Intelligent Engineering Systems* (*INES 2007*), Budapest, Hungary, June 29–July 1, 2007, pp. 13–19.

[W09] B. M. Wilamowski, Neural network architectures and learning algorithms. *IEEE Industrial Electronics Magazine* 3(4):56–63.

[W89] P. D. Wasserman, *Neural Computing Theory and Practice*. Van Nostrand Reinhold, New York, 1989.

[W96] B. M. Wilamowski, Neural networks and fuzzy systems, Chaps. 124.1–124.8. In *The Electronic Handbook*. CRC Press, 1996, pp. 1893–1914.

[WB01] B. M. Wilamowski and J. Binfet, Microprocessor implementation of fuzzy systems and neural networks. *International Joint Conference on Neural Networks* (*IJCNN'01*), Washington, DC, July 15–19, 2001, pp. 234–239.

[WCKD07] B. M. Wilamowski, N. J. Cotton, O. Kaynak, and G. Dundar, Method of computing gradient vector and Jacobian matrix in arbitrarily connected neural networks. *IEEE International Symposium on Industrial Electronics* (*ISIE 2007*), Vigo, Spain, June 4–7, 2007, pp. 3298–3303.

[WCKD08] B. M. Wilamowski, N. J. Cotton, O. Kaynak, and G. Dundar, Computing gradient vector and Jacobian matrix in arbitrarily connected neural networks. *IEEE Transactions on Industrial Electronics* 55(10):3784–3790, October 2008.

[WH10] B. M. Wilamowski and H. Yu, Improved computation for Levenberg Marquardt training. *IEEE Transactions on Neural Networks* 21:930–937, 2010.

[WHM03] B. M. Wilamowski, D. Hunter, and A. Malinowski, Solving parity-N problems with feedforward neural network. *Proceedings of the International Joint Conference on Neural Networks* (*IJCNN'03*), Portland, OR, July 20–23, 2003, pp. 2546–2551.

[WJ96] B. M. Wilamowski and R. C. Jaeger, Implementation of RBF type networks by MLP networks. *IEEE International Conference on Neural Networks*, Washington, DC, June 3–6, 1996, pp. 1670–1675.

[WJK99] B. M. Wilamowski, R. C. Jaeger, and M. O. Kaynak, Neuro-fuzzy architecture for CMOS implementation. *IEEE Transactions on Industrial Electronics* 46(6):1132–1136, December 1999.

[WJPM96] B. M. Wilamowski, R. C. Jaeger, M. L. Padgett, and L. J. Myers, CMOS implementation of a pulse-coupled neuron cell. *IEEE International Conference on Neural Networks*, Washington, DC, June 3–6, 1996, pp. 986–990.

[WPJ96] B. M. Wilamowski, M. L. Padgett, and R. C. Jaeger, Pulse-coupled neurons for image filtering. *World Congress of Neural Networks*, San Diego, CA, September 15–20, 1996, pp. 851–854.

[Z92] J. Zurada, *Introduction to Artificial Neural Systems*. West Publishing Company, St. Paul, MN, 1992.

[Z95] J. M. Zurada, *Artificial Neural Systems*. PWS Publishing Company, St. Paul, MN, 1995.

7

Radial-Basis-Function Networks

Åge J. Eide
Ostfold University College

Thomas Lindblad
Royal Institute of Technology

Guy Paillet
General Vision Inc.

7.1 Introduction

Neural networks are characterized by being massively parallel in their architecture and that they use a learning paradigm rather than being programmed. These make neural networks very useful in several areas where they can "learn from examples" and then process new data in a redundant way. In order to take advantage of these features, novel silicon architectures must be conceived. While "computer simulation" of neural networks brings interesting results, they fail to achieve the promise of natively parallel implementations.

7.2 Radial-Basis-Function Networks

Artificial neural networks process sets of data, classifying the sets according to similarities between the sets. A very simple neural network is shown in Figure 7.1. There are more than a dozen different architectures of such neural networks, as well as many ways to process the input data sets. The two inputs in Figure 7.1 may or may not be weighted before reaching a node (the circle) for specific processing. The radial basis function (RBF) represents one such processing method.

For an RBF neural network, the weights following the input layer are all set like one. After the processing, φ_1 and φ_2, the results are weighted, w_1 and w_2, before entering the final summing node in the output layer. Also note the fixed input, +1, with weight b, the use of which will become clear below.

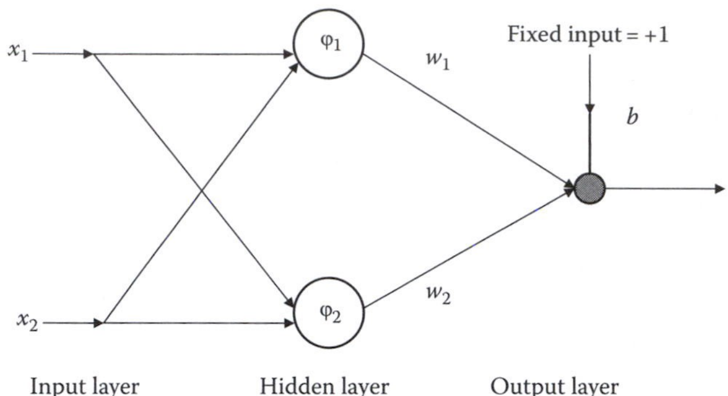

FIGURE 7.1 A simple RBF network. Observe that in most cases other than for RBF networks, the inputs are weighted before reaching the nodes in the hidden layer.

7.2.1 Radial Basis Function

As indicated in Figure 7.1, the RBFs (φ_1 and φ_2) are the hidden layer in the neural network. We define the RBF as a real valued function of the form

$$\varphi(\mathbf{x}) = \varphi(||\,\mathbf{x}\,||) \tag{7.1}$$

It thus depends only on the distance from the origin, or from a point $\mathbf{m}_i$,

$$\varphi(\mathbf{x},\ \mathbf{m}_i) = \varphi(||\,\mathbf{x} - \mathbf{m}_i\,||) \tag{7.2}$$

The norm $||\cdot||$ is usually Euclidean distance. An RBF is typically used to build up function approximations like

$$\varphi(x) = \sum_{i=1}^{N} w_i \varphi(||\,\mathbf{x} - \mathbf{m}_i\,||) \tag{7.3}$$

The RBFs are each associated with different centers, $\mathbf{m}_i$, and weighted by a coefficient, w_i.

Commonly used types of RBFs include ($r = ||\,\mathbf{x} - \mathbf{m}_i||$)

Gaussian:

$$\varphi(r) = \exp\left(-\frac{r^2}{2\sigma}\right) \quad \text{for some } \sigma > 0 \quad \text{and} \quad r \in R \tag{7.4}$$

Multiquadrics:

$$\varphi(r) = \left(\sqrt{r^2 + c^2}\right) \quad \text{for some } c > 0 \quad \text{and} \quad r \in R \tag{7.5}$$

Inverse multiquadrics:

$$\varphi(r) = \frac{1}{\sqrt{r^2 + c^2}} \quad \text{for some } c > 0 \quad \text{and} \quad r \in R \tag{7.6}$$

7.2.2 Use of the RBF: An Example

A typical use of an RBF is to separate complex patterns, that is, the elements of the pattern are not linearly separable. The use of RBFs may solve the problem in one, or if needed, in two steps:

a. Performing a nonlinear transformation, $\varphi_i(\mathbf{x})$ on the input vectors **x**, to a hidden space where the $\varphi_i(\mathbf{x})$s may be linearly separable. The dimension of the hidden space is equal to the number of nonlinear functions $\varphi_i(\mathbf{x})$ taking part in the transformations. The dimension of the hidden space may, as a first try, be set equal to the dimension of the input space.
   If the $\varphi_i(\mathbf{x})$s of the hidden space do not become linearly separable, then one may.
b. Choose a higher dimension on the hidden space, than that of the input space. The "Cover's theorem on the separability of patterns" [19] suggests this will increase the probability of achieving linear separability of the elements in the hidden layer.

An RBF network consists then of an input layer, a hidden layer of nonlinear transfer functions, the $\varphi_i(\mathbf{x})$s, which then are input to an ordinary feed-forward perceptron. Figure 7.1 shows a two input RBF network, with two nodes also in the hidden layer.

In the process of separating complex pattern, the use of an RBF network will consist of choosing a suitable radial function for the hidden layer, here φ_1 and φ_2, and furthermore finding a working set of weights, here w_1, w_2, and b.

Below is an example of this procedure applied on the called XOR problem [19]. In the XOR problem there are four points or patterns in the two dimensional $x_1 - x_2$ plane: (1,1), (0,1), (0,0), and (1,0). The requirement is to construct an RBF network that produces a binary output 0 in response to the input pattern (1,1) or (0,0), and a binary output 1 to the two other input patterns (0,1) or (1,0).

In Figure 7.2, the four points are marked on the $x_1 - x_2$ plane. One easily observes the impossibility of using a straight line to divide the points according to the requirement.

In this example, the RBF used will be a Gaussian

$$\varphi(\mathbf{x},\ \mathbf{t}_i) = G(\|\mathbf{x} - \mathbf{t}_i\|) = \exp(-\|\mathbf{x} - \mathbf{t}_i\|^2), \quad i = 1,\ 2$$

The Gaussian centers $\mathbf{t}_1$ and $\mathbf{t}_2$ are

$\mathbf{t}_1 = [1, 1]$
$\mathbf{t}_2 = [0, 0]$

The relationships between inputs $\mathbf{x}_j$ and output d_j (j = 1, 2, 3, 4) are

$\mathbf{x}_j$	d_j
(1, 1)	0
(0, 1)	1
(0, 0)	0
(1, 0)	1

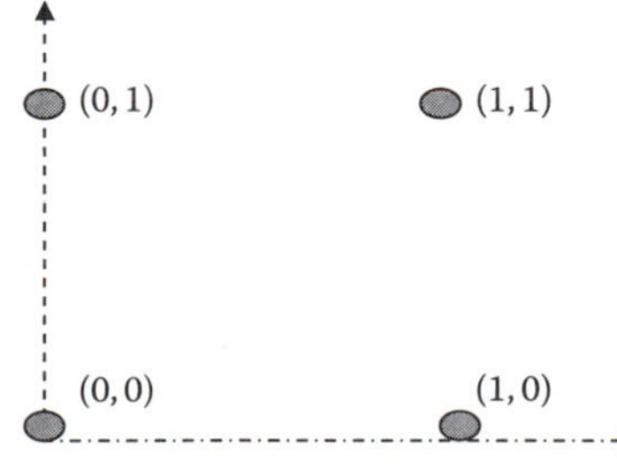

FIGURE 7.2 The four patterns of the XOR problem.

The outputs from the hidden layer will then be modified by the weights w_1 and w_2, such that

$$\sum_{i=1}^{2} w_i G(\|\mathbf{x}_j - \mathbf{t}_i\|) + b = d_j$$

where

i = 1, 2 refers to the two nodes (φ_i) in the hidden layer
j = 1, 2, 3, 4 refers to the four input patterns

The detail in finding the weights is demonstrated in the MATLAB® program below.

In MATLAB notation, the input value x, the corresponding output value *d*, and the centre values *t*:

```
% The input values
x = [1 1;0 1; 0 0;1 0];
% The target values
d = [0 1 0 1]';
% Two neurons with Gaussian centres
t = [1 1; 0 0];
```

With the elements in the *G*-matrix written as

$$g_{ji} = G\left(\|\mathbf{x}_j - \mathbf{t}_i\|\right), \quad j = 1, 2, 3, 4; \quad i = 1, 2$$

that are elements in a 4 by 2 matrix. One has in detail MATLAB notation for the g_{ji}

```
% The Gaussian matrix (4 by 2)
for i=1:2
for j=1:4
    a=(x(j,1)-t(i,1))^2 + (x(j,2)-t(i,2))^2;
    g(j,i)=exp(-a);
end
end
```

with the result

```
g =
1.0000    0.1353
0.3679    0.3679
0.1353    1.0000
0.3679    0.3679
```

where

g_{11} is the output from φ_1 (Figure 7.1) based on input $\mathbf{x}_1 = (1,1)$

g_{12} is the output from φ_2, based on the same input $\mathbf{x}_1 = (1,1)$

Thus, the **g**-matrix displays the transformation from the input plane to the $(\varphi_1\varphi_2)$ plane. A simple plot may show, by inspection, it is possible to separate the two groups by a straight line:

```
for j=1:4
      plot(g(j,1),g(j,2),'or')
      hold on
end
axis([-0.2 1.2 -0.2 1.2])
hold off
grid
```

Now, before adjusting the w_is, the bias b has to be included in the contribution to the output node:

```
% Prepare with bias
b=[1 1 1 1]';
```

so the new weight matrix is

```
% The new Gaussian
G = [g b]
```

with the result

```
G =
1.0000  0.1353  1.0000
0.3679  0.3679  1.0000
0.1353  1.0000  1.0000
0.3679  0.3679  1.0000
```

The perceptron part of the network, the output from the two RBFs together with the bias, may now be presented as

$$\mathbf{Gw} = \mathbf{d}$$

where $\mathbf{w} = [w_1\ w_2\ b]^T$ is the weight vector, which is to be calculated, and

$$\mathbf{d} = [0\ 1\ 0\ 1]^T$$

is the desired output. T indicates a transpose vector. One observes that the matrix **G** is not square (an overdetermined problem) and there is no unique inverse matrix **G**. To overcome this, one may use the pseudoinverse solution [19]:

$$\mathbf{w} = \mathbf{G}^{+}\mathbf{d} = (\mathbf{G}^T\mathbf{G})^{-1}\mathbf{G}^T\mathbf{d}$$

So, the following needs to be calculated in MATLAB:

```
% The transpose: G^T
gt = G';

%An inverse: (G^TG)^-1
gg = gt*G;
gi = inv(gg);

%The new "Gaussian" G+
gp = gi*gt;
```

which gives

```
gp =

 1.8296  -1.2513   0.6731  -1.2513
 0.6731  -1.2513   1.8296  -1.2513
-0.9207   1.4207  -0.9207   1.4207
```

```
%The weights

w = gp*d
```

which gives

```
w =
-2.5027
-2.5027
 2.8413
```

and which may be checked

```
cd=G*w
```

which gives

```
cd =
-0.0000
 1.0000
 0.0000
 1.0000
```

and is identical to the original desired output d.

7.2.3 Radial-Basis-Function Network

The approximation by the sum, mentioned above (7.3), can also be interpreted as a rather simple single-layer type of artificial neural network. Here, the RBFs taking on the role of the activation functions of the neural network. It can be shown that any continuous function on a compact interval can be interpolated with arbitrary accuracy by a sum of this form, given that a sufficiently large number N of RBFs is used.

The strong point of the RBF-net is its capability to separate entangled classes. While the RBF-net will do quite well for the so-called double spiral (Figure 7.3), conventional nets will not do so well (as will be shown later).

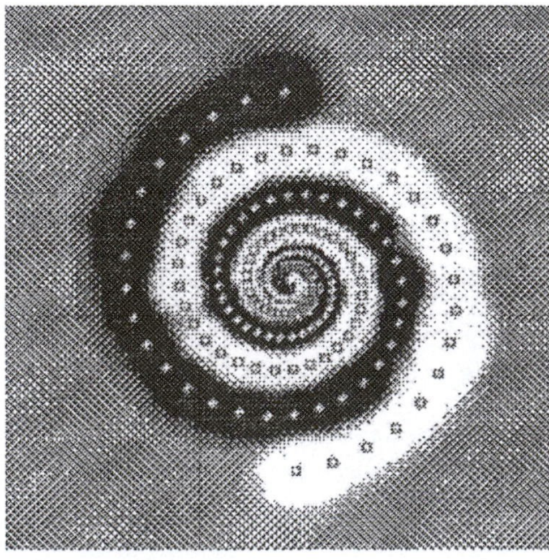

FIGURE 7.3 The twin spiral problem is hard to solve with a conventional feed-forward neural network but easily solved for the RBF-net.

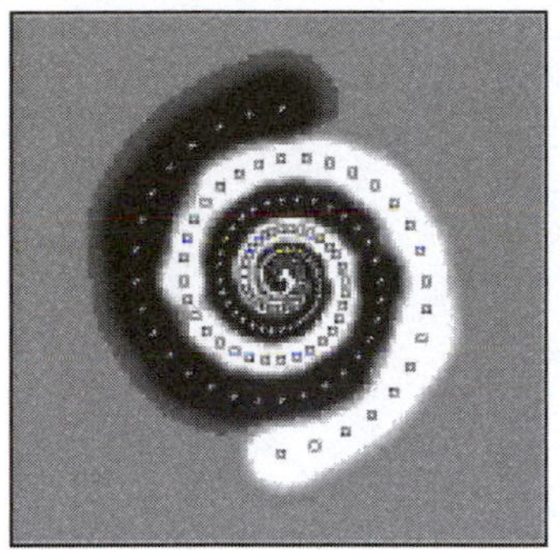
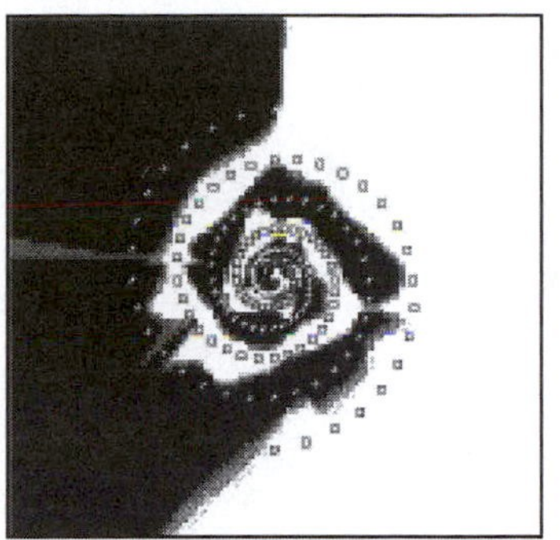
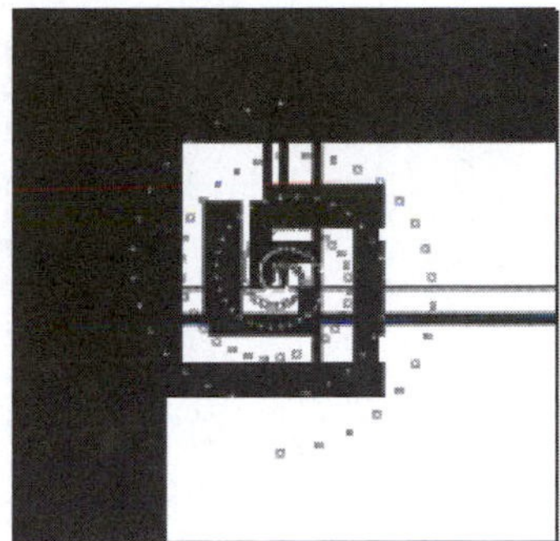

FIGURE 7.4 Results of tests using the double-spiral input. The results are shown for an RBF-DDA case (left), MLP-PROP (middle), and a decision-tree type of neural network (right). (From Ref. [15].)

7.2.4 Learning Paradigms

The most well-known learning paradigm for feed-forward neural networks is the backpropagation (BP) paradigm. It has several salient pros and cons. For the RCC networks, however, the RCC or restricted Coulomb energy (RCE-P) algorithms are generally used. They have the feature that one does not need to fix the number of hidden neuron beforehand. Instead, these are added during training "as needed." The big disadvantage with these paradigms is that the standard deviation is adjusted with one single global parameter. The RCE training algorithm was introduced by Reilly, Cooper, and Elbaum (hence the RCE [18]). The RCE and its probabilistic extension the RCE-P algorithm take advantage of a growing structure in which hidden units, as mentioned, are only introduced when necessary. The nature of both algorithms allows training to reach stability much faster than most other algorithms. But, again, RCE-P networks do not adjust the standard deviation of their prototypes individually (using only one global value for this parameter).

This latter disadvantage is taken care of in the Dynamic Decay Adjustment-algorithm or the DDA-algorithm. It was introduced by Michael R. Berthold and Jay Diamond [13] and yields more efficient networks at the cost of more computer mathematics. It is thus of importance to consider this when implementing any neural network system and considering the task to solve. In a task where the groups are more or less interlaced, the DDA is clearly superior as shown in the "double-spiral test," in Figure 7.4. It could also be mentioned that the DDA took only four epochs to train, while the RPROP (short for resilient backpropagation [20]), is a learning heuristics for supervised learning in artificial neural networks. It is similar to the Manhattan update rule. (Cf., for example, Ref. [20] for details.) The neural network was trained for 40,000 epochs, still with worse result [15]. In spite of the superior results for the RBF-net, the multilayer perceptrons (or MLPs for short) are still the most prominent and well-researched class of neural networks.

7.3 Implementation in Hardware

Generally speaking, very few implementations of neural networks have been done outside the university world, where such implementations have been quite popular. A few chips are or have been available, of which the most well known are the Intel Electrically Trainable Neural Network (ETANN), which used a feed-forward architecture, and the IBM ZISC [3,4], which used an RBF type of architecture as described above. In 2008, an evolution of the ZISC, CogniMem (for cognitive memory) has been introduced by Recognetics Inc.

The ETANN chip and 64 neurons, which could be time multiplexed, that is, used two times. It was possible to connect up to eight ETANN chips, which thus limited the number of neurons. The ETANN was probably developed by Intel in collaboration with the U.S. Navy at China Lake and never entered production. However, it was used in several applications in high-energy physics.

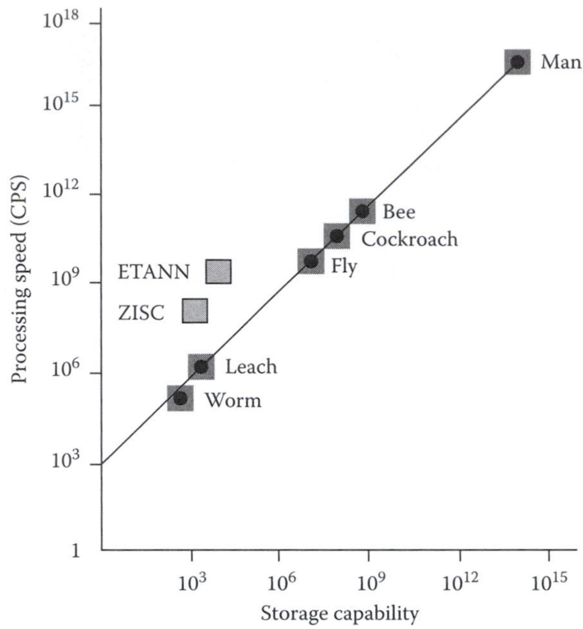

FIGURE 7.5 Comparison between two neural network chips, the ETANN and the ZISC to the number of neurons and connections in the brains of some animals.

The ZISC chip was developed by a team of IBM France, based on a hardware architecture devised by Guy Paillet. This hardware architecture is at the intersection of parallel processing technology and the RBF model (Figure 7.5).

Comparing the ETANN and ZISC to biological species may be wrong, but anyhow sets the silicon in perspective to various animals (Figure 7.6).

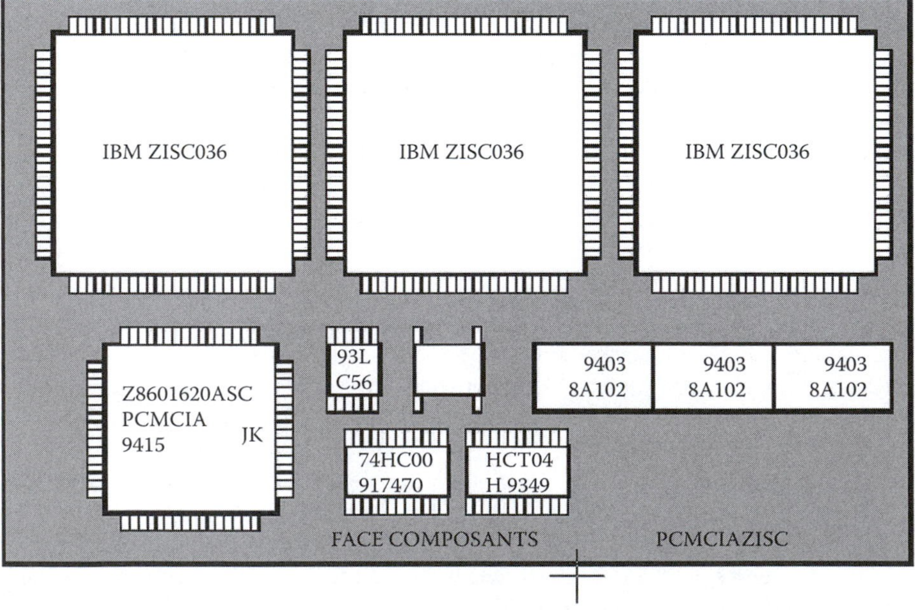

FIGURE 7.6 Three ZISC chips were conveniently mounted on a PCMCIA card.

7.3.1 Implementation for High-Energy Physics

Although the number of cascaded-ZISC chips may be application or bus dependent, there should be no problem using up to 10 chips. Building larger networks is a matter of grouping small networks together by placing re-powering devices where needed [1]. Bus and CPU interface examples are found in the *ZISC036 Data Book* [1].

The VMEbus and VME modules are used by researchers in physics. A multipurpose, experimental VMEbus card was used in a demonstration (cf. Figure 7.7 and also Ref. [5] for details on an IBM/ISA implementation) The VME-board holds four piggy-back PCBs with one chip each (cf. Figure 7.7). The PCBs, holding the ZISCs, are made to carry another card on top using Euro-connectors. Hence up to 40 ZlSCs could, in principle, be mounted in four "ZISC-towers" on the VME-card.

In an early study of the ZISC036 using a PC/ISA-board [5], the computer codes were written in Borland C++ under DOS and Windows. In the VME implementation, we rely on a VME to SBus hardware interface and pertinent software. This software is written using the GNU C++ and the VMIC SBus interface library.

As mentioned previously, a neural network of the RBF-type [9–11] is somewhat different from more conventional NNW architectures. In very general terms, the approach is to map an N-dimensional space by prototypes. Each of these prototypes is associated with a category and an influence field representing a part of the N-dimensional space around the prototype. Input vectors within that field are assigned the category of that prototype. (In the ZISC implementation, the influence fields are represented by

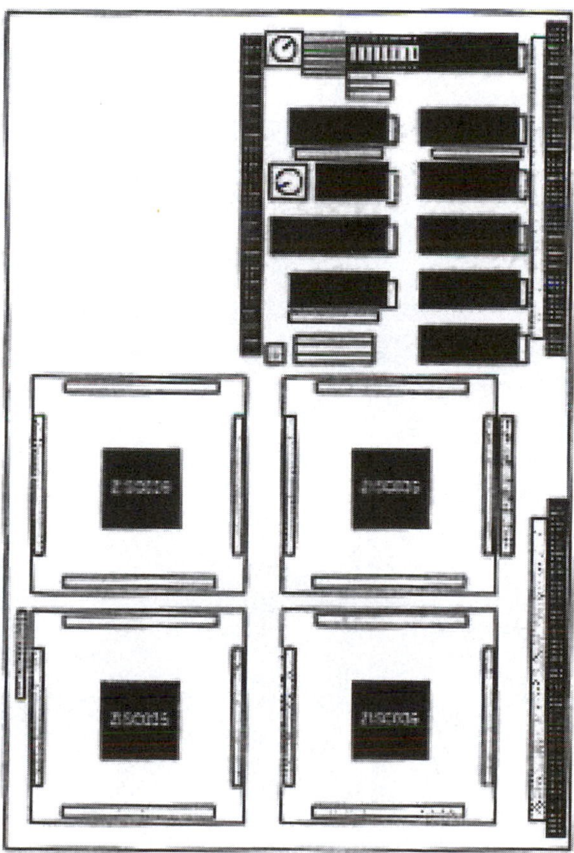

FIGURE 7.7 Schematic layout of the VME/ZISC036 board. The lower part shows the piggy-back area, which can hold 4–40 ZISC chips.

hyper-polygons rather than hyper-spheres as in a more theoretical model. Two user-selectable distance norms are supported by the chip [1]). Several prototypes can be associated with one category and influence fields may overlap.

There are several learning algorithms associated with the RBF-architecture, but the most common ones are the RCE [9] and RCE-like ones. The one used by the ZISC chip is "RCE-like." A nearest-neighbor evaluation is also available.

We have added the Intel ETANN and the Bellcore CLNN32/64 neural nets [12] to the ZISC036 with a LHC physics "benchmark test." The inputs in this test are the moments and transverse moments of the four leading particles, obtained in a simulation of a LHC search for a heavy Higgs (cf. Ref. [12] for details). Two-dimensional plots of these moments (p versus p_t), for the leading particle, are shown in Figure 7.8.

Although only some preliminary results have been obtained, it is fair to say that a system with eight inputs and just 72 RBF-neurons could recognize the Higgs to a level of just above 70% and the background to about 85%. This is almost as good as the CLNN32/64 chips discussed in Ref. [12]. Further details and results will be presented at the AIHENP-95 [13].

In 2008, a powerful evolution of the ZISC, CM1K (CogniMem 1024 neurons), was released by CogniMem Ltd. based on a new implementation by Anne Menendez and Guy Paillet. The neuron density as well as additional features such as low power, high speed, and small size allow CM1K to operate both as "near sensor trainable pattern recognition" and as a pattern recognition server, taking especially advantage or virtually unlimited expendability. A single CM1K chip achieves 2.6 Billion CUPS

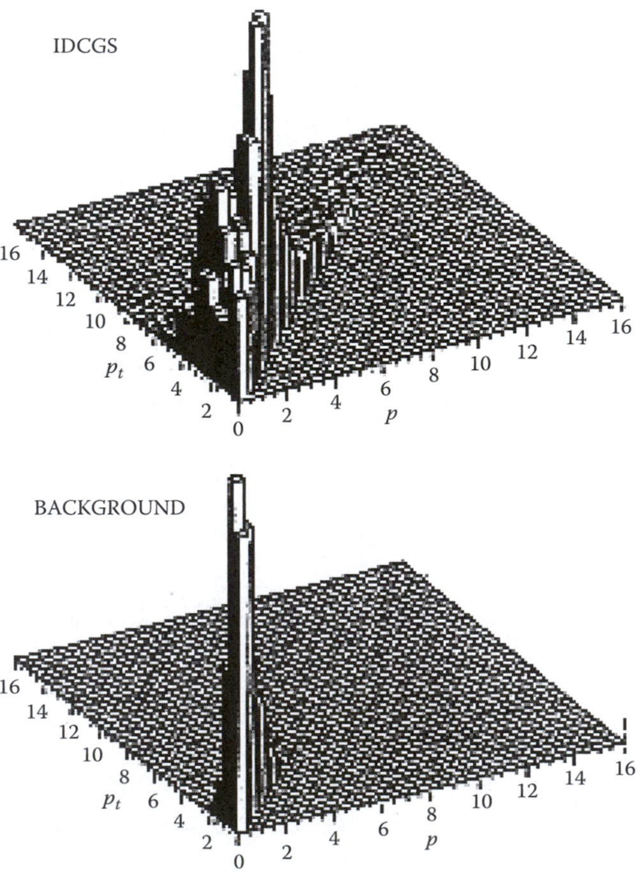

FIGURE 7.8 Examples of inputs. The input neurons 1 and 2 get their values from either of the two histograms shown here (there are three additional histograms for inputs 3 and 8).

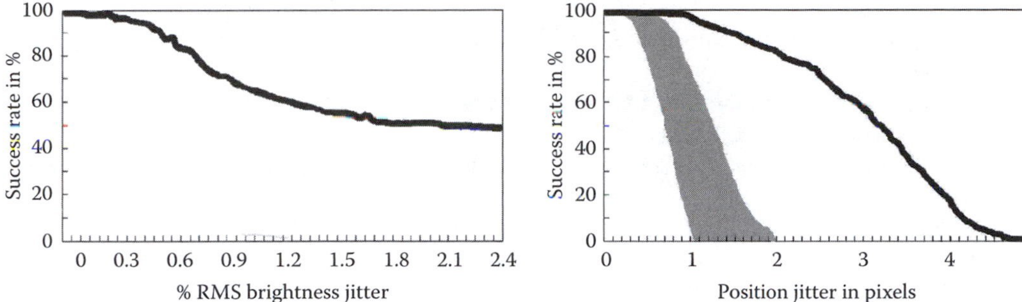

FIGURE 7.9 Star tracker results as discussed in the text. The shaded area in the figure to the right indicates where most star tracker data falls, showing that the RBF-net does very well when it comes to position jitter robustness.

with a 256kB storage capacity. An architecture featuring 1024 CM1K will reach 2.6 Trillion CUPS with 256MB storage capacity. Practically, such a system will be able to find the nearest neighbor of one 256 bytes vector versus one millions in less than 10μs. Finding the subsequent closest neighbor will take an additional 1.3μs per neighbor vector.

7.3.2 Implementing a Star Tracker for Satellites

A star tracker is an optical device measuring the direction to one or more stars, using a photocell or solid-state camera to observe the star. One may use a single star, and of the most used are Sirius (the brightest) and Canoponus. However, for more complex missions, entire star field databases are used to identify orientation.

Using the ZISC chip to implement a star tracker turned out to be very successful [17]. The RMS brightness jitter as well as the position jitter in pixels turned out to be equally good or even better than most conventional star tracker of same complexity. This is shown in Figure 7.9.

7.3.3 Implementing an RBF network in VHDL

This may seem as a very good idea and has been done in several university projects [14]. Here, we simply refer one diploma report [16]. This report uses the VHDL and Xilinx ISE 4.1i. Explanation about how components were created in Xilinx and generated in CoreGenerator. It also contains VHDL code, wiring diagram, and explanation how to work with Xilinx in creating new projects and files. All solutions are described and presented together with component diagrams to show what the various kinds of components do and how they connect to each other.

References

1. *ZISC036 Data Book*, IBM Essonnes Components Development Laboratory, IBM Microelectronics, Corbeil-Essonnes, France.
2. J.-P. LeBouquin, *IBM Microelectronics ZISC, Zero Instruction Set Computer Preliminary Information*, Poster Show WCNN, San Diego, CA, 1994 and addendum to conference proceedings.
3. J.-P. LeBouquin, IBM Essones Components Development Laboratory, Corbeil-Essonnes, France, private communications.
4. J.P. LeBouquin, *ZISC036 Data Book*, version 2.1 (11/1994) IBM Microelectronics ZISC Neural Network Series, ZISC/ISA Accelerator for PC (Users Manual), Prel., ver. 0 (11/1994).
5. M.J.D. Powell, Radial basis functions for multivariable interpolation: A review, In: D.F. Griffiths and G.A. Watson (eds.), *Numerical Analysis*, Longman Scientific and Technical, Harlow, U.K., 1987, pp. 223–241.

6. M.J.D. Powell, Radial basis functions for multivariable interpolation: A review. *IMA Conference on Algorithms for the Approximation of Functions and Data*, RMCS Shrivenham, U.K., 1985, pp. 143–167.
7. J.E. Moody and C.J. Darken, Fast learning in networks of locally-tuned processing units, *Neural Computation*, 1, 281–294, 1989.
8. D.-L. Reilly, L.N. Cooper, and C. Elbaum, A neural model for category learning, *Biological Cybernetics*, 45, 35–41, 1982.
9. S. Renals, Radial basis function for speech pattern classification, *Electronics Letters*, 25, 437–439, 1989.
10. S. Haykin, *Neural Networks, A Comprehensive Foundation*, IEEE Press, Washington, DC, ISBN 0-02-352761-1.
11. A. Eide, C.S. Lindsey, Th. Lindblad, M. Minerskjld, G. Szekely, and G. Sekhiniadze, An implementation of the Zero Instruction Set Computers (ZISC036) on a PC/ISA-bus card, Invited talk, 1994 WNN/FNN WDC (December 1994) and to be published by SPIE.
12. Th. Lindblad, C.S. Lindsey, F. Block, and A. Jayakumar, Using software and hardware neural networks in a Higgs search, *Nuclear Instruments and Methods A*, 356, 498–506, February 15, 1995.
13. M.R. Berthold and J. Diamond, *Boosting the Performance of the RBF Networks with Dynamic Decay Adjustment*, in G. Tesauro, D.S. Touretsky, and T.K. Leen (eds), Advances in Neural Information Processing Systems 7, MIT Press, Cambridge, MA, 1995.
14. Th. Lindblad, G. Székely, M.L. Padgett, Å.J. Eide, and C.S. Lindsey, Implementing the dynamic decay algorithm in a CNAPS parallel computer system, *Nuclear Instruments and Methods A*, 381, 502–207, 1996.
15. D. Berggren and A. Erlandsson, ZISC78—Hardware implementation of neural networks, a comparison, Diploma work, KTH, Stockholm, Sweden, TRITA-FYS 2002:24 ISSN 0280-316X ISRN KTH/FYS/2002:24-SE.
16. S. Kurjakovic and A. Svennberg, Implementering av RBF i VHDL (in Swedish), Diploma work, KTH, Stockholm, Sweden, TRITA-FYS 2002:45 ISSN 0280-316X ISRN KTH/FYS/-02:45-SE.
17. C.S. Lindsey, A.J. Eide, and Th. Lindblad, unpublished.
18. D.L. Reilly, L.N. Cooper, and C. Elbaum, A neural model for category learning, *Biological Cybernetics*, 45, 35–41, 1982.
19. S. Haykin, *Neural Networks*, Prentice Hall, Upper Saddle River, NJ, 1999.
20. M. Riedmiller, Rprop-description and implementation detail, University of Karlsruhe, Technical Report 1994. http://citeseer.ist.psu.edu/cache/papers/cs2/20/http:zSzzSzamy.informatik.uos.dezSzriedmillerzSzpublicationszSzrprop.details.pdf/riedmiller94rprop.pdf

8

GMDH Neural Networks

Marcin Mrugalski
University of Zielona Góra

Józef Korbicz
University of Zielona Góra

8.1 Introduction

The scope of applications of mathematical models in contemporary industrial systems is very broad and includes system design [12], control [5,6,27], and diagnosis [15,16,19,22–24]. The models are usually created on the basis of the physical laws describing the system behavior. Unfortunately, in the case of most industrial systems, these laws are too complex or unknown. Thus, the phenomenological models are often not available. In order to solve this problem, the system identification approach can be applied [22,25].

One of the most popular nonlinear system identification approaches is based on the application of artificial neural networks (ANNs) [3,11,21]. These can be most adequately characterized as computational models with particular properties such as generalization abilities, the ability to learn, parallel data processing, and good approximation of nonlinear systems. However, ANNs, despite the small number of assumptions in comparison to analytical methods [4,9,15,22], still require a significant amount of a priori information about the model's structure. Moreover, there are no efficient algorithms for selecting structures of classical ANNs, and hence many experiments should be carried out to obtain an appropriate configuration. Experts should decide on the quality and quantity of inputs, the number of layers and neurons, as well as the form of their activation function. The heuristic approach that follows the determination of the network architecture corresponds to a subjective choice of the final model, which, in the majority of cases, will not approximate with the required quality.

To tackle this problem, the Group Method of Data Handling (GMDH) approach can be employed [14,19]. The concept of this approach is based on iterative processing of an operation defined as a sequence leading to the evolution of the resulting neural network structure. The GMDH approach also allows developing the formula of the GMDH model due to inclusion of the additional procedures, which can be used to extend the scope of the application. GMDH neural models can be used in the identification of static and dynamic systems, both single-input single-output (SISO) and multi-input multi-output (MIMO).

The main objective of this chapter is to present the structure and properties of GMDH neural networks. In particular, the chapter is organized as follows: Section 8.2 outlines the fundamentals of GMDH neural networks and presents the processes of the synthesis of structure and parameters estimation of GMDH network. Section 8.3 is devoted to generalizations of GMDH approach. In particular, the problem of the modeling of dynamical systems is considered. Moreover, the processes of the synthesis of single-input and multi-output GMDH neural networks are described. The subsequent Section 8.4 shows an application of the GMDH neural model to the robust fault detection. Finally, Section 8.5 concludes the chapter.

8.2 Fundamentals of GMDH Neural Networks

The concept of the GMDH approach relies on replacing the complex neural model by the set of hierarchically connected partial models. The model is obtained as a result of neural network structure synthesis with the application of the GMDH algorithm [7,14]. The synthesis process consists of partial model structure selection and parameter estimation. The parameters of each partial model (a neuron) are estimated separately. In the next step of process synthesis, the partial models are evaluated, selected, and included to the newly created neuron layers (Figure 8.1). During the network synthesis, new layers are added to the network. The process of network synthesis leads to the evolution of the resulting model structure to obtain the best quality approximation of real system output signals. The process is completed when the optimal degree of network complexity is achieved.

8.2.1 Synthesis of the GMDH Neural Network

Based on the kth measurement of the system inputs $u(k) \in \mathbb{R}'^{n_u}$, the GMDH network grows its first layer of neurons. It is assumed that all the possible couples of inputs from $u_1^{(l)}(k),\ldots,u_{n_u}^{(l)}(k)$, belonging to the training data set T, constitute the stimulation, which results in the formation of the neurons outputs $\hat{y}_n^{(l)}(k)$:

$$\hat{y}_n^{(l)}(k) = f(\boldsymbol{u}(k)) = f(u_1^{(l)}(k),\ldots,u_{n_u}^{(l)}(k)), \tag{8.1}$$

where

l is the layer number of the GMDH network

n is the neuron number in the lth layer

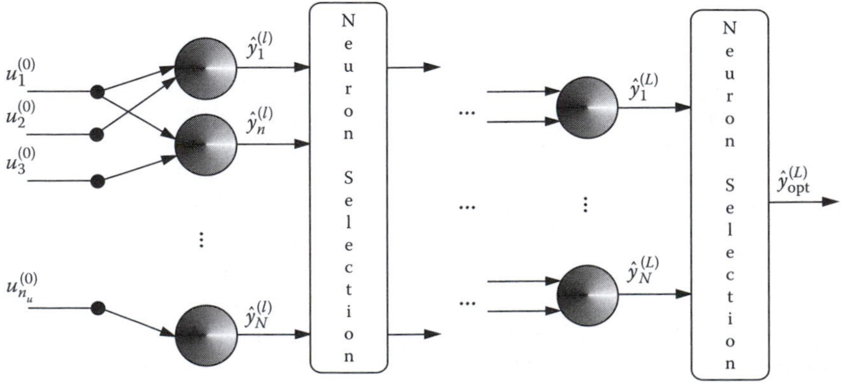

FIGURE 8.1 Synthesis of the GMDH-type neural network.

The GMDH approach allows much freedom in defining an elementary model transfer function f (e.g., tangent or logarithmic functions) [14,20]. The original GMDH algorithm developed by Ivakhnenko [13] is based on linear or second-order polynomial transfer functions, such as

$$f(u_i(k), u_j(k)) = p_0 + p_1 u_i(k) + p_2 u_j(k) + p_3 u_i(k) u_j(k) + p_4 u_i^2(k) + p_5 u_j^2(k). \tag{8.2}$$

In this case, after network synthesis, the general relation between the model inputs and the output $\hat{y}(k)$ can be described in the following way:

$$\hat{y} = f(\boldsymbol{u}(k)) = p_0 + \sum_{i=1}^{n_u} p_i u_i(k) + \sum_{i=1}^{n_u} \sum_{j=1}^{n_u} p_{ij} u_i(k) u_j(k) + ,\ldots, \tag{8.3}$$

From the practical point of view, (8.3) should be not too complex because it may complicate the learning process and extend the computation time. In general, in the case of the identification of static nonlinear systems, the partial model can be described as follows:

$$\hat{y}_n^{(l)}(k) = \xi\left(\left(\boldsymbol{r}_n^{(l)}(k)\right)^T \boldsymbol{p}_n^{(l)}\right), \tag{8.4}$$

where $\xi(\cdot)$ denotes a nonlinear invertible activation function, that is, there exists $\xi^{-1}(\cdot)$. Moreover, $\boldsymbol{r}_n^{(l)}(k) = \boldsymbol{f}\left([u_i^{(l)}(k), u_j^{(l)}(k)]^T\right)$, $i, j = 1, \ldots, n_u$ and $\boldsymbol{p}_n^{(l)} \in \mathbb{R}^{n_p}$ are the regressor and parameter vectors, respectively, and $\boldsymbol{f}(\cdot)$ is an arbitrary bivariate vector function, for example, $\boldsymbol{f}(\boldsymbol{x}) = [x_1^2, x_2^2, x_1 x_2, x_1, x_2, 1]^T$, that corresponds to the bivariate polynomial of the second degree.

The number of neurons in the first layer of the GMDH network depends on the number of the external inputs n_u:

$$\begin{cases} \hat{y}_1^{(1)}(k) = f(u_1^{(1)}(k), u_2^{(1)}(k), \hat{\boldsymbol{p}}_{1,2}) \\ \hat{y}_2^{(1)}(k) = f(u_1^{(1)}(k), u_3^{(1)}(k), \hat{\boldsymbol{p}}_{1,3}) \\ \ldots \\ \hat{y}_{n_y}^{(1)}(k) = f(u_{n_u-1}^{(1)}(k), u_{n_u}^{(1)}(k), \hat{\boldsymbol{p}}_{n_u-1,n_u}) \end{cases},$$

where $\hat{\boldsymbol{p}}_{1,2}, \hat{\boldsymbol{p}}_{1,3}, \ldots, \hat{\boldsymbol{p}}_{n_u-1,n_u}$ are estimates of the network parameters and should be obtained during the identification process.

At the next stage of GMDH network synthesis, a validation data set v, not employed during the parameter estimation phase, is used to calculate a processing error of each partial model in the current lth network layer. The processing error can be calculated with the application of the evaluation criterion such as: the final prediction error (FPE), the Akaike information criterion (AIC) or the F-test. Based on the defined evaluation criterion it is possible to select the best-fitted neurons in the layer. Selection methods in GMDH neural networks play the role of a mechanism of structural optimization at the stage of constructing a new layer of neurons. During the selection, neurons that have too large a value of the evaluation criterion $Q(\hat{y}_n^{(l)}(k))$ are rejected.

A few methods of performing the selection procedure can be applied [20]. One of the most-often used is the *constant population method*. It is based on a selection of g neurons, whose evaluation criterion $Q(\hat{y}_n^{(l)}(k))$ reaches the least values. The constant g is chosen in an empirical way and the most important advantage of this method is its simplicity of implementation. Unfortunately, constant population method has very restrictive structure evolution possibilities. One way out of this problem is the application of the *optimal population method*. This approach is based on rejecting the neurons whose value of the evaluation criterion is larger than an arbitrarily determined threshold e_h. The threshold is usually selected for each layer in an empirical way depending on the task considered. The difficulty with the selection of the threshold results in the fact that the optimal population method is not applied too often. One of the most interesting ways of performing the selection procedure is the application of the method based on the soft selection approach. An outline of the *soft selection method* [18] is as follows:

Input: The set of all n_y neurons in the lth layer, n_j—the number of opponent neurons, n_w—the number of winnings required for nth neuron selection.

Output: The set of neurons after selection.

1. Calculate the evaluation criterion $Q(\hat{y}_n^{(l)}(k))$ for $n = 1, \ldots, n_y$ neurons.
2. Conduct a series of n_y competitions between each nth neuron in the layer and n_j randomly selected neurons (the so-called opponent) from the same layer. The nth neuron is the so-called winner neuron when

$$Q(\hat{y}_n^{(l)}(k)) \leq Q(\hat{y}_j^{(l)}(k)), \quad j = 1, \ldots, n_j,$$

   where $\hat{y}_j^{(l)}(k)$ denotes a signal generated by the opponent neuron.
3. Select the neurons for the $(l+1)$-th layer with the number of winnings bigger than n_w (the remaining neurons are removed).

The property of soft selection follows from the specific series of competitions. It may happen that the potentially unfitted neuron is selected. Everything depends on its score in the series of competition. The main advantage of such an approach in comparison with other selection methods is that it is possible to use potentially unfitted neurons, which in the next layers may improve the quality of the model. Moreover, if the neural network is not fitted perfectly to the identification data set, it is possible to achieve a network that possesses better generalization abilities. One of the most important parameters that should be chosen in the selection process is the number of n_j opponents. A bigger value of n_j makes the probability of the selection of a neuron with a small quality index lower. In this way, in an extreme situation, when $n_j \gg n_y$, the soft selection method will behave as the constant population method, which is based on the selection only of the best-fitted neurons. Some experimental results performed on a number of selected examples indicate that the soft selection method makes it possible to obtain a more flexible network structure. Another advantage, in comparison to the optimal population method, is that an arbitrary selection of the threshold is avoided. Instead, we have to select a number of winnings n_w. It is, of course, a less sophisticated task.

After the selection procedure, the outputs of the selected neurons become the inputs to other neurons in the next layer:

$$\begin{cases} u_1^{(l+1)}(k) = \hat{y}_1^{(l)}(k), \\ u_2^{(l+1)}(k) = \hat{y}_2^{(l)}(k), \\ \quad \ldots \\ u_{n_u}^{(l+1)}(k) = \hat{y}_{n_y}^{(l)}(k). \end{cases} \tag{8.5}$$

In an analogous way, the new neurons in the next layers of the network are created. During the synthesis of the GMDH network, the number of layers suitably increases. Each time a new layer is added, new neurons are introduced. The synthesis of the GMDH network is completed when the optimum criterion is achieved. The idea of this criterion relies on the determination of the quality

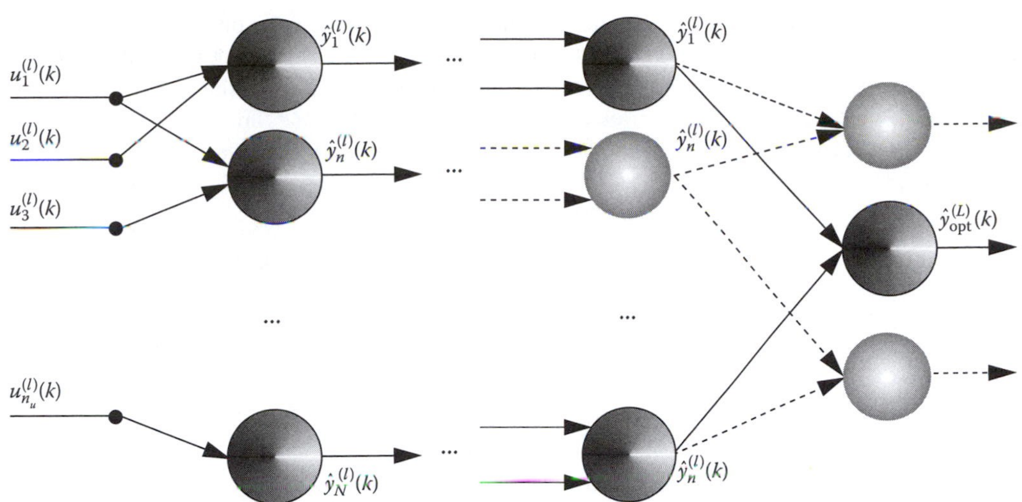

FIGURE 8.2 Final structure of the GMDH neural network.

index $Q(\hat{y}_n^{(l)}(k))$ for all N neurons included in the l layer. $Q_{min}^{(l)}$ represents the processing error for the best neuron in this layer:

$$Q_{min}^{(l)} = \min_{n=1,\ldots,N} Q(\hat{y}_n^{(l)}(k)). \tag{8.6}$$

The values $Q(\hat{y}_n^{(l)}(k))$ can be determined with the application of the defined evaluation criterion, which was used in the selection process. The values $Q_{min}^{(l)}$ are calculated for each layer in the network. The optimum criterion is achieved when the following condition occurs:

$$Q_{opt}^{(L)} = \min_{l=1,\ldots,L} Q_{min}^{(l)}. \tag{8.7}$$

$Q_{opt}^{(L)}$ represents the processing error for the best neuron in the network, which generates the model output. In other words, when additional layers do not improve the performance of the network, the synthesis process is stopped.

To obtain the final structure of the network (Figure 8.2), all unnecessary neurons are removed, leaving only those that are relevant to the computation of the model output. The procedure of removing unnecessary neurons is the last stage of the synthesis of the GMDH neural network.

8.2.2 Parameters Estimation of the GMDH Neural Network

The application of the GMDH approach during neural network synthesis allows us to apply parameter estimation of linear-in-parameters model algorithms, for example, the least mean square (LMS) [7,14]. This follows from the fact that the parameters of each partial model are estimated separately and the neuron's activation function $\xi(\cdot)$ fulfills the following conditions:

1. $\xi(\cdot)$ is continuous and bounded, that is, $\forall u \in \mathbb{R}: a < \xi(u) < b$.
2. $\xi(\cdot)$ is monotonically increasing, that is, $\forall u, y \in \mathbb{R}: u \leq y$ iff $\xi(u) \leq \xi(y)$.
3. $\xi(\cdot)$ is invertible, that is, there exists $\xi^{-1}(\cdot)$.

The advantage of the LMS approach is the simple computation algorithm that gives good results even for small sets of measuring data. Unfortunately, the usual statistical parameter estimation framework assumes that the data are corrupted by errors, which can be modeled as realizations of independent random variables with a known or parameterized distribution. A more realistic approach is to assume that the errors lie between given prior bounds. It leads directly to the bounded error set estimation class of algorithms, and one of them, called the outer bounding ellipsoid (OBE) algorithm [17,26], can be employed to solve the parameter estimation problem considered.

The OBE algorithm requires the system output to be described in the form

$$y_n^{(l)}(k) = \left(\boldsymbol{r}_n^{(l)}(k)\right)^T \boldsymbol{p}_n^{(l)} + \varepsilon_n^{(l)}(k). \tag{8.8}$$

Moreover, it is assumed that the output error, $\varepsilon_n^{(l)}(k)$, can be defined as

$$\varepsilon_n^{(l)}(k) = y_n^{(l)}(k) - \hat{y}_n^{(l)}(k), \tag{8.9}$$

where

$y_n^{(l)}(k)$ is the kth scalar measurement of the system output
$\hat{y}_n^{(l)}(k)$ is the corresponding neuron output

The problem is to estimate the parameter vector $\boldsymbol{p}_n^{(l)}$, that is, to obtain $\hat{\boldsymbol{p}}_n^{(l)}$, as well as an associated parameter uncertainty in the form of the admissible parameter space $\mathbb{E}$. In order to simplify the notation, the index $_n^{(l)}$ is omitted. As has been already mentioned, it is possible to assume that $\varepsilon(k)$ lies between given prior bounds. In this case, the output error is assumed to be bounded as follows:

$$\varepsilon^m(k) \le \varepsilon(k) \le \varepsilon^M(k), \tag{8.10}$$

where the bounds $\varepsilon^m(k)$ and $\varepsilon^m(k)$ ($\varepsilon^m(k) \neq \varepsilon^M(k)$) can be estimated [29] or are known a priori [17]. An example can be provided by data collected with an analogue-to-digital converter or for measurements performed with a sensor of a given type. Based on the measurements $\{\mathbf{r}(k), y(k)\}$, $k = 1\ldots n_T$ and the error bounds (8.10), a finite number of linear inequalities is defined. Each inequality associated with the kth measurement can be put in the following standard form:

$$-1 \le \bar{y}(k) - \bar{\hat{y}}(k) \le 1, \tag{8.11}$$

where

$$\bar{y}(k) = \frac{2y(k) - \varepsilon^M(k) - \varepsilon^m(k)}{\varepsilon^M(k) - \varepsilon^m(k)}, \tag{8.12}$$

$$\bar{\hat{y}}(k) = \frac{2}{\varepsilon^M(k) - \varepsilon^m(k)}\hat{y}(k). \tag{8.13}$$

The inequalities (8.11) define two parallel hyperplanes for each kth measurement:

$$\mathbb{H}^+ = \left\{\mathbf{p} \in \mathbb{R}^{n_p}: y(k) - \boldsymbol{r}^T(k-1)\boldsymbol{p} = 1\right\}, \tag{8.14}$$

$$\mathbb{H}^- = \left\{\mathbf{p} \in \mathbb{R}^{n_p}: y(k) - \boldsymbol{r}^T(k-1)\boldsymbol{p} = -1\right\}, \tag{8.15}$$

and bounding a strip $\mathbb{S}(k)$ containing a set of $\mathbf{p}$ values, which satisfy the constraints with $y(k)$:

$$\mathbb{S}(k) = \left\{\mathbf{p} \in \mathbb{R}^{n_p}: -1 \le \bar{y}(k) - \bar{\hat{y}}(k) \le 1\right\}. \tag{8.16}$$

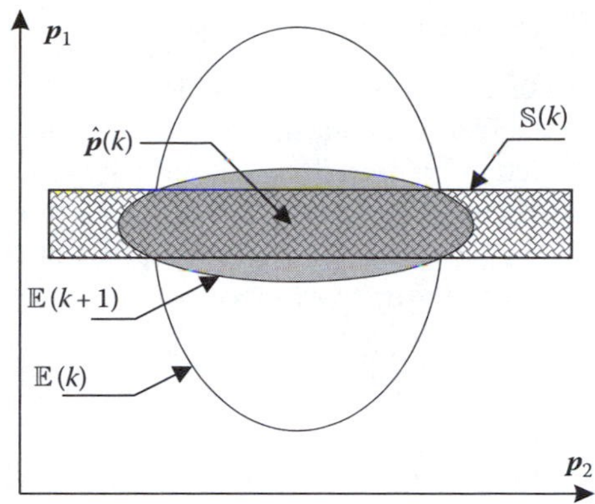

FIGURE 8.3 Recursive determination of the outer ellipsoid.

By the intersection of the strips $\mathbb{S}(k)$ for the $k = 1 \ldots n_T$ measurements of the parameters, a feasible region $\mathbb{E}$ is obtained and its center is chosen as the parameter estimate. Unfortunately, the polytopic region $\mathbb{E}$ becomes very complicated when the number of measurements and parameters is significant, and hence its determination is time consuming. An easier solution relies on the approximation of the convex polytopes $\mathbb{S}(k)$ by simpler ellipsoids. In a recursive OBE algorithm, which is based on this idea, the measurements are taken into account one after another to construct a succession of ellipsoids containing all values of **p** consistent with all previous measurements. After the first k observations, the set of feasible parameters is characterized by the ellipsoid:

$$\mathbb{E}(\hat{\boldsymbol{p}}(k), \boldsymbol{P}(k)) = \{\boldsymbol{p} \in \mathbb{R}^{n_p} : (\boldsymbol{p} - \hat{\boldsymbol{p}}(k))^T \boldsymbol{P}^{-1}(k)(\boldsymbol{p} - \hat{\boldsymbol{p}}(k)) \leq 1\}, \tag{8.17}$$

where

$\hat{\boldsymbol{p}}(k)$ is the center of the ellipsoid constituting kth parameter estimate
$\boldsymbol{P}(k)$ is a positive-definite matrix that specifies its size and orientation

By means of the intersection of the strip (8.16) and the ellipsoid (8.17), a region of possible parameter estimates is obtained. This region is outer bounded by a new $\mathbb{E}(k + 1)$ ellipsoid. The OBE algorithm provides rules for computing $\boldsymbol{p}(k)$ and $\boldsymbol{P}(k)$ in such a way that the volume of $\mathbb{E}(\hat{\boldsymbol{p}}(k + 1), \boldsymbol{P}(k + 1))$ is minimized (cf. Figure 8.3). The center of the last n_Tth ellipsoid constitutes the resulting parameter estimate, while the ellipsoid itself represents the feasible parameter set. However, any parameter vector $\hat{\boldsymbol{p}}$ contained in $\mathbb{E}(n_T)$ is a valid estimate of $\boldsymbol{p}$. A detailed structure of the OBE recursive algorithm is described in [26].

8.3 Generalizations of the GMDH Algorithm

The assumptions of GMDH networks presented in Section 8.2 give a lot of freedom in defining the particular elements of the algorithm of synthesis. The mentioned possibilities relate to, for example, the definition of the transition function, evaluation criteria of the processing accuracy, or selection methods. The concept of the GMDH also allows developing the formula of the GMDH network, through the application of additional procedures, which can be used to extend the scope of the application.

8.3.1 Dynamics in GMDH Neural Networks

The partial model described in Section 8.2.1 can be used for the identification of nonlinear static systems. Unfortunately, most industrial systems are dynamic in nature [15,22]. The application of the static neural network will result in a large model uncertainty. Thus, during system identification, it seems desirable to employ models, which can represent the dynamics of the system. In the case of the classical neural network, for example, the multi-layer perceptron (MLP), the modeling problem of the dynamics is solved by the introduction of additional inputs. The input vector consists of suitably delayed inputs and outputs:

$$\hat{y}(k) = f(\mathbf{u}(k), \mathbf{u}(k-1), \ldots, \mathbf{u}(k-n_u), \hat{y}(k-1), \ldots, \hat{y}(k-n_y)), \tag{8.18}$$

where n_u and n_y represent the number of delays. Unfortunately, the described approach cannot be applied in the GMDH neural network easily, because such a network is constructed through gradual connection of the partial models. The introduction of global output feedback lines complicates the synthesis of the network. On the other hand, the behavior of each partial model should reflect the behavior of the identified system. It follows from the rule of the GMDH algorithm that the parameters of each partial model are estimated in such a way that their output is the best approximation of the real system output. In this situation, the partial model should have the ability to represent the dynamics. One way out of this problem is to use dynamic neurons [16].

Due to the introduction of different local feedbacks to the classical neuron model, it is possible to obtain several types of dynamic neurons. The most well-known architectures are the so-called neurons with local activation feedback [8], neurons with local synapse feedback [1], and neurons with output feedback [10]. The main advantage of networks constructed with the application of dynamic neurons is the fact that their stability can be proved relatively easily. As a matter of the fact, the stability of the network only depends on the stability of neurons. The feed-forward structure of such networks seems to make the training process easier. On the other hand, the introduction of dynamic neurons increases the parameter space significantly. This drawback together with the nonlinear and multi-modal properties of an identification index implies that parameter estimation becomes relatively complex.

In order to overcome this drawback, it is possible to use another type of a dynamic neuron model [16]. Dynamic in such an approach is realized by the introduction of a linear dynamic system—an infinite impulse response (IIR) filter. In this way, each neuron in the network reproduces the output signal based on the past values of its inputs and outputs. Such a neuron model (Figure 8.4) consists of two submodules: the filter module and the activation module.

The filter module is described by the following equation:

$$\tilde{y}_n^{(l)}(k) = \left(\boldsymbol{r}_n^{(l)}(k)\right)^T \hat{\boldsymbol{p}}_n^{(l)}, \tag{8.19}$$

where

$\boldsymbol{r}_n^{(l)}(k) = [-\tilde{y}_n^{(l)}(k-1), \ldots, -\tilde{y}_n^{(l)}(k-n_a), \boldsymbol{u}_n^{(l)}(k), \boldsymbol{u}_n^{(l)}(k-1), \ldots, \boldsymbol{u}_n^{(l)}(k-n_b)]$ is the regressor

$\hat{\boldsymbol{p}}_n^{(l)} = [a_1, \ldots, a_{n_a}, \boldsymbol{v}_0, \boldsymbol{v}_1, \ldots, \boldsymbol{v}_{n_b}]$ is the filter parameters

The filter output is used as the input for the activation module:

$$\hat{y}_n^{(l)}(k) = \xi\left(\tilde{y}_n^{(l)}(k)\right). \tag{8.20}$$

The application of dynamic neurons in the process of GMDH network synthesis can improve the model quality. To additionally reduce the uncertainty of the dynamic neural model, it is necessary to assume the appropriate order of the IIR filter. This problem can be solved by the application of the Lipschitz index approach based on the so-called Lipschitz quotients [21].

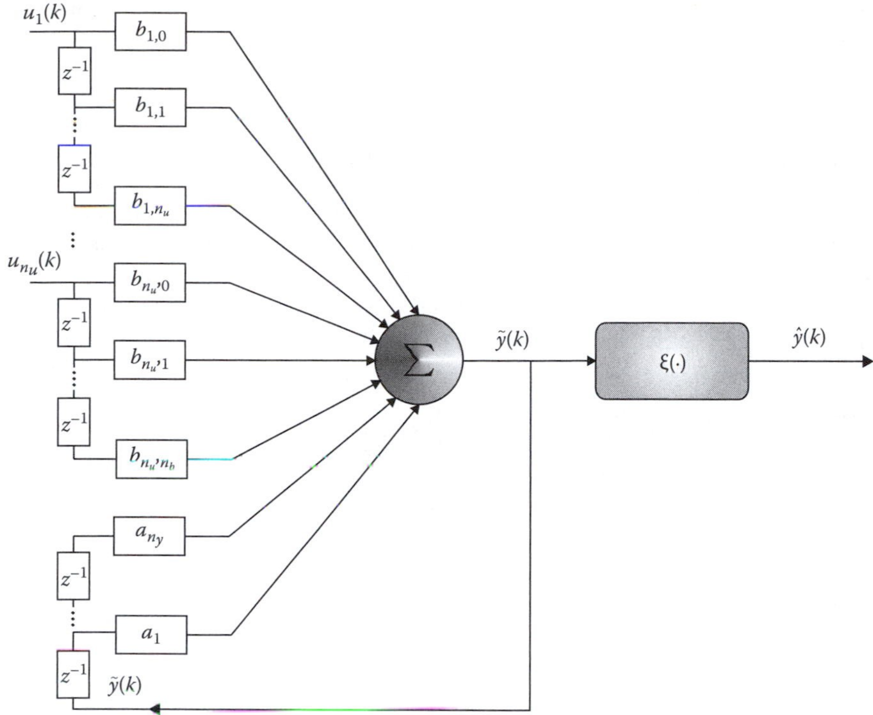

FIGURE 8.4 Dynamic neuron model.

8.3.2 Synthesis of the Single-Input Dynamic GMDH Neural Network

The main advantage of the GMDH neural network is its application to systems with a large number of inputs. Unfortunately, this method is not deprived of weakness consisting in the impossibility of conducting network synthesis, when the number of inputs is less than three. For example, such a situation takes place during system identification of SISO systems.

To overcome this problem, it is possible to decompose IIR filters in the way shown in Figure 8.5. A network of n_u inputs is built from p input neurons with IIR filters. N new elements are formed in the following way:

$$N = \frac{(n_u + n_d n_u)!}{p!(n_u + n_d n_u - p)!}. \tag{8.21}$$

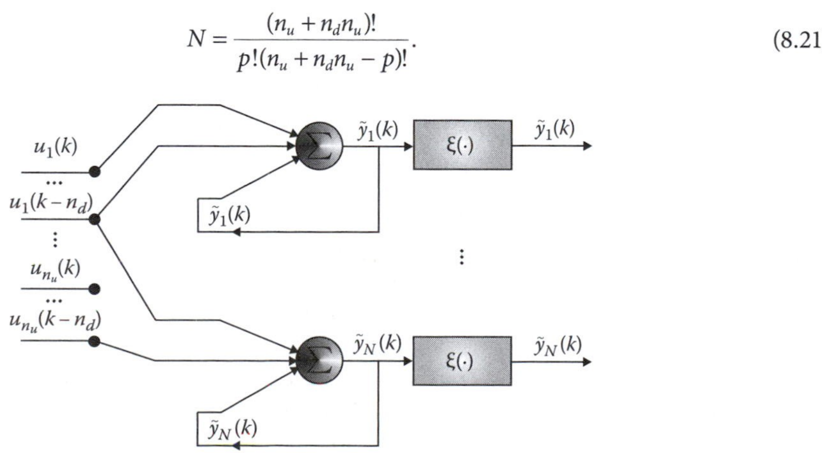

FIGURE 8.5 Decomposition of IIR filters.

As can be observed, as a result of such a decomposition, a greater number of inputs can be obtained in the next layer of the network. The further process of network synthesis is carried out in accordance with the manner described in Section 8.2.

8.3.3 Synthesis of the Multi-Output GMDH Neural Network

The assumptions of the GMDH approach presented in Section 8.2 lead to the formation of a neural network of many inputs and one output. The constructed structure approximates the dependence $y(k) = f(u_1(k),\ldots,u_{n_u}(k))$. However, systems of many inputs and many outputs $y_1(k)$, …, $y_R(k)$ are found in practical applications most often. The synthesis of this model can be realized similarly as in the case of multi-input and single-output models. In the first step of network synthesis, based on all the combinations of inputs, the system output $\hat{y}_1^{(1)}(k)$ is obtained. Next, based on the same combinations of inputs, the remaining outputs $\hat{y}_2^{(1)}(k),\ldots,\hat{y}_R^{(1)}(k)$ are obtained:

$$\begin{cases} \hat{y}_{1,1}^{(1)}(k) = f(u_1^{(1)}(k), u_2^{(1)}(k)) \\ \quad\cdots \\ \hat{y}_{1,N}^{(1)}(k) = f(u_{m-1}^{(1)}(k), u_{n_u}^{(1)}(k)) \\ \quad\vdots \\ \hat{y}_{R,1}^{(1)}(k) = f(u_1^{(1)}(k), u_2^{(1)}(k)) \\ \quad\cdots \\ \hat{y}_{R,N}^{(1)}(k) = f(u_{m-1}^{(1)}(k), u_{n_u}^{(1)}(k)) \end{cases} \tag{8.22}$$

Selection of best performing neurons, for their processing accuracy in the layer, is realized with application of selection methods described in Section 8.2. The independent evaluation of any of the processing errors $Q_1, Q_2, \ldots, Q_R$ is performed after the generation of each layer of neurons.

$$\begin{cases} Q_1(\hat{y}_{1,1}^{(l)}(k)),\ldots,Q_R(\hat{y}_{1,1}^{(l)}(k)) \\ \quad\cdots \\ Q_1(\hat{y}_{1,N}^{(1)}(k)),\ldots,Q_R(\hat{y}_{1,N}^{(1)}(k)) \\ \quad\vdots \\ Q_1(\hat{y}_{R,1}^{(l)}(k)),\ldots,Q_R(\hat{y}_{R,1}^{(l)}(k)) \\ \quad\cdots \\ Q_1(\hat{y}_{R,N}^{(1)}(k)),\ldots,Q_R(\hat{y}_{R,N}^{(1)}(k)). \end{cases} \tag{8.23}$$

According to the chosen selection method, elements that introduce too big a processing error of each output $y_1(k)$, …, $y_R(k)$ are removed (Figure 8.6). The effectiveness of a neuron in the processing of at least one output signal is sufficient to leave the neuron in the network.

Based on all the selected neurons, a new layer is created. In an analogous way, new layers of the network are introduced. During the synthesis of next layers, all outputs from the previous layer must be used to generate each output $y_1(k)$, …, $y_R(k)$. This follows from the fact that in real industrial systems, outputs are usually correlated, so the output $\hat{y}_{r,n}^{(l)}(k)$ should be obtained based on all the potential outputs $\hat{y}_{1,1}^{(l-1)}(k),\ldots,\hat{y}_{R,N}^{(l-1)}(k)$.

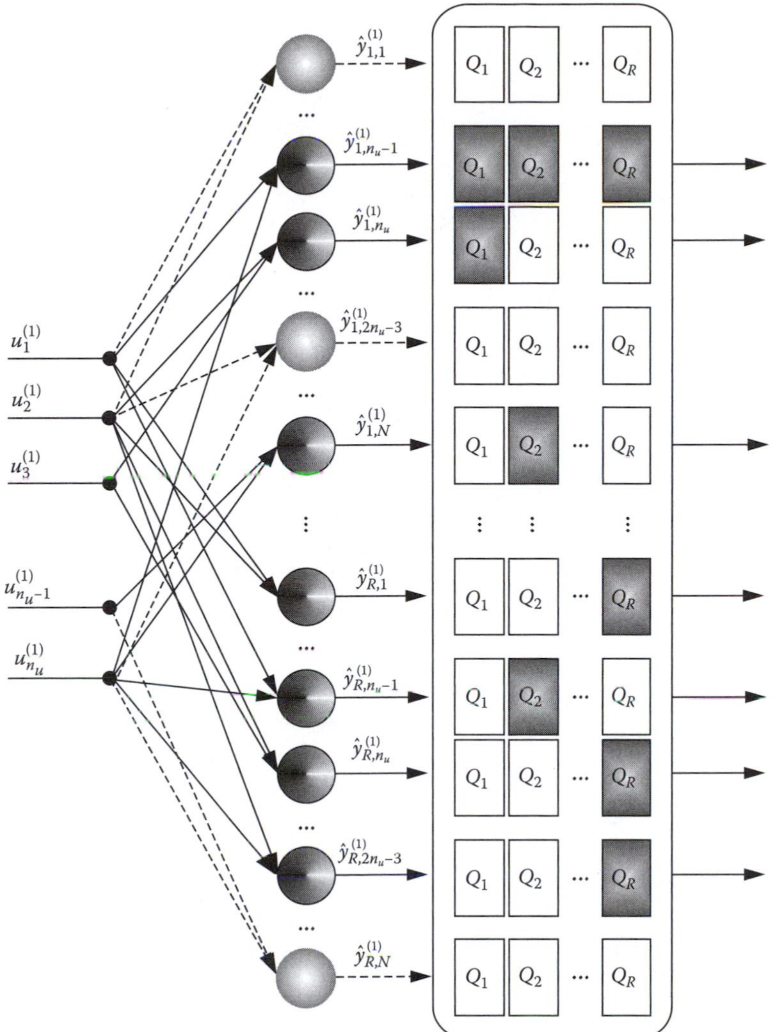

FIGURE 8.6 Selection in the MIMO GMDH network.

The termination of the synthesis of the GMDH network in the presented solution appears independently of all values of the partial processing error $Q_1, Q_2, \ldots, Q_R$ (Figure 8.7):

$$Q_{r,\min}^{(l)} = \min_{n=1,\ldots,N} Q_r(\hat{y}_{r,n}^{(l)}(k)) \quad \text{for} \quad r = 1,\ldots,R. \tag{8.24}$$

The synthesis of the network is completed when each of the calculated criteria values reaches the minimum:

$$Q_{\text{opt}}^{(L_r)} = \min_{l=1,\ldots,L_r} Q_{r,\min}^{(l)} \quad \text{for} \quad r = 1,\ldots,R. \tag{8.25}$$

The output $\hat{y}_r(k)$ is connected to the output of this neuron, for which $Q_{r,\min}^{(l)}$ achieves the least value. The particular minimum could occur at different stages of network synthesis. This is why in the multi-output network, outputs of the resulting structure are usually in different layers (Figure 8.8).

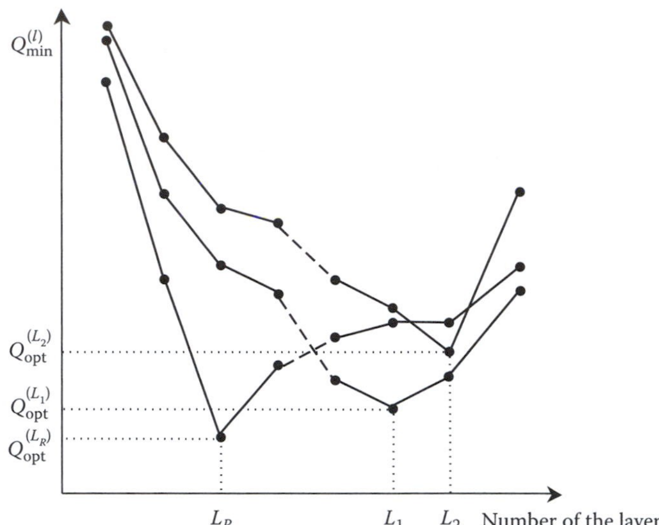

FIGURE 8.7 Termination of the synthesis of the multi-output GMDH network.

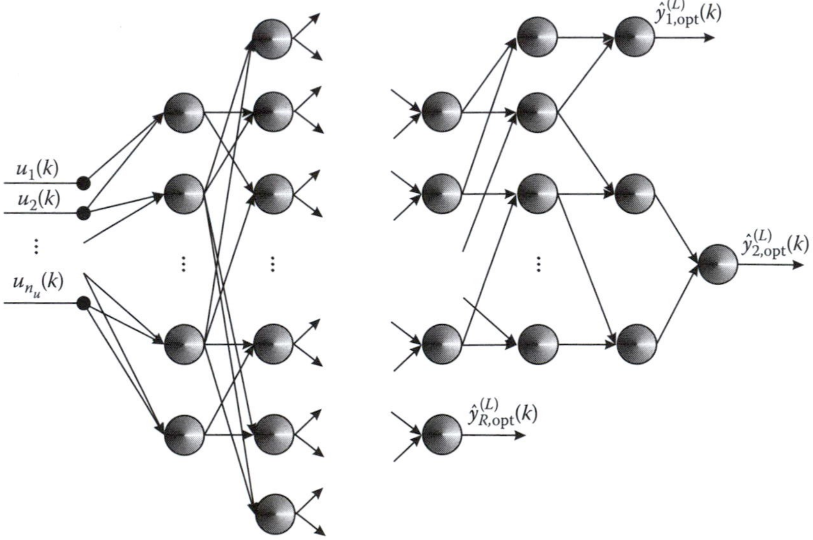

FIGURE 8.8 Final structure of the MIMO GMDH neural network.

8.4 Application of GMDH Neural Networks

The main objective of this section is to present the application of the GMDH neural model to robust fault detection. A fault can be generally defined as an unexpected change in a system of interest. Model-based fault diagnosis can be defined as the detection, isolation, and identification of faults in the system based on a comparison of system available measurements with information represented by the system mathematical model [15,22].

8.4.1 Robust GMDH Model-Based Fault Detection

The comparison of the system $y(k)$ and the model response $\hat{y}(k)$ leads to the generation of the residual:

$$\varepsilon(k) = y(k) - \hat{y}(k), \tag{8.26}$$

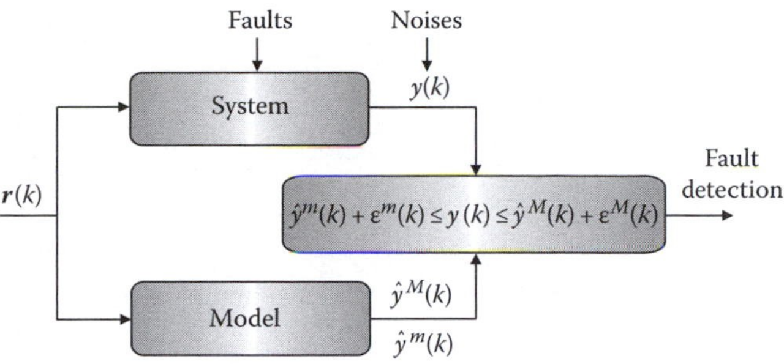

FIGURE 8.9 Robust fault detection with the adaptive time-variant threshold.

which is a source of information about faults for further processing. In the model-based fault detection approach, it is assumed that $\varepsilon(k)$ should be close to zero in the fault-free case, and it should be distinguishably different from zero in the case of a fault. Under such an assumption, the faults are detected by setting a fixed threshold on the residual. In this case, the fault can be detected when the absolute value of the residuum $|\varepsilon(k)|$ is larger than an arbitrarily assumed threshold value δ_y. The difficulty with this kind of residual evaluation is that the measurement of the system output $y(k)$ is usually corrupted by noise and disturbances $\varepsilon^m(k) \le \varepsilon(k) \le \varepsilon^M(k)$, where $\varepsilon^m(k) \le 0$ and $\varepsilon^M(k) \ge 0$. Another difficulty follows from the fact that the model obtained during system identification is usually uncertain [19,29]. Model uncertainty can appear during model structure selection and also parameters estimation. In practice, due to modeling uncertainty and measurement noise, it is necessary to assign wider thresholds in order to avoid false alarms, which can imply a reduction of fault detection sensitivity.

To tackle this problem, the adaptive time-variant threshold that is adapted according to system behavior can be applied. Indeed, knowing the model structure and possessing knowledge regarding its uncertainty, it is possible to design a robust fault detection scheme. The idea behind the proposed approach is illustrated in Figure 8.9.

The proposed technique relies on the calculation of the model output uncertainty interval based on the estimated parameters whose values are known at some confidence level:

$$\hat{y}^m(k) \le \hat{y}(k) \le \hat{y}^M(k). \tag{8.27}$$

Additionally, as the measurement of the controlled system response $y(k)$ is corrupted by noise, it is necessary to add the boundary values of the output error $\varepsilon^m(k)$ and $\varepsilon^M(k)$ to the model output uncertainty interval. A system output interval defined in this way should contain the real system response in the fault-free mode. The occurrence of a fault is signaled when the system output $y(k)$ crosses the system output uncertainty interval:

$$\hat{y}^m(k) + \varepsilon^m(k) \le y(k) \le \hat{y}^M(k) + \varepsilon^M(k). \tag{8.28}$$

The effectiveness of the robust fault detection method requires determining of a mathematical description of model uncertainty and knowing maximal and minimal values of disturbances ε.

To solve this problem, the GMDH model can be applied, which is constructed according to the procedure described in Section 8.2.1. At the beginning, it is necessary to adapt the OBE algorithm to parameter estimation of the partial models (8.8) with the nonlinear activation function $\xi(\cdot)$. In order to avoid the noise additivity problem, it is necessary to transform the relation

$$\varepsilon^m(k) \le y(k) - \xi\left(\left(\boldsymbol{r}_n^{(l)}(k)\right)^T \boldsymbol{p}_n^{(l)}\right) \le \varepsilon^M(k) \tag{8.29}$$

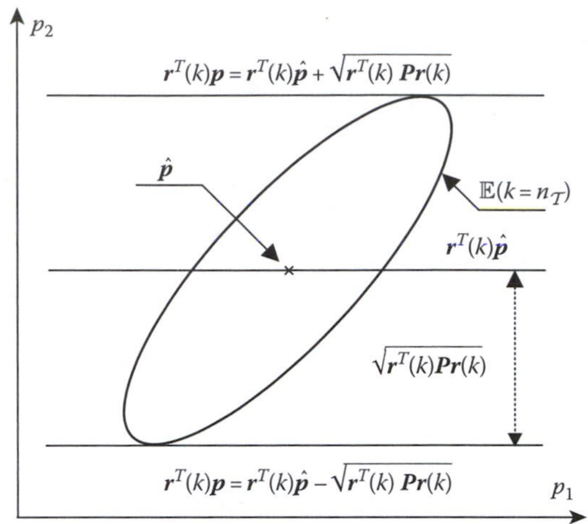

FIGURE 8.10 Relation between the size of the ellipsoid and model output uncertainty.

using $\xi^{-1}(\cdot)$, and hence

$$\xi^{-1}\left(y(k) - \varepsilon^{M}(k)\right) \le \left(r_n^{(l)}(k)\right)^T p_n^{(l)} \le \xi^{-1}\left(y(k) - \varepsilon^{m}(k)\right). \tag{8.30}$$

The transformation (8.30) is appropriate if the conditions concerning the properties of the nonlinear activation function $\xi(\cdot)$ are fulfilled. The methodology described in Section 8.2.2 makes it possible to obtain $\hat{p}$ and $\mathbb{E}$. But from the point of view of applications of the GMDH model to fault detection, it is important to obtain the model output uncertainty interval. The range of this interval for the partial model output depends on the size and the orientation of the ellipsoid $\mathbb{E}$ (cf. Figure 8.10).

Taking the minimal and maximal values of the admissible parameter set $\mathbb{E}$ into consideration, it is possible to determine the minimal and maximal values of the model output uncertainty interval for each partial model of the GMDH neural network:

$$r^T(k)\hat{p} - \sqrt{r^T(k)Pr(k)} \le r^T(k)p \le r^T(k)\hat{p} + \sqrt{r^T(k)Pr(k)}. \tag{8.31}$$

The partial models in the lth ($l > 1$) layer of the GMDH neural network are based on outputs incoming from the (l – 1)-th layer. Since (8.31) describes the model output uncertainty interval in the (l – 1)-th layer, parameters of the partial models in the next layers have to be obtained with an approach that solves the problem of an uncertain regressor. Let us denote an unknown "true" value of the regressor $r_n(k)$ by a difference between a known (measured) value of the regressor $r(k)$ and the error in the regressor $e(k)$:

$$r_n(k) = r(k) - e(k), \tag{8.32}$$

where the regressor error $e(k)$ is bounded as follows:

$$-e_i \le e_i(k) \le \varepsilon_i, \quad i = 1, \ldots, n_p. \tag{8.33}$$

Substituting (8.32) into (8.31), it can be shown that the partial models output uncertainty interval has the following form:

$$\hat{y}^{m}(k) \le r^T(k)p \le \hat{y}^{M}(k), \tag{8.34}$$

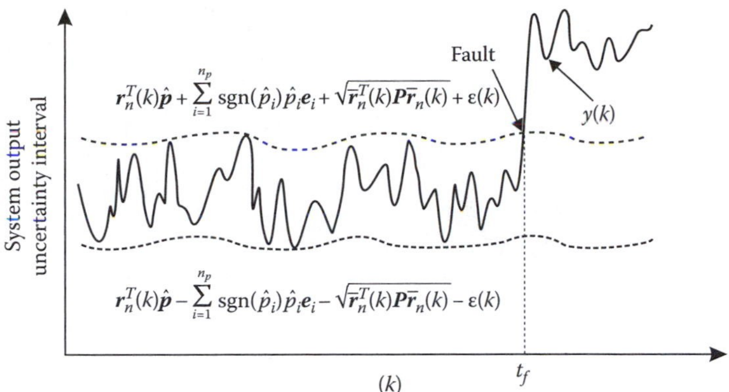

FIGURE 8.11 Fault detection via the system output uncertainty interval.

where

$$\hat{y}^m(k) = \boldsymbol{r}_n^T(k)\hat{\boldsymbol{p}} + \boldsymbol{e}^T(k)\hat{\boldsymbol{p}} - \sqrt{(\boldsymbol{r}_n(k)+\boldsymbol{e}(k))^T \boldsymbol{P}(\boldsymbol{r}_n(k)+\boldsymbol{e}(k))}, \tag{8.35}$$

$$\hat{y}^M(k) = \boldsymbol{r}_n^T(k)\hat{\boldsymbol{p}} + \boldsymbol{e}^T(k)\hat{\boldsymbol{p}} + \sqrt{(\boldsymbol{r}_n(k)+\boldsymbol{e}(k))^T \boldsymbol{P}(\boldsymbol{r}_n(k)+\boldsymbol{e}(k))}. \tag{8.36}$$

In order to obtain the final form of the expression (8.34), it is necessary to take into consideration the bounds of the regressor error (8.33) in the expressions (8.35) and (8.36):

$$\hat{y}^m(k) = \boldsymbol{r}_n^T(k)\hat{\boldsymbol{p}} + \sum_{i=1}^{n_p} \text{sgn}(\hat{p}_i)\hat{p}_i e_i - \sqrt{\bar{\boldsymbol{r}}_n^T(k)\boldsymbol{P}\bar{\boldsymbol{r}}_n(k)}, \tag{8.37}$$

$$\hat{y}^M(k) = \boldsymbol{r}_n^T(k)\hat{\boldsymbol{p}} + \sum_{i=1}^{n_p} \text{sgn}(\hat{p}_i)\hat{p}_i e_i + \sqrt{\bar{\boldsymbol{r}}_n^T(k)\boldsymbol{P}\bar{\boldsymbol{r}}_n(k)}, \tag{8.38}$$

where

$$\bar{r}_{n,i}(k) = r_{n,i}(k) + \text{sgn}\left(r_{n,i}(k)\right)\boldsymbol{e}_i. \tag{8.39}$$

The GMDH model output uncertainty interval (8.37) and (8.38) should contain the real system response in the fault-free mode. As the measurements of the system response are corrupted by noise, it is necessary to add the boundary values of the output error (8.10) to the model output uncertainty interval (8.37) and (8.38). The newly defined interval (Figure 8.11) is called the system output uncertainty interval and it is calculated for the partial model in the last GMDH neural network layer, which generates the model output. The occurrence of a fault is signaled when the system output signal crosses the system output uncertainty interval.

8.4.2 Robust Fault Detection of the Intelligent Actuator

In order to show the effectiveness of the GMDH model-based fault detection system, the actuator model from the Development and Application of Methods for Actuator Diagnosis in Industrial Control System (DAMADICS) benchmark [2] was employed (Figure 8.12), where V_1, V_2, and V_3 denote the bypass valves, *ACQ* and *CPU* are the data acquisition and positioner central processing units, respectively. *E/P* and *FT* are the electro-pneumatic and value flow transducers. Finally, *DT* and *PT* represent the displacement and the pressure transducers. On the ground of process analysis and taking into account the expert

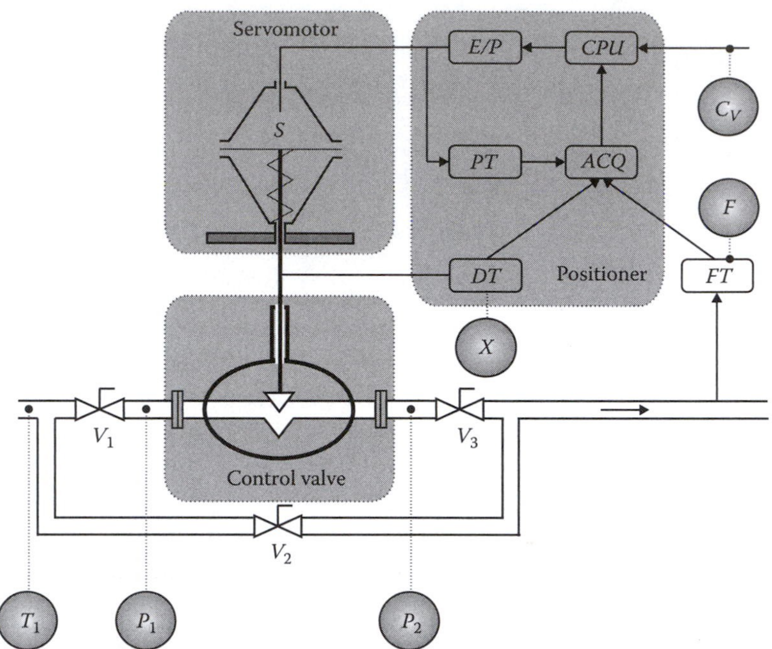

FIGURE 8.12 Diagram of the actuator.

process knowledge, the following model of the juice flow at the outlet of the valve $F = r_F(X, P_1, P_2, T_1)$ and the servomotor rod displacement $X = r_X(C_V, P_1, P_2, T_1)$ were considered, where $r_F(\cdot)$ and $r_X(\cdot)$ denote the modeled relationships, C_v is the control valve, P_1 and P_2 are the pressures at the inlet and the outlet of the valve, respectively, and T_1 represents the juice temperature at the inlet of the valve.

The DAMADICS benchmark makes it possible to generate data for 19 different faults. In the benchmark scenario, the abrupt A and incipient I faults are considered. Furthermore, the abrupt faults can be regarded as small S, medium M, and big B, according to the benchmark descriptions. The synthesis process of the GMDH neural network proceeds according to the steps described in Section 8.2.1. During the synthesis of the GMDH networks, dynamic neurons with the IIR filter (8.19) were applied. The selection of best performing neurons in each layer of the GMDH network in terms of their processing accuracy was realized with the application of the soft selection method based on the following evaluation criterion:

$$Q_V = \frac{1}{n_V}\sum_{k=1}^{n_V}\left|\left(\hat{y}^M(k)+\varepsilon^M(k)\right)-\left(\hat{y}^m(k)+\varepsilon^m(k)\right)\right|. \tag{8.40}$$

The values of this criterion were calculated separately for each neuron in the GMDH network, whereas $\hat{y}^m(k)$ and $\hat{y}^M(k)$ in (8.40) were obtained with (8.31) for the neurons in the first layer of the network and with (8.34) for the subsequent ones. Table 8.1 presents the results for the subsequent layers, that is, these values were obtained for the best performing partial models in a particular layer.

The results show that the gradual decrease of the value of the evaluation criteria occurs when a new layer of the GMDH network is introduced. It follows from the increasing of the model complexity as well as its modeling abilities. However, when the model is too complex, the quality index Q_V increases. This situation occurs when the fifth layer of the network is added. It means that the model corresponding to $F = r_F(\cdot)$ and $X = r_X(\cdot)$ should have only four layers. The final structures of GMDH neural networks are presented in Figures 8.13 and 8.14.

After the synthesis of the $F = r_F(\cdot)$ and $X = r_X(\cdot)$ GMDH models, it is possible to employ them for robust fault detection. This task can be realized with the application of the system output interval

TABLE 8.1 Evolution of Q_V for the Subsequent Layers

Layer	1	2	3	4	5
$r_F(\cdot)$	1.1034	1.0633	1.0206	0.9434	1.9938
$r_X(\cdot)$	0.3198	0.2931	0.2895	0.2811	0.2972

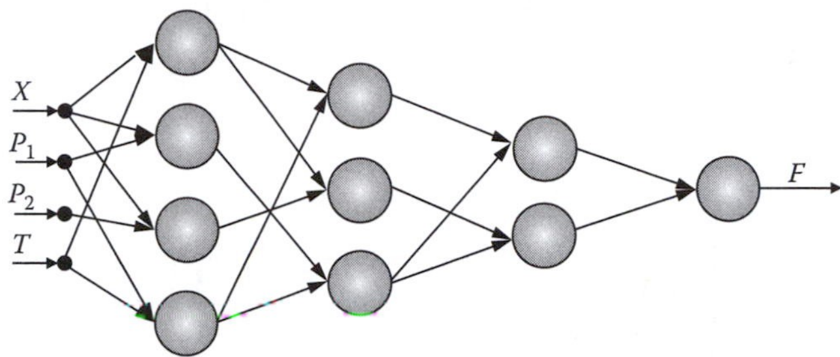

FIGURE 8.13 Final structure of $F = r_F(\cdot)$.

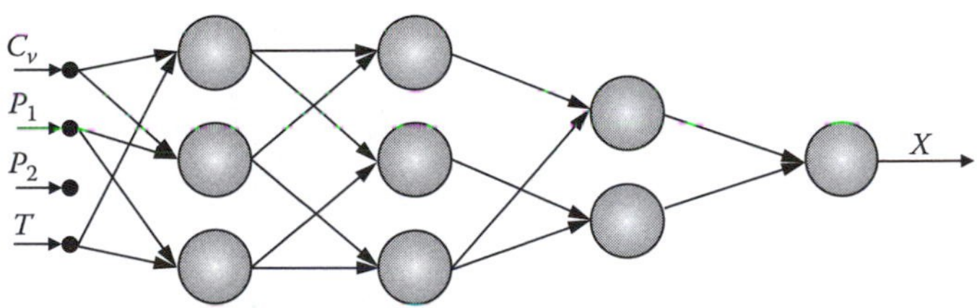

FIGURE 8.14 Final structure of $X = r_X(\cdot)$.

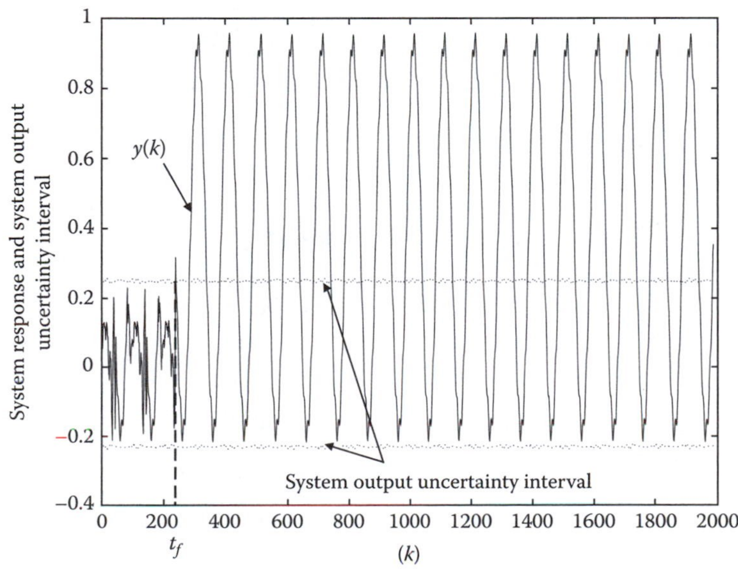

FIGURE 8.15 System response and the system output uncertainty interval for the big abrupt fault f_1.

defined in Section 8.4.1. Figures 8.15 through 8.18 present the system responses and the corresponding system output uncertainty intervals for the faulty data, where t_f denotes the moment of fault occurrence.

Table 8.2 shows the results of fault detection of all the faults considered. The notation given in Table 8.2 can be explained as follows: *ND* means that it is impossible to detect a given fault, D_F or D_X means that it is possible to detect a fault with $r_F(\cdot)$ or $r_X(\cdot)$, respectively, while D_{FX} means that a given fault can be detected with both $r_F(\cdot)$ or $r_X(\cdot)$. From the results presented in Table 8.2, it can be seen that it is impossible to detect the faults f_5, f_9, and f_{14}. Moreover, some small and medium faults cannot be detected, that is, f_8 and f_{12}. This situation can be explained by the fact that the effect of these faults is at the same level as the effect of noise.

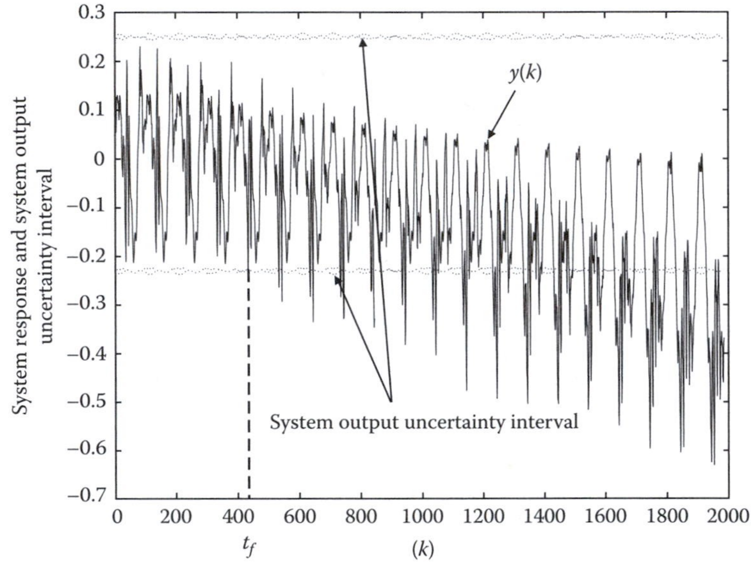

FIGURE 8.16 System response and the system output uncertainty interval for the incipient fault f_2.

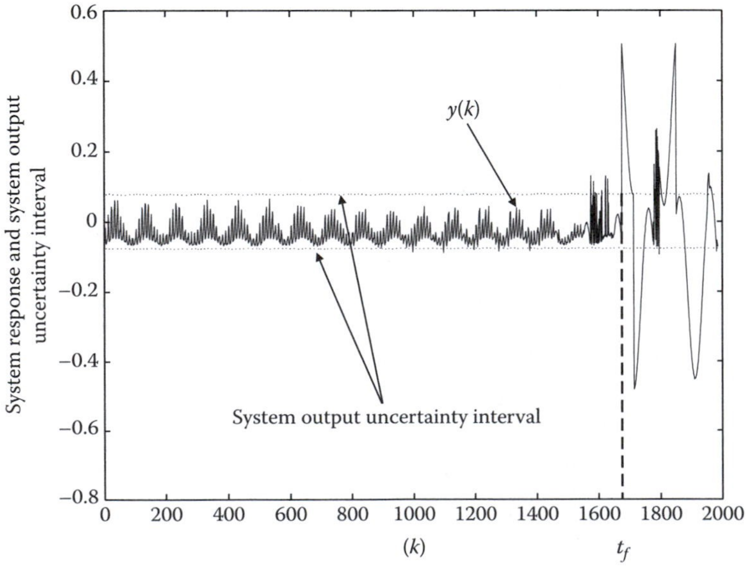

FIGURE 8.17 System response and the system output uncertainty interval for the incipient fault f_4.

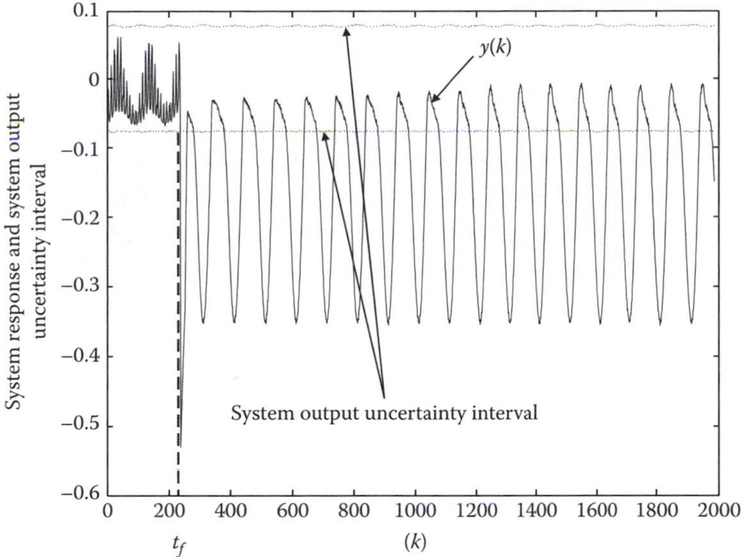

FIGURE 8.18 System response and the system output uncertainty interval for the abrupt medium fault f_7.

TABLE 8.2 Results of Fault Detection

f	Faults Description	S	M	B	I
f_1	Valve clogging	ND	$D_{F,X}$	$D_{F,X}$	
f_2	Valve plug or valve seat sedimentation			D_F	D_F
f_3	Valve plug or valve seat erosion				D_F
f_4	Increase of valve or busing friction				D_X
f_5	External leakage				ND
f_6	Internal leakage (valve tightness)				D_F
f_7	Medium evaporation or critical flow	$D_{F,X}$	D_X	D_X	
f_8	Twisted servomotor's piston rod	ND	ND	D_X	
f_9	Servomotors housing or terminals tightness				ND
f_{10}	Servomotor's diaphragm perforation	D_X	D_{FX}	D_{FX}	
f_{11}	Servomotor's spring fault			D_X	D_{FX}
f_{12}	Electro-pneumatic transducer fault	ND	ND	D_X	
f_{13}	Rod displacement sensor fault	D_F	D_F	D_F	D_{FX}
f_{14}	Pressure sensor fault	ND	ND	ND	
f_{15}	Positioner feedback fault			D_X	
f_{16}	Positioner supply pressure drop	D_F	D_X	D_X	
f_{17}	Unexpected pressure change across the valve			D_{FX}	D_{FX}
f_{18}	Fully or partly opened bypass valves	D_F	D_F	D_F	D_F
f_{19}	Flow rate sensor fault	D_F	D_F	D_F	

Note: S, small; M, medium; B, big; I, incipient.

8.5 Conclusions

One of the crucial problems occurring during system identification with the application of the ANNs is the choice of an appropriate neural model architecture. The GMDH approach solves this problem and allows us to chose such an architecture directly only on the basis of measurements data. The application of this approach leads to the evolution of the resulting neural model architecture in such a way so as to obtain the best quality approximation of the identified system. It is worth emphasizing that the original GMDH algorithm can be easily extended and applied to the identification of dynamical SISO and MIMO systems.

Unfortunately, irrespective of the identification method used, there is always the problem of model uncertainty, that is, the model–reality mismatch. Even though the application of the GMDH approach to the model structure selection can improve the quality of the model, the resulting structure is not the same as that of the system. The application of the OBE algorithm to parameters estimation of the GMDH model also allows us to obtain a neural model uncertainty.

In the illustrative part of the chapter, an example concerning a practical application of GMDH neural model was presented. The calculation of the GMDH model uncertainty in the form of the system output uncertainty interval allowed performing robust fault detection of industrial systems. In particular, the proposed approach was tested on fault detection of the intelligent actuator connected with an evaporation station. The obtained results show that almost all faults can be detected, except for a few incipient or small ones.

References

1. A.D. Back and A.C. Tsoi, FIR and IIR synapses. A new neural network architectures for time series modelling, *Neural Computation*, 3, 375–385, 1991.
2. M. Bartyś, R.J. Patton, M. Syfert, S. Heras, and J. Quevedo, Introduction to the DAMADICS actuator FDI benchmark study, *Control Engineering Practice*, 14, 577–596, 2006.
3. T. Bouthiba, Fault location in EHV transmission lines using artificial neural networks, *International Journal of Applied Mathematics and Computer Science*, 14, 69–78, 2004.
4. J. Chen and R.J. Patton, *Robust Model Based Fault Diagnosis for Dynamic Systems*, London, U.K.: Kluwer Academic Publishers, 1999.
5. E. Delaleau, J.P. Louis, and R. Ortega, Modeling and control of induction motors, *International Journal of Applied Mathematics and Computer Science*, 11, 105–129, 2001.
6. E. Etien, S. Cauet, L. Rambault, and G. Champenois, Control of an induction motor using sliding mode linearization, *International Journal of Applied Mathematics and Computer Science*, 12, 523–531, 2001.
7. S.J. Farlow, *Self-Organizing Methods in Modelling—GMDH Type Algorithms*, New York: Marcel Dekker, 1984.
8. P. Fasconi, M. Gori, and G. Soda, Local feedback multilayered networks, *Neural Computation*, 4, 120–130, 1992.
9. J. Gertler, *Fault Detection and Diagnosis in Engineering Systems*, New York: Marcel Dekker, 1998.
10. M. Gori, Y. Bengio, and R.D. Mori, BPS: A learning algorithm for capturing the dynamic nature of speech, *Neural Computation*, 2, 417–423, 1989.
11. M.M. Gupta, J. Liang, and N. Homma, *Static and Dynamic Neural Networks*, Hoboken, NJ: John Wiley & Sons, 2003.
12. J. Herskovits, P. Mappa, E. Goulart, and C.M. Mota Soares, Mathematical programming models and algorithms for engineering design optimization, *Computer Methods in Applied Mechanics and Engineering*, 194, 3244–3268, 2005.
13. A.G. Ivakhnenko, Polynominal theory of complex systems, *IEEE Transactions on System, Man and Cybernetics*, SMC-1(4), 44–58, 1971.
14. A.G. Ivakhnenko and J.A. Mueller, Self-organizing of nets of active neurons, *System Analysis Modelling Simulation*, 20, 93–106, 1995.
15. J. Korbicz, J.M. Kościelny, Z. Kowalczuk, and W. Cholewa, Eds., *Fault Diagnosis. Models, Artificial Intelligence, Applications*, Berlin, Germany: Springer-Verlag, 2004.
16. J. Korbicz and M. Mrugalski, Confidence estimation of GMDH neural networks and its application in fault detection systems, *International Journal of System Science*, 39(8), 783–800, 2008.

17. M. Milanese, J. Norton, H. Piet-Lahanier, and E. Walter, Eds., *Bounding Approaches to System Identification*, New York: Plenum Press, 1996.
18. M. Mrugalski, E. Arinton, and J. Korbicz, Systems identification with the GMDH neural networks: A multi-dimensional case, in *Computer Science: Artificial Neural Nets and Genetic Algorithms*, D.W. Pearson, N.C. Steel, and R.F. Albrecht, Eds., Wien, New York: Springer-Verlag, 2003.
19. M. Mrugalski, M. Witczak, and J. Korbicz, Confidence estimation of the multi-layer perceptron and its application in fault detection systems, *Engineering Applications of Artificial Intelligence*, 21, 895–906, 2008.
20. J.E. Mueller and F. Lemke, *Self-Organising Data Mining*, Hamburg, Germany: Libri, 2000.
21. O. Nelles, *Non-Linear Systems Identification. From Classical Approaches to Neural Networks and Fuzzy Models*, Berlin, Germany: Springer, 2001.
22. R.J. Patton, P.M. Frank, and R.N. Clark, *Issues of Fault Diagnosis for Dynamic Systems*, Berlin, Germany: Springer-Verlag, 2000.
23. R.J. Patton, J. Korbicz, M. Witczak, and J.U. Faisel, Combined computational intelligence and analytical methods in fault diagnosis, in *Intelligent Control Systems Using Computational Intelligence Techniques*, A.E. Ruano, Ed., London, U.K.: The Institution of Electrical Engineers, IEE, pp. 349–392, 2005.
24. H.R. Scola, R. Nikoukah, and F. Delebecque, Test signal design for failure detection: A linear programming approach, *International Journal of Applied Mathematics and Computer Science*, 13, 515–526, 2003.
25. T. Soderstrom and P. Stoica, *System Identification*, Hemel Hempstead, U.K.: Prentice-Hall International, 1989.
26. E. Walter and L. Pronzato, *Identification of Parametric Models from Experimental Data*, Berlin, Germany: Springer-Verlag, 1997.
27. C. Wang and D.J. Hill, Learning from neural control, *IEEE Transactions on Neural Networks*, 17, 130–146, 2006.
28. M. Witczak, Advances in model-based fault diagnosis with evolutionary algorithms and neural networks, *International Journal of Applied Mathematics and Computer Science*, 16, 85–99, 2006.
29. M. Witczak, J. Korbicz, M. Mrugalski, and R.J. Patton, A GMDH neural network based approach to robust fault detection and its application to solve the DAMADICS benchmark problem, *Control Engineering Practice*, 14, 671–683, 2006.

9

Optimization of Neural Network Architectures

Andrzej Obuchowicz
University of Zielona Góra

The construction process of an artificial neural network (ANN), which is required to solve a given problem, usually consists of four steps [40]. In the first step, a set of pairs of input and output patterns, which should represent characteristics of a problem as well as possible, is selected. In the next step, an architecture of the ANN, the number of units, their ordering into layers or modules, synaptic connections, and other structure parameters are defined. In the third step, free parameters of the ANN (e.g., weights of synaptic connections, slope parameters of activation functions) are automatically trained using a set of training patterns (a learning process). Finally, the obtained ANN is evaluated in accordance with a given quality measure. The process is repeated until the quality measure of the ANN is satisfied. Recently, in practice, there have appeared many effective methods of ANN training and quality estimation. But researchers usually choose an ANN architecture and select a representative set of input and output patterns based rather on their intuition and experience than on an automatic procedure. In this chapter, algorithms of automatic ANN structure selection are presented and analyzed.

The chapter is ordered in the following way: First, the problem of ANN architecture selection is defined. Next, some theoretical aspects of this problem are considered. Chosen methods of automatic ANN structure selection are presented in a systematic way in the following sections.

9.1 Problem Statement

Before the problem of ANN architecture optimization is formulated, some useful definitions will be introduced. The ANN is represented by an ordered pair $NN = (NA, \boldsymbol{v})$ [7,40,41]. NA denotes the ANN architecture:

$$NA = (\{V_i \mid i = 0, \ldots, M\}, \mathcal{E}). \quad (9.1)$$

$\{V_i \mid i = 0, \ldots, M\}$ is a family of $M + 1$ sets of neurons, called *layers*, including at least two nonempty sets V_0 and V_M that define $s_0 = \text{card}(V_0)$ input and $s_M = \text{card}(V_M)$ output units, respectively, $\mathbb{E}$ is a set of connections between neurons in the network. The vector $\boldsymbol{v}$ contains all free parameters of the network, among which there is the set of weights of synaptic connections $\omega: \mathcal{E} \to \mathbb{R}$. In general, sets $\{V_i \mid i = 0, \ldots, M\}$ do not have to be disjunctive, thus there can be input units that are also outputs of the *NN*. Units that do not belong to either V_0 or V_M are called *hidden* neurons. If there are cycles of synaptic connections in the set $\mathcal{E}$, then we have a dynamic network.

The most popular type of neural networks is a feed-forward neural network, called also a multilayer perceptron (MLP), whose architecture possesses the following properties:

$$\forall i \neq j \quad V_i \cap V_j = \varnothing, \tag{9.2}$$

$$\mathcal{E} = \bigcup_{i=0}^{M-1} V_i \times V_{i+1}. \tag{9.3}$$

Layers in the MLP are disjunctive. The main task of the input units of the layer V_0 is preliminary input data processing $\boldsymbol{u} = \{u_p \mid p = 1, 2, \ldots, P\}$ and passing them into units of the hidden layer. Data processing can comprise, for example, scaling, filtering, or signal normalization. Fundamental neural data processing is carried out in hidden and output layers. It is necessary to notice that links between neurons are designed in such a way that each element of the previous layer is connected with each element of the next layer. There are no feedback connections. Connections are assigned suitable weight coefficients, which are determined, for each separate case, depending on the task the network should solve.

In this chapter, for simplicity of presentation, the attention will be focused on the methods of MLP architecture optimization based on neural units with monotonic activation functions. The wide class of problems connected with the RBF, recurrent, or cellular networks will be passed over.

The fundamental learning algorithm for the MLP is the BP algorithm [48,58]. This algorithm is of iterative type and it is based on the minimization of a sum-squared error utilizing optimization gradient-descent method. Unfortunately, the standard BP algorithm is slowly convergent; however, it is widely used and in recent years its numerous modifications and extensions have been proposed. Currently, the Levenberg–Marquardt algorithm [16] seems to be most often applied by researchers.

Neural networks with the MLP architecture owe their popularity to many effective applications, for example, in the pattern recognition problems [30,53] and the approximation of nonlinear functions [22]. It has been proved that using the MLP with only one hidden layer and a suitable number of neurons, it is possible to approximate any nonlinear static relation with arbitrary accuracy [6,22]. Thus, taking relatively simple algorithms applied to MLP learning into consideration, this type of network becomes a very attractive tool for building models of static systems.

Let us consider the network that has to approximate a given function $f(\boldsymbol{u})$. Let $\Phi = \{(\boldsymbol{u}, \boldsymbol{y})\}$ be a set of all possible (usually uncountable) pairs of vectors from the domain $\boldsymbol{u} \in \mathbb{D} \subset \mathbb{R}^{s_0}$ and from the range $\boldsymbol{y} \in \mathbb{D}' \subset \mathbb{R}^{s_M}$, which realize the relation $\boldsymbol{y} = f(\boldsymbol{u})$. The goal is to construct an *NN* with an architecture NA^{opt} and a set of parameters $\boldsymbol{v}^{opt}$, which fulfills the relation $\boldsymbol{y}_{NA,\boldsymbol{v}} = f_{NA,\boldsymbol{v}}(\boldsymbol{u})$ that a given cost function $\sup_{(\boldsymbol{u},\boldsymbol{y})\in\Phi} J_T(\boldsymbol{y}_{NA,\boldsymbol{v}}, \boldsymbol{y})$ will be minimized. So, the following pair has to be found:

$$(NA^{opt}, \boldsymbol{v}^{opt}) = \arg\min \left[\sup_{(\boldsymbol{u},\boldsymbol{y})\in\Phi} J_T(\boldsymbol{y}_{NA,\boldsymbol{v}}, \boldsymbol{y}) \right]. \tag{9.4}$$

Practically, a solution of the above problem is not possible to be obtained because of the infinite cardinality of the set Φ. Thus, in order* to estimate the solution, two finite sets $\Phi_L, \Phi_T \subset \Phi$ are selected. The set Φ_L is used in a learning process of an ANN of a given architecture NA:

$$\boldsymbol{v}^{*} = \arg\min_{\boldsymbol{v}\in\mathcal{V}} \left[\max_{(\boldsymbol{u},\boldsymbol{y})\in\Phi_L} J_L(\boldsymbol{y}_{NA,\boldsymbol{v}}, \boldsymbol{y}) \right], \tag{9.5}$$

where $\mathcal{V}$ is the space of network parameters. In general, cost functions of learning, $J_L(\boldsymbol{y}_{NA,\boldsymbol{v}}, \boldsymbol{y})$, and testing, $J_T(\boldsymbol{y}_{NA,\boldsymbol{v}}, \boldsymbol{y})$, processes can have different definitions. The set Φ_T is used in the searching process of the NA^*, for which

$$NA^* = \arg\min_{NA \in \mathcal{A}} \left[\max_{(\boldsymbol{u},\boldsymbol{y}) \in \Phi_T} J_T(\boldsymbol{y}_{NA,\boldsymbol{v}^*}, \boldsymbol{y}) \right], \tag{9.6}$$

where $\mathcal{A}$ is the space of neural-network architectures. Obviously, the solutions of both tasks (9.5) and (9.6) need not necessarily be unique. Then, a definition of an additional criterion is needed.

There are many definitions of the selection of the best neural-network architecture. The most popular ones are [41]

- Minimization of the number of network free parameters. In this case, the subset

$$\mathcal{A}_\delta = \{NA: J_T(\boldsymbol{y}_{NA,\boldsymbol{v}^*}, \boldsymbol{y}) \le \delta\} \subset \mathcal{A} \tag{9.7}$$

  is looked for. The network with the architecture $NA \in \mathcal{A}_\delta$ and the smallest number of training parameters is considered to be optimal. This criterion is crucial when VLSI implementation of the neural network is planned.
- Maximization of the network generalization ability. The sets of training Φ_L and testing Φ_T patterns have to be disjunctive, $\Phi_L \cap \Phi_T = \emptyset$. Then, J_T is the conformity measure between the network reply on testing patterns and desired outputs. Usually, both quality measures J_L and J_T are similarly defined:

$$J_{L(T)}(\boldsymbol{y}_{NA,\boldsymbol{v}}, \boldsymbol{y}) = \sum_{k=1}^{\text{card}(\Phi_{L(T)})} (\boldsymbol{y}_{NA,\boldsymbol{v}} - \boldsymbol{y})^2. \tag{9.8}$$

  The restriction of the number of training parameters is the minor criterion in this case. The above criterion is important for approximating networks or neural models.
- Maximization of the noise immunity. This criterion is used in networks applied in classification or pattern recognition problems. The quality measure is the maximal noise level of the pattern that is still recognized by the network.

Two first criterions are correlated. Gradually decreasing the number of hidden neurons and synaptic connections causes the drop of nonlinearity level of the network mapping, and then the network generalization ability increases. The third criterion needs some redundancy of the network parameters. This fact usually clashes with previous criterions. In most publications, the second criterion is chosen.

The quality of the estimates obtained with neural networks strongly depends on selection finite learning Φ_L and testing Φ_T sets. Small network structures may not be able to approximate the desired relation between inputs and outputs with a satisfying accuracy. On the other hand, if the number of network free parameters is too large (in comparison with card(Φ_L)), then the function $f_{NA^*,\boldsymbol{v}^*}(\boldsymbol{u})$ realized by the network strongly depends on the actual set of learning patterns (*the bias/variance dilemma* [12]).

It is very important to note that the efficiency of the method of neural-network architecture optimization strongly depends on the learning algorithm used. In the case of a multimodal topology of the network error function, the effectiveness of the classical learning algorithms based on the gradient-descent method (e.g., the BP algorithm and its modifications) is limited. These methods usually localize some local optimum and the superior algorithm searching for the optimal architecture receives wrong information about the trained network quality.

9.2 MLP as a Canonical Form Approximator of Nonlinearity

Let us consider the MLP network with two hidden layers and units with a sigmoid activation function (Figure 9.1). Four spaces can be distinguished: the input space U and its successive patterns $Y_{h1} = R_{h1}(U)$, $Y_{h2} = R_{h2}(Y_{h1})$, and $Y = R_0(Y_{h2})$, where R_{h1}, R_{h2}, and R_0 are mappings realized by both hidden and output layers, respectively. Numbers of input and output units are defined by dimensions of input and output spaces. The number of hidden units in both hidden layers depends on an approximation problem solved by a network. Further deliberations are based on the following theorem [56]:

Theorem 9.2.1 Let Φ_L be a finite set of training pairs associated with finite and compact manifolds. Let f be some continuous function. Taking into account the space of three-level MLPs, there exists an unambiguous approximation of the canonical decomposition of the function f, if and only if the number of hidden neurons in each hidden layer is equal to the dimension of the subspace of the canonical decomposition of the function f.

Theorem 9.2.1 gives necessary and sufficient conditions for the existence of MLP approximation of the canonical decomposition of any continuous function. These conditions are as follows: U and Y must be fully represented by the learning set Φ_L. The network contains more than two hidden layers, which are enough for implementing the discussed approximation of the canonical decomposition of any continuous function. The goal of the first hidden layer is to map the n-dimensional input space U into the space $Y_{h1} = R_{h1}(U)$, which is an inverse image of the output space in the sense of the function f. Thus, the mapping $Y_{h1} \rightarrow Y$ is invertible. The number of units in the first hidden layer card(V_1) is equal to the dimension of the minimal space, which still fully represents input data and is, in general, lower than the dimension of input vectors.

Theorem 9.2.1 guarantees that an approximation of the canonical form of the function f exists and is unambiguous. If card(V_1) is higher than the dimension of the canonical decomposition space of the function f, the network does not approximate the canonical decomposition but can still be the best approximation of the function f. However, such an approximation is not unambiguous and depends on the initial condition of the learning process. On the other hand, if the number card(V_1) is too low, the obtained approximation is not optimal. So, both the deficiency and excess of neurons in the first hidden layer lead to poor approximation.

As has been pointed out above, the first layer reduces the dimension of the actual input space to the level sufficient for optimal approximation. The next two layers, the second hidden one and the output

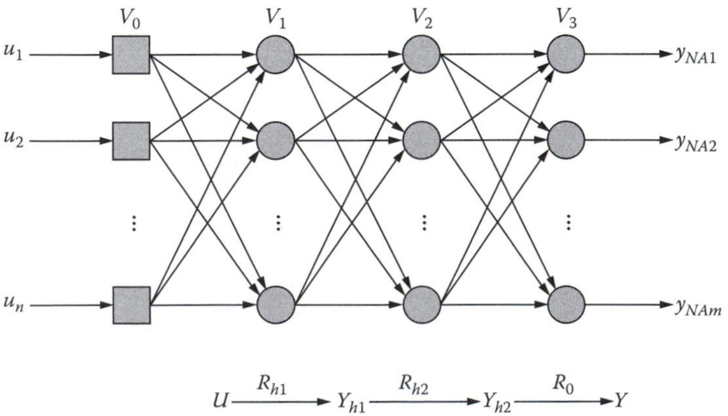

FIGURE 9.1 MLP network with three layers.

one, are sufficient for the realization of such an approximation [6,22]. The number of units in the second hidden layer card(V_2) is determined by an assumed error of approximation. The lowest error needs a higher card(V_2). The crucial tradeoff that one has to make is between the learning capability of the MLP and fluctuations due to the finite sample size. If card(V_2) is too small, the network might not be able to approximate the functional relationship between the input and target output well enough. If card(V_2) is too great (compared to the number of learning samples), the realized network function will depend too much on the actual realization of the learning set [12].

The above consideration suggests that the MLP can be used for the approximation of the canonical decomposition of any function specified on the compact topological manifold. The following question comes to mind: Why is the canonical decomposition needed? Usually, essential variables, which fully describe the input–output relation, are not precisely defined. Thus, the approximation of this relation can be difficult. The existence of the first layer allows us to transform real data to the form of the complete set of variables of an invertible mapping. If the input space agrees with the inverse image of the approximated mapping, the first hidden layer is unnecessary.

9.3 Methods of MLP Architecture Optimization

9.3.1 Methods Classification

Procedures that search for the optimal ANN architecture have been studied for a dozen or so years, particularly intensively in the period between 1989 and 1991. At that time, almost all standard solutions were published. In subsequent years, the number of publications significantly decreased. Most of the proposed methods were dedicated to specific types of neural networks. But new results are still needed.

There is a big collection of bibliography items and various methods available to solve this problem. Recently, a variety of architecture optimization algorithms have been proposed. They can be divided into three classes [7,40,41]:

- Bottom-up approaches
- Top-down (pruning) approaches
- Discrete optimization methods

Starting with a relatively small architecture, bottom-up procedures increase the number of hidden units and thus increase the power of the growing network. Bottom-up methods [2,9,10,21,33,51,55] prove to be the most flexible approach, though computationally expensive (complexity of all known algorithms is exponential). Several bottom-up methods have been reported to learn even hard problems with a reasonable computational effort. The resulting network architectures can hardly be proven to be optimal. But further criticism concerns the insertion of hidden neurons as long as elements of the learning set are misclassified. Thus, the resulting networks exhibit a poor generalization performance and are disqualified for many applications.

Most neural-network applications use the neural model of binary, bipolar, sigmoid, or hyperbolic tangent activation function. A single unit of this type represents a hyperplane, which separates its input space into two subspaces. Through serial–parallel unit connections in the network, the input space is divided into subspaces that are polyhedral sets. The idea of the top-down methods is gradual reduction of the hidden unit number in order to simplify the shapes of the division of the input space. In this way, the generalization property can be improved. Top-down approaches [1,4,5,11,18,23,28,29, 35,39,45,46,57] inherently assume the knowledge of a sufficiently complex network architecture that can always be provided for finite size learning samples. Because the algorithms presented up to now can only handle special cases of redundancy reduction in a network architecture, they are likely to result in a network that is still oversized. In this case, the cascade-reduction method [39], where the obtained architecture using a given top-down method is an initial architecture for the next searching process, can be a good solution.

The space of ANN architectures is infinite discrete, and there are many bibliography items dealing with the implementation of discrete optimization methods to solve the ANN architecture optimization problem. In particular, evolutionary algorithms, especially genetic algorithms (GA), seem to have gained a strong attraction within this context (c.f. [3,15,17,19,25–27,32,34,36,44,50,54]). Nevertheless, implementations of the A* algorithm [7,38,41,43], the simulating annealing [31,40,41] and the tabu search [31,41,43] deserve an attention.

9.3.2 Bottom-Up Approaches

One of the first bottom-up methods was proposed by Mezard and Nadal [33]. Their *tiling algorithm* is dedicated for the MLP that has to map Boolean functions of binary inputs. Creating subsequent layers neuron by neuron, the tiling algorithm successively reduces the number of learning patterns, which are not linearly separable. A similar approach was introduced by Frean [10]. Both algorithms give MLP architectures in a finite time, and these architectures aspire to be almost optimal. In [21], an extension of the back-propagation (BP) algorithm was proposed. This algorithm allows us to add or reduce hidden units depending on the actual position of the training process. Ash [2] and Setiono and Hui [51] stated that the training process of sequentially created networks is initiated using values of parameters from previously obtained networks. Wang and coworkers [55] built an algorithm based on their Theorem 9.2.1 [56], which describes necessary and sufficient conditions under which there exists neural-network approximation of the canonical decomposition of any continuous function. The cascade-correlation algorithm [9] builds an ANN of an original architecture.

9.3.2.1 Tiling Algorithm

The tiling algorithm [33] was proposed for feed-forward neural networks with one output and binary function activation of all neurons. Using such a network, any Boolean function of n inputs or some approximation of such a function (if the number of learning patterns $p < 2^n$) can be realized. The authors propose a strategy in which neurons are added to the network in the following order (Figure 9.2): The first neuron of each layer fulfils a special role and is called a *master unit*. An output of the master unit of the latest added layer is used for the calculation of a recent network quality measure. In the best case, the output of the network is faultless and the algorithm run is finished. Otherwise, *auxiliary nodes* are introduced into the last layer until the layer outputs become a "suitable representation" of the problem, that is, for two different learning patterns (with different desired outputs) the output vectors of the layer are different. If the "suitable representation" is achieved, then the layer construction process is finished and a new master unit of a new output layer is introduced and trained.

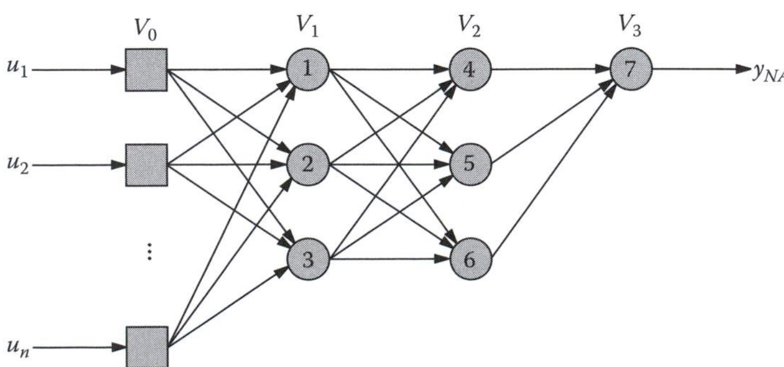

FIGURE 9.2 Order of node adding to the network in the tiling algorithm.

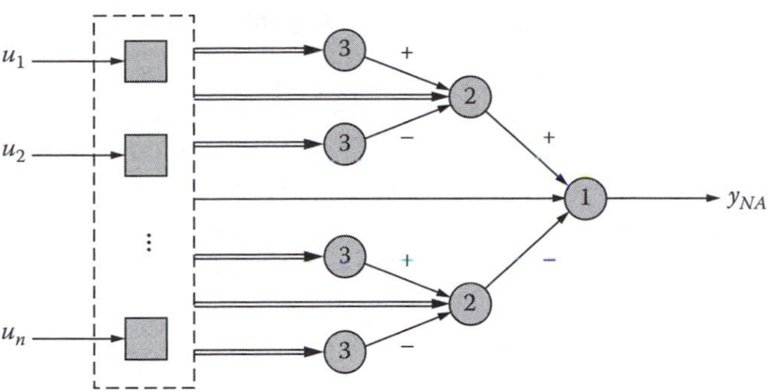

FIGURE 9.3 Upstart network.

In detail, the layer construction process is as follows: Let the layer V_{L-1} be "suitable"; then the set of the learning patterns can be divided into classes $\{C^\nu|\nu = 1\ldots p_{L-1}\}$, which are attributed to different activation vectors (so-called prototypes) $\{\tau^\nu|\nu = 1\ldots p_{L-1}\}$. Each learning pattern of C^ν is attributed to the same desired output y^ν. The master unit of the layer V_L is trained by Rosenblatt's algorithm [47] using the learning pairs $\{(\tau^\nu, y^\nu)|\nu = 1\ldots p_{L-1}\}$. If the output error is equal to zero, then the patterns $\{(\tau^\nu, y^\nu)|\nu = 1\ldots p_{L-1}\}$ are linearly separated and the master unit of V_L is the output unit of the network. If the output error of the master unit of V_L is not equal to zero, then auxiliary nodes have to be introduced. Two nodes are just in the layer V_L: the bias and master unit. Thus, the learning patterns belong to two classes of prototypes: $(\tau'_0 = 1, \tau'_1 = 0)$ and $(\tau'_0 = 1, \tau'_1 = 1)$. Because the output error of the master unit of V_L is not faultless, there exists at least one "unsuitable" class C^μ. A new auxiliary unit is trained the relationship $\tau^\mu \rightarrow y^\mu$, using only patterns from the "unsuitable" class C^μ. After training, the class C^μ can be divided into two "suitable" classes. In the other case, the "unsuitable" subclass of C^μ is used to train a new auxiliary unit. Such a consistent procedure leads to the creation of the "suitable" layer V_L. Tiling algorithm convergence is proved in [33].

9.3.2.2 Upstart Algorithm

The upstart algorithm was proposed by Frean [10]. This method is also dedicated to feed-forward neural networks with one output and binary function activation of all neurons. Unlike the tiling algorithm, the upstart method does not build a network layer by layer from the input to output, but new units are introduced between the input and output layers, and their task is to correct the error of the output unit.

Let the network have to learn some binary classification. The basic idea is as follows: A given unit Z generates another one, which corrects its error. Two types of errors can be distinguished: the switch on fault (1 instead of 0) and the *switch off fault* (0 instead of 1). Let us consider the switch on fault case. The answer of Z can be corrected by a big negative weight of the synaptic connection from a new unit X, which is active only in the case of the switch on fault of Z. If Z is in the state of the switch off fault, then this state can be corrected by a big positive weight of connection from a new unit Y, which is active in suitable time. In order to fit weights of the input connections to the units X and Y using Rosenblatt's algorithm, their desired outputs are specified based on the activity of Z. The X and Y units are called *daughters* and Z is the *parent*. The scheme of the building process is shown in Figure 9.3, and the desired outputs of the X and Y units are presented in Table 9.1.

The network architecture generated by the upstart algorithm is not conventional. It possesses the structure of a hierarchical tree and each unit is connected with network input nodes.

TABLE 9.1 Desired Outputs of (a) X and (b) Y Depending on Current o_Z and Desired t_Z Outputs of Z

	t_Z	
o_Z	0	1
(a)		
0	0	0
1	1	0
(b)		
0	0	1
1	0	0

9.3.2.3 BP Algorithm That Varies the Number of Hidden Units

The BP algorithm seems to be the most popular procedure of MLP training. Unfortunately, this algorithm is not devoid of problems. The learning process sometimes gets stuck in a local, unsatisfied optimum of the weight space. Hirose and coworkers [21] found that the increase of the weight space dimension by adding hidden neurons allows the BP algorithm to escape from the local optimum trap. The phase of MLP growth is finished when the result of the learning process is satisfied. The obtained network is usually too big, described by too many free parameters. Thus, the next cyclic phase is started, in which hidden units are in turn removed until the learning algorithm convergency is lost.

The proposed algorithm effectively reduces the time of MLP design. But the generalization ability is limited. Usually, the obtained network fits the learning set of patterns too exactly.

9.3.2.4 Dynamic Node Creation

The dynamic node creation algorithm was proposed by Ash [2]. This procedure is dedicated to MLP design with one hidden layer and is similar to the algorithm described in the previous paragraph. The initial architecture contains a small number of hidden units (usually two). Next, neurons are added to the network until the desired realization is fulfilled with a given accuracy. After adding a given unit, the network is trained using the BP algorithm. The crucial novelty is the fact that the learning process, started after adding a new neuron, is initialized from the weights obtained in the previous learning process of the smaller network. Only new free parameters (connected with the new node) of the network are randomly chosen. The main success of the dynamic node creation algorithm is the construction of the network with six hidden units, which solve the problem of the parity of six bits. This solution is very difficult to obtain when the network with six hidden neurons is trained by the BP algorithm initialized from a set of randomly chosen weights.

Setiono and Chi Kwong Hui [51] developed the dynamic node creation algorithm by implementing a learning process based on the BCFG optimization algorithm rather than on the BP algorithm. The application of this modification accelerates the convergence of the learning process. Taking into account the fact that, in the initial phase, the BCFG algorithm trains a network of a small size, the problem of the memory space needed for the implementation of the Hessian matrix estimate is negligible.

9.3.2.5 Canonical Form Method

Theorem 9.2.1 as well as the definition of δ-linear independence of vectors presented below are the basis of the canonical form method [55].

Definition 9.3.1 Let E be an n-dimensional vector space and $\{\mathbf{a}_i\}_{i=1}^{r}$ be vectors from E. The set of vectors $\{\mathbf{a}_i\}_{i=1}^{r}$ is called δ-linear independent if the matrix $[\mathbf{a}_i]_{i=1}^{r}$ fulfills the following relationship:

$$\det\left(\mathbb{A}^T\mathbb{A}\right) > \delta, \tag{9.9}$$

where δ is a given positive constant.

Let $\mathbb{U}$ be a matrix whose columns are created by the successive input learning patterns. Let $\mathbb{Y}_1 = R_1\mathbb{U}$ and $\mathbb{Y}_2 = R_1\mathbb{Y}_1$ be matrices whose columns are vectors of replays of first and second hidden layer, respectively, on input patterns. Based on Theorem 9.2.1 and Definition 9.3.1, Wang and coworkers [55] prove that the optimal numbers of units in the first p_1 and second p_2 layers have to fulfill the following inequalities:

$$\det\left(\mathbb{Y}_1^T\mathbb{Y}_1\right)_{p_1} \geq \delta, \quad \det\left(\mathbb{Y}_1^T\mathbb{Y}_1\right)_{p_1+1} < \delta, \tag{9.10}$$

$$\det\left(\mathbb{Y}_2^T\mathbb{Y}_2\right)_{p_2} \geq \delta, \quad \det\left(\mathbb{Y}_2^T\mathbb{Y}_2\right)_{p_2+1} < \delta, \tag{9.11}$$

where $\det\left(\mathbb{Y}_i^T\mathbb{Y}_i\right)_{p_i}$ means that the matrix $\mathbb{Y}_i$ has p_i rows.

The p_1 and p_2 searching process is initialized from a relatively small number of units in both hidden layers. The inequalities (9.10) and (9.11) (or some of their modifications [55]) are checked for each analyzed MLP architecture after the learning process. If the inequalities are not met, then a new node is added to an appropriate hidden layer and the network is trained again.

9.3.2.6 Cascade-Correlation Method

The cascade-correlation method [9] is similar to the upstart method, but it is applied to networks of continuous activation functions of nodes. The algorithm iteratively reduces the error generated by output units. In order to obtain this effect, the hidden nodes, which correlate or anti-correlate the quality measure based on the output response, are introduced into the network. The cascade-correlation algorithm is initialized with a structure that contains only output processing units. Free parameters of these units are trained using a simple gradient-descent method. If the response of the network is satisfied, then algorithm processing is finished. Otherwise, a candidate node to be a hidden unit is introduced. This candidate unit receives signals from all network inputs and previously introduced hidden units. The output of the candidate unit is not connected with the network in this stage. Parameters tuning of the candidate is based on the maximization of the following quality measure:

$$J_c = \sum_{i=1}^{m} \left| \sum_{(\boldsymbol{u},y)\in\Phi_L} (y_c(\boldsymbol{u}) - \overline{y}_c)(E_i(\boldsymbol{u}) - \overline{E}_i) \right|, \tag{9.12}$$

where

- m is the number of output units
- Φ_L is a set of learning patterns
- $y_c(\boldsymbol{u})$ is the response of the candidate on inputs $\boldsymbol{u}$
- $E_i(\boldsymbol{u})$ is the output error of the ith output unit generated by inputs $\boldsymbol{u}$

$$\overline{y}_c = \left(\frac{1}{\text{card}(\Phi_L)}\right) \sum_{(\boldsymbol{u},y)\in\Phi_L} y_c(\boldsymbol{u}),$$

$$\overline{E}_i = \left(\frac{1}{\text{card}(\Phi_L)}\right) \sum_{(\boldsymbol{u},y)\in\Phi_L} E_i(\boldsymbol{u}).$$

For J_c maximization, the gradient-ascent method is applied. When the learning process of the candidate is finished, it is included in the network, its parameters are frozen and all free parameters of the output units are trained again. This cycle is repeated until the output error is acceptable. A sample of the cascade-correlation network is presented in Figure 9.4. The black dots represent connection weights

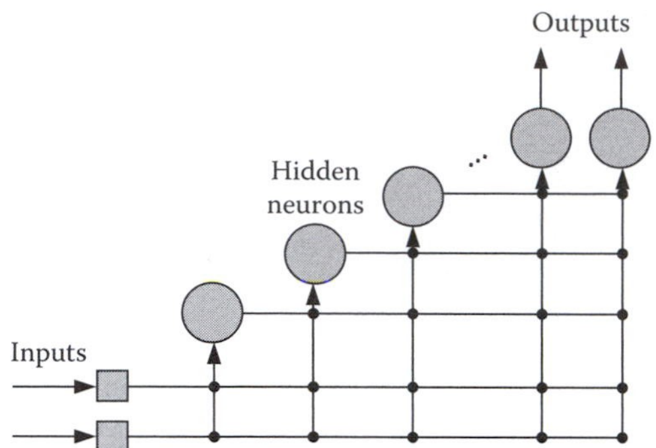

FIGURE 9.4 Sample of the cascade-correlation network with two inputs and two outputs.

between units. It is important to notice that the obtained network is not optimal in the sense of the number of network free parameters but in the sense of modeling quality.

9.3.3 Top-Down (Pruning) Methods

An ANN architecture that is able to be trained with an acceptably small accuracy has to be initially created. Such an architecture is characterized by some redundancy. Top-down methods try to reduce the number of free parameters of the network, preserving the leaning error at an acceptable level. Three classes of top-down methods can be distinguished:

- Penalty function methods
- Sensitivity methods
- Methods based on information analysis

In the first class, a penalty function, which punishes too big architectures, is added to the network quality criterion. In the second class, synaptic connections, for which the weight influence on the quality measure J_T (9.6) is negligibly small, are eliminated. The third class can be treated as an expanded version of the second one. The decision regarding given node pruning is made after an analysis of the covariance matrix (or its estimation) of hidden units outputs. The number of significantly large eigenvalues of this matrix is the necessary number of hidden units.

9.3.3.1 Penalty Function Methods

The idea behind penalty function methods is the modification of the learning criterion, J_L, by adding a component $\Gamma(\boldsymbol{v})$, which punishes for redundancy architecture elements:

$$J_L'(\boldsymbol{y}_{NA,\boldsymbol{v}},\boldsymbol{y}) = J_L(\boldsymbol{y}_{NA,\boldsymbol{v}},\boldsymbol{y}) + \gamma\Gamma(\boldsymbol{v}), \tag{9.13}$$

where γ is a penalty coefficient. Usually, the correction of network free parameters is conducted in two stages. First, new values of $\boldsymbol{v}'$ are calculated using standard learning methods (e.g., the BP algorithm), next, these values are corrected as follows:

$$\boldsymbol{v} = \boldsymbol{v}'(1-\eta\gamma\Gamma_K(\boldsymbol{v}')), \tag{9.14}$$

where η is a learning factor.

There are two attitudes to penalty function $\Gamma(\boldsymbol{v})$ design in MLP networks:

- Penalty for redundancy synaptic connections
- Penalty for redundancy hidden units

In the first case [20], the penalty function can be defined in a different form:

$$\Gamma(\mathbf{w}) = \|\mathbf{w}\|^2, \quad \Gamma_{K,ij} = 1; \tag{9.15}$$

$$\Gamma(\mathbf{w}) = \frac{1}{2}\sum_{i,j}\frac{w_{ij}^2}{1+w_{ij}^2}, \quad \Gamma_{K,ij} = \frac{1}{(1+w_{ij}'^2)^2}; \tag{9.16}$$

$$\Gamma(\mathbf{w}) = \frac{1}{2}\sum_{i,j}\frac{w_{ij}^2}{1+\sum_k w_{ik}^2}, \quad \Gamma_{K,ij} = \frac{1+2\sum_{k\neq j} w_{ij}'^2}{\left(1+\sum_k w_{ik}'^2\right)^2}; \tag{9.17}$$

where $\Gamma_{K,ij}$ is a function that corrects the weight of the connection (j, i) (9.14). The first of the functions $\Gamma(\mathbf{w})$, (9.15), is a penalty for too high values of weights. A disadvantage of this method is that it corrects all weights to the same extent, even when the problem solution specification needs weights of high values. In the case (9.16), this problem is avoided. It is easy to see that the expression (9.16) is similar to (9.15) for low values of weights and is negligibly small for high values. The expression (9.17) preserves the properties of (9.16), and, additionally, it eliminates units whose norm of the weight vector is near zero.

Units whose activation changes to a small extent during the learning process can be considered as redundant [5]. Let $\Delta_{i,p}$ be the activation change of the ith hidden unit after the presentation of the pth learning pattern; thus the penalty function can be chosen in the following form:

$$\Gamma(\mathbf{w}) = \sum_i \sum_p e(\Delta_{i,p}^2), \tag{9.18}$$

where the internal summation is conducted over all learning patterns and the external summation over all hidden units. There are many possibilities of $e(\Delta_{i,p}^2)$ definition. One looks for such a function $e(\Delta_{i,p}^2)$ that a small activation change forces big parameter corrections, and vice versa. Because the weight corrections corresponding to the penalty function are proportional to its partial derivatives over individual $\Delta_{i,p}^2$, the above property is met when

$$\frac{\partial e(\Delta_{i,p}^2)}{\partial \Delta_{i,p}^2} = \frac{1}{(1+\Delta_{i,p}^2)^n}. \tag{9.19}$$

The exponent n controls the penalty process. The higher it is, the better the elimination of a hidden unit with low activation.

9.3.3.2 Sensitivity Methods

The sensitivity of the synaptic connection (j, i) is defined as its elimination influence on the value of J_T. There are some different sensitivity measures in the literature. Mozer and Smolensky [35] introduce into the model of the ith neuron an additional set of coefficients $\{\alpha_{ij}\}$:

$$y_i = f\left(\sum_j w_{ij}\alpha_{ij}u_j\right), \tag{9.20}$$

where

- the summation j is done over all input connections of the ith unit
- w_{ij} is a weight of the jth input of the unit considered
- y_i is its output signal
- u_j is the jth input signal
- $f()$ represents the activation function

If $\alpha_{ij} = 0$, then the connection is removed, while for $\alpha_{ij} = 1$ this connection exists in a normal sense. The sensitivity measure is defined as follows:

$$s_{ij}^{MS} = -\left.\frac{\partial J_T}{\partial \alpha_{ij}}\right|_{\alpha_{ij}=1}. \tag{9.21}$$

Karnin [23] proposes a simpler definition. Sensitivity is a difference between quality measures of a full network J_T and after the connection (j, i) removing J_T^{ij}:

$$s_{ij}^{K} = -(J_T - J_T^{ij}). \tag{9.22}$$

This idea was analyzed and developed in [45].

One of the most well-known sensitivity methods is the optimal brain damage (OBD) algorithm [29]. If all weights of the network are described by one vector $\boldsymbol{w}$, then the Taylor series of the network quality measure around the current solution $\boldsymbol{w}^*$ has the following form:

$$J_T(\boldsymbol{w}) - J_T(\boldsymbol{w}^*) = \nabla^T J_T(\boldsymbol{w}^*)(\boldsymbol{w}-\boldsymbol{w}^*) + \frac{1}{2}(\boldsymbol{w}-\boldsymbol{w}^*)^T H(\boldsymbol{w}^*)(\boldsymbol{w}-\boldsymbol{w}^*) + O(\|\boldsymbol{w}-\boldsymbol{w}^*\|^2), \tag{9.23}$$

where $\mathbb{H}(\boldsymbol{w}^*)$ is the Hessian matrix. Because the weight reduction follows the learning process, it can be assumed that $\boldsymbol{w}^*$ is an argument of the J_T local minimum and $\nabla J_T(\boldsymbol{w}^*) = 0$. If it is assumed that the value of $O(\|\boldsymbol{w} - \boldsymbol{w}^*\|^2)$ is negligibly small, then

$$J_T(\boldsymbol{w}) - J_T(\boldsymbol{w}^*) \approx \frac{1}{2}(\boldsymbol{w}-\boldsymbol{w}^*)^T H(\boldsymbol{w}^*)(\boldsymbol{w}-\boldsymbol{w}^*). \tag{9.24}$$

The Hessian matrix is usually of great dimension, because even an ANN of a medium size has hundreds of free parameters, and its calculation is time consuming. In order to simplify this problem, it is assumed [29] that the diagonal elements of $\mathbb{H}(\boldsymbol{w}^*)$ are predominant. Thus

$$s_{ij}^{OBD} = \frac{1}{2}\frac{\partial^2 J_T}{\partial w_{ij}^2} w_{ij}^2. \tag{9.25}$$

The calculation of the diagonal elements of the Hessian matrix shown in (9.25) is based on the application of the back-propagation technique to the second derivatives. The activation $u_{L,i}$ of the ith unit of the layer V_L has the form

$$u_{L,i} = f\left(\sum_{j=0}^{n_{L-1}} w_{L,ij} u_{L-1,j}\right), \tag{9.26}$$

where $n_{L-1} = \text{card}(V_{L-1})$ is the number of units in the previous layer. In this notation, $u_{0,i}$ describes the ith input signal or the whole network, and $u_{M,i} = y_{NN,i}$ is the ith output signal of the network. It is easy to see that (see 9.26)

$$\frac{\partial^2 J_T}{\partial w_{ij}^2} = \frac{\partial^2 J_T}{\partial u_{L,i}^2} u_{L-1,j}^2. \tag{9.27}$$

Finally, second derivatives of the quality criterion over input signals to individual neurons have the following form:

- For the output layer V_M

$$\frac{\partial^2 J_T}{\partial u_{M,i}^2} = \frac{\partial^2 J_T}{\partial y_{NA,i}^2} (f')^2 + \left(\frac{\partial J_T}{\partial y_{NA,i}}\right)^2 f'' \tag{9.28}$$

- For other layers $(V_L | L = 1 \ldots M-1)$

$$\frac{\partial^2 J_T}{\partial u_{L,i}^2} = (f')^2 \sum_{k=1}^{n_{L+1}} w_{L+1,ki}^2 \frac{\partial^2 J_T}{\partial u_{L+1,k}^2} + f'' \sum_{k=1}^{n_{L+1}} w_{L+1,ki}^2 \frac{\partial J_T}{\partial u_{L+1,k}}. \tag{9.29}$$

The improved version of the OBD method is the optimal brain surgeon (OBS) algorithm [18]. In this method, the elimination of the ith weight vector $\boldsymbol{w}^*$ (9.24) is treated as a step of the learning process, in which a newly obtained weight vector $\boldsymbol{w}$ differs from $\boldsymbol{w}^*$ by only one element—the removed weight, w_i. Thus, the following relation is fulfilled:

$$\boldsymbol{e}_i^T (\boldsymbol{w} - \boldsymbol{w}^*) + w_i = 0, \tag{9.30}$$

where $\boldsymbol{e}_i$ is a unit vector with one on the ith location. The problem is to find the weight that meets the following condition:

$$w_i = \arg\min_{\boldsymbol{w}} \left(\frac{1}{2} (\boldsymbol{w} - \boldsymbol{w}^*)^T \mathbb{H}(\boldsymbol{w}^*)(\boldsymbol{w} - \boldsymbol{w}^*) \right) \tag{9.31}$$

and the relation (9.30). It is proved in [18] that the weight w_i, which meets the condition (9.31) is also an argument of the minimum of the following criterion:

$$s_i^{OBS} = \frac{1}{2} \frac{w_i^2}{[\mathbb{H}^{-1}]_{ii}}. \tag{9.32}$$

The connection (j, i), for which the value of s_{ij} (9.21), (9.22), (9.25) or (9.32) is the smallest, is pruned. Apart from the OBS method, the network is trained again after pruning. In the OBS method, the weight vector is corrected (see (9.31) and (9.30)):

$$\Delta \boldsymbol{w} = \frac{w_i}{[\mathbb{H}^{-1}]_{ii}} \mathbb{H}^{-1} \boldsymbol{e}_i. \tag{9.33}$$

Another solution was proposed in [4], where units are pruned also without subsequent retraining. The method is based on the simple idea of iteratively removing hidden units and then adjusting the remaining weights while maintaining the original input–output behavior. If a chosen hidden unit (kth) is pruned, then all its input and output connections are also removed. Let the ith unit be a successor of the kth unit. We want to keep the value of the ith unit output. In order to achieve this goal, weights of others input synaptic connections have to be corrected so that the following relation holds:

$$\sum_{j\in K_{V_h}} w_{ji}u_j^{(\mu)} = \sum_{j\in K_{V_h}-\{k\}} (w_{ji}-\delta_{ji})u_j^{(\mu)} \tag{9.34}$$

for all learning patterns μ. K_{V_h} denotes the set of indices of units preceding the ith unit (before pruning), and δ_{ji} is the correction of the weight w_{ji}. Equation 9.34 can be reduced to a set of linear equations:

$$\sum_{j\in K_{V_h}-\{k\}} \delta_{ji}u_j^{(\mu)} = w_{hi}u_h^{(\mu)}. \tag{9.35}$$

Castellano and coworkers [4] solved this set of equations in the least-square sense using an efficient preconditioned conjugate gradient procedure.

9.3.3.2.1 Methods Based on Information Analysis

Let us consider an ANN of the regression type: one hidden layer of units with a sigmoid activation function and one output linear unit. Generalization to a network with many outputs is simple. Let us assume that the network has learned some input–output relation with a given accuracy. Let $u_{p,i}^h$ be an input signal of the ith hidden unit from the Pth learning pattern. The covariance matrix C connected with the outputs of the hidden layer, and calculated over whole learning set, has the form

$$\mathbb{C} = \left[c_{ij} = \frac{1}{\text{card}(\Phi_L)} \sum_{p=1}^{\text{card}(\Phi_L)} \left(u_{p,i}^h - \overline{u}_i^h\right)\left(u_{p,j}^h - \overline{u}_j^h\right)\right], \tag{9.36}$$

where $\overline{u}_i^h = \frac{1}{\text{card}(\Phi_L)}\sum_{p=1}^{\text{card}(\Phi_L)} u_{p,i}^h$. The covariance matrix is symmetric and positive semi-defined; thus it can be transformed to a diagonal form using some orthonormal matrix $\mathbb{U}$:

$$\mathbb{C} = \mathbb{U}\,\text{diag}(\lambda_i | i = 1,\ldots,n)\mathbb{U}^T, \tag{9.37}$$

where n is the number of hidden units.

Topological optimization at the neural level can be done using the analysis of the $\mathbb{C}$ eigenvalues ($\lambda_i | i = 1, \ldots, n$), where one assumes that the number of pruned units is equal to the number of negligible low eigenvalues [57]. However, the network has to be retrained after units pruning, and there is no direct relation between pruned units and pointed negligible eigenvalues of $\mathbb{C}$.

Alippi and coworkers [1] propose different optimization procedure, during which the network generalization ability increases without retraining. This goal is achieved by introducing a *virtual layer*. This layer is located between the hidden layer and the output unit, and it possesses the same number of units n as the hidden layer (rys. 5). Weights of connections between the hidden and virtual layers are chosen in the form of the matrix $\mathbb{U}$, and between the virtual layer and output unit they are equal to $\mathbb{U}^T\boldsymbol{w}$, where $\boldsymbol{w}$ is the weight vector, obtained during the training process, of connections between the hidden layer and the output unit. In this way the network output does not change: $y_{NA,v} = \boldsymbol{w}^T\mathbb{U}(\mathbb{U}^T\boldsymbol{u}^h) = \boldsymbol{w}^T\boldsymbol{u}^h$. It is easy to see that the covariance matrix $\mathbb{C}_v$ corresponding to outputs of virtual units is diagonal (9.37): $\mathbb{C}_v = \text{diag}(\lambda_i | i = 1, \ldots, n)$. It means that the outputs of virtual units are independent. If the variance λ_i is negligibly small, then one can assume that $\lambda_i = 0$; thus, the ith virtual unit, independently off an input signal, has a constant output $\bar{u}_i^h$. This value can be added to the *bias* of the output unit and the virtual unit considered can be pruned. Such a process is repeated until the generalization ability of the network increases, that is, until J_T decreases. The above method finds the optimal network architecture in the sense of the generalization ability. In the sense of the minimization of the number of network free parameters, such a network is still redundant.

The OBD and OBS methods use the second order Taylor expansion of the error function to estimate its changes when the weights are perturbed. It is assumed that the first derivative is equal to zero (pruning after learning—locally optimal weights) and the error function around the optimum can be treated as a quadratic function. The OBD method assumes additionally that the off-diagonal terms of the Hessian matrix are zero.

Engelbrecht [8] showed that objective function sensitivity analysis, which is the main idea of the OBD, can be replaced with output sensitivity analysis. His pruning algorithm based on output sensitivity analysis involves a first order Taylor expansion of the ANN output. The basic idea is that a parameter with low average sensitivity and with negligible low sensitivity variance taken over all learning patterns has a negligible effect on the ANN output. Lauret et al. [28] also propose an output sensitivity analysis, but theirs is based on the Fourier amplitude sensitivity test (FAST) method described in [49].

The ANN considered in [11,46] is also of regression type (Figure 9.5a). The method proposed in [46] consists of two phases. In the first phase (the so-called additive phase), the procedure starts with the smallest architecture—with one hidden unit, and subsequent units are added iteratively until the biggest possible network for the problem considered is known. For each structure, the condition number $K(\mathbb{Z})$ of the Jacobian matrix

$$\mathbb{Z} = \frac{\partial \boldsymbol{f}(\boldsymbol{x},\boldsymbol{v})}{\partial \boldsymbol{v}} \tag{9.38}$$

of the approximated function $\boldsymbol{f}(\boldsymbol{x},\boldsymbol{v})$ over network parameters v is calculated using singular value decomposition (SVD). If the condition number $K(\mathbb{Z}) > 10^8$, the procedure is stopped. Moreover, the

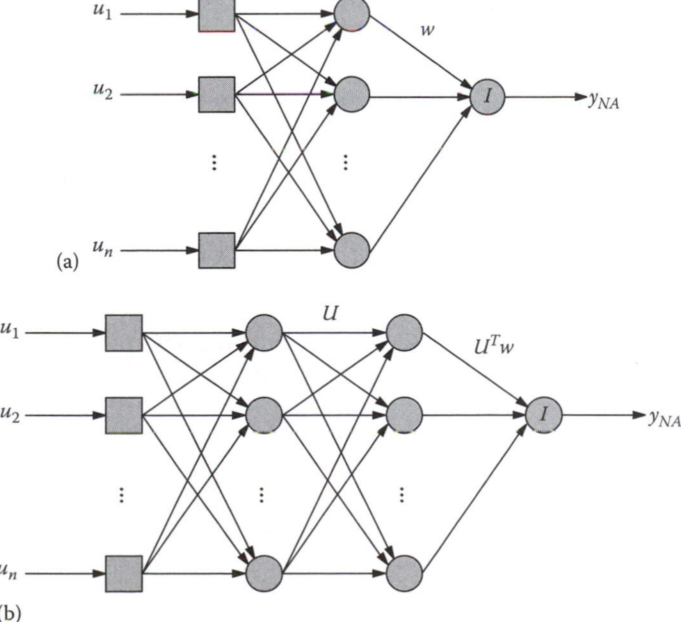

FIGURE 9.5 (a) Regression type network and (b) its version with the virtual layer.

multi-start Levenberg–Marquardt method is used in order to estimate network free parameters, and the residual signal and the estimated generalization error are remembered. This phase is similar to that presented by Fukumizu [11], where the procedure is theoretically based on the theory of statistical learning and the theory of measurement optimization. If the condition number in [46] exceeds 10^8, then the inverse $(\mathbb{Z}^T\mathbb{Z})^{-1}$, which is an estimation of the Fisher matrix, is less then 10^{-16} (under the floating-point precision in standard computers). When the first phase is stopped and the second is started [46], Fukumizu [11] carries out integrate tests for each hidden node and prunes redundant unit in accordance with the proposed sufficient conditions of Fisher matrix singularity. The aim of the second phase [46] is removing first redundant units and next redundant connections between input and hidden layers. The statistical hypothesis for each architecture obtained in the first phase is tested. This hypothesis checks whether the family of functions represented by the neural model contains the approximated function. In order to do it, the estimator of Fisher distribution based on (9.8) is calculated. If the Gaussian noise is assumed and the null hypothesis cannot be rejected, then we have the proof that the residual signal contains only random disturbances. The ANN architecture that fulfils the above test is still simplified by pruning redundant connections between the input and hidden layer. This task is also solved using statistical hypothesis testing described in detail in [46].

9.3.4 Discrete Optimization Methods

The space of ANN architectures is infinite discrete. The main problem is to choose a representation of each architecture and to order them in a structure comfortable for searching. The most popular method is encoding the network architecture in the sequence of symbols from a finite alphabet. Also, the graph, tree, and matrix representations are implemented by researchers in order to find the best possible ANN architecture.

9.3.4.1 Evolutionary Algorithms

The application of evolutionary algorithms to the ANN design process has about 20 years of history. These algorithms as global optimization algorithms can be used in neural networks in three tasks:

- Selection of parameters of the ANN with a fitted structure (learning process)
- Searching for the optimal ANN architecture—the learning process is conducted using other methods
- Application of evolutionary algorithms to both of the above tasks simultaneously

The last two tasks are the subject of this chapter. The most popular class of evolutionary algorithms is GA, which seems to be the most natural tool for a discrete space of ANN architectures. This fact results from the classical chromosome structure—a string of bits. Such a representation is used in many applications (c.f. [3,17,19,50]). Initially, an ANN architecture, NA_{max}, sufficient to represent an input–output relation is selected. This architecture defines the upper limit of ANN architecture complexity. Next, all units from the input, hidden, and output layers of NA_{max} are numbered from 1 to N. In this way, the searching space of architectures is reduced to a class of digraphs of N nodes. An architecture NA (a digraph) is represented by an incidence matrix $\mathbb{V}$ of N^2 elements. Each element is equal to 0 or 1. If $V_{ij} = 1$, then the synaptic connection between the ith and the jth node exists. The chromosome is created by rewriting the matrix $\mathbb{V}$ row by row to one binary string of length N^2. If the initial population of chromosomes is randomly generated, then the standard GA can be applied to search for the optimal ANN architecture.

It is easy to see that the chromosome created by the above procedure can represent any ANN architecture, also with backward connections. If one wants to delimit searching to MLP architectures, then the matrix $\mathbb{V}$ contains many elements equal to 0, which cannot be changed during the searching process. In this case, the definition of genetic operators is complicated and a large space of memory is unnecessarily occupied. It is sensible to omit such elements in the chromosome [44].

In practice, the neural networks applied contain hundreds to thousands of synaptic connections. Standard genetic operators working on such long chromosomes are not effective. Moreover, when the complexity of the ANN architecture increases, the convergence of the evolutionary process decreases. Thus, many researchers look for representations of ANN architectures that simplify the evolutionary process. In [34], the architecture of the ANN is directly represented by the incidence matrix $\mathbb{V}$. The crossover operator is defined as a random exchange of rows or columns between two matrices of population. In the mutation operator, each bit of the matrix is diverted with some (very low) probability.

The above method of genotypic representation of the ANN architecture is called *direct encoding* [25]. It means that there is a possibility to interpret each bit of the chromosome directly as to whether or not a concrete synaptic connection exists. The disadvantage of these methods is too low convergence in the case of a complex ANN architecture searched for or a complete loss of convergence in the limit. Moreover, if the initial architecture is of great size, then the searching process does not find an optimal solution but is only characterized by some reduction level of the network. In these cases, the quality measure of the searching method can be defined by the so-called compression factor [44], defined as

$$\kappa = \frac{\eta^*}{\eta_{max}} \times 100\%, \tag{9.39}$$

where

η^* is the number of synaptic connections of the architecture obtained by the evolutionary process

η_{max} is the maximal number of synaptic connections permissible for the selected representation of the ANN architecture

Methods of *indirect encoding* were proposed in papers [25,27,32]. In [32], an individual of the population contains the binary code of network architecture parameters (the number of hidden layers, the number of units in each hidden layer, etc.) and parameters of the back-propagation learning process (the learning factor, the momentum factor, the desired accuracy, the maximum number of iterations, etc.). Possible values of each parameter belong to a discrete, finite set whose numerical force is determined by a fixed number of bits of this parameter representation. In this way, the evolutionary process searches for the optimal ANN architecture and the optimal learning process simultaneously.

Another proposal [25] is encoding based on the graph creation system. Let the searching space be limited to ANN architectures of 2^{h+1} units at the very most. Then the incidence matrix can be represented by a tree even of high h, where each element either possesses four descendants or is a leaf. Each leaf is one of 16 possible binary matrices of size 2×2. The new type of individual representation needs new definitions of crossover and mutation operators, both of which are explained in Figure 9.6.

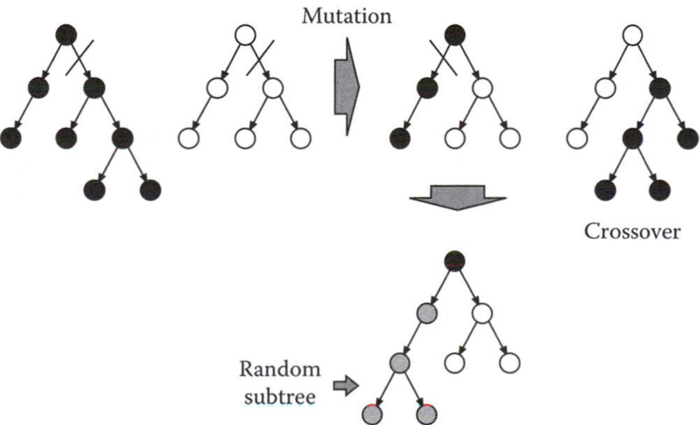

FIGURE 9.6 Genetic operators for tree representations of chromosomes—crossover and mutation.

Koza and Rice [27] propose quite a different approach to the MLP creation process based on genetic programming (GP). The single output of the network can be described as follows:

$$y_{NA,v} = f\left(\sum_j w_{ij}^M u_j^{M-1}\right). \tag{9.40}$$

The output of each network processing unit can be expressed by the formulae

$$u_i^L = f\left(\sum_j w_{ij}^m u_j^{L-1}\right) \quad L = 1,2,\ldots,M-1; \quad i = 1,2,\ldots,n_L. \tag{9.41}$$

The expressions (9.40) and (9.41) can be represented by a hierarchical tree of operators (nodes):

$$F = \{f, W, +, -, *, \%\} \tag{9.42}$$

and terms (leaves):

$$T = \{u_1, \ldots, u_n, R\}, \tag{9.43}$$

where

f is a nonlinear activation function

W is a weight function (the product of the signal and weight)

Both functions can have various numbers of arguments. Other elements of *F* are the ordinary operators of addition (+), subtraction (–), and multiplication (*); the operator (%) is the ordinary division apart from the division by 0—in this case the output is equal to 0. The set of terms (9.43) contains network input signals and some atomic floating-point constant *R*. The MLP is represented by a tree with nodes selected from the set *F* (9.42), and leaves selected from the set *T* (9.43). The population of trees is exposed to the evolutionary process, which uses the genetic operators presented in Figure 9.6.

*A** algorithm. The *A** algorithm [37] is a way to implement the best-first search to a problem graph. The algorithm will operate by searching a directed graph in which each node n_i represents a point in the problem space. Each node will contain, in addition to a description of the problem state it represents, an indication of how promising it is, a parent link that points back to the best node from which it came, and a list of the nodes that were generated from it. The parent link will make it possible to recover the path to the goal once the goal is found. The list of successors will make it possible, if a better path is found to an already existing node, to propagate the improvement down to its successors.

A heuristic function $f(n_i)$ is needed that estimates the merits of each generated node. In the *A** algorithm, this cost function is defined as a sum of two components:

$$f(n_i) = g(n_i) + h(n_i), \tag{9.44}$$

where $g(n_i)$ is the cost of the best path from the start node n_0 to the node n_i, and it is known exactly to be the sum of the cost of each of the rules that were applied along the best path from n_0 to n_i, and $h(n_i)$ is the estimation of the addition cost getting from the node n_i to the nearest goal node. The function $h(n_i)$ contains the knowledge about the problem.

In order to implement the A^* algorithm to the MLP architecture optimization process [7], the following components have to be defined:

- The set, G, of goal architectures
- The expansion operator Ξ: $A \to 2^A$, which determines the set of network architectures being successors of the architecture $NA \in A$
- The cost function $g(NA, NA')$ connected with each expansion operation
- The heuristic function $h(NA)$

The goal MLP architecture is obtained if the learning process is finished with a given accuracy:

$$G = \left\{ NA \in A \middle| \max_{\Phi_T} J_T(y_{NA,v}, y) \le \eta_0 \right\}, \tag{9.45}$$

where η_0 is a chosen nonnegative constant.

The expansion operator Ξ generates successors in a twofold way:

- By adding a hidden layer—the successor NA' of the NA is obtained by adding a new hidden layer directly in front of the output layer with the number of hidden units equal to the number of output units.
- By adding a hidden unit—it is assumed that the current architecture NA possesses at least one hidden layer, the successor NA' is created by adding a new hidden unit to the selected hidden layer.

In this way, the space of the MLP architecture is ordered into the digraph presented in Figure 9.7.

It can be proved that there exist sets of free parameters for successors NA' of NA created such that [7]

$$\forall NA' \in \Xi(NA) \quad \exists v': \quad J_T(y_{NA',v'}, y) \le J_T(y_{NA,v^*}, y). \tag{9.46}$$

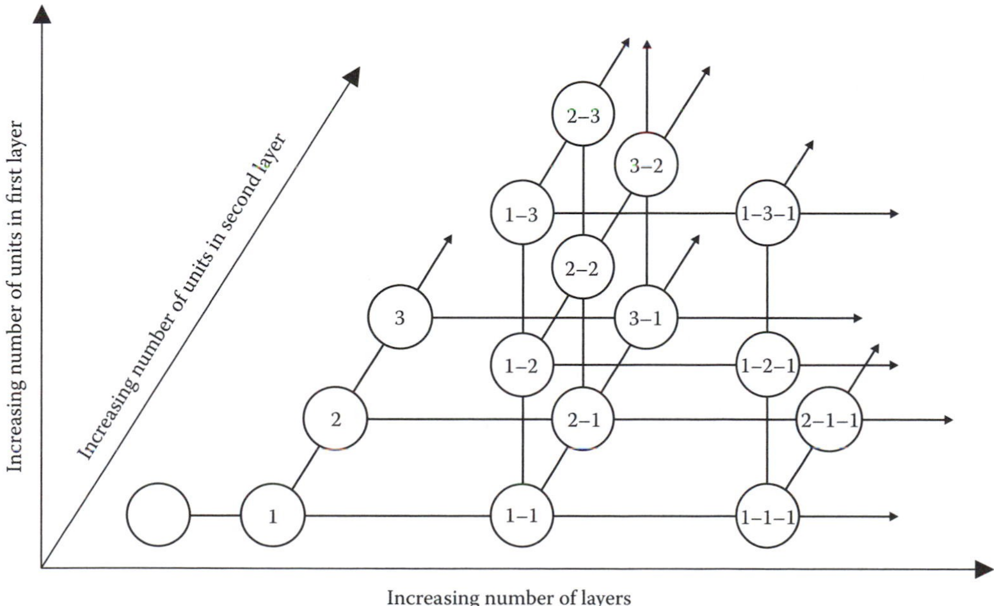

FIGURE 9.7 Digraph of MLP architectures for single input and a single output problem. p–q–r describes the architecture with p units in the first hidden layer, r units in the second one, and q units in the third one.

Each expansion $NA' \in \Xi(NA)$ is connected with the increasing of the cost vector:

$$g(NA,NA') = \begin{bmatrix} \vartheta_h(NA') - \vartheta_h(NA) \\ \vartheta_l(NA') - \vartheta_l(NA) \end{bmatrix}, \tag{9.47}$$

where

$\vartheta_h(NA) = \sum_{i=1}^{M-1} \text{card}(V_i)$ is the number of hidden units

$\vartheta_l = M - 1$ is the number of hidden layers of the NA

The effectiveness of the A^* algorithm strongly depends on the chosen definition of the heuristic function $h(NA)$. For the MLP architecture optimization problem, Doering and co-workers [7] propose it as follows:

$$h(NA) = \frac{1}{\alpha e_L(NA_0) + \beta e_T(NA_0)} \begin{bmatrix} \alpha e_L(NA) + \beta e_T(NA) \\ 0 \end{bmatrix}, \tag{9.48}$$

where

$$e_{L(T)} = \frac{1}{\text{card}\,\Phi_{L(T)}} \sum_{\Phi_{L(T)}} J_{L(T)}(y_{NA,v^*}, y)$$

is the mean error obtained for the learning (testing) set; moreover, $\alpha + \beta = 1$ and $\alpha, \beta \geq 0$. A big learning error significantly influences the heuristic function in the case of architectures near the initial architecture NA_0. This error decreases during the searching process and the selection of successor is dominated by the generalization component.

In order to compare different goal functions (9.44), the relation of the linear order $\dot{\leq}$ of two vectors a, $b \in R^2$ must be defined:

$$(\mathbf{a} \dot{\leq} \mathbf{b}) \Leftrightarrow (a_1 \leq b_1) \vee ((a_1 = b_1) \wedge (a_2 \leq b_2)). \tag{9.49}$$

The A^* algorithm is a very effective tool for MLP architecture optimization. The advantage of this algorithm over the cascade-correlation algorithm has been shown [7]. However, its computational complexity is very large in the case of complex goal architectures, because the number of successors increases very fast with the current architecture complexity. This problem is especially visible in the case of networks of dynamic units [38].

9.3.4.2 Simulated Annealing and Tabu Search

Simulated annealing (SA) [24] is based on the observation of the crystal annealing process, which has to reduce crystal defects. The system state is represented by a point S in the space of feasible solutions of a given optimization problem. The neighboring state S' of the state S differs from S only in one parameter. The minimized objective function E is called the energy by the physical analogy, and the control parameter T is called the temperature. The SA algorithm states the following steps:

1. Choose the initial state $S = S_0$ and the initial temperature $T = T_0$.
2. If the stop condition is satisfied, then stop with the solution S, else go to 3.
3. If the equilibrium state is achieved, go to 8, else go to 4.
4. Randomly choose a new neighboring state S' of the state S.
5. Calculate $\Delta E = E(S') - E(S)$.
6. If $\Delta E < 0$ or $\chi < \exp(-\Delta E/T)$, where χ is a uniformly distributed random number from the interval $[0,1)$, then $S = S'$.
7. Go to 3.
8. Update T and go to 2.

The nonnegative temperature ($T > 0$) allows us to choose the state S', whose energy is higher than the energy of the actual state S, as a base state for the further search, and then there is a chance to avoid getting stuck in a local optimum. Dislocations, which deteriorate the system energy, are controlled by the temperature T. Their range and occurring frequency decrease with $T \to 0$. As the equilibrium state can be chosen a state, in which the energy almost does not change (with a given accuracy, which is a function of temperature) in a given time interval. This criterion is relatively strong and cannot be accomplished. So, usually, the number of iterations is fixed for a given temperature. The initial temperature is the measure of the maximal "thermal" fluctuations in the system. Usually, it is assumed that the chance of achieving any system energy should be high at the beginning of the searching process. The linear decreasing of the temperature is not recommended. The linear *annealing strategy* causes the exponential decrease of "thermal" fluctuations, and the searching process usually gets stuck in a local optimum. Two *annealing strategies* are recommended:

$$T(t_n) = \begin{cases} \dfrac{T_0}{1+\ln t_n}, \\ \alpha T(t_{n-1}) \end{cases} \tag{9.50}$$

where

t_n is the number of temperature updating

$\alpha \in [0,1]$ is a given constant

The annealing strategy determines the stop condition. If the strategy (9.50) is used, the process is stopped when the temperature is almost equal to 0 ($T < \varepsilon$).

The tabu search metaheuristic was proposed by Glover [13]. This algorithm models processes existing in the human memory. This memory is implemented as a simple list of solutions explored recently. The algorithm starts from a given solution, x_0, which is treated as actually the best solution $x^* \leftarrow x_0$. The tabu list is empty: $T = \varnothing$. Next, the set of neighboring solutions are generated, excluding solutions noted in the tabu list, and the best solution of this set is chosen as a new base point. If x' is better than x^*, then $x^* \leftarrow x'$. The actual base point x' is added to the tabu list. This process is iteratively repeated until a given criterion is satisfied. There are many implementations of the tabu search idea, which differ between each other in the method of tabu list managing, for example, the tabu navigation method (TNM), the cancellation sequence method (CSM), and the reverse elimination method (REM). A particular description of these methods can be found in [14].

First, the SA algorithm was implemented for neural-network learning. But, in most cases, the evolutionary approaches performed better than SA (e.g., [52,54]). In the case of ANN architecture optimization, the SA and tabu search algorithms were applied in two ways. In the first one, the architecture is represented by a binary string, in which each position shows whether or not a given synaptic connection exists. The energy function for SA is the generalization criterion (9.8). Such a defined SA algorithm has given better results than the GA, based on the same chromosome representation [41]. A similar representation, enlarged by an additional real vector of network free parameters, was implemented for SA and the tabu search in [28] for simultaneous weight and architecture adjusting. Both algorithms are characterized by slow convergence to optimal solution. Thus, Lauret et al. [28] propose a very interesting hybrid algorithm that exploits the ideas of both algorithms. In this approach, a set of new solutions is generated, and the best one is selected according to the cost function, as performed by the tabu search. But the best solution is not always accepted since the decision is guided by the Boltzmann probability distribution as it is done in SA. Such a methodology performed better than the standard SA and tabu search algorithms. The other idea is the cascade reduction [39]. One starts with a network structure that is supposed to be sufficiently complex and reduces it using a given algorithm. Thus, we obtain a network with $\eta^*(0)$ parameters from $\eta_{max}(0)$. In the next step, we assume that $\eta_{max}(1) = \eta^*(0)$ and apply reduction again. This process is repeated until $\eta^*(k) = \eta_{max}(k)(=\eta^*(k-1))$.

In the second approach of the SA and tabu search algorithms to ANN architecture representation, both algorithms search the graph of network architectures, which has been used for the A^* algorithm (Figure 9.7) [41–43]. The neighboring solutions of a given architecture are all of its predecessors and successors.

9.4 Summary

The ANN architecture optimization problem is one of the most basic and important subtasks of neural application design. Both insufficiency and redundancy of network processing units lead to an unsatisfactory quality of the ANN model. Although the set of solutions proposed in the literature is very rich, especially in the case of feed-forward ANNs, there is still no procedure that is fully satisfactory for researchers. Two types of methods have been exploited in the last years: methods based on information analysis and discrete optimization algorithms. The methods of the first class are mathematically well grounded, but they are usually dedicated to simple networks (like regression networks), and their applicability is limited. Moreover, because most of these methods are grounded on the statistical analysis approach, rich sets of learning patterns are needed. The methods of discrete optimization seem to be most attractive for ANN structure design, especially in the case of dynamic neural networks, which still expect efficient architecture optimization methods.

References

1. Alippi, C., R. Petracca, and V. Piuri. 1995. Off-line performance maximisation in feed-forward neural networks by applying virtual neurons and covariance transformations. In *Proc. IEEE Int. Symp. Circuits and Systems ISCAS'95*, pp. 2197–2200, Seattle, WA.
2. Ash, T. 1989. Dynamic node creation. *Connect. Sci.*, 1(4):365–375.
3. Bornholdt, S. and D. Graudenz. 1991. General asymmetric neural networks and structure design by genetic algorithms. Deutsches Elektronen-Synchrotron 91–046, Hamburg, Germany.
4. Castellano, G., A.M. Fanelli, and M. Pelillo. 1997. An iterative pruning algorithm for feedforward neural networks. *IEEE Trans. Neural Netw.*, 8(3):519–531.
5. Chauvin. Y. 1989. A back-propagation algorithm with optimal use of hidden units. In *Advances in NIPS1*, ed. D. Touretzky, pp. 519–526. San Mateo, CA: Morgan Kaufmann.
6. Cybenko, G. 1989. Approximation by superpositions of a sigmoidal function. *Math. Control Signals Syst.*, 2:303–314.
7. Doering, A., M. Galicki, and H. Witte. 1997. Structure optimization of neural networks with the A*-algorithm. *IEEE Trans. Neural Netw.*, 8(6):1434–1445.
8. Engelbrecht, A.P. 2001. A new pruning heuristic based on variance analysis of sensitivity information. *IEEE Trans. Neural Netw.*, 12(6):1386–1399.
9. Fahlman, S.E. and C. Lebiere. 1990. The cascade-correlation learning architecture. In *Advances in NIPS2*, ed. D. Touretzky, pp. 524–532. San Mateo, CA: Morgan Kaufmann.
10. Frean, M. 1990. The upstart algorithm: A method for constructing and training feedforward neural networks. *Neural Comput.*, 2:198–209.
11. Fukumizu, K. 1998. Dynamic of batch learning in multilayer networks—Overrealizability and overtraining, http://www.ism.ac.jp/fukumizu/research.html
12. Geman, S., E. Bienenstack, and R. Doursat. 1992. Neural networks and the bias/variance dilemma. *Nat. Comput.*, 4(1):1–58.
13. Glover, F. 1986. Future paths for integer programming and links to artificial intelligence. *Comput. Operat. Res.*, 13(5):533–549.
14. Glover, F. and M. Laguna. 1997. *Tabu Search.* Norwell, MA: Kluwer Academic Publishers.

15. Goh, Ch.-K. and E.-J. Teoh. 2008. Hybrid multiobjective evolutionary design for artificial neural networks. *IEEE Trans. Neural Netw.*, 19(9):1531–1547.
16. Hagan, M.T. and M.B. Menhaj. 1994. Training feedforward networks with the Marquardt algorithm. *IEEE Trans. Neural Netw.*, 5:989–993.
17. Harp, S.A., T. Samad, and A. Guha. 1989. Designing application-specific neural networks using genetic algorithms. In *Advances in NIPS1*, ed. D. Touretzky, pp. 447–454. San Mateo, CA: Morgan Kaufmann.
18. Hassibi, B. and D. Stork. 1993. Second order derivatives for network pruning: Optimal brain surgeon. In *Advances in NIPS5*, ed. D. Touretzky, pp. 164–171. San Mateo, CA: Morgan Kaufmann.
19. Heo, G.-S. and I.-S. Oh. 2008. Simultaneous node pruning of input and hidden layers using genetic algorithms. In *Proc. 7th Int. Conf. Machine Learning and Cybernetics*, pp. 3428–3432, Kunming, China.
20. Hertz, J., A. Krogh, and R.G. Palmeer. 1991. *Introduction to the Theory of Neural Computation*. Reading, MA: Addison-Wesley Publishing Company, Inc.
21. Hirose, Y., K. Yamashita, and S. Hijiya. 1991. Back-propagation algorithm which varies the number of hidden units. *Neural Netw.*, 4(1):61–66.
22. Hornik, K., M. Stinchcombe, and H. White. 1989. Multilayer feedforward networks are universal approximators. *Neural Netw.*, 2:359–366.
23. Karnin, E.D. 1990. A simple procedure for pruning back-propagation trained neural networks. *IEEE Trans. Neural Netw.*, 1:239–242.
24. Kirkpatrick, S.C., D. Gellat, and M.P. Vecchi. 1983. Optimization by simulated annealing. *Science*, 220:671–680.
25. Kitano, H. 1990. Designing neural networks using genetic algorithms with graph generation system. *Complex Syst.*, 4:461–476.
26. Kottathra K. and Y. Attikiouzel. 1996. A novel multicriteria optimization algorithm for the structure determination of multilayer feedforward neural networks. *J. Netw. Comput. Appl.*, 19:135–147.
27. Koza, J.R. and J.P. Rice. 1991. Genetic generation of both the weights and architecture for a neural network. In *Proc. Int. Joint Conf. Neural Networks (IJCNN-91)*, pp. 397–404, Seattle, WA.
28. Lauret, P., E. Fock, and T.A. Mara. 2006. A node pruning algorithm based on a Fourier amplitude sensitivity test method. *IEEE Trans. Neural Netw.*, 17(2):273–293.
29. LeCun, Y., J. Denker, and S. Solla. 1990. Optimal brain damage. In *Advances in NIPS2*, ed. D. Touretzky, pp. 598–605. San Mateo, CA: Morgan Kaufmann.
30. Looney, C.G. 1997. *Pattern Recognition Using Neural Networks*. Oxford, U.K: Oxford University Press.
31. Ludemir, T.B., A. Yamazaki, and C. Zanchettin. 2006. An optimization methodology for neural network weight and architectures. *IEEE Trans. Neural Netw.*, 17(6):1452–1459.
32. Marshall, S.J. and R.F. Harrison. 1991. Optimization and training of feedforward neural networks by genetic algorithms. In *Proc 2nd Int. Conf. Artificial Neural Networks*, pp. 39–42, Boumemouth, U.K.
33. Mezard, M. and J.-P. Nadal. 1989. Learning feedforward layered networks: The tiling algorithm. *J. Phys. A*, 22:2191–2204.
34. Miller, G., P. Todd, and S. Hedge. 1989. Designing neural networks using genetic algorithms. In *Proc. Int. Conf. Genetic Algorithms*, Fairfax, VA, pp. 379–384.
35. Mozer, M. and P. Smolensky. 1989. Skeletonization—A technique for trimming the fat from a network via relevance assessment. In *Advances in NIPS1*, ed. D. Touretzky, pp. 107–115. San Mateo, CA: Morgan Kaufmann.
36. Nagao, T., T. Agui, and H. Nagahashi. 1993. Structural evolution of neural networks having arbitrary connections by a genetic method. *IEICE Trans. Inf. Syst.*, E76-D(6):689–697.
37. Nilson, N. 1980. *Principles of Artificial Intelligence*. New York: Springer-Verlag.
38. Obuchowicz, A. 1999. Architecture optimization of a network of dynamic neurons using the A*-algorithm. In *Proc. 7th European Congress on Intelligent Techniques and Soft Computing (EUFIT'99)*, Aachen, Germany (published on CD-ROM).
39. Obuchowicz, A. 1999. Optimization of a neural network structure with the cascade-reduction method. In *Proc. 4th Int. Conf. Neural Networks and Their Applications*, pp. 437–442, Zakopane, Poland.

40. Obuchowicz, A. 2000. Optimization of neural network architectures. In *Biocybernetics and Biomedical Engineering 2000, Neural Networks*, eds. W. Duch, J. Korbicz, L. Rutkowski, and R. Tadeusiewicz, pp. 323–368. Warsaw, Poland: Academic Publishing House EXIT (in Polish).
41. Obuchowicz, A. 2003. *Evolutionary Algorithms in Global Optimization and Dynamic System Diagnosis*. Lubuskie Scientific Society, Zielona Góra, Poland.
42. Obuchowicz, A. and K. Patan. 1998. Network of dynamic neurons as a residual generator. The architecture optimization. In *Proc. 3rd Conf. Diagnostics of Industrial Processes*, pp. 101–106, Jurata, Poland (in Polish). Technical University of Gdańsk Press.
43. Obuchowicz, A. and K. Patan. 2003. Heuristic search for optimal architecture of a locally recurrent neural network. In *Intelligent Information Systems*, eds. M. Kłopotek and S.T. Wierzchoń, pp. 285–292. Heidelberg, Germany: Physica-Verlag.
44. Obuchowicz, A. and K. Politowicz. 1997. Evolutionary algorithms in optimization of a multilayer feedforward neural network architecture. In *Proc. 4th Int. Symp. Methods and Models in Automation and Robotics*, pp. 739–743, Miedzyzdroje, Poland.
45. Ponnapalli, P.V.S., K.C. Ho, and M. Thomson. 1999. A formal selection and pruning algorithm for feedforward artificial neural network optimization. *IEEE Trans. Neural Netw.*, 10(4):964–968.
46. Rivals, I. and L. Personnaz. 2003. Neural-network construction and selection in nonlinear modeling. *IEEE Trans. Neural Netw.*, 14(4):804–819.
47. Rosenblatt, F. 1962. *Principles of Neurodynamics*. New York: Spartan.
48. Rumelhart, D.E., G.E. Hinton, and R.J. Williams. 1986. Learning representations by back-propagation errors. *Nature*, 323:533–536.
49. Saltelli, A., K.-S. Chan, and E.M. Scott. 2000. *Sensitivity Analysis*. New York: Wiley.
50. Senfa Ch. and T. Changbao. 2007. Neural network structure optimization and its application for passenger flow predicting of comprehensive transportation between cities. In *Proc. IEEE Int. Conf. Grey Systems and Intelligent Services*, pp. 1087–1091, Nanjing, China.
51. Setiono, R. and L. Chi Kwong Hui. 1995. Use of a quasi-Newton method in feedforward neural network construction algorithm. *IEEE Trans. Neural Netw.*, 6(1):273–277.
52. Sexton, R.S., R.E. Dorsey, and J.D. Johnson. 1999. Optimization of neural networks: A comparative analysis of the genetic algorithm and simulated annealing. *Eur. J. Operat. Res.*, 114:589–601.
53. Sharkey, A.J.C. (ed.). 1999. *Combining Artificial Neural Nets*. London, U.K.: Springer-Verlag.
54. Tsai J.-T., J.-H. Chou, and T.K. Liu. 2006. Tuning the structure and parameters of a neural network by using hybrid Taguchi-genetic algorithm. *IEEE Trans. Neural Netw.*, 17(1):69–80.
55. Wang, Z., Ch. Di Massimo, M.T. Tham, and A.J. Morris. 1994. A procedure for determining the topology of multilayer feedforward neural networks. *Neural Netw.*, 7(2):291–300.
56. Wang, Z., M. Tham, and A.J. Morris. 1992. Multilayer feedforward neural networks: A canonical form approximation of nonlinearity. *Int. J. Control*, 56:665–672.
57. Weigend, A.S. and D.E. Rumelhart. 1991. The effective dimension of the space of hidden units. In *Proc. IJCNN*, Singapore.
58. Werbos, P.J. 1974. Beyond regression: New tools for prediction and analysis in the behavioral sciences. PhD thesis, Harvard University, Cambridge, MA.

10

Parity-*N* Problems as a Vehicle to Compare Efficiencies of Neural Network Architectures

Bogdan M. Wilamowski
Auburn University

Hao Yu
Auburn University

Kun Tao Chung
Auburn University

10.1 Introduction

Parity-*N* problems have been studied deeply in many literatures [WT93,AW95,HLS99,WH10]. The *N*-bit parity function can be interpreted as a mapping (defined by 2^N binary vectors) that indicates whether the sum of the *N* elements of every binary vector is odd or even. It is shown that threshold networks with one hidden layer require *N* hidden threshold units to solve the parity-*N* problem [M61,HKP91,WH03]. If the network has bridged connections across layers, then the number of hidden threshold units can be reduced by half. In this case, only *N*/2 neurons are required in the hidden layer for the parity-*N* problem [M61]. After that, Paturi and Saks [PS90] showed that only $N/\log_2 N$ neurons are required. Siu, Roychowdhury, and Kailath [SRT91] showed that when one more hidden layer is introduced, the total number of hidden units could be only $2\sqrt{N}$.

In this chapter, the parity-*N* problem is solved by different networks, so as to compare the efficiency of neural architecture.

One may notice that, in parity problems, the same value of sum of all inputs results with the same outputs. Therefore, considering all the weights on network inputs as "1," the number of training patterns of parity-*N* problem can be reduced from 2^N to $N+1$.

Figure 10.1 shows both the original eight training patterns and the simplified four training patterns, which are identical.

Based on this pattern simplification, a linear neuron (with slope equal to 1) can be used as the network input (see Figure 10.2b). This linear neuron works as a summator. It does not have bias input and does not need to be trained.

Input	Sum of inputs	Output
0 0 0	0	0
0 0 1	1	1
0 1 0	1	1
0 1 1	2	0
1 0 0	1	1
1 0 1	2	0
1 1 0	2	0
1 1 1	3	1

(a)

Input	Output
0	0
1	1
2	0
3	1

(b)

FIGURE 10.1 Training simplification for the parity-3 problem: (a) original patterns and (b) simplified patterns.

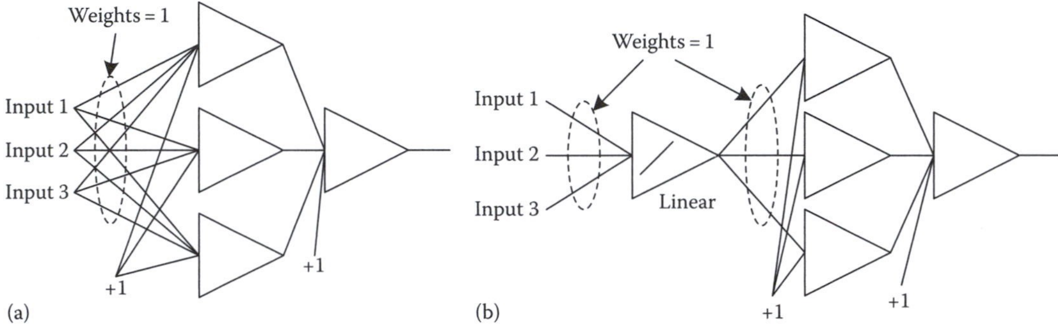

FIGURE 10.2 Two equivalent networks for the parity-3 problem: (a) parity-3 inputs and (b) linear neuron inputs.

10.2 MLP Networks with One Hidden Layer

Multilayer perceptron (MLP) networks are the most popular networks, because they are regularly formed and easy for programming. In MLP networks, neurons are organized layer by layer and there are no connections across layers.

Both parity-2 (XOR) and parity-3 problems can be visually illustrated in two and three dimensions respectively, as shown in Figure 10.3.

Similarly, using MLP networks with one hidden layer to solve the parity-7 problem, there could be at least seven neurons in the hidden layer to separate the eight training patterns (using a simplification described in introduction), as shown in Figure 10.4a.

In Figure 10.4a, eight patterns {0, 1, 2, 3, 4, 5, 6, 7} are separated by seven neurons (bold line). The thresholds of the hidden neurons are {0.5, 1.5, 2.5, 3.5, 4.5, 5.5, 6.5}. Then summing the outputs of hidden neurons weighted by {1, −1, 1, −1, 1, −1, 1}, the net inputs at the output neurons could be only {0, 1}, which can be separated by the neuron with threshold 0.5. Therefore, parity-7 problem can be solved by the architecture shown in Figure 10.4b.

Generally, if there are n neurons in MLP networks with a single hidden layer, the largest possible parity-N problem that can be solved is

$$N = n - 1 \tag{10.1}$$

where

n is the number of neurons

N is the parity index

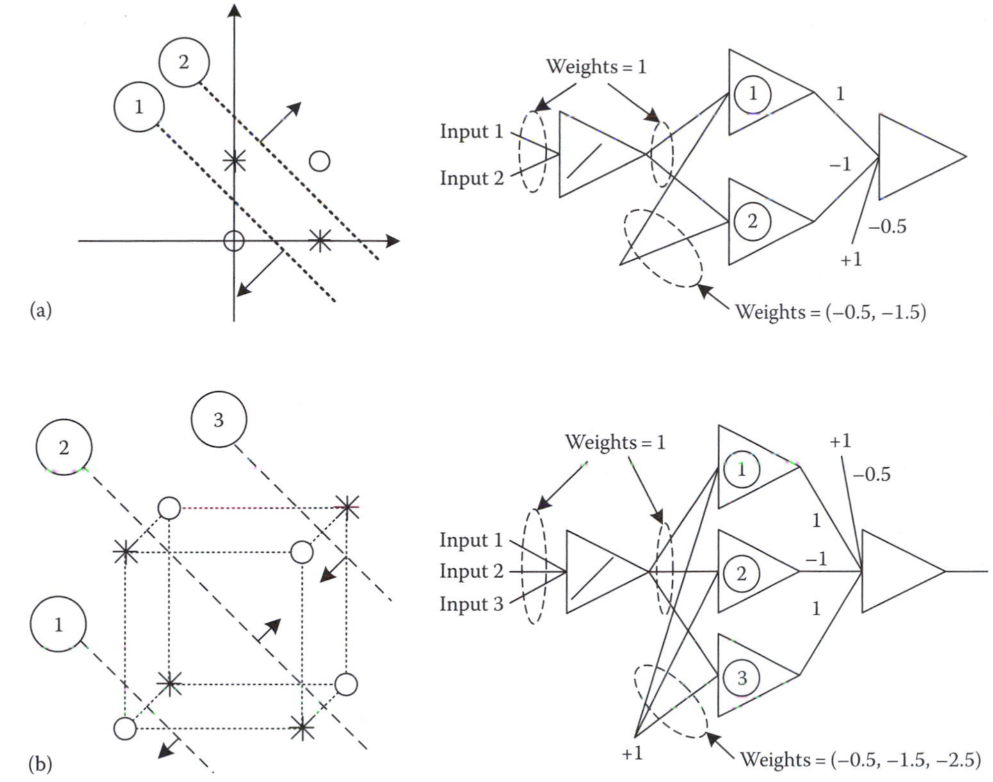

FIGURE 10.3 Graphical interpretation of pattern separation by hidden layer and network implementation using unipolar neurons for (a) XOR problem and (b) parity-3 problem.

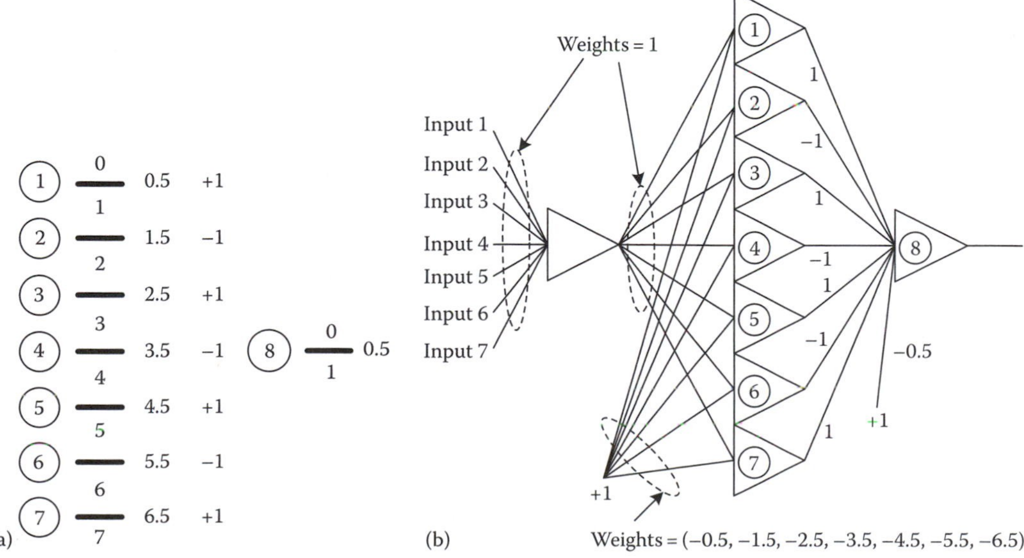

FIGURE 10.4 Solving the parity-7 problem using MLP network with one hidden layer: (a) analysis and (b) architecture.

10.3 BMLP Networks

In MLP networks, if connections across layers are permitted, then networks have bridged multilayer perceptron (BMLP) topologies. BMLP networks are more powerful than traditional MLP networks.

10.3.1 BMLP Networks with One Hidden Layer

Considering BMLP networks with only one hidden layer, all network inputs are also connected to the output neuron or neurons.

For the parity-7 problem, the eight simplified training patterns can be separated by three neurons to four subpatterns {0, 1}, {2, 3}, {4, 5}, and {6, 7}. The threshold of the hidden neurons should be {1.5, 3.5, 5.5}. In order to transfer all subpatterns to the unique pattern {0, 1} for separation, patterns {2, 3}, {4, 5}, and {6, 7} should be reduced by 2, 4, and 6 separately, which determines the weight values on connections between hidden neurons and output neurons. After pattern transformation, the unique pattern {0, 1} can be separated by the output neuron with threshold 0.5. The design process is shown in Figure 10.5a and the corresponding solution architecture is shown in Figure 10.5b.

For the parity-11 problem, similar analysis and related BMLP networks with single hidden layer solution architecture are presented in Figure 10.6.

Generally, for n neurons in BMLP networks with one hidden layer, the largest parity-N problem that can be possibly solved is

$$N = 2n - 1 \tag{10.2}$$

10.3.2 BMLP Networks with Multiple Hidden Layers

If BMLP networks have more than one hidden layer, then the further reduction of the number of neurons are possible, for solving the same problem.

For the parity-11 problem, using 4 neurons, in both 11 = 2 = 1 = 1 and 11 = 1 = 2 = 1 architectures, can find solutions.

Considering the 11 = 2 = 1 = 1 network, the 12 simplified training patterns would be separated by two neurons at first, into {0, 1, 2, 3}, {4, 5, 6, 7}, and {8, 9, 10 11}; the thresholds of the two neurons are 3.5 and 7.5, separately. Then, subpatterns {4, 5, 6, 7} and {8, 9, 10, 11} are transformed to {0, 1, 2, 3} by subtracting −4

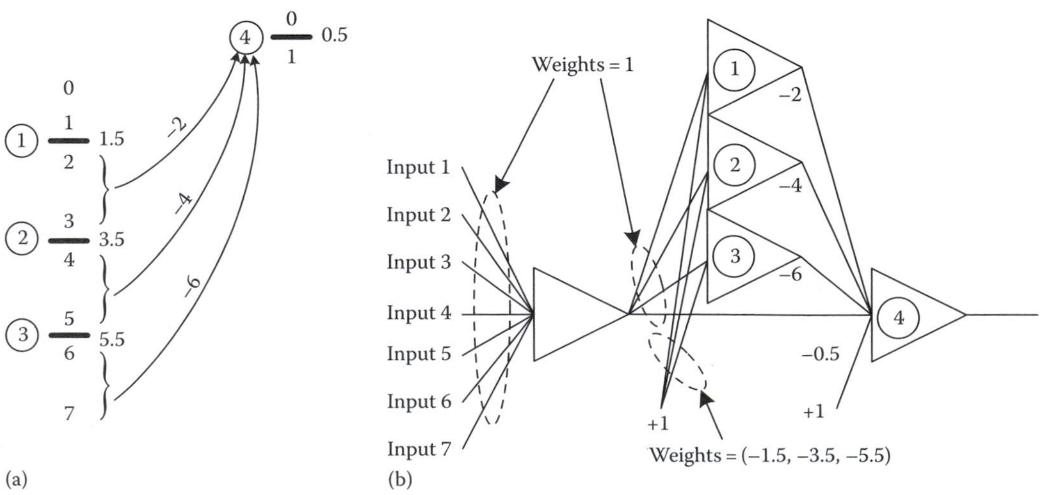

FIGURE 10.5 Solving the parity-7 problem using BMLP networks with one hidden layer: (a) analysis and (b) architecture.

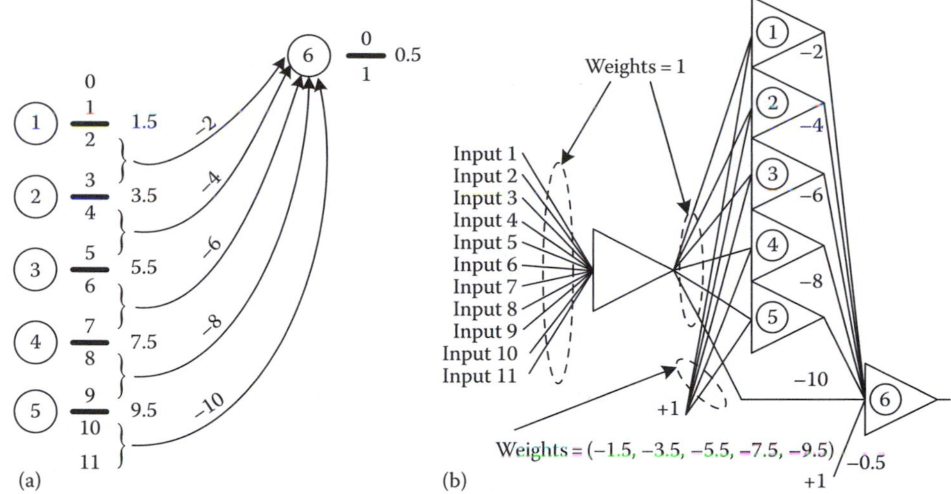

FIGURE 10.6 Solving the parity-11 problem using BMLP networks with single hidden layer: (a) analysis and (b) architecture.

and −8 separately, which determines the weight values on connections between the first hidden layer and followed layers. In the second hidden layer, one neuron is introduced to separate {0, 1, 2, 3} into {0, 1} and {2, 3}, with threshold 1.5. After that, subpattern {2, 3} is transferred to {0, 1} by setting weight value as −2 on the connection between the second layer and the output layer. At last, output neuron with threshold 0.5 separates the pattern {0, 1}. The whole procedure is presented in Figure 10.7.

Figure 10.8 shows the 11 = 1 = 2 = 1 BMLP network with two hidden layers, for solving the parity-11 problem.

Generally, considering the BMLP network with two hidden layers, the largest parity-N problem can be possibly solved is

$$N = 2(m+1)(n+1) - 1 \tag{10.3}$$

where m and n are the numbers of neurons in the two hidden layers, respectively.

For further derivation, one may notice that if there are k hidden layers and n_i is the number of neurons in related hidden layer, where i is ranged from 1 to k, then

$$N = 2(n_1+1)(n_2+1)\cdots(n_{k-1}+1)(n_k+1) - 1 \tag{10.4}$$

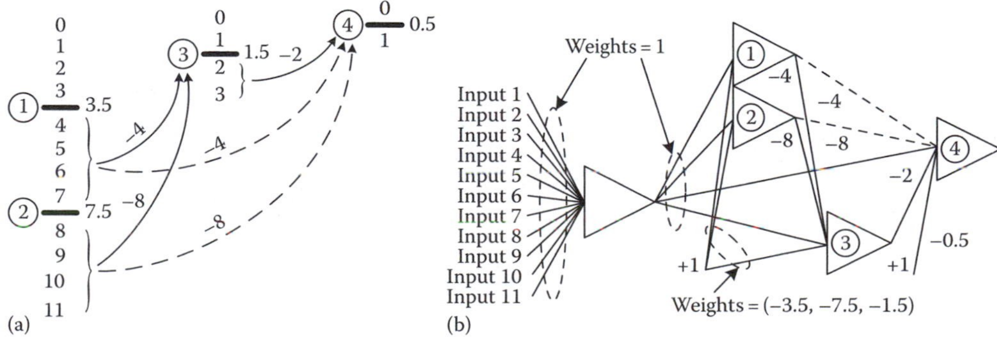

FIGURE 10.7 Solving the parity-11 problem using BMLP networks with two hidden layers, 11=2=1=1: (a) analysis and (b) architecture.

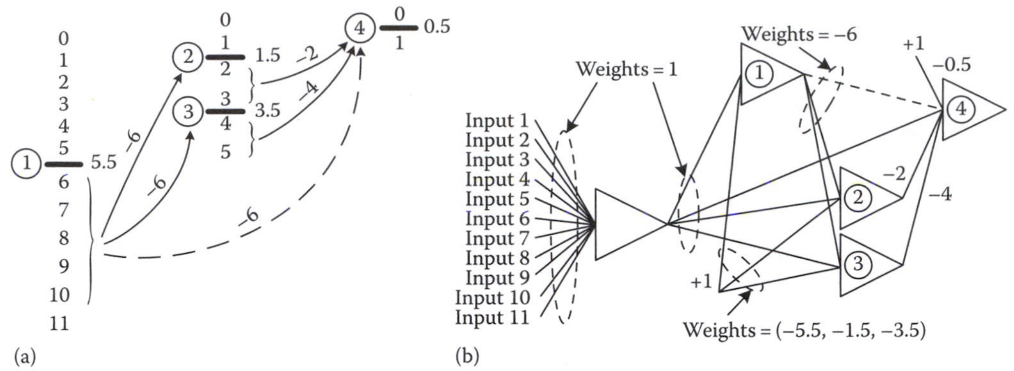

FIGURE 10.8 Solving the parity-11 problem using BMLP networks with two hidden layers, 11=1=2=1: (a) analysis and (b) architecture.

10.4 FCC Networks

Fully connected cascade (FCC) networks can solve problems using the smallest possible number of neurons [W09]. In the FCC networks, all possible routines are weighted, and each neuron contributes to a layer.

For parity-7 problem, the simplified eight training patterns are divided by one neuron at first, as {0, 1, 2, 3} and {4, 5, 6, 7}; the threshold of the neuron is 3.5. Then the subpattern {4, 5, 6, 7} is transferred to {0, 1, 2, 3} by weights equal to −4, connected to the followed neurons. Again, by using another neuron, the patterns in the second hidden layer {0, 1, 2, 3} can be separated as {0, 1} and {2, 3}; the threshold of the neuron is 1.5. In order to transfer the subpattern {2, 3} to {1, 2}, 2 should be subtracted from subpattern {2, 3}, which determines that the weight between the second layer and the output layer is −2. At last, output neurons with threshold 0.5 is used to separate the pattern {0, 1}, see Figure 10.9.

Figure 10.10 shows the solution of parity-15 problem using FCC networks.

Considering the FCC networks as special BMLP networks with only one neuron in each hidden layer, for n neurons in FCC networks, the largest N for parity-N problem can be derived from Equation 10.4 as

$$N = 2\underbrace{(1+1)(1+1)\cdots(1+1)(1+1)}_{n-1} - 1 \tag{10.5}$$

or

$$N = 2^n - 1 \tag{10.6}$$

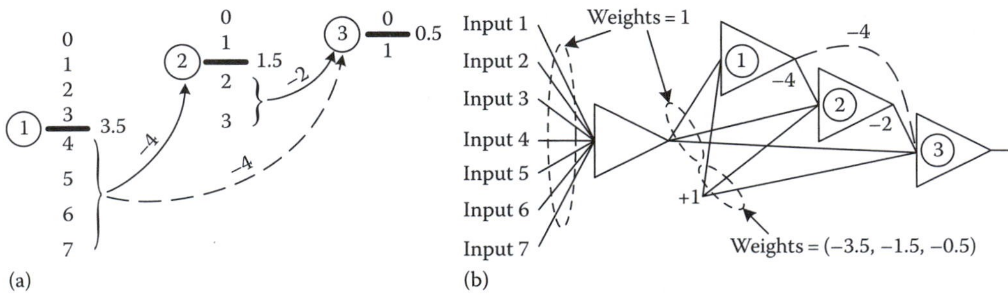

FIGURE 10.9 Solving the parity-7 problem using FCC networks: (a) analysis and (b) architecture.

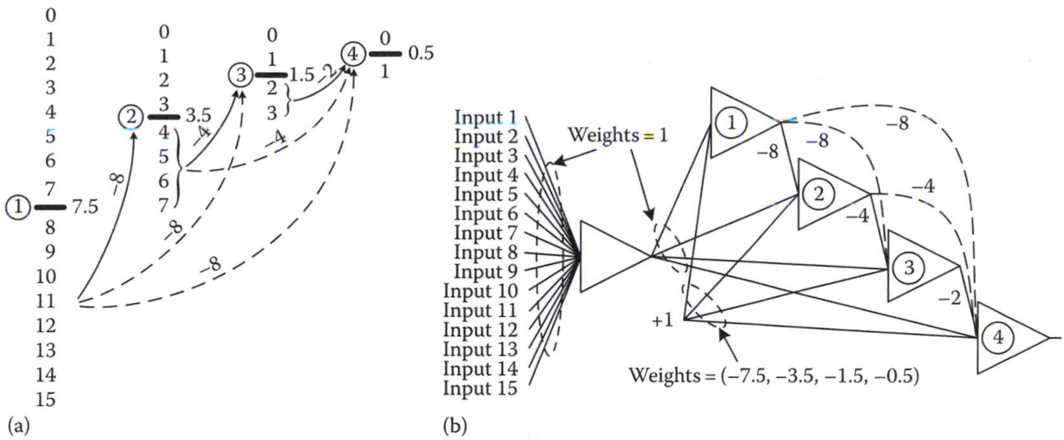

FIGURE 10.10 Solving the parity-15 problem using FCC networks: (a) analysis and (b) architecture.

TABLE 10.1 Different Architectures for Solving the Parity-*N* Problem

Network Structure	Parameters	Parity-*N* Problem
MLP with single hidden layer	*n* neurons	$n - 1$
BMLP with one hidden layer	*n* neurons	$2n + 1$
BMLP with multiple hidden layers	*h* hidden layers, each with n_i neurons	$2(n_1 + 1)(n_2 + 1)\cdots(n_{h-1} + 1)(n_h + 1) - 1$
FCC	*n* neurons	$2^n - 1$

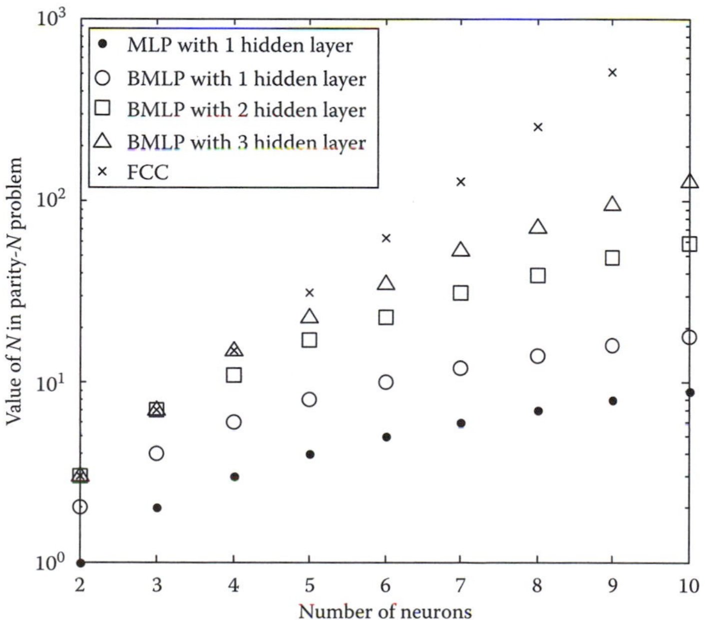

FIGURE 10.11 Efficiency comparison among various neural network architectures.

10.5 Comparison of Topologies

Table 10.1 concludes the analysis above, for the largest parity-*N* problem that can be solved with a given network structure.

Figure 10.11 shows comparisons of the efficiency of various neural network architectures.

10.6 Conclusion

This chapter analyzed the efficiency of different network architectures, using parity-*N* problems. Based on the comparison in Table 10.1 and Figure 10.11, one may notice that, for the same number of neurons, FCC networks are able solve parity-*N* problems with the least number of neurons than other architectures.

However, FCC networks also have the largest number of layers and this makes them very difficult to be trained. For example, few algorithms can be so powerful to train the parity-*N* problem with the given optimal architectures, such as four neurons for the parity-15 problem. So, the reasonable architecture would be the BMLP network with couple hidden layers.

References

[AW95] T. J. Andersen, and B. M. Wilamowski, A modified regression algorithm for fast one layer neural network training, *World Congress of Neural Networks*, vol. 1, pp. 687–690, Washington, DC, July 17–21, 1995.

[HKP91] J. Hertz, A. Krogh, and R. Palmer, *Introduction to the Theory of Neural Computation*, Addison-Wesley, Reading, MA, 1991.

[HLS99] M. E. Hohil, D. Liu, and S. H. Smith, Solving the N-bit parity problem using neural networks, *Neural Networks*, 12, 1321–1323, 1999.

[M61] R. C. Minnick, Linear-input logic, *IRE Transactions on Electronic Computers*, EC-10, 6–16, March 1961.

[PS90] R. Paturi and M. Saks, On threshold circuits for parity, *IEEE Symposium on Foundations of Computer Science*, vol. 1, pp. 397–404, October 1990.

[SRT91] K. Y. Siu, V. Roychowdhury, and T. Kailath, Depth size trade off for neural computation, *IEEE Transactions on Computers*, 40, 1402–1412, December 1991.

[W09] B. M. Wilamowski, Neural network architectures and learning algorithms, *IEEE Industrial Electronics Magazine*, 3(4), 56–63, 2009.

[WH03] B. M. Wilamowski, D. Hunter, and A. Malimowski Solving parity-N problems with feedforward neural network, *Proceedings of the IJCNN'03 International Joint Conference on Neural Networks*, pp. 2546–2551, Portland, OR, July 20–23, 2003.

[WH10] B. M. Wilamowski and H. Yu, Improved computation for Levenberg Marquardt training, *IEEE Transactions on Neural Networks*, 21(6), 930–937, 2010.

[WT93] B. M. Wilamowski and L. Torvik, Modification of gradient computation in the back-propagation algorithm, Presented at *ANNIE'93—Artificial Neural Networks in Engineering*, St. Louis, MO, November 14–17, 1993.

11

Neural Networks Learning

Bogdan M. Wilamowski
Auburn University

11.1 Introduction

The concept of systems that can learn was well described over half a century ago by Nilsson [N65] in his book *Learning Machines* where he summarized many developments of that time. The publication of the Mynsky and Paper [MP69] book slowed down artificial neural network research, and the mathematical foundation of the back-propagation algorithm by Werbos [W74] went unnoticed. A decade later, Rumelhart et al. [RHW86] showed that the error back-propagation (EBP) algorithm effectively trained neural networks [WT93,WK00,W07,FAEC02,FAN01]. Since that time many learning algorithms have been developed and only a few of them can efficiently train multilayer neuron networks. But even the best learning algorithms currently known have difficulty training neural networks with a reduced number of neurons.

Similar to biological neurons, the weights in artificial neurons are adjusted during a training procedure. Some use only local signals in the neurons, others require information from outputs; some require a supervisor who knows what outputs should be for the given patterns, and other unsupervised algorithms need no such information. Common learning rules are described in the following sections.

11.2 Foundations of Neural Network Learning

Neural networks can be trained efficiently only if networks are transparent so small changes in weights' values produce changes on neural outputs. This is not possible if neurons have hard-activation functions. Therefore, it is essential that all neurons have soft activation functions (Figure 11.1).

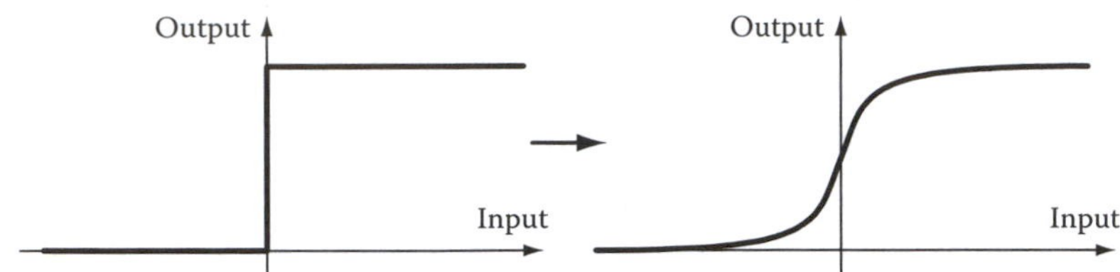

FIGURE 11.1 Neurons in the trainable network must have soft activation functions.

Neurons strongly respond to input patterns if weights' values are similar to incoming signals. Let us analyze the neuron shown in Figure 11.2 with five inputs, and let us assume that input signal is binary and bipolar (−1 or +1). For example, inputs **X** = [1, −1, 1, −1, −1] and also weights **W** = [1, −1, 1, −1, −1] then the *net* value

$$net = \sum_{i=1}^{5} x_i w_i = \mathbf{X}\mathbf{W}^T = 5 \tag{11.1}$$

This is maximal *net* value, because for any other input signals the net value will be smaller. For example, if input vector differs from the weight vector by one bit (it means the Hamming distance HD = 1), then the *net* = 3. Therefore,

$$net = \sum_{i=1}^{n} x_i w_i = \mathbf{X}\mathbf{W}^T = n - 2HD \tag{11.2}$$

where
  n is the size of the input
  HD is the Hamming distance between input pattern **X** and the weight vector **W**

This is true for binary bipolar values, but this concept can be extended to weights and patterns with analog values, as long as both lengths of the weight vector and input pattern vectors are the same. Therefore, the weights' changes should be proportional to the input pattern

$$\Delta\mathbf{W} \sim \mathbf{X} \tag{11.3}$$

In other words, the neuron receives maximum excitation if input pattern and weight vector are equal. The learning process should continue as long as the network produces wrong answers. Learning may stop if there are no errors on the network outputs. This implies the rule that weight change should be

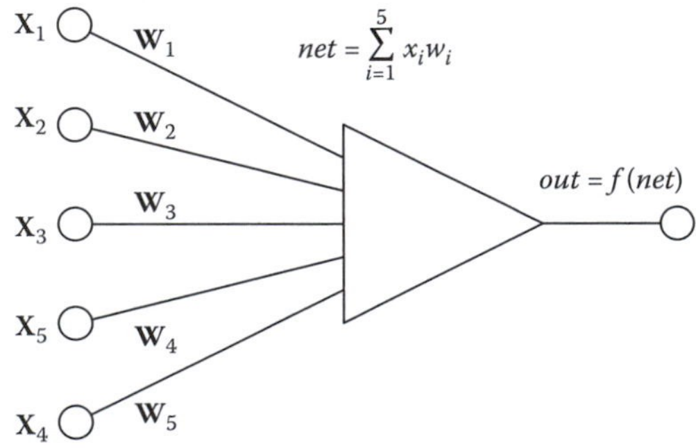

FIGURE 11.2 Neuron as the Hamming distance classifier.

proportional to the error. This rule is used in the popular EBP algorithm. Unfortunately, when errors become smaller, then weight corrections become smaller and the training process has very slow asymptotic character. Therefore, this rule is not used in advanced fast-learning algorithms such as LM [HM94] or NBN [WCKD08,W09,WY10].

11.3 Learning Rules for Single Neuron

11.3.1 Hebbian Learning Rule

The Hebb [H49] learning rule is based on the assumption that if two neighbor neurons must be activated and deactivated at the same time, then the weight connecting these neurons should increase. For neurons operating in the opposite phase, the weight between them should decrease. If there is no signal correlation, the weight should remain unchanged. This assumption can be described by the formula

$$\Delta w_{ij} = c x_i o_j \tag{11.4}$$

where
- w_{ij} is the weight from *i*th to *j*th neuron
- c is the learning constant
- x_i is the signal on the *i*th input
- o_j is the output signal

The training process usually starts with values of all weights set to zero. This learning rule can be used for both soft- and hard-activation functions. Since desired responses of neurons are not used in the learning procedure, this is the unsupervised learning rule. The absolute values of the weights are usually proportional to the learning time, which is undesired.

11.3.2 Correlation Learning Rule

The correlation learning rule is based on a similar principle as the Hebbian learning rule. It assumes that weights between simultaneously responding neurons should be largely positive, and weights between neurons with opposite reaction should be largely negative.

Contrary to the Hebbian rule, the correlation rule is the supervised learning. Instead of actual response, o_j, the desired response, d_j, is used for the weight-change calculation

$$\Delta w_{ij} = c x_i d_j \tag{11.5}$$

where d_j is the desired value of output signal. This training algorithm usually starts with initialization of weights to zero.

11.3.3 Instar Learning Rule

If input vectors and weights are normalized, or if they have only binary bipolar values (−1 or +1), then the *net* value will have the largest positive value when the weights and the input signals are the same. Therefore, weights should be changed only if they are different from the signals

$$\Delta w_i = c(x_i - w_i) \tag{11.6}$$

Note that the information required for the weight is taken only from the input signals. This is a very local and unsupervised learning algorithm [G69].

11.3.4 Winner Takes All

The winner takes all (WTA) is a modification of the instar algorithm, where weights are modified only for the neuron with the highest *net* value. Weights of remaining neurons are left unchanged. Sometimes this algorithm is modified in such a way that a few neurons with the highest *net* values are modified at the same time. Although this is an unsupervised algorithm because we do not know what desired outputs are, there is a need for a "judge" or "supervisor" to find a winner with a largest *net* value. The WTA algorithm, developed by Kohonen [K88], is often used for automatic clustering and for extracting statistical properties of input data.

11.3.5 Outstar Learning Rule

In the outstar learning rule, it is required that weights connected to a certain node should be equal to the desired outputs for the neurons connected through those weights

$$\Delta w_{ij} = c(d_j - w_{ij}) \tag{11.7}$$

where

d_j is the desired neuron output

c is the small learning constant, which further decreases during the learning procedure

This is the supervised training procedure, because desired outputs must be known. Both instar and outstar learning rules were proposed by Grossberg [G69].

11.3.6 Widrow–Hoff LMS Learning Rule

Widrow and Hoff [WH60] developed a supervised training algorithm that allows training a neuron for the desired response. This rule was derived so the square of the difference between the *net* and output value is minimized. The $Error_j$ for jth neuron is

$$Error_j = \sum_{p=1}^{P} \left(net_{jp} - d_{jp} \right)^2 \tag{11.8}$$

where

P is the number of applied patterns

d_{jp} is the desired output for jth neuron when pth pattern is applied

net is given by

$$net = \sum_{i=1}^{n} w_i x_i \tag{11.9}$$

This rule is also known as the least mean square (LMS) rule. By calculating a derivative of Equation 11.8 with respect to w_{ij} to find the gradient, the formula for the weight change can be found:

$$\frac{\partial Error_j}{\partial w_{ij}} = 2x_{ij} \sum_{p=1}^{P} \left(d_{jp} - net_{jp} \right) \tag{11.10}$$

so

$$\Delta w_{ij} = c x_{ij} \sum_{p=1}^{P} \left(d_{jp} - net_{jp} \right) \tag{11.11}$$

Note that weight change, Δw_{ij}, is a sum of the changes from each of the individual applied patterns. Therefore, it is possible to correct the weight after each individual pattern is applied. This process is known as *incremental updating*. The *cumulative updating* is when weights are changed after all patterns have been applied once. Incremental updating usually leads to a solution faster, but it is sensitive to the order in which patterns are applied. If the learning constant c is chosen to be small, then both methods give the same result. The LMS rule works well for all types of activation functions. This rule tries to enforce the *net* value to be equal to desired value. Sometimes, this is not what the observer is looking for. It is usually not important what the *net* value is, but it is important if the *net* value is positive or negative. For example, a very large *net* value with a proper sign will result in a correct output and in a large error as defined by Equation 11.8, and this may be the preferred solution.

11.3.7 Linear Regression

The LMS learning rule requires hundreds of iterations, using formula (11.11), before it converges to the proper solution. If the linear regression is used, the same result can be obtained in only one step [W02,AW95]. Considering one neuron and using vector notation for a set of the input patterns **X** applied through weight vector **w**, the vector of *net* values **net** is calculated using

$$\mathbf{X}\mathbf{w}^{\mathrm{T}} = \mathbf{net} \tag{11.12}$$

where
- **X** is the rectangular array $(n + 1) \times p$ of input patterns
- n is the number of inputs
- p is the number of patterns

Note that the size of the input patterns is always augmented by one, and this additional weight is responsible for the threshold (see Figure 11.3).

This method, similar to the LMS rule, assumes a linear activation function, and so the *net* values should be equal to desired output values **d**

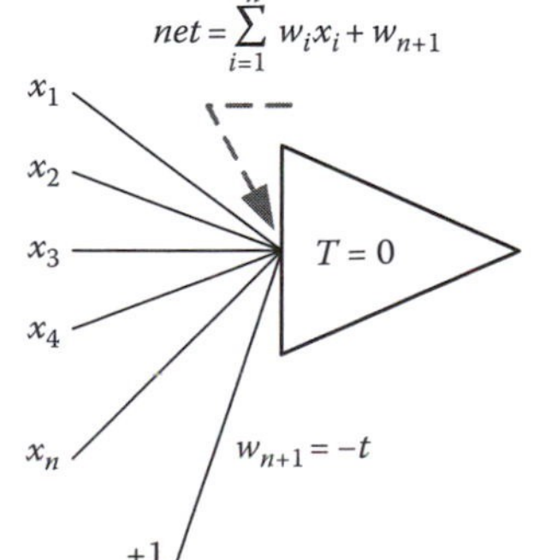

FIGURE 11.3 Single neuron with the threshold adjusted by additional weight w_{n+1}.

$$\mathbf{X}\mathbf{w}^{\mathrm{T}} = \mathbf{d} \tag{11.13}$$

Usually, $p > n + 1$, and the preceding equation can be solved only in the least mean square error sense. Using the vector arithmetic, the solution is given by

$$\mathbf{W} = \left(\mathbf{X}^{\mathrm{T}} \mathbf{X} \right)^{-1} \mathbf{X}^{\mathrm{T}} \mathbf{d} \tag{11.14}$$

The linear regression that is an equivalent of the LMS algorithm works correctly only for linear activation functions. For typical sigmoidal activation functions, this learning rule usually produces a wrong answer. However, when it is used iteratively by computing $\Delta\mathbf{W}$ instead of **W**, correct results can be obtained (see Figure 11.4).

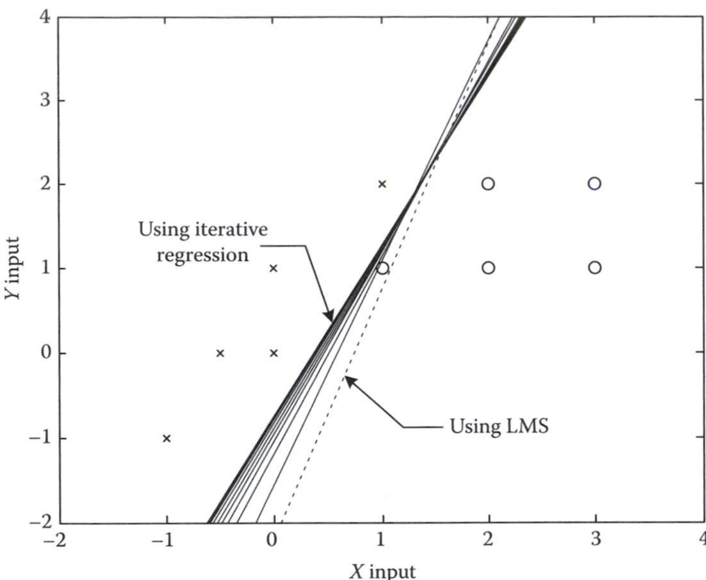

FIGURE 11.4 Single neuron training to separate patterns using LMS rule and using iterative regression with sigmoidal activation function.

11.3.8 Delta Learning Rule

The LMS method assumes linear activation function ***net*** = ***o***, and the obtained solution is sometimes far from optimum as it is shown in Figure 11.4 for a simple two dimensional case, with four patterns belonging to two categories. In the solution obtained using the LMS algorithm, one pattern is misclassified. The most common activation functions and its derivatives are for bipolar neurons:

$$o = f(net) = \tanh\left(\frac{k\,net}{2}\right) \qquad f'(o) = 0.5k\left(1 - o^2\right) \tag{11.15}$$

and for unipolar neurons

$$o = f(net) = \frac{1}{1 + \exp\left(-k\,net\right)} \qquad f'(o) = ko\left(1 - o\right) \tag{11.16}$$

where k is the slope of the activation function at $net = 0$. If error is defined as

$$Error_j = \sum_{p=1}^{P} \left(o_{jp} - d_{jp}\right)^2 \tag{11.17}$$

then the derivative of the error with respect to the weight w_{ij} is

$$\frac{dError_j}{dw_{ij}} = 2\sum_{p=1}^{P} \left(o_{jp} - d_{jp}\right)\frac{df(net_{jp})}{dnet_{jp}} x_i \tag{11.18}$$

where $o = f(net)$ are given by (11.15) or (11.16) and the *net* is given by (11.10). Note that this derivative is proportional to the derivative of the activation function $f'(net)$.

In case of the incremental training for each applied pattern

$$\Delta w_{ij} = cx_i f_j'(d_j - o_j) = cx_i \delta_j \tag{11.19}$$

Using the cumulative approach, the neuron weight, w_{ij}, should be changed after all patterns are applied:

$$\Delta w_{ij} = cx_i \sum_{p=1}^{P} (d_{jp} - o_{jp}) f_{jp}' = cx_i \sum_{p=1}^{P} \delta_{pj} \tag{11.20}$$

The weight change is proportional to input signal x_i, to the difference between desired and actual outputs $d_{jp} - o_{jp}$, and to the derivative of the activation function f_{jp}'. Similar to the LMS rule, weights can be updated using both ways: incremental and cumulative methods. One-layer neural networks are relatively easy to train [AW95,WJ96,WCM99]. In comparison to the LMS rule, the delta rule always leads to a solution close to the optimum. When the delta rule is used, then all patterns on Figure 11.4 are classified correctly.

11.4 Training of Multilayer Networks

The multilayer neural networks are more difficult to train. The most commonly used feed-forward neural network is multilayer perceptron (MLP) shown in Figure 11.5. Training is difficult because signals propagate by several nonlinear elements (neurons) and there are many signal paths. The first algorithm for multilayer training was error back-propagation algorithm [W74,RHW86,BUD09,WJK99, FFJC09,FFN01], and it is still often used because of its simplicity, even though the training process is very slow and training of close-to-optimal networks seldom produces satisfying results.

11.4.1 Error Back-Propagation Learning

The delta learning rule can be generalized for multilayer networks [W74,RHW86]. Using a similar approach, the gradient of the global error can be computed with respect to each weight in the network, as was described for the delta rule. The difference is that on top of a nonlinear activation function of a neuron, there is another nonlinear term $F\{z\}$ as shown in Figure 11.6. The learning rule for EBP can be derived in a similar way as for the delta learning rule:

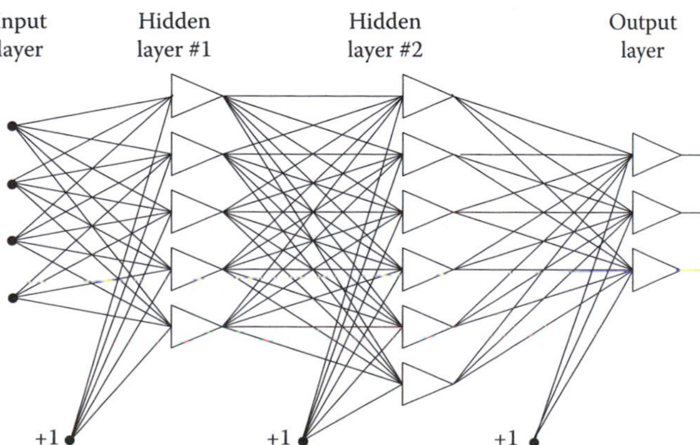

FIGURE 11.5 An example of the four layer (4-5-6-3) feed-forward neural network, which is sometimes known also as multilayer perceptron (MLP) network.

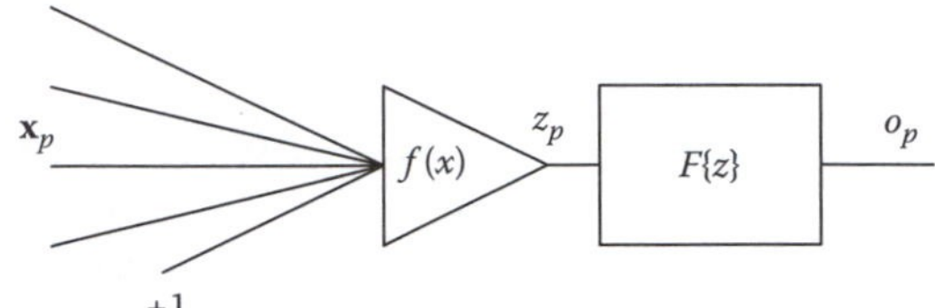

FIGURE 11.6 Error back propagation for neural networks with one output.

$$o_p = F\left\{f\left(w_1 x_{p1} + w_2 x_{p2} + \cdots + w_n x_n\right)\right\} \tag{11.21}$$

$$TE = \sum_{p=1}^{np}\left[d_p - o_p\right]^2 \tag{11.22}$$

$$\frac{d(TE)}{dw_i} = -2\sum_{p=1}^{np}\left[\left(d_p - o_p\right)F'\{z_p\}f'\left(net_p\right)x_{pi}\right] \tag{11.23}$$

The weight update for a single pattern p is

$$\Delta \mathbf{w}_p = \alpha\left(d_p - o_p\right)F_o'\{z_p\}f'\left(net_p\right)\mathbf{x}_p \tag{11.24}$$

In the case of batch training (weights are changed once all patterns are applied),

$$\Delta \mathbf{w} = \alpha\sum_{p=1}^{np}\left(\Delta \mathbf{w}_p\right) = \alpha\sum_{p=1}^{np}\left[\left(d_p - o_p\right)F_o'\{z_p\}f''\left(net_p\right)\mathbf{x}_p\right] \tag{11.25}$$

The main difference is that instead of using just derivative of activation function f' as in the delta learning rule, the product of $f'F'$ must be used. For multiple outputs as shown in Figure 11.7, the resulted weight change would be the sum of all the weight changes from all outputs calculated separately for each output using Equation 11.24. In the EBP algorithm, the calculation process is organized in such a way that error signals, Δ_p, are being propagated through layers from outputs to inputs as it is shown in Figure 11.8. Once the delta values on neurons inputs are found, then weights for this neuron are updated using a simple formula:

$$\Delta \mathbf{w}_p = \alpha \mathbf{x}_p \Delta_p \tag{11.26}$$

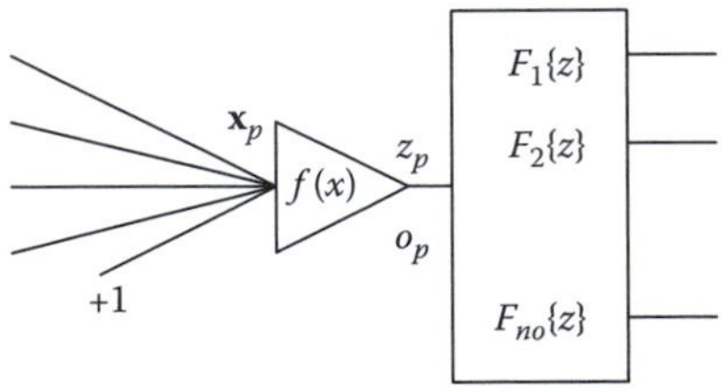

$\mathbf{x}_p$ $f(x)$ z_p o_p $+1$ $F_1\{z\}$ $F_2\{z\}$ $F_{no}\{z\}$

FIGURE 11.7 Error back propagation for neural networks with multiple outputs.

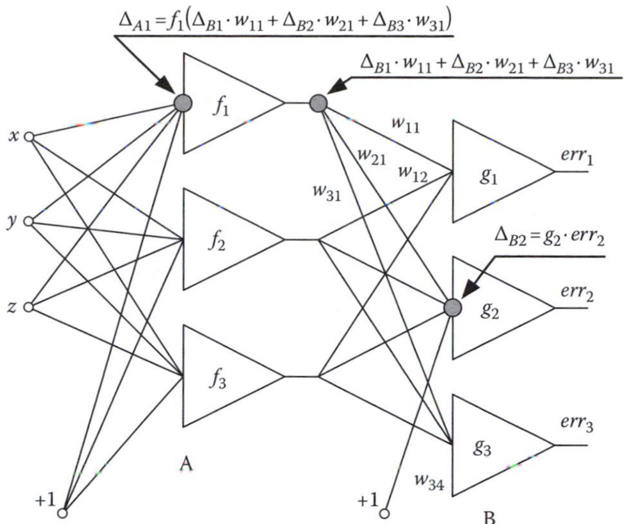

FIGURE 11.8 Calculation errors in neural network using error back-propagation algorithm. The symbols f_i and g_i represent slopes of activation functions.

where Δ was calculated using the error back-propagation process for all outputs K:

$$\Delta_p = \sum_{k=1}^{K} \left[\left(d_{pk} - o_{pk}\right) F_k'\{z_{pk}\} f'\left(net_{pk}\right) \right] \tag{11.27}$$

The calculation of the back-propagating error is kind of artificial to the real nervous system. Also, the error back-propagation method is not practical from the point of view of hardware realization. Instead, it is simpler to find signal gains $A_{j,k}$ from the input of the jth neuron to each of the network output k (Figure 11.9). For each pattern, the Δ value for a given neuron, j, can be now obtained for each output k:

$$\Delta_{j,k} = A_{j,k}\left(o_k - d_k\right) \tag{11.28}$$

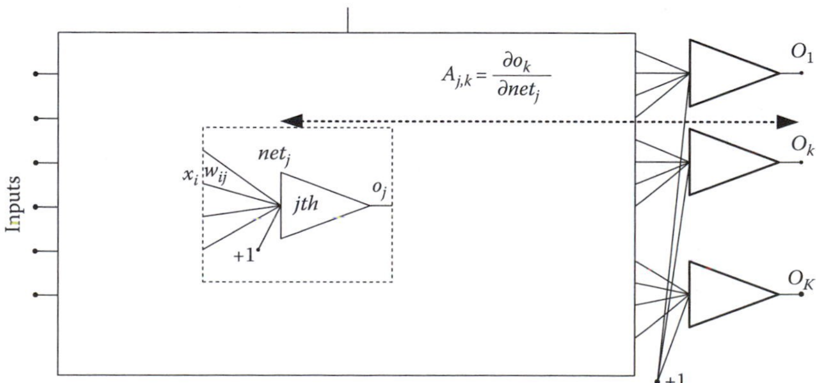

FIGURE 11.9 Finding gradients using evaluation of signal gains, $A_{j,k}$.

and for all outputs for neuron j

$$\Delta_j = \sum_{k=1}^{K} \Delta_{j,k} = \sum_{k=1}^{K} \left[A_{jk} \left(o_k - d_k \right) \right] \tag{11.29}$$

Note that the above formula is general, no matter whether neurons are arranged in layers or not. One way to find gains $A_{j,k}$ is to introduce an incremental change on the input of the *jth* neuron and observe the change in the *kth* network output. This procedure requires only forward signal propagation, and it is easy to implement in a hardware realization. Another possible way is to calculate gains through each layer and then find the total gains as products of layer gains. This procedure is equally or less computation intensive than a calculation of cumulative errors in the error back-propagation algorithm.

11.4.2 Improvements of EBP

11.4.2.1 Momentum

The back-propagation algorithm has a tendency for oscillation (Figure 11.10) [PS94]. In order to smooth up the process, the weights increment, Δw_{ij}, can be modified according to Rumelhart et al. [RHW86]:

$$w_{ij}(n+1) = w_{ij}(n) + \Delta w_{ij}(n) + \eta \Delta w_{ij}(n-1) \tag{11.30}$$

or according to Sejnowski and Rosenberg [SR87]

$$w_{ij}(n+1) = w_{ij}(n) + \left(1 - \alpha\right) \Delta w_{ij}(n) + \eta \Delta w_{ij}(n-1) \tag{11.31}$$

11.4.2.2 Gradient Direction Search

The back-propagation algorithm can be significantly accelerated, when after finding components of the gradient, weights are modified along the gradient direction until a minimum is reached. This process can be carried on without the necessity of computational intensive gradient calculation at each step. The new gradient components are calculated once a minimum on the direction of the previous gradient is obtained. This process is only possible for cumulative weight adjustment. One method to find a minimum along the gradient direction is the three step process of finding error for three points along gradient direction and then, using a parabola approximation, jump directly to the minimum (Figure 11.11).

11.4.2.3 Elimination of Flat Spots

The back-propagation algorithm has many disadvantages that lead to very slow convergence. One of the most painful is that in the back-propagation algorithm, it has difficulty to train neurons with the maximally wrong answer. In order to understand this problem, let us analyze a bipolar activation function

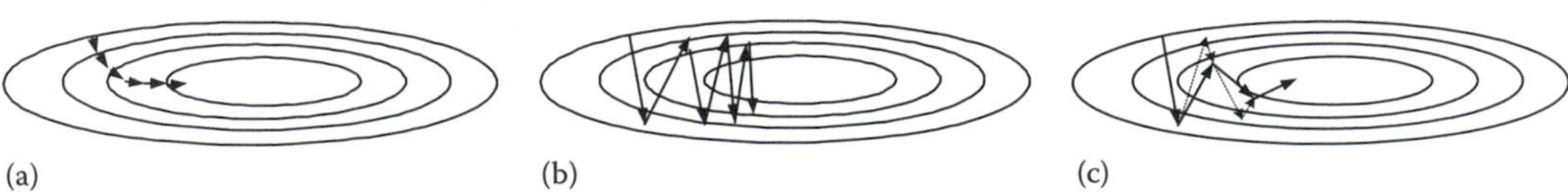

FIGURE 11.10 Illustration of convergence process for (a) too small learning constant, (b) too large learning constant, and (c) large learning constant with momentum.

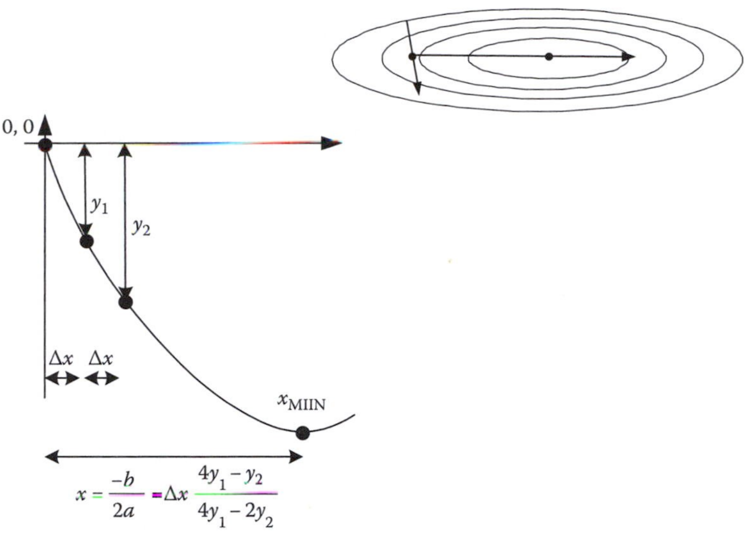

FIGURE 11.11 Search on the gradient direction before a new calculation of gradient components.

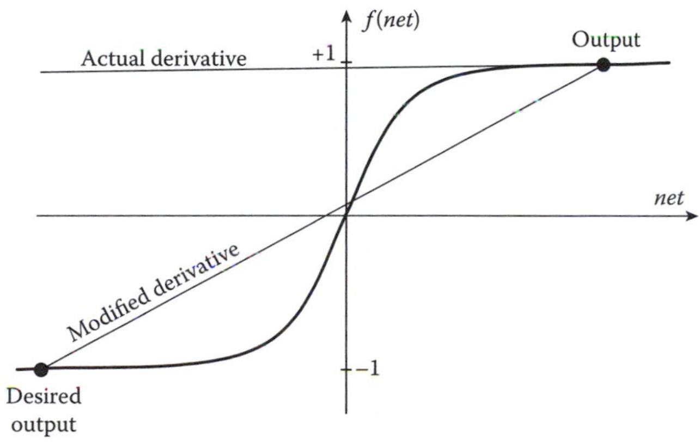

FIGURE 11.12 Lack of error back propagation for very large errors.

shown in Figure 11.12. Maximum error equal 2 exists if the desired output is −1 and actual output is +1. At this condition, the derivative of the activation function is close to zero so the neuron is not transparent for error propagation. In other words, this neuron with large output error will not be trained or it will be trained very slowly. In the mean time, other neurons will be trained and weights of this neuron would remain unchanged.

To overcome this difficulty, a modified method for derivative calculation was introduced by Wilamowski and Torvik [WT93]. The derivative is calculated as the slope of a line connecting the point of the output value with the point of the desired value as shown in Figure 11.11:

$$f_{modif} = \frac{o_{desired} - o_{actual}}{net_{desired} - net_{actual}} \tag{11.32}$$

If the computation of the activation derivative as given by

$$f'(net) = k\left[1 - o^2\right] \tag{11.33}$$

is replaced by

$$f'(net) = k\left[1 - o^2\left(1 - \left(\frac{err}{2}\right)^2\right)\right] \tag{11.34}$$

then for small errors

$$f'(net) = k\left[1 - o^2\right] \tag{11.35}$$

and for large errors ($err = 2$)

$$f'(net) = k \tag{11.36}$$

Note that for small errors, the modified derivative would be equal to the derivative of activation function at the point of the output value. Therefore, modified derivative is used only for large errors, which cannot be propagated otherwise.

11.4.3 Quickprop Algorithm

The fast-learning algorithm using the approach below was proposed by Fahlman [F88], and it is known as the *quickprop*:

$$\Delta w_{ij}(t) = -\alpha S_{ij}(t) + \gamma_{ij}\Delta w_{ij}(t-1) \tag{11.37}$$

$$S_{ij}(t) = \frac{\partial E(\mathbf{w}(t))}{\partial w_{ij}} + \eta w_{ij}(t) \tag{11.38}$$

where

- α is the learning constant
- γ is the memory constant (small 0.0001 range) leads to reduction of weights and limits growth of weights
- η is the momentum term selected individually for each weight

$$\begin{aligned} 0.01 < \alpha < 0.6 \quad &\text{when} \quad \Delta w_{ij} = 0 \text{ or sign of } \Delta w_{ij} \\ \alpha = 0 \quad &\text{otherwise} \end{aligned} \tag{11.39}$$

$$S_{ij}(t)\Delta w_{ij}(t) > 0 \tag{11.40}$$

$$\Delta w_{ij}(t) = -\alpha S_{ij}(t) + \gamma_{ij}\Delta w_{ij}(t-1) \tag{11.41}$$

The momentum term selected individually for each weight is a very important part of this algorithm. Quickprop algorithm sometimes reduces computation time hundreds of times.

11.4.4 RPROP-Resilient Error Back Propagation

Very similar to EBP, but weights are adjusted without using values of the propagated errors but only its sign. Learning constants are selected individually to each weight based on the history:

$$\Delta w_{ij}(t) = -\alpha_{ij}\,\mathrm{sgn}\left(\frac{\partial E(\mathbf{w}(t))}{\partial w_{ij}(t)}\right) \tag{11.42}$$

$$S_{ij}(t) = \frac{\partial E(\mathbf{w}(t))}{\partial w_{ij}} + \eta w_{ij}(t) \tag{11.43}$$

$$\alpha_{ij}(t) = \begin{cases} \min\left(a\cdot\alpha_{ij}(t-1), \alpha_{\max}\right) & \text{for } S_{ij}(t)\cdot S_{ij}(t-1) > 0 \\ \max\left(b\cdot\alpha_{ij}(t-1), \alpha_{\min}\right) & \text{for } S_{ij}(t)\cdot S_{ij}(t-1) < 0 \\ \alpha_{ij}(t-1) & \text{otherwise} \end{cases}$$

11.4.5 Back Percolation

Error is propagated as in EBP and then each neuron is "trained" using the algorithm to train one neuron such as pseudo inversion. Unfortunately, pseudo inversion may lead to errors, which are sometimes larger than 2 for bipolar or larger than 1 for unipolar.

11.4.6 Delta-Bar-Delta

For each weight, the learning coefficient is selected individually. It was developed for quadratic error functions

$$\Delta\alpha_{ij}(t) = \begin{cases} a & \text{for } S_{ij}(t-1)D_{ij}(t) > 0 \\ -b\cdot\alpha_{ij}(t-1) & \text{for } S_{ij}(t-1)D_{ij}(t) < 0 \\ 0 & \text{otherwise} \end{cases} \tag{11.44}$$

$$D_{ij}(t) = \frac{\partial E(t)}{\partial w_{ij}(t)} \tag{11.45}$$

$$S_{ij}(t) = (1-\xi)D_{ij}(t) + \xi S_{ij}(t-1) \tag{11.46}$$

11.5 Advanced Learning Algorithms

Let us introduce basic definitions for the terms used in advanced second-order algorithms. In the first-order methods such as EBP, which is the steepest descent method, the weight changes are proportional to the gradient of the error function:

$$\mathbf{w}_{k+1} = \mathbf{w}_k - \alpha\mathbf{g}_k \tag{11.47}$$

where **g** is gradient vector.

$$\text{gradient } \mathbf{g} = \begin{matrix} \dfrac{\partial E}{\partial w_1} \\ \dfrac{\partial E}{\partial w_2} \\ \vdots \\ \dfrac{\partial E}{\partial w_n} \end{matrix} \tag{11.48}$$

For the second-order Newton method, the Equation 11.42 is replaced by

$$\mathbf{w}_{k+1} = \mathbf{w}_k - \mathbf{A}_k^{-1}\mathbf{g}_k \tag{11.49}$$

where $\mathbf{A}_k$ is Hessian.

$$\mathbf{A} = \begin{matrix} \dfrac{\partial^2 E}{\partial w_1^2} & \dfrac{\partial^2 E}{\partial w_2 \partial w_1} & \cdots & \dfrac{\partial^2 E}{\partial w_n \partial w_1} \\ \dfrac{\partial^2 E}{\partial w_1 \partial w_2} & \dfrac{\partial^2 E}{\partial w_2^2} & \cdots & \dfrac{\partial^2 E}{\partial w_n \partial w_2} \\ \vdots & \vdots & \ddots & \vdots \\ \dfrac{\partial^2 E}{\partial w_1 \partial w_n} & \dfrac{\partial^2 E}{\partial w_2 \partial w_n} & \cdots & \dfrac{\partial^2 E}{\partial w_n^2} \end{matrix} \tag{11.50}$$

Unfortunately, it is very difficult to find Hessians so in the Gauss–Newton method the Hessian is replaced by product of Jacobians:

$$\mathbf{A} = 2\mathbf{J}^T\mathbf{J} \tag{11.51}$$

where

$$\mathbf{J} = \begin{bmatrix} \dfrac{\partial e_{11}}{\partial w_1} & \dfrac{\partial e_{11}}{\partial w_2} & \cdots & \dfrac{\partial e_{11}}{\partial w_n} \\ \dfrac{\partial e_{21}}{\partial w_1} & \dfrac{\partial e_{21}}{\partial w_2} & \cdots & \dfrac{\partial e_{21}}{\partial w_n} \\ \vdots & \vdots & & \vdots \\ \dfrac{\partial e_{M1}}{\partial w_1} & \dfrac{\partial e_{M1}}{\partial w_2} & \cdots & \dfrac{\partial e_{M1}}{\partial w_N} \\ \vdots & \vdots & & \vdots \\ \dfrac{\partial e_{1P}}{\partial w_1} & \dfrac{\partial e_{1P}}{\partial w_2} & \cdots & \dfrac{\partial e_{1P}}{\partial w_N} \\ \dfrac{\partial e_{2P}}{\partial w_1} & \dfrac{\partial e_{2P}}{\partial w_2} & \cdots & \dfrac{\partial e_{2P}}{\partial w_N} \\ \vdots & \vdots & & \vdots \\ \dfrac{\partial e_{MP}}{\partial w_1} & \dfrac{\partial e_{MP}}{\partial w_2} & \cdots & \dfrac{\partial e_{MP}}{\partial w_N} \end{bmatrix} \tag{11.52}$$

Knowing Jacobian, **J**, the gradient can be calculated as

$$\mathbf{g} = 2\mathbf{J}^T\mathbf{e} \tag{11.53}$$

Therefore, in Gauss–Newton algorithm, the weight update is calculated as

$$\mathbf{w}_{k+1} = \mathbf{w}_k - \left(\mathbf{J}_k^T\mathbf{J}_k\right)^{-1}\mathbf{J}_k^T\mathbf{e} \tag{11.54}$$

This method is very fast, but it works well only for systems that are almost linear and this is not true for neural networks.

11.5.1 Levenberg–Marquardt Algorithm

Levenberg and Marquardt, in order to secure convergence, modified Equation (11.46) to the form

$$\mathbf{w}_{k+1} = \mathbf{w}_k - \left(\mathbf{J}_k^T\mathbf{J}_k + \mu\mathbf{I}\right)^{-1}\mathbf{J}_k^T\mathbf{e} \tag{11.55}$$

where the μ parameter changed during the training process. If $\mu = 0$ algorithm works as Gauss–Newton method and for large values of μ algorithm works as steepest decent method.

The Levenberg–Marquardt algorithm was adopted for neural network training by Hagan and Menhaj [HM94], and then Demuth and Beale [DB04] adopted the LM algorithm in MATLAB® Neural Network Toolbox.

The LM algorithm is very fast, but there are several problems:

1. It was written only for MLP networks, which are not the best architectures for neural networks.
2. It can handle only problems with relatively small patterns because the size of Jacobian is proportional to the number of patterns.

11.5.2 Neuron by Neuron

The recently developed neuron by neuron (NBN) algorithm [WCKD08,CWD08,W09,YW09] is very fast. Figures 11.13 and 11.14 show speed comparison of EBP and NBN algorithms to solve the parity-4 problem.

The NBN algorithm eliminates most deficiencies of the LM algorithm. It can be used to train neural networks with arbitrarily connected neurons (not just MLP architecture). It does not require to compute and to store large Jacobians, so it can train problems with basically unlimited number of patterns [WH10,WHY10]. Error derivatives are computed only in forward pass, so back-propagation process is not needed. It is equally fast, but in the case of networks with multiple outputs faster than LM

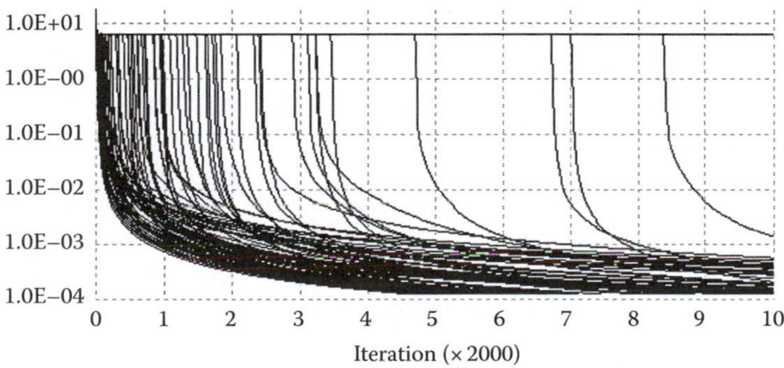

FIGURE 11.13 Sum of squared errors as a function of number of iterations for the parity-4 problem using EBP algorithm, and 100 runs.

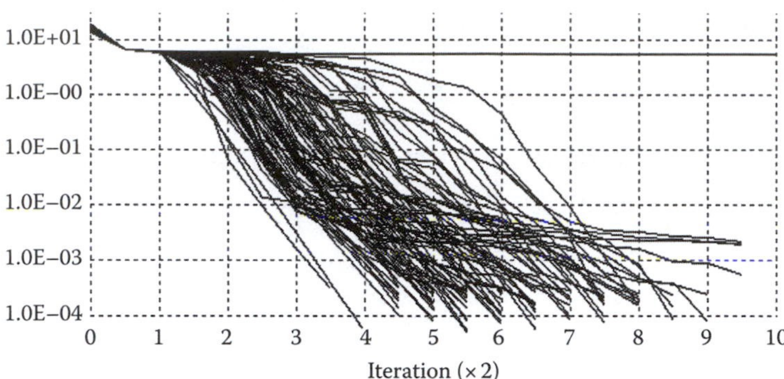

FIGURE 11.14 Sum of squared errors as a function of number of iterations for the parity-4 problem using NBN algorithm and 100 runs.

algorithm. It can train networks that are impossible to train with other algorithms. A more detailed description of the NBN algorithm is given in Chapter 13.

11.6 Warnings about Neural Network Training

It is much easier to train neural networks where the number of neurons is larger than required. But, with a smaller number of neurons the neural network has much better generalization abilities. It means it will respond correctly for patterns not used for training. If too many neurons are used, then the network can be overtrained on the training patterns, but it will fail on patterns never used in training. With a smaller number of neurons, the network cannot be trained to very small errors, but it may produce much better approximations for new patterns. The most common mistake made by many researchers is that in order to speed up the training process and to reduce the training errors, they use neural networks with a larger number of neurons than required. Such networks would perform very poorly for new patterns not used for training [W09,ISIE,PE10].

11.7 Conclusion

There are several reasons for the frustration of people trying to adapt neural networks for their research:

1. In most cases, the relatively inefficient MLP architecture is used instead of more powerful topologies [WHM03] where connections across layers are allowed.
2. When a popular learning software is used, such as EBP, the training process is not only very time consuming, but frequently the wrong solution is obtained. In other words, EBP is often not able to find solutions for neural network with the smallest possible number of neurons.
3. It is easy to train neural networks with an excessive number of neurons. Such complex architectures for a given pattern can be trained to very small errors, but such networks do not have generalization abilities. Such networks are not able to deliver a correct response to new patterns, which were not used for training [W09,HW09]. In other words, the main purpose of using neural networks is missed. In order to properly utilize neural networks, its architecture should be as simple as possible to perform the required function.
4. In order of find solutions for close-to-optimal architectures, second-order algorithms such as NBN or LM should be used [WCKD07,WCKD08]. Unfortunately, the LM algorithm adopted in the popular MATLAB NN Toolbox can handle only MLP topology without connections across layers and these topologies are far from optimal.

The importance of the proper learning algorithm was emphasized, since with an advanced learning algorithm we can train those networks, which cannot be trained with simple algorithms. The software used in this work, which implements the NBN algorithm, can be downloaded from [WY09].

References

[AW95] T. J. Andersen and B. M. Wilamowski, A modified regression algorithm for fast one layer neural network training, in *World Congress of Neural Networks*, vol. 1, pp. 687–690, Washington, DC, July 17–21, 1995.

[BUD09] B. M. Wilamowski, Neural Networks or Fuzzy Systems, Workshop on Intelligent Systems, Budapest, Hungary, August 30, 2009, pp. 1–12.

[CWD08] N. J. Cotton, B. M. Wilamowski, and G. Dundar, A neural network implementation on an inexpensive eight bit microcontroller, in *12th International Conference on Intelligent Engineering Systems (INES 2008)*, pp. 109–114, Miami, FL, February 25–29, 2008.

[DB04] H. Demuth and M. Beale, *Neural Network Toolbox, User's Guide, Version 4*, The MathWorks, Inc., Natick, MA, revised for version 4.0.4 edition, October 2004, http://www.mathworks.com

[F88] S. E. Fahlman, Faster-learning variations on back-propagation: An empirical study, in *Connectionist Models Summer School*, eds. T. J. Sejnowski G. E. Hinton, and D. S. Touretzky, Morgan Kaufmann, San Mateo, CA, 1988.

[FAEC02] F. Fnaiech, S. Abid, N. Ellala, and M. Cheriet, A comparative study of fast neural networks learning algorithms, *IEEE International Conference on Systems, Man and Cybernetics*, Hammamet, Tunisia, October 6–9, 2002.

[FAN01] F. Fnaiech, S. Abid, and M. Najim, A fast feed-forward training algorithm using a modified form of standard back-propagation algorithm, *IEEE Transactions on Neural Network*, 12(2), 424–430, March 2001.

[FFJC09] N. Fnaiech, F. Fnaiech, B.W. Jervis, and M. Cheriet, The combined statistical stepwise and iterative neural network pruning algorithm, *Intelligent Automation and Soft Computing*, 15(4), 573–589, 2009.

[FFN01] F. Fnaiech, N. Fnaiech, and M. Najim, A new feed-forward neural network hidden layer's neurons pruning algorithm, *IEEE International Conference on Acoustic Speech and Signal Processing (ICASSP'2001)*, Salt Lake City, UT.

[G69] S. Grossberg, Embedding fields: A theory of learning with physiological implications, *Journal of Mathematical Psychology*, 6, 209–239, 1969.

[H49] D. O. Hebb, *The Organization of Behavior, a Neuropsychological Theory*, John Wiley, New York, 1949.

[HM94] M. T. Hagan and M. Menhaj, Training feedforward networks with the Marquardt algorithm, *IEEE Transactions on Neural Networks*, 5(6), 989–993, 1994.

[HW09] H. Yu and B. M. Wilamowski, C++ implementation of neural networks trainer, in *13th International Conference on Intelligent Engineering Systems (INES-09)*, Barbados, April 16–18, 2009.

[ISIE10] B. M. Wilamowski Efficient Neural Network Architectures and Advanced Training Algorithms, *ICIT10 3rd International Conference on Information Technologies*, Gdansk, Poland June 28–30, 2010, pp. 345–352.

[K88] T. Kohonen, The neural phonetic typewriter, *IEEE Computer*, 27(3), 11–22, 1988.

[MP69] M. Minsky and S. Papert, *Perceptrons*, MIT Press, Cambridge, MA, 1969.

[RHW86] D. E., Rumelhart, G. E. Hinton, and R. J. Wiliams, Learning representations by back-propagating errors, *Nature*, 323, 533–536, 1986.

[N65] N. J. Nilson, *Learning Machines: Foundations of Trainable Pattern Classifiers*, McGraw Hill, New York, 1965.

[PE10] B. M. Wilamowski, Special Neural Network Architectures for Easy Electronic Implementations POWERENG.2009, Lisbon, Portugal, March 18–20, 2009, pp. 17–22.

[PS94] V. V. Phansalkar and P. S. Sastry, Analysis of the back-propagation algorithm with momentum, *IEEE Transactions on Neural Networks*, 5(3), 505–506, March 1994.

[SR87] T. J. Sejnowski and C. R. Rosenberg, Parallel networks that learn to pronounce English text. *Complex Systems*, 1, 145–168, 1987.

[W02] B. M. Wilamowski, Neural networks and fuzzy systems, Chap. 32 in *Mechatronics Handbook*, ed. R. R. Bishop, pp. 33-1–32-26, CRC Press, Boca Raton, FL, 2002.

[W07] B. M. Wilamowski, Neural networks and fuzzy systems for nonlinear applications, in *11th International Conference on Intelligent Engineering Systems (INES 2007)*, pp. 13–19, Budapest, Hungary, June 29, 2007–July 1, 2007.

[W09] B. M. Wilamowski, Neural network architectures and learning algorithms, *IEEE Industrial Electronics Magazine*, 3(4), 56–63.

[W74] P. Werbos, Beyond regression: New tools for prediction and analysis in behavioral sciences, PhD diss., Harvard University, Cambridge, MA, 1974.

[WCKD07] B. M. Wilamowski, N. J. Cotton, O. Kaynak, and G. Dundar, Method of computing gradient vector and Jacobian matrix in arbitrarily connected neural networks, *IEEE International Symposium on Industrial Electronics (ISIE 2007)*, pp. 3298–3303, Vigo, Spain, 4–7 June 2007.

[WCKD08] B. M. Wilamowski, N. J. Cotton, O. Kaynak, and G. Dundar, Computing gradient vector and Jacobian matrix in arbitrarily connected neural networks, *IEEE Transactions on Industrial Electronics*, 55(10), 3784–3790, October 2008.

[WCM99] B. M. Wilamowski, Y. Chen, and A. Malinowski, Efficient algorithm for training neural networks with one hidden layer, in *International Joint Conference on Neural Networks (IJCNN'99)*, pp. 1725–1728, Washington, DC, July 10–16, 1999. #295 Session: 5.1.

[WH10] B. M. Wilamowski and H. Yu Improved computation for Levenberg Marquardt training *IEEE Transactions on Neural Networks*, 21, 930–937, 2010.

[WH60] B. Widrow and M. E. Hoff, Adaptive switching circuits. 1960 IRE Western Electric Show and Convention Record, Part 4 (Aug. 23) pp. 96–104, 1960.

[WHM03] B. Wilamowski, D. Hunter, and A. Malinowski, Solving parity-*n* problems with feedforward neural network, in *Proceedings of the International Joint Conference on Neural Networks (IJCNN'03)*, pp. 2546–2551, Portland, OR, July 20–23, 2003.

[WHY10] B. M. Wilamowski and H. Yu, Neural Network Learning without Backpropagation, *IEEE Transactions on Neural Networks*, 21(11), 1793-1803, November 2010.

[WJ96] B. M. Wilamowski and R. C. Jaeger, Implementation of RBF type networks by MLP networks, in *IEEE International Conference on Neural Networks*, pp. 1670–1675, Washington, DC, June 3–6, 1996.

[WJK99] B. M. Wilamowski, R. C. Jaeger, and M. O. Kaynak, Neuro-fuzzy architecture for CMOS implementation, *IEEE Transaction on Industrial Electronics*, 46(6), 1132–1136, December 1999.

[WK00] B. M. Wilamowski and O. Kaynak, Oil well diagnosis by sensing terminal characteristics of the induction motor, *IEEE Transactions on Industrial Electronics*, 47(5), 1100–1107, October 2000.

[WT93] B. M. Wilamowski and L. Torvik, Modification of gradient computation in the back-propagation algorithm, in *Artificial Neural Networks in Engineering (ANNIE'93)*, St. Louis, MO, November 14–17, 1993; also in C. H. Dagli, ed., *Intelligent Engineering Systems through Artificial Neural Networks*, vol. 3, pp. 175–180, ASME Press, New York, 1993.

[WY09] NNT—Neural Network Trainer, http://www.eng.auburn.edu/~wilambm/nnt/

[YW09] H. Yu and B. M. Wilamowski, C++ implementation of neural networks trainer, in *13th International Conference on Intelligent Engineering Systems (INES-09)*, Barbados, April 16–18, 2009.

12

Levenberg–Marquardt Training

Hao Yu
Auburn University

Bogdan M. Wilamowski
Auburn University

12.1 Introduction

The Levenberg–Marquardt algorithm [L44,M63], which was independently developed by Kenneth Levenberg and Donald Marquardt, provides a numerical solution to the problem of minimizing a nonlinear function. It is fast and has stable convergence. In the artificial neural-networks field, this algorithm is suitable for training small- and medium-sized problems.

Many other methods have already been developed for neural-networks training. The steepest descent algorithm, also known as the error backpropagation (EBP) algorithm [EHW86,J88], dispersed the dark clouds on the field of artificial neural networks and could be regarded as one of the most significant breakthroughs for training neural networks. Many improvements have been made to EBP [WT93,AW95,W96,WCM99], but these improvements are relatively minor [W02,WHM03,YW09,W09,WH10]. Sometimes instead of improving learning algorithms special neural network architectures are used, which are easy to train [WB99,WB01,WJ96,PE10,BUD90]. The EBP algorithm is still widely used today; however, it is also known as an inefficient algorithm because of its slow convergence. There are two main reasons for the slow convergence [ISIE10,WHY10]: the first reason is that its step sizes should be adequate to the gradients (Figure 12.1). Logically, small step sizes should be taken where the gradient is steep so as not to rattle out of the required minima (because of oscillation). So, if the step size is a constant, it needs to be chosen small. Then, in the place where the gradient is gentle, the training process would be very slow. The second reason is that the curvature of the error surface may not be the same in all directions, such as the Rosenbrock function, so the classic "error valley" problem [O92] may exist and may result in the slow convergence.

The slow convergence of the steepest descent method can be greatly improved by the Gauss–Newton algorithm [O92]. Using second-order derivatives of error function to "naturally" evaluate the curvature of error surface, the Gauss–Newton algorithm can find proper step sizes for each direction and converge very fast; especially, if the error function has a quadratic surface, it can converge directly in the first iteration. But this improvement only happens when the quadratic approximation of error function is reasonable. Otherwise, the Gauss–Newton algorithm would be mostly divergent.

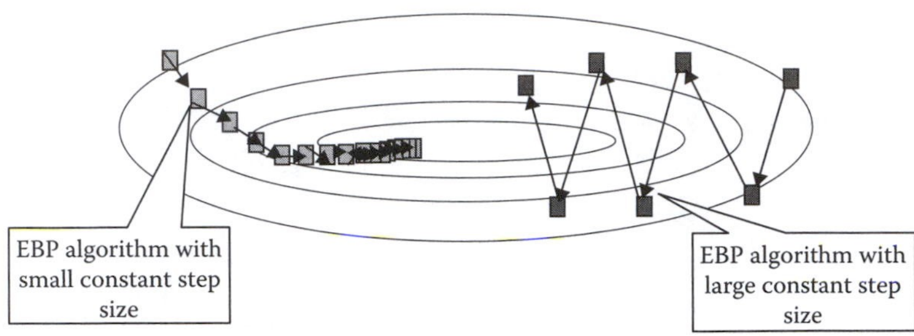

FIGURE 12.1 Searching process of the steepest descent method with different learning constants: trajectory on the left is for small learning constant that leads to slow convergence; trajectory on the right is for large learning constant that causes oscillation (divergence).

The Levenberg–Marquardt algorithm blends the steepest descent method and the Gauss–Newton algorithm. Fortunately, it inherits the speed advantage of the Gauss–Newton algorithm and the stability of the steepest descent method. It's more robust than the Gauss–Newton algorithm, because in many cases it can converge well even if the error surface is much more complex than the quadratic situation. Although the Levenberg–Marquardt algorithm tends to be a bit slower than Gauss–Newton algorithm (in convergent situation), it converges much faster than the steepest descent method.

The basic idea of the Levenberg–Marquardt algorithm is that it performs a combined training process: around the area with complex curvature, the Levenberg–Marquardt algorithm switches to the steepest descent algorithm, until the local curvature is proper to make a quadratic approximation; then it approximately becomes the Gauss–Newton algorithm, which can speed up the convergence significantly.

12.2 Algorithm Derivation

In this part, the derivation of the Levenberg–Marquardt algorithm will be presented in four parts: (1) steepest descent algorithm, (2) Newton's method, (3) Gauss–Newton's algorithm, and (4) Levenberg–Marquardt algorithm.

Before the derivation, let us introduce some commonly used indices:

- p is the index of patterns, from 1 to P, where P is the number of patterns.
- m is the index of outputs, from 1 to M, where M is the number of outputs.
- i and j are the indices of weights, from 1 to N, where N is the number of weights.
- k is the index of iterations.

Other indices will be explained in related places.

Sum square error (SSE) is defined to evaluate the training process. For all training patterns and network outputs, it is calculated by

$$E(\boldsymbol{x},\boldsymbol{w}) = \frac{1}{2}\sum_{p=1}^{P}\sum_{m=1}^{M} \boldsymbol{e}_{p,m}^2 \tag{12.1}$$

where

$\boldsymbol{x}$ is the input vector

$\boldsymbol{w}$ is the weight vector

$\boldsymbol{e}_{p,m}$ is the training error at output m when applying pattern p and it is defined as

$$\boldsymbol{e}_{p,m} = \boldsymbol{d}_{p,m} - \boldsymbol{o}_{p,m} \tag{12.2}$$

where

$\boldsymbol{d}$ is the desired output vector

$\boldsymbol{o}$ is the actual output vector

12.2.1 Steepest Descent Algorithm

The steepest descent algorithm is a first-order algorithm. It uses the first-order derivative of total error function to find the minima in error space. Normally, gradient $\boldsymbol{g}$ is defined as the first-order derivative of total error function (12.1):

$$\boldsymbol{g} = \frac{\partial E(\boldsymbol{x},\boldsymbol{w})}{\partial \boldsymbol{w}} = \begin{bmatrix} \dfrac{\partial E}{\partial w_1} & \dfrac{\partial E}{\partial w_2} & \cdots & \dfrac{\partial E}{\partial w_N} \end{bmatrix}^T \tag{12.3}$$

With the definition of gradient $\boldsymbol{g}$ in (12.3), the update rule of the steepest descent algorithm could be written as

$$\boldsymbol{w}_{k+1} = \boldsymbol{w}_k - \alpha \boldsymbol{g}_k \tag{12.4}$$

where α is the learning constant (step size).

The training process of the steepest descent algorithm is asymptotic convergence. Around the solution, all the elements of gradient vector would be very small and there would be a very tiny weight change.

12.2.2 Newton's Method

Newton's method assumes that all the gradient components $g_1, g_2, \ldots, g_N$ are functions of weights.

$$\begin{cases} g_1 = F_1(w_1, w_2 \cdots w_N) \\ g_2 = F_2(w_1, w_2 \cdots w_N) \\ \cdots \\ g_N = F_N(w_1, w_2 \cdots w_N) \end{cases} \tag{12.5}$$

where $F_1, F_2, \ldots, F_N$ are nonlinear relationships between weights and related gradient components.

Unfold each g_i $(i = 1, 2, \ldots, N)$ in Equations 12.5 by Taylor series and take the first-order approximation:

$$\begin{cases} g_1 \approx g_{1,0} + \dfrac{\partial g_1}{\partial w_1}\Delta w_1 + \dfrac{\partial g_1}{\partial w_2}\Delta w_2 + \cdots + \dfrac{\partial g_1}{\partial w_N}\Delta w_N \\ g_2 \approx g_{2,0} + \dfrac{\partial g_2}{\partial w_1}\Delta w_1 + \dfrac{\partial g_2}{\partial w_2}\Delta w_2 + \cdots + \dfrac{\partial g_2}{\partial w_N}\Delta w_N \\ \cdots \\ g_N \approx g_{N,0} + \dfrac{\partial g_N}{\partial w_1}\Delta w_1 + \dfrac{\partial g_N}{\partial w_2}\Delta w_2 + \cdots + \dfrac{\partial g_N}{\partial w_N}\Delta w_N \end{cases} \tag{12.6}$$

By combining the definition of gradient vector $\boldsymbol{g}$ in (12.3), it could be determined that

$$\frac{\partial g_i}{\partial w_j} = \frac{\partial \left(\dfrac{\partial E}{\partial w_j} \right)}{\partial w_j} = \frac{\partial^2 E}{\partial w_i \partial w_j} \tag{12.7}$$

By inserting Equation 12.7 to 12.6:

$$\begin{cases} g_1 \approx g_{1,0} + \dfrac{\partial^2 E}{\partial w_1^2}\Delta w_1 + \dfrac{\partial^2 E}{\partial w_1 \partial w_2}\Delta w_2 + \cdots + \dfrac{\partial^2 E}{\partial w_1 \partial w_N}\Delta w_N \\ g_2 \approx g_{2,0} + \dfrac{\partial^2 E}{\partial w_2 \partial w_1}\Delta w_1 + \dfrac{\partial^2 E}{\partial w_2^2}\Delta w_2 + \cdots + \dfrac{\partial^2 E}{\partial w_2 \partial w_N}\Delta w_N \\ \cdots \\ g_N \approx g_{N,0} + \dfrac{\partial^2 E}{\partial w_N \partial w_1}\Delta w_1 + \dfrac{\partial^2 E}{\partial w_N \partial w_2}\Delta w_2 + \cdots + \dfrac{\partial^2 E}{\partial w_N^2}\Delta w_N \end{cases} \tag{12.8}$$

Comparing with the steepest descent method, the second-order derivatives of the total error function need to be calculated for each component of gradient vector.

In order to get the minima of total error function E, each element of the gradient vector should be zero. Therefore, left sides of the Equations 12.8 are all zero, then

$$\begin{cases} 0 \approx g_{1,0} + \dfrac{\partial^2 E}{\partial w_1^2}\Delta w_1 + \dfrac{\partial^2 E}{\partial w_1 \partial w_2}\Delta w_2 + \cdots + \dfrac{\partial^2 E}{\partial w_1 \partial w_N}\Delta w_N \\ 0 \approx g_{2,0} + \dfrac{\partial^2 E}{\partial w_2 \partial w_1}\Delta w_1 + \dfrac{\partial^2 E}{\partial w_2^2}\Delta w_2 + \cdots + \dfrac{\partial^2 E}{\partial w_2 \partial w_N}\Delta w_N \\ \cdots \\ 0 \approx g_{N,0} + \dfrac{\partial^2 E}{\partial w_N \partial w_1}\Delta w_1 + \dfrac{\partial^2 E}{\partial w_N \partial w_2}\Delta w_2 + \cdots + \dfrac{\partial^2 E}{\partial w_N^2}\Delta w_N \end{cases} \tag{12.9}$$

By combining Equation 12.3 with 12.9

$$\begin{cases} -\dfrac{\partial E}{\partial w_1} = -g_{1,0} \approx \dfrac{\partial^2 E}{\partial w_1^2}\Delta w_1 + \dfrac{\partial^2 E}{\partial w_1 \partial w_2}\Delta w_2 + \cdots + \dfrac{\partial^2 E}{\partial w_1 \partial w_N}\Delta w_N \\ -\dfrac{\partial E}{\partial w_2} = -g_{2,0} \approx \dfrac{\partial^2 E}{\partial w_2 \partial w_1}\Delta w_1 + \dfrac{\partial^2 E}{\partial w_2^2}\Delta w_2 + \cdots + \dfrac{\partial^2 E}{\partial w_2 \partial w_N}\Delta w_N \\ \cdots \\ -\dfrac{\partial E}{\partial w_N} = -g_{N,0} \approx \dfrac{\partial^2 E}{\partial w_N \partial w_1}\Delta w_1 + \dfrac{\partial^2 E}{\partial w_N \partial w_2}\Delta w_2 + \cdots + \dfrac{\partial^2 E}{\partial w_N^2}\Delta w_N \end{cases} \tag{12.10}$$

There are N equations with N unknowns so that all Δw_n can be calculated. With the solutions, the weight space can be updated iteratively.

Equations 12.10 can be also written in matrix form

$$\begin{bmatrix} -g_1 \\ -g_2 \\ \cdots \\ -g_N \end{bmatrix} = \begin{bmatrix} -\dfrac{\partial E}{\partial w_1} \\ -\dfrac{\partial E}{\partial w_2} \\ \cdots \\ -\dfrac{\partial E}{\partial w_N} \end{bmatrix} = \begin{bmatrix} \dfrac{\partial^2 E}{\partial w_1^2} & \dfrac{\partial^2 E}{\partial w_1 \partial w_2} & \cdots & \dfrac{\partial^2 E}{\partial w_1 \partial w_N} \\ \dfrac{\partial^2 E}{\partial w_2 \partial w_1} & \dfrac{\partial^2 E}{\partial w_2^2} & \cdots & \dfrac{\partial^2 E}{\partial w_2 \partial w_N} \\ \cdots & \cdots & \cdots & \cdots \\ \dfrac{\partial^2 E}{\partial w_N \partial w_1} & \dfrac{\partial^2 E}{\partial w_N \partial w_2} & \cdots & \dfrac{\partial^2 E}{\partial w_N^2} \end{bmatrix} \times \begin{bmatrix} \Delta w_1 \\ \Delta w_2 \\ \cdots \\ \Delta w_N \end{bmatrix} \tag{12.11}$$

where the square matrix is Hessian matrix:

$$H = \begin{bmatrix} \frac{\partial^2 E}{\partial w_1^2} & \frac{\partial^2 E}{\partial w_1 \partial w_2} & \cdots & \frac{\partial^2 E}{\partial w_1 \partial w_N} \\ \frac{\partial^2 E}{\partial w_2 \partial w_1} & \frac{\partial^2 E}{\partial w_2^2} & \cdots & \frac{\partial^2 E}{\partial w_2 \partial w_N} \\ \cdots & \cdots & \cdots & \cdots \\ \frac{\partial^2 E}{\partial w_N \partial w_1} & \frac{\partial^2 E}{\partial w_N \partial w_2} & \cdots & \frac{\partial^2 E}{\partial w_N^2} \end{bmatrix} \tag{12.12}$$

Equation 12.11 can be written in matrix form as:

$$-\boldsymbol{g} = \boldsymbol{H}\Delta \boldsymbol{w} \tag{12.13}$$

So

$$\Delta \boldsymbol{w} = -\boldsymbol{H}^{-1}\boldsymbol{g} \tag{12.14}$$

Therefore, the update rule for Newton's method is

$$\boldsymbol{w}_{k+1} = \boldsymbol{w}_k - \boldsymbol{H}_k^{-1}\boldsymbol{g}_k \tag{12.15}$$

As the second-order derivatives of total error function, the Hessian matrix $\boldsymbol{H}$ gives the proper evaluation on the change of gradient vector. By comparing Equations 12.4 and 12.15, one may notice that well-matched step sizes are given by the inverted Hessian matrix.

12.2.3 Gauss–Newton Algorithm

If Newton's method is applied for weight updating, in order to get the Hessian matrix $\boldsymbol{H}$, the second-order derivatives of total error function have to be calculated and it could be very complicated. In order to simplify the calculating process, Jacobian matrix $\boldsymbol{J}$ is introduced as

$$J = \begin{bmatrix} \frac{\partial e_{1,1}}{\partial w_1} & \frac{\partial e_{1,1}}{\partial w_2} & \cdots & \frac{\partial e_{1,1}}{\partial w_N} \\ \frac{\partial e_{1,2}}{\partial w_1} & \frac{\partial e_{1,2}}{\partial w_2} & \cdots & \frac{\partial e_{1,2}}{\partial w_N} \\ \cdots & \cdots & \cdots & \cdots \\ \frac{\partial e_{1,M}}{\partial w_1} & \frac{\partial e_{1,M}}{\partial w_2} & \cdots & \frac{\partial e_{1,M}}{\partial w_N} \\ \cdots & \cdots & \cdots & \cdots \\ \frac{\partial e_{P,1}}{\partial w_1} & \frac{\partial e_{P,1}}{\partial w_2} & \cdots & \frac{\partial e_{P,1}}{\partial w_N} \\ \frac{\partial e_{P,2}}{\partial w_1} & \frac{\partial e_{P,2}}{\partial w_2} & \cdots & \frac{\partial e_{P,2}}{\partial w_N} \\ \cdots & \cdots & \cdots & \cdots \\ \frac{\partial e_{P,M}}{\partial w_1} & \frac{\partial e_{P,M}}{\partial w_2} & \cdots & \frac{\partial e_{P,M}}{\partial w_N} \end{bmatrix} \tag{12.16}$$

By integrating Equations 12.1 and 12.3, the elements of gradient vector can be calculated as

$$g_i = \frac{\partial E}{\partial w_i} = \frac{\partial\left(\frac{1}{2}\sum_{p=1}^{P}\sum_{m=1}^{M} e_{p,m}^2\right)}{\partial w_i} = \sum_{p=1}^{P}\sum_{m=1}^{M}\left(\frac{\partial e_{p,m}}{\partial w_i} e_{p,m}\right) \tag{12.17}$$

Combining Equations 12.16 and 12.17, the relationship between the Jacobian matrix $\boldsymbol{J}$ and the gradient vector $\boldsymbol{g}$ would be

$$\boldsymbol{g} = \boldsymbol{J}\boldsymbol{e} \tag{12.18}$$

where error vector $\boldsymbol{e}$ has the form

$$\boldsymbol{e} = \begin{bmatrix} e_{1,1} \\ e_{1,2} \\ \cdots \\ e_{1,M} \\ \cdots \\ e_{P,1} \\ e_{P,2} \\ \cdots \\ e_{P,M} \end{bmatrix} \tag{12.19}$$

Inserting Equation 12.1 into 12.12, the element at ith row and jth column of the Hessian matrix can be calculated as

$$h_{i,j} = \frac{\partial^2 E}{\partial w_i \partial w_j} = \frac{\partial^2\left(\frac{1}{2}\sum_{p=1}^{P}\sum_{m=1}^{M} e_{p,m}^2\right)}{\partial w_i \partial w_j} = \sum_{p=1}^{P}\sum_{m=1}^{M}\frac{\partial e_{p,m}}{\partial w_i}\frac{\partial e_{p,m}}{\partial w_j} + S_{i,j} \tag{12.20}$$

where $S_{i,j}$ is equal to

$$S_{i,j} = \sum_{p=1}^{P}\sum_{m=1}^{M}\frac{\partial^2 e_{p,m}}{\partial w_i \partial w_j} e_{p,m} \tag{12.21}$$

As the basic assumption of the Newton's method is that $S_{i,j}$ is closed to zero [TM94], the relationship between the Hessian matrix $\boldsymbol{H}$ and the Jacobian matrix $\boldsymbol{J}$ can be rewritten as

$$\boldsymbol{H} \approx \boldsymbol{J}^T\boldsymbol{J} \tag{12.22}$$

By combining Equations 12.15, 12.18, and 12.22, the update rule of the Gauss–Newton algorithm is presented as

$$\boldsymbol{w}_{k+1} = \boldsymbol{w}_k - \left(\boldsymbol{J}_k^T\boldsymbol{J}_k\right)^{-1}\boldsymbol{J}_k\boldsymbol{e}_k \tag{12.23}$$

Obviously, the advantage of the Gauss–Newton algorithm over the standard Newton's method (Equation 12.15) is that the former does not require the calculation of second-order derivatives of the total error function, by introducing Jacobian matrix $\boldsymbol{J}$ instead. However, the Gauss–Newton algorithm still faces the same convergent problem like the Newton algorithm for complex error space optimization. Mathematically, the problem can be interpreted as the matrix $\boldsymbol{J}^T\boldsymbol{J}$ may not be invertible.

12.2.4 Levenberg–Marquardt Algorithm

In order to make sure that the approximated Hessian matrix $\boldsymbol{J}^T\boldsymbol{J}$ is invertible, the Levenberg–Marquardt algorithm introduces another approximation to the Hessian matrix:

$$\boldsymbol{H} \approx \boldsymbol{J}^T\boldsymbol{J} + \mu\boldsymbol{I} \tag{12.24}$$

where

μ is always positive, called combination coefficient
$\boldsymbol{I}$ is the identity matrix

From Equation 12.24, one may notice that the elements on the main diagonal of the approximated Hessian matrix will be larger than zero. Therefore, with this approximation (Equation 12.24), it can be sure that matrix $\boldsymbol{H}$ is always invertible.

By combining Equations 12.23 and 12.24, the update rule of the Levenberg–Marquardt algorithm can be presented as

$$\boldsymbol{w}_{k+1} = \boldsymbol{w}_k - \left(\boldsymbol{J}_k^T\boldsymbol{J}_k + \mu\boldsymbol{I}\right)^{-1}\boldsymbol{J}_k\boldsymbol{e}_k \tag{12.25}$$

As the combination of the steepest descent algorithm and the Gauss–Newton algorithm, the Levenberg–Marquardt algorithm switches between the two algorithms during the training process. When the combination coefficient μ is very small (nearly zero), Equation 12.25 is approaching to Equation 12.23 and the Gauss–Newton algorithm is used. When combination coefficient μ is very large, Equation 12.25 approximates to Equation 12.4 and the steepest descent method is used.

If the combination coefficient μ in Equation 12.25 is very big, it can be interpreted as the learning coefficient in the steepest descent method (12.4):

$$\alpha = \frac{1}{\mu} \tag{12.26}$$

Table 12.1 summarizes the update rules for various algorithms.

TABLE 12.1 Specifications of Different Algorithms

Algorithms	Update Rules	Convergence	Computation Complexity
EBP algorithm	$\boldsymbol{w}_{k+1} = \boldsymbol{w}_k - \alpha\boldsymbol{g}_k$	Stable, slow	Gradient
Newton algorithm	$\boldsymbol{w}_{k+1} = \boldsymbol{w}_k - \boldsymbol{H}_k^{-1}\boldsymbol{g}_k$	Unstable, fast	Gradient and Hessian
Gauss–Newton algorithm	$\boldsymbol{w}_{k+1} = \boldsymbol{w}_k - \left(\boldsymbol{J}_k^T\boldsymbol{J}_k\right)^{-1}\boldsymbol{J}_k\boldsymbol{e}_k$	Unstable, fast	Jacobian
Levenberg–Marquardt algorithm	$\boldsymbol{w}_{k+1} = \boldsymbol{w}_k - \left(\boldsymbol{J}_k^T\boldsymbol{J}_k + \mu\boldsymbol{I}\right)^{-1}\boldsymbol{J}_k\boldsymbol{e}_k$	Stable, fast	Jacobian
NBN algorithm [08WC][a]	$\boldsymbol{w}_{k+1} = \boldsymbol{w}_k - \boldsymbol{Q}_k^{-1}\boldsymbol{g}_k$	Stable, fast	Quasi Hessian[a]

[a] Reference Chapter 12.

12.3 Algorithm Implementation

In order to implement the Levenberg–Marquardt algorithm for neural network training, two problems have to be solved: how does one calculate the Jacobian matrix, and how does one organize the training process iteratively for weight updating.

In this section, the implementation of training with the Levenberg–Marquardt algorithm will be introduced in two parts: (1) calculation of the Jacobian matrix; (2) training process design.

12.3.1 Calculation of the Jacobian Matrix

In this section j and k are used as the indices of neurons ranging from 1 to nn, where nn is the number of neurons contained in a topology; i is the index of neuron inputs ranging from 1 to ni, where ni is the number of inputs and it may vary for different neurons.

As an introduction of basic concepts of neural network training, let us consider a neuron j with ni inputs, as shown in Figure 12.2. If neuron j is in the first layer, all its inputs would be connected to the inputs of the network, otherwise, its inputs can be connected to outputs of other neurons or to networks inputs if connections across layers are allowed.

Node y is an important and flexible concept. It can be $y_{j,i}$, meaning the ith input of neuron j. It also can be used as y_j to define the output of neuron j. In the following derivation, if node y has one index then it is used as a neuron output node, but if it has two indices (neuron and input), it is a neuron input node.

The output node of neuron j is calculated using

$$y_j = f_j(net_j) \tag{12.27}$$

where f_j is the activation function of neuron j and net value net_j is the sum of weighted input nodes of neuron j:

$$net_j = \sum_{i=1}^{ni} w_{j,i} y_{j,i} + w_{j,0} \tag{12.28}$$

where

$y_{j,i}$ is the ith input node of neuron j, weighted by $w_{j,i}$

$w_{j,0}$ is the bias weight of neuron j

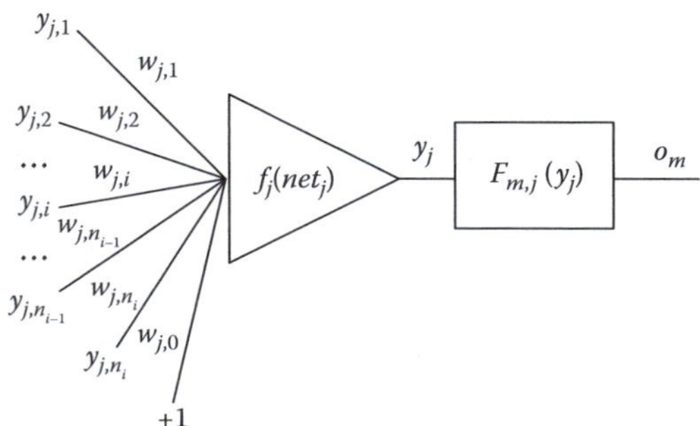

FIGURE 12.2 Connection of a neuron j with the rest of the network. Nodes $y_{j,i}$ could represent network inputs or outputs of other neurons. $F_{m,j}(y_j)$ is the nonlinear relationship between the neuron output node y_j and the network output o_m.

Using Equation 12.28, one may notice that derivative of net_j is

$$\frac{\partial net_j}{\partial w_{j,i}} = y_{j,i} \tag{12.29}$$

and slope s_j of activation function f_j is

$$s_j = \frac{\partial y_j}{\partial net_j} = \frac{\partial f_j(net_j)}{\partial net_j} \tag{12.30}$$

Between the output node y_j of a hidden neuron j and network output o_m, there is a complex nonlinear relationship (Figure 12.2):

$$o_m = F_{m,j}(y_j) \tag{12.31}$$

where o_m is the mth output of the network.

The complexity of this nonlinear function $F_{m,j}(y_j)$ depends on how many other neurons are between neuron j and network output m. If neuron j is at network output m, then $o_m = y_j$ and $F'_{mj}(y_j) = 1$, where F'_{mj} is the derivative of nonlinear relationship between neuron j and output m.

The elements of the Jacobian matrix in Equation 12.16 can be calculated as

$$\frac{\partial e_{p,m}}{\partial w_{j,i}} = \frac{\partial \left(d_{p,m} - o_{p,m}\right)}{\partial w_{j,i}} = -\frac{\partial o_{p,m}}{\partial w_{j,i}} = -\frac{\partial o_{p,m}}{\partial y_j}\frac{\partial y_j}{\partial net_j}\frac{\partial net_j}{\partial w_{j,i}} \tag{12.32}$$

Combining with Equations 12.28 through 12.30, 12.31 can be rewritten as

$$\frac{\partial e_{p,m}}{\partial w_{j,i}} = -F'_{mj} s_j y_{j,i} \tag{12.33}$$

where F'_{mj} is the derivative of nonlinear function between neuron j and output m.

The computation process for the Jacobian matrix can be organized according to the traditional backpropagation computation in first-order algorithms (like the EBP algorithm). But there are also differences between them. First of all, for every pattern, in the EBP algorithm, only one backpropagation process is needed, while in the Levenberg–Marquardt algorithm the backpropagation process has to be repeated for every output separately in order to obtain consecutive rows of the Jacobian matrix (Equation 12.16). Another difference is that the concept of backpropagation of δ parameter [N89] has to be modified. In the EBP algorithm, output errors are parts of the δ parameter:

$$\delta_j = s_j \sum_{m=1}^{M} F'_{mj} e_m \tag{12.34}$$

In the Levenberg–Marquardt algorithm, the δ parameters are calculated for each neuron j and each output m, separately. Also, in the backpropagation process, the error is replaced by a unit value [TM94]:

$$\delta_{m,j} = s_j F'_{mj} \tag{12.35}$$

By combining Equations 12.33 and 12.35, elements of the Jacobian matrix can be calculated by

$$\frac{\partial e_{p,m}}{\partial w_{j,i}} = -\delta_{m,j} y_{j,i} \tag{12.36}$$

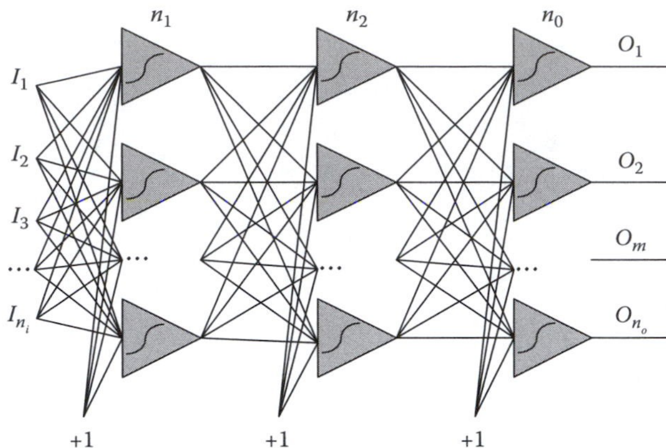

FIGURE 12.3 Three-layer multilayer perceptron network: the number of inputs is n_i, the number of outputs is n_0, and n_1 and n_2 are the numbers of neurons in the first and second layers separately.

There are two unknowns in Equation 12.36 for the Jacobian matrix computation. The input node, $y_{j,i}$, can be calculated in the forward computation (signal propagating from inputs to outputs); while $\delta_{m,j}$ is obtained in the backward computation, which is organized as errors backpropagating from output neurons (output layer) to network inputs (input layer). At output neuron m ($j = m$), $\delta_{m,j} = s_m$.

For better interpretation of forward computation and backward computation, let us consider the three-layer multilayer perceptron network (Figure 12.3) as an example.

For a given pattern, the forward computation can be organized in the following steps:

a. Calculate net values, slopes, and outputs for all neurons in the first layer:

$$net_j^1 = \sum_{i=1}^{ni} I_i w_{j,i}^1 + w_{j,0}^1 \tag{12.37}$$

$$y_j^1 = f_j^1\left(net_j^1\right) \tag{12.38}$$

$$s_j^1 = \frac{\partial f_j^1}{\partial net_j^1} \tag{12.39}$$

where

I_i are the network inputs
the superscript "1" means the first layer
j is the index of neurons in the first layer

b. Use the outputs of the first layer neurons as the inputs of all neurons in the second layer, do a similar calculation for net values, slopes, and outputs:

$$net_j^2 = \sum_{i=1}^{n1} y_i^1 w_{j,i}^2 + w_{j,0}^2 \tag{12.40}$$

$$y_j^2 = f_j^2\left(net_j^2\right) \tag{12.41}$$

$$s_j^2 = \frac{\partial f_j^2}{\partial net_j^2} \quad (12.42)$$

c. Use the outputs of the second layer neurons as the inputs of all neurons in the output layer (third layer), do a similar calculation for net values, slopes, and outputs:

$$net_j^3 = \sum_{i=1}^{n_2} y_i^2 w_{j,i}^3 + w_{j,0}^3 \quad (12.43)$$

$$o_j = f_j^3\left(net_j^3\right) \quad (12.44)$$

$$s_j^3 = \frac{\partial f_j^3}{\partial net_j^3} \quad (12.45)$$

After the forward calculation, node array $\boldsymbol{y}$ and slope array $\boldsymbol{s}$ can be obtained for all neurons with the given pattern.

With the results from the forward computation, for a given output j, the backward computation can be organized as

d. Calculate error at the output j and initial δ as the slope of output j:

$$e_j = d_j - o_j \quad (12.46)$$

$$\delta_{j,j}^3 = s_j^3 \quad (12.47)$$

$$\delta_{j,k}^3 = 0 \quad (12.48)$$

where

d_j is the desired output at output j

o_j is the actual output at output j obtained in the forward computation

$\delta_{j,j}^3$ is the self-backpropagation

$\delta_{j,k}^3$ is the backpropagation from other neurons in the same layer (output layer)

e. Backpropagate δ from the inputs of the third layer to the outputs of the second layer

$$\delta_{j,k}^2 = w_{j,k}^3 \delta_{j,j}^3 \quad (12.49)$$

where k is the index of neurons in the second layer, from 1 to n_2.

f. Backpropagate δ from the outputs of the second layer to the inputs of the second layer

$$\delta_{j,k}^2 = \delta_{j,k}^2 s_k^2 \quad (12.50)$$

where k is the index of neurons in the second layer, from 1 to n_2.

g. Backpropagate δ from the inputs of the second layer to the outputs of the first layer

$$\delta_{j,k}^1 = \sum_{i=1}^{n_2} w_{j,i}^2 \delta_{j,i}^2 \quad (12.51)$$

where k is the index of neurons in the first layer, from 1 to n_1.

h. Backpropagate δ from the outputs of the first layer to the inputs of the first layer

$$\delta^1_{j,k} = \delta^1_{j,k} s^1_k \tag{12.52}$$

where k is the index of neurons in the second layer, from 1 to n_1.

For the backpropagation process of other outputs, the steps (d)–(h) are repeated.

By performing the forward computation and backward computation, the whole **δ** array and ***y*** array can be obtained for the given pattern. Then related row elements (*no* rows) of the Jacobian matrix can be calculated by using Equation 12.36.

For other patterns, by repeating the forward and backward computation, the whole Jacobian matrix can be calculated.

The pseudo code of the forward computation and backward computation for the Jacobian matrix in the Levenberg–Marquardt algorithm is shown in Figure 12.4.

12.3.2 Training Process Design

With the update rule of the Levenberg–Marquardt algorithm (Equation 12.25) and the computation of the Jacobian matrix, the next step is to organize the training process.

According to the update rule, if the error goes down, which means it is smaller than the last error, it implies that the quadratic approximation on total error function is working and the combination coefficient μ could be changed smaller to reduce the influence of gradient descent part (ready to speed up). On the other hand, if the error goes up, which means it's larger than the last error, it shows that it's necessary to follow the gradient more to look for a proper curvature for quadratic approximation and the combination coefficient μ is increased.

```
for all patterns
%Forward computation
   for all layers
      for all neurons in the layer
          calculate net;         % Equation (29)
          calculate output;      % Equation (28)
          calculate slope;       % Equation (31)
      end;
   end;
%Backward computation
  initial delta as slope;
  for all outputs
      calculate error;
      for all layers
          for all neurons in the previous layer
              for all neurons in the current layer
                  multiply delta through weights
                  sum the backpropagated delta at proper nodes
              end;
              multiply delta by slope;
          end;
      end;
  end;
end;
```

FIGURE 12.4 Pseudo code of forward computation and backward computation implementing the Levenberg–Marquardt algorithm.

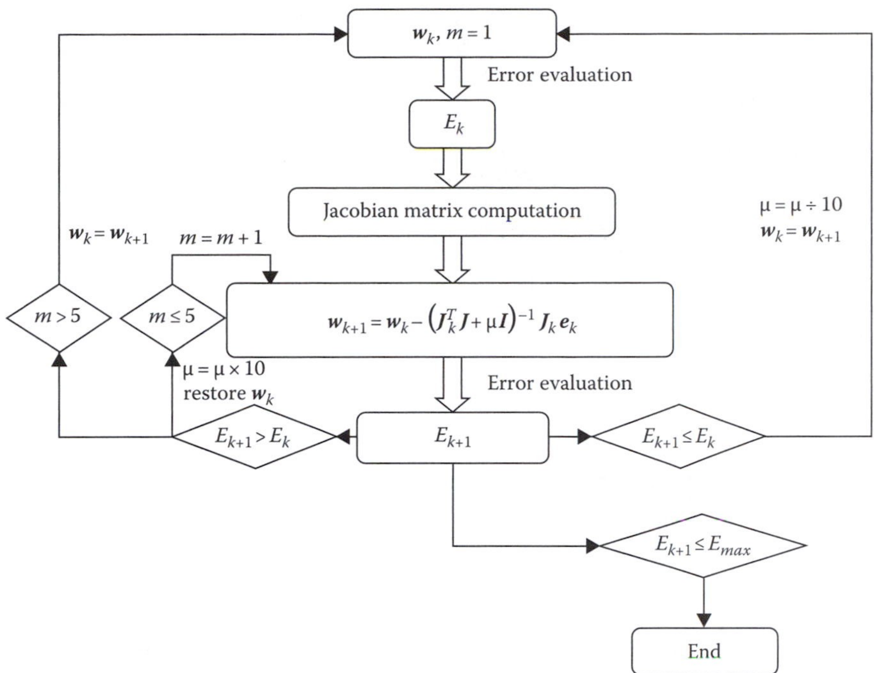

FIGURE 12.5 Block diagram for training using the Levenberg–Marquardt algorithm: w_k is the current weight, w_{k+1} is the next weight, E_{k+1} is the current total error, and E_k is the last total error.

Therefore, the training process using the Levenberg–Marquardt algorithm could be designed as follows:

i. With the initial weights (randomly generated), evaluate the total error (SSE).
ii. Do an update as directed by Equation 12.25 to adjust weights.
iii. With the new weights, evaluate the total error.
iv. If the current total error is increased as a result of the update, then retract the step (such as reset the weight vector to the previous value) and increase combination coefficient μ by a factor of 10 or by some other factors. Then go to step ii and try an update again.
v. If the current total error is decreased as a result of the update, then accept the step (such as keep the new weight vector as the current one) and decrease the combination coefficient μ by a factor of 10 or by the same factor as step iv.
vi. Go to step ii with the new weights until the current total error is smaller than the required value.

The flowchart of the above procedure is shown in Figure 12.5.

12.4 Comparison of Algorithms

In order to illustrate the advantage of the Levenberg–Marquardt algorithm, let us use the parity-3 problem (see Figure 12.6) as an example and make a comparison among the EBP algorithm, the Gauss–Newton algorithm, and the Levenberg algorithm [WCKD07].

Three neurons in multilayer perceptron network (Figure 12.7) are used for training, and the required training error is 0.01. In order to compare the convergent rate, for each algorithm, 100 trials are tested with randomly generated weights (between −1 and 1).

Inputs			Outputs
−1	−1	−1	−1
−1	−1	1	1
−1	1	−1	1
−1	1	1	−1
1	−1	−1	1
1	−1	1	−1
1	1	−1	−1
1	1	1	1

FIGURE 12.6 Training patterns of the parity-3 problem.

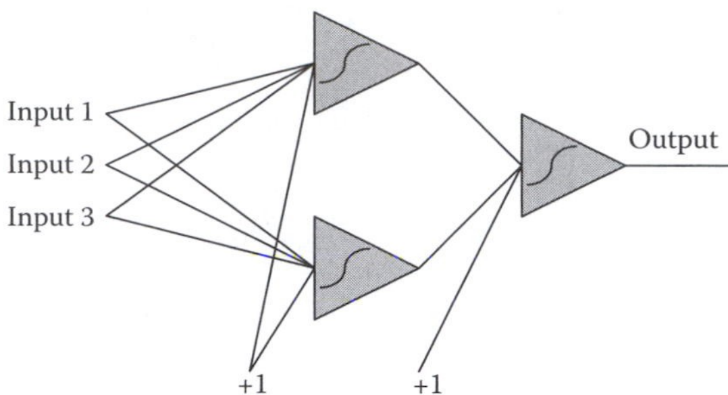

FIGURE 12.7 Three neurons in multilayer perceptron network.

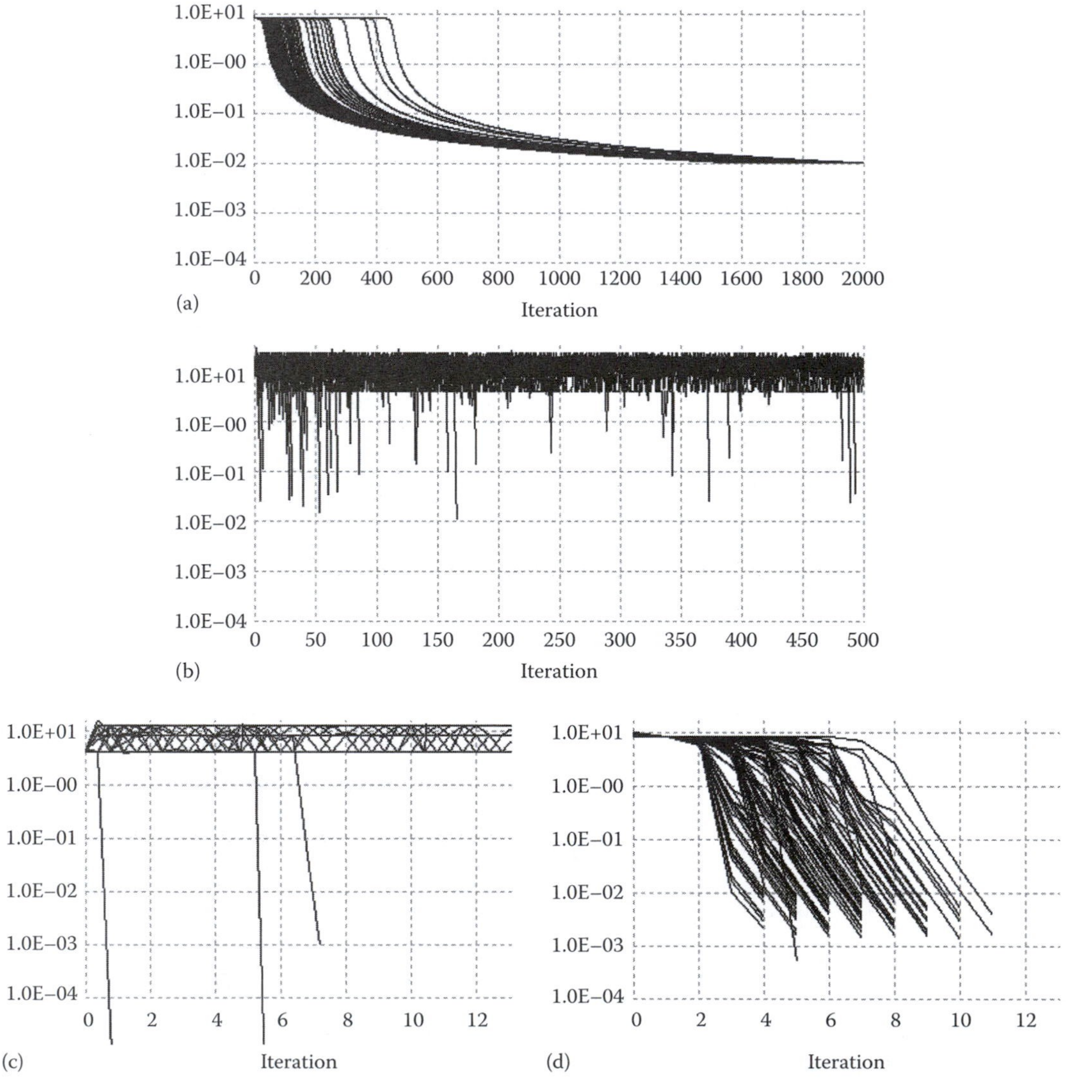

FIGURE 12.8 Training results of the parity-3 problem: (a) the EBP algorithm ($\alpha = 1$), (b) the EBP algorithm ($\alpha = 100$), (c) the Gauss–Newton algorithm, and (d) the Levenberg–Marquardt algorithm.

TABLE 12.2 Comparison among Different Algorithms for Parity-3 Problem

Algorithms	Convergence Rate (%)	Average Iteration	Average Time (ms)
EBP algorithm ($\alpha = 1$)	100	1646.52	320.6
EBP algorithm ($\alpha = 100$)	79	171.48	36.5
Gauss–Newton algorithm	3	4.33	1.2
Levenberg–Marquardt algorithm	100	6.18	1.6

The training results are shown in Figure 12.8 and the comparison is presented in Table 12.2. One may notice that: (1) for the EBP algorithm, the larger the training constant α is, the faster and less stable the training process will be; (2) the Levenberg–Marquardt is much faster than the EBP algorithm and more stable than the Gauss–Newton algorithm.

For more complex parity-N problems, the Gauss–Newton method cannot converge at all, and the EBP algorithm also becomes more inefficient to find the solution, while the Levenberg–Marquardt algorithm may lead to successful solutions.

12.5 Summary

The Levenberg–Marquardt algorithm solves the problems existing in both the gradient descent method and the Gauss–Newton method for neural-networks training, by the combination of those two algorithms. It is regarded as one of the most efficient training algorithms [TM94].

However, the Levenberg–Marquardt algorithm has its flaws. One problem is that the Hessian matrix inversion needs to be calculated each time for weight updating and there may be several updates in each iteration. For small size networks training, the computation is efficient, but for large networks, such as image recognition problems, this inversion calculation is going to be a disaster and the speed gained by second-order approximation may be totally lost. In that case, the Levenberg–Marquardt algorithm may be even slower than the steepest descent algorithm. Another problem is that the Jacobian matrix has to be stored for computation, and its size is $P \times M \times N$, where P is the number of patterns, M is the number of outputs, and N is the number of weights. For large-sized training patterns, the memory cost for the Jacobian matrix storage may be too huge to be practical. Only very recently Levenberg–Marquardt algorithm was implemented on other than MLP (multilayer preceptron) networks [WCKD08,WH10,WHY10,ISIE10].

Even though there are still some problems not solved for the Levenberg–Marquardt training, for small- and medium-sized networks and patterns, the Levenberg–Marquardt algorithm is remarkably efficient and strongly recommended for neural network training.

References

[AW95] T. J. Andersen and B. M. Wilamowski, A modified regression algorithm for fast one layer neural network training, *World Congress of Neural Networks*, vol. 1, pp. 687–690, Washington, DC, July 17–21, 1995.

[BUD09] B. M. Wilamowski, Neural Networks or Fuzzy Systems, Workshop on Intelligent Systems, Budapest, Hungary, August 30, 2009, pp. 1–12.

[EHW86] D. E. Rumelhart, G. E. Hinton, and R. J. Williams, Learning representations by back-propagating errors, *Nature*, 323, 533–536, 1986.

[ISIE10] B. M. Wilamowski, Efficient Neural Network Architectures and Advanced Training Algorithms, *ICIT10 3rd International Conference on Information Technologies,* Gdansk, Poland, June 28–30, 2010, pp. 345–352.

[J88] P. J. Werbos, Back-propagation: Past and future, in *Proceedings of International Conference on Neural Networks*, vol. 1, pp. 343–354, San Diego, CA, 1988.

[L44] K. Levenberg, A method for the solution of certain problems in least squares, *Quarterly of Applied Mathematics*, 5, 164–168, 1944.

[M63] D. Marquardt, An algorithm for least-squares estimation of nonlinear parameters, *SIAM Journal on Applied Mathematics*, 11(2), 431–441, June 1963.

[N89] Robert Hecht Nielsen, Theory of the back propagation neural network in *Proceedings 1989 IEEE IJCNN*, pp. 1593–1605, IEEE Press, New York, 1989.

[O92] M. R. Osborne, Fisher's method of scoring, *International Statistical Review*, 86, 271–286, 1992.

[PE10] B. M. Wilamowski, Special Neural Network Architectures for Easy Electronic Implementations, POWERENG.2009, Lisbon, Portugal, March 18–20, 2009, pp. 17–22.

[TM94] M. T. Hagan and M. Menhaj, Training feedforward networks with the Marquardt algorithm, *IEEE Transactions on Neural Networks*, 5(6), 989–993, 1994.

[W02] B. M. Wilamowski, Neural networks and fuzzy systems, Chap. 32 in *Mechatronics Handbook*, ed. R. R. Bishop, CRC Press, Boca Raton, FL, pp. 33-1–32-26, 2002.

[W09] B. M. Wilamowski, Neural network architectures and learning algorithms, *IEEE Industrial Electronics Magazine*, 3(4), 56–63, 2009.

[W96] B. M. Wilamowski, Neural networks and fuzzy systems, Chaps. 124.1 to 124.8 in *The Electronic Handbook*, CRC Press, Boca Raton, FL, pp. 1893–1914, 1996.

[WB01] B. M. Wilamowski and J. Binfet Microprocessor implementation of fuzzy systems and neural networks, in *International Joint Conference on Neural Networks (IJCNN'01)*, pp. 234–239, Washington, DC, July 15–19, 2001.

[WB99] B. M. Wilamowski and J. Binfet, Do fuzzy controllers have advantages over neural controllers in microprocessor implementation in *Proceedings of 2nd International Conference on Recent Advances in Mechatronics (ICRAM'99)*, pp. 342–347, Istanbul, Turkey, May 24–26, 1999.

[WCKD07] B. M. Wilamowski, N. J. Cotton, O. Kaynak, and G. Dundar, Method of computing gradient vector and Jacobian matrix in arbitrarily connected neural networks, in *IEEE International Symposium on Industrial Electronics (ISIE 2007)*, pp. 3298–3303, Vigo, Spain, June 4–7, 2007.

[WCKD08] B. M. Wilamowski, N. J. Cotton, O. Kaynak, and G. Dundar, Computing gradient vector and Jacobian matrix in arbitrarily connected neural networks, *IEEE Transactions on Industrial Electronics*, 55(10), 3784–3790, October 2008.

[WCM99] B. M. Wilamowski. Y. Chen, and A. Malinowski, Efficient algorithm for training neural networks with one hidden layer, in *1999 International Joint Conference on Neural Networks (IJCNN'99)*, pp. 1725–1728, Washington, DC, July 10–16, 1999. #295 Session: 5.1.

[WH10] B. M. Wilamowski and H. Yu, Improved computation for Levenberg Marquardt training, *IEEE Transactions on Neural Networks*, 21, 930–937, 2010.

[WHM03] B. Wilamowski, D. Hunter, and A. Malinowski, Solving parity-n problems with feedforward neural network, in *Proceedings of the IJCNN'03 International Joint Conference on Neural Networks*, pp. 2546–2551, Portland, OR, July 20–23, 2003.

[WHY10] B. M. Wilamowski and H. Yu, Neural Network Learning without Backpropagation, *IEEE Transactions on Neural Networks*, 21(11), 1793–1803, November 2010.

[WJ96] B. M. Wilamowski and R. C. Jaeger, Implementation of RBF type networks by MLP networks, *IEEE International Conference on Neural Networks*, pp. 1670–1675, Washington, DC, June 3–6, 1996.

[WT93] B. M. Wilamowski and L. Torvik, Modification of gradient computation in the back-propagation algorithm, in *Artificial Neural Networks in Engineering (ANNIE'93)*, St. Louis, MO, November 14–17, 1993.

[YW09] H. Yu and B. M. Wilamowski, C++ implementation of neural networks trainer, in *13th International Conference on Intelligent Engineering Systems (INES-09)*, Barbados, April 16–18, 2009.

13 NBN Algorithm

Bogdan M. Wilamowski
Auburn University

Hao Yu
Auburn University

Nicholas Cotton
Naval Surface Warfare Centre

13.1 Introduction

Since the development of EBP—error backpropagation—algorithm for training neural networks, many attempts have been made to improve the learning process. There are some well-known methods like momentum or variable learning rate and there are less known methods which significantly accelerate learning rate [WT93,AW95,WCM99,W09,WH10,PE10,BUD09,WB01]. The recently developed NBN (neuron-by-neuron) algorithm [WCHK07,WCKD08,YW09,WHY10] is very efficient for neural network training. Compared to with the well-known Levenberg–Marquardt (LM) algorithm (introduced in Chapter 12) [L44,M63], the NBN algorithm has several advantages: (1) the ability to handle arbitrarily connected neural networks; (2) forward-only computation (without backpropagation process); and (3) the direct computation of quasi-Hessian matrix (no need to compute and store Jacobian matrix). This chapter is organized around the three advantages of the NBN algorithm.

13.2 Computational Fundamentals

Before the derivation, let us introduce some commonly used indices in this chapter:

- p is the index of patterns, from 1 to np, where np is the number of patterns.
- m is the index of outputs, from 1 to no, where no is the number of outputs.

- j and k are the indices of neurons, from 1 to nn, where nn is the number of neurons.
- i is the index of neuron inputs, from 1 to ni, where ni is the number of inputs and it may vary for different neurons.

Other indices will be explained in related places.

Sum square error (SSE) E is defined to evaluate the training process. For all patterns and outputs, it is calculated by

$$E = \frac{1}{2}\sum_{p=1}^{np}\sum_{m=1}^{no} e_{p,m}^2 \tag{13.1}$$

where $e_{p,m}$ is the error at output m defined as

$$e_{p,m} = o_{p,m} - d_{p,m} \tag{13.2}$$

where $d_{p,m}$ and $o_{p,m}$ are desired output and actual output, respectively, at network output m for training pattern p.

In all algorithms, besides the NBN algorithm, the same computations are being repeated for one pattern at a time. Therefore, in order to simplify notations, the index p for patterns will be skipped in following derivations, unless it is essential.

13.2.1 Definition of Basic Concepts in Neural Network Training

Let us consider neuron j with ni inputs, as shown in Figure 13.1. If neuron j is in the first layer, all its inputs would be connected to the inputs of the network; otherwise, its inputs can be connected to outputs of other neurons or to network inputs if connections across layers are allowed.

Node y is an important and flexible concept. It can be $y_{j,i}$, meaning the ith input of neuron j. It also can be used as y_j to define the output of neuron j. In this chapter, if node y has one index (neuron), then it is used as a neuron output node; while if it has two indices (neuron and input), it is a neuron input node.

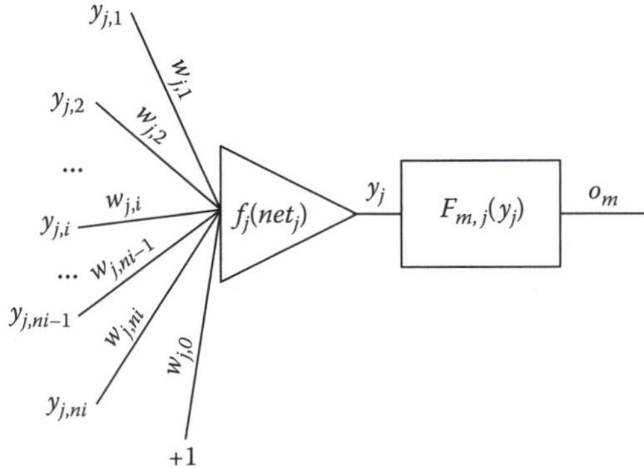

FIGURE 13.1 Connection of a neuron j with the rest of the network. Nodes $y_{j,i}$ could represent network inputs or outputs of other neurons. $F_{m,j}(y_j)$ is the nonlinear relationship between the neuron output node y_j and the network output o_m.

Output node of neuron j is calculated using

$$y_j = f_j(net_j) \tag{13.3}$$

where

f_j is the activation function of neuron j

net value net_j is the sum of weighted input nodes of neuron j:

$$net_j = \sum_{i=1}^{ni} w_{j,i} y_{j,i} + w_{j,0} \tag{13.4}$$

where

$y_{j,i}$ is the ith input node of neuron j

weighted by $w_{j,i}$, and $w_{j,0}$ is the bias weight of neuron j

Using (13.4) one may notice that derivative of net_j is

$$\frac{\partial net_j}{\partial w_{j,i}} = y_{j,i} \tag{13.5}$$

and slope s_j of activation function f_j is

$$s_j = \frac{\partial y_j}{\partial net_j} = \frac{\partial f_j(net_j)}{\partial net_j} \tag{13.6}$$

Between the output node y_j of a hidden neuron j and network output o_m, there is a complex nonlinear relationship (Figure 13.1):

$$o_m = F_{m,j}(y_j) \tag{13.7}$$

where o_m is the mth output of the network.

The complexity of this nonlinear function $F_{m,j}(y_j)$ depends on how many other neurons are between neuron j and network output m. If neuron j is at network output m, then $o_m = y_j$ and $F'_{m,j}(y_j) = 1$, where $F'_{m,j}$ is the derivative of nonlinear relationship between neuron j and output m.

13.2.2 Jacobian Matrix Computation

The update rule of the LM algorithm is [TM94]

$$\mathbf{w}_{n+1} = \mathbf{w}_n - \left(\mathbf{J}_n^T \mathbf{J}_n + \mu \mathbf{I}\right)^{-1} \mathbf{J}_n \mathbf{e}_n \tag{13.8}$$

where

n is the index of iterations

μ is the combination coefficient

$\mathbf{I}$ is the identity matrix

$\mathbf{J}$ is the Jacobian matrix (Figure 13.2)

From Figure 13.2, one may notice that, for every pattern p, there are no rows of the Jacobian matrix where no is the number of network outputs. The number of columns is equal to number of weights in the networks and the number of rows is equal to $np \times no$.

$$\mathbf{J} = \begin{array}{c} \quad \text{neuron 1} \qquad \cdots \qquad \text{neuron } j \qquad \cdots \\ \left[\begin{array}{cccccc}
\frac{\partial e_{1,1}}{\partial w_{1,1}} & \frac{\partial e_{1,1}}{\partial w_{1,2}} & \cdots & \frac{\partial e_{1,1}}{\partial w_{j,1}} & \frac{\partial e_{1,1}}{\partial w_{j,2}} & \cdots \\
\frac{\partial e_{1,2}}{\partial w_{1,1}} & \frac{\partial e_{1,2}}{\partial w_{1,2}} & \cdots & \frac{\partial e_{1,2}}{\partial w_{j,1}} & \frac{\partial e_{1,2}}{\partial w_{j,2}} & \cdots \\
\cdots & \cdots & \cdots & \cdots & \cdots & \cdots \\
\frac{\partial e_{1,no}}{\partial w_{1,1}} & \frac{\partial e_{1,no}}{\partial w_{1,2}} & \cdots & \frac{\partial e_{1,no}}{\partial w_{j,1}} & \frac{\partial e_{1,no}}{\partial w_{j,2}} & \cdots \\
\cdots & \cdots & \cdots & \cdots & \cdots & \cdots \\
\frac{\partial e_{p,1}}{\partial w_{1,1}} & \frac{\partial e_{p,1}}{\partial w_{1,2}} & \cdots & \frac{\partial e_{p,1}}{\partial w_{j,1}} & \frac{\partial e_{p,1}}{\partial w_{j,2}} & \cdots \\
\cdots & \cdots & \cdots & \cdots & \cdots & \cdots \\
\frac{\partial e_{p,m}}{\partial w_{1,1}} & \frac{\partial e_{p,m}}{\partial w_{1,2}} & \cdots & \frac{\partial e_{p,m}}{\partial w_{j,1}} & \frac{\partial e_{p,m}}{\partial w_{j,2}} & \cdots \\
\cdots & \cdots & \cdots & \cdots & \cdots & \cdots \\
\frac{\partial e_{np,1}}{\partial w_{1,1}} & \frac{\partial e_{np,1}}{\partial w_{1,2}} & \cdots & \frac{\partial e_{np,1}}{\partial w_{j,1}} & \frac{\partial e_{np,1}}{\partial w_{j,2}} & \cdots \\
\frac{\partial e_{np,1}}{\partial w_{1,1}} & \frac{\partial e_{np,2}}{\partial w_{1,2}} & \cdots & \frac{\partial e_{np,2}}{\partial w_{j,1}} & \frac{\partial e_{np,2}}{\partial w_{j,2}} & \cdots \\
\cdots & \cdots & \cdots & \cdots & \cdots & \cdots \\
\frac{\partial e_{np,no}}{\partial w_{1,1}} & \frac{\partial e_{np,no}}{\partial w_{1,2}} & \cdots & \frac{\partial e_{np,no}}{\partial w_{j,1}} & \frac{\partial e_{np,no}}{\partial w_{j,2}} & \cdots
\end{array}\right] \end{array}
\begin{array}{l} \\ \left.\begin{array}{l} m=1 \\ m=2 \\ \cdots \\ m=no \end{array}\right\} p=1 \\ \cdots \qquad\quad \cdots \\ \left.\begin{array}{l} m=1 \\ \cdots \\ m=m \end{array}\right\} p=p \\ \cdots \qquad\quad \cdots \\ \left.\begin{array}{l} m=1 \\ m=2 \\ \cdots \\ m=no \end{array}\right\} p=np \end{array}$$

FIGURE 13.2 Structure of the Jacobian matrix: (1) the number of columns is equal to the number of weights and (2) each row corresponds to a specified training pattern p and output m.

The elements of the Jacobian matrix can be calculated by

$$\frac{\partial e_m}{\partial w_{j,i}} = \frac{\partial e_m}{\partial y_j}\frac{\partial y_j}{\partial net_j}\frac{\partial net_j}{\partial w_{j,i}} \tag{13.9}$$

By combining with (13.2), (13.5), (13.6), and (13.7), (13.9) can be written as

$$\frac{\partial e_m}{\partial w_{j,i}} = y_{j,i}s_jF'_{m,j} \tag{13.10}$$

In second-order algorithms, the parameter δ [N89,TM94] is defined to measure the EBP process, as

$$\delta_{m,j} = s_jF'_{m,j} \tag{13.11}$$

By combining (13.10) and (13.11), elements of the Jacobian matrix can be calculated by

$$\frac{\partial e_m}{\partial w_{j,i}} = y_{j,i}\delta_{m,j} \tag{13.12}$$

Using (13.12), in backpropagation process, the error can be replaced by a unit value "1."

13.3 Training Arbitrarily Connected Neural Networks

The NBN algorithm introduced in this chapter is developed for training arbitrarily connected neural networks using the LM update rule. Instead of layer-by-layer computation (introduced in Chapter 12), the NBN algorithm does the forward and backward computation based on NBN routings [WCHK07], which makes it suitable for arbitrarily connected neural networks.

13.3.1 Importance of Training Arbitrarily Connected Neural Networks

The traditional implementation of the LM algorithm [TM94], like being adopted in MATLAB® neural network toolbox (MNNT), was developed only for standard multilayer perceptron (MLP) networks, it turns out that the MLP networks are not efficient.

Figure 13.3 shows the smallest structures to solve parity-7 problem. The standard MLP network with one hidden layer (Figure 13.3a) needs at least eight neurons to find the solution. The BMLP (bridged multiplayer perceptron) network (Figure 13.3b) can solve the problem with four neurons. The FCC (fully connected cascade) network (Figure 13.3c) is the most powerful one, and it only requires three neurons to get the solutions. One may notice that the last two types of networks are better choices for efficient training, but they also require more challenging computation.

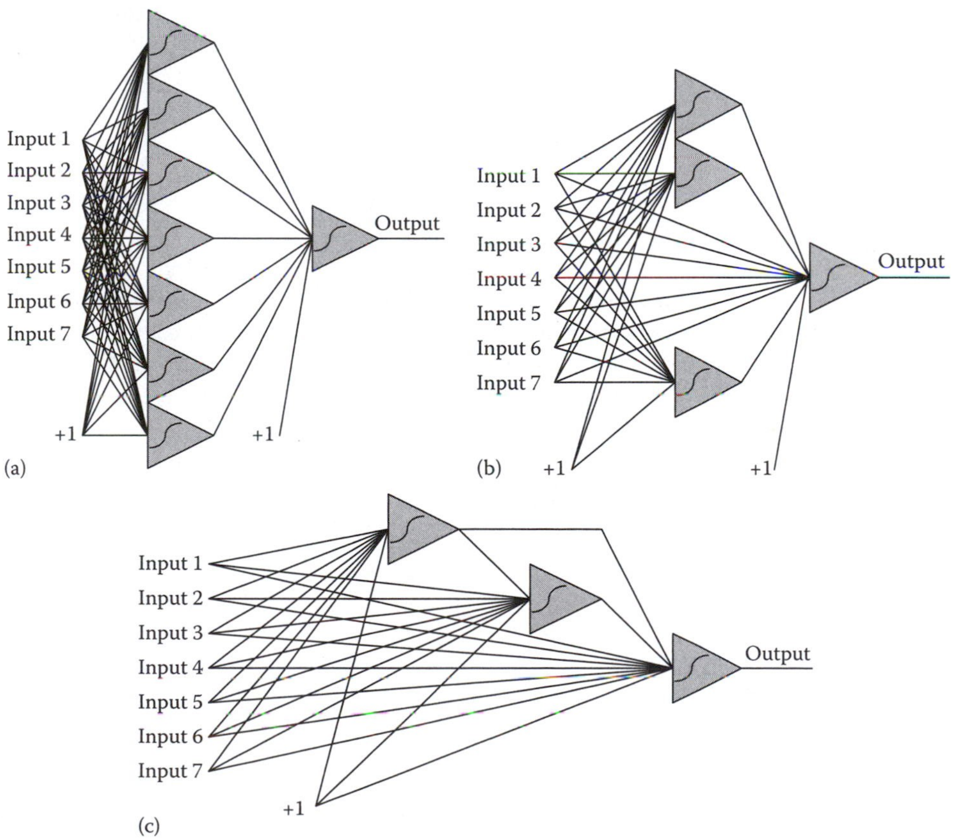

FIGURE 13.3 Smallest structures for solving parity-7 problem: (a) standard MLP network (64 weights), (b) BMLP network (35 weights), and (c) FCC network (27 weights).

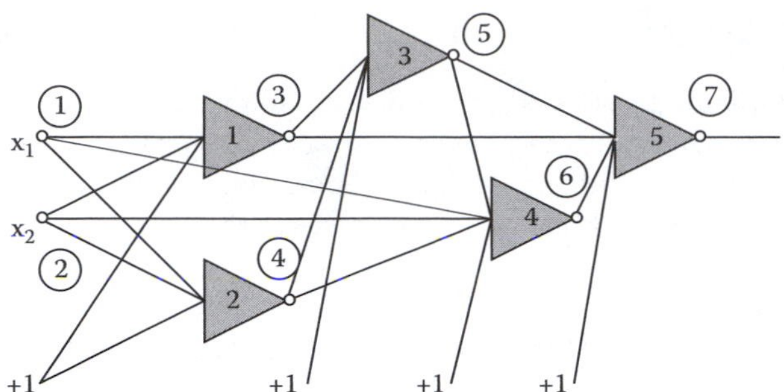

FIGURE 13.4 Five neurons in arbitrarily connected network.

13.3.2 Creation of Jacobian Matrix for Arbitrarily Connected Neural Networks

In this section, the NBN algorithm for calculating the Jacobian matrix for arbitrarily connected feedforward neural networks is presented. The rest of the computations for weight updating follow the LM algorithm, as shown in Equation 13.9.

In the forward computation, neurons are organized according to the direction of signal propagation, while in the backward computation, the analysis will follow the backpropagation procedures.

Let us consider the arbitrarily connected network with one output, as shown in Figure 13.4.

For the network in Figure 13.4, using the NBN algorithm, the network topology can be described similarly in the SPICE program:

```
n1 [model] 3 1 2
n2 [model] 4 1 2
n3 [model] 5 3 4
n4 [model] 6 1 2 4 5
n5 [model] 7 3 5 6
```

Notice that each line corresponds to one neuron. The first part (n_1–n_5) is the neuron name (Figure 13.4). The second part "*[model]*" is the neuron models, such as bipolar, unipolar, and linear. Models are declared in separate lines where the types of activation functions and the neuron gains are specified. The first digit in each line after the neuron model indicates the network nodes starting with the output node of the neuron, followed with its input nodes.

Please notice that neurons must be ordered from inputs to neuron outputs. It is important that, for each given neuron, the neuron inputs must have smaller indices than its output.

The row elements of the Jacobian matrix for a given pattern are being computed in the following three steps [WCKD08]:

1. Forward computation
2. Backward computation
3. Jacobian element computation

13.3.2.1 Forward Computation

In the forward computation, the neurons connected to the network inputs are first processed so that their outputs can be used as inputs to the subsequent neurons. The following neurons are then processed as their input values become available. In other words, the selected computing sequence has to follow the concept of feedforward signal propagation. If a signal reaches the inputs of several neurons at the same time, then these neurons can be processed in any sequence. In the example in Figure 13.4, there are two possible ways in which neurons can be processed in the forward direction: $n_1n_2n_3n_4n_5$ or $n_2n_1n_3n_4n_5$. The two procedures will lead to different computing processes but with exactly the same results. When the forward pass is concluded, the following two temporary vectors are stored: the first vector **y** with the values of the signals on the neuron output nodes and the second vector **s** with the values of the slopes of the neuron activation functions, which are signal dependent.

13.3.2.2 Backward Computation

The sequence of the backward computation is opposite to the forward computation sequence. The process starts with the last neuron and continues toward the input. In the case of the network in Figure 13.4, the following are two possible sequences (backpropagation paths): $n_5n_4n_3n_2n_1$ or $n_5n_4n_3n_1n_2$, and also they will have the same results. To demonstrate the case, let us use the $n_5n_4n_3n_2n_1$ sequence. The vector δ represents signal propagation from a network output to the inputs of all other neurons. The size of this vector is equal to the number of neurons.

For the output neuron n_5, its sensitivity is initialed using its slope $\delta_{1,5} = s_5$. For neuron n_4, the delta at n_5 will be propagated by w_{45}—the weight between n_4 and n_5, then by the slope of neuron n_4. So the delta parameter of n_4 is presented as $\delta_{1,4} = \delta_{1,5}w_{45}s_4$. For neuron n_3, the delta parameters of n_4 and n_5 will be propagated to the output of neuron n_3 and summed, then multiplied by the slope of neuron n_3, as $\delta_{1,3} = (\delta_{1,5}w_{35} + \delta_{1,4}w_{34})s_3$. For the same procedure, it could be obtained that $\delta_{1,2} = (\delta_{1,3}w_{23} + \delta_{1,4}w_{24})s_2$ and $\delta_{1,1} = (\delta_{1,3}w_{13} + \delta_{1,5}w_{15})s_1$. After the backpropagation process is done at neuron $N1$, all the elements of array δ are obtained.

13.3.2.3 Jacobian Element Computation

After the forward and backward computation, all the neuron outputs **y** and vector δ are calculated. Then using Equation 13.12, the Jacobian row for a given pattern can be obtained.

By applying all training patterns, the whole Jacobian matrix can be calculated and stored.

For arbitrarily connected neural networks, the NBN algorithm for the Jacobian matrix computation can be organized as shown in Figure 13.5.

```
for all patterns (np)
% Forward computation
 for all neurons (nn)
  for all weights of the neuron (nx)
   calculate net;          % Eq. (4)
  end;
  calculate neuron output; % Eq. (3)
  calculate neuron slope;  % Eq. (6)
 end;
 for all outputs (no)
  calculate error;         % Eq. (2)
%Backward computation
  initial delta as slope;
  for all neurons starting from output neurons (nn)
   for the weights connected to other neurons (ny)
    multiply delta through weights
    sum the backpropagated delta at proper nodes
   end;
   multiply delta by slope (for hidden neurons);
  end;
  related Jacobian row computation; %Eq. (12)
 end;
end;
```

FIGURE 13.5 Pseudo code using NBN algorithm for Jacobian matrix computation.

13.3.3 Solve Problems with Arbitrarily Connected Neural Networks

13.3.3.1 Function Approximation Problem

Function approximation is usually used in nonlinear control realm of neural networks, for control surface prediction. In order to approximate the function shown below, 25 points are selected from 0 to 4 as the training patterns. With only four neurons in FCC networks (as shown in Figure 13.6), the training result is presented in Figure 13.7.

$$z = 4exp\left(-0.15(x-4)^2 - 0.5(y-3)^2\right) + 10^{-9} \quad (13.13)$$

13.3.3.2 Two-Spiral Problem

Two-spiral problem is considered a good evaluation of both training algorithms and training architectures [AS99]. Depending on the neural network architecture, different numbers of neurons are required for successful training. For example, using standard MLP networks with one hidden layer, 34 neurons are required for two-spiral problem [PLI08]; while with the FCC architecture, it can be solved with only eight neurons using the NBN algorithm. NBN algorithms are not only much faster but also can train reduced size networks which cannot be handled by the traditional EBP algorithm (see Table 13.1).

For the EBP algorithm, learning constant is 0.005 (largest possible to avoid oscillation) and momentum is 0.5; maximum iteration is 1,000,000 for EBP algorithm and 1000 for the LM algorithm; desired error = 0.01; all neurons are in FCC networks; there are 100 trials for each case.

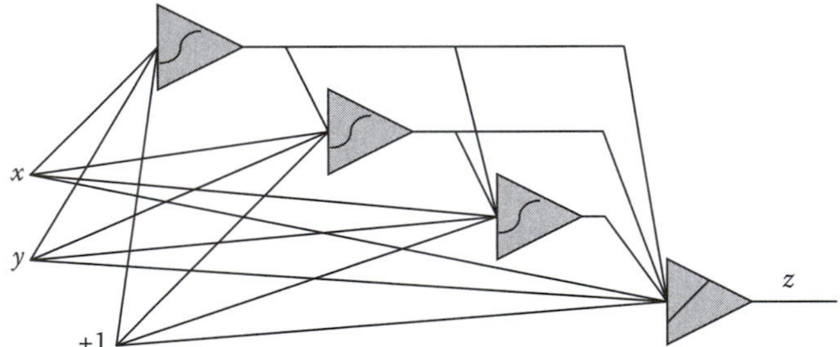

FIGURE 13.6 Network used for training the function approximation problem; notice the output neuron is a linear neuron with gain = 1.

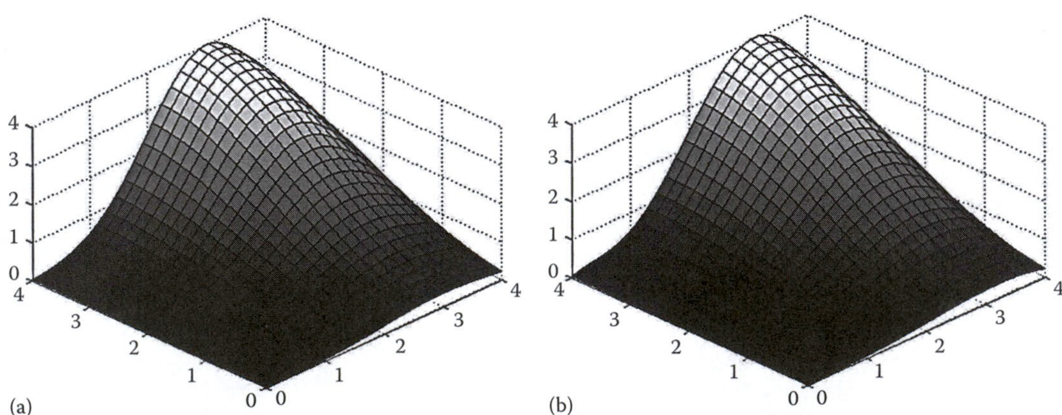

FIGURE 13.7 Averaged SSE between desired surface (a) and neural prediction (b) is 0.0025.

TABLE 13.1 Training Results of Two-Spiral Problem

	Success Rate (%)		Average Number of Iterations		Average Time (s)	
Neurons	EBP	NBN	EBP	NBN	EBP	NBN
8	0	13	Failing	287.7	Failing	0.88
9	0	24	Failing	261.4	Failing	0.98
10	0	40	Failing	243.9	Failing	1.57
11	0	69	Failing	231.8	Failing	1.62
12	63	80	410,254	175.1	633.91	1.70
13	85	89	335,531	159.7	620.30	2.09
14	92	92	266,237	137.3	605.32	2.40

13.4 Forward-Only Computation

The NBN procedure introduced in Section 3 requires both forward and backward computation. Especially, as shown in Figure 13.5, one may notice that for networks with multiple outputs, the backpropagation process has to be repeated for each output.

In this section, an improved NBN computation is introduced to overcome the problem, by removing the backpropagation process in the computation of the Jacobian matrix.

13.4.1 Derivation

The concept of $\delta_{m,j}$ was described in Section 13.2. One may notice that $\delta_{m,j}$ can be interpreted also as a signal gain between net input of neuron j and the network output m. Let us extend this concept to gain coefficients between all neurons in the network (Figures 13.8 and 13.10). The notation of $\delta_{k,j}$ is an extension of Equation 13.11 and can be interpreted as signal gain between neurons j and k, and it is given by

$$\delta_{k,j} = \frac{\partial F_{k,j}(y_j)}{\partial net_j} = \frac{\partial F_{k,j}(y_j)}{\partial y_j}\frac{\partial y_j}{\partial net_j} = F'_{k,j}s_j \qquad (13.14)$$

where

k and j are indices of neurons

$F_{k,j}(y_j)$ is the nonlinear relationship between the output node of neuron k and the output node of neuron j

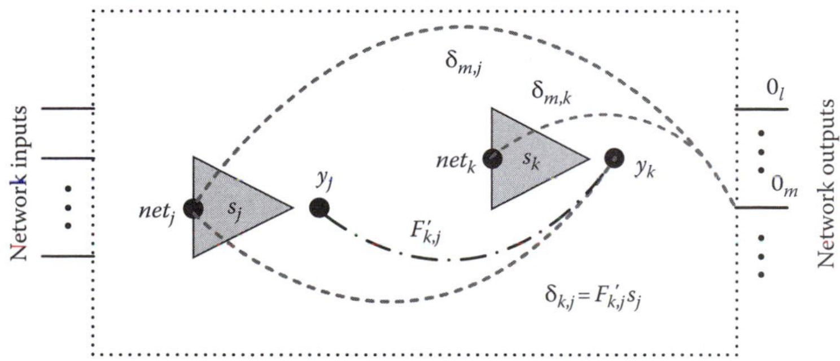

FIGURE 13.8 Interpretation of $\delta_{k,j}$ as a signal gain, where in feedforward network neuron j must be located before neuron k.

Naturally in feedforward networks, $k \geq j$. If $k = j$, then $\delta_{k,k} = s_k$, where s_k is the slope of activation function calculated by Equation 13.6. Figure 13.8 illustrates this extended concept of $\delta_{k,j}$ parameter as a signal gain.

The matrix $\boldsymbol{\delta}$ has a triangular shape and its elements can be calculated in the forward-only process. Later, elements of the Jacobian can be obtained using Equation 13.12, where only last rows of matrix $\boldsymbol{\delta}$ associated with network outputs are used. The key issue of the proposed algorithm is the method of calculating of $\delta_{k,j}$ parameters in the forward calculation process, and it will be described in the next part of this section.

13.4.2 Calculation of δ Matrix for FCC Architectures

Let us start our analysis with fully connected neural networks (Figure 13.9). Any other architecture could be considered as a simplification of fully connected neural networks by eliminating connections (setting weights to zero). If the feedforward principle is enforced (no feedback), fully connected neural networks must have cascade architectures.

Slopes of neuron activation functions s_j can be also written in the form of δ parameter as $\delta_{j,j} = s_j$. By inspecting Figure 13.10, δ parameters can be written as follows:

For the first neuron, there is only one δ parameter

$$\delta_{1,1} = s_1 \tag{13.15}$$

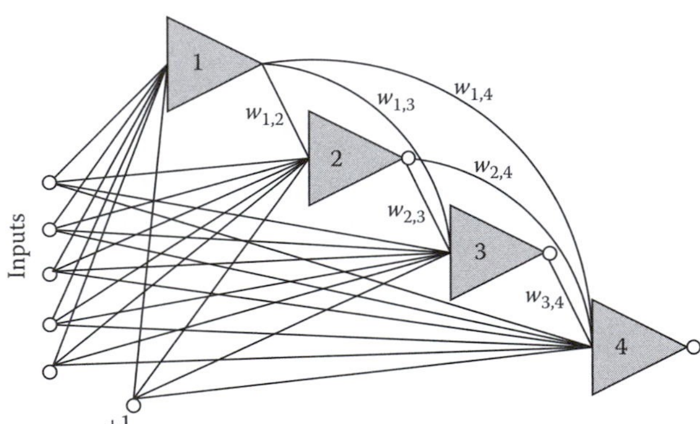

FIGURE 13.9 Four neurons in fully connected neural network, with five inputs and three outputs.

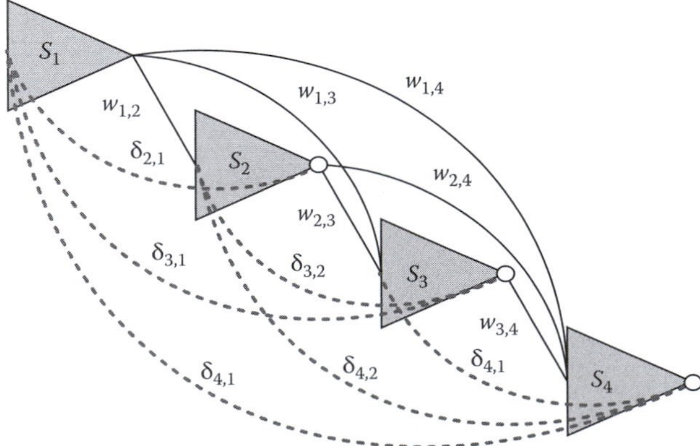

FIGURE 13.10 The $\delta_{k,j}$ parameters for the neural network of Figure 13.9. Input and bias weights are not used in the calculation of gain parameters.

For the second neuron, there are two δ parameters

$$\begin{aligned}\delta_{2,2} &= s_2 \\ \delta_{2,1} &= s_2 w_{1,2} s_1\end{aligned} \tag{13.16}$$

For the third neuron, there are three δ parameters

$$\begin{aligned}\delta_{3,3} &= s_3 \\ \delta_{3,2} &= s_3 w_{2,3} s_2 \\ \delta_{3,1} &= s_3 w_{1,3} s_1 + s_3 w_{2,3} s_2 w_{1,2} s_1\end{aligned} \tag{13.17}$$

One may notice that all δ parameters for the third neuron can be also expressed as a function of δ parameters calculated for previous neurons. Equations 13.17 can be rewritten as

$$\begin{aligned}\delta_{3,3} &= s_3 \\ \delta_{3,2} &= \delta_{3,3} w_{2,3} \delta_{2,2} \\ \delta_{3,1} &= \delta_{3,3} w_{1,3} \delta_{1,1} + \delta_{3,3} w_{2,3} \delta_{2,1}\end{aligned} \tag{13.18}$$

For the fourth neuron, there are four δ parameters

$$\begin{aligned}\delta_{4,4} &= s_4 \\ \delta_{4,3} &= \delta_{4,4} w_{3,4} \delta_{3,3} \\ \delta_{4,2} &= \delta_{4,4} w_{2,4} \delta_{2,2} + \delta_{4,4} w_{3,4} \delta_{3,2} \\ \delta_{4,1} &= \delta_{4,4} w_{1,4} \delta_{1,1} + \delta_{4,4} w_{2,4} \delta_{2,1} + \delta_{4,4} w_{3,4} \delta_{3,1}\end{aligned} \tag{13.19}$$

The last parameter $\delta_{4,1}$ can be also expressed in a compacted form by summing all terms connected to other neurons (from 1 to 3)

$$\delta_{4,1} = \delta_{4,4} \sum_{i=1}^{3} w_{i,4} \delta_{i,1} \tag{13.20}$$

The universal formula to calculate $\delta_{k,j}$ parameters using already calculated data for previous neurons is

$$\delta_{k,j} = \delta_{k,k} \sum_{i=j}^{k-1} w_{i,k} \delta_{i,j} \tag{13.21}$$

where in feedforward network, neuron j must be located before neuron k, so $k \geq j$; $\delta_{k,k} = s_k$ is the slope of activation function of neuron k; $w_{j,k}$ is the weight between neuron j and neuron k; and $\delta_{k,j}$ is a signal gain through weight $w_{j,k}$ and through other part of network connected to $w_{j,k}$.

In order to organize the process, the $nn \times nn$ computation table is used for calculating signal gains between neurons, where nn is the number of neurons (Figure 13.11). Natural indices (from 1 to nn) are given for each neuron according to the direction of signal propagation. For signal gain computation, only connections between neurons need to be concerned, while the weights connected to network inputs and biasing weights of all neurons will be used only at the end of the process. For a given pattern, a sample of the $nn \times nn$ computation table is shown in Figure 13.11. One may notice that the indices of rows and columns are the same as the indices of neurons. In the following derivation, let us use k and j

Neuron Index	1	2	...	j	...	k	...	nn
1	$\delta_{1,1}$	$w_{1,2}$	...	$w_{1,j}$	...	$w_{1,k}$	...	$w_{1,nn}$
2	$\delta_{2,1}$	$\delta_{2,2}$	...	$w_{2,j}$	...	$w_{2,k}$	...	$w_{2,nn}$
...	...	...	...	...	...	...	...	...
j	$\delta_{j,1}$	$\delta_{j,2}$	...	$\delta_{j,j}$	...	$w_{j,k}$	...	$w_{j,nn}$
...	...	...	...	...	...	...	...	...
k	$\delta_{k,1}$	$\delta_{k,2}$	...	$\delta_{k,j}$	...	$\delta_{k,k}$	...	$w_{k,nn}$
...	...	...	...	...	...	...	...	...
nn	$\delta_{nn,1}$	$\delta_{nn,2}$	...	$\delta_{nn,j}$	...	$\delta_{nn,k}$	...	$\delta_{nn,nn}$

FIGURE 13.11 The $nn \times nn$ computation table; gain matrix $\boldsymbol{\delta}$ contains all the signal gains between neurons; weight array $\boldsymbol{w}$ presents only the connections between neurons, while network input weights and biasing weights are not included.

used as neuron indices to specify the rows and columns in the computation table. In feedforward network, $k \geq j$ and matrix $\boldsymbol{\delta}$ has a triangular shape.

The computation table consists of three parts: weights between neurons in the upper triangle, vector of slopes of activation functions in main diagonal, and signal gain matrix $\boldsymbol{\delta}$ in lower triangle. Only the main diagonal and lower triangular elements are computed for each pattern. Initially, elements on main diagonal $\delta_{k,k} = s_k$ are known as slopes of activation functions and values of signal gains $\delta_{k,j}$ are being computed subsequently using Equation 13.21.

The computation is being processed NBN starting with the neuron closest to network inputs. At first, the row number one is calculated and then elements of subsequent rows. Calculation on row below is done using elements from above rows using Equation 13.21. After completion of forward computation process, all elements of $\boldsymbol{\delta}$ matrix in the form of the lower triangle are obtained.

In the next step, elements of the Jacobian matrix are calculated using Equation 13.12. In the case of neural networks with one output, only the last row of $\boldsymbol{\delta}$ matrix is needed for the gradient vector and Jacobian matrix computation. If networks have more outputs *no*, then last *no* rows of $\boldsymbol{\delta}$ matrix are used. For example, if the network shown in Figure 13.9 has three outputs, the following elements of $\boldsymbol{\delta}$ matrix are used

$$\begin{bmatrix} \delta_{2,1} & \delta_{2,2} = s_2 & \delta_{2,3} = 0 & \delta_{2,4} = 0 \\ \delta_{3,1} & \delta_{3,2} & \delta_{3,3} = s_3 & \delta_{3,4} = 0 \\ \delta_{4,1} & \delta_{4,2} & \delta_{4,3} & \delta_{4,4} = s_4 \end{bmatrix} \tag{13.22}$$

and then for each pattern, the three rows of Jacobian matrix, corresponding to three outputs, are calculated in one step using Equation 13.12 without additional propagation of δ

$$\begin{bmatrix} \delta_{2,1} \times \{y_1\} & s_2 \times \{y_2\} & 0 \times \{y_3\} & 0 \times \{y_4\} \\ \delta_{3,1} \times \{y_1\} & \delta_{3,2} \times \{y_2\} & s_3 \times \{y_3\} & 0 \times \{y_4\} \\ \underbrace{\delta_{4,1} \times \{y_1\}}_{\text{neuron 1}} & \underbrace{\delta_{4,2} \times \{y_2\}}_{\text{neuron 2}} & \underbrace{\delta_{4,3} \times \{y_3\}}_{\text{neuron 3}} & \underbrace{s_4 \times \{y_4\}}_{\text{neuron 4}} \end{bmatrix} \tag{13.23}$$

where neurons' input vectors $\boldsymbol{y}_1$ through $\boldsymbol{y}_4$ have 6, 7, 8, and 9 elements respectively (Figure 13.9), corresponding to number of weights connected. Therefore, each row of the Jacobian matrix has 6 + 7 + 8 + 9 = 30 elements. If the network has three outputs, then from six elements of $\boldsymbol{\delta}$ matrix and 3 slopes,

90 elements of the Jacobian matrix are calculated. One may notice that the size of newly introduced **δ** matrix is relatively small, and it is negligible in comparison with other matrices used in calculation.

The improved NBN procedure gives all the information needed to calculate the Jacobian matrix (13.12), without backpropagation process; instead, δ parameters are obtained in relatively simple forward computation (see Equation 13.21).

13.4.3 Training Arbitrarily Connected Neural Networks

The proposed computation above was derived for fully connected neural networks. If the network is not fully connected, then some elements of the computation table are zero. Figure 13.12 shows computation

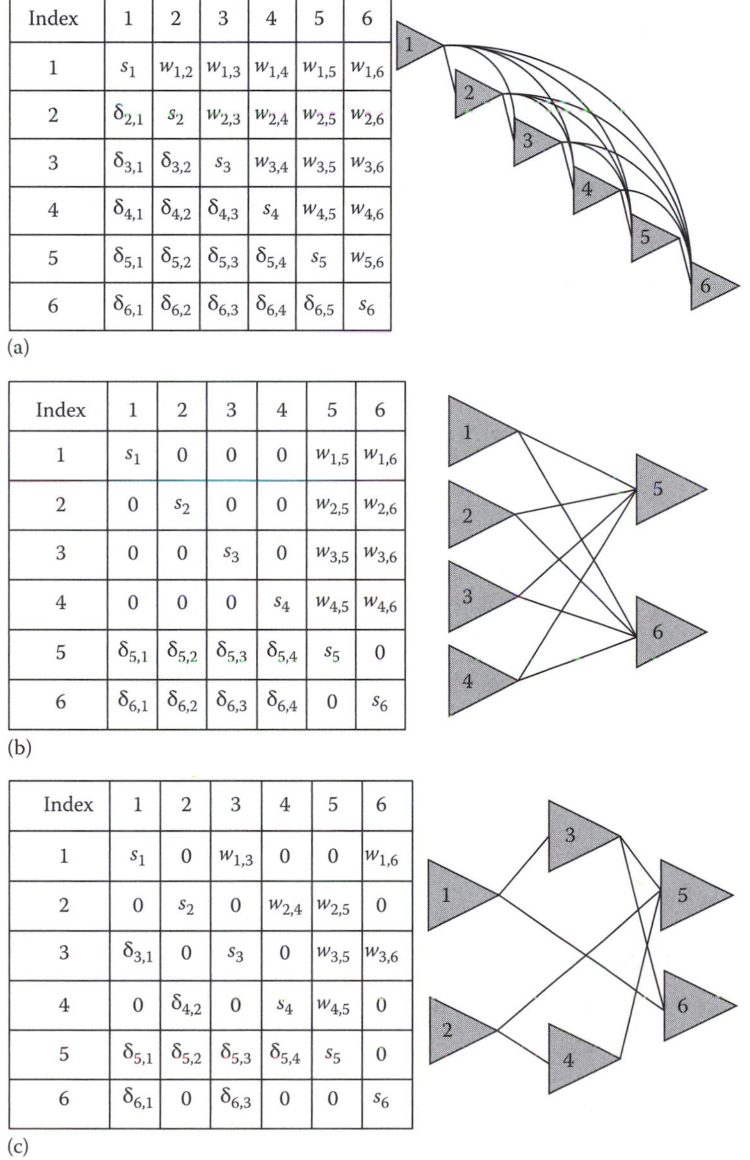

Index	1	2	3	4	5	6
1	s_1	$w_{1,2}$	$w_{1,3}$	$w_{1,4}$	$w_{1,5}$	$w_{1,6}$
2	$\delta_{2,1}$	s_2	$w_{2,3}$	$w_{2,4}$	$w_{2,5}$	$w_{2,6}$
3	$\delta_{3,1}$	$\delta_{3,2}$	s_3	$w_{3,4}$	$w_{3,5}$	$w_{3,6}$
4	$\delta_{4,1}$	$\delta_{4,2}$	$\delta_{4,3}$	s_4	$w_{4,5}$	$w_{4,6}$
5	$\delta_{5,1}$	$\delta_{5,2}$	$\delta_{5,3}$	$\delta_{5,4}$	s_5	$w_{5,6}$
6	$\delta_{6,1}$	$\delta_{6,2}$	$\delta_{6,3}$	$\delta_{6,4}$	$\delta_{6,5}$	s_6

(a)

Index	1	2	3	4	5	6
1	s_1	0	0	0	$w_{1,5}$	$w_{1,6}$
2	0	s_2	0	0	$w_{2,5}$	$w_{2,6}$
3	0	0	s_3	0	$w_{3,5}$	$w_{3,6}$
4	0	0	0	s_4	$w_{4,5}$	$w_{4,6}$
5	$\delta_{5,1}$	$\delta_{5,2}$	$\delta_{5,3}$	$\delta_{5,4}$	s_5	0
6	$\delta_{6,1}$	$\delta_{6,2}$	$\delta_{6,3}$	$\delta_{6,4}$	0	s_6

(b)

Index	1	2	3	4	5	6
1	s_1	0	$w_{1,3}$	0	0	$w_{1,6}$
2	0	s_2	0	$w_{2,4}$	$w_{2,5}$	0
3	$\delta_{3,1}$	0	s_3	0	$w_{3,5}$	$w_{3,6}$
4	0	$\delta_{4,2}$	0	s_4	$w_{4,5}$	0
5	$\delta_{5,1}$	$\delta_{5,2}$	$\delta_{5,3}$	$\delta_{5,4}$	s_5	0
6	$\delta_{6,1}$	0	$\delta_{6,3}$	0	0	s_6

(c)

FIGURE 13.12 Three different architectures with six neurons: (a) FCC network, (b) MLP network, and (c) arbitrarily connected neural network.

```
for all patterns (np)
% Forward computation
 for all neurons (nn)
  for all weights of the neuron (nx)
   calculate net;              % Eq. (4)
  end;
  calculate neuron output;       % Eq. (3)
  calculate neuron slope;        % Eq. (6)
  set current slope as delta;
  for weights connected to previous neurons (ny)
   for previous neurons (nz)
     multiply delta through weights then sum; % Eq. (24)
    end;
    multiply the sum by the slope;          % Eq. (25)
  end;
  related Jacobian elements computation;   % Eq. (12)
 end;
 for all outputs (no)
  calculate error;    % Eq. (2)
 end;
end;
```

FIGURE 13.13 Pseudo code of the forward-only computation, in second-order algorithms.

tables for different neural network topologies with six neurons each. Please notice zero elements are for not connected neurons (in the same layers). This can further simplify the computation process for popular MLP topologies (Figure 13.12b).

Most of the used neural networks have many zero elements in the computation table (Figure 13.12). In order to reduce the storage requirements (do not store weights with zero values) and to reduce computation process (do not perform operations on zero elements), a part of the NBN algorithm in Section 13.3 was adopted for forward computation.

In order to further simplify the computation process, Equation 13.21 is completed in two steps

$$x_{k,j} = \sum_{i=j}^{k-1} w_{i,k}\delta_{i,j} \tag{13.24}$$

and

$$\delta_{k,j} = \delta_{k,k}x_{k,j} = s_k x_{k,j} \tag{13.25}$$

The complete algorithm with forward-only computation is shown in Figure 13.13. By adding two additional steps using Equations 13.24 and 13.25 (highlighted in bold in Figure 13.13), all computations can be completed in the forward-only computing process.

13.4.4 Experimental Results

Several problems are presented to test the computing speed of two different NBN algorithms—with and without backpropagation process.

The testing of time costs for both the backpropagation computation and the forward-only computation are divided into forward part and backward part separately.

13.4.4.1 ASCII Codes to Image Conversion

This problem is to associate 256 ASCII codes with 256 character images, each of which is made up of 7 × 8 pixels (Figure 13.14). So there are 8 bit inputs (inputs of parity-8 problem), 256 patterns, and

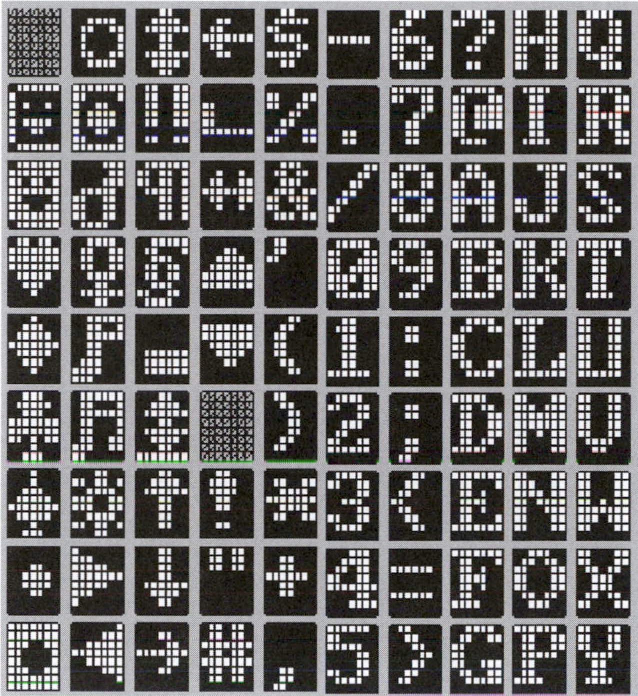

FIGURE 13.14 The first 90 images of ASCII characters.

TABLE 13.2 Comparison for ASCII Character Recognition Problem

Computation Methods	Time Cost (ms/Iteration)		Relative Time (%)
	Forward	Backward	
Backpropagation	8.24	1,028.74	100
Forward-only	61.13	0.00	5.9

56 outputs. In order to solve the problem, the structure, 112 neurons in 8-56-56 MLP network, is used to train those patterns using the NBN algorithms. The computation time is presented in Table 13.2.

13.4.4.2 Parity-7 Problem

Parity-*N* problems are aimed to associate *n*-bit binary input data with their parity bits. It is also considered to be one of the most difficult problems in neural network training, although it has been solved analytically [BDA03].

Parity-7 problem is trained with the NBN algorithms, using both the forward-only computation and traditional computation separately. Two different network structures are used for training: eight neurons in 7-7-1 MLP network (64 weights) and three neurons in FCC network (27 weights). Time cost comparison is shown in Table 13.3.

13.4.4.3 Error Correction Problems

Error correction is an extension of parity-*N* problems for multiple parity bits. In Figure 13.15, the left side is the input data, made up of signal bits and their parity bits, while the right side is the related corrected signal bits and parity bits as outputs, so number of inputs is equal to the number of outputs.

TABLE 13.3 Comparison for Parity-7 Problem

Networks	Computation Methods	Time Cost (μs/Iteration) Forward	Backward	Relative Time (%)
MLP	Backpropagation	158.57	67.82	100
	Forward-only	229.13	0.00	101.2
FCC	Backpropagation	54.14	31.94	100
	Forward-only	86.30	0.00	100.3

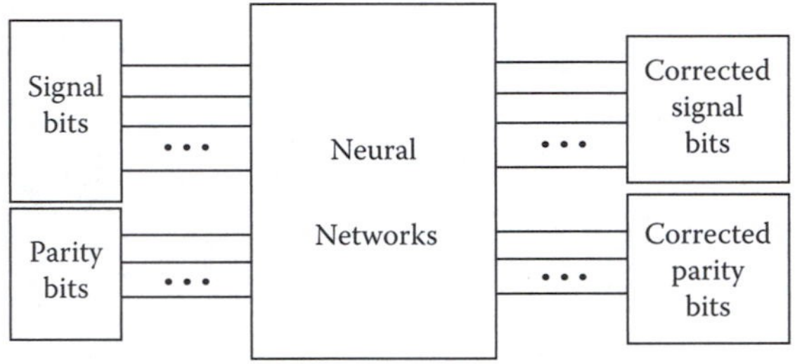

FIGURE 13.15 Using neural networks to solve error correction problem; errors in input data can be corrected by well-trained neural networks.

Two error correction experiments are presented, one has 4 bit signal with its 3 bit parity bits as inputs, 7 outputs, and 128 patterns (16 correct patterns and 112 patterns with errors), using 23 neurons in 7-16-7 MLP network (247 weights); the other has 8 bit signal with its 4 bit parity bits as inputs, 12 outputs, and 3328 patterns (256 correct patterns and 3072 patterns with errors), using 42 neurons in 12-30-12 MLP network (762 weights). Error patterns with one incorrect bit must be corrected. Both backpropagation computation and the forward-only computation were performed with the NBN algorithms. The testing results are presented in Table 13.4.

13.4.4.4 Encoders and Decoders

Experiment results on 3-to-8 decoder, 8-to-3 encoder, 4-to-16 decoder, and 16-to-4 encoder, using the NBN algorithms, are presented in Table 13.5. For 3-to-8 decoder and 8-to-3 encoder, 11 neurons are used in 3-3-8 MLP network (44 weights) and 8-83 MLP network (99 weights) respectively; while for 4-to-16 decoder and 16-to-4 encoder, 20 neurons are used in 4-4-16 MLP network (100 weights) and 16-16-4 MLP network (340 weights) separately.

In the encoder and decoder problems, one may notice that for the same number of neurons, the more outputs the networks have, the more efficiently the forward-only computation works.

From the presented experimental results, one may notice that, for networks with multiple outputs, the forward-only computation is more efficient than the backpropagation computation; while for single output situation, the forward-only computation is slightly worse.

TABLE 13.4 Comparison for Error Correction Problem

Problems	Computation Methods	Time Cost (ms/Iteration) Forward	Backward	Relative Time (%)
4 bit signal	Backpropagation	0.43	2.82	100
	Forward-only	1.82	0.00	56
8 bit signal	Backpropagation	40.59	468.14	100
	Forward-only	175.72	0.00	34.5

TABLE 13.5 Comparison for Encoders and Decoders

Problems	Computation Methods	Time Cost (μs/Iteration)		Relative Time (%)
		Forward	Backward	
3-to-8	Traditional	10.14	55.37	100
decoder	Forward-only	27.86	0.00	42.5
8-to-3	Traditional	7.19	26.97	100
encoder	Forward-only	29.76	0.00	87.1
4-to-16	Traditional	40.03	557.51	100
decoder	Forward-only	177.65	0.00	29.7
16-to-4	Traditional	83.24	244.20	100
encoder	Forward-only	211.28	0.00	62.5

13.5 Direct Computation of Quasi-Hessian Matrix and Gradient Vector

Using Equation 13.8 for weight updating, one may notice that the matrix multiplication $\mathbf{J}^T\mathbf{J}$ and $\mathbf{J}^T\mathbf{e}$ have to be calculated

$$\mathbf{H} \approx \mathbf{Q} = \mathbf{J}^T\mathbf{J} \tag{13.26}$$

$$\mathbf{g} = \mathbf{J}^T\mathbf{e} \tag{13.27}$$

where

matrix $\mathbf{Q}$ is the quasi-Hessian matrix
$\mathbf{g}$ is the gradient vector [YW09]

Traditionally, the whole Jacobian matrix $\mathbf{J}$ is calculated and stored [TM94] for further multiplication operation using Equations 13.26 and 13.27. The memory limitation may be caused by the Jacobian matrix storage, as described below.

In the NBN algorithm, quasi-Hessian matrix $\mathbf{Q}$ and gradient vector $\mathbf{g}$ are calculated directly, without Jacobian matrix computation and storage. Therefore, the NBN algorithm can be used in training the problems with unlimited number of training patterns.

13.5.1 Memory Limitation in the LM Algorithm

In the LM algorithm, Jacobian matrix $\mathbf{J}$ has to be calculated and stored for the Hessian matrix computation [TM94]. In this procedure, as shown in Figure 13.2, at least $np \times no \times nn$ elements (Jacobian matrix) have to be stored. For small and median size pattern training, this method may work smoothly. However, it would be a huge memory cost for training large-sized patterns, since the number of elements of Jacobian matrix $\mathbf{J}$ is proportional to the number of patterns.

For example, the pattern recognition problem in MNIST handwritten digit database [CKOZ06] consists of 60,000 training patterns, 784 inputs, and 10 outputs. Using only the simplest possible neural network with 10 neurons (one neuron per each output), the memory cost for the entire Jacobian matrix storage is nearly 35 Gb. This huge memory requirement cannot be satisfied by any Windows compliers, where there is a 3 Gb limitation for single-array storage. Therefore, the LM algorithm cannot be used for problems with large number of patterns.

13.5.2 Review of Matrix Algebra

There are two ways to multiply rows and columns of two matrices. If the row of the first matrix is multiplied by the column of the second matrix, then we obtain a scalar, as shown in Figure 13.16a. When the column of the first matrix is multiplied by the row of the second matrix then the result is a partial matrix **q** (Figure 13.16b) [L05]. The number of scalars is $nn \times nn$, while the number of partial matrices **q**, which later have to be summed, is $np \times no$.

When $\mathbf{J}^T$ is multiplied by **J** using routine shown in Figure 13.16b, at first, partial matrices **q** (size: $nn \times nn$) need to be calculated $np \times no$ times, then all of $np \times no$ matrices **q** must be summed together. The routine of Figure 13.16b seems complicated; therefore, almost all matrix multiplication processes use the routine of Figure 13.16a, where only one element of resulted matrix is calculated and stored each time.

Even the routine of Figure 13.16b seems to be more complicated; after detailed analysis (see Table 13.6), one may conclude that the computation time for matrix multiplication of the two ways is basically the same.

In a specific case of neural network training, only one row of Jacobian matrix **J** (column of $\mathbf{J}^T$) is known for each training pattern, so if routine from Figure 13.16b is used then the process of creation of the quasi-Hessian matrix can start sooner without the necessity of computing and storing the entire Jacobian matrix for all patterns and all outputs.

Table 13.7 roughly estimates the memory cost in two multiplication methods separately.

The analytical results in Table 13.7 show that the column-row multiplication (Figure 13.16b) can save a lot of memory.

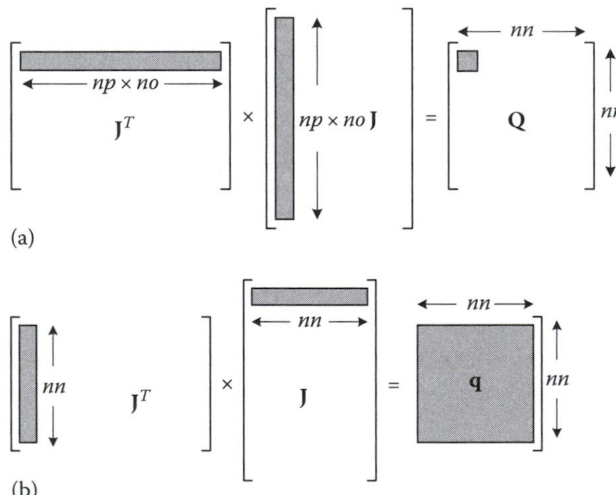

FIGURE 13.16 Two ways of multiplying matrices: (a) row–column multiplication results in a scalar and (b) column–row multiplication results in a partial matrix **q**.

TABLE 13.6 Computation Analysis

Multiplication Methods	Addition	Multiplication
Row–column	$(np \times no) \times nn \times nn$	$(np \times no) \times nn \times nn$
Column–row	$nn \times nn \times (np \times no)$	$nn \times nn \times (np \times no)$

np is the number of training patterns, no is the number of outputs, and nn is the number of weights.

TABLE 13.7 Memory Cost Analysis

Multiplication Methods	Elements for Storage
Row–column	$(np \times no) \times nn + nn \times nn + nn$
Column–row	$nn \times nn + nn$
Difference	$(np \times no) \times nn$

13.5.3 Quasi-Hessian Matrix Computation

Let us introduce the quasi-Hessian submatrix $\mathbf{q}_{p,m}$ (size: $nn \times nn$)

$$\mathbf{q}_{p,m} = \begin{bmatrix} \left(\frac{\partial e_{p,m}}{\partial w_1}\right)^2 & \frac{\partial e_{p,m}}{\partial w_1}\frac{\partial e_{p,m}}{\partial w_2} & \cdots & \frac{\partial e_{p,m}}{\partial w_1}\frac{\partial e_{p,m}}{\partial w_{nn}} \\ \frac{\partial e_{p,m}}{\partial w_2}\frac{\partial e_{p,m}}{\partial w_1} & \left(\frac{\partial e_{p,m}}{\partial w_2}\right)^2 & \cdots & \frac{\partial e_{p,m}}{\partial w_2}\frac{\partial e_{p,m}}{\partial w_{nn}} \\ \cdots & \cdots & \cdots & \cdots \\ \frac{\partial e_{p,m}}{\partial w_{nn}}\frac{\partial e_{p,m}}{\partial w_1} & \frac{\partial e_{p,m}}{\partial w_{nn}}\frac{\partial e_{p,m}}{\partial w_2} & \cdots & \left(\frac{\partial e_{p,m}}{\partial w_{nn}}\right)^2 \end{bmatrix} \tag{13.28}$$

Using the procedure in Figure 13.5b, the $nn \times nn$ quasi-Hessian matrix $\mathbf{Q}$ can be calculated as the sum of submatrices $\mathbf{q}_{p,m}$

$$\mathbf{Q} = \sum_{p=1}^{np}\sum_{m=1}^{no}\mathbf{q}_{p,m} \tag{13.29}$$

By introducing $1 \times nn$ vector $\mathbf{j}_{p,m}$

$$\mathbf{j}_{p,m} = \begin{bmatrix} \frac{\partial e_{p,m}}{\partial w_1} & \frac{\partial e_{p,m}}{\partial w_2} & \cdots & \frac{\partial e_{p,m}}{\partial w_{nn}} \end{bmatrix} \tag{13.30}$$

submatrices $\mathbf{q}_{p,m}$ in Equation 13.13 can be also written in the vector form (Figure 13.5b)

$$\mathbf{q}_{p,m} = \mathbf{j}_{p,m}^T\mathbf{j}_{p,m} \tag{13.31}$$

One may notice that for the computation of submatrices $\mathbf{q}_{p,m}$, only N elements of vector $\mathbf{j}_{p,m}$ need to be calculated and stored. All the submatrices can be calculated for each pattern p and output m separately, and summed together, so as to obtain the quasi-Hessian matrix $\mathbf{Q}$.

Considering the independence among all patterns and outputs, there is no need to store all the quasi-Hessian submatrices $\mathbf{q}_{p,m}$. Each submatrix can be summed to a temporary matrix after its computation. Therefore, during the direct computation of the quasi-Hessian matrix $\mathbf{Q}$ using (13.29), only memory for nn elements is required, instead of that for the whole Jacobian matrix with $(np \times no) \times nn$ elements (Table 13.7).

From (13.13), one may notice that all the submatrices $\mathbf{q}_{p,m}$ are symmetrical. With this property, only upper (or lower) triangular elements of those submatrices need to be calculated. Therefore, during the improved quasi-Hessian matrix $\mathbf{Q}$ computation, multiplication operations in (13.31) and sum operations in (13.29) can be both reduced by half approximately.

13.5.4 Gradient Vector Computation

Gradient subvector $\boldsymbol{\eta}_{p,m}$ (size: $nn \times 1$) is

$$\boldsymbol{\eta}_{p,m} = \begin{bmatrix} \frac{\partial e_{p,m}}{\partial w_1} e_{p,m} \\ \frac{\partial e_{p,m}}{\partial w_2} e_{p,m} \\ \cdots \\ \frac{\partial e_{p,m}}{\partial w_{nn}} e_{p,m} \end{bmatrix} = \begin{bmatrix} \frac{\partial e_{p,m}}{\partial w_1} \\ \frac{\partial e_{p,m}}{\partial w_2} \\ \cdots \\ \frac{\partial e_{p,m}}{\partial w_{nn}} \end{bmatrix} \times e_{p,m} \tag{13.32}$$

With the procedure in Figure 13.16b, gradient vector $\mathbf{g}$ can be calculated as the sum of gradient subvector $\boldsymbol{\eta}_{p,m}$

$$\mathbf{g} = \sum_{p=1}^{np} \sum_{m=1}^{no} \boldsymbol{\eta}_{p,m} \tag{13.33}$$

Using the same vector $\mathbf{j}_{p,m}$ defined in (13.30), gradient subvector can be calculated using

$$\boldsymbol{\eta}_{p,m} = \mathbf{j}_{p,m} e_{p,m} \tag{13.34}$$

Similarly, the gradient subvector $\boldsymbol{\eta}_{p,m}$ can be calculated for each pattern and output separately, and summed to a temporary vector. Since the same vector $\mathbf{j}_{p,m}$ is calculated during the quasi-Hessian matrix computation above, there is only an extra scalar $e_{p,m}$ need to be stored.

With the improved computation, both the quasi-Hessian matrix $\mathbf{Q}$ and gradient vector $\mathbf{g}$ can be computed directly, without the Jacobian matrix storage and multiplication. During the process, only a temporary vector $\mathbf{j}_{p,m}$ with nn elements needs to be stored; in other words, the memory cost for the Jacobian matrix storage is reduced by $(np \times no)$ times. In the MINST problem mentioned in Section 13.5.1, the memory cost for the storage of Jacobian elements could be reduced from more than 35 GB to nearly 30.7 kB.

13.5.5 Jacobian Row Computation

The key point of the improved computation above for quasi-Hessian matrix $\mathbf{Q}$ and gradient vector $\mathbf{g}$ is to calculate vector $\mathbf{j}_{p,m}$ defined in (13.30) for each pattern and output. This vector is equivalent to one row of the Jacobian matrix $\mathbf{J}$.

By combining Equations 13.12 and 13.30, the elements of vector $\mathbf{j}_{p,m}$ can be calculated by

$$\mathbf{j}_{p,m} = \left[\delta_{p,m,1}\left[y_{p,1,1} \quad \cdots \quad y_{p,1,i} \quad \cdots\right] \cdots \delta_{p,m,j}\left[y_{p,j,1} \quad \cdots \quad y_{p,j,i} \quad \cdots\right] \cdots\right] \tag{13.35}$$

where $y_{p,j,i}$ is the ith input of neuron j, when training pattern p.

Using the NBN procedure introduced in Section 13.3, all elements $y_{p,j,i}$ in Equation 13.35 can be calculated in the forward computation, while vector $\boldsymbol{\delta}$ is obtained in the backward computation; or, using the improved NBN procedure in Section 13.4, both vectors $\mathbf{y}$ and δ can be obtained in the improved forward computation. Again, since only one vector $\mathbf{j}_{p,m}$ needs to be stored for each pattern

```
% Initialization
Q=0;
g=0
% Improved computation
for p=1:np          % Number of patterns
 % Forward computation
 ...
 for m=1:no         % Number of outputs
  % Backward computation
  ...
  calculate vector j_p,m;          % Eq. (35)
  calculate sub matrix q_p,m;      % Eq. (31)
  calculate sub vector η_p,m;      % Eq. (34)
  Q=Q+q_p,m;                       % Eq. (29)
  g=g+η_p,m;                       % Eq. (33)
 end;
end;
```

FIGURE 13.17 Pseudo code of the improved computation for the quasi-Hessian matrix and gradient vector in NBN algorithm.

and output in the improved computation, the memory cost for all those temporary parameters can be reduced by ($np \times no$) times. All matrix operations are simplified to vector operations.

Generally, for the problem with np patterns and no outputs, the NBN algorithm without the Jacobian matrix storage can be organized as the pseudo code shown in Figure 13.17.

13.5.6 Comparison on Memory and Time Consumption

Several experiments are designed to test the memory and time efficiencies of the NBN algorithm, comparing with the traditional LM algorithm. They are divided into two parts: (1) memory comparison and (2) time comparison.

13.5.6.1 Memory Comparison

Three problems, each of which has huge number of patterns, are selected to test the memory cost of both the traditional computation and the improved computation. LM algorithm and NBN algorithm are used for training, and the test results are shown in Tables 13.8 and 13.9. In order to make a more precise comparison, memory cost for program code and input files were not used in the comparison.

TABLE 13.8 Memory Comparison for Parity Problems

Parity-N problems	$N = 14$	$N = 16$
Patterns	16,384	65,536
Structures[a]	15 neurons	17 neurons
Jacobian matrix sizes	5,406,720	27,852,800
Weight vector sizes	330	425
Average iteration	99.2	166.4
Success rate (%)	13	9
Algorithms	**Actual Memory Cost (Mb)**	
LM algorithm	79.21	385.22
NBN algorithm	3.41	4.3

[a] All neurons are in FCC networks.

TABLE 13.9 Memory Comparison for MINST Problem

Problem	MINST
Patterns	60,000
Structures	784 = 1 single layer network[a]
Jacobian matrix sizes	47,100,000
Weight vector sizes	785
Algorithms	**Actual Memory Cost (Mb)**
LM algorithm	385.68
NBN algorithm	15.67

[a] In order to perform efficient matrix inversion during training, only one of ten digits is classified each time.

TABLE 13.10 Time Comparison for Parity Problems

Parity-N Problems	$N = 9$	$N = 11$	$N = 13$	$N = 15$
Patterns	512	2,048	8,192	32,768
Neurons	10	12	14	16
Weights	145	210	287	376
Average iterations	38.51	59.02	68.08	126.08
Success rate (%)	58	37	24	12
Algorithms	**Averaged Training Time (s)**			
Traditional LM	0.78	68.01	1508.46	43,417.06
Improved LM	0.33	22.09	173.79	2,797.93

From the test results in Tables 13.8 and 13.9, it is clear that memory cost for training is significantly reduced in the improved computation.

13.5.6.2 Time Comparison

Parity-N problems are presented to test the training time for both traditional computation and the improved computation using the LM algorithm. The structures used for testing are all FCC networks. For each problem, the initial weights and training parameters are the same.

From Table 13.10, one may notice that the NBN computation cannot only handle much larger problems, but also computes much faster than the LM algorithm, especially for large-sized pattern training. The larger the pattern size is, the more time efficient the improved computation will be.

13.6 Conclusion

In this chapter, the NBN algorithm is introduced to solve the structure and memory limitation in the LM algorithm. Based on the specially designed NBN routings, the NBN algorithm can be used not only for traditional MLP networks, but also other arbitrarily connected neural networks.

The NBN algorithm can be organized in two procedures—with backpropagation process and without backpropagation process. Experimental results show that the former one is suitable for networks with single output, while the latter one is more efficient for networks with multiple outputs.

The NBN algorithm does not require to store and to multiply large Jacobian matrix. As a consequence, memory requirement for the quasi-Hessian matrix and gradient vector computation is decreased by $(P \times M)$ times, where P is the number of patterns and M is the number of outputs. An additional benefit of memory reduction is also a significant reduction in computation time. Therefore, the training speed of the NBN algorithm becomes much faster than the traditional LM algorithm [W09,ISIE10].

In the NBN algorithm, the quasi-Hessian matrix can be computed on fly when training patterns are applied. Moreover, it has the special advantage for applications which require dynamically changing the number of training patterns. There is no need to repeat the entire multiplication of $\mathbf{J}^T\mathbf{J}$, but only add to or subtract from the quasi-Hessian matrix. The quasi-Hessian matrix can be modified as patterns are applied or removed.

There are two implementations of the NBN algorithm on the website: http://www.eng.auburn.edu/~wilambm/nnt/index.htm. MATLAB® version can handle arbitrarily connected networks, but the Jacobian matrix is computed and stored [WCHK07]. In the C++ version [YW09], all new features of the NBN algorithm mentioned in this chapter are implemented.

References

[AS99] J. R. Alvarez-Sanchez, Injecting knowledge into the solution of the two-spiral problem. *Neural Compute and Applications*, 8, 265–272, 1999.

[AW95] T. J. Andersen and B. M. Wilamowski, A modified regression algorithm for fast one layer neural network training, *World Congress of Neural Networks*, Washington DC, July 17–21, 1995, Vol. 1, 687–690.

[BDA03] B. M. Wilamowski, D. Hunter, and A. Malinowski, Solving parity-N problems with feedforward neural networks. *Proceedings of the 2003 IEEE IJCNN*, pp. 2546–2551, IEEE Press, 2003.

[BUD09] B. M. Wilamowski, Neural Networks or Fuzzy Systems, Workshop on Intelligent Systems, Budapest, Hungary, August 30, 2009, pp. 1–12.

[CKOZ06] L. J. Cao, S. S. Keerthi, Chong-Jin Ong, J. Q. Zhang, U. Periyathamby, Xiu Ju Fu, and H. P. Lee, Parallel sequential minimal optimization for the training of support vector machines, *IEEE Transactions on Neural Networks*, 17(4), 1039–1049, April 2006.

[ISIE10] B. M. Wilamowski, Efficient Neural Network Architectures and Advanced Training Algorithms, *ICIT10 3rd International Conference on Information Technologies*, Gdansk, Poland June 28–30, 2010, pp. 345–352.

[L05] David C. Lay, *Linear Algebra and its Applications*, Addison-Wesley Publishing Company, 3rd version, pp. 124, July, 2005.

[L44] K. Levenberg, A method for the solution of certain problems in least squares. *Quarterly of Applied Mathematics*, 5, 164–168, 1944.

[M63] D. Marquardt, An algorithm for least-squares estimation of nonlinear parameters. *SIAM J. Appl. Math.*, 11(2), 431–441, June 1963.

[N89] Robert Hecht Nielsen, Theory of the back propagation neural network. *Proceedings of the 1989 IEEE IJCNN*, pp. 1593–1605, IEEE Press, New York, 1989.

[PE10] B. M. Wilamowski, Special Neural Network Architectures for Easy Electronic Implementations, POWERENG.2009, Lisbon, Portugal, March 18–20, 2009, pp. 17–22.

[PLI08] Jian-Xun Peng, Kang Li, and G. W. Irwin, A new Jacobian matrix for optimal learning of single-layer neural networks, *IEEE Transactions on Neural Networks*, 19(1), 119–129, January 2008.

[TM94] M. T. Hagan and M. Menhaj, Training feedforward networks with the Marquardt algorithm. *IEEE Transactions on Neural Networks*, 5(6), 989–993, 1994.

[W09] B. M. Wilamowski, Neural network architectures and learning algorithms, *IEEE Industrial Electronics Magazine*, 3(4), 56–63.

[WB01] B. M. Wilamowski and J. Binfet, Microprocessor implementation of fuzzy systems and neural networks, *International Joint Conference on Neural Networks (IJCNN'01)*, Washington DC, July 15–19, 2001, pp. 234–239.

[WCHK07] B. M. Wilamowski, N. Cotton, J. Hewlett, and O. Kaynak, Neural network trainer with second order learning algorithms, *Proceedings of the International Conference on Intelligent Engineering Systems*, June 29, 2007–July 1, 2007, pp. 127–132.

[WCKD08] B. M. Wilamowski, N. J. Cotton, O. Kaynak, and G. Dundar, Computing gradient vector and Jacobian matrix in arbitrarily connected neural networks, *IEEE Transactions on Industrial Electronics*, 55(10), 3784–3790, October 2008.

[WCM99] B. M. Wilamowski, Y. Chen, and A. Malinowski, Efficient algorithm for training neural networks with one hidden layer, presented at *1999 International Joint Conference on Neural Networks (IJCNN'99)*, Washington, DC, July 10–16, 1999, pp. 1725–1728.

[WH10] B. M. Wilamowski and H. Yu, Improved computation for Levenberg Marquardt training, *IEEE Transactions on Neural Networks*, 21(6), 930–937, June 2010.

[WHY10] B. M. Wilamowski and H. Yu, Neural Network Learning without Backpropagation, *IEEE Transactions on Neural Networks*, 21(11), 1793–1803, November 2010.

[WT93] B. M. Wilamowski and L. Torvik, Modification of gradient computation in the back-propagation algorithm, presented at *Artificial Neural Networks in Engineering (ANNIE'93)*, St. Louis, MI, November 14–17, 1993.

[YW09] Hao Yu and B. M. Wilamowski, C++ implementation of neural networks trainer, *13th International Conference on Intelligent Engineering Systems (INES-09)*, Barbados, April 16–18, 2009.

14

Accelerating the Multilayer Perceptron Learning Algorithms

Sabeur Abid
University of Tunis

Farhat Fnaiech
University of Tunis

Barrie W. Jervis
Sheffield Hallam University

14.1 Introduction

Feedforward neural networks such as the multilayer perceptron (MLP) are some of the most popular artificial neural network structures being used today. MLPs have been the subject of intensive research efforts in recent years because of their interesting learning and generalization capacity and applicability to a variety of classification, approximation, modeling, identification, and control problems.

The classical, the simplest, and the most used method for training an MLP is the steepest descent backpropagation algorithm, known also as the standard backpropagation (SBP) algorithm. Unfortunately, this algorithm suffers from a number of shortcomings, mainly the slow learning rate. Therefore, many researchers have been interested in accelerating the learning with this algorithm, or in proposing new, fast learning algorithms.

Fast learning algorithms constituted an appealing area of research in 1988 and 1989. Since then, there have been many attempts to find fast training algorithms, and consequently a number of learning algorithms have been proposed with significant progress being made on these and related issues.

In this chapter, a survey of different neural network fast training procedures is presented. More than a decade of progress in accelerating the feedforward neural network learning algorithms is reviewed. An overview of further up-to-date new techniques is also discussed. Different algorithms and techniques are presented in unified forms and are discussed with particular emphasis on their corresponding behavior, including the reduction of the iteration number, and their computational complexities, generalization capacities, and other parameters. Experimental results on benchmark applications are delivered, which allows a comparison of the performances of some algorithms with respect to others.

14.2 Review of the Multilayer Perceptron

Many models of neurons have been proposed in the literature [1,2]. The one most used is called the perceptron, and is given in Figure 14.1.

The subscript j stands for the number of the neurons in the network while s is the number of the corresponding layers. Based on this model, the MLP is constructed, as indicated in Figure 14.2.

Each layer consists of n_s neurons (n_s:1,…,L) and $n_{s-1}+1$ inputs. The first input of each layer is a bias input (typical values can be equal to 0.5 or 1). The first layer with n_0+1 inputs is the input layer. The Lth layer with n_L nodes is the output layer.

MLP rules:

For any neuron j in a layer s, the output signal is defined by

$$y_j^{[s]} = f\left(u_j^{[s]}\right) \tag{14.1}$$

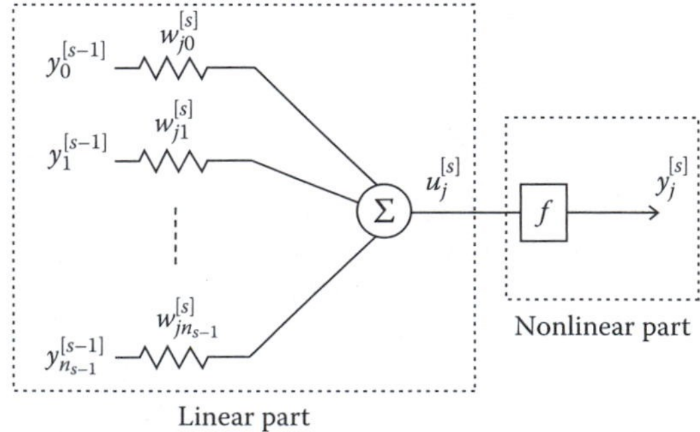

FIGURE 14.1 A model for a single neuron (perceptron) in an MLP.

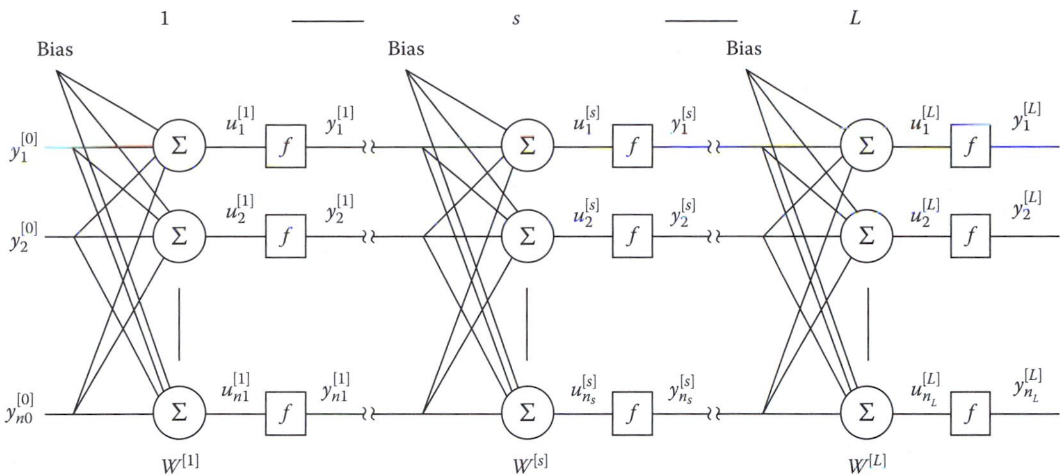

FIGURE 14.2 Fully connected feedforward MLP.

where $u_j^{[s]}$ is the linear output signal and f is an activation function assumed to be differentiable over $\mathfrak{R}$:

$$u_j^{[s]} = \sum_{i=0}^{n_{s-1}} w_{ji}^{[s]} y_i^{[s-1]} = \left(W_j^{[s]}\right)^T Y^{[s-1]} \tag{14.2}$$

where

$W_j^{[s]} = [w_{j0}^{[s]}, w_{j1}^{[s]}, \ldots, w_{jn_{s-1}}^{[s]}]^T$ for $j = 1 \ldots n_s$; $s = 1 \ldots L$

$Y^{[s]} = \left[y_0^{[s]}\ y_1^{[s]} \ldots y_{n_s}^{[s]}\right]^T$ for $s = 0 \ldots L - 1$

The terms $y_0^{[s]}$ depict the bias terms.

14.3 Review of the Standard Backpropagation Algorithm

The backpropagation algorithm is the standard algorithm used for training an MLP. It is a generalized least mean square (LMS) algorithm that minimizes a cost function equal to the sum of the squares of the error signals between the actual and the desired outputs.

Let us define the total error function for all output neurons and for the current pattern as

$$E_p = \sum_{j=1}^{n_L} \frac{1}{2}\left(e_j^{[L]}\right)^2 \tag{14.3}$$

where $e_j^{[L]}$ depicts the nonlinear error of the jth output unit and it is given by

$$e_j^{[L]} = d_j^{[L]} - y_j^{[L]} \tag{14.4}$$

where $d_j^{[L]}$ and $y_j^{[L]}$ are the desired and the current output signals of the corresponding unit in the output layer, respectively.

In the literature, E_p is called the performance index and is expressed by the more general form [2,40]:

$$E_p = \sum_{j=1}^{n_L} \sigma(e_j) \tag{14.5}$$

In this case, $\sigma(.)$ becomes an error output function typically convex called the loss or weighting function. In the case of $\sigma(e_{jp}) = 1/2e_{jp}^2$, we retrieve the ordinary L_2-norm criterion [2].

The criterion to be minimized is the sum of all error over all the training patterns:

$$E = \sum_{p=1}^{N} E_p \tag{14.6}$$

The SBP algorithm is derived using the gradient descent method expressed by

$$\Delta w_{ji}^{[s]}(k) = w_{ji}^{[s]}(k) - w_{ji}^{[s]}(k-1) = -\mu \left.\frac{\partial E}{\partial w_{ji}^{[s]}}\right|_{w_{ji}^{[s]} = w_{ji}^{[s]}(k)} \tag{14.7}$$

where

μ is a positive real constant called the learning rate.
Note that minimizing E is equivalent to minimizing E_p.
In what follows, the subscript k will be omitted for simplification

In vector notation, we can express Equation 14.7 for the neuron j (i.e., for all synaptic weights of the jth neuron) in a layer s as follows:

$$\Delta W_j^{[s]} = -\mu\left(\nabla_j^{[s]}(E_p)\right) \tag{14.8}$$

where

$$\nabla_j^{[s]}(E_p) = \left[\frac{\partial E_p}{\partial w_{j0}^{[s]}}, \frac{\partial E_p}{\partial w_{j1}^{[s]}}, \ldots, \frac{\partial E_p}{\partial w_{jn_{s-1}}^{[s]}}\right]^T \tag{14.9}$$

The evaluation of this gradient vector changes according to whether the derivative is computed with respect to the synaptic weights of the output layer or any of the hidden layers.

For a neuron j in the output layer L,

$$\nabla_j^{[L]}(E_p) = \left[f'\left(u_j^{[L]}\right)\ e_j^{[L]}\ \ y_0^{[L-1]}, f'\left(u_j^{[L]}\right)\ e_j^{[L]}\ \ y_1^{[L-1]}, \ldots, \left(u_j^{[L]}\right)\ e_j^{[L]}\ \ y_{n_{L-1}}^{[L-1]}\right]^T \tag{14.10}$$

For the entire number of the output neurons, the gradient signals vector is given by

$$\nabla^{[L]}(E_p) = \left[\left(\nabla_1^{[L]}\right)^T, \left(\nabla_2^{[L]}\right)^T, \ldots, \left(\nabla_{n_L}^{[L]}\right)^T\right]^T \tag{14.11}$$

For a neuron j in a hidden layer s, Equation 14.9 becomes

$$\nabla_j^{[s]}(E_p) = \left[f'\left(u_j^{[s]}\right)\ e_j^{[s]}\ \ y_0^{[s-1]}, f'\left(u_j^{[s]}\right)\ e_j^{[s]}\ \ y_1^{[s-1]}, \ldots, f'\left(u_j^{[s]}\right)\ e_j^{[s]}\ \ y_{n_{s-1}}^{[s-1]}\right]^T \tag{14.12}$$

where $s = (L-1), \ldots, 1$ and

$$e_j^{[s]} = \sum_{r=1}^{n_{s+1}} \left(e_r^{[s+1]} w_{rj}^{[s+1]}\right) \tag{14.13}$$

$e_j^{[s]}$ is assumed to be the estimated error in the hidden layer. Note that these estimated errors are computed in a backward direction from layer $(L-1)$ to 1.

For all the neurons in the sth layer,

$$\nabla^{[s]}(E_p) = \left[\left(\nabla_1^{[s]}\right)^T, \left(\nabla_2^{[s]}\right)^T, \ldots, \left(\nabla_{n_L}^{[s]}\right)^T\right]^T \tag{14.14}$$

Let us define the general gradient vector of E_p for all neural network synaptic weights as

$$\nabla E_p = \nabla E_p(W) = \left[\left(\nabla^{[1]}(E_p)\right)^T, \left(\nabla^{[2]}(E_p)\right)^T, \ldots, \left(\nabla^{[L]}(E_p)\right)^T\right]^T \tag{14.15}$$

where

$$W = \left[\left(W_1^{[1]}\right)^T, \left(W_2^{[1]}\right)^T, \ldots, \left(W_{n_1}^{[1]}\right)^T, \left(W_1^{[2]}\right)^T, \left(W_2^{[2]}\right)^T, \ldots,\right.$$
$$\left.\left(W_{n_2}^{[2]}\right)^T, \left(W_1^{[L]}\right)^T, \left(W_2^{[L]}\right)^T, \ldots, \left(W_{n_L}^{[L]}\right)^T\right]^T \tag{14.16}$$

The general updating rule for all the synaptic weights in the network becomes

$$\Delta W(k) = \mu \nabla E_p(W) \tag{14.17}$$

As mentioned above, the learning of the MLP using the SBP algorithm is plagued by slow convergence. Eight different approaches for increasing the convergence speed are summarized as follows.

14.4 Different Approaches for Increasing the Learning Speed

14.4.1 Weight Updating Procedure

We distinguish the online (incremental) method and the batch method. In the first, a pattern (a learning example) is presented at the input and then all weights are updated before the next pattern is presented. In the batch method, the weight changes Δw are accumulated over some number (usually all) of the learning examples before the weights are actually changed.

Practically, we have found that the convergence speeds of these two methods are similar.

14.4.2 Principles of Learning

In the learning phase, in each iteration, patterns can be selected arbitrarily or in a certain order. The order of presenting data during the learning phase affects the learning speed most. In general, presenting data in a certain order yields slightly faster training.

14.4.3 Estimation of Optimal Initial Conditions

In the SBP algorithm, the user always starts with random initial weight values. Finding optimal initial weights to start the learning phase can considerably improve the convergence speed.

14.4.4 Reduction of the Data Size

In many applications (namely, in signal processing or image processing), one has to deal with data of huge dimensions. The use of these data in their initial forms becomes intractable. Preprocessing data, for example, by extracting features or by using projection onto a new basis speeds up the learning process and simplifies the use of the NN.

14.4.5 Estimation of the Optimal NN Structure

Usually, the NN structure is evaluated by a trial-and-error approach. Starting with the optimal NN structure, i.e., the optimal number of the hidden layers and their corresponding number of neurons, would considerably reduce the training time needed. The interested reader can find a chapter on MLP pruning algorithms used to determine the optimal structure in this handbook. However it was recently shown that MLP architectures, as shown in Figure 14.2, are not as powerful as other neural network architectures such as FCC and BMLP networks [6,13,20].

14.4.6 Use of Adaptive Parameters

The use of an adaptive slope of the activation function or a global adaptation of the learning rate and/or momentum rate can increase the convergence speed in some applications.

14.4.7 Choice of the Optimization Criterion

In order to improve the learning speed or the generalization capacities, many other sophisticated optimization criteria can be used. The standard (L_2-norm) least squares criterion is not the only cost function to be used for deriving the synaptic weights. Indeed, when signals are corrupted with non-Gaussian noise, the standard L_2-norm cost function performs badly. This will be discussed later.

14.4.8 Application of More Advanced Algorithms

Numerous heuristic optimization algorithms have been proposed to improve the convergence speed of the SBP algorithm. Unfortunately, some of these algorithms are computationally very costly and time consuming, i.e., they require a large increase of storage and computational cost, which can become unmanageable even for a moderate size of neural network.

As we will see later, the first five possibilities depend on the learning algorithm, and despite the multiple attempts to develop theories that help to find optimal initial weights or initial neural network structures etc., there exist neither interesting results nor universal rules or theory allowing this.

The three remaining possibilities are related to the algorithm itself. The search for new algorithms or new optimization functions has made good progress and has yielded good results.

In spite of the big variations in the proposed algorithms, they fall roughly into two categories.

The first category involves the development of algorithms based on first-order optimization methods (FOOM). This is the case for the SBP algorithm developed above.

Assume that $E(\underline{w})$ is a cost function to minimize with respect to the parameter vector $\underline{w}$, and $\nabla E(\underline{w})$ is the gradient vector of $E(\underline{w})$ with respect to $\underline{w}$. The FOOMs are based on the following rule:

$$\Delta\underline{w} = -\mu\frac{\partial E(\underline{w})}{\partial\underline{w}} = -\mu\nabla E(\underline{w}) \qquad (14.18)$$

which is known in the literature as the steepest descent algorithm or the gradient descent method [1–4]. μ is a positive constant that governs the amplitude of the correction applied to w in each iteration and thus governs the convergence speed.

The second category involves the development of algorithms based on the second-order optimization methods (SOOM). This aims to accelerate the learning speed of the MLP, too. All of these methods are based on the following rule:

$$\Delta(\underline{w}) = -\left[\nabla^2 E(\underline{w})\right]^{-1}\nabla E(\underline{w}) \qquad (14.19)$$

where $\nabla^2 E(\underline{w})$ is the matrix of second-order derivative of $E(\underline{w})$ with respect to $\underline{w}$, known as the Hessian matrix. This method is known in the literature as Newton's method [21,22]. It is known by its high convergence speed but it needs the computation of the Hessian matrix inverse $[\nabla^2 E(\underline{w})]^{-1}$. Dimensions of this matrix grow with those of the network size, and in practice it is a very difficult task to find the exact value of this matrix. For this reason, a lot of research has been focused on finding an approximation to this matrix (i.e. the Hessian matrix inverse). The most popular approaches have used quasi-Newton methods (i.e., the conjugate gradient of secant methods). All of them proceed by approximating $[\nabla^2 E(\underline{w})]^{-1}$, and are considered to be more efficient, but their storage and their computational requirements go up as the square of the network size [7]. As mentioned above, these algorithms are faster than those based on FOOM, but because of their complexity, neural networks researchers and users prefer using the SBP algorithm to benefit from its simplicity and are very interested in finding a modified version of this algorithm that is faster than the SBP version.

14.5 Different Approaches to Speed Up the SBP Algorithm

Several parameters in the SBP algorithm can be updated during the learning phase for the purpose of accelerating the convergence speed. These parameters are the learning coefficient μ, the momentum term Ω, and even the activation function slope. We summarize below all the suggested ideas in this regard.

14.5.1 Updating the Learning Rate

From about 1988 several authors have been interested in improving the convergence speed of the SBP algorithm by updating the learning rate [2,3,5,12]. Several rules have been proposed for this purpose. Some of them are effectively interesting but others do not differ too much from the algorithm with constant step.

We shall review the multiple suggestions for the updating rules of the learning rate μ and/or the momentum term Ω in the order of their appearance:

- In 1988, two approaches were published, namely, the quickprop (QP) algorithm by Fahlman in [24] and the delta bar delta (DBD) rule by Jacobs in [10].
- In the QP algorithm, the activation function is equal to

$$f(u_j) = \frac{1}{(1+e^{-au_j})} + 0.1u_j \tag{14.20}$$

where u_j is given by (14.2).

The simplified QP algorithm can be summarized in the following rules [2]:

$$\Delta w_{ji}(k) = \begin{cases} \gamma_{ji}^{(k)} \Delta w_{ji}(k-1), & \text{if } \Delta w_{ji}(k-1) \neq 0 \\ \mu_0 \dfrac{\partial E}{\partial w_{ji}}, & \Delta w_{ji}(k-1) = 0 \end{cases} \tag{14.21}$$

where

$$\gamma_{ji}^{(k)} = \min \left\{ \frac{\dfrac{\partial E(w^{(k)})}{\partial w_{ji}}}{\dfrac{\partial E(w^{(k-1)})}{\partial w_{ji}} - \dfrac{\partial E(w^{(k)})}{\partial w_{ji}}}, \gamma_{\max} \right\} \tag{14.22}$$

and the parameters $\gamma_{\max}$ and μ_0 are typically chosen equal to $0.01 \leq \mu_0 \leq 0.6$ and $\gamma_{\max} = 1.75$.

- The DBD algorithm assumes that each synaptic coefficient has its own learning rate. The updating rule of each learning coefficient $\mu_{ji}^{[s]}$ depends on the sign of the local gradient $\delta_{ji}^{[s]}(k) = \partial E/\partial w_{ji}^{[s]}$ in each iteration:

$$\mu_{ji}^{[s]}(k) = \begin{cases} \mu_{ji}^{[s]}(k-1)+\alpha & \text{if } \nu_{ji}^{[s]}(k-1)\delta_{ji}^{[s]}(k) > 0 \\ \mu_{ji}^{[s]}(k-1) & \text{if } \nu_{ji}^{[s]}(k-1)\delta_{ji}^{[s]}(k) < 0 \\ 0 & \text{otherwise} \end{cases} \tag{14.23}$$

where
  α and β are arbitrarily parameters
  $\nu_{ji}^{[s]}(k) = (1-\nu)\delta_{ji}^{[s]}(k) + \lambda\nu_{ji}^{[s]}(k-1)$
  λ is a positive real smaller than 1 [2]

- In 1990, the Super SAB algorithm was proposed by Tollenaere in Ref. [25]. It represents a slight modification of the DBD one. Each synaptic weight w_{ji} has its own learning rate such that

$$\Delta w_{ji}(k) = -\mu_{ji}^{(k)}\frac{\partial E}{\partial w_{ji}} + \gamma \Delta w_{ji}(k-1) \tag{14.24}$$

where

$$\mu_{ji}^{(k)} = \begin{cases} \alpha\mu_{ji}^{(k-1)} & \text{if } \dfrac{\partial E(w^{(k)})}{\partial w_{ji}}\dfrac{\partial E(w^{(k-1)})}{\partial w_{ji}} > 0 \\ \beta\mu_{ji}^{(k-1)} & \text{otherwise} \end{cases} \tag{14.25}$$

and $\alpha \approx 1/\beta$.

- In 1991, Darken and Moody in [8] have suggested a modification of the learning rate along the training phase such that

$$\mu(k) = \mu_0 \frac{1}{1+k/k_0} \tag{14.26}$$

or

$$\mu(k) = \mu_0 \frac{1+\dfrac{c}{\mu_0}\dfrac{k}{k_0}}{1+\dfrac{c}{\mu_0}\dfrac{k}{k_0}+k_0\left(\dfrac{k}{k_0}\right)^2} \tag{14.27}$$

where
  c and k_0 are positive constants
  μ_0 is the initial value of the learning coefficient μ

These are called the "search-then-converge strategy: STCS" algorithms.

The plots of the evolution of μ versus time for formula (14.26) for two different values of k_0 are shown in Figure 14.3.

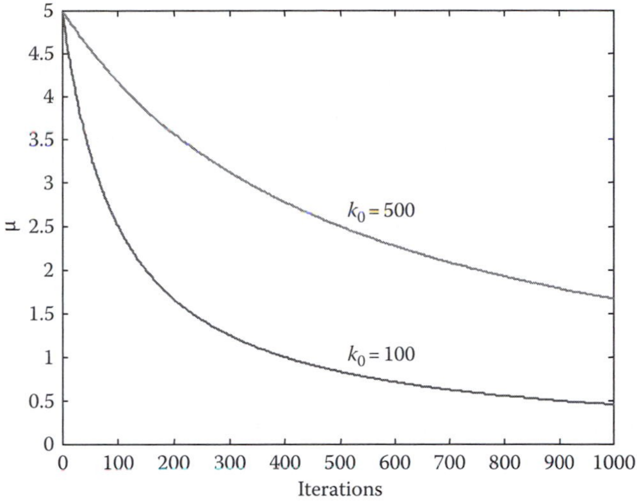

FIGURE 14.3 Evolution of μ versus time for different values of k_0.

Note that at the beginning of the learning phase (search phase) μ is relatively large. This ensures the rapidity of the algorithm when starting. As the training progresses, μ becomes smaller (the convergence phase). This avoids oscillations at the end of the learning phase and ensures a smooth convergence, but it decreases the speed of convergence.

- In 1992, the famous conjugate gradient method was developed separately by Johansson [11] and Battiti [26]. The major advantage of the method lies in the use of the second-order derivative of the error signal. In the case of a quadratic error surface, the method needs only *N* iterations for convergence (*N* is the number of synaptic weights of the neural network). The algorithm can be summarized into [11]

$$\Delta w(k) = \mu(k)dr(k) \tag{14.28}$$

where $dr(k)$ is the descent direction. Many expressions have been proposed to update the descent direction. The most widely used formula is given by

$$dr(k) = -\nabla E(w(k)) + \beta(k)dr(k-1) \tag{14.29}$$

where
$\nabla E(w(k)) = \partial E/\partial w(k)$
$\beta(k)$ is in general computed using the Polak–Ribiere formula:

$$\beta(k) = \frac{[\nabla E(w(k)) - \nabla E(w(k-1))]^T \nabla E(w(k))}{\nabla E(w(k-1))^T \nabla E(w(k-1))} \tag{14.30}$$

- In 1994, two other approaches were developed, the RPROP algorithm, Riedmiller [27] and the accelerated backpropagation algorithm (ABP algorithm), Parlos et al. [23].

The RPROP algorithm is characterized by the use of the sign of the gradient instead of its numeric value in the updating equations:

$$\Delta w_{ji}(k) = -\Delta_{ji}(k)sign(\nabla E(w_{ji}(k))) \tag{14.31}$$

where

$$\Delta_{ji}(k)=\begin{cases}1.2\Delta_{ji}(k-1) & \text{if } \nabla E(w_{ji}(k))\nabla E(w_{ji}(k-1))>0\\ 0.5\Delta_{ji}(k-1) & \text{if } \nabla E(w_{ji}(k))\nabla E(w_{ji}(k-1))<0\\ 0 & \text{otherwise}\end{cases} \tag{14.32}$$

In the ABP algorithm, the batch update principle is suggested, and weights are changed according to

$$\Delta w_{ji}(k)=-\rho(E)\frac{1}{\left\|\partial E/\partial w_{ji}\right\|^2}\frac{\partial E}{\partial w_{ji}} \tag{14.33}$$

where $\rho(E)$ is a function of the error signal. Different expressions were suggested for this, such as

$$\rho(E)=\mu;\quad \rho(E)=\mu E;\quad \text{or}\quad \rho(E)=\mu\tanh\left(\frac{E}{E_0}\right)$$

where μ and E_0 are constant, nonnegative numbers. The choice of $\rho(E)$ defines the decay mode in the search space.

- In 1996, another algorithm called SASS was proposed by Hannan et al. [28]. The updating equations are the same as for RPROP (14.31), but $\Delta_{ji}(k)$ is given by

$$\Delta_{ji}(k)=\begin{cases}2\Delta_{ji}(k-1) & \text{if } g_{ji}(k)g_{ji}(k-1)\geq 0 \quad\text{and}\quad g_{ji}(k)g_{ji}(k-2)\geq 0\\ 0.5\Delta_{ji}(k-1) & \text{otherwise}\end{cases} \tag{14.34}$$

This approach seems to be similar to that for RPROP, but experimental results show that it behaves differently in several cases.

Referring to the works of Alpsan [29], Hannan [31], and Smagt [30], we can conclude that the RPROP algorithm is the fastest among the eight described above. In [31], Hannan et al. state that *conclusions about the general performance of the algorithms cannot be drawn*. Indeed, generalized conclusions about algorithm properties and speed can neither be based on some simulation results nor on inspecting some statistics.

- In 2006, a new approach was proposed by L. Behera et al. [12] using the Lyapunov function. This approach seems to be more effective than those mentioned above. The following optimization criterion

$$E=\frac{1}{2}\sum_{p=1}^{n_L}\left(d_j^{[L]}-y_j^{[L]}\right)^2 \tag{14.35}$$

can be expressed as a Lyapunov function such that

$$V_1=\frac{1}{2}(\tilde{y}^T\tilde{y}) \tag{14.36}$$

where

$$\tilde{y}=\left[d_1^{[L]}-y_1^{[L]},\ldots,d_{n_L}^{[L]}-y_{n_L}^{[L]}\right]^T$$

Then, the learning rate is updated during the learning phase according to

$$\eta_a = \frac{\mu \|\tilde{y}\|^2}{\|J^T \tilde{y}\|^2} \tag{14.37}$$

where μ is a starting constant to be chosen arbitrarily at the beginning of the learning phase and J is defined as

$$J = \frac{\partial y^{[L]}}{\partial W} \in \Re^{1\times m} \tag{14.38}$$

and represents the instantaneous value of the Jacobian matrix.

This method has a significantly high convergence speed, owing to the updating of the learning rate using information provided by the Jacobian matrix.

14.5.2 Updating the Activation Function Slope

In [9], Krushke et al. have shown that the speed of the SBP algorithm can be increased by updating the slope of the activation function, a.

To find the updating rule for the slope, we apply the gradient method with respect to a:

$$\Delta a(k) = -\mu_a \frac{\partial E}{\partial a} \tag{14.39}$$

14.6 Some Simulation Results

To compare the performance of the different algorithms presented, it is first necessary to define the sensitivity to the initialization of the synaptic weights and the generalization capacity of a neural network.

14.6.1 Evaluation of the Sensitivity to the Initialization of the Synaptic Weights

It is known that the convergence of all iterative search methods depends essentially on the starting point chosen. In our case, these are the initial synaptic coefficients (weights). In the case of poor initialization, the iterative algorithm may diverge. As we will see below, there is no rule for defining or choosing a priori the initial departure point or even the range in which it lies. There are many algorithms that are very sensitive to the initialization while others are less sensitive. To study the sensitivity to the initialization of the weights, we have to test the convergence of the algorithm for a huge number of different, randomly initialized trials (Monte Carlo test). By a trial, we mean one training phase with one random weight initialization. The ending criterion is equal to the mean squared error for all the output neurons and for all the training patterns:

$$E = \frac{1}{N} \sum_{p=1}^{N} \sum_{i=1}^{n_L} \left(e_i^{(L)}\right)^2 \tag{14.40}$$

where N is the total number of training patterns.

Each training phase is stopped if E reaches a threshold fixed beforehand. This threshold is selected depending on the application, and it will be denoted in what follows as *ending_threshold*. Each learning trial must be stopped if the ending threshold is not reached after an iteration number fixed a priori. The choice of this number also depends on the application. This number is denoted by *iter_number*. The convergence is assumed to have failed if *iter_number* is reached before the value of *ending_threshold*.

The sensitivity to weight initialization is evaluated via the proposed formula:

$$S_w(\%) = 100 \cdot \left(1 - \frac{\text{Number of convergent trials}}{\text{Total number of trials}}\right) \tag{14.41}$$

Thus, the smaller S_w, the less sensitive is the algorithm to weight initialization.

14.6.2 Study of the Generalization Capability

To study the generalization capability (G_c) after the training phase, we should present new patterns (testing patterns), whose desired outputs are known, to the network and compare the actual neural outputs with the desired ones. If the norm of the error between these two inputs is smaller than a threshold chosen beforehand (denoted in what follows as *gen_threshold*), then the new pattern is assumed to be recognized with success.

The generalization capability is evaluated via the following proposed formula:

$$G_c(\%) = 100 \cdot \frac{\text{Recognized pattern number}}{\text{Total testing patterns number}} \tag{14.42}$$

Thus, the larger G_c, the better the generalization capability of the network.

14.6.3 Simulation Results and Performance Comparison

In these simulation tests, we have compared the performances of the presented algorithms with that of the conventional SBP algorithm. For this purpose, all algorithms are used to train networks for the same problem. We present two examples here, the 4-b parity checker (logic problem), and the circle-in-the-square problem (analog problem). For all algorithms, learning parameters (such as μ, λ, β,…) are selected after many trials (100) to maximize the performance of each algorithm. However, an exhaustive search for the best possible parameters is beyond the scope of this work, and optimal values may exist for each algorithm. In order to make suitable comparisons, we kept the same neural network size for testing all the training algorithms.

The problem of choosing the learning parameters

Like all optimization methods that are based on the steepest descent of the gradient, the convergence of the algorithms is strongly related to

- The choice of the learning parameters such as μ, λ, β, …
- The initial conditions, namely, the initial synaptic coefficients.

The learning parameters govern the amplitudes of the correction terms and consequently affect the stability of the algorithm. To date, there is no practical guideline that allows the computation or even the choice of these parameters in an optimal way. In the literature, we can only find some attempts (even heuristic) to give formulae which contribute to speeding up the convergence or to stabilizing the algorithm [44–48].

The same applies to the initial choice of the synaptic coefficients. They are generally chosen in an arbitrarily manner, and there are no laws for defining their values a priori.

Note that the initial synaptic coefficients fix the point from which the descent will start in the opposite direction of the gradient. Consequently, the algorithm will converge in reasonable time if this starting point is near an acceptable local minimum or preferably a global minimum. Otherwise the algorithm will be trapped in the nearest local minimum or will not converge in a reasonable time.

In conclusion, like many other neural network parameters (such as the hidden layer number, the number of neurons per hidden layer, the nature of the activation function etc.), adequate values of these parameters can only be found by experiment assisted by expertise.

14.6.3.1 4-b Parity Checker

The aim of this application is to determine the parity of a 4-bit binary number. The neural network inputs are logic values (0.9 for the higher level and −0.9 for the lower level). At each iteration, we present to the network the 16 input combinations with their desired outputs (0.1 for the lower level and 0.9 for the higher one). The network size is (4,8,2,1), i.e., 4 inputs, two hidden layers with 8 and 2 neurons, and one output neuron. The synaptic coefficients are initialized randomly in the range [−3, +3].

To evaluate the sensitivity to initialization of the weights S_w, we have chosen *iter_number* = 500 and *ending_threshold* = 10^{-3}.

For the generalization test G_c, we have presented 4-bit distorted numbers to the network. The distortion rate with respect to the exact 4-bit binary numbers is about 30%, and *gen_threshold* = 0.1.

Note that

- We have followed the same procedure to determine the performances of each algorithm for all the simulation examples.
- Although these results were obtained after several experiments (in order to maximize the performance of each algorithm), they may be slightly changed depending on the values of *ending_threshold*, *iter_number*, *gen_threshold*, and even on the range of the initial synaptic coefficients and their statistical distribution.

TABLE 14.1 Comparison of the Performance of the Different Algorithms with respect to the SBP Algorithm for the 4-b Parity Checker

	S_w (%)	G_c (%)
SBP	74	88
QP	76	88
DBD	76	89
Super SAB	75	87
STCS	74	88
CG	78	96
RPROP	78	94
ABP	75	90
SASS	76	89
LSBP	79	97
VAFS	72	81

Tables 14.1 and 14.2 summarize the performance of all the algorithms for this application. From these results, we note that the new MLMSF network is less affected by the choice of the initial weights and has a good generalization capability with respect to the SBP algorithm.

14.6.3.1.1 Conclusion

From these results, one concludes that all the proposed algorithms have almost the same performance in the sensitivity to the initialization, the generalization capacity, and the time gain. The CG, RPROP, and the LSBP algorithms have a slight superiority with respect to the other algorithms. However, the speed of convergence remains less than expected.

TABLE 14.2 Improvement Ratios with respect to the SBP Algorithm

	In Iteration	In Time
QP	1.18	1.10
DBD	1.21	1.20
Super SAB	1.25	1.13
STCS	1.23	1.21
CG	1.31	1.25
RPROP	1.28	1.24
ABP	1.19	1.11
SASS	1.20	1.11
LSBP	1.28	1.23
VAFS	1.13	1.07

14.6.3.2 Circle-in-the-Square Problem

In this application, the neural network has to decide whether a point with coordinates (x, y), varying from −0.5 to +0.5, is in the circle of radius equal to 0.35 [1]. Training patterns, which alternate between the two classes (inside and the outside the circle), are presented to the network. In each iteration, we present 100 input/output patterns to the network. The networks size is (2,8,2,1) and the synaptic coefficients are initialized randomly in the range [−1,+1].

To evaluate S_w, we have chosen *iter_number* = 200, *ending_threshold* = 10^{-2}. For G_c, we have presented to the network a new coordinate (x, y), and we have chosen *gen_threshold* = 0.1.

Tables 14.3 and 14.4 summarize the performances of the algorithms for this application.

14.6.3.2.1 Conclusion

For this analog problem, one notes that all the algorithms have almost the same performance. For this reason, several algorithms were proposed to increase the convergence of the learning algorithms for the multilayer perceptron networks.

TABLE 14.3 Comparison of the Performances of the Different Algorithms with respect to the SBP Algorithm for the Circle-in-the-Square Problem

	S_w (%)	G_c (%)
SBP	66	84
QP	68	83
DBD	68	83
Super SAB	68	85
STCS	71	86
CG	75	91
RPROP	74	91
ABP	69	82
SASS	68	84
LSBP	77	89
VAFS	68	76

14.7 Backpropagation Algorithms with Different Optimization Criteria

The choice of the training algorithm determines the rate of convergence, the time required to reach the solution, and the optimality of the latter.

In the field of neural networks, training algorithms may differ from that of the SBP algorithm by the optimization criterion and/or by the method with which updating equations are derived.

Different forms of optimization criterion have been proposed in order to increase the convergence speed and/or to improve the generalization capability, and two algorithms using more advance optimization procedures have been published.

In [14] Karayiannis et al. have developed the following generalized criterion for training an MLP:

$$E_p = (1-\lambda)\sum_{j=1}^{n_L}\sigma_1(e_{jp}) + \lambda\sum_{j=1}^{n_L}\sigma_2(e_{jp}) \tag{14.43}$$

where $\sigma_1(e_{jp})$ and $\sigma_2(e_{jp})$ are known as loss functions that must be convex and differentiable, and $\lambda \in [0,1]$.

Inspired by this equation we have proposed [1] a new learning algorithm that is remarkably faster than the SBP algorithm. It is based on the following criterion:

$$E_p = \sum_{j=1}^{n_L}\frac{1}{2}e_{1jp}^{[L]^2} + \sum_{j=1}^{n_L}\frac{1}{2}\lambda e_{2jp}^{[L]^2} \tag{14.44}$$

where e_1 and e_2 are the nonlinear and the linear output errors, respectively. To work with a system of linear equations, the authors in [16–19] used an inversion of the output layer nonlinearity and an estimation of the desired output the hidden layers. Then they applied the recursive least-squares (RLS) algorithm at each layer yielding a fast training algorithm. To avoid the inversion of the output nonlinearity, the authors in [15] used the standard threshold logic-type nonlinearity as an approximation of the sigmoid. These approaches yield fast training with respect to the SBP algorithm but still are approximation dependant.

At the same time, a real-time learning algorithm based on the EKF technique was developed [32–36]. In these works, a Kalman filter is assigned to each connected weight. Parameter-free tuning is the major advantage. In the following, we provide a summary of some of these algorithms.

TABLE 14.4 Improvement Ratios with respect to the SBP Algorithm

	In Iteration	In Time
QP	1.10	1.03
DBD	1.13	1.10
Super SAB	1.12	1.06
STCS	1.12	1.12
CG	1.18	1.20
RPROP	1.20	1.20
ABP	1.13	1.05
SASS	1.14	1.04
LSBP	1.18	1.17
VAFS	1.12	1.03

14.7.1 Modified Backpropagation Algorithm

Based on Equation 14.43, we have developed [1] a new error function using the linear and nonlinear neuronal outputs.

Recall that the output error (called the nonlinear output error signal) is now given by

$$e_{1j}^{[s]} = d_j^{[s]} - y_j^{[s]} \tag{14.45}$$

Now, the linear output error signal can easily be found by

$$e_{2j}^{[s]} = ld_j^{[s]} - u_j^{[s]} \tag{14.46}$$

where $ld_j^{[s]}$ is given by

$$ld_j^{[s]} = f^{-1}\left(d_j^{[s]}\right) \tag{14.47}$$

This is illustrated in Figure 14.4.

The proposed optimization criterion (14.44) is given by

$$E_p = \sum_{j=1}^{n_L} \frac{1}{2} e_{1jp}^{[L]^2} + \sum_{j=1}^{n_L} \frac{1}{2} \lambda e_{2jp}^{[L]^2}$$

Applying the gradient descent method to E_p, we obtain the following updating equations for the output layer $[L]$ and the hidden layers $[s]$ from $(L-1)$ to 1, respectively:

$$\Delta w_{ji}^{[L]}(k) = \mu f'\left(u_j^{[L]}\right) e_{1jp}^{[L]} y_i^{[L-1]} + \mu\lambda e_{2jp}^{[L]} y_i^{[L-1]} \tag{14.48}$$

$$\Delta w_{ji}^{[s]} = \mu y_{ip}^{[s-1]} f'\left(u_{jp}^{[s]}\right) e_{1jp}^{[s]} + \mu\lambda y_{ip}^{[s-1]} e_{2jp}^{[s]} \tag{14.49}$$

where

$$e_{1jp}^{[s]} = \sum_{r=1}^{n_{s+1}} e_{1rp}^{[s+1]} f'(u_{rp}^{[s+1]}) w_{rp}^{[s]} \tag{14.50}$$

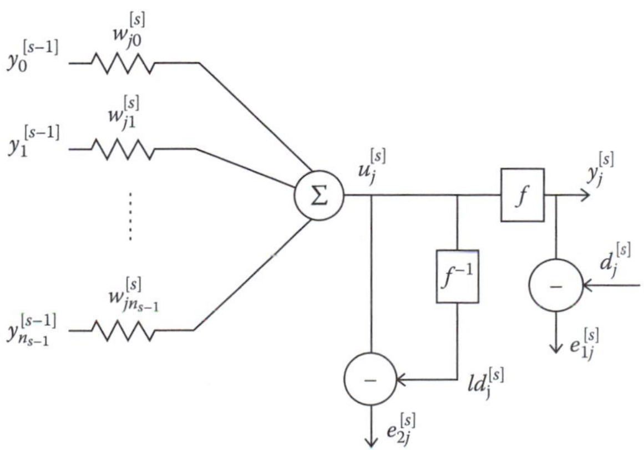

FIGURE 14.4 Finding the nonlinear and the linear error signal in a neuron.

and

$$e_{2jp}^{[s]} = f'(u_{jp}^{[s]}) \sum_{r=1}^{n_{s+1}} e_{2rp}^{[s+1]} w_{rj}^{[s+1]} \tag{14.51}$$

are assumed to be the estimates of the nonlinear and linear error signals for the hidden layers, respectively.

14.7.2 Least Squares Algorithms for Neural Network Training

The existence of the nonlinearity in the activation function makes the backpropagation algorithm a nonlinear one. If this nonlinearity can be avoided in one way or another, one can make use of all least-squares adaptive filtering techniques for solving this problem. These techniques are known to have rapid convergence properties.

The development of a training algorithm using RLS methods for NNs was first introduced in [16] and then later extended in [15,17–19].

For the purpose of developing a system of linear equations, two approaches can be used. The first is to invert the sigmoidal output node function, as in the case of Figure 14.4 [19]. The second is the use of a standard threshold logic-type nonlinearity, as shown in Figure 14.5 [15].

14.7.2.1 Linearization by Nonlinearity Inversion

Scalero et al. in [19] have proposed a new algorithm, which modifies weights based on the minimization of the *MSE* between the linear desired output and the actual linear output of a neuron.

It is shown in Figure 14.4 that a neuron is formed by a linear part (a scalar product) and a nonlinear part, and then it is possible to separate the linear and nonlinear parts to derive a linear optimization problem.

The optimization problem is then based on the following optimization criterion:

$$E(k) = \frac{1}{2} \sum_{t=1}^{k} \rho(k,t) \sum_{j=1}^{n_L} \left(le_j^{[L]} \right)^2 (t) \tag{14.52}$$

where $\rho(k,t)$ is the variable weighting sequence which satisfies

$$\rho(k,t) = \lambda(k,t)\rho(k-1,t) \tag{14.53}$$

and $\rho(k,k) = 1$.

This means that we may write

$$\rho(k,t) = \prod_{j=t+1}^{k} \lambda(j) \tag{14.54}$$

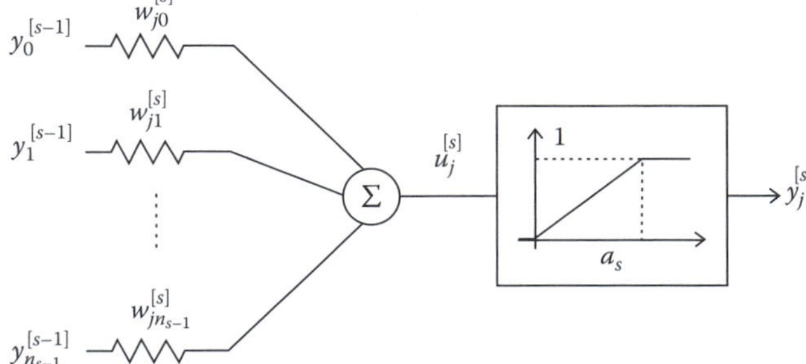

FIGURE 14.5 Standard logic-type nonlinearity.

In the most applications, a constant exponential weighting is used, i.e., $\rho(k,t) = \lambda^{(k-t)}$ where λ is a positive number, less than, but close to 1, which is called the "forgetting factor."

This error (14.52) can be minimized by taking partial derivatives of $E(k)$ with respect to each weight and equating them to zero. The result will be a set of $n_s + 1$ linear equations where $n_s + 1$ is the number of weights in the neuron.

Minimizing $E(k)$ with respect to a weight $w_{ji}^{[L]}$ produces

$$w_j^L(k) = R^{-1}(k)p(k) \tag{14.55}$$

where

$$R(k) = \sum_{p=1}^{k} y_p^{L-1}(y_p^{L-1})^T \tag{14.56}$$

and

$$p(k) = \sum_{p=1}^{k} ld_{jp}^L y_p^{L-1} \tag{14.57}$$

Except for a factor of $1/k$, (14.56) and (14.57) are estimates of the correlation matrix and correlation vec tor, respectively, and they improve with increasing k.

Both (14.56) and (14.57) can be written in a recursive form as

$$R(k) = \lambda R(k-1) + y_k^{[L-1]} y_k^{[L-1]^T} \tag{14.58}$$

and

$$p(k) = \lambda p(k-1) + ld_{jk}^L y_k^{L-1} \tag{14.59}$$

Although (14.58) and (14.59) are in a recursive form, what one needs is the recursive equation for the inverse autocorrelation matrix $R^{-1}(k)$, as required by (14.55). This can be achieved by using either the matrix inversion lemma [37] or what may be viewed as its compact form, the Kalman filter [32–36].

14.7.2.2 Linearization by Using Standard Threshold Logic-Type Nonlinearity

In this approach, the standard threshold logic-type nonlinearity shown in Figure 14.5 is used.

The neuronal outputs will be expressed by the following system:

$$y_j^{[s]} = f\left(u_j^{[s]}\right) = \begin{cases} 0 & si \quad u_j^{[s]} \le 0 \\ \dfrac{1}{a_s} u_j^{[s]} & si \quad 0 < u_j^{[s]} < a_s \\ 1 & si \quad u_j^{[s]} \ge a_s \end{cases} \tag{14.60}$$

The optimization criterion at the iteration k which incorporates a limited memory is given by

$$E(k) = \frac{1}{2}\sum_{t=1}^{k}\rho(k,t)\sum_{j=1}^{n_L}\left(e_j^{[L]}(t)\right)^2 \tag{14.61}$$

where $e_j^{[L]}(t) = d_j^{[L]} - y_j^{[L]}$.

For the weight vectors, $W_j^{[L]}(k)$, in the final (output) layer, since the desired outputs are specified $E_p(k)$ can be minimized for $W_j^{[L]}(k)$ by taking their partial derivatives with respect to $W_j^{[L]}(k)$ and setting it equal to zero, thus

$$\frac{\partial E_p(k)}{\partial w_j^{[L]}(k)} = 0 \tag{14.62}$$

This leads to the following deterministic, normal equation:

$$R_L(k)W_j^{[L]} = P_L(k) \tag{14.63}$$

where

$$R_L(k) = \sum_{t=1}^{k}\lambda^{k-t}\left(\frac{1}{a_L}\right)^2 Y^{[L-1]}(t)Y^{[L-1]^T}(t) \tag{14.64}$$

and

$$P_L(k) = \sum_{t=1}^{k}\lambda^{k-t}\left(\frac{1}{a_L}\right)Y^{[L-1]}(t)d_j^{[L]}(t) \tag{14.65}$$

This equation can be solved efficiently using the weighted RLS algorithm [37,38].

Note that when the total input to node j does not lie on the ramp region of the threshold logic nonlinearity function, the derivative in (14.62) is always zero. This implies that the normal equation in (14.63) will only be solved when the input to the relevant nodes lies within the ramp region; otherwise no updating is required. This is the case for all the other layers.

Similar normal equations for the other layers can be obtained by taking the partial derivatives of $E_p(k)$ with respect to the weight vectors in these layers and setting the results equal to zero.

For the weight vector in layer [s], we have

$$\frac{\partial E_p(k)}{\partial w_j^{[s]}(k)} = 0 \tag{14.66}$$

which leads, using the chain rule, to

$$\sum_{t=1}^{k}\lambda^{k-t}\sum_{i=0}^{n_s}\frac{\partial y_i^{[s]}(t)}{\partial w_r^{[s]}(k)}e_i^{[s]}(t) = 0 \tag{14.67}$$

where

$$e_i^{[s]}(t) = \frac{1}{a_{s+1}} \sum_{j=1}^{n_{s+1}} w_{ji}^{[s+1]}(k) e_j^{[s+1]}(t) \tag{14.68}$$

by defining the hidden desired output of the jth neuron layer $[s]$ as

$$d_i^{[s]}(t) = y_i^{[s]}(t) + e_i^{[s]}(t) \tag{14.69}$$

The following normal equation can be obtained for the layer $[s]$:

$$\sum_{t=1}^{n} \lambda^{n-t} \frac{1}{a_s} y^{[s-1]}(t) \left(d_i^{[s]} - \frac{1}{a_s} y^{[s-1]^T} w_i^{[s]}(k) \right) = 0 \tag{14.70}$$

The equations for updating the weight vectors $W_j^{[L]}(k)$'s can be derived using the matrix inversion lemma as

$$K^{[s]}(k) = \frac{R^{[s]^{-1}}(k-1) Y^{[s-1]}(k)}{\lambda + Y^{[s-1]^T}(k) R^{[s]^{-1}}(k-1) Y^{[s-1]}(k)} \tag{14.71}$$

$$R^{[s]^{-1}}(k) = \lambda^{-1} \left(I - K^{[s]}(k) Y^{[s-1]}(k) \right) R^{[s]^{-1}}(k-1) \quad \forall L \in [1, L] \tag{14.72}$$

$$\Delta w_j^{[s]}(k) = K^{[s]}(k) \left(d_j^{[s]}(k) - \frac{1}{a_s} Y^{[s-1]^T} w_j^{[s]}(k-1) \right) \quad \forall j \in [1, n_s] \tag{14.73}$$

14.8 Kalman Filters for MLP Training

The learning algorithm of an MLP can be regarded as parameter estimation for such a nonlinear system. A lot of estimation methods of general nonlinear systems have been reported so far. For linear dynamic systems with white input and observation noise, the Kalman algorithm [49] is known to be an optimum algorithm.

In [35], the classical Kalman filter method has been proposed to train MLPs, and better results have been shown compared with the SBP algorithm. In order to work with a system of linear equations, an inversion of the output layer nonlinearity and an estimation of the desired output summation in the hidden layers are used as in Section 14.7.2.

Extended versions of the Kalman filter algorithm can be applied to nonlinear dynamic systems by linearizing the system around the current estimate of the parameters. Although it is computationally complex, this algorithm updates parameters consistent with all previously seen data and usually converges in a few iterations.

14.8.1 Multidimensional Kalman Filter Algorithm (FKF)

To solve the nonlinear network problem, one should portion it into linear and nonlinear parts [19]. Since the fast Kalman filter is only applied to solve linear filtering problems, the same linearization approach as in [19] is considered.

Looking at Figure 14.1, the linear part of a neuron may be viewed as a multidimensional input linear filter.

Therefore, the MLP training problem is transformed into a filtering problem by using fast Kalman filter [37].

The algorithm works as follows:

- FKF procedure

  For each layer s from 1 to L, compute the following quantities:

  - *Forward prediction error*

$$e_s^a = y_{s-1}(k) - A_s^T(k-1)y_{s-1}(k-1) \tag{14.74}$$

  - *Forward prediction matrix*

$$A_s(k) = A_s(k-1) + G_s(k-1)\left(e_s^a(k)\right)^T \tag{14.75}$$

  - *A posterior forward prediction error*

$$\varepsilon_s^a(k) = y_{s-1}(k) - A_s^T(k)y_{s-1}(k-1) \tag{14.76}$$

  - *Energy matrix*

$$E_s^a(k) = \lambda E_s^a(k-1) + e_s^a(k)\left(\varepsilon_s^a(k)\right) \tag{14.77}$$

  - *Augmented Kalman gain*

$$G_1^{(s)}(k) = \begin{vmatrix} 0 \\ \cdots \\ G_s(k-1) \end{vmatrix} + \begin{vmatrix} I_s \\ \cdots \\ -A_s(k) \end{vmatrix} \left(E_s^a(k)\right)^T \varepsilon_s^a(k)$$

$$= \begin{vmatrix} M_s(k) \\ \cdots \\ m_s(k) \end{vmatrix} \tag{14.78}$$

  - *Backward prediction error*

$$e_s^b(k) = y_{s-1}(k-1) - B_s^T(k-1)y_{-1}(k) \tag{14.79}$$

  - *Kalman gain*

$$G_s(k) = \frac{1}{1-\left(e_s^b(k)\right)^T m_s(k)}\left(M_s(k) + B_s(k-1)m_s(k)\right) \tag{14.80}$$

  - *Backward prediction matrix*

$$B_s(k) = B_s(k-1) + G_s(k)\left(e_s^b(k)\right)^T \tag{14.81}$$

It follows that the updating rules for the different synaptic weights are given as follows:

- *For the weight of each neuron j in the output layer L*

$$w_j^{[L]}(k) = w_j^{[L]}(k-1) + G_L(k)\left(d_j(k) - y_j^{[L]}(k)\right) \quad (14.82)$$

- *For each hidden layer s = 1 through L − 1, the weight vectors are updated by*

$$w_j^{[s]}(k) = w_j^{[s]}(k-1) + G_s(k)e_j^{[s]}\mu \quad (14.83)$$

where $e_j^{[s]}$ is given by (14.50).

It should be noted that to avoid the inversion of the energy matrix $E_s^a(k)$ at each layer, we can use a recursive form of the inversion matrix lemma [50].

For the different steps of the algorithm, see Appendix 14.A.

14.8.2 Extended Kalman Filter Algorithm

The extended Kalman filter is well known as a state estimation method for a nonlinear system, and can be used as a parameter estimation method by augmenting the state with unknown parameters. Since the EKF-based algorithm approximately gives the minimum variance estimate of the link weights, it is expected that it converges in fewer iterations than the SBP algorithm.

Since mathematical derivations for the EKF are widely available in the literature [51,52], we shall briefly outline the EKF applied to a discrete time system.

Consider a nonlinear finite dimensional discrete time system of the form

$$\begin{cases} x(k+1) = f_k(x(k)) + w(k) \\ y(k) = h_k(x(k)) + v(k) \end{cases} \quad (14.84)$$

where

$x(k)$ is the state vector

$y(k)$ is the observation vector

f_k and h_k are time-variant nonlinear functions

Also, $w(k)$ and $v(k)$ are assumed to be zero mean, independent, Gaussian white noise vectors with known covariance $Q(k)$ and $R(k)$, respectively.

The initial state $x(0)$ is assumed to be a Gaussian random vector with mean x_0 and covariance P_0.

Defining the current estimated state vector based on the observations up to time $k-1$ as $\hat{x}(k/k-1)$, the EKF updates the state vector as each new pattern is available.

The final results are given by the following equations:

$$\hat{x}\left(k+\frac{1}{k}\right) = f_k\left(\hat{x}\left(\frac{k}{k}\right)\right) \quad (14.85)$$

$$\hat{x}\left(\frac{k}{k}\right) = \hat{x}\left(\frac{k}{k-1}\right) + K(k)\left[y(k) - h_k\left(\hat{x}\left(\frac{k}{k-1}\right)\right)\right] \quad (14.86)$$

$$K(k) = P\left(\frac{k}{k-1}\right)H^T(k)\left[H(k)P\left(\frac{k}{k-1}\right)H^T(k) + R(k)\right]^{-1} \quad (14.87)$$

$$P\left(k+\frac{1}{k}\right)=F(k)P\left(\frac{k}{k}\right)F^T(k)+Q(k) \tag{14.88}$$

$$P\left(\frac{k}{k}\right)=P\left(\frac{k}{k-1}\right)-K(k)H(k)P\left(\frac{k}{k-1}\right) \tag{14.89}$$

This algorithm is initialized by $\hat{x}(0/-1) = x_0$ and $P(0/-1) = P_0$. The matrix $K(k)$ is the Kalman gain.

$F(k)$ and $H(k)$ are defined by

$$F(k)=\left(\frac{\partial f_k}{\partial x}\right)_{x=\hat{x}(k/k)} \tag{14.90}$$

$$H(k)=\left(\frac{\partial h_k}{\partial x}\right)_{x=\hat{x}(k/k-1)} \tag{14.91}$$

The standard Kalman filter for the linear system, in which $f_k(x(k)) = A_k x(k)$ and $h_k(x(k)) = B_k x(k)$ gives the minimum variance estimate of $x(k)$. In other words, $\hat{x}(k/k-1)$ is optimal in the sense that the trace of the error covariance defined by

$$p\left(\frac{k}{k-1}\right)=E\left\{\left(x(k)-\hat{x}\left(\frac{k}{k-1}\right)\right)\left(x(k)-\hat{x}\left(\frac{k}{k-1}\right)\right)^T\right\} \tag{14.92}$$

is minimized, where $E(.)$ denotes here the expectation operator. On the other hand, the EKF for the nonlinear system is no longer optimal, and $\hat{x}(k/k-1)$ and $P(k/k-1)$ express approximate conditional mean and covariance, respectively. Because the EKF is based on the linearization of $f_k(x(k))$ and $h_k(x(k))$ around $\hat{x}(k/k)$ and $\hat{x}(k/k-1)$, respectively, and on the use of the standard Kalman filter, it is also noted that the EKF may get stuck at a local minimum if the initial estimates are not appropriate [54]. Nevertheless, a lot of successful applications have been reported because of its excellent convergence properties.

We will show now how a real-time learning algorithm for the MLP can be derived from the EKF [32,53]. Since the EKF is a method of estimating the state vector, we shall put the unknown linkweights as the state vector

$$M=\left[(W^1)^T,(W^2)^T,\ldots,(W^L)^T\right]^T \tag{14.93}$$

The MLP is then expressed by the following nonlinear system equations:

$$\begin{cases} M(k+1)=M(k) \\ d(k)=h_k(M(k))+v(k) \\ \qquad\;\; =y^{[L]}(k)+v(k) \end{cases} \tag{14.94}$$

The input to the MLP for a pattern k combined with the structure of the MLP is expressed by a nonlinear time-variant function h_k. The observation vector is expressed by the desired output vector $d(t)$, and $v(k)$ is assumed to be a white noise vector with covariance matrix $R(k)$ regarded as a modeling error.

The application of the EKF to this system gives the following real-time learning algorithm

$$\hat{M}(k) = \hat{M}(k-1) + K(k)\left[d(k) - h_k(M(k-1))\right] \tag{14.95}$$

$$K(k) = P(k-1)H^T(k)\left[H(k)P(k-1)H^T(k) + R(k)\right]^{-1} \tag{14.96}$$

$$P(k) = P(k-1) - K(k)H(k)P(k-1) \tag{14.97}$$

We note that we put here $P(t) = P(t/t)$ and $\hat{M}(k) = \hat{M}(k/k)$ since $P(t) = P(t+1/t)$ and $\hat{M}(k/k) = \hat{M}(k+1/k)$. Also $\hat{y}^{[L]}(k)$ denotes the estimate of $y^{[L]}(k)$ based on the observations up to time $k - 1$, which is computed by $\hat{y}^{[L]}(k) = h_k(\hat{M}(k-1))$. According to (14.91), $H(k)$ is expressed by

$$H(k) = \left(\frac{\partial Y^L(k)}{\partial W}\right)_{M=\hat{M}(k-1)} \tag{14.98}$$

$$H(k) = [H_1^1(k),\ldots,H_{n_1}^1(k),H_1^2(k),\ldots,H_{n_2}^2(k),H_1^L(k),\ldots,H_{n_L}^L(k)] \tag{14.99}$$

with the definition of

$$H_j^s(k) = \left(\frac{\partial Y^L(k)}{\partial W_j^s}\right)_{M=\hat{M}(k-1)} \tag{14.100}$$

The different steps of the algorithm are given in Appendix 14.B.

14.9 Davidon–Fletcher–Powell Algorithms

In [41,42], a quasi-Newton method called Broyden, Fletcher Goldfarb, and Shanno (BFGS) method have been applied to train the MLP and it's found that this algorithm converges much faster than the SBP algorithm, but the potential drawback of the BFGS method lies on the huge size of memory needed to store the Hessian matrix. In [7], the Marquardt–Levenberg algorithm was applied to train the feedforward neural network and simulation results on some problems showed that the algorithm is very faster than the conjugate gradient algorithm and the variable learning rate algorithm. The great drawback of this algorithm is its high computational complexity and its sensitivity for the initial choice of the parameter μ in the update of the Hessian matrix.

In this part, we present a new fast training algorithm based on the Davidon–Fletcher–Powell (DFP) method. The DFP algorithm consists of approximating the Hessian matrix (as is the case of all quasi-Newton methods).

The new Hessian matrix is approximated also by using only the gradient vector $\nabla E(W)$ as in ML method but the approximation in this case uses more information provided by the descent direction $\underline{d}_r$ and the step length λ obtained by minimizing the cost function $E(W + \lambda \underline{d}_r)$ that governs the amplitude of descent and updating.

Let $H(k)$ be the inverse of the Hessian matrix. The DFP algorithm is based on updating $H(k)$ iteratively by

$$H(k+1) = H(k) + \frac{\delta(k)\delta^T(k)}{\delta^T(k)\gamma(k)} - \frac{H(k)\gamma(k)\gamma^T(k)H(k)}{\gamma^T(k)H(k)\gamma(k)} \tag{14.101}$$

where $\delta(k)$ and $\gamma(k)$ are the different parameters used in DFP algorithm.

The DFP algorithm for mathematical programming works as indicated in Appendix 14.C.

14.9.1 Davidon–Fletcher–Powell Algorithm for Training MLP

In this section, we apply the DFP algorithm first to train a single output layer perceptron, and then we extend all equations to MLP.

14.9.1.1 DFP Algorithm for Training a Single Output Layer Perceptron

First, let us develop the new algorithm for a neuron j in the output layer $[L]$ of the network. The optimization criterion $E(W_j^{[L]})$ for the current pattern is defined as

$$E(W_j^{[L]}) = \frac{1}{2}\left(e_j^{[L]}\right)^2 \tag{14.102}$$

where $e_j^{[L]}$ is the neuron output error defined by Equation 14.4. Note that the difficult task in the application of the DFP algorithm is how to find the optimal value of λ that minimizes $E\left(W_j^{[L]} + \lambda d_{rj}^{[L]}\right)$.

The first method to search λ that minimize $E\left(W_j^{[L]} + \lambda d_{rj}^{[L]}\right)$ is to solve the following equation:

$$\frac{\partial E\left(W_j^{[L]} + \lambda d_{rj}^{[L]}\right)}{\partial \lambda} = 0 \tag{14.103}$$

Recall that

$$d_{rj}^{[L]}(k) = -H_j^{[L]}(k)\nabla E\left(W_j^{[L]}(k)\right) \tag{14.104}$$

For this purpose, we have to derivate E with respect to λ.

$$\begin{aligned}\frac{\partial E\left(W_j^{[L]} + \lambda d_{rj}^{[L]}\right)}{\partial \lambda} &= \frac{1}{2}\frac{\partial e_j^2\left(W_j^{[L]} + \lambda d_{rj}^{[L]}\right)}{\partial \lambda} \\ &= 2f'\left(W_j^{[L]} + \lambda d_{rj}^{[L]}\right)\left(f\left(W_j^{[L]} + \lambda d_{rj}^{[L]}\right) - d_j^{[L]}\right)\left(y_j^{[L-1]}\right)^T d_{rj}^{[L]}(k)\end{aligned} \tag{14.105}$$

$E\left(W_j^{[L]} + \lambda d_{rj}^{[L]}\right)$ is minimum when $\left(\partial E\left(W_j^{[L]} + \lambda d_{rj}^{[L]}\right)\right)\big/(\partial\lambda) = 0$, which leads to the optimal value of λ:

$$\lambda^* = -\frac{-\left(\dfrac{1}{a}\log\left(\dfrac{1 - d_j^{[L]}}{d_j^{[L]}}\right)\right) + \left(Y^{[L-1]}\right)^T W_j^{[L]}(k)}{y_j^{[L-1]} d_{rj}^{[L]}} \tag{14.106}$$

We can apply this search method of λ^* in the case of a single-layer perceptron but this becomes a very difficult task when we going up to train an MLP.

To avoid this hard computation to find an optimal step length $\lambda^*(k)$ for a given descent direction $d_r(k)$, we can use Wolfe's line search [41,42].

This line search procedure satisfies the Wolfe linear search conditions:

$$E(\underline{w}(k) + \lambda(k)\underline{d}_r(k)) - E(\underline{w}(k)) \le 10^{-4}\lambda(k)\underline{d}_r{}^T(k)\nabla E(\underline{w}(k)) \tag{14.107}$$

$$\underline{d}_r^T \nabla E(\underline{w}(k) + \lambda(k)d(k)) \ge 0.9\underline{d}_r^T \nabla E(\underline{w}(k)) \tag{14.108}$$

In Appendixes 14.D and 14.E, we give the different steps of Wolfe's line search method.

Before extending this method to train an MLP, let us give a practical example for a single neuron trained with ML, and the two versions of DFP algorithm.

14.9.1.2 DFP Algorithm for Training an MLP

In the previous section, we have applied the new algorithm to a single-layer perceptron with tow techniques to find a step length λ.

To train an MLP, considered with a single hidden layer, we need to minimize the error function defined by

$$E(\underline{w}) = \sum_{j=1}^{n_L} \left(d_j - f\left(\sum_{r=0}^{n_1} \left(f\left(W_r^{[1]^T} y^{[0]} \right) \cdot W_{jr}^{[2]} \right) \right) \right)^2 \qquad (14.109)$$

The problem here is how to determine the weights of the network that minimize the error function $E(\underline{w})$. This is can be considered as an unconstrained optimization problem [43]. For this reason, we will extend the equations of the DFP method on the entire network to optimize the cost function presented in (14.109).

The learning procedure with DFP algorithm needs to find an optimal value λ^*, which minimizes the function $E(\underline{w} + \lambda \underline{d}_r)$.

To evaluate the coefficient λ, we will use the Wolfe line search because the exact method is very difficult since we must drive $E(\underline{w} + \lambda \underline{d}_r)$ with respect to λ.

When we have applied the DFP algorithm to train a neural network with a single hidden layer, we have found that this algorithm is faster than the ML one. Different steps of the algorithm are given in Appendix 14.E.

TABLE 14.5 Performance Comparison of the Fast Algorithms with respect to the SBP One for the 4-b Parity Checker

	S_w (%)	G_c (%)
MBP	59	97
RLS	56	98
FKF	49	98
EKF	52	98
DFP	46	98

TABLE 14.6 Improvement Ratios with respect to SBP

	In Iteration	In Time
MBP	2.8	2.6
RLS	3.4	2.7
FKF	3.3	2.1
EKF	3.4	2.3
DFP	3.5	2.6

14.10 Some Simulation Results

In this section, we present some simulation results of the last five algorithms. Some of them are based on the Newton method. To perform a good comparison, we have used the same problems as in Section 14.6: the 4-b parity checker and the circle-in-the-square problem. The comparison is done by the comparison of S_w, G_c, and the improvement ratios and always with respect to the performance of the SBP algorithm.

14.10.1 For the 4-b Parity Checker

Tables 14.5 and 14.6 present the different simulation results for this problem.

14.10.2 For the Circle-in-the-Square Problem

Tables 14.7 and 14.8 present the different simulation results for this problem.

From these results, one can conclude that the performance of these fast algorithms is quite similar; the choice of the best algorithm for training an MLP is still always dependent on the problem, the MLP structure, and the expertise of the user.

TABLE 14.7 Performance Comparison of the Fast Algorithms with respect to the SBP One for the Circle-in-the-Square Problem

	S_w (%)	G_c (%)
MBP	55	99
RLS	55	98
FKF	52	99
EKF	52	99
DFP	50	99

TABLE 14.8 Improvement Ratios with respect to SBP

	In Iteration	In Time
MBP	3.1	2.9
RLS	3.2	2.8
FKF	3.3	2.3
EKF	3.4	2.6
DFP	3.4	2.5

14.11 Conclusion

This work constitutes a brief survey related to the different techniques and algorithms that have been proposed in the literature to speed up the learning phase in an MLP. We have shown that the learning speed depends on several factors; some are related to the network and its use and others are related to the algorithm itself.

Despite the variety of the proposed methods, one isn't capable of giving a forward confirmation on the best algorithm that is suitable in a given application. However, the algorithms based on the SOOM are more rapid than those based on the gradient method. Recently developed NBN algorithm [39] described in Chapter 13 is very fast but its success rate is 100% not only for the provided here benchmark of Parity-4 but also for the Parity-5 and Parity-6 problems using the same MLP architecture (N,8,2,1) [55].

Appendix 14.A: Different Steps of the FKF Algorithm for Training an MLP

1. Initialization
   - From layer s=1 to L, equalize all $y_0^{[s-1]}$ to a value different from 0 (e.g., 0.5).
   - Randomize all the weights $w_{ji}^{[s]}$ at random values between ±0.5.
   - Initialize the matrix inverse $E_s^a(0)$.
   - Initialize the forgetting factor λ.
2. Select training pattern
   - Select an input/output pair to be processed into the network.
   - The input vector is $y_p^{[0]}$ and corresponding output is $d_p^{[L]}$.
3. Run selected pattern through the network for each layer s from 1 to L and calculate the summation output:

$$u_j^{[s]} = \sum_{i=0}^{n_{s-1}} w_{ji} y_i^{[s-1]}$$

   and the nonlinear output

$$y_j^{[s]} = f\left(u_j^{[s]}\right) = sigmoide\left(u_j^{[s]}\right)$$

4. FKF procedure

   For each layer s from 1 to L, compute the following quantities:
   - Forward prediction error

$$e_s^a = y_{s-1}(k) - A_s^T(k-1)y_{s-1}(k-1)$$

   - Forward prediction matrix

$$A_s(k) = A_s(k-1) + G_s(k-1)\left(e_s^a(k)\right)^T$$

   - A posterior forward prediction error

$$\varepsilon_s^a(k) = y_{s-1}(k) - A_s^T(k)y_{s-1}(k-1)$$

   - Energy matrix

$$E_s^a(k) = \lambda E_s^a(k-1) + e_s^a(k)\left(\varepsilon_s^a(k)\right)^T$$

- Augmented Kalman gain

$$G_1^{(s)}(k) = \begin{vmatrix} 0 \\ L \\ G_s(k-1) \end{vmatrix} + \begin{vmatrix} I_s \\ L \\ -A_s(k) \end{vmatrix} \left(E_s^a(k)\right)^T \varepsilon_s^a(k)$$

$$= \begin{vmatrix} M_s(k) \\ L \\ m_s(k) \end{vmatrix}$$

- Backward prediction error

$$e_s^b(k) = y_{s-1}(k-1) - B_s^T(k-1)y_{-1}(k)$$

- Kalman gain

$$G_s(k) = \frac{1}{1-\left(e_s^b(k)\right)^T m_s(k)}$$

$$\left[M_s(k) + B_s(k-1)m_s(k)\right]$$

- Backward prediction matrix

$$B_s(k) = B_s(k-1) + G_s(k)\left(e_s^b(k)\right)^T$$

5. Backpropagate the signal error
   - Compute $f'(u_j^{[s]}) = f(u_j^{[s]})\left(1 - f(u_j^{[s]})\right)$
   - Compute the error signal of the output layer ($s = L$): $e_j^{[L]} = f'(u_j^{[L]})\left(d_j^{[L]} - y_j^{[L]}\right)$
   - For each node j of the hidden layer, start from $s = L-1$ down $s = 1$ and calculate

$$e_j^{[s]} = f'(u_j^{[s]})\sum_i \left(e_i^{[s+1]} w_{ji}^{[s+1]}\right)$$

6. Calculate the desired summation output at the Lth layer using the inverse of the sigmoid

$$ld_j^{[L]} = -\text{Log}\left(\frac{1-d_j^{[L]}}{d_j^{[L]}}\right)$$

   for each neuron j.
7. Calculate the weight vectors in the output layer L

$$w_j^{[L]}(k) = w_j^{[L]}(k-1) + G_L(k)\left(d_j(k) - y_j^{[L]}(k)\right)$$

   for each neuron j.
   For each hidden layer $s = 1$ through L^{-1}, the weight vectors are updated by

$$w_j^{[s]}(k) = w_j^{[s]}(k-1) + G_s(k)e_j^{[s]}\mu$$

   for each neuron j.
8. Test for ending the running

Appendix 14.B: Different Steps of the EKF for Training an MLP

For $k = 1,2,\ldots$

$$\hat{y}_j^{s+1} = f(\hat{W}_j^s(k-1)^T \hat{y}^s)$$

$$\hat{\lambda}(k) = \hat{\lambda}(k-1) + \mu(k) \left[\frac{(d(k) - \overset{L}{y}(k))^T (d(k) - \overset{L}{y}(k))}{n_L} - \hat{\lambda}(k-1) \right]$$

$$\hat{y}_j^L(k) = \hat{x}^L(k)$$

For $s = L$ at 1:

$$\Delta_j^s(k) = \begin{cases} y_j^{[L]}(k)(1 - y_j^{[L]}(k))[0,\ldots,0,\overset{j}{1},0,\ldots,0]^T \\ \hat{y}_j^{s+1}(k)\left(1 - \hat{y}_j^{s+1}(k)\right) \sum_{j=1}^{n_{s+2}} w_{ji}^{s+1}(k-1)\Delta_j^{s+1}(k) \end{cases}$$

$$\psi_j^s(k) = P_j^s(k-1)y^s(k)$$

$$\alpha_j^s(k) = y_j^s(k)^T \psi_j^s(k)$$

$$\beta_j^s(k) = \Delta_j^{sT} \Delta_j^s$$

$$W_j^s(k) = W_j^s(k-1) + \frac{\Delta_j^s(k)^T (d(k) - y_j^s(k))}{\lambda(k) + \alpha_j^s(k)\beta_j^s(k)} \psi_j^i(k)$$

$$P_j^s(k) = P_j^s(k-1) - \frac{\beta_j^s(k)}{\lambda(k) + \alpha_j^s \beta_j^s} \psi_j^s \psi_j^{sT}$$

$$y_{j+1}^s(k) = y_j^s(k) + \Delta_j^s(k) y^{sT}(k)(y_j^s(k) - y_j^s(k-1))$$

Appendix 14.C: Different Steps of the DFP Algorithm for Mathematical Programming

1. Initializing the vector $W(0)$ and
   a positive definite initial Hessian inverse matrix $H(0)$.
   Select a convergence threshold: ct
2. Compute the descent direction $d_r(k)$

$$\underline{d}_r(k) = -H(k)\nabla E(W(k))$$

3. Search the optimal value $\lambda^*(k)$ such as

$$E(W(k)+\lambda^*(k)\underline{d}_r(k))=\min_{\lambda\geq 0}E(W(k)+\lambda(k)\underline{d}_r(k))$$

4. Update $W(k)$:

$$W(k+1)=W(k)+\lambda^*(k)\underline{d}_r(k)$$

5. Compute

$$\delta(k)=W(k+1)-W(k)$$

$$\gamma(k)=\nabla E\big(W(k+1)\big)-\nabla E\big(W(k)\big)$$

$$\Delta(k)=\frac{\delta(k)\delta^T(k)}{\delta^T(k)\gamma(k)}-\frac{H(k)\gamma(k)\gamma^T(k)H(k)}{\gamma^T(k)H(k)\gamma(k)}$$

6. Update the inverse matrix $H(k)$:

$$H(k+1)=H(k)+\Delta(k)$$

7. Compute the cost function value $E(W(k))$
   If $E(W(k)) > ct$. Go to step 2.

Appendix 14.D: Different Steps of Wolfe's Line Search Algorithm

For a given descent direction $d_r(k)$, we will evaluate an optimal value λ^* that satisfies the two Wolfe's conditions.

1. Set $\lambda_0 = 0$; choose $\lambda_1 > 0$ and $\lambda_{\max}$; $i = 1$
   Repeat
2. Evaluate $E(\underline{w} + \lambda_i \underline{d}_r) = \Phi(\lambda_i)$ and check the 1st Wolfe condition:
   If $\Phi(\lambda_i)\Phi(0) + 10^{-4}\lambda_i\Phi'(0)\cdot d_r$
   Then $\lambda^* \leftarrow$ Zoom $(\lambda_{i-1},\lambda_i)$ and stop
3. Evaluate $\Phi'(\lambda_i)d_r$ and check the 2nd Wolfe condition:
   If $|\Phi'(\lambda_i)\cdot d_r| \leq 0.9\Phi'(0)\cdot d_r$
   Then $\lambda^* = \lambda_i$ and stop
   If $\Phi'(\lambda_i)\cdot d_r \geq 0$
   Then $\lambda^* \leftarrow$ Zoom $(\lambda_i,\lambda_{i-1})$ and stop
4. Choose $\lambda_{i+1} \infty (\lambda_i,\lambda_{\max})$
   $i = i+1$;
   End (repeat)

   "Zoom" phase
   Repeat
1. Interpolate to find a trial step length λ_j between λ_{lo} and λ_{hi}
2. Evaluate $\Phi(\lambda_j)$
   If $[\Phi(\lambda_j) > \Phi(0) + 10^{-4}\lambda_j\Phi'(0)\cdot d_r]$ or $[\Phi(\lambda_j) > \Phi(\lambda_{lo})]$

Then $\lambda_{hi} = \lambda_j$
$\Phi(\lambda_{hi}) = \Phi(\lambda_j)$
Else, compute $\Phi'(\lambda_j) \cdot d_r$
If $|\Phi'(\lambda_j) \cdot d_r| \leq 0.9\Phi'(0) \cdot d_r$
Then $\lambda^* = \lambda_j$ and stop
If $\Phi'(\lambda_j)d_r \cdot (\lambda_{hi} - \lambda_{lo}) \geq 0$
Then $\lambda_{hi} = \lambda_{lo}$
$\Phi(\lambda_{hi}) = \Phi(\lambda_{lo})$
$\lambda_{lo} = \lambda_j$
$\Phi(\lambda_{lo}) = \Phi(\lambda_j)$

3. Evaluate $\Phi'(\lambda_j)$
   End (repeat)

Appendix 14.E: Different Steps of the DFP Algorithm for Training an MLP

1. Initializing randomly the synaptic coefficients $\underline{w}(0)$ and the definite positive Hessian inverse matrix $H(0)$.
2. Select an input/output pattern and compute

$$u_j^{[L]} = \left(W_j^{[L]}\right)^T y_j^{[L-1]} \quad \text{and} \quad f(u_j^{[L]}) = y_j^{[L]}$$

3. Compute the error function:

$$E(\underline{w}) = \sum_{j=1}^{n_L} e_j^2 = \sum_{j=1}^{n_L} \left(d_j^{[L]} - y_j^{[L]}\right)^2$$

4. Evaluate $\nabla E(\underline{w}) = (\partial E(\underline{w}))/(\partial \underline{w})$ where $\underline{w} = (W^{[L-1]}, W^{[L]})$
5. Compute the descent direction $d_r(k)$

$$d_r(k) = -H(k)\nabla E(\underline{w})$$

6. Compute the optimal value $\lambda^*(k)$ that satisfies the two Wolfe's conditions:

$$E(\underline{w}(k) + \lambda(k)\underline{d}_r(k)) - E(\underline{w}(k)) \leq 10^{-4}\lambda(k)\underline{d}_r^T(k)\nabla E(\underline{w}(k))$$

$$\underline{d}_r^T \nabla E(\underline{w}(k) + \lambda(k)d(k)) \geq 0.9\underline{d}_r^T \nabla E(\underline{w}(k))$$

7. Update $\underline{w}(k)$ for the output layer:

$$\underline{w}(k+1) = \underline{w}(k) + \lambda^*(k)\underline{d}_r(k)$$

8. Compute:

$$\delta(k) = \underline{w}(k+1) - \underline{w}(k)$$

$$\gamma(k) = \nabla E\left(\underline{w}(k+1)\right) - \nabla E\left(\underline{w}(k)\right)$$

$$\Delta(k) = \frac{\delta(k)\delta^T(k)}{\delta^T(k)\gamma(k)} - \frac{H(k)\gamma(k)\gamma^T(k)H(k)}{\gamma^T(k)H(k)\gamma(k)}$$

9. Update the inverse matrix $H(k)$:

$$H(k+1) = H(k) + \Delta(k)$$

10. Compute the global error: $E = \sum_p E_p$

    If $E >$ threshold then return to 2.

References

1. S. Abid, F. Fnaiech, and M. Najim, A fast feed-forward training algorithm using a modified form of the standard back propagation algorithm, *IEEE Trans. Neural Network*, 12(2), 424–430, March 2001.
2. A. Cichocki and R. Unbehauen, *Neural Network for Optimization and Signal Processing*, John Wiley & Sons Ltd. Baffins Lane, Chichester, England, 1993.
3. C. Charalambous, Conjugate gradient algorithm for efficient training of artificial neural networks, *IEEE Proc.*, 139(3), 301–310, 1992.
4. F. Fnaiech, S. Abid, N. Ellala, and M. Cheriet, A comparative study of fast neural networks learning algorithms, *IEEE International Conference on Systems, Man and Cybernetics*, Hammamet, Tunisia, October 6–9, 2002.
5. T.P. Vogl, J.K. Mangis, A.K. Zigler, W.T. Zink, and D.L. Alkon, Accelerating the convergence of the backpropagation method, *Biol. Cybern.*, 59, 256–264, September 1988.
6. B.M. Wilamowski, Neural Network Architectures and Learning algorithms, *IEEE Ind. Electro Mag.*, 3(4), 56–63, November 2009.
7. M.T. Hagan and M.B. Menhaj, Training feedforward networks with the Marquardt algorithm, *IEEE Trans. Neural Netw.*, 5, 989, November 1994.
8. C. Darken and J. Moody, Towards faster stochastic gradient search, *Advances in Neural Information Processing Systems 4*, Morgan Kaufman, San Mateo, CA, 1991, pp. 1009–1016.
9. J.K. Kruschke and J.R. Movellan, Benefits of gain: Speeded learning and minimal layers in back-propagation networks, *IEEE Trans. Syst. Cybern.*, 21, 273–280, 1991.
10. R.A. Jacobs, Increased rates of convergence through learning rate adaptation, *Neural Netw.*, 1, 295–307, 1988.
11. E.M. Johansson, F.V. Dowla, and D.M. Goodman, Backpropagation learning for multilayer feed-forward networks using the conjugate gradient methods, *Int. J. Neural Syst.*, 2, pp. 291–301, 1992.
12. L. Behera, S. Kumar, and A. Patnaik, On adaptive learning rate that guarantees convergence in feed-forward networks, *IEEE Trans. Neural Netw.*, 17(5), 1116–1125, September 2006.
13. B.M. Wilamowski, D. Hunter, and A. Malinowski, Solving parity-N problems with feedforward neural network, *Proceedings of the IJCNN'03 International Joint Conference on Neural Netw.*, Portland, OR, July 20–23, 2003, pp. 2546–2551.
14. N.B. Karayiannis and A.N. Venetsanopoulos, Fast learning algorithm for neural networks, *IEEE Trans. Circuits Syst.*, 39, 453–474, 1992.
15. M.R. Azimi-Sadjadi and R.J. Liou, Fast learning process of multilayer neural network using recursive least squares method, *IEEE Trans. Signal Process.*, 40(2), 443–446, February 1992.
16. M.R. Azimi-Sadjadi and S. Citrin, Fast learning process of multilayer neural nets using recursive least squares technique. In *Proceedings of the IEEE International Conference on Neural Networks*, Washington DC, May 1989.

17. M.R. Azimi-Sadjadi, S. Citrin, and S. Sheedvash, Supervised learning process of multilayer perceptron neural networks using fast recursive least squares. In *Proceedings of the IEEE International Conference on Acoustics, Speech, Signal Processing (ICASSP'90)*, New Mexico, April 1990, pp. 1381–1384.
18. F.B. Konig and F. Barmann, A learning algorithm for multilayered neural networks based on linear squares problems, *Neural Netw.*, 6, 127–131, 1993.
19. R.S. Scalero and N. Tepedelenglioglu, A fast new algorithm for training feedforward neural networks, *IEEE Trans. Signal Process.*, 40(1), 202–210, January 1992.
20. B. M. Wilamowski, Challenges in Applications of Computational Intelligence in Industrial Electronics, *ISIE10 - International Symposium on Industrial Electronics*, Bari, Italy, July 4–7, 2010, pp. 15–22.
21. H.S.M. Beigi and C.J. Li, Learning algorithms for neural networks based on quasi-Newton methods with self-scaling, *ASME Trans. J. Dyn. Syst. Meas. Contr.*, 115, 38–43, 1993.
22. D.B. Parker, Optimal algorithms for adaptive networks: Second order backpropagation, second-order direct propagation, and second-order Hebbian learning, in *Proceedings of the First International Conference on Neural Networks*, San Diego, CA, Vol. 2, November 1987, pp. 593–600.
23. A.G. Parlos, B. Fernandez, A.F. Atiya, J. Muthusami, and W.K. Tsai, An accelerated learning algorithm for multilayer perceptron networks, *IEEE Trans. Netw.*, 5(3), 493–497, May 1994.
24. S. Fahlman, An empirical study of learning speed in backpropagation networks, Technical Report CMCU-CS-162, Computer Science Department, Carnegie-Mellon University, Pittsburgh, PA, 1988.
25. T. Tollenaere, SuperSab: Fast adaptive BP with good scaling properties, *Neural Netw.*, 3, 561–573, 1990.
26. R. Battiti, First and second order methods for learning, *Neural Comput.*, 4, 141–166, 1992.
27. M. Riedmiller, Advanced supervised learning in MLPs, *Comput. Stand. Interfaces*, 16, 265–278, 1994.
28. J. Hannan and M. Bishop, A class of fast artificial NN training algorithms, Technical Report, JMH-JMB 01/96, Department of Cybernetics, University of Reading, Reading, U.K., 1996.
29. D. Alpsan et al., Efficacy of modified BP and optimisation methods on a real-word medical problem, *Neural Netw.*, 8(6), 945–962, 1995.
30. P. Van Der Smagt, Minimisation methods for training feedforward NNs, *Neural Netw.*, 7(1), 1–11, 1994.
31. J.M. Hannan and J.M. Bishop, A comparison of fast training algorithms over two real problems, *Fifth International Conference on Artificial Neural Networks*, Cambridge, MA, July, 7–9, 1997, pp. 1–6.
32. Y. Liguni, H. Sakai, and H. Tokumaru, A real-time algorithm for multilayered neural network based on the extended Kalman filter, *IEEE Trans. Signal Process.*, 40(4), 959–966, April, 1992.
33. A. Kimura, I. Arizono, and H. Ohta, A back-propagation algorithm based on the extended Kalman filter, *Proceedings of International Joint Conference on Neural Networks*, Washington, DC, October 25–29, 1993, Vol. 2, pp. 1669–1672.
34. F. Fnaiech, D. Bastard, V. Buzenac, R. Settineri, and M. Najim, A fast Kalman filter based new algorithm for training feedforward neural networks, *Proceedings of European Signal Processing Conference (EUSIPCO'94)*, Eidumbrg, Scottland, U.K., September 13–16, 1994.
35. F. Heimes, Extended Kalman filter neural network training: experimental results and algorithm improvements, *IEEE International Conference on Systems, Man, and Cybernetics*, San Diego, CA, October 11–14, 1998, Vol. 2, pp. 1639–1644.
36. Y. Zhang and X.R. Li, A fast U-D factorization based learning algorithm with applications to nonlinear system modelling and identification, *IEEE Trans. Neural Netw.*, 10, 930–938, July 1999.
37. S.S. Haykin, *Adaptive Filter Theory*, Englewood Cliffs, NJ: Prentice-Hall, 1986.
38. G.C. Goodwin and K.S. Sin, *Adaptive Filtering, Prediction, and Control*, Englewood Cliffs, NJ: Prentice-Hall, 1984.
39. B.M. Wilamowski and H. Yu, Improved Computation for Levenberg Marquardt Training, *IEEE Trans. Neural Netw.*, 21(6), 930–937, June 2010.
40. S. Abid, F. Fnaiech, and M. Najim, Evaluation of the feedforward neural network covariance matrix error, *IEEE International Conference on Acoustic Speech and Signal Processing (ICASSP'2000)*, Istanbul, Turkey, June 5–9, 2000.

41. R. Fletcher, *Practical Methods of Optimization: Unconstrained Optimization*, Vol. 1, New York: John Wiley & Sons, 1980.
42. R. Fletcher, *Practical Methods of Optimization: Constrained Optimization*, Vol. 2, New York: John Wiley & Sons, 1981.
43. O.L. Mangasarian, Mathematical programming in neural network, Technical Report 1129, Computer Sciences Department, University of Wisconsin-Madison, Madison, WI, 1992.
44. A. Atiya and C. Ji, How initial conditions affect generalization performance in large networks, *IEEE Trans. Neural Netw.*, 2(2), 448–451, March 1997.
45. N. Weymaere and J.P. Martens, On the initialization and optimisation of multilayer perceptrons, *IEEE Trans. Neural Netw.*, 5(5), 738–751, September 1994.
46. C.L. Chen and R.S. Nutter, Improving the training speed of three layer feedforward neural nets by optimal estimation of the initial weights, in *Proceedings of the International Joint Conference on Neural Networks*, Seattle, WA, 1991, Vol. 3, pp. 2063–2068.
47. B.M. Wilamowski and L. Torvik, Modification of gradient computation in the back-propagation algorithm, *Presented at ANNIE'93—Artificial Neural Networks in Engineering*, St. Louis, MO, November 14–17, 1993.
48. T.J. Andersen and B.M. Wilamowski, A modified regression algorithm for fast one layer neural network training, *World Congress of Neural Networks*, vol. 1, pp. 687–690, Washington, DC, July 17–21, 1995.
49. R.E. Kalman, A new approach to linear filtering and prediction problems, *J. Basic Eng. Trans. ASME, Series D*, 82(1), 35–45, 1960.
50. M.G. Bellanger, *Adaptive Digital Filters and Signal Analysis*, New York: Marcel Dekker Inc., 1987.
51. B.D.O. Anderson and J.B. Moore, *Optimal Filtering*, Englewood Cliffs, NJ: Prentice Hall, 1979.
52. C.K. Chui and G. Chen, *Kalman Filtering*, New York: Springer Verlag, 1987.
53. S. Singhal and L. Wu, Training feed-forward networks with the extended Kalaman algorithm, Bell Communications Research, Inc., Morristown, NJ, 1989, IEEE.
54. L. Ljung, Asymptotic behavior of the extended Kalman filter as a parameter estimator for linear systems, *IEEE Trans. Automat. Contr.*, AC-24(1), 36–50, February 1979.
55. B.M. Wilamowski, N.J. Cotton, O. Kaynak, and G. Dundar, Computing Gradient Vector and Jacobian Matrix in Arbitrarily Connected Neural Networks, *IEEE Trans. Ind. Electron.*, 55(10), 3784–3790, October 2008.

15

Feedforward Neural Networks Pruning Algorithms

Nader Fnaiech
University of Tunis

Farhat Fnaiech
University of Tunis

Barrie W. Jervis
Sheffield Hallam University

15.1 Introduction

Since the 1990s, the feedforward neural network (FNN) has been universally used to model and simulate complex nonlinear problems, including the supposedly unknown mathematical relationship existing between the input and output data of applications. The implementation of an FNN requires the implementation of the different steps below to ensure proper representation of the function to be approximated:

Step 1: Identification of the structure of the multilayer neural network:
First, we must choose the network architecture, i.e., the neural network structure, by determining the number of layers and the number of neurons per layer required. To date, the choice remains arbitrary and intuitive, which makes the identification of the structure a fundamental problem to be solved, and has a huge effect on the remaining steps. The work of Funahashi (1989) and Cybenko (1989) shows that any continuous function can be approximated by an FNN with three layers, using a sigmoid activation function for neurons in the hidden layer, and linear activation functions for neurons in the output layer. This work shows that there is a neural network to approximate a nonlinear function, but does not specify

the number of neurons in the hidden layer. The number of existing neurons in the input layer and output layer is fixed by the number of inputs and outputs, respectively, of the system to be modeled. The power of neural network architectures strongly depends on the used architecture [WY10]. For example using 10 neurons in popular MLP architecture with one hidden layer only Parity-9 problem can be solved [WHM03,W10]. However if FCC (Fully connected architecture) is used then as big problem as Parity 1023 can be solved with the same 10 neurons. Generally, if connections across layers are allowed then neural networks are becoming more powerful [W09]. Similarly, the choice of the number of neurons per layer is another important problem: the choices of the initial synaptic coefficients are arbitrary, so there must be an effective and convergent learning phase.

Step 2: Learning phase:
Second, this learning phase consists of updating the parameters of nonlinear regression by minimizing a cost function applied to all the data, so that the network achieves the desired function. There are many learning methods which depend on several factors including the choice of cost function, the initialization of the weights, the criterion for stopping the learning, etc. (For further details see the related chapters in this handbook.)

Step 3: Generalization phase or testing phase:
Finally, we must test the quality of the network obtained by presenting examples not used in the learning phase. To do so, it is necessary to divide the available data into a set of learning patterns and another of generalization patterns. At this stage, we decide whether the neural network obtained is capable of achieving the desired function within an acceptable error; if it is not, we need to repeat the steps 1, 2, and 3 by changing one (or more) of the following:

- The structure
- The initial synaptic coefficients
- The parameters of the learning phase: the learning algorithm, stopping criteria, etc.

In general, the advantages of a multilayer neural network can be summed up in its adjustable synaptic coefficients, and a feedback propagation learning algorithm, which is trained on the data. The effectiveness of the latter depends on the complex structure of the neural network used. This makes the structural identification stage of "optimizing the number of layers necessary in a multilayer neural network as well as the number of neurons per layer, in order to improve its performance" an essential step for guaranteeing the best training and generalization. To this end, much research has been devoted to solve this problem of choosing the optimal structure for a multilayer neural network, from the point of view of the number of layers and the number of neurons per layer: such as the pruning and growing algorithms. In this chapter, we will focus on the pruning algorithms.

15.2 Definition of Pruning Algorithms

Commencing with a multilayer neural network of large structure, the task for the pruning algorithm is to optimize the number of layers and the number of neurons needed to model the desired function or application. After pruning, the FNN retains the necessary number of layers, and numbers of neurons in each layer to implement the application. The resulting FNN is regarded as having an optimal structure.

15.3 Review of the Literature

An FFN should ideally be optimized to have a small and compact structure with good learning and generalization capabilities. Many researchers have proposed pruning algorithms to reduce the network size. These algorithms are mainly based on

- The Iterative Pruning Algorithm [CF97]
- Statistical methods [CGGMM95]

- The combined statistical stepwise and iterative neural network pruning algorithm [FFJC09]
  - Pruning heuristics using sensitivity analysis [E01]
  - Nonlinear ARMAX models [FFN01]
  - Adaptive training and pruning in feed-forward networks [WCSL01]
  - Optimal brain damage [LDS90]
  - Pruning algorithm based on genetic algorithm [MHF06]
  - Pruning algorithm based on fuzzy logic [J00]

In this chapter, two existing, published techniques used in pruning algorithms are reviewed. A new pruning algorithm proposed by the authors in [FFJC09] will also be described and discussed.

15.4 First Method: Iterative-Pruning (IP) Algorithm

This method was proposed first by Castellano et al. [CF97] in 1997. The unnecessary neurons in a large FNN of arbitrary structure and parameters are removed, yielding a less complex neural network with a better performance. In what follows, we give some details to help the user to understand the procedure to implement the IP algorithm.

15.4.1 Some Definitions and Notations

An FNN can be represented by the following graph, $N = (V, E, w)$, this notation is known as an "acyclic weighted directed graph" as reported in [CF97] with

$V = \{1, 2, \ldots, n\}$: Set of n neurons
$E \subseteq V*V$: The set of connections between the different neurons V
$w: E \to R$: The function that associates a real value w_{ij} for each connection $(i, j) \in E$

- Each neuron $i \in V$ is associated with two specific sets:
  - The own set of "Projective Field": The set of neurons j fed by the neuron i.

$$P_i = \{j \in V : (j, i) \in E\} \tag{15.1}$$

  - The own set of "Receptive field": The set of neurons j directed to neuron i.

$$R_i = \{j \in V : (i, j) \in E\} \tag{15.2}$$

- We define by p_i and r_i the cardinals respectively of the sets P_i and R_i.
- In the case of a multilayer neural network, the projective and receptive sets of any neuron i belonging to layer l are simply the sets of neurons in layers $(l + 1)$ and $(l - 1)$, respectively.
- The set of neurons V can be divided into three subsets:
  - V_I: The input set neurons of the neural network
  - V_O: The output set neurons of the neural network
  - V_H: The set of hidden neurons of the neural network
- A multilayer neural network works as follows: each neuron receives input information from the external environment as a pattern, and it spreads to the neurons belonging to the projective sets. Similarly, each neuron $i \in V_H \cup V_O$ receives its own receptive set and its input is calculated as follows:

$$u_i = \sum_{j \in R_i} w_{ij} y_j \tag{15.3}$$

where
u_i is the linear output of neuron i
y_j represents the nonlinear output of the neuron j

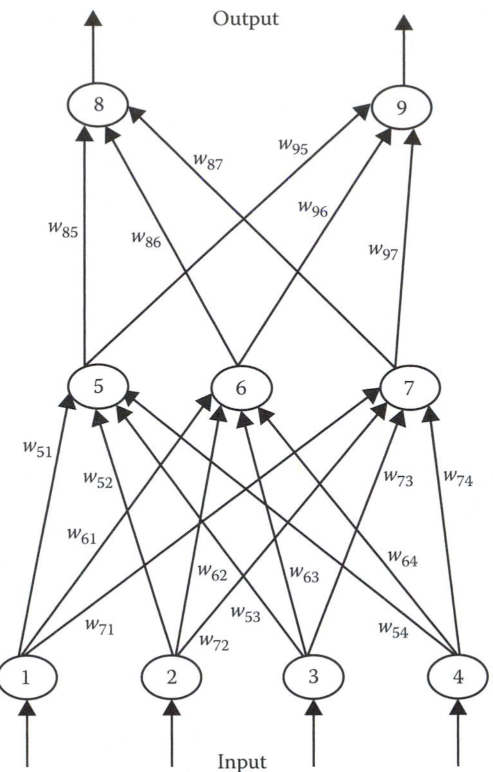

FIGURE 15.1 An example of a multilayer neural network: $V_I = \{1, 2, 3, 4\}$, $V_H = \{5, 6, 7\}$. The own projective set of neuron 7 is $P_7 = \{8, 9\}$.

This neuron then sends the nonlinear output signal y_i to all the neurons belonging to its own projective set P_i:

$$y_i = f(u_i) \tag{15.4}$$

where $f(.)$ is the activation function of the neuron i.

This procedure continues until the output neurons correspond to the input example.

Figure 15.1 shows an example of a multilayer neural network and illustrates the notations introduced above.

15.4.2 Formulation of the Pruning Problem

The pruning procedure (IP) consists of a series of steps of elimination of neurons in the hidden layer of the FNN. The elimination of a neuron directly implies the elimination of all connections associated with this neuron. So, the main question which arises is how is the neuron to be eliminated chosen?

Suppose that a neuron h was chosen for elimination according to a pre-specified criterion, which will be detailed further in the following paragraph. Consider the new set of neural network connections:

$$E_{new} = E_{old} - \left(\{h\} * P_h \cup R_h * \{h\}\right) \tag{15.5}$$

with E_{old} the set of connections before the elimination of the neuron h.

This elimination is followed directly by an adjustment of all the connections related to each neuron belonging to all projective P_h in order to improve (or maintain) the performance of the initial neural network during the learning phase.

Consider the neuron $i \in P_h$. The corresponding linear output of the neuron for a pattern $p \in \{1, \ldots, M\}$ is given by

$$u_i^{(p)} = \sum_{j \in R_i} w_{ij} y_j^{(p)} \tag{15.6}$$

where $y_j^{(p)}$ is the nonlinear output neuron j corresponding to the pattern p.

After removing the neuron h, one has to update the remaining synaptic coefficients w_{ij} in the same layer by a specified value δ in order to maintain the initial performance of the neural network. This idea is illustrated in Figure 15.2.

The linear summation at the output of the pruned layer becomes

$$\sum_{j \in R_i} w_{ij} y_j^{(p)} = \sum_{j \in R_i - \{h\}} (w_{ij} + \delta_{ij}) y_j^{(p)} \tag{15.7}$$

with

$p = 1 \ldots M$, $i \in P_h$

δ_{ij} is the adjustment factor to be determined later.

A simple mathematical development gives

$$\sum_{j \in R_i - \{h\}} \delta_{ij} y_j^{(p)} = w_{ih} y_h^{(p)} \tag{15.8}$$

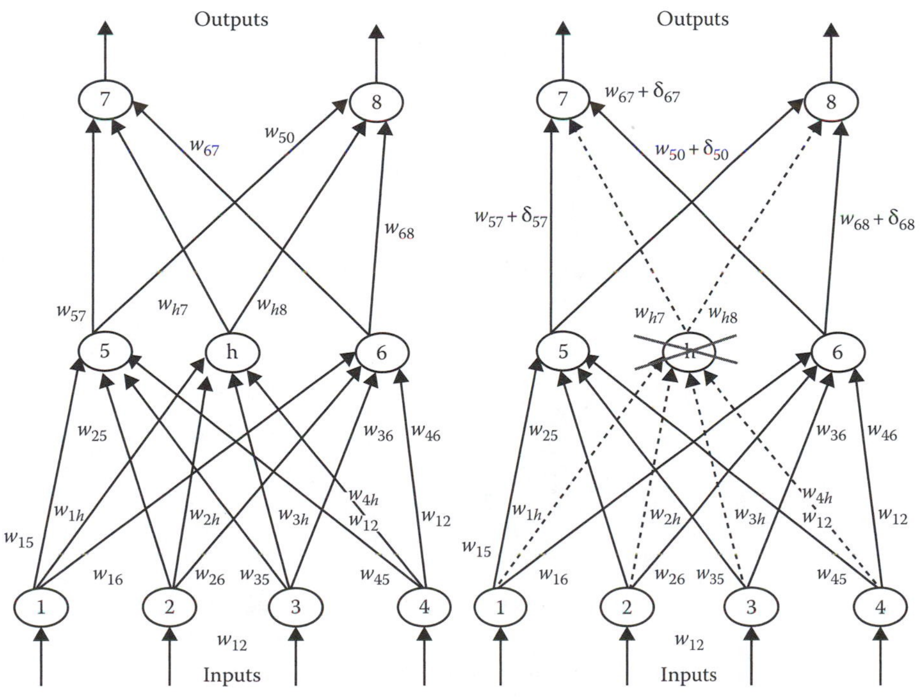

FIGURE 15.2 The procedure of the IP algorithm. The neuron h was chosen for elimination. All connections associated with this neuron are eliminated and all related connections P_h are adjusted with the parameter $\bar{\delta} = [\delta_{57}\ \delta_{58}\ \delta_{67}\ \delta_{68}]$.

This leads to a set of Mp_i linear equations with $k_h = \sum_{i \in P_h} (r_i - 1)$ unknown parameters $\{\delta_{ij}\}$. Similarly, we can observe that k_h represents the total number of connections on their way to the set P_h. It is interesting to write these equations in a matrix form. Thus, consider the following vectors:

- For each $i \in P_h$, the M vector $\overline{y}_i$ contains all values of the output neuron i for each example p:

$$\overline{y}_i = (y_1^{(1)}, \ldots, y_i^{(M)})^T \quad (15.9)$$

- Define $Y_{i,h}$: the $(M*(r_i - 1))$ matrix whose columns are the new releases of neurons $j \in R_i - \{h\}$:

$$Y_{i,h} = [\overline{y}_{j1} \ \ \overline{y}_{j2} \ \ \ldots. \ \ \overline{y}_{j(r_i-1)}] \quad (15.10)$$

with the index jk, for $k = 1, \ldots, (r_i - 1)$ belonging to $R_i - \{h\}$.

Now solve the system in its matrix form:

$$Y_{i,h}\overline{\delta}_i = \overline{z}_{i,h} \quad (15.11)$$

For each $i \in P_h$, with $\overline{\delta}_i$ the unknown parameters, and

$$\overline{z}_{i,h} = w_{ih}\overline{y}_h \quad (15.12)$$

Finally, consider all the set of equations, one gets

$$Y_h\overline{\delta} = \overline{z}_h \quad (15.13)$$

with

$$Y_h = diag(Y_{i1,h}, Y_{i2,h}, ..Y_{ik,h} ..., Y_{ip_h,h}) \quad (15.14)$$

$$\overline{\delta} = (\overline{\delta}_{i1}^T, \overline{\delta}_{i2}^T, \ldots \overline{\delta}_{ik}^T \ldots, \overline{\delta}_{ip_h}^T)^T \quad (15.15)$$

$$\overline{z}_h = (\overline{z_{i1,h}^T}, \overline{z_{i2,h}^T}, \ldots, \overline{z_{ik,h}^T} \ldots, \overline{z_{p_h,h}^T})^T \quad (15.16)$$

Here, indexed $ik(k = 1, \ldots, p_h)$ vary in P_h.

One can conclude by solving this set of equations by minimizing the following criterion:

$$\text{Minimize } \|\overline{z}_h - Y_h\overline{\delta}\|_2 \quad (15.17)$$

Several methods can be used to solve this optimization problem. These include the conjugate gradient preconditioned normal equation (CGPCNE) algorithm (see Appendix 15.A).

15.4.3 How to Choose the Neuron to Be Removed?

The adopted strategy is to associate each neuron in the hidden layer with an indicator giving its contribution to the neural network. After solving Equation 15.17, the choice of a neuron h must ensure a minimum effect on the outputs of the neural network.

The CGPCNE method has an objective to reduce the next *term* at each iteration:

$$\rho_h(\bar{\delta}_k) = \|\bar{z} - Y_h\bar{\delta}_k\|_2 \tag{15.18}$$

where $\|.\|_2$ is the Euclidian norm.

But the choice of the neuron h is made before applying the algorithm CGPCNE. So, this choice is made at iteration $k = 0$:

$$\rho_h(\bar{\delta}_0) = \|\bar{z}_h - Y_h\bar{\delta}_0\|_2 \tag{15.19}$$

Therefore, the neuron h can be chosen according to the following criterion:

$$h = \arg\min_{h \in V_H} \rho_h(\bar{\delta}_0) \tag{15.20}$$

where V_H is the set of hidden neurons in the neural network.

In general, the initial value of $\bar{\delta}_0$ is typically set to zero. A general formulation of the criterion may be written as

$$h = \arg\min \sum_{i \in P_h} w_{hi}^2 \left\|\bar{y}_h\right\|_2 \tag{15.21}$$

A summary of the iterative-pruning (IP) algorithm is given in Table 15.1.

TABLE 15.1 Summary of the IP Algorithm

Step 0: Choose an oversized $NN(k = 0)$ then apply the standard back propagation (SBP) training algorithm

Repeat

Step 1: Identify excess unit h from network $NN(k)$:

$$h = \arg\min \sum_{i \in P_h} w_{hi}^2 \sqrt{\left(y_h^{(1)}\right)^2 + \cdots + \left(y_h^{(\mu)}\right)^2 + \cdots + \left(y_h^{(M)}\right)^2}$$

where

P_h represents the set of units that are fed by unit h (called projective field)

w_{hi} represents weights connected from neuron h to neuron i

$y_h^{(\mu)}$ represents the output of unit h corresponding to patterns $\mu \in \{1,\ldots,M\}$

Step 2: Apply the CGPCNE algorithm to determine $\bar{\delta}$ (refer to appendix)

Step 3: Remove the unit h with all its incoming and outgoing connections and build a new network $NN(k + 1)$ as follows:

$$w_{ji}(k+1) = \begin{cases} w_{ji}(k) & \text{if } i \notin P_h \\ w_{ji}(k) + \delta_{ji} & \text{if } i \in P_h \end{cases}$$

Step 4: $k := k + 1$

continue a until deterioration performance of $NN(k)$ appears

15.5 Second Method: Statistical Stepwise Method (SSM) Algorithm

Unlike the previous approach, Cottrel et al. [CGGMM95] proposed another algorithm that seeks the unknown synaptic coefficients, which must be removed to ensure the neural network performs better. This idea is based on statistical studies such as the Akaike information criterion (AIC) and the Bayesian information criterion (BIC).

15.5.1 Some Definitions and Notations

A neural network can be defined by the existing set of non-zero synaptic coefficients. Therefore we can associate each neural network (NN) model with its set of synaptic coefficients. In general, the quality of the neural network is evaluated by calculating its performance, such as by the quadratic error $S(NN)$:

$$S(NN) = \sum_{p=1}^{M} (y_p - \hat{y}_p)^2 \tag{15.22}$$

where

NN indicate the set of all synaptic coefficients of the initial starting neural network, i.e., $NN = \left\{ w_{11}^{(0)}, w_{12}^{(0)}, \ldots, w_{ij}^{(l)}, \ldots \right\}$

y_p, $\hat{y}_p$ are, respectively, the desired output and current actual output of the neural network for an input example p

This criterion may be sufficient if one is interested only in the examples used during the learning phase. In this case, one can use the following information criteria

- AIC:

$$\text{AIC} = \ln \frac{S(NN)}{M} + \frac{2m}{M} \tag{15.23}$$

- BIC:

$$\text{BIC} = \ln \frac{S(NN)}{M} + \frac{m \log M}{M} \tag{15.24}$$

where

m is the total number of all synaptic coefficients in the neural network

M is the total number of examples used in the learning phase

These criteria give good results when M is large and tends to infinity.

15.5.2 General Idea

The "Statistical Stepwise method" proposed by Cottrell may be considered to successively eliminate the synaptic coefficients w_{l_1}, w_{l_2}, ...w_{l_L}, which is equivalent to the successive implementation of the following neural networks:

$$NN_{l_1} (\text{with}\, w_{l_1} = 0) \quad NN_{l_1 l_2} (\text{with}\, w_{l_1} = w_{l_2} = 0), \quad \text{and so on.}$$

Figure 15.3 shows an example of removing synaptic connection in a couple of iterations.

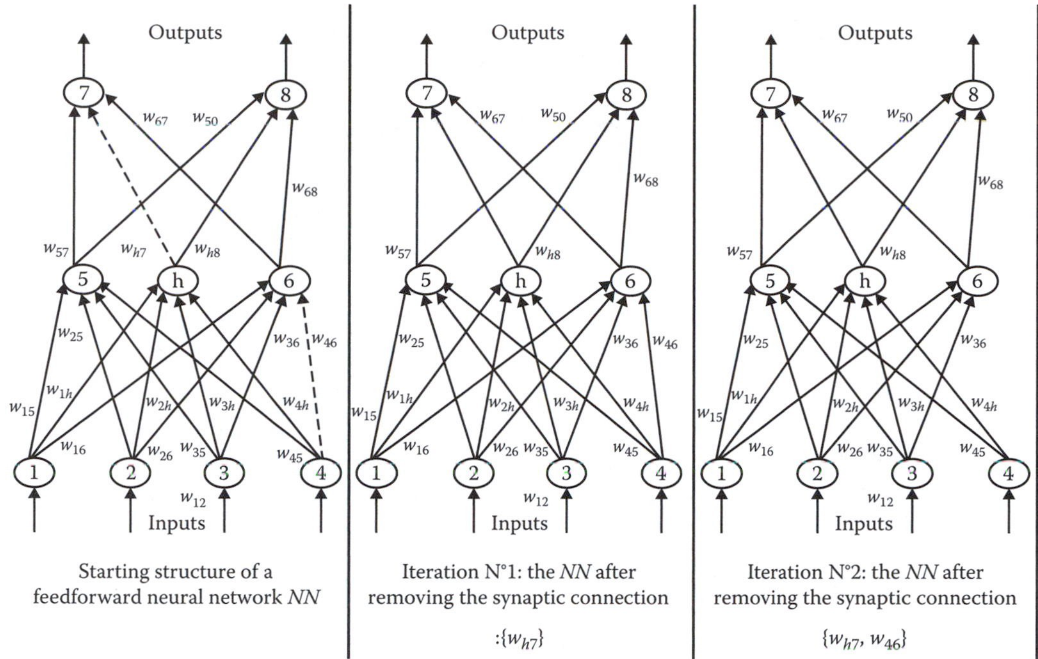

FIGURE 15.3 Example of an initial NN and the removal process using the SSM algorithm over two iterations.

15.5.3 Summary of the Steps in the SSM Algorithm

After training an initially large neural network *NN*, one can apply the following steps to eliminate the insignificant synaptic coefficients in order to ensure better performance (Table 15.2).

15.6 Third Method: Combined Statistical Stepwise and Iterative Neural Network Pruning (SSIP) Algorithm [FFJC09]

The algorithms IP and SSIP are used to simultaneously remove unnecessary neurons or weight connections from a given FNN *NN* in order to "optimize" its structure. Some modifications to the previous pruning algorithms published in [CF97] and [CGGMM95] can be reported. As indicated in the previous pruning algorithms IP and SSM, the stop criterion used in the pruning phase is aimed at finishing the removal process when the first deterioration occurs in the generalization step. However, due to the complexity and nonlinear character of the neural network, if the pruning process is continued without stopping the algorithm when the first deterioration occurs, another pruned network, which gives a better performance, may be formed. Consequently, the pruning was extended beyond the first deterioration to determine any network which yielded a better generalization. In the next section, we shall start to add some additional modifications to improve the pruning capabilities of the standard IP and SSM algorithms, and consequently modified IP and modified SSM algorithms are defined as follows:

1. **Modified IP algorithm**

   The modified IP algorithm for the elimination of one neuron is

   - Apply the steps (1, 2, and 3) of (IP) for the elimination of one neuron.
   - Compute the performance of the new *NN* after removal of the neuron.
   - If the performance of the new *NN* is better than that of $NN_{optimal}(k)$, save the new optimal structure.
   - Decrease the number of neurons.

TABLE 15.2 Summary of the SSM Algorithm

Step 0: Choose an oversized *NN*, then apply the SBP training algorithm.

Repeat

Step 1: For each weight connection $w_l = w_{ij}$, compute the following ratio:

$$Q_{ij} = \sqrt{\frac{2MS_{ij}}{\frac{S(NN)}{M-m}}}$$

where

$S_{ij} = \frac{1}{2M}\left(S(NN_{ij}) - S(NN)\right)$ is the saliency of weight w_{ij} is the increase of the residual error resulting from the elimination of w_{ij}

$S(NN) = \sum_{t=1}^{M} (Y_t - \hat{Y}_{NN})^2$ is the sum of squared residuals of *NN*

$S(NN_{ij}) = \sum_{t=1}^{M} (Y_t - \hat{Y}_{NN_{ij}})^2$ the sum of squared residuals of NN_{ij} (without the weight w_{ij})

M is the number of training patterns

$\hat{Y}_{NN}$ is the output of the *NN* model

m is the total number of remaining weights (before the elimination of w_{ij})

Y_t is the desired output

$\hat{Y}_{NNij}$ is the NN_{ij} output with weight w_{ij} removed

Step 2: Determine $l_{\min}$ corresponding to the minimum value of these ratio

$$w_{l_{\min}} = w_{ij} = \arg\min Q_{ij} = \arg\min \sqrt{\frac{2MS_{ij}}{\frac{S(NN)}{M-m}}}$$

Step 3: If $Q_{l_{\min}} < \tau$ (with $1 < \tau < 1.96$ [CF97]), one can

- Eliminate the weight $w_{l_{\min}}$ corresponding to $l_{\min}$
- Retrain the new $NN = NN_{ij}$ with SBP

Else, do not eliminate the weight $w_{l_{\min}}$

Until the performance of the *NN* deteriorates or no insignificant weights remain

2. **Modified SSM algorithm**

   The modified SSM algorithm for the elimination of one weight connection is

   - Apply the steps (1, 2, and 3) of SSM algorithm for the elimination of one weight.
   - If the weight is to be eliminated
     a. Compute the performance of the new *NN* after weight removal.
     b. If the performance of the new *NN* is better than that of $NN_{optimal}(k)$, save the new optimal structure.
     c. Decrease the number of weights.

With these modifications, the pruning algorithm SSIP is based on the idea of successive elimination of unnecessary neurons and of insignificant weights. It may be regarded as a combination of the modified IP algorithm (to remove an excess neuron) and the modified statistical stepwise algorithm (to prune insignificant links). Two versions of the new pruning algorithm SSIP, namely $SSIP_1$ and $SSIP_2$, are proposed.

TABLE 15.3 First Version of the New Pruning Algorithm: $SSIP_1$

Step 0: Choose an oversized *NN* then apply the SBP training algorithm.
Step 1:
Comments: Apply the modified IP algorithm (removal of neuron) to all the NN
Repeat
Apply the modified IP algorithm to remove one insignificant neuron from all the NN
Until
One neuron remains in each layer
Conclusion: We have the $NN_{(IP)} = NN_{optimal}(k)$ *after removing the unnecessary neurons from NN*
Step 2:
Comments: Apply the modified SSM algorithm (removal of weight) to all the $NN_{(IP)}$
Repeat
Apply the modified SSM algorithm to eliminate one insignificant weight from all the $NN_{(IP)}$ *the latest optimal structure*
Until
There are no more insignificant weights to be removed or there is only at least one weight between any two layers
Conclusion: The final optimal structure obtained is labeled: $NN_{optimal} = NN_{(SSIP_1)}$

15.6.1 First Version: $SSIP_1$

The modified IP algorithm is firstly applied to remove insignificant neurons, and then the modified SSM algorithm is applied to remove any insignificant weight.

This version $SSIP_1$ is given in Table 15.3.

15.6.2 Second Version: $SSIP_2$

The two modified algorithms are applied separately to each layer, while retaining the previously optimized structure in each previous layer. The general algorithmic steps are given in Table 15.4.

TABLE 15.4 Second Version of the New Pruning Algorithm: $SSIP_2$

Step 0: Choose an oversized *NN*(*L* layers), then apply the SBP training algorithm
Step 1: Start from the input layer
Step 2:
Comments: Apply the modified IP algorithm (removal of neuron) only to this layer
Repeat
Apply the modified IP algorithm to remove only one neuron from this layer
Until
All insignificant nodes have been removed or there is only one neuron in this layer
Comments: Apply the modified SSM algorithm (removal of weight) only to this layer
Repeat
Apply the modified SSM algorithm to eliminate one insignificant weight from this layer
Until
No insignificant weights are left or there is only one weight in this layer
Step 3: Go on to the following layer (2, 3 …) and repeat step 2 until layer $= L - 1$
Step 4:
Comments: Apply the modified SSM algorithm (removal of weight) only to the output layer
Repeat
Apply the modified SSM algorithm to eliminate one insignificant weight from this layer
Until
There are no remaining insignificant weights or there is only one weight in this layer
Conclusion: The final optimal structure obtained is labeled $NN_{optimal} = NN_{(SSIPI_2)}$

15.7 Comments

All the later algorithms, SSM, IP, and $SSIP_{1,2}$ have been applied in real-world applications. In [FF02], the authors have applied the SSM algorithm to circle in the square problem while in [M93] another method of pruning has been proposed. In [FAN01,SJF96], the authors applied SSM, IP, and $SSIP_{1,2}$ to medical brain diseases classification problems. In [P97], the authors present another pruning method near to SSM algorithm to prune connections between neurons.

15.8 Simulations and Interpretations

To test the effectiveness of the later two versions of $SSIP_{1,2}$ algorithm, as compared to the SSM and IP algorithms, three parameters are used to decide whether or not the pruned network is suitable, namely:

- The complexity: the total number of links necessary for modeling each application.
- The sensitivity of learning: the percentage of the number of patterns used during learning, which are perfectly trained.
- The sensitivity of generalization of the patterns, which were not used during the learning phase.

Two applications are selected to justify the performance of the proposed method. These include

1. Brain Disease Detection [SJF96]: A *NN* is used to differentiate between Schizophrenic (SH), Parkinson's disease (PD), Huntington's disease (HD) patients, and normal control (NS) subjects. Each individual is characterized by 17 different variables.
2. Texture classification [FSN98]: The *NN* is used to classify some images that are randomly chosen from eight initial textures. The least mean square filter coefficients of each texture image are used to form the input vector (8 variables) to the neural network.

For the two problems, several structures (1 input layer, 2 hidden layers, 1 output layer) with different initial conditions are examined by the IP, SSM, and $SSIP_{1,2}$ algorithms in order to remove the insignificant parameters (units or weights connections) for each structure. A statistical study has been performed to evaluate both the two new versions of SSIP1 and SSIP2. Hence, 100 realizations of different initial structures of weighting coefficients have been used and tested. Tables 15.5 and 15.6 illustrate the average performance of each pruning algorithm.

TABLE 15.5 Summary of Results of Texture Classification

Algorithm	Percentage of Pruned Unit (%)	Percentage of Pruned Links in Layer 1 (%)	Percentage of Pruned Links in Layer 2 (%)	Percentage of Pruned Links in Layer 3 (%)	Complexity (Number of Links)	Sensitivity of Learning (%)	Sensitivity of Generalization (%)
NN without pruning	—	—	—	—	163	79.64 ± 10	79.37 ± 7
NN with IP [CF97]	26.29 ± 11	—	—	—	114 ± 35	78.74 ± 8	77.79 ± 5
NN with SSM [CGGMM95]	—	65.87 ± 21	55.47 ± 18	28.24 ± 13	75 ± 25	87.62 ± 3	81.06 ± 4
NN with $SSIP_1$ [FFJC09]	26.90 ± 7	25 ± 10	24.72 ± 12	6 ± 4	90.25 ± 10	91.63 ± 6	86.75 ± 4

TABLE 15.6 Summary of Results of Detection of Brain Diseases

Algorithm	Percentage of Pruned Unit (%)	Percentage of Pruned Links in Layer 1 (%)	Percentage of Pruned Links in Layer 2 (%)	Percentage of Pruned Links in Layer 3 (%)	Complexity (Number of Links)	Sensitivity of Learning (%)	Sensitivity of Generalization (%)
NN without pruning	—	—	—	—	298.5	47.30 ± 20	52.77 ± 10
NN with IP [CF97]	68.55 ± 15	—	—	—	68.5 ± 15	43.16 ± 12	50 ± 14
NN with SSM [CGGMM95]	—	75.36 ± 5	79.27 ± 12	4.80 ± 3	114 ± 20	52.95 ± 25	43.15 ± 18
NN with $SSIP_2$ [FFJC09]	39.21 ± 18	65.14 ± 12	44.28 ± 7	22.18 ± 5	77.5 ± 12	77.15 ± 15	76.39 ± 12

The first thing that can be seen from these tables is that the new proposed pruning algorithms, $SSIP_{1,2}$, indeed offer good learning and generalization capabilities.

- Note that during the pruning process, the SSIP algorithm eliminates around 59% of links from the total number in order to achieve an improvement of +39% for the sensitivity of learning and +26% for the sensitivity of generalization.
  Whereas for the SSM and IP algorithms, we have
- For the SSM algorithm, there is an elimination of about 57.5% of links, which corresponds to an improvement of about +10.95% for the sensitivity of learning and of −8% for the sensitivity of generalization.
- For the IP algorithm, there is an elimination of around 53.5% of the links, which corresponds to about a −4.94% degradation in the sensitivity of learning and a −3.64% degradation in the sensitivity of generalization.

These results highlight the superiority of the $SSIP_{1,2}$ versus SSM and IP used separately.

15.9 Conclusions

In this chapter, we have discussed several approaches of pruning algorithms in order to obtain a multilayer neural network "optimal" in terms of number of neurons or synaptic coefficients per layer.

We have detailed three pruning algorithms such as IP, SSM, and SSIP, from what one can understand the pruning procedure to be applied in order to handle in practice an optimal structure of a neural network. From the simulation results of these algorithms, we can conclude the following:

- The effectiveness of any pruning algorithm depends not only on the efficiency rules of the pruning but also on (1) the manner of applying these rules and (2) the criteria used for stopping these pruning algorithms.
- The pruning results depend greatly on the initialization of the starting *NN* weighting coefficients.
- Sometimes and when one stars the pruning, the user should pay attention to not stop the running of the algorithm at the first performance degradation because this may be tricky and he should continue the pruning iterations.
- As a concluding remark, in this chapter, we have presented some ideas for pruning *NN*, while we are sure that the pruning problem is an open area of research and we believe that still many things remain to be done in this subject.

Appendix 15.A: Algorithm of CGPCNE—Conjugate Gradient Preconditioned Normal Equation

This is the procedure used in the pruning algorithm (IP) proposed in [CF97] to solve the following problem:

To determine the parameter δ (as shown in Figure 15.A.1) to Minimize $\left\|\overline{z}_h - Y_h\overline{\delta}\right\|_2$ with:

$$\overline{z}_h = (\overline{z_{i1,h}^T}, \overline{z_{i2,h}^T}, \ldots, \overline{z_{ik,h}^T}, \ldots, \overline{z_{p_{h,h}}^T})^T$$

$$\overline{z}_{i,h} = w_{ih}\overline{y}_h = w_{hi}(y_h^{(1)} \quad y_h^{(2)} \quad \ldots \quad y_h^{(M)})^T$$

$$Y_h = diag(Y_{i1,h}, Y_{i2,h}, \ldots, Y_{ip_{h,h}})$$

$$Y_{i,h} = [\overline{y}_{j1} \ \ \overline{y}_{j2} \ \ \ldots \ \ \overline{y}_{jr_i-1}]$$

where

h is the index of the neuron to be detected and removed

$\{j_1, j_2, \ldots, r\}$ are the indexes for the set of neurons that feed the neuron i (h not considered in this set)

Let us define:

D: a diagonal matrix whose non-zero elements are calculated as follows $(D)_{jj} = \left\|Y(:,j)\right\|_2^2$ (with the notation $Y(:, j)$ indicating the jth column of the matrix Y)

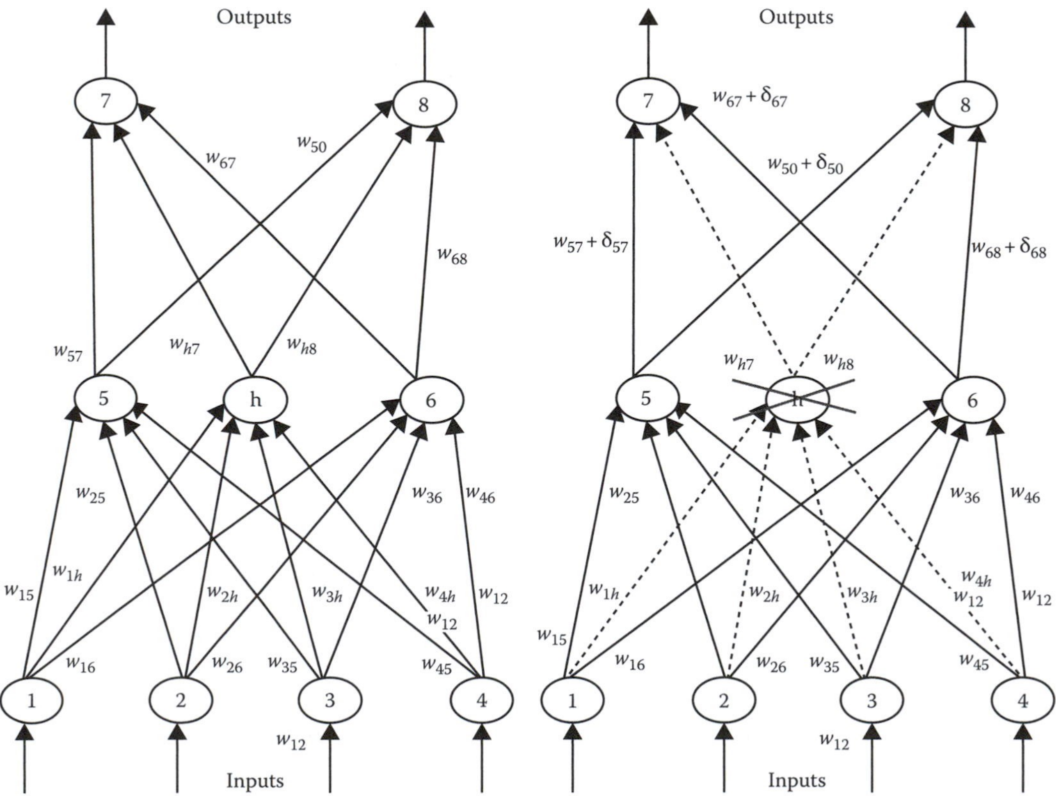

FIGURE 15.A.1 The procedure for the IP algorithm.

L: Triangular matrix defined as follows $(L)_{jk} = Y(:,j)^T\, Y(:,k), j > k$.

C_w: A matrix of "preconditioning" is calculated by $C_w = (D + wL)D^{-1/2}$ where w is a relaxation parameter lying between 0 and 2.

Hence, the steps in the following procedure CGPCNE are

1. Initialization: $\bar{r}_0 := \bar{z} - Y\,\bar{\delta}_0$; $\bar{s}_0 := C_w^{-1} Y^T \bar{r}_0$; $\bar{p}_0 := \bar{s}_0$; $k := 0$
2. Repeat

a. $\bar{q}_k := Y C_w^{-T} \bar{p}_k$	e. $\beta_k := \left\|\bar{s}_{k+1}\right\|_2^2 / \left\|\bar{s}_k\right\|_2^2$
b. $\alpha_k = \left\|\bar{s}_k\right\|_2^2 / \left\|\bar{q}_k\right\|_2^2$	f. $\bar{p}_{k+1} := \bar{s}_{k+1} + \beta_k \bar{p}_k$
c. $\bar{r}_{k+1} := \bar{r}_k - \alpha_k \bar{q}_k$	g. $\bar{\delta}_{k+1} := \bar{\delta}_k + \alpha_k C_w^{-T} \bar{p}_k$
d. $\bar{s}_{k+1} := C_w^{-1} Y^T \bar{r}_{k+1}$	h. $k := k + 1$

$$\text{until } \left\|\bar{\delta}_k - \bar{\delta}_{k-1}\right\|_2 < \varepsilon,$$

where ε is a small predetermined constant.

References

[CF97] G. Castellano and A.M. Fanelli, An iterative pruning algorithm for feed-forward neural networks, *IEEE Transactions on Neural Networks*, 8(3), 519–531, May 1997.

[CGGMM95] M. Cottrell, B. Girard, Y. Girard, M. Mangeas, and C. Muller, Neural modelling for time series: Statistical stepwise method for weight elimination, *IEEE Transactions on Neural Networks*, 6(6), 1355–1363, November 1995.

[E01] A.P. Engelbrecht, A new pruning heuristic based on variance analysis of sensitivity information, *IEEE Transactions on Neural Networks*, 12(6), 1386–1399, November 2001.

[FAN01] F. Fnaiech, S. Abid, and M. Najim, A fast feed-forward training algorithm using a modified form of standard back-propagation algorithm, *IEEE Transactions on Neural Network*, 12(2), 424–430, March 2001.

[FF02] N. Fnaiech, F. Fnaiech, and M. Cheriet, A new feed-forward neural network pruning algorithm: Iterative-SSM pruning, *IEEE International Conference on Systems, Man and Cybernetics*, Hammamet, Tunisia, October 6–9, 2002.

[FFJC09] N. Fnaiech, F. Fnaiech, B.W. Jervis, and M. Cheriet, The combined statistical stepwise and iterative neural network pruning algorithm, *Intelligent Automation and Soft Computing*, 15(4), 573–589, 2009.

[FFN01] F. Fnaiech, N. Fnaiech, and M. Najim, A new feed-forward neural network hidden layer's neurons pruning algorithm, *IEEE International Conference on Acoustic Speech and Signal Processing (ICASSP'2001)*, Salt Lake City, UT.

[FSN98] F. Fnaiech, M. Sayadi, and M. Najim, Texture characterization based on two dimensional lattice coefficients, *IEEE ICASSP'98*, Seattle, WA and *IEEE ICASSP'99*, Phoenix, AZ.

[J00] J.-G. Juang, Trajectory synthesis based on different fuzzy modeling network pruning algorithms, *Proceeding of the 2000 IEEE, International Conference on Control Applications*, Anchorage, AK, 2000.

[LDS90] Y. LeCun, J.S. Denker, and S.A. Solla, Optimal brain damage. In D. Touretzky, ed. *Advances in Neural Information Processing Systems*, Vol. 2, pp. 598–605. Morgan Kaufmann, Palo Alto, CA, 1990.

[M93] J.E. Moody, Prediction risk and architecture selection for neural networks, *Form Statistic to Neural Networks: Theory and Pattern Recognition Application*. Springer-Verlag, Berlin, Germany, 1993.

[MHF06] S. Mei, Z. Huang, and K. Fang, A neural network controller based on genetic algorithms, *International Conference on intelligent Processing Systems*, Beijing, China, October 28–31, 1997.

[P97] L. Prechelt, Connection pruning with static and adaptative pruning schedules, *Neuro-Comput.*, 16(1), 49–61, 1997.

[SJF96] M. Sayadi, B.W. Jervis, and F. Fnaiech, Classification of brain conditions using multilayer perceptrons trained by the recursive least squares algorithm, *Proceeding of the 2nd International Conference on Neural Network and Expert System in Medicine and Healthcare*, Plymouth, U.K., pp. 5–13, 28–30, August 1996.

[W09] B.M. Wilamowski, Neural Network Architectures and Learning algorithms, *IEEE Ind. Electron. Mag.*, 3(4), 56–63, November 2009.

[W10] B.M. Wilamowski, Challenges in Applications of Computational Intelligence in Industrial Electronics *ISIE10 - International Symposium on Industrial Electronics,* Bari, Italy, July 4–7, 2010, pp. 15–22.

[WCSL01] K.-W. Wong, S.-J. Chang, J. Sun, and C.S. Leung, Adaptive training and pruning in feed-forward networks, *Electronics Letters*, 37(2), 106–107, 2001.

[WHM03] B.M. Wilamowski, D. Hunter, and A. Malinowski, Solving parity-N problems with feedforward neural network, *Proceedings of the IJCNN'03 International Joint Conference on Neural Networks*, pp. 2546–2551, Portland, OR, July 20–23, 2003.

[WY10] B.M. Wilamowski and H. Yu, Improved Computation for Levenberg Marquardt Training, *IEEE Trans. Neural Netw.* 21(6), 930–937, June 2010.

16

Principal Component Analysis

Anastasios Tefas
Aristotle University of Thessaloniki

Ioannis Pitas
Aristotle University of Thessaloniki

16.1 Introduction

Principal component analysis (PCA) is a classical statistical data analysis technique that is widely used in many real-life applications for dimensionality reduction, data compression, data visualization, and more usually for feature extraction [6]. In this chapter, the theory of PCA is explained in detail and practical implementation issues are presented along with various application examples. The mathematical concepts behind PCA such as mean value, covariance, eigenvalues, and eigenvectors are also briefly introduced.

The history of PCA starts in 1901 from the work of Pearson [9], who proposed a linear regression method in N dimensions using least mean squares (LMS). However, Hotelling is considered to be the founder of PCA [4], since he was the first to propose PCA for analyzing the variance of multidimensional random variables. PCA is equivalent to the Karhunen–Loève transform for signal processing [2].

The principal idea behind PCA is that, in many systems that are described by many, let us assume N, random variables, the degrees of freedom M are less than N. Thus, although the dimensionality of the observation vectors is N, the system can be described by $M < N$ uncorrelated but hidden random variables. These hidden random variables are usually called *factors* or *features* of the observation vectors. In the following, the term *sample vectors* will refer to the observation vectors. In statistics, the term *superficial dimensionality* refers to the dimensionality N of the sample vector, whereas the term *intrinsic dimensionality* refers to the dimensionality M of the feature vectors. Obviously, if the N dimensions of the sample vectors are uncorrelated, then the intrinsic dimensionality is also $M = N$. As the correlation between the random variables increases, less features are needed in order to represent the sample vectors and thus $M < N$. In the limit, only one feature ($M = 1$) is enough for representing the sample vectors.

In the following, we will use lower case bold roman letters to represent column vectors (e.g., $\mathbf{x} \in \mathcal{R}^N$) and uppercase bold roman letters to represent matrices (e.g., $\mathbf{W} \in \mathcal{R}^{M \times N}$). The transpose of a vector or a matrix is denoted using the superscript T, so that $\mathbf{x}^T$ will be a row vector. The notation $(x_1, x_2, \ldots, x_N)$ is used for representing a row vector and $\mathbf{x} = (x_1, x_2, \ldots, x_N)^T$ is used for representing a column vector.

The PCA algorithm can be defined as the orthogonal projection of the sample vectors onto a lower dimension linear subspace such that the variance of the projected sample vectors (i.e., the feature vectors) is maximized in this subspace [4]. This definition is equivalent to the orthogonal projection, such that the mean squared distance between the sample vectors and their projections in the lower dimensional space is minimized [9].

Let us begin the detailed PCA description by considering that the sample vectors are represented by a random vector $\mathbf{x}_i = (x_1, x_2,\ldots, x_N)^T$, $i = 1 \ldots K$, having expected value $E\{\mathbf{x}\} = 0$ and autocorrelation matrix $\mathbf{R}_x = E\{\mathbf{x}\mathbf{x}^T\}$. According to the PCA transform, the feature vector $\hat{\mathbf{x}}_i \in \mathcal{R}^M$ is a linear transformation of the sample vector $\mathbf{x}_i$:

$$\hat{\mathbf{x}}_i = \mathbf{W}^T \mathbf{x}_i \tag{16.1}$$

where $\mathbf{W} \in \mathcal{R}^{N\times M}$ is a matrix having less columns than rows.

We can consider that if we want to reduce the dimension of the sample vectors from N to 1, we should find an appropriate vector $\mathbf{w} \in \mathcal{R}^N$, such that for each sample vector $\mathbf{x}_i \in \mathcal{R}^N$, the projected scalar value will be given by the inner product, $\hat{x} = \mathbf{w}^T\mathbf{x}$. If we want to reduce the sample vectors dimension from N to M, then we should find M projection vectors $\mathbf{w}_i \in \mathcal{R}^N$, $i = 1 \ldots M$. The projected vector $\hat{\mathbf{x}} \in \mathcal{R}^M$ will be given by $\hat{\mathbf{x}} = \mathbf{W}^T\mathbf{x}$, where the matrix $\mathbf{W} = (\mathbf{w}_1, \mathbf{w}_2,\ldots, \mathbf{w}_M)$.

16.2 Principal Component Analysis Algorithm

The basic idea in PCA is to reduce the dimensionality of a set of multidimensional data using a linear transformation and retain as much data variation as possible. This is achieved by transforming the initial set of N random variables to a new set of M random variables that are called *principal components* using projection vectors. The principal components are ordered in such way that the principal components retain information (variance) in descending order. That is, the projection of the sample vectors to $\mathbf{w}_1$ should maximize the variance of the projected samples.

Let us denote by $\mathbf{m}_x$ the mean vector of the sample data:

$$\mathbf{m}_x = \frac{1}{K}\sum_{i=1}^{K}\mathbf{x}_i, \tag{16.2}$$

where K is the number of sample vectors in the data set. The mean value of the data after the projection to the vector $\mathbf{w}_1$ is given by

$$m_x = \frac{1}{K}\sum_{i=1}^{K}\mathbf{w}_1^T\mathbf{x}_i = \mathbf{w}_1^T\mathbf{m}_x \tag{16.3}$$

and the variance of the projected data to the vector $\mathbf{w}_1$ is given by

$$s_x = \frac{1}{K}\sum_{i=1}^{K}\left(\mathbf{w}_1^T\mathbf{x}_i - \mathbf{w}_1^T\mathbf{m}_x\right)^2 = \mathbf{w}_1^T\mathbf{S}_x\mathbf{w}_1, \tag{16.4}$$

where $\mathbf{S}_x$ is the covariance matrix of the sample vectors defined by

$$\mathbf{S}_x = \frac{1}{K}\sum_{i=1}^{K}(\mathbf{x}_i - \mathbf{m}_x)(\mathbf{x}_i - \mathbf{m}_x)^T. \tag{16.5}$$

Without loss of generality, we can consider the principal components to be normalized vectors (i.e., $\mathbf{w}_1^T\mathbf{w}_1 = 1$), since we need to find only the direction of the principal component. In order to find the first principal component $\mathbf{w}_1$ we seek the vector $\mathbf{w}_1$ that maximizes the projected data variance s_x in (16.4) and has unit magnitude:

$$\text{maximize } \mathbf{w}_1^T \mathbf{S}_x \mathbf{w}_1 \tag{16.6}$$

$$\text{subject to } \mathbf{w}_1^T \mathbf{w}_1 = 1. \tag{16.7}$$

The above optimization problem can be solved using Langrange multipliers [3]:

$$J = \mathbf{w}_1^T \mathbf{S}_x \mathbf{w}_1 + \lambda_1 (1 - \mathbf{w}_1^T \mathbf{w}_1). \tag{16.8}$$

By setting the derivative of the Langrangian with respect to the projection vector equal to zero we obtain:

$$\frac{\vartheta J}{\vartheta \mathbf{w}_1} = 0 \Rightarrow \mathbf{S}_x \mathbf{w}_1 = \lambda_1 \mathbf{w}_1, \tag{16.9}$$

which implies that the variance s_x is maximized if the sample data are projected using an eigenvector $\mathbf{w}_1$ of their covariance matrix $\mathbf{S}_x$. The corresponding Langrange multiplier is the eigenvalue that corresponds to the eigenvector $\mathbf{w}_1$. Moreover, if we multiply both sides of (16.9) by the vector $\mathbf{w}_1$, we get:

$$\mathbf{w}_1^T \mathbf{S}_x \mathbf{w}_1 = \lambda_1 \mathbf{w}_1^T \mathbf{w}_1 = \lambda_1. \tag{16.10}$$

That is, the solution to the PCA problem is to perform eigenanalysis to the covariance matrix of the sample data. All the eigenvectors and their corresponding eigenvalues are solutions of the optimization problem. The value of the corresponding eigenvalue is equal to the resulting variance after the projection and thus, if we want to maximize the variance of the projected samples we should choose as projection vector the eigenvector that corresponds to the largest eigenvalue. In the literature, either the projection values or the eigenvectors are called principal components [6]. An example of PCA on a two-dimensional data set is illustrated in Figure 16.1. It is obvious that

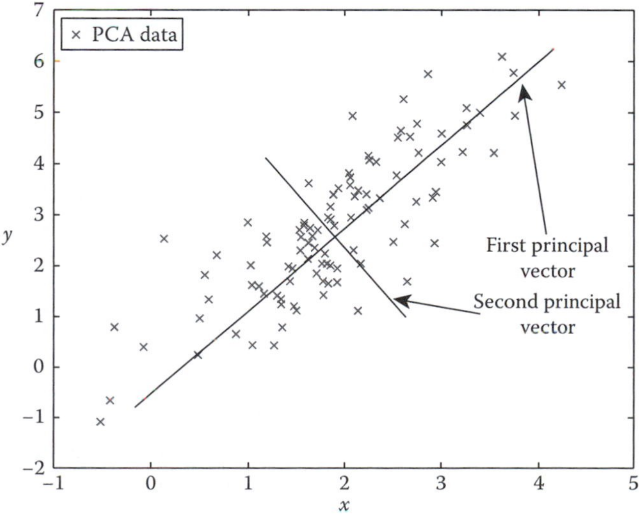

FIGURE 16.1 PCA example.

the first principal component is the one that maximizes the variance of the projected samples. The second principal component is orthogonal to the first one.

If we want to proceed in finding the next principal component, we should seek for another projection vector $\mathbf{w}_2$ that maximizes the variance of the projected samples and it is orthonormal to $\mathbf{w}_1$. It is straightforward to form the corresponding Langrangian and maximize it with respect to $\mathbf{w}_2$. The projection vector that maximizes the variance of the projected samples is the eigenvector of the covariance matrix that corresponds to the second larger eigenvalue.

Similarly, we can extract M eigenvectors of the covariance matrix that will form the M principal components. These column vectors form the PCA projection matrix $\mathbf{W} = (\mathbf{w}_1, \mathbf{w}_2, \ldots, \mathbf{w}_M)$ that can be used to reduce the dimensionality of the sample vectors from N to M. Let us also note that if an eigenvalue of the covariance matrix is zero, then the variance of the projected samples in the corresponding eigenvector is zero according to (16.9). That is, all the sample vectors are projected to the same scalar value when using this eigenvector as projection vector. Thus, we can conclude that there is no information (i.e., data variation) in the directions specified by the null eigenvectors of the covariance matrix. Moreover, we can discard all these directions that correspond to zero eigenvalues, without any loss of information. The number of nonzero eigenvalues is equal to the covariance matrix rank. The procedure of discarding the dimensions that correspond to zero eigenvalues is usually called removal of the null subspace.

Another interpretation of the PCA algorithm is that it extracts the directions that minimize the mean square error between the sample data and the projected sample data. That is, in terms of reconstruction error, PCA finds the optimal projections for representing the data (i.e., minimizes the representation error). It is straightforward to prove that PCA minimizes the mean square error of the representation [1].

16.3 Computational Complexity and High-Dimensional Data

Let us proceed now in estimating the computational complexity of the PCA algorithm to a data set of K samples $\mathbf{x}_i$ of dimensionality N. To do so, we should firstly consider that we need to calculate the mean vector $\mathbf{m}_x$ and the covariance matrix $\mathbf{S}_x$ of the data set and, afterward, to perform eigenanalysis to the data covariance matrix. The computational complexity will be given in terms of the number of basic operations, such as additions, multiplications, and divisions. The complexity of calculating the mean vector of the data set in (16.2) is $O(KN)$, since we need $K-1$ additions for each of the N dimensions. Similarly, the complexity of calculating the covariance matrix in (16.5) is $O(KN^2)$. Finally, we have to calculate the eigenvectors of the covariance matrix which is a procedure with complexity $O(N^3)$ [3]. If we want to calculate only the first M eigenvectors, then the complexity is reduced to $O(MN^2)$.

It is obvious from the above analysis that, if the data have high-dimensionality, the computational complexity of PCA is large. Moreover, the memory needed for storing the $N \times N$ covariance matrix is also very large. In many real applications, such as image processing, the dimensionality N of the sample vectors is very large resulting in demanding implementations of the PCA. For example, if we try to use PCA for devising an algorithm for face recognition, even if we use images of relatively small size (e.g., 32×32), the resulting sample vectors will have 1024 dimensions. Straightforward application of the PCA algorithm requires the eigenanalysis of the 1024×1024 covariance matrix. This is a computationally intensive task. The problem is even worse, if we consider full-resolution images of many megapixels where applying PCA is practically computationally infeasible.

In many applications, however, that use high-dimensional data, the number of sample vectors K is much smaller than their dimensionality $N \gg K$. In this case, the covariance matrix is not full-rank and the eigenvectors of the covariance matrix that correspond to nonzero eigenvalues and, thus, constitute the principal components are at most $K-1$. So, there is no sense in calculating more than $K-1$ principal components, since these eigenvectors project the data samples to scalar values with zero variance. As we have already noted, applying PCA in very high-dimensional data is computationally infeasible and, thus, we should follow a different approach for calculating the $M < K$ principal components.

Let us define by $\mathbf{X} = (\mathbf{x}_1 - \mathbf{m}_x, \mathbf{x}_2 - \mathbf{m}_x, \ldots, \mathbf{x}_K - \mathbf{m}_x)$ the centered data matrix, with dimensions $N \times K$ that represents the sample vectors. Each column of the matrix $\mathbf{X}$ is a sample vector after subtracting the data mean vector. The covariance matrix of the sample data is given by

$$\mathbf{S}_x = \frac{1}{K}\mathbf{X}\mathbf{X}^T. \tag{16.11}$$

Let us define by $\hat{\mathbf{S}}_x = \frac{1}{K}\mathbf{X}^T\mathbf{X}$ an auxiliary matrix of dimension $K \times K$. By performing eigenanalysis to $\hat{\mathbf{S}}_x$ we can calculate the K-dimensional eigenvectors $\hat{\mathbf{w}}_i$ that correspond to the K largest eigenvalues λ_i, $i = 1,\ldots, K$ of $\hat{\mathbf{S}}_x$. For these eigenvectors, we have

$$\hat{\mathbf{S}}_x\hat{\mathbf{w}}_i = \lambda_i\hat{\mathbf{w}}_i \Rightarrow \frac{1}{K}\mathbf{X}^T\mathbf{X}\hat{\mathbf{w}}_i = \lambda\hat{\mathbf{w}}_i. \tag{16.12}$$

Multiplying (16.12) from both sides by $\mathbf{X}$, we get

$$\frac{1}{K}\mathbf{X}\mathbf{X}^T(\mathbf{X}\hat{\mathbf{w}}_i) = \lambda_i(\mathbf{X}\hat{\mathbf{w}}_i) \Rightarrow \mathbf{S}_x\mathbf{u}_i = \lambda_i\mathbf{u}_i \tag{16.13}$$

with $\mathbf{u}_i = \mathbf{X}\hat{\mathbf{w}}_i$. That is, the vectors $\mathbf{u}_i$ of dimension N are eigenvectors of the covariance matrix $\mathbf{S}_x$ of the initial sample vectors. Following this procedure, we can calculate efficiently the $K - 1$ eigenvectors of the covariance matrix that correspond to nonzero eigenvalues. The remaining $N - K + 1$ eigenvectors of the covariance matrix correspond to zero eigenvalues and are not important. We should also note that these eigenvectors are not normalized and thus, should be normalized in order to have unit length.

16.4 Singular Value Decomposition

In many cases, PCA is implemented using the singular value decomposition (SVD) of the covariance matrix. SVD is an important tool for factorizing an arbitrary real or complex matrix, with many applications in various research areas, such as signal processing and statistics. SVD is widely used for computing the pseudoinverse of a matrix, for least-squares data fitting, for matrix approximation and rank, null space calculation. SVD is closely related to PCA, since it gives a general solution to matrix decomposition and, in many cases, SVD is more stable numerically than PCA. Let $\mathbf{X}$ be an arbitrary $N \times M$ matrix and $\mathbf{C} = \mathbf{X}^T\mathbf{X}$ be a rank R, square, symmetric $M \times M$ matrix. The objective of SVD is to find a decomposition of the matrix $\mathbf{X}$ to three matrices $\mathbf{U}$, $\mathbf{S}$, $\mathbf{V}$ of dimensions $N \times M$, $M \times M$, and $M \times M$, respectively, such that

$$\mathbf{X} = \mathbf{U}\mathbf{S}\mathbf{V}^T, \tag{16.14}$$

where $\mathbf{U}^T\mathbf{U} = \mathbf{I}$, $\mathbf{V}^T\mathbf{V} = \mathbf{I}$ and $\mathbf{S}$ is a diagonal matrix. That is, the matrix $\mathbf{X}$ can be expressed as the product of a matrix with orthonormal columns, a diagonal, and an orthogonal matrix.

There is a direct relation between PCA and SVD when principal components are calculated using the covariance matrix in (16.5). If we consider data samples that have zero mean value (e.g., by centering), then, PCA can be implemented using SVD. To do so, we perform SVD to the data matrix $\mathbf{X}$ used for the definition of the covariance matrix in (16.11). Thus, we find the matrices $\mathbf{U}$, $\mathbf{S}$, and $\mathbf{V}$, such that $\mathbf{X} = \mathbf{U}\mathbf{S}\mathbf{V}^T$. According to (16.11) the covariance matrix is given by

$$\mathbf{S}_x = \frac{1}{K}\mathbf{X}\mathbf{X}^T = \frac{1}{K}\mathbf{U}\mathbf{S}^2\mathbf{U}^T. \tag{16.15}$$

In this case, $\mathbf{U}$ is an $N \times M$ matrix. The eigenvectors of the covariance matrix that correspond to nonzero eigenvalues are stored in the first N columns of $\mathbf{U}$, if $N < M$, or in the first M columns if $N \leq M$. The corresponding eigenvalues are stored in $\mathbf{S}^2$. That is, diagonalization of the covariance matrix using SVD yields the principal components. In practice, PCA is considered as a special case of SVD and in many cases SVD has better numerical stability.

16.5 Kernel Principal Component Analysis

The standard PCA algorithm can be extended to support nonlinear principal components using nonlinear kernels [10]. The idea is to substitute the inner products in the space of the sample data with kernels that transform the scalar products to a higher, even infinite, dimensional space. PCA is then applied to this higher dimensional space resulting in nonlinear principal components. This method is called kernel principal component analysis (KPCA).

Let us assume that we have subtracted the mean value and the modified sample vectors $\mathbf{x}_i$ have zero mean. We can consider nonlinear transformations $\phi(\mathbf{x}_i)$ that map the sample vectors to a high-dimensional space. Let us also assume that the transformed data have also zero mean. That is, $\sum_{i=1}^{K} \phi(\mathbf{x}_i) = 0$. The covariance matrix in the nonlinear space is given by

$$\mathbf{S}_x^{\phi} = \frac{1}{K} \sum_{i=1}^{K} \phi(\mathbf{x}_i)\phi(\mathbf{x}_i)^T \tag{16.16}$$

The next step is to perform eigenanalysis to $\mathbf{S}_x^{\phi}$, whose dimension is very high (even infinite), and direct eigenanalysis is infeasible. Thus, a procedure similar to the one described for high-dimensional data in Section 16.3 can be followed [10], in order to perform eigenanalysis to the auxiliary matrix of dimension $K \times K$, as described in (16.12), considering nonlinear kernels.

That is, in order to perform KPCA the auxiliary kernel matrix $\mathbf{K}_x$ should be computed as follows:

$$[\mathbf{K}_x]_{i,j} = \phi(\mathbf{x}_i)^T \phi(\mathbf{x}_j) = \mathcal{K}(\mathbf{x}_i, \mathbf{x}_j), \tag{16.17}$$

where $\mathcal{K}$ is an appropriate nonlinear kernel satisfying Mercer's conditions [10].

The matrix $\mathbf{K}_x$ is positive semidefinite of dimension $K \times K$, where K the number of data samples. Afterward, eigenanalysis is performed to the matrix $\mathbf{K}_x$ in order to calculate its eigenvectors. Once again the null space is discarded, by eliminating the eigenvectors that correspond to zero eigenvalues. The remaining vectors are ordered according to their eigenvalues and they are normalized such that $\mathbf{v}_i^T \mathbf{v}_i = 1$ for all i that correspond to nonzero eigenvalues. The normalized vectors $\mathbf{v}_i$ form the projection matrix $\mathbf{V} = (\mathbf{v}_1, \mathbf{v}_2, \ldots, \mathbf{v}_L)$.

The nonlinear principal components $\tilde{\mathbf{x}}$ of a test sample $\mathbf{x}$ are calculated as follows:

$$[\tilde{\mathbf{x}}]_j = \left(\mathbf{v}_j^T \phi(\mathbf{x})\right) = \sum_{i=1}^{K} [\mathbf{v}_j]_i \phi(\mathbf{x}_i)^T \phi(\mathbf{x}) = \sum_{i=1}^{K} [\mathbf{v}_j]_i \mathcal{K}(\mathbf{x}_i, \mathbf{x}_j), \tag{16.18}$$

where $[\cdot]_i$ denotes the ith element of the corresponding vector or matrix. The kernels $\mathcal{K}$ that are most commonly used in the literature are the Gaussian $\mathcal{K}(\mathbf{x}_i, \mathbf{x}_j) = \exp(p\|\mathbf{x}_i - \mathbf{x}_j\|^2)$ and the polynomial $\mathcal{K}(\mathbf{x}_i, \mathbf{x}_j) = (\mathbf{x}_i^T \mathbf{x}_j + 1)^p$ ones [13]. We should also note that an appropriate centralization [10] should be used in the general case, since the data samples do not have zero mean in the high-dimensional space. Another remark is that KPCA applies eigenanalysis in square matrices having dimensions equal to the number of samples and, thus, it may be computationally intensive, if there are many sample vectors in the data set.

16.6 PCA Neural Networks

PCA can also be performed without eigenanalysis. This can be achieved by using neural networks that can be trained to extract the principal components of the training set [2]. These neural networks are based on Hebbian learning and have a single output layer. The initial algorithm for training these networks has been proposed in [7]. The PCA neural networks are unsupervised. Compared to the standard approach of diagonalizing the covariance matrix, they have an advantage when the data are nonstationary. Furthermore, they can be implemented incrementally. Another advantage of the PCA neural networks is that they can be used to construct nonlinear variants of PCA by adding nonlinear activation functions. However, the major disadvantage is the slow convergence and numerical instability.

Another way to perform PCA using neural networks is by constructing a multilayer perceptron having a hidden layer with M neurons, where M is smaller than the dimension N of the input data. The neural network is trained in order to produce the input vector at the output level. Thus, the input vector is reduced in dimension in the hidden layer and it is reconstructed in the output layer. In this case, the neurons in the hidden layer perform PCA. Nonlinear extensions of this neural network can be considered for performing nonlinear PCA. The optimization problem solved by this network is hard and convergence is slow [2].

16.7 Applications of PCA

PCA has been successfully used in many applications such as dimensionality reduction and feature extraction for pattern recognition and data mining, lossy data compression, and data visualization [1,2]. Pattern representation is very critical in pattern recognition applications and PCA is a good solution for preprocessing the data prior to classification.

In face recognition, which is one of the most difficult classification tasks, PCA has been used to develop the EigenFaces algorithm [12], which is considered as the baseline algorithm for comparison. Features with good approximation quality, such as the ones produced by PCA, however, are not always good discriminative features. Thus, they cannot usually be used in classification tasks, whenever class-dependent information is available. A solution to this problem is to use the PCA as a preprocessing step only. Discriminant analysis is applied in a second step, in order to compute linear projections that are useful for extracting discriminant features [8,11]. An example of PCA versus linear discriminant analysis for classification is shown in Figure 16.2. It is obvious that, although PCA finds the projection that

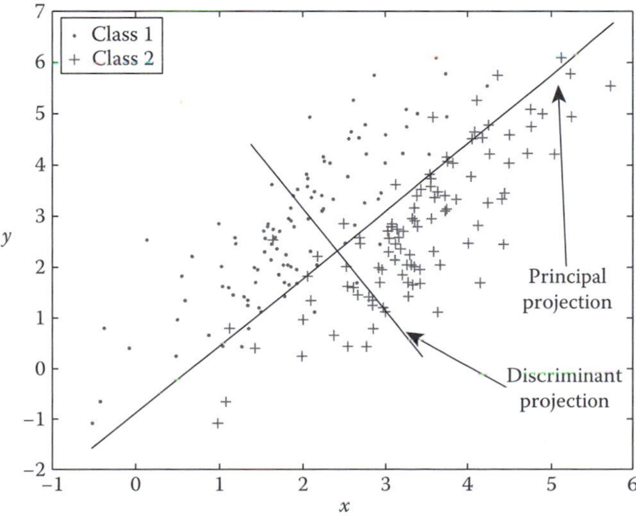

FIGURE 16.2 PCA versus linear discriminant analysis in classification problems.

better represents the entire data set that comprises of the samples of both classes, it fails in separating the two classes samples after the projection. Instead, linear discriminant analysis can find the projection that better separates the samples of the two classes.

Kernel PCA provides a drop-in replacement of PCA which, besides second order correlations, takes also into account higher order correlations. KPCA has been successfully used as a preprocessing step in the KPCA plus linear discriminant analysis algorithm that has been proven very efficient for many demanding classification tasks [5]. KPCA has been also used as a preprocessing step to support vector machines in order to give a powerful classification algorithm in [14].

PCA can also be used for lossy data compression. Compression is achieved by using a small number of principal components instead of the full-dimensional data. For example, data samples that lie in $\mathcal{R}^N$ can be compressed by representing them in $\mathcal{R}^M$, with $M \ll N$, using the M principal components of each data sample. An example is given in Figure 16.3, where the compression result of an image using PCA is illustrated. The original image is split in subimages of size 16 × 16 and then PCA is applied in order to extract the most representative information. From the 256 dimensions only 16 are retained and the

(a) Original image

(b) Compressed image

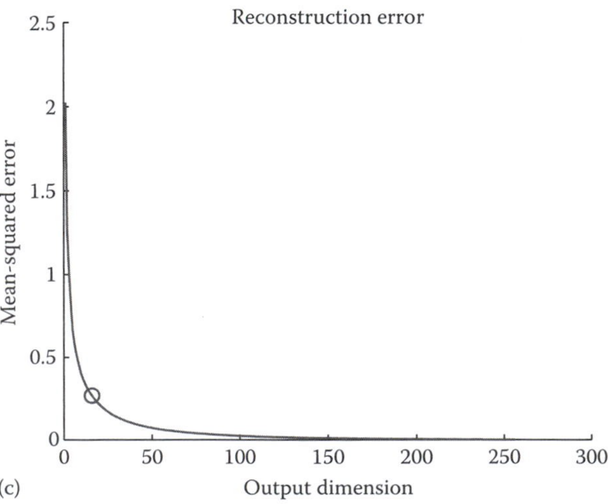

(c)

FIGURE 16.3 Image compression using PCA. The original image in (a) is reconstructed in (b) using only 16 eigenvectors from the 256 of the original image. The mean-squared error is plotted in (c).

reconstructed image is shown in Figure 16.3b. The mean-squared error of the reconstruction is plotted in Figure 16.3c, where it can be observed that as the number of the retained eigenvectors increases, the mean square error of the image reconstruction decreases rapidly. The compression ratio can be controlled by the number of the principal components that should be retained.

16.8 Conclusions

PCA is a very powerful tool for the statistical analysis of a data set. It provides low-dimensional representations of the sample data, by retaining as much data variation as possible. It is extensively used in many applications for dimensionality reduction, feature extraction, lossy data compression, and visualization. Many variants of PCA have been proposed for efficient and stable extraction of the principal components, extension to nonlinear components and combination with powerful data classification algorithms.

References

1. C.M. Bishop. *Pattern Recognition and Machine Learning*. Information Science and Statistics. Springer-Verlag, Berlin, Germany, 2006.
2. K.I. Diamantaras and S.Y. Kung. *Principal Component Neural Networks: Theory and Applications*. Wiley-Interscience, New York, 1996.
3. G.H. Golub and C.F. Van Loan. *Matrix Computations*, 3rd edn. Studies in Mathematical Sciences. The Johns Hopkins University Press, Baltimore, MD, October 15, 1996.
4. H. Hotelling. Analysis of a complex of statistical variables into principal components. *Journal of Educational Psychology*, 24:498–520, 1933.
5. J. Yang, A.F. Frangi, J.Y. Yang, D. Zhang, and Z. Jin. KPCA plus LDA: A complete kernel Fisher discriminant framework for feature extraction and recognition. *IEEE Transactions on Pattern Analysis and Machine Intelligence*, 27(2):230–244, 2005.
6. I.T. Jolliffe. *Principal Component Analysis*, 2nd edn. Series in Statistics. Springer, Berlin Germany, October 1, 2002.
7. E. Oja. A simplified neuron model as a principal component analyzer. *Journal of Mathematical Biology*, 15:267–273, 1982.
8. J. Hespanha, P. Belhumeur, and D. Kriegman. Eigenfaces vs. Fisherfaces: Recognition using class specific linear projection. *IEEE Transactions on Pattern Analysis and Machine Intelligence*, 19(7):711–720, 1997.
9. K. Pearson. On line and planes of closest fit systems of points in space. *Philosophical Magazine*, 2(6):559–572, 1901.
10. B. Schölkopf, A. Smola, and K.-R. Müller. Nonlinear component analysis as a kernel eigenvalue problem. *Neural Computation*, 10(5):1299–1319, July 1998.
11. D.L. Swets and J. Weng. Using discriminant eigenfeatures for image retrieval. *Transactions on Pattern Analysis and Machine Intelligence*, 18:831–836, 1996.
12. M. Turk and A. Pentland. Eigenfaces for recognition. *Journal of Cognitive Neuroscience*, 3(1):71–86, 1991.
13. V. Vapnik. *The Nature of Statistical Learning Theory*. Springer, Berlin, Germany, 1999.
14. S. Zafeiriou, A. Tefas, and I. Pitas. Minimum class variance support vector machines. *IEEE Transactions on Image Processing*, 16(10):2551–2564, 2007.

17

Adaptive Critic Neural Network Control

Gary Yen
Oklahoma State University

17.1 Introduction

This chapter introduces a family of neural network (NN) control architectures known as adaptive critic controllers as a natural extension of simpler architectures. The merits of each architecture are discussed and their shortcomings exposed, which in turn becomes the motivation for the next. The first architecture is an application of a single NN with a classical training algorithm, which implies the requirement of full knowledge of the plant's dynamics at all times. The controller is then improved by the addition of a second NN capable of generating online a map of the plant's dynamics; however, the training algorithm remains fundamentally the same.

The addition of a third NN and a change in the training paradigm leads to the adaptive critic architecture known as heuristic dynamic programming (HDP), followed by dual heuristic programming (DHP). Finally, the developments culminate in a full description of the most advanced adaptive critic architecture so far, known as globalized dual heuristic programming (GDHP). Presented in great detail, the particular GDHP training algorithm contained in this chapter was developed for application in the demanding field of fault tolerant control (FTC) [1]. In order to better comprehend the needs that drive many researchers to seek the great potential adaptive power of GDHP and also to give perspective to some examples presented in the end of this chapter, a short introduction to FTC is available for the readers.

The ultimate goal of this chapter is to provide the readers with basic motivation, background, and description of different adaptive critic controllers by presenting a series of NN adaptive controller architectures ranging from a single NN adaptive controller to GDHP.

17.2 Background

Increased performance requirements are often achieved at the cost of plant and control simplicity. As overall complexity rises, so does the chance of occurrence, diversity, and severity of faults. Therefore, availability, defined as the probability that a system or equipment will operate satisfactory and effectively at any point of time [2], becomes a factor of great importance. For automated production processes, for example, availability is now considered to be the single factor with highest impact on profitability [3].

FTC is a field of research that aims to increase availability and reduce the risk of safety hazards by specifically designing control algorithms capable of maintaining stability and/or performance despite the occurrence of faults [4]. As complex systems suffer from faults, the original model parameters or even its own dynamic structure may change in a multitude of unpredictable ways. Even if the system has a satisfactory linearization around the nominal operation point, nonlinearities may become of paramount importance after a fault occurs [5]. When the stochastic nature of faults is taken into consideration and to even predict all fault scenarios is made impossible, it becomes clear that the problem of interest of FTC cannot be dealt with without an online nonlinear adaptive control strategy. Successful applications of adaptive critic architecture controllers to FTC problems [6] have been credited to the controllers' great flexibility and known effectiveness to work in noisy, nonlinear environments while making minimal assumptions regarding the nature of that environment [7].

It is important to state here that, for the benefit of the discussion in this chapter, the required redundancy is assumed to exist in the system. Hardware redundancy requires two or more independent instruments that perform the same function, while analytical redundancy uses two components based on different principles to measure a variable, where at least one of them uses a mathematical model in analytical form. In either case, from the theoretical point of view, this assumption matches the requirement for sustained observability and controllability (or global reachability for nonlinear systems) through fault scenarios.

17.3 Single NN Control Architecture

The goal of this approach is to use a NN to generate a nonlinear map connecting the states of the plant $x(t)$, previous inputs $u(t-1)$, and current target $x^t(t)$ to an input $u(t)$ that will minimize the utility function $U(t)$ defined by Equation 17.1:

$$U(t) = \frac{1}{2}\left(x(t) - x^t(t)\right)^T Q\left(x(t) - x^t(t)\right) + \frac{1}{2}\rho u(t)^T \tag{17.1}$$

where

Q is a diagonal square matrix that can be used to assign different degrees of importance to each state

R is the equivalent matrix that penalizes the amount of control action used

ρ is a scalar used to balance the minimization of the tracking error and the energy use during the process

In order to differentiate it from the other NNs that will be introduced in later architectures, this NN is named action neural network (AcNN). Figure 17.1 depicts such architecture. When performing the training of the AcNN, the information of how its weights affect the states of the plant is required. However, backpropagation through the AcNN only provides information on how the inputs $u(t)$ are affected by its weights. Therefore, this approach requires the availability of a differential model of the

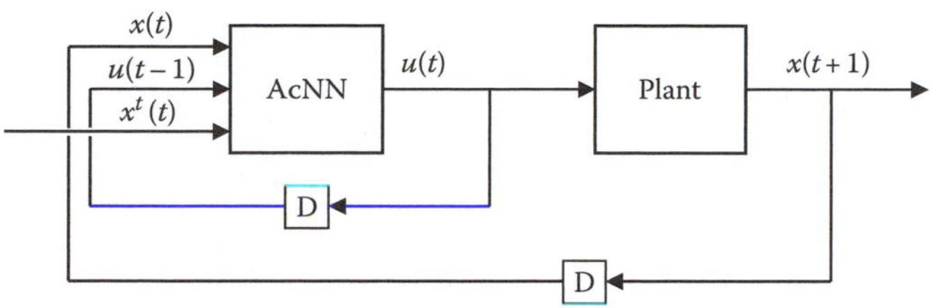

FIGURE 17.1 Single NN control architecture.

dynamics of the plant from which the information on how the states $x(t)$ are affected by the inputs $u(t)$ can be extracted. As a result, this architecture is not suitable for FTC application, since faults are assumed to modify the dynamics of the plant in unpredictable ways, making it impossible to design models beforehand.

17.4 Adaptive Control Architecture Using Two NNs

Since it is not possible to offline design models of the plant dynamics for all fault scenarios, in this architecture a second NN is introduced with the goal of performing online plant identification. Once this network has converged to represent a map of the dynamics of the plant, the derivative of the states with respect to the inputs can be extracted through standard backpropagation. Such network will be referred to as the identification NN (IdNN). Figure 17.2 displays this second approach.

Although no critical restrictions prevent this architecture to be used as a solution to the FTC problem, its performance can still be largely improved if the training algorithm for the AcNN is reevaluated. In these first two architectures, the AcNN is trained at each iteration with the goal of reducing the current value of the utility function $U(t)$. This is performed under the assumption that this process will ultimately lead to a set of weights that minimize the utility function for all times. However, this training approach provides no mechanisms to minimize the values that $U(t)$ assumes *during* training (or the time it takes). Clearly, it is of the interest of FTC to provide a new control solution to a fault scenario as quick as possible and with minimum performance impact.

17.5 Heuristic Dynamic Programming

Seeking to overcome the limitations of the previous approaches, the first adaptive critic controller is introduced. Adaptive critic architectures have a much greater potential to achieve the required degrees of reconfiguration and stability because more than the simple instantaneous difference between desired and actual states is available to be used as performance index. Due to the continuous interaction between

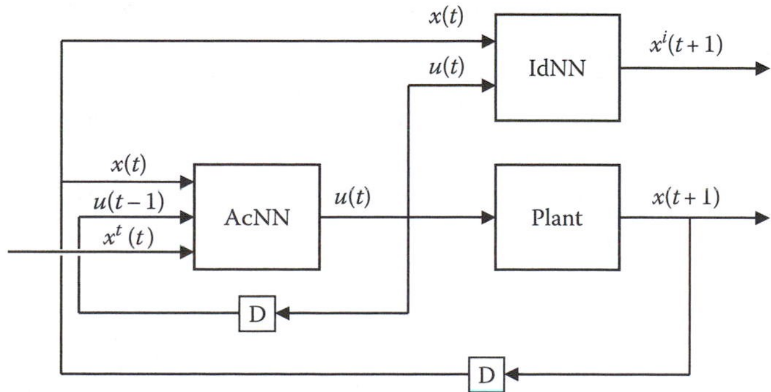

FIGURE 17.2 Direct adaptive control architecture using two NNs.

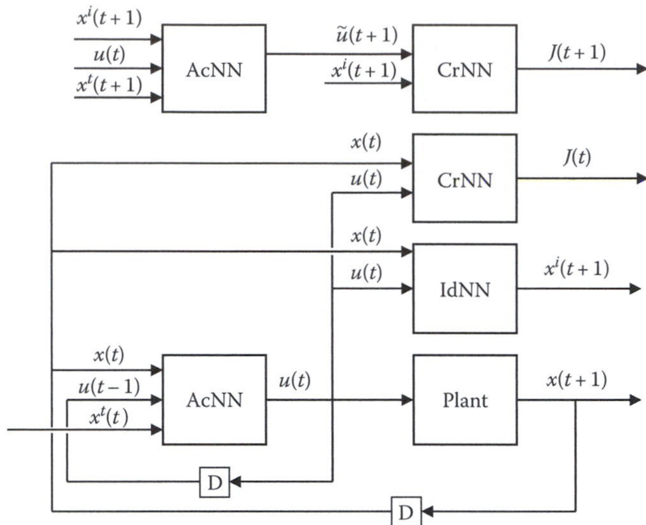

FIGURE 17.3 Heuristic dynamic programming.

the controller and the plant, the quality of a certain control strategy can only be fully measured after analyzing all future effects it has on the control mission, in our case trajectory tracking.

Therefore, HDP trains the AcNN to minimize not only the present utility function, but also the sum of all future values of $U(t)$ with a decaying factor γ ($0 < \gamma < 1$). Such quantity is referred to as the cost-to-go $J(t)$, as defined by the Hamilton–Jacobi–Bellman Equation 17.2, and represents the core of dynamic programming [10].

$$J(t) = \sum_{k=0}^{\infty} \gamma^k U(t+k) \tag{17.2}$$

Problems formulated in this form are the main focus of dynamic programming, which solves it through a backward search from the final step [8]. To make the problem tractable to an online learning approach, adaptive critic designs require an estimate of the actual cost-to-go to be constantly determined [9,10]. Although ACDs can be implemented with any differentiable structure [11], NNs have been widely used [12] due to their generalization and nonlinear mapping capabilities as well as having suitable methods for online learning. Given the complexity of FTC systems, dynamic or recurrent NNs were chosen due to their more efficient handling of dynamic nonlinear mapping [13]. It is in this context that we introduce a third NN, denominated the critic neural network (CrNN), responsible for approximating $J(t)$. The resulting block diagram is shown in Figure 17.3.

In other words, the training of the AcNN is done in the direction of the minimization of the cost-to-go approximation. In HDP, this is accomplished by starting the training path of the AcNN with the information of how the inputs and states will affect the current cost-to-go $J(t)$. Since the CrNN is trained to estimate it, such information can be easily extracted from the NN via backpropagation though time [14,15].

17.6 Dual Heuristic Programming

DHP reevaluates the purpose of the CrNN and redesigns it. Although in HDP the CrNN is trained to estimate $J(t)$, its true purpose is to provide the AcNN with the partial derivatives of $J(t)$ with respect to the states and inputs (usually referred to as $\lambda^x(t)$ and $\lambda^u(t)$, respectively). In DHP architecture, as shown

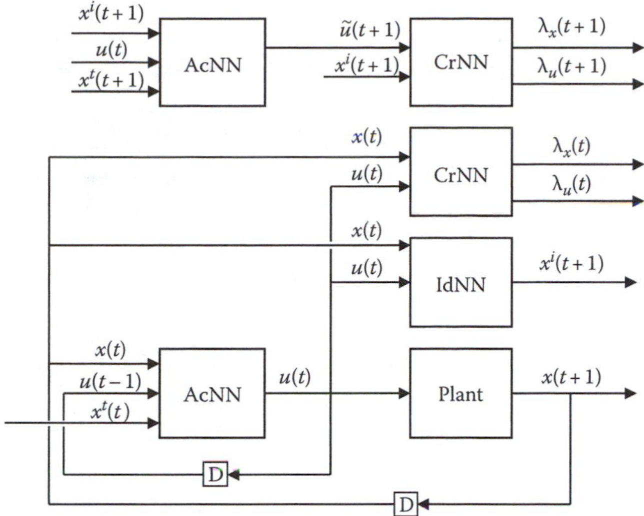

FIGURE 17.4 Dual heuristic programming.

in Figure 17.4, the CrNN is trained to output such derivatives directly. Using this direct approach, DHP is capable of generating smoother derivatives and has shown improved performance when compared to HDP. Those results were presented in Ref. [11], where both methods were applied to a turbogenerator, characterized as a highly complex, nonlinear, fast-acting, multivariable system with dynamic characteristics that vary as operating conditions change. Also, results from the application of DHP to the FTC challenge from early stages of the presented work can be found in Ref. [6]. These benefits come with the tradeoff of a more complex training algorithm for the CrNN as shown in Ref. [16].

17.7 Globalized Dual Heuristic

17.7.1 Introduction

The adaptive critic GDHP algorithm combines the HDP and DHP approaches to generate the most complete and powerful adaptive critic design [7]. In GDHP, $\lambda^x(t)$ and $\lambda^u(t)$ are determined with the precision and smoothness of DHP, while improving the CrNN training by also estimating $J(t)$ as in HDP [17]. Figure 17.5 depicts the block diagram of this approach.

In this section, the adaptive critic architecture of GDHP is presented in detail. Following this introduction, the adaptive control problem of interest to FTC is stated mathematically and the adopted notation introduced. The next three subsections are focused each on one of the NNs that composes the GDHP architecture: identifier, action, and critic. Each NN has its structure presented, followed by a discussion on its training algorithm and the ways through which information required by other networks is extracted. Finally, all information contained in this section is summarized in the complete GDHP algorithm presented in a manner that can be readily applied.

17.7.2 Preliminaries

The first step is to define $x(t)$ in Equation 17.3 and $u(t)$ in Equation 17.4, column vectors of the nx states and nu inputs at time t, and the Tap Delay Line (TDL) vectors $\bar{x}(t)$ in Equation 17.5 and $\bar{u}(t)$ in Equation 17.6 that combine information of TDL_x and TDL_u sampling times, respectively.

$$x(t) = \begin{bmatrix} x_1(t) & x_2(t) & \cdots & x_{nx}(t) \end{bmatrix}^T \tag{17.3}$$

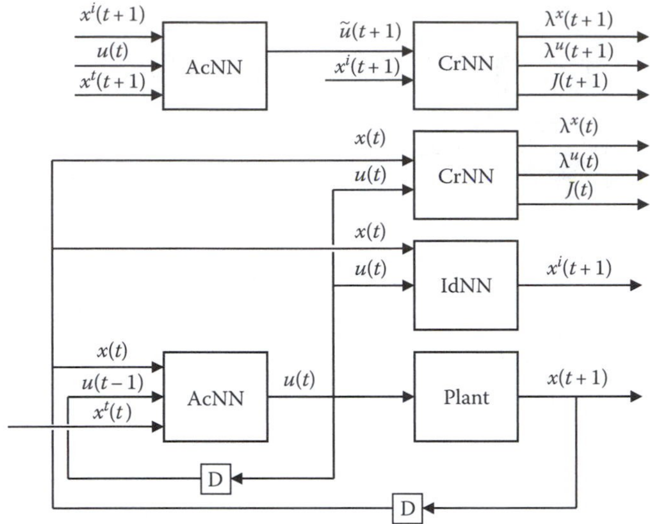

FIGURE 17.5 Globalized dual heuristic programming.

$$u(t) = \begin{bmatrix} u_1(t) & u_2(t) & \cdots & u_{nu}(t) \end{bmatrix}^T \quad (17.4)$$

$$\bar{x}(t) = \begin{bmatrix} x(t)^T & x(t-1)^T & \cdots & x(t-TDL_x+1)^T \end{bmatrix}^T \quad (17.5)$$

$$\bar{u}(t) = \begin{bmatrix} u(t)^T & u(t-1)^T & \cdots & u(t-TDL_u+1)^T \end{bmatrix}^T \quad (17.6)$$

Given the causal plant described in Equation 17.7 with nonlinear $f(\cdot)$ subject to abrupt faults characterized by discontinuous changes in its parameters or structure, the primary goal of the controller (17.8) is to make the states track the desired trajectory $x^t(t)$. Since particular fault scenarios may render regions of the state space unreachable to the plant, the controller is not required to reduce the tracking error to zero, but rather minimize it under the constrains of each particular fault. In the controller, $g(\cdot)$ is a nonlinear continuously differentiable approximator composed of three NNs: identification, action, and critic. The way each NN is trained online and how they interact in the GDHP architecture is explained in detail in the following sections.

$$x(t) = f\left(\bar{x}(t-1), \bar{u}(t-1)\right) \quad (17.7)$$

$$u(t) = g\left(\bar{x}(t), \bar{u}(t-1), x^t(t)\right) \quad (17.8)$$

17.7.3 Identification Neural Network

The IdNN (shown in Figure 17.6) is responsible for generating a differentiable map that matches the dynamics of the plant. Note that, in the used notation, all variables related specifically to the IdNN receive the superscript i. Designed as a two-layered recurrent NN [18,19] with input $p^i(t)$ in Equation 17.9, *nhi* neurons in the hidden layer and a tangent sigmoid transfer function in Equation 17.10, the IdNN outputs a vector of the estimated states $x^i(t)$ (11).

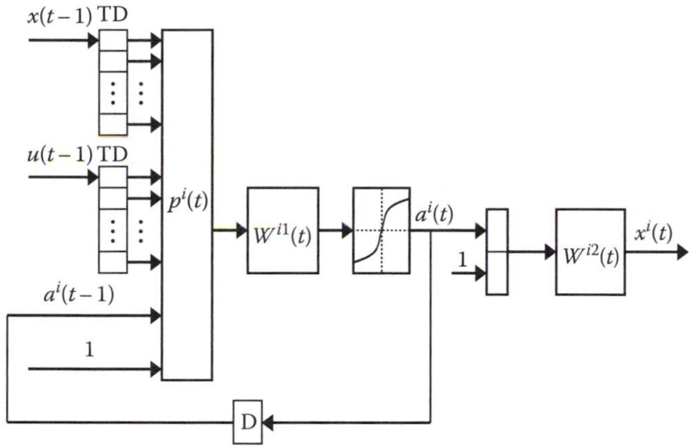

FIGURE 17.6 IdNN RNN architecture.

$$p^i(t) = \begin{bmatrix} \bar{x}(t-1) \\ \bar{u}(t-1) \\ a^i(t-1) \\ 1 \end{bmatrix} \tag{17.9}$$

$$a^i(t) = \tan sig\left(W^{i1}(t)p^i(t)\right) \tag{17.10}$$

$$x^i(t) = W^{i2}(t)\begin{bmatrix} a^i(t) \\ 1 \end{bmatrix} \tag{17.11}$$

The network is trained online with the goal of minimizing the identification error $E^i(t)$ subject to the relative importance matrix S in Equation 17.12. Generally, the matrix S is set as the identity; however, by adjusting the magnitude of the diagonal elements, the IdNN can be made to focus more on the reduction of the identification error of certain states. By applying the steepest descent training algorithm, the weight update (valid for both layers) is given by Equation 17.13.

$$E^i(t) = \frac{1}{2}\left(x^i(t) - x(t)\right)^T S\left(x^i(t) - x(t)\right) \tag{17.12}$$

$$w^i(t+1) = w^i(t) - \beta^i\left(\frac{dx^i(t)}{dw^i}\right)^T S\left(x^i(t) - x(t)\right) \tag{17.13}$$

where

w^i is a column vector of the elements of the corresponding weight matrix w^i

β^i is the learning rate

Equations 17.14 through 17.16 show how the required derivatives are calculated.

$$\frac{da^i(t)}{dw^{i1}} = \left(I - diag\left(a^i(t)\right)^2\right)\left(\begin{bmatrix} p^i(t)^T & 0 & 0 \\ 0 & \ddots & 0 \\ 0 & 0 & p^i(t)^T \end{bmatrix} + W^{i1}_{(:,end-nhi:end-1)}(t)\frac{da^i(t-1)}{dw^{i1}}\right) \tag{17.14}$$

$$\frac{dx^i(t)}{dw^{i1}} = W^{i2}_{(:,1:nhi)}(t)\frac{da^i(t)}{dw^{i1}} \tag{17.15}$$

$$\frac{dx^i(t)}{dw^{i2}} = \begin{bmatrix} \left[a^i(t)^T \;\; 1\right] & 0 & 0 \\ 0 & \ddots & 0 \\ 0 & 0 & \left[a^i(t)^T \;\; 1\right] \end{bmatrix} \tag{17.16}$$

where

w^{i1} corresponds to the weights of the first layer

w^{i2} to those of the second

The standard MATLAB® notation for the indication of rows and columns within a matrix is used. In such notation, when used by itself, the colon indicates all entities in a particular dimension (e.g., all rows or all columns), while when used between two numbers or variables it indicates the range between and containing such values in the corresponding dimension (e.g., all rows from 5 to *nhi*)

In order to train both the AcNN and the CrNN, information on the plant dynamics is required. Once the IdNN has converged to an estimator of the plant, the derivative of the output with respect to the input calculated by Equations 17.17 through 17.19 can be used as an approximation to part of the plant dynamics. Equations 17.20 and 17.21 show how the previous derivative is used to build the complete dynamic description when TDL_x is greater than 1.

$$\frac{da^i(t)}{d\bar{u}(t)} = \left[0_{(nhi,nu)} \quad \vdots \quad \frac{da^i(t)}{d\bar{u}(t-1)}_{(:,1:end-nu)} \right] \tag{17.17}$$

$$\frac{da^i(t+1)}{d\bar{u}(t)} = \left(I - diag\left(a^i(t+1)\right)^2\right) W^{i1}_{(:,1:end-1)}(t+1) \begin{bmatrix} \dfrac{d\bar{x}(t)}{d\bar{u}(t)} \\ \dfrac{d\bar{u}(t)}{d\bar{u}(t)} \\ \dfrac{da^i(t)}{d\bar{u}(t)} \end{bmatrix} \tag{17.18}$$

$$\frac{dx(t+1)}{d\bar{u}(t)} \approx \frac{dx^i(t+1)}{d\bar{u}(t)} = W^{i2}_{(:,1:nhi)}(t+1)\frac{da^i(t+1)}{d\bar{u}(t)} \tag{17.19}$$

$$\frac{d\bar{x}(t)}{d\bar{u}(t)} = \left[0_{(nx*TDLx,nu)} \quad \vdots \quad \frac{d\bar{x}(t)}{d\bar{u}(t-1)}_{(:,1:end-nu)} \right] \tag{17.20}$$

$$\frac{d\bar{x}(t+1)}{d\bar{u}(t)} = \begin{bmatrix} \dfrac{dx(t+1)}{d\bar{u}(t)} \\ \cdots\cdots\cdots \\ \dfrac{d\bar{x}(t)}{d\bar{u}(t)}_{(1:end-nx,:)} \end{bmatrix} \tag{17.21}$$

The information on the plant dynamics is completed with the knowledge of how the current and past states affect the state on the next step. Therefore, there is also need to use the differential map of the

IdNN to calculate the derivative of $x(t+1)$ with respect to $\bar{x}(t)$. The process through which such derivative is obtained, detailed in Equations 17.22 through 17.26, is analogous to the one performed in (17.17) through (17.21). Note that in the process of calculating both derivatives, the causality of the plant is taken into consideration. However, while causality restricts $d\bar{x}(t+1)/d\bar{u}(t)$ to a block upper triangular matrix, $d\bar{x}(t+1)/d\bar{x}(t)$ is an upper triangular matrix with ones in the diagonal.

$$\frac{da^i(t)}{d\bar{x}(t)} = \left[0_{(nhi,nx)} \quad \vdots \quad \frac{da^i(t)}{d\bar{x}(t-1)}_{(:,1:end-nx)} \right] \tag{17.22}$$

$$\frac{da^i(t+1)}{d\bar{x}(t)} = \left(I - diag\left(a^i(t+1)\right)^2 \right) W^{i1}_{(:,1:end-1)}(t+1) \begin{bmatrix} \dfrac{d\bar{x}(t)}{d\bar{x}(t)} \\ \dfrac{d\bar{u}(t)}{d\bar{x}(t)} \\ \dfrac{da^i(t)}{d\bar{x}(t)} \end{bmatrix} \tag{17.23}$$

$$\frac{dx(t+1)}{d\bar{x}(t)} \approx \frac{dx^i(t+1)}{d\bar{x}(t)} = W^{i2}_{(:,1:nhi)}(t+1)\frac{da^i(t+1)}{d\bar{x}(t)} \tag{17.24}$$

$$\frac{d\bar{x}(t)}{d\bar{x}(t)} = \begin{bmatrix} I_{(nx)} & \vdots & \\ \cdots\cdots\cdots & \vdots & \dfrac{d\bar{x}(t)}{d\bar{x}(t-1)}_{(:,1:end-nx)} \\ 0_{(nx*(TDLx-1),nx)} & \vdots & \end{bmatrix} \tag{17.25}$$

$$\frac{d\bar{x}(t+1)}{d\bar{x}(t)} = \begin{bmatrix} \dfrac{dx(t+1)}{d\bar{x}(t)} \\ \cdots\cdots\cdots \\ \dfrac{d\bar{x}(t)}{d\bar{x}(t)}_{(1:end-nx,:)} \end{bmatrix} \tag{17.26}$$

17.7.4 Action Neural Network

The core of the GDHP adaptive controller, the AcNN is responsible for the generation of the control input $u(t)$. Similar to the IdNN, the AcNN is also built on a two-layered architecture, as can be seen in Figure 17.7 and in the network description in Equations 17.27 through 17. 29. Equivalently, the superscript a is used over all variables specifically related to the AcNN.

$$p^a(t) = \begin{bmatrix} \bar{x}(t) \\ \bar{u}(t-1) \\ a^a(t-1) \\ x^t(t) \\ 1 \end{bmatrix} \tag{17.27}$$

$$a^a(t) = \tan sig\left(W^{a1}(t)p^a(t)\right) \tag{17.28}$$

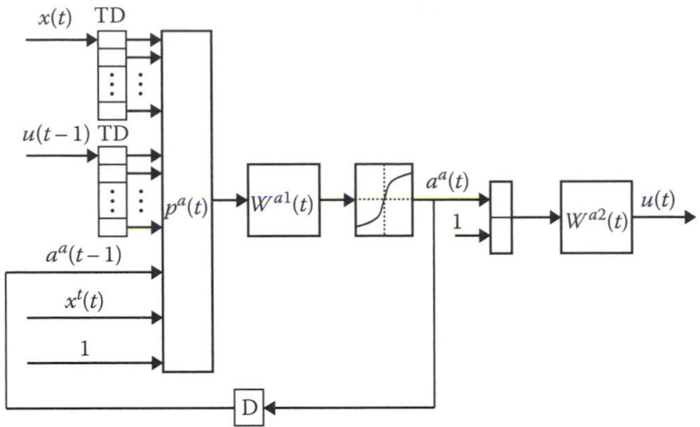

FIGURE 17.7 AcNN RNN architecture.

$$u(t) = W^{a2}(t)\begin{bmatrix} a^a(t) \\ 1 \end{bmatrix} \tag{17.29}$$

The training of the AcNN has the goal of producing the control sequence $u(t)$ that minimizes the cost function $J(t)$, defined in Equation 17.2 as the sum of all future values of the utility function $U(t)$ (1) with a decaying factor γ $(0 < \gamma < 1)$. The diagonal matrices Q and R have the same purpose as S in the IdNN while ρ adjusts the degree at which the amount of energy spent in the control effort is penalized relative to the tracking error.

As in the IdNN, a steepest descent training algorithm was applied, resulting in the update Equation 17.30. For reasons that will become clear in the description of the critic, the differentiation of $J(t)$ with respect to the weights of the AcNN is not performed directly from the infinite sum (17.2). The relationship (17.31) is used instead, resulting in Equation 17.32.

$$w^a(t+1) = w^a(t) - \beta^a\left(\frac{dJ(t)}{dw^a}\right)^T \tag{17.30}$$

$$J(t) = U(t) + \gamma * J(t+1) \tag{17.31}$$

$$\frac{dJ(t)}{dw^a} = \left(\frac{dU(t)}{d\bar{u}(t)} + \gamma * \lambda^x(t+1)\frac{d\bar{x}(t+1)}{d\bar{u}(t)} + \gamma * \lambda^u(t+1)\frac{d\bar{u}(t+1)}{d\bar{u}(t)}\right)\frac{d\bar{u}(t)}{dw^a} \tag{17.32}$$

where

β^a is the learning rate of the AcNN

$\lambda^x(t) = \dfrac{\partial J(t)}{\partial \bar{x}(t)}$ and $\lambda^u(t) = \dfrac{\partial J(t)}{\partial \bar{u}(t)}$ are outputs of the CrNN

The next step is the calculation of the derivative of the input with respect to the weights of the AcNN. Equations 17.33 through 17.34 for the first layer and Equations 17.35 through 17.36 for the second layer were derived in the same fashion as Equations 17.14 through 17.16 of the IdNN. Equation 17.37 describes the way the full temporal derivative is obtained for both layers. It is important to call to attention that, different from the IdNN, the AcNN is positioned in a closed loop with the plant. Therefore, in Equations 17.33 and 17.35, the AcNN derivation path extends to include information on the dynamics of the plant, approximated by the IdNN.

$$\frac{da^a(t)}{dw^{a1}} = \left(I - diag\left(a^a(t)\right)^2\right)\left(\begin{bmatrix} p^a(t)^T & 0 & 0 \\ 0 & \ddots & 0 \\ 0 & 0 & p^a(t)^T \end{bmatrix} + W^{a1}_{(:,1:end-nx-1)}(t)\begin{bmatrix} \frac{d\bar{x}(t)}{d\bar{u}(t-1)}\frac{d\bar{u}(t-1)}{dw^{a1}} \\ \frac{d\bar{u}(t-1)}{dw^{a1}} \\ \frac{da^a(t-1)}{dw^{a1}} \end{bmatrix}\right) \tag{17.33}$$

$$\frac{du(t)}{dw^{a1}} = W^{a2}_{(:,1:nha)}(t)\frac{da^a(t)}{dw^{a1}} \tag{17.34}$$

$$\frac{da^a(t)}{dw^{a2}} = \left(I - diag\left(a^a(t)\right)^2\right)W^{a1}_{(:,1:end-nx-1)}(t)\begin{bmatrix} \frac{d\bar{x}(t)}{d\bar{u}(t-1)}\frac{d\bar{u}(t-1)}{dw^{a2}} \\ \frac{d\bar{u}(t-1)}{dw^{a2}} \\ \frac{da^a(t-1)}{dw^{a2}} \end{bmatrix} \tag{17.35}$$

$$\frac{du(t)}{dw^{a2}} = \begin{bmatrix} [a^a(t)^T \ \ 1] & 0 & 0 \\ 0 & \ddots & 0 \\ 0 & 0 & [a^a(t)^T \ \ 1] \end{bmatrix} + W^{a2}_{(:,1:nha)}(t)\frac{da^a(t)}{dw^{a2}} \tag{17.36}$$

$$\frac{d\bar{u}(t)}{dw^a} = \begin{bmatrix} \frac{du(t)}{dw^a} \\ \frac{d\bar{u}(t-1)}{dw^a_{(1:end-nu,:)}} \end{bmatrix} \tag{17.37}$$

On Equations 17.18 and 17.32, the derivative of the tap delayed input with respect to itself was required. Equations 17.38 through 17.42 display how those are calculated. Since $W^a(t+1)$ is not yet available at this stage, the terms with superscript tilde are obtained using $W^a(t)$ as an approximation. Note that $p^a(t+1)$ used for the calculation of $\tilde{a}^a(t+1)$ can be generated by using the IdNN to estimate the future states of the plant assuming $x^t(t+1)$ available.

$$\frac{da^a(t)}{d\bar{u}(t)} = \begin{bmatrix} 0_{(nha,nu)} & \vdots & \frac{da^a(t)}{d\bar{u}(t-1)}_{(:,1:end-nu)} \end{bmatrix} \tag{17.38}$$

$$\frac{da^a(t+1)}{d\bar{u}(t)} = \left(I - diag\left(\tilde{a}^a(t+1)\right)^2\right)\tilde{W}^{a1}_{(:,1:end-nx-1)}(t+1)\begin{bmatrix} \frac{d\bar{x}(t+1)}{d\bar{u}(t)} \\ \frac{d\bar{u}(t)}{d\bar{u}(t)} \\ \frac{da^a(t)}{d\bar{u}(t)} \end{bmatrix} \tag{17.39}$$

$$\frac{du(t+1)}{d\bar{u}(t)} = \tilde{W}^{a2}_{(:,1:end-1)}(t+1)\frac{da^a(t+1)}{d\bar{u}(t)} \tag{17.40}$$

$$\frac{d\bar{u}(t)}{d\bar{u}(t)} = \begin{bmatrix} I_{(nu)} & \vdots & \frac{d\bar{u}(t)}{d\bar{u}(t-1)}_{(:,1:end-nu)} \\ \cdots\cdots\cdots & \vdots & \\ 0_{(nu*(TDLu-1),nu)} & \vdots & \end{bmatrix} \tag{17.41}$$

$$\frac{d\bar{u}(t+1)}{d\bar{u}(t)} = \begin{bmatrix} \frac{du(t+1)}{d\bar{u}(t)} \\ \cdots\cdots\cdots \\ \frac{d\bar{u}(t)}{d\bar{u}(t)}_{(1:end-nu,:)} \end{bmatrix} \tag{17.42}$$

In later developments, the information on how the future input is affected by the present states of the plant is required. For such purpose, Equations 17.43 through 17.47 are provided.

$$\frac{da^a(t)}{d\bar{x}(t)} = \begin{bmatrix} 0_{(nha,nx)} & \vdots & \frac{da^a(t)}{d\bar{x}(t-1)}_{(:,1:end-nx)} \end{bmatrix} \tag{17.43}$$

$$\frac{da^a(t)}{d\bar{x}(t)} = \left(I - diag\left(a^a(t)\right)^2\right) W^{a1}_{(:,1:end-nx-1)}(t) \begin{bmatrix} \frac{d\bar{x}(t)}{d\bar{x}(t)} \\ \frac{d\bar{u}(t-1)}{d\bar{x}(t)} \\ \frac{da^a(t-1)}{d\bar{x}(t)} \end{bmatrix} \tag{17.44}$$

$$\frac{du(t)}{d\bar{x}(t)} = W^{a2}_{(:,1:end-1)}(t) \frac{da^a(t)}{d\bar{x}(t)} \tag{17.45}$$

$$\frac{d\bar{u}(t-1)}{d\bar{x}(t)} = \begin{bmatrix} 0_{(nu*TDLu,nx)} & \vdots & \frac{d\bar{u}(t-1)}{d\bar{x}(t-1)}_{(:,1:end-nx)} \end{bmatrix} \tag{17.46}$$

$$\frac{d\bar{u}(t)}{d\bar{x}(t)} = \begin{bmatrix} \frac{du(t)}{d\bar{x}(t)} \\ \cdots\cdots\cdots \\ \frac{d\bar{u}(t-1)}{d\bar{x}(t)}_{(1:end-nu,:)} \end{bmatrix} \tag{17.47}$$

17.7.5 Critic Neural Network

The third and final NN, the critic is responsible for the estimation of the cost function $J(t)$ and of its derivatives with respect to the inputs and states ($\lambda^u(t)$ and $\lambda^x(t)$, respectively). Consistent with the notation of the other two NNs, all variables specifically related to the CrNN are marked by a superscript c. As shown from the network description in Figure 17.8 and Equations 17.48 through 17.50, the beforementioned derivatives are obtained directly as outputs of the network, instead of through backpropagation from the cost function.

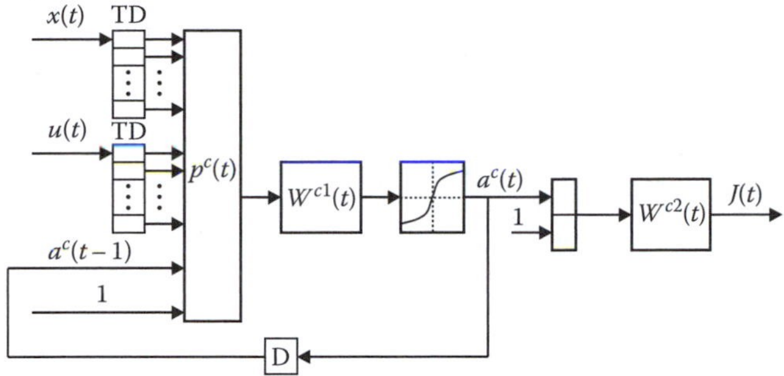

FIGURE 17.8 CrNN RNN architecture.

$$p^c(t) = \begin{bmatrix} \bar{x}(t) \\ \bar{u}(t) \\ a^c(t-1) \\ 1 \end{bmatrix} \tag{17.48}$$

$$a^c(t) = \tan sig\left(W^{c1}(t)p^c(t)\right) \tag{17.49}$$

$$\begin{bmatrix} \lambda^x(t)^T \\ \lambda^u(t)^T \\ J(t) \end{bmatrix} = W^{c2}(t)\begin{bmatrix} a^c(t) \\ 1 \end{bmatrix} \tag{17.50}$$

The GDHP critic's weight update Equation 17.51 is a combination of the training algorithms of HDP (minimizing the estimation error of $J(t)$) and DHP (minimizing the estimation error of $\lambda(t)$). Although the influence of the HDP and DHP algorithms can be decoupled in the update of the weights of the second layer, both terms equally affect all the weights of the first layer. This superposition of training approaches in the first layer of the CrNN is the main source of the synergy of GDHP [20].

$$w^c(t+1) = w^c(t) - \beta^c(1-\eta)\left(\frac{dJ(t)}{dw^c}\right)^T \left(J(t) - J^o(t)\right) - \beta^c\eta \begin{bmatrix} \dfrac{d\lambda^x(t)}{dw^c} \\ \dfrac{d\lambda^u(t)}{dw^c} \end{bmatrix}^T \left(\begin{bmatrix} \lambda^x(t) \\ \lambda^u(t) \end{bmatrix} - \begin{bmatrix} \lambda^{x^o}(t) \\ \lambda^{u^o}(t) \end{bmatrix}\right) \tag{17.51}$$

where

β^c is the learning rate of the CrNN

$\eta \in [0, 1]$ is a parameter that adjusts how HDP and DHP are combined in GDHP

For $\eta = 0$, the training of the CrNN reduces to a pure HDP, while $\eta = 1$ does the same for DHP.

Since the cost function $J(t)$ is a weighted sum of present and future variables, the targets, $J^o(t)$, $\lambda^{x^o}(t)$, and $\lambda^{u^o}(t)$, are not analytically available when performing online learning. In order to generate values that will in time converge to the true targets, relationship (31) is used, resulting in Equations 17.52 through 17.54.

$$J^o(t) = U(t) + \gamma * J(t+1) \tag{17.52}$$

$$\lambda^{x^o}(t) = \frac{\partial J^o(t)}{\partial x(t)} = \frac{\partial U(t)}{\partial x(t)} + \gamma\left(\lambda^x(t+1)\frac{dx(t+1)}{dx(t)} + \lambda^u(t+1)\frac{du(t+1)}{dx(t)}\right) \tag{17.53}$$

$$\lambda^{u^o}(t) = \frac{\partial J^o(t)}{\partial u(t)} = \frac{\partial U(t)}{\partial u(t)} + \gamma\left(\lambda^x(t+1)\frac{dx(t+1)}{du(t)} + \lambda^u(t+1)\frac{du(t+1)}{du(t)}\right) \tag{17.54}$$

The next step is the calculation of the partial derivatives of the critic's outputs with respect to its weights. Equations 17.55 through 17.57 demonstrate how those are obtained.

$$\frac{da^c(t)}{dw^{c1}} = \left(I - diag\left(a^c(t)\right)^2\right)\left(\begin{bmatrix} p^c(t)^T & 0 & 0 \\ 0 & \ddots & 0 \\ 0 & 0 & p^c(t)^T \end{bmatrix} + W^{c1}_{(:,end-nhc:end-1)}(t)\frac{da^c(t-1)}{dw^{c1}}\right) \tag{17.55}$$

$$\begin{bmatrix} \dfrac{d\lambda^x(t)^T}{dw^{c1}} \\ \dfrac{d\lambda^u(t)^T}{dw^{c1}} \\ \dfrac{dJ(t)^T}{dw^{c1}} \end{bmatrix} = W^{c2}_{(:,1:end-1)}(t)\frac{da^c(t)}{dw^{c1}} \tag{17.56}$$

$$\begin{bmatrix} \dfrac{d\lambda^x(t)^T}{dw^{c2}} \\ \dfrac{d\lambda^u(t)^T}{dw^{c2}} \\ \dfrac{dJ(t)^T}{dw^{c2}} \end{bmatrix} = \begin{bmatrix} [a^a(t)^T \;\; 1] & 0 & 0 \\ 0 & \ddots & 0 \\ 0 & 0 & [a^a(t)^T \;\; 1] \end{bmatrix} \tag{17.57}$$

Completing the requirements of Equations 17.53 through 17.54, the partial derivatives of the utility function with respect to the states and inputs are provided in Equations 17.58 through 17.59. Equation 17.60 shows how the full derivative of the utility function with respect to the inputs is calculated, as required in (17.32).

$$\frac{\partial U(t)}{\partial x(t)} = \left(x(t) - x^t(t)\right)^T S \tag{17.58}$$

$$\frac{\partial U(t)}{\partial u(t)} = \rho u(t)^T R \tag{17.59}$$

$$\frac{dU(t)}{d\bar{u}(t)} = \frac{\partial U(t)}{\partial x(t)}\frac{dx(t)}{d\bar{u}(t)} + \frac{\partial U(t)}{\partial u(t)}\frac{du(t)}{d\bar{u}(t)} \tag{17.60}$$

17.7.6 Complete GDHP Algorithm

A key issue in all adaptive critic designs implementation is how to coordinate the online training of the three NNs. While the IdNN is trained independently since it uses information of the plant alone, the training of each AcNN and CrNN depends on the weights of the other. If no provisions are made, both

TABLE 17.1 Pseudocode for the Presented GDHP Controller

1. Set $t = 1$, $e = 1$. Initialize NNs weights and network derivatives. Estimate $x^i(1)$
2. Sample the plant states $x(t)$ and desired trajectory $x^t(t)$
3. Update the weights of the IdNN by generating $w^i(t + 1)$—Equations 17.13 through 17.16
4. Feedforward through all 3 NNs (AcNN and CrNN twice) to generate in this order: $u(t)$, $x^i(t + 1)$, $\bar{u}(t + 1)$, $\lambda^x(t)$, $\lambda^u(t)$, $J(t)$, $\lambda^x(t + 1)$, $\lambda^u(t + 1)$ and $J(t + 1)$—Equations 17.9 through 17.11, 17.31 through 17.32, and 17.48 through 17.50
5. Calculate $U(t)$—Equation 17.1
6. Backpropagate to generate $\frac{d\bar{x}(t+1)}{d\bar{u}(t)}$, $\frac{d\bar{x}(t+1)}{d\bar{x}(t)}$, $\frac{d\bar{u}(t+1)}{d\bar{u}(t)}$ and $\frac{d\bar{u}(t)}{d\bar{x}(t)}$—Equations 17.17 through 17.26 and 17.38 through 17.47
7. Calculate $\frac{dU(t)}{d\bar{u}(t)}$—Equations 17.58 through 17.59
8. Update the weights of the AcNN by generating $w^a(t + 1)$—Equations 17.30 through 17.37
9. Update the weights of the CrNN by generating $w^c(t + 1)$—Equations 17.51 through 17.57
10. If e = epoch, copy the weights of CrNN#1 to CrNN#2 and set $e = 1$
11. $t = t + 1$, $e = e + 1$. Return to 2

networks are forced to follow a moving target, making the whole process potentially slower and likely unstable. In Ref. [20], four different strategies were discussed and compared through the application on two different test beds, demonstrating the superior performance, stability, and reduced training time of a particular one that we choose to implement. Although the original work was developed for the DHP architecture, the extension to GDHP is straightforward. The strategy of interest differs from others by the fact that it utilizes two distinct NNs to implement the critic. The first (CrNN#1) outputs $J(t)$ and $\lambda(t)$ and is trained at every iteration whereas the second (CrNN#2) outputs $J(t + 1)$ and $\lambda(t + 1)$ and is updated with a copy of the first only once at a given period of iterations (i.e., epoch). With such training approach, it is possible to train both AcNN and CrNN continuously allowing the adaptive critic controller to start responding to a fault as soon as it occurs.

With all the mathematical content of GDHP already available in Equations 17.1 through 17.60, a pseudocode version of the actual algorithm is presented in a condensed format in Table 17.1.

17.8 Fault Tolerant Control

Failure prevention is not a new concept in theory or practice of engineering. The components or machinery that forms a system are often built with safety protections such as fuses or limit switches. Continued operation or start-up is prevented if sensors like those inform that conditions are met to enter a local shutdown mode. This local safety approach though, does not guarantee global fail-safe operation for the complete system. A ship propulsion system depicts an example where the application of local safety with the lack of analysis of the global implications resulted in many events where consequences vary from irregularity to major economic loss and casualties [3].

Another failure prevention approach derives from the use of direct hardware redundancy. If three or more independent sensors are used to directly measure the same variable, a majority voting can be used not only to detect a fault, but also to isolate the faulty sensor. When only two redundant sensors are available, isolation is not necessarily achievable, but fault detection is still guaranteed. The remedial action to be taken is then simply ignoring the isolated sensor or generating an alarm when no trustable signal is available.

The same principle is applied to components and actuators, though it is possible in those cases that more than one output from different elements operating at only a fraction of its total capability is used at the same time. After a fault is isolated in one of the elements, the failure prevention approach then becomes one of energy redistribution among the healthy set.

FTC's goal is to prevent failures at system level through proper actions in the programmable parts of a control loop. In this approach, analytical redundancy can be used in place of its hardware counterpart. Analytical redundancy helps not only to reduce the cost involved in using extra elements, but also delivers greater design freedom to avoid the loss of performance that may result from direct hardware redundancy implementation. When sensors are considered, the use of analytical relations united with the actual measurements also increases the degree of confidence of the considered variable. Since FTC focus on the overall mission goal and aims for continuous system availability, different from the other failure prevention approaches mentioned earlier, a loss of performance is allowed after a fault occurs. As a matter of fact, given the specific redundancies available in a given system, a reconfiguration to a state of inferior performance might be an optimal solution when the mission objective, such as stability, is preferred.

17.8.1 Passive versus Active Approaches

One possible way to implement fault tolerance is to design static control laws capable to compensate for some plant uncertainties such as disturbances and noise [5]. If the effects of a fault are small enough to be in the range covered by the robustness of the controller, no specific reconfiguration is required. Since no information about the faults is typically utilized by the control system, this type of approach is often referred to as "passive FTC."

By utilizing fault information extracted from the system, it becomes possible to design a reconfigurable controller that modifies the control function (parameters or structure) in response to faults, characterizing an "active FTC." This approach is preferable over the passive one when tolerance to a wider range of faults is intended since the required increase in robustness has a negative effect on the performance, even under nominal operation. As depicted in the generic active FTC diagram in Figure 17.9, it is common to separate the control algorithm into two distinct blocks: a baseline controller and a supervisor system. While the baseline controller focuses on the maintenance of the immediate control objectives, the supervisor extracts fault information, determines remedial action, and executes them by modifying the baseline controller.

17.8.2 Active FTC Methods

Active FTC systems compensate for the effects of a fault either by selecting a new precomputed control law (projection-based methods) or by synthesizing a new control law online (online automatic controller redesign methods) [21].

Gain scheduling (GS) [22], fuzzy decision logic [23], and structural analysis [3] are some of the possible ways to implement projection-based active FTC. Models and precomputed controllers for the system under nominal conditions and under the effect of the faults of interest are used during the design phase to grant the controller quick and correct responses to the envisioned scenarios. However,

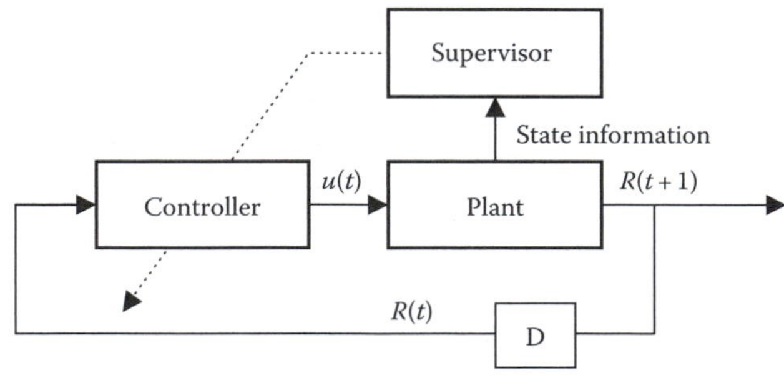

FIGURE 17.9 Generic active fault tolerant architecture depicting the base line controller and the supervisory system. In the figure, D represents a delay block, $u(t)$ is the controlled input, and $R(t)$ is output of the plant.

since fault information at least to the level of isolation is essential, it is necessary for the models of the faulty scenarios to be accurate enough to be distinguishable under the effect of noise and disturbances. Even when precision is not taken into account, the mere task of offline design of characteristic models for all possible fault scenarios is by itself a challenging one, especially if complex nonlinear plants are considered.

Online automatic redesign methods are of particular interest in light of the goal of the proposed work due to its capability of providing specific control actions even to fault scenarios that had not been necessarily anticipated during the design phase. Reconfigurable control can be used to implement online redesign requiring only the residuals generated by fault detection. Nevertheless, the flexibility gained by this approach comes at the expense of slower response since the controller must be allowed time to learn the new dynamics and modify itself. Since the reconfigurable controller does not require knowledge of the dynamics of the system under the effect of each specific fault, it is inherently immune to modeling errors and possesses a greater potential to deal with unmeasured disturbances and noise-corrupted data.

A reconfiguration approach in which the eigenstructure can be directly assigned to the close-loop system to achieve the desired system stability and dynamic performance is known as eigenstructure assignment (EA) [24]. The conditions for exact assignment are the existence of a sufficient number of actuators and measurements available and that the desired eigenvectors reside in the achievable subspaces. The limitations of EA are that the system performance may not be optimal in any sense, and that the system requirements are often not easily specified in terms of the eigenstructure [25].

The pseudo-inverse method (PIM), on the other hand, is a reconfiguration method that is optimal in the sense that it minimizes the Frobenius norm of the difference matrix between the original and the impaired closed-loop system transition matrices. Since in its initial formulation stability cannot be guaranteed, a modified pseudo-inverse method (MPIM) was proposed [26]. In its initial formulation, however, MPIM required full state feedback and relied on stability bounds that could give very conservative results. Those limitations were the focus of Ref. [27], where the problem was reevaluated from an optimization point of view while focusing on FTC application. Although the state feedback constraint was relaxed to output feedback, the method still requires residuals for each parameter of the model to be generated (comparison between transition matrices), limiting in the reconfigurable fault scenarios to those with the same dynamic structure than the nominal mode.

However, both EA and PIM-based controllers are restricted to implementation on linear models. When a fixed dynamical nonlinear structure is available and only the parameters are unknown, adaptive control can be used. Even to this restricted case, the assumptions that have to be made concerning the unknown plant to develop a stable adaptive controller were established only in the 1980s [28]. The problem becomes truly formidable when the plant is nonlinear and the input–output characteristics are unknown and time varying.

From a system theoretic point of view, artificial NNs can be considered as practically implementable parametrizations of nonlinear maps from one finite dimension space to another. Theoretical works by several researchers have proven that, even with one hidden layer, NNs can uniformly approximate at any degree of precision any piecewise continuous function over a compact domain, provided the network has a sufficient number of units, or neurons. Therefore, NN can, by their very nature, cope with complexity, uncertainty, and nonlinearity, and NN have been used successfully to identify and control nonlinear dynamic systems [29].

Multilayer neural networks (MNN) and radial basis functions networks (RBFN) have proven extremely successful in pattern recognition problems, while recurrent neural networks (RNN) have been used in associative memories as well as optimization problems [30]. From the theoretic point of view MNN and RBFN represent static nonlinear maps while RNN are represented by nonlinear dynamic feedback systems [13].

In Ref. [31], a recurrent high-order neural network (RHONN) was developed with the goal of identification of dynamical systems displaying similar convergence properties of classical adaptive and robust adaptive schemes. A Lyapunov-based approach is used to prove the convergence property of the learning algorithm that ensures that the identification error converges to zero exponentially and that, if it is initially zero, it remains in zero during the whole identification process. Later, in Ref. [13], the identifications capabilities of the RHONN were used to provide state information to a sliding mode controller to solve a tracking problem. However, the RHONN displays serious restrictions to its applicability to complex systems due to a lack of scalability in its heavily connected architecture.

In Ref. [32], a simplified RNN is used to identify the system and its parameters used as input to a controller based on feedback linearization and pole placement. Stability though, is only assured if the controlled system remains stable, a limitation that greatly decreases the applicability of the method to the FTC problem.

A RNN-based adaptive controller specially developed to deal with nonlinear systems with unknown dynamics is presented in Ref. [33]. In the proposed configuration, the output from the RNN adaptive controller was applied to the system summed with the output of a linearizing controller designed offline to deal with the nonlinearities in the nominal model. The proposed learning algorithm was stable in the Lyapunov sense, but the restrictions applied to achieve such proof make this approach capable only to deal with incipient faults.

In order to achieve semi-global boundedness of all signals in a control loop of a MIMO system, a backstepping approach is used in Ref. [34] to divide the MIMO nonlinear model into a series of SISO nonlinear models and design controllers separately using RBFNs. However, in order to achieve such degree of decouplability, it must be possible to describe the system in block-triangular form. Even if true for the nominal model, a fault may increase relationships between states that could previously be ignored, making it impossible for the system to fit in a block-triangular form again.

Taking inspiration in a PID controller, a modified RNN architecture is applied in a model reference adaptive control framework to control an automotive engine in Ref. [35]. Although identification and control are performed by RNNs, the identification is performed offline while only the controller is trained online. Therefore, direct application of this method to systems which dynamics may be affected by faults in unexpected ways is not possible.

17.8.3 Multiple Model as a Framework

Even though a reconfigurable adaptive controller is a key element without which solutions for unknown faults cannot be designed online, if used as a FTC architecture alone, it displays two major limitations. The first involves the fact that a reconfigurable controller makes it impossible for any available fault knowledge to be incorporated during design time. Although an ideal reconfigurable controller will always reach a solution (given its existence) for a given fault scenario, the amount of time it must be allowed to learn the new dynamics and modify itself accordingly could be greatly reduced by the direct application of a known solution. The second major limitation is caused by the known tradeoff between adaptability and long-term memory. As a reconfigurable controller is optimized to deal with a broader scope of faults with minimum reconfiguration time, previously configured controllers are forgotten and the reconfiguration process has to be repeated even when returning to the healthy condition from an intermittent fault scenario.

Multiple models architecture (MMA) [23,29] presents a framework in which projection-based methods and online redesign can be synergistically integrated to provide the fast and specific response of the first combined with the flexibility and robustness of the second. More specifically, in Refs. [9,36] it was shown that implementing a reconfigurable controller in a MMA has the potential to overcome the cited limitations for the tracking of complex nonlinear plants. Since then, MMA has been applied to FTC by combining fault scenarios and their respective control solutions in model banks coordinated by a supervisor. However, most publications so far are based on fixed model banks built offline and therefore are incapable of improving the controller response in the reoccurrence of faults that were unexpected during design time. In Ref. [37], a dynamic model bank (DMB) is used to allow the insertion of new plant dynamics as they were identified online, but the use of a linear controller and the lack of a complete fault detection and diagnosis (FDD) scheme significantly limit its applicability.

To better understand the MMA approach, its simplest implementation, GS, will first be introduced and discussed. GS is a technique that aims to provide control over nonlinear systems without requiring the design of nonlinear controllers. The first step in GS is to linearize the model about one or more operating points. Then linear design methods are applied to the linearized model at each operating point in order to arrive at a set of linear feedback control laws that perform satisfactorily when the closed-loop system is operating near the respective operating points. The zone of the state space where a controller still performs satisfactory is denoted operating region. The final step is the actual GS, which is intended to handle the nonlinear aspects of the design problem. The basic idea involves interpolating in some way the linear control law designs at intermediate operating conditions. It is usual in GS applications to choose a particular structure for the linear controllers (e.g., PID) and therefore its parameters (gains) are modified (scheduled) according to the states of the closed-loop system.

In addition to the evident simplicity brought by the design of the controllers for linear approximations instead of the global nonlinear models, GS also provides the potential to respond rapidly to changing operating conditions and its real-time computational burden is light [22]. However, since the design process of GS in its original formulation is based only on local information of a limited set of operation points, no global characteristic (stability, performance, robustness, etc.) can be guaranteed. In the same way that a well-designed set of linear controllers does not necessarily result in even a globally stable control law for the nonlinear system, reachable nonlinear systems may provide uncontrollable linearized models, preventing GS to be applied at all.

Advanced MMAs make use of local nonlinear models to design its controllers, resulting in potentially bigger operating regions for controller. Given enough information of the system, this property allows the main components of a system dynamics to be represented in a finite set of nonlinear models, making it possible to incorporate global stability, performance, and robustness requirements in the design phase of multiple models. Model predictive control, feedback linearization, and sliding mode [38] are examples of such methods. Another benefit from the use of nonlinear models and controllers is that it provides the possibility to dramatically reduce the total number of models, making it feasible to apply the MMA concept to systems with widely diversified complex dynamics.

Nevertheless, independent from the linearity of the models used to generate the set of controllers, the quality of the end result of the application of a MMA approach is still largely affected by a wide range of design choices concerning how many to create, where to position, and how to interpolate the controllers designed at each operating point or region.

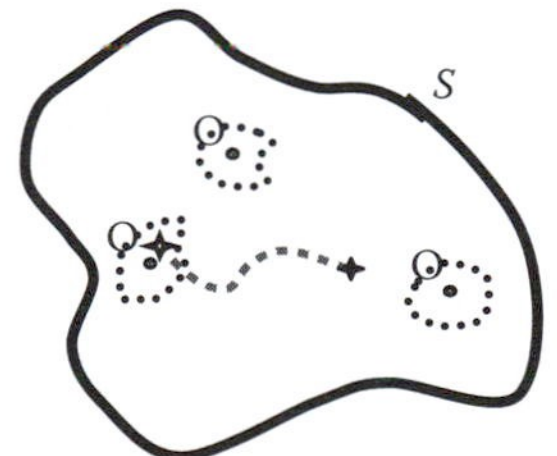

FIGURE 17.10 Performing multiple model control with sparsely distributed operating regions. O_1–O_3 are operating regions around each operating point. The system is originally in the position of the parameter spaced marked by the white star and follows the depicted trajectory.

For a better understanding and comparison between different approaches, the parameter space representation presented in Ref. [9] will be used. The parameter space (S) is an augmented version of the state space representation that includes "states" of the environment that contain information of sensors present in the plant used solely to extract fault information. Temperature, for example, can be considered an environmental state if the model of the plant does not take it into account directly, but as the temperature deviates from the nominal condition the dynamics of the plant are altered. In the examples that follow, the parametric space is a bounded region that encompasses the physically achievable values of each state.

For the sake of visualization, the following discussion will be held with examples using two-dimensional parameter spaces. The conclusions however are not limited to this particular case, being possible to apply all the discussed methods in higher dimensional spaces directly.

Figure 17.10 shows a basic MMA setting where a set of controllers is devised for some specific operating regions sparsely distributed in the parameter space. Each of the operating regions (O_1, O_2, and O_3) is generated around an operating point and limited by the range of the state space of the plant in which the corresponding controller performs with a satisfactory performance. In the dimensions of the parameter space that do not represent states of the plant, the operating regions represent the robustness of the controller.

If the plant is in a position in the parameter space that is close enough to an operating point to be inside its operating region, it is reasonable to apply the respective precomputed control law. This is the case of the original position (white star) of the trajectory shown in Figure 17.10. However, variations in the set-point or the occurrence of faults may take the system to a point away from all operating points that were considered offline (black star) and the question of what control law to use is raised. As a matter of fact, since precise description of the operating regions is not often available in practice, such question may arise even while the plant is still inside the respective operating region.

Perhaps the most intuitive approach, one of the ways to generate control laws for in-between operating points, is to assign a mean of the parameters of the controllers at each operating point to the parameters of the active controller, weighting it by their geometrical distance with respect to the present position in the parameter space. The main critic to this method is that it does not take into account the nonlinear characteristic of the system that creates a heterogeneous parameter space. In Figure 17.10, for example, since the plant finds itself closer to O_2, weighting the sum by the geometrical distance alone would result in a control law more similar to the one devised for that operating point. However, if a strong nonlinearity existed between O_2 and the present system position, the ideal control law may be more similar to those created for O_1 and O_3.

Among the techniques that have been researched aiming to overcome this limitation, some are of special interest to this study as they were specifically designed for FTC applications. In Ref. [23], a set of IF–THEN rules was used in a fuzzy logic framework to compare the present position in the parameter space with the symptoms of known faults. The degree of similarity with each fault scenario was then used to weigh the mean that adjusts the parameters of the controller. Assuming that knowledge is available regarding the status of the system that make it prone to develop each expected fault, in Ref. [5] this approach was improved by applying the fuzzy algorithm only to the set of possible faults at a given position in the parameter space.

A different approach was taken in Ref. [25] where the probability of occurrence of each expected fault was modeled in a finite-state Markov chain with known transition probabilities. With this information at hand, the mean of control parameters was weighted favorably to the most probable fault scenarios.

Regardless of the weighting scheme chosen, it is still an approximation of the behavior of the system outside the considered regions of operation and as such it is inevitably susceptible to nonlinearities active outside those regions. One way to solve this deficiency is to generate closely connected models by dividing the whole parameter space into evenly spaced operating regions as shown in Figure 17.11a. The natural tradeoff of this method is that increased control performance tends to require controllers

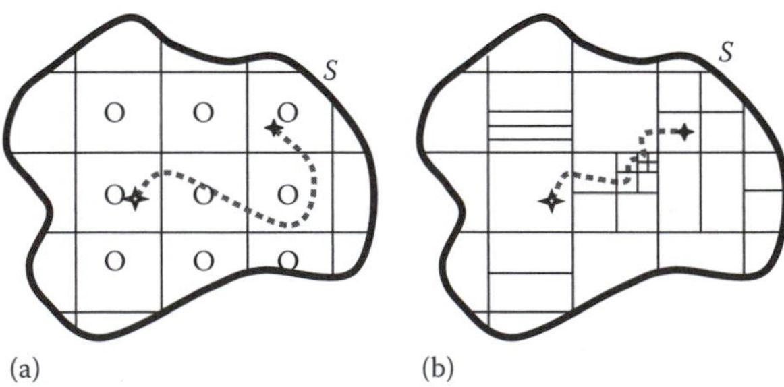

FIGURE 17.11 Closely connected multiple model implementations: (a) fixed size operating regions and (b) plant–dynamics-dependent operating regions. In the figure, each rectangular section represents an operating region. The system is originally in the position of the parameter spaced marked by the white star and follows the depicted trajectories.

designed for smaller, and therefore less complex, regions. This in turn causes the final number of models needed to cover the whole space to grow, requiring extensive design work since a control law has to be designed for each model. This relationship can be clearly seen in a series of simulation results performed in Ref. [36]. If made small enough, each region can be represented by a linear model given by the linearization of the nonlinear plant on the center of the operating region, making it possible to apply GS.

Since all parameter space is covered by previously designed controllers, a basic closely connected MMA algorithm would be composed only of two steps: determine the present position of the plant in the parameter space and apply the corresponding controller. However, even though each controller is designed to provide a desirable behavior for the plant while inside its respective operating region, if no special procedure is performed, switching directly from one control law to another may cause all kinds of unwanted responses as the plant navigates from one operating region to another. In Ref. [9], a minimum time (or number of iterations) was set for permanence inside an operating region before switching takes place, creating in this way a time-based hysteresis in an effort to prevent oscillations between adjacent operating regions. Another approach, requiring all controllers to possess the same structure, is to create an area on the border of adjacent operating regions in which the parameters of both controllers are combined causing one to gradually change to another. However, both methods are solely heuristic solutions and no proof of their efficiency, let alone deterministic way to configure their design parameters, is available. A method that guarantees stability of systems when perform control switching has been presented in Ref. [39]. The referenced paper describes a way to compute a pretransition sub-region inside an operating region from which stability is assured when switching to another specific operating region. In Ref. [36], an adaptive controller that operates in parallel with the MMA is used to assure stability of the system during the transient behavior generated by switching controllers.

If complete information on the dynamics of the nonlinear system is available beforehand, it is possible to divide the parameter space taking into account the sensibility of different areas (as shown in Figure 17.11b) and produce a combination of controllers with good performance based on a compact set of operating regions. It is important to notice that, independent of the number and uniformity of the regions, because no interpolation is fundamentally necessary, different control structures or strategies can be used for each region. From the point of view of FTC applications that consider the occurrence of unexpected faults, model weighting is not an attractive technique since there is no reason to assume that a new fault dynamic will hold any relationship with those previously known. When closely connected multiple models are considered, the quality of the response depends on the robustness of the design of each controller and the way the control laws are switched from one to another. Although in this formulation an active FTC is being performed for expected faults, no direct action can be defined for unexpected dynamics. At the same time that the requirement for robustness increases, since the areas of sensitivity are no longer available at design time, a large number of evenly spaced operating regions have to be created making the memory requirement and design effort increase greatly.

It is therefore interesting to explore yet another way to apply the MMA concept in which controllers are designed online as new operating regions are reached [36]. Since no information about the parameter space is supposed to be available at design time, nonlinear online identification is required in order to learn new operating regions (models) and recognize the ones to which a controller has already been designed. In this way, different from the previously discussed methods that adjust the controller based on the position of the plant in the parameter space, the online building of models achieves the same in an indirect manner by the identification error of the models designed so far. Therefore, if at a given moment the identification error of every known model (contained in a dynamic database) is high, the plant is considered to be in an unknown region of the parameter space, while if the error of one of the models is low, it indicates that the plant is inside the previously designed operating region. What is considered to be "high" or "low" depends on an identification threshold selected by the user. By reducing this threshold, the operating region of each model shrinks, causing a greater number of models to be generated. In this sense, a parallel can be traced between the setting of the identification threshold and the choice of how many fixed models to have in the closely connected operating regions approach.

It is interesting to notice that, due to the indirect measuring through the dynamics of the plant, the operating regions now span in the space of the identifier, not in the parameter space. If a NN is used to approximate the plant dynamics, for example, the operating regions span in the dimension defined by its weights. Operating regions described in the parameter space possess all its dimensions with direct physical meaning since they came from sensor readings. Although this property can be highly desirable in certain operations such as translating expert knowledge to the model database, changes in the dynamics are not always directly linked to the position on the environmental space. For example, a high temperature in a certain part of a system may not instantly incur in a fault, but may increase the probability of its occurrence. Since the identifier focus on the change in the dynamics and not on the secondary symptoms of faults, it does not suffer from such drawback. On the other hand, in order to extract information from expert sources it is necessary to duplicate the described conditions in simulation so that the identifier is able to produce a model in its own space.

As with the identification models, the control laws must also be devised online. A single control strategy that modifies itself based on the identified models, such as approximate feedback linearization [28], is a valid approach for plants which dynamics do not present extreme nonlinearities. When it is not the case, highly flexible nonlinear adaptive controllers [29] may be applicable.

If a new model is added to the database every time the identification threshold is exceeded, the area of the parameter space to which the system is exposed will be filled with closely connected models and therefore there is no need to use the same control structure for every operating region. Particular solutions previously known to exist to particular regions of the parameter space can then be directly introduced. For example, fuzzy logic can be used to extract expert knowledge on the solution of a particular fault, while NNs are used to generate novel control laws to cope with unexpected fault scenarios.

Sparsely connected model distribution can also be attained by the online MMA approach if a second threshold to measure model dissimilarity is created. The dissimilarity threshold, always greater than the identification one, indicates the regions in which the present dynamics of the plant are considered to be different enough from all the models in the database to justify the addition of a new model. Such scheme was implemented in Ref. [37] where the parameters of the controllers for regions not covered by the models in the database were adjusted by a mean of the known controllers weighted by the inverse of the identification error of their respective models. In this way, the control laws for regions between models hold more similarity to the ones devised for similar plant dynamics.

Apart from the above-mentioned concerns involving the transient behavior of the system when switching is performed, the application of MMA to FTC harbors two other points that require careful consideration. The first of them is the fact that the task to link either the present location on the parameter space or the prediction errors of identification models to the occurrence of a particular fault represents a FDI process and as such is vulnerable in all issues outlined. The second point focus on FTC applications that require new models to be designed online in a continuously growing database. In such a scenario, the nonlinear adaptive controller is required to be at the same time: quick to converge, highly flexible, and possess guaranteed stability, often conflicting characteristics in practice.

17.9 Case Studies

In order to demonstrate the capabilities of the identification network and provide a better understanding of the fine interrelations between the supervisor and DHP controller, two numerical examples are exploited. In both examples, faults are simulated by instantly or gradually changing the model of the plant. To give a better insight to the challenge of each fault scenario, linear models of fixed order similar to those employed in Ref. [37] are used here. This information, however, is not used in any way during the design of the fault tolerant controller that continues to take the plant as possessing a generic nonlinear model.

17.9.1 Identification on Using an RNN

The goal of the following example is to display the capabilities of the single-layered recurrent network to perform the identification of linear difference systems. An input signal is supplied in the form of a fixed frequency sine wave that changes mean and amplitude only once during the simulation. Since in the final application the input to the plant generated by the actor network is not necessarily composed of a large range of frequencies, the input with a limited spectrum represents a challenging but possible scenario in practice.

Four systems are presented in the sequence displayed in Table 17.2. The network is allowed 50 s for the identification of the first model, 30 s for the second, and 20 s for the third. The fourth and final model is unstable and the applied sinusoidal input steeply drives the output to positive infinity. A variable learning rate with maximum value of 0.004 is used.

In Figure 17.12, the performance of the identifier can be seen. The small learning rate applied generates a slow initial reaction, but the identification signal remains close enough to the true plant output throughout the simulation in spite of the changes in the range of input and plant dynamics from model 1 to model 3. As the fourth dynamic causes the output of the plant to grow steadily at increasing rates, it is not feasible for an identifier with a maximum learning rate to produce true identification indefinitely. Still the RNN-based identifier fulfils its goal until the output becomes 45 times larger than the normal range of operation. In the complete scheme, this would allow the actor network more than 70 iterations to restructure itself in any way that would at least decrease the rate of divergence.

TABLE 17.2 Sequence of Changes in the Dynamics of the Plant Applied for the Identification Example

Start Time (ds)	Plant Dynamics
0	$y(t) = 1.810\,y(t-1) - 0.8187\,y(t-2) + 0.00566\,u(t-1) + 0.00566\,u(t-2)$
500	$y(t) = 1.810\,y(t-1) - 0.9000\,y(t-2) + 0.00566\,u(t-1) + 0.00566\,u(t-2)$
800	$y(t) = 1.810\,y(t-1) - 0.9048\,y(t-2) + 0.00242\,u(t-1) + 0.00234\,u(t-2)$
1000	$y(t) = 1.919\,y(t-1) - 0.9048\,y(t-2) - 0.00242\,u(t-1) + 0.00234\,u(t-2)$

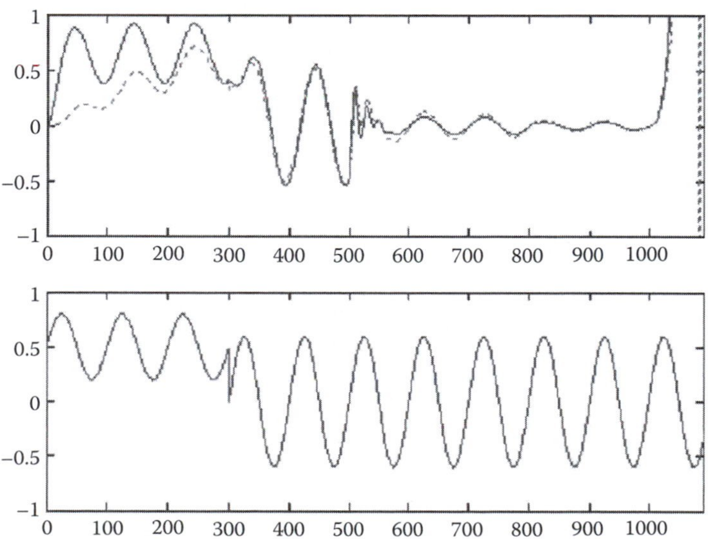

FIGURE 17.12 Results of the identification simulation. Plant signals are displayed in solid lines and the identification network output in dashed lines.

17.9.2 FTC Using a GDHP Controller

In this section, the GDHP adaptive critic controller is presented with the challenge of making a nonlinear MIMO plant (61) follow the sinusoidal trajectories described by (17.62). As faults will be introduced latter, the original dynamics will also be referred to as the nominal dynamics. This plant, as well as the tracking trajectories, was suggested in Ref. [9] as a simulation test bed for online nonlinear adaptive control. In the original paper, results were only shown after the controller was allowed to adapt for 500,000 iterations.

$$\begin{aligned}
x_1(t+1) &= 0.9x_1(t)\sin(x_1(t)) + \left(2 + \frac{1.5x_1(t)u_1(t)}{1+x_1^2(t)u_1^2(t)}\right)u_1(t) + \left(x_1(t) + 2\frac{x_1(t)}{1+x_1^2(t)}\right)u_2(t)\\
x_2(t+1) &= x_3(t)\big(1+\sin\big(4x_3(t)\big)\big) + \frac{x_3(t)}{1+x_3^2(t)}\\
x_3(t+1) &= \big(3+\sin\big(2x_1(t)\big)\big)u_2(t)
\end{aligned} \tag{17.61}$$

$$\begin{aligned}
x_1^t(t) &= 0.5\sin\left(\frac{2\pi t}{50}\right) + 0.5\sin\left(\frac{2\pi t}{25}\right)\\
x_2^t(t) &= 0.25\sin\left(\frac{2\pi t}{50}\right) + 0.75\sin\left(\frac{2\pi t}{25}\right)
\end{aligned} \tag{17.62}$$

Note that although in this example we assume that all states are available (and therefore will be mapped), we are only interested to track two of the states. Therefore, the matrices for the utility function and identification goal are adjusted as shown in (17.63).

$$S = \begin{bmatrix} 1 & 0 & 0\\ 0 & 1 & 0\\ 0 & 0 & 1 \end{bmatrix},\quad R = \begin{bmatrix} 1 & 0\\ 0 & 1 \end{bmatrix},\quad Q = \begin{bmatrix} 1 & 0 & 0\\ 0 & 1 & 0\\ 0 & 0 & 0 \end{bmatrix} \tag{17.63}$$

For all the results shown in this section, the learning rates for the IdNN, AcNN, and CrNN were set, respectively, to $\beta^i = 0.01$, $\beta^a = 0.001$, and $\beta^c = 0.04$. For the CrNN training, the GDHP algorithm is set to combine HDP and DHP with equal weights, i.e., $\eta = 0.5$. For both inputs and states, tap delay lines of size 10 were used. The future horizon for the calculation of $J(t)$ was set to approximately 50 iterations by having $\gamma = 0.9$.

Figure 17.13 shows a performance comparison of the GDHP programming algorithm running with 20 hidden neurons and with 50 hidden neurons in each of the three NNs. The results indicate that the

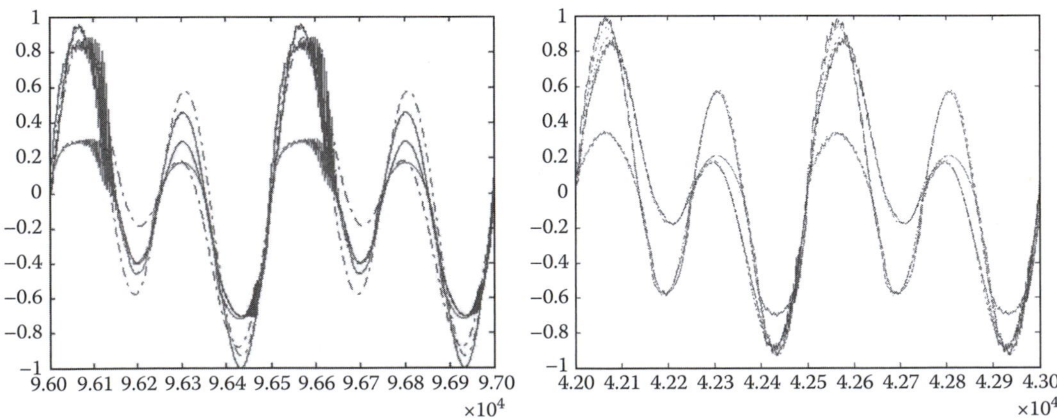

FIGURE 17.13 Results of the application of GDHP with different number of hidden neurons for the online adaptation of the nominal plant. The desired trajectories are plotted in dash-dotted lines, while the actual plant outputs are in solid lines.

GDHP algorithm managed to adapt the weights of the NNs so as to best follow the trajectories, given the power granted by the number of hidden neurons available.

This first experiment demonstrates that GDHP is capable of achieving excellent results, provided that enough hidden neurons are available and that enough time is provided. It is also important to point out that the result shown in Figure 17.13 was achieved after 42,000 iterations, more than 10 times faster than the original paper that used a classical neural control design.

In the next experiment, a fault is introduced at the iteration 32,000 by changing the nominal dynamics of the first state of the plant to the following:

$$x_1(t+1) = 0.5x_1(t)\sin(x_1(t)) + \left(4 + \frac{1.5x_1(t)u_1(t)}{1+x_1^2(t)u_1^2(t)}\right)u_1(t) + \left(x_1(t) + 2\frac{x_1(t)}{1+x_1^2(t)}\right)u_2(t) \tag{17.64}$$

As it can be seen in Figure 17.14, after a short transient where large spikes are generate, the GDHP controller manages to generate a control sequence that once again tracks the desired trajectories closely. Figure 17.15 gives an indirect idea of the amount of reconfiguration performed by depicting how different the control inputs for the nominal and fault scenarios had to be to achieve the same tracking.

As pointed previously, the future horizon for $J(t)$ is adjusted directly γ. This fact allows for a very simple way to compare the performance of the GDHP adaptive control algorithm with the classical neural control approach, with only IdNN and AcNN discussed. As it can be seen in Equation 17.32, setting $\gamma = 0$ reduces the training of the AcNN to a minimization of $U(t)$ only, while Equations 17.52 through 17.54 show that the CrNN is reduced to an estimator of the utility function and its derivatives only.

Without changing any other parameter, the experiment in which the fault is introduced at iteration 32,000 was run again for $\gamma = 0$. The sum of the absolute tracking error over two complete periods (1000 iterations) was used as a performance indicator. The comparison of the evolution of such indicator during training for $\gamma = 0.9$ and $\gamma = 0$ is brought in Figure 17.16. The faster convergence of the GDHP controller clearly demonstrates the advantage of possessing a functional CNN over the classical approach. As a matter of fact, the classical approach with the same parameters could not even maintain stability of the plant after the fault is introduced. Although it is reasonable to argue that different parameters (in special different learning rates) might have produced better results for the classical approach, the presented result stands as an indication that GDHP might be a more stable (i.e., less affected by design parameters) NN control paradigm.

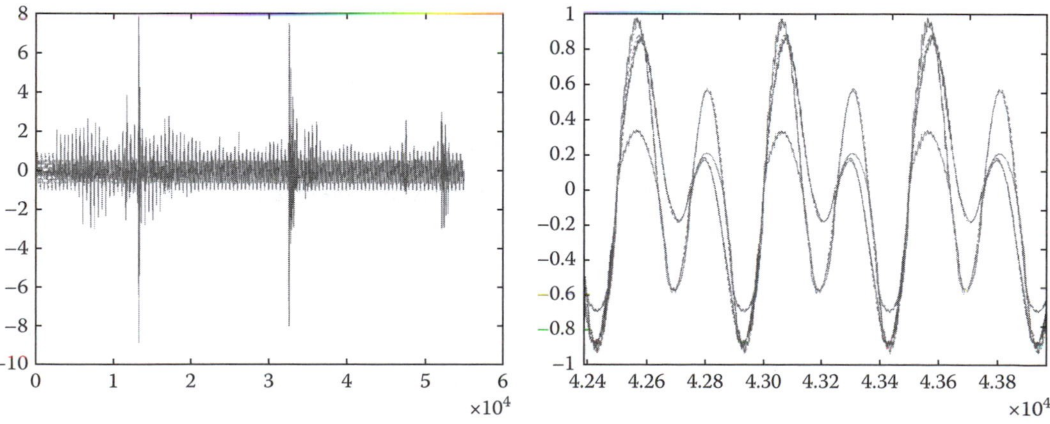

FIGURE 17.14 Performance of the GDHP controller as a fault changes the plant dynamic at iteration 32,000. A plot of the general development (left) and the tracking performance after 10,000 iterations following the fault occurrence (right).

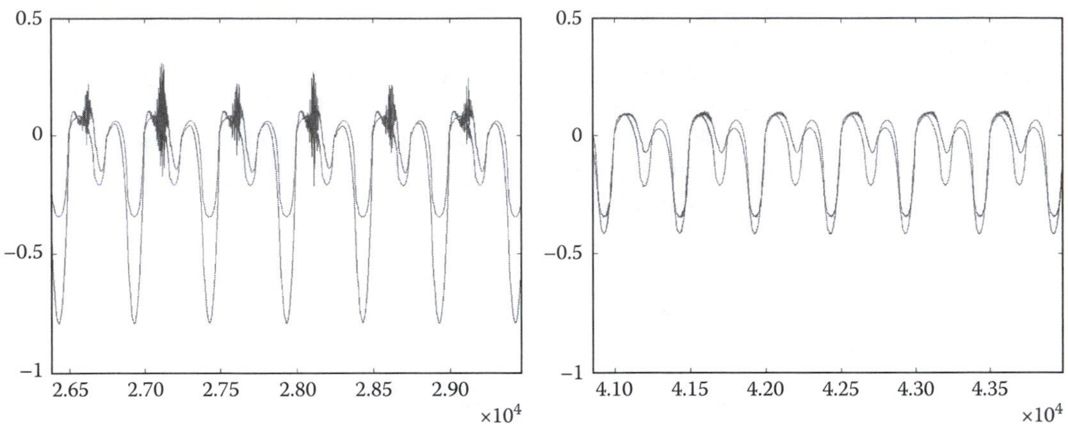

FIGURE 17.15 Input sequences developed by the GDHP controller to track the desired states under nominal (left) and fault (right) conditions.

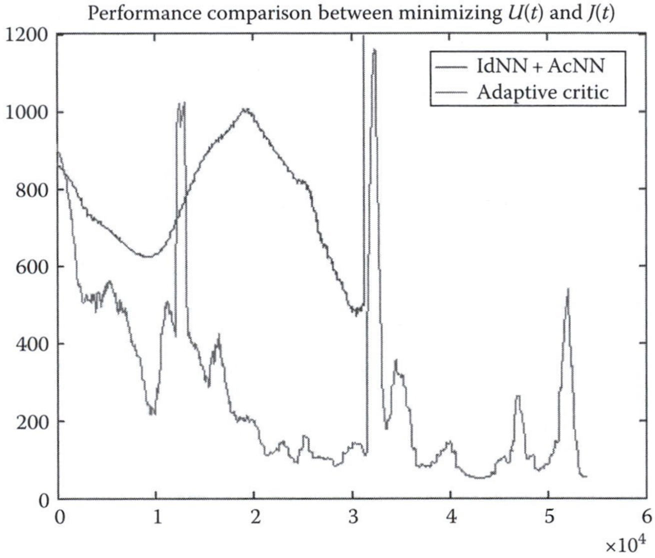

FIGURE 17.16 Tracking performance comparison between GDHP adaptive critic and classical NN adaptive control design.

Finally, to demonstrate the capabilities of the GDHP to deal with a plant that varies its dynamics under several different fault scenarios, the following dynamics were presented:

Nominal—iterations 1–30,000

$$
\begin{aligned}
x_1(t+1) &= 0.9x_1(t)\sin(x_1(t)) + \left(2 + \frac{1.5x_1(t)u_1(t)}{1+x_1^2(t)u_1^2(t)}\right)u_1(t) + \left(x_1(t) + 2\frac{x_1(t)}{1+x_1^2(t)}\right)u_2(t) \\
x_2(t+1) &= x_3(t)\big(1+\sin\big(4x_3(t)\big)\big) + \frac{x_3(t)}{1+x_3^2(t)} \\
x_3(t+1) &= \big(3+\sin\big(2x_1(t)\big)\big)u_2(t)
\end{aligned}
\tag{17.65}
$$

Fault 1—iterations 30,001–40,000

$$\begin{aligned}
x_1(t+1) &= 0.5x_1(t)\sin(x_1(t)) + \left(4 + \frac{1.5x_1(t)u_1(t)}{1+x_1^2(t)u_1^2(t)}\right)u_1(t) + \left(x_1(t) + 2\frac{x_1(t)}{1+x_1^2(t)}\right)u_2(t) \\
x_2(t+1) &= x_3(t)\big(1+\sin\big(4x_3(t)\big)\big) + \frac{x_3(t)}{1+x_3^2(t)} \\
x_3(t+1) &= \big(3+\sin\big(2x_1(t)\big)\big)u_2(t)
\end{aligned} \tag{17.66}$$

Fault 2—iterations 40,001–50,000

$$\begin{aligned}
x_1(t+1) &= 0.5x_1(t)\sin(x_1(t)) + 4u_1(t) + \left(x_1(t) + 2\frac{x_1(t)}{1+x_1^2(t)}\right)u_2(t) \\
x_2(t+1) &= x_3(t)\big(1+\sin\big(4x_3(t)\big)\big) + \frac{x_3(t)}{1+x_3^2(t)} \\
x_3(t+1) &= \big(5+\sin\big(2x_1(t)\big)\big)u_2(t)
\end{aligned} \tag{17.67}$$

Fault 3—iterations 50,001–60,000

$$\begin{aligned}
x_1(t+1) &= 0.5x_1(t)\sin(x_1(t)) + \left(4 + \frac{1.5x_1(t)u_1(t)}{1+x_1^2(t)u_1^2(t)}\right)u_1(t) + \left(x_1(t) + 2\frac{x_1(t)}{1+x_1^2(t)}\right)u_2(t) \\
x_2(t+1) &= x_3(t) \\
x_3(t+1) &= \big(3+\sin\big(2x_1(t)\big)\big)u_2(t)
\end{aligned} \tag{17.68}$$

Return to Nominal—iterations 60,001–70,000

$$\begin{aligned}
x_1(t+1) &= 0.9x_1(t)\sin(x_1(t)) + \left(2 + \frac{1.5x_1(t)u_1(t)}{1+x_1^2(t)u_1^2(t)}\right)u_1(t) + \left(x_1(t) + 2\frac{x_1(t)}{1+x_1^2(t)}\right)u_2(t) \\
x_2(t+1) &= x_3(t)\big(1+\sin\big(4x_3(t)\big)\big) + \frac{x_3(t)}{1+x_3^2(t)} \\
x_3(t+1) &= \big(3+\sin\big(2x_1(t)\big)\big)u_2(t)
\end{aligned} \tag{17.69}$$

As depicted in Figures 17.17 and 17.18, the GDHP adaptive controller was capable of devising new nonlinear controllers as it adapted online so as to maintain the trajectory as close as the desired as possible under different fault scenarios that modified the plant dynamics several times.

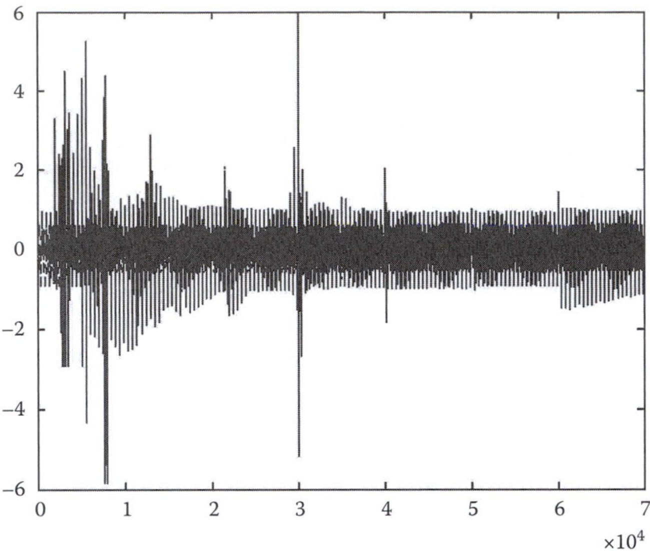

FIGURE 17.17 Trajectory tracking results as the dynamics are changed several times during the experiment.

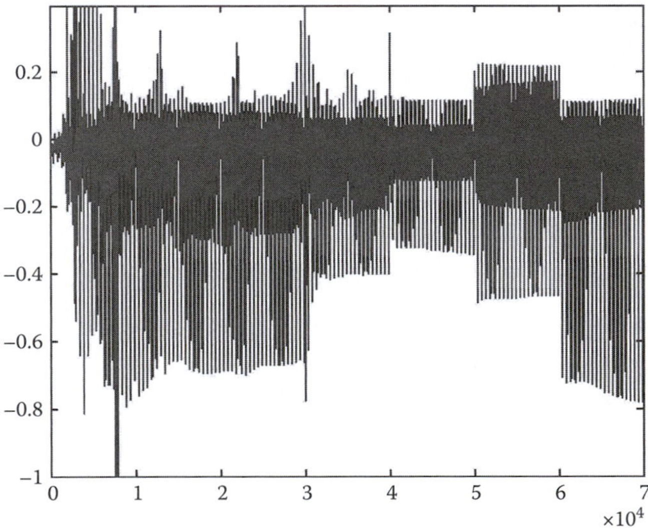

FIGURE 17.18 Different control input sequences developed as each fault scenario was presented.

17.10 Concluding Remarks

In this chapter, a particular implementation of GDHP adaptive critic controller architecture was introduced as a solution to the FTC problem that composed the initial motivation. A series of computer simulations provided favorable results indicating that the GDHP architecture might hold the potential to be the superior ADC architecture. A complete and detailed mathematical description of the proposed GDHP algorithm used in the experiments was introduced in a format that should lead to easy implementation in any matrix-oriented compiler. Finally, the applicability of the GDHP controller to the FTC problem was demonstrated in a comprehensive example where the plant dynamics were modified several times by the occurrence of different faults.

References

1. G.G. Yen and P.G. DeLima, Improving the performance of globalized dual heuristic programming for fault tolerant control through an online learning supervisor, *IEEE Transactions on Automation Science and Engineering*, 2, 121–131, 2005.
2. K. Åström, P. Albertos, M. Blanke, A. Isidori, W. Schaufelberger, and R. Sanz (Eds.), *Control of Complex Systems*, Springer: London, U.K., 2001.
3. M. Blanke, R. Izadi-Zamanabadi, S. Bøgh, and C. Lunau, Fault-tolerant control systems—A holistic view, *Control Engineering Practice*, 5, 693–702, 1997.
4. M. Blanke, M. Staroswiecki, and N. Wu, Concepts and methods in fault-tolerant control, *Proceedings of the American Control Conference*, Arlington, VA, pp. 2606–2620, 2001.
5. Y. Diao, Fault tolerant systems design using adaptive estimation and control, PhD dissertation, The Ohio State University, Columbus, OH, 2000.
6. P.G. DeLima and G.G. Yen, Accommodating controller malfunctions through fault tolerant control architecture, *IEEE Transactions on Aerospace and Electronics Systems*, 43, 706–722, 2007.
7. D. Prokhorov, R. Santiago, and D. Wunsch, Adaptive critic designs: A case study for neurocontrol, *Neural Networks*, 8, 1367–1372, 1995.
8. D. Prokhorov and D. Wunsch, Adaptive critic designs, *IEEE Transactions on Neural Networks*, 8, 997–1007, 1997.
9. K. Narendra, J. Balakrishnan, and M. Ciliz, Adaptation and learning using multiple models, switching and tuning, *IEEE Control Systems Magazine*, 15, 37–51, 1995.
10. J. Murray, C. Cox, R. Saeks, and G. Lendaris, Globally convergent approximate dynamic programming applied to an autolander, *Proceedings of the American Control Conference*, Arlington, VA, pp. 2901–2906, 2001.
11. G. Venayagamoorthy, R. Harley, and D. Wunsch, Comparison of a heuristic programming and a dual heuristic programming based adaptive critics neurocontroller for a turbogenerator, *Proceedings of the International Joint Conference on Neural Networks*, Como, Italy, pp. 233–238, 2000.
12. P. Werbos, Stable adaptive control using new critic designs, available at xxx.lanl.gov/abs/adap-org/9810001, 1998.
13. E. Sanchez and M. Bernal, Adaptive recurrent neural control for nonlinear systems tracking, *IEEE Transactions on Systems, Man and Cybernetics*, 30, 886–889, 2000.
14. P. Werbos, Backpropagation though time: What it does and how to do it, *Proceedings of the IEEE*, 78, 1550–1560, 1990.
15. O. Jesús and M. Hagan, Backpropagation through time for a general class of recurrent network, *Proceedings of the International Joint Conference on Neural Networks*, Washington, DC, pp. 2638–2643, 2001.
16. G. Lendaris and C. Paintz, Training strategies for critic and action neural networks in dual heuristic programming method, *Proceedings of the International Joint Conference on Neural Networks*, Atlanta, GA, pp. 712–717, 1997.
17. P.G. DeLima and G.G. Yen, Multiple model fault tolerant control using globalized dual heuristic programming, *Proceedings of the International Symposium on Intelligent Control*, Houston, TX, pp. 523–528, 2003.
18. R. Williams, Training recurrent networks using the extended Kalman filter, *Proceedings of the International Joint Conference on Neural Networks*, Baltimore, MD, pp. 241–246, 1992.
19. M. Livstone, J. Farrell, and W. Baker, A computationally efficient algorithm for training recurrent connectionists networks, *Proceedings of the American Control Conference*, Chicago, IL, pp. 555–561, 1992.
20. G. Lendaris, C. Paintz, and T. Shannon, More on training strategies for critic and action neural networks in dual heuristic programming method, *Proceedings of the IEEE International Conference on Systems, Man and Cybernetics*, Orlando, FL, pp. 3067–3072, 1997.
21. M. Mahmoud, J. Jiang, and Y. Zhang, Analysis of the stochastic stability of fault tolerant control systems, *Proceedings of the Conference on Decision and Control*, Phoenix, AZ, pp. 3188–3193, 1999.

22. W. Rugh, Analytical framework for gain scheduling, *IEEE Control Systems Magazine*, 11, 79–84, 1991.
23. C. Lopez-Toribio and R. Patton, Takagi-Sugeno fuzzy fault-tolerant control of a non-linear system, *Proceedings of the Conference on Decision and Control*, Phoenix, AZ, pp. 4368–4373, 1999.
24. I. Konstantopoulos and P. Antsaklis, An eigenstructure assignment approach to control reconfiguration, *Proceedings of the IEEE Mediterranean Symposium of Control and Automation*, Chania, Greece, pp. 328–333, 1996.
25. Y. Zhang and J. Jiang, An interacting multiple-model based fault tolerant detection, diagnosis and fault-tolerant control approach, *Proceedings of the Conference on Decision and Control*, Phoenix, AZ, pp. 3593–3598, 1999.
26. S. Kanev and M. Verhaegen, A bank of reconfigurable LQG controllers for linear systems subjected to failures, *Proceedings of the Conference on Decision and Control*, 4, 3684–3689, 2000.
27. I. Konstantopoulos and P. Antsaklis, An optimization approach to control reconfiguration, *Dynamics and Control*, 9, 255–270, 1999.
28. K. Narendra and S. Mukhopadhyay, Adaptive control of nonlinear multivariable systems using neural networks, *Neural Networks*, 7, 737–752, 1994.
29. K. Narendra and O. Driollet, Adaptive control using multiple models, switching, and tuning, *Proceedings of the Adaptive Systems for Signal Processing, Communications, and Control Symposium*, Lake Louise, Canada, pp. 159–164, 2000.
30. K. Narendra and K. Parthasarathy, Identification and control of dynamical systems using neural networks, *IEEE Transactions on Neural Networks*, 1, 4–27, 1990.
31. E. Kosmatopoulos, M. Christodoulou, and P. Ioannou, Dynamical neural networks that ensure exponential identification error convergence, *Neural Networks*, 10, 299–314, 1997.
32. G. Kulawski and M. Brdyś, Stable adaptive control with recurrent networks, *Automatica*, 36, 5–22, 2000.
33. C. Hwang and C. Lin, A discrete-time multivariable neuro-adaptive control for nonlinear unknown dynamic systems, *IEEE Transactions on Systems, Man and Cybernetics*, 30, 865–877, 2000.
34. S. Ge, C. Wang and Y. Tan, Adaptive control of partially known nonlinear multivariable systems using neural networks, *Proceedings of the IEEE International Symposium on Intelligent Control*, Mexico City , Mexico, pp. 292–297, 2001.
35. G. Puskorius and L. Feldkamp, Automotive engine idle speed control with recurrent neural networks, *Proceedings of the American Control Conference*, San Francisco, CA, pp. 311–316, 1993.
36. J. Boskovic, S. Liu, and R. Mehra, On-line failure detection and identification (FDI) and adaptive reconfiguration control (ARC) in aerospace applications, *Proceedings of the American Control Conference*, Arlington, VA, pp. 2625–2626, 2001.
37. D. Filev and T. Larsson, Intelligent adaptive control using multiple models, *Proceedings of the IEEE International Symposium on Intelligent Control*, Mexico City, Mexico, pp. 314–319, 2001.
38. R. Murray-Smith and T. Johansen (Eds.), *Multiple Model Approaches to Modeling Control*, Taylor & Francis: Basingstoke, U.K., 1997.
39. H. Pei and B. Krogh, Stability regions for systems with mode transitions, *Proceedings of the American Control Conference*, Arlington, VA, pp. 4834–4839, 2001.

18

Self-Organizing Maps

Gary Yen
Oklahoma State University

18.1 Introduction

The self-organizing map (SOM) is a neural network paradigm for exploratory data analysis. The idea of the SOM was originally motivated by localized regions of activities in the human cortex, where similar regions react to similar stimuli. This model stems from Kohonen's work [1] and builds upon earlier work of Willshaw and von der Malsburg [2]. As a data analysis tool, the SOM can be used at the same time both to reduce the amount of data by clustering and for projecting the data nonlinearly onto a lower-dimensional display [3]. Because of its benefits, the SOM has been used in a wide variety of scientific and industrial applications, such as image recognition, signal processing, and natural language processing. In the research community, it has received significant attention in the context of clustering, data mining, topology preserving, vector projection, and data visualization.

The SOM is equipped with an unsupervised and competitive learning algorithm. It consists of an array of neurons placed in a regular, usually two-dimensional (2D) grid. Each neuron is associated with a weight vector (or prototype vector). Similar to other competitive networks, the learning rule is based on weight adaptations. In the original algorithm of SOM, only one neuron (winner) at a time is activated corresponding to each input. The presentation of each input pattern consists of a localized region of activity in the SOM network. During the learning process, a sufficient number of different realizations of the input patterns are fed to the neurons so that the neurons become tuned to various input patterns in an orderly fashion. The principal goal of the SOM is to adaptively transform an incoming pattern of arbitrary dimension into the low-dimensional SOM grid. The locations of the responses in the array tend to become ordered in the learning process as if some meaningful nonlinear coordinate system for the different input features were being created over the network. This projection can be visualized in numerous ways in order to reveal the characteristics of the underlying input data or to analyze the quality of the obtained mapping [4].

18.1.1 Structure

The neurons in a SOM are usually placed in a regularly spaced one-, two-, or higher dimensional grid. The 2D grid is most commonly used because it provides more information than the one-dimensional (1D) and is less problematic than the higher dimensional ones. The positions of the neurons in the grid are fixed, so they won't move during the training phase of the SOM.

The neurons are connected to adjacent neurons by a neighborhood relation, which dictates the structure or topology of the map. The neurons most often are connected to each other via a rectangular or hexagonal grid structure. The grid structures are illustrated in Figure 18.1, where neurons are marked with black dots. Each neuron has neighborhoods of increasing diameter surrounding it. The neighborhood size controls the smoothness and generalization of the mapping. Neighborhoods of different sizes in both topologies are also illustrated in Figure 18.1. Neighborhood 1, the neighborhood of diameter 1, includes the center neuron itself and its immediate neighbors. The neighborhood of diameter 2 includes the neighborhood 1 neurons and their immediate neighbors. The map topology is usually planar but toroidal topologies [5] have also been used. Figure 18.2 illustrates these two types of topologies.

18.1.2 Initialization

In the basic SOM algorithm, the layout and number of neurons are determined before training. They are fixed from the beginning. The number of neurons determines the resolution of the resulting map. A sufficiently high number of neurons should be chosen to obtain a map with decent resolution. Yet, this number should not be too high, as the computational complexity grows quadratically with the number of neurons [6].

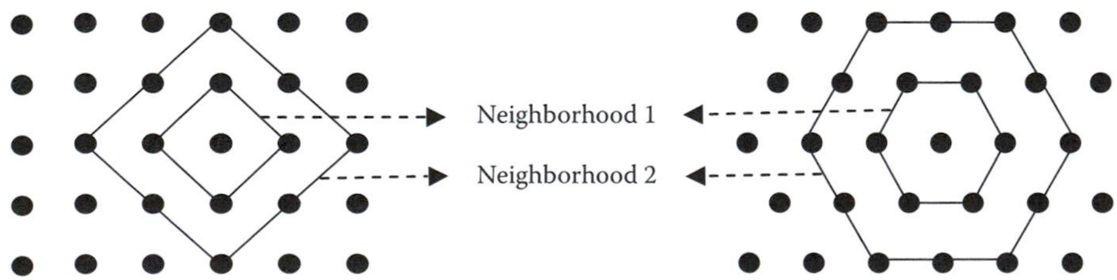

FIGURE 18.1 The SOM grid structure: (a) rectangular grid and (b) hexagonal grid.

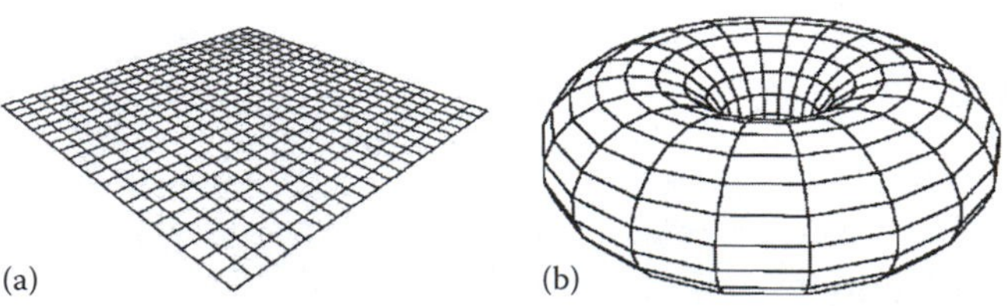

FIGURE 18.2 Two types of SOM topologies: (a) planar topology and (b) toroidal topology.

Each neuron in the SOM is associated with an n-dimensional weight vector

$$w_i = [w_{i1}, w_{i2}, \ldots, w_{in}]^T$$

where
- n is the dimension of the input vectors
- T denotes the matrix transpose

The weight vector is often referred to as the prototype vector. In this chapter, the terms weight vector and prototype vector are used interchangeably. Before the training phase, initial values are assigned to the weight vectors. Three types of network initializations are proposed by Kohonen [3]:

1. Random initialization, where simply random values are given to weight vectors. This is the case if little is known about the input data at the time of the initialization.
2. Initialization using initial samples, which has the advantage that the initial locations of the weight vectors lie in the same part of the input space as the data points.
3. Linear initialization, where the weight vectors are initialized to lie in the linear subspace spanned by two largest eigenvectors of the input data. This helps to stretch the SOM to the orientation in which the input data set has the most significant amount of information.

18.1.3 Training

The SOM is an unsupervised neural network, which means the training of a SOM is completely data driven. No external supervisor is available to provide target outputs. The SOM learns only from the input vectors through repetitive adaptations of the weight vectors of the neurons.

The training of the SOM is an iterative process. At each time step, one input vector x is drawn randomly from the input data set and presented to the network. The training consists of two essential steps:

1. Winner selection
   This step is often called competition. For each input pattern, a similarity measure is calculated between it and all the weight vectors of the map. The neuron with the greatest similarity with the input vector will be chosen as the winning neuron, also called the best-match unit (BMU). Usually the similarity is defined by a distance measure, typically Euclidean distance. Therefore the winner, denoted as c, is the neuron whose weigh vector is the closest to the data sample in the input space. This can be defined mathematically as the neuron for which

$$c = \arg\min_i \{\|x - w_i\|\} \tag{18.1}$$

2. Updating weight vectors
   After the winner is determined, the winning unit and its neighbors are adjusted by modifying their weight vectors toward the current input according to the learning rule formulated as

$$w_i(t+1) = w_i(t) + \alpha(t)h_{ci}(t)[x(t) - w_i(t)] \tag{18.2}$$

   where
   - $x(t)$ is the input vector randomly drawn from input set at time t
   - $\alpha(t)$ is the learning rate function
   - $h_{ci}(t)$ is the neighborhood function centered on the winner at time t

This adaptation rule of the weights is closely related to the k-means clustering. The weight vector of each neuron represents a cluster center. Like the k-means, the weight of the best matching neuron (cluster center) is updated in a small step in the direction of the input vector x. However, unlike k-means, the winner and the neurons surrounding it are updated instead of the winner alone. The size of the surrounding region is specified by $h_{ci}(t)$, which is a non-increasing function of time and of the distance of neuron *i* from the winner *c*.

As a result of the update rule, the neuron whose weight vector is the closest to the input vector is updated to be even closer. Consequently, the winning unit is more likely to win the competition the next time a similar input sample is presented, while less likely to win when a very different input sample is presented. As more input samples are presented to the network, the SOM gradually learns to recognize groups of similar input patterns in such a way that neurons physically close together on the map respond to similar input vectors.

18.1.4 Analysis of the Updating Rule

The update rule in Equation 18.2 can be rewritten as

$$w_i(t+1) = [1 - \alpha(t)h_{ci}(t)]w_i(t) + \alpha(t)h_{ci}(t)x(t) \tag{18.3}$$

This equation characterizes the influence of data samples during training and directly shows how the parameters, $\alpha(t)$ and $h_{ci}(t)$, affect the motion of w_i. Every time a data sample $x(t)$ is presented to the network, the value of $x(t)$, scaled down by $\alpha(t)*h_{ci}(t)$, is superimposed on w_i and all previous values $x(t')$, $t' = 0, 1, \ldots, t-1$, are scaled down by the factor $[1 - \alpha(t)*h_{ci}(t)]$, which we assume is less than 1. The contribution of the data samples can be shown more clearly by rewriting Equation 18.3 into a non-iterative form.

Given $w_i(0)$ as the initial condition, Equation 18.3 can be transformed into the following form by iteratively substituting $w_i(t')$ with $w_i(t'-1)$, $t' = t, t-1, \ldots, 1$,

$$w_i(t+1) = A(t+1)w_i(0) + \sum_{n=1}^{t} B(t+1,n)x(n) \tag{18.4}$$

The coefficient $A(t)$ describes the effect of the initial weight value on $w_i(t)$ and $B(t,n)$ describes the effect of the data point presented at time *n* on $w_i(t)$. Both $A(t)$ and $B(t,n)$ are functions of $\alpha(t)*h_{ci}(t)$ and decrease with *t*. Equation 18.4 shows that $w_i(t+1)$, the weight vector at time $t+1$, depends on a weighted sum of the initial condition and every data points presented to the network. $w_i(t+1)$ can be therefore considered as a "memory" of all the values of $x(t')$, $t' = 0, 1, \ldots, t$. As the weight function $B(t,n)$ is a function of $\alpha(t)$ and $h_{ci}(t)$, the influence of a training sample on the final weight vector depends on the specific learning rate and neighborhood function used during the self-organizing process.

18.1.5 Neighborhood Function

The neighborhood function is a non-increasing function of time and of the distance of unit *i* from the winner neuron *c*. The form of the neighborhood function determines the rate of change around the winner neuron. The simplest neighborhood function is the bubble function, which is constant over the defined neighborhood of the winner unit and zero elsewhere. Using the bubble neighborhood function, every neuron in the neighborhood is updated the same proportion of the difference between the unit and the presented sample vector.

Another widely applied, smooth neighborhood function is the Gaussian neighborhood function

$$h_{ci}(t) = \exp\left[\frac{-\|r_c - r_i\|^2}{2\sigma(t)^2}\right] \tag{18.5}$$

where

$\sigma(t)$ is the width of the Gaussian kernel

$\|r_c - r_i\|^2$ is the distance between the winner *c* and the neuron *i* with r_c and r_i representing the 2D positions of neurons *c* and *i* on the SOM grid

Usually, the radius of the neighborhood is large at first and decreases during the training. One commonly used form of $\sigma(t)$ is given by

$$\sigma(t) = \sigma(0)\left(\frac{\sigma(f)}{\sigma(0)}\right)^{\frac{t}{t_{max}}} \tag{18.6}$$

where
$\sigma(0)$ is the initial neighborhood radius
$\sigma(f)$ is the final neighborhood radius
t_{max} is the number of training iterations

Therefore, $\sigma(t)$ is a monotonically decreasing function of time. The decreasing neighborhood radius ensures that the global order is obtained at the beginning, whereas toward the end, the local corrections of the weight vectors of the map will be more specific.

18.1.6 Learning Rate

The learning rate $\alpha(t)$ is a function decreasing with time. It can be linear, exponential, or inversely proportional to time. The linear learning rate function can be defined as

$$\alpha(t) = \alpha(0) * \left(1 - \frac{t}{t_{max}}\right) \tag{18.7}$$

where $\alpha(0)$ is the initial learning rate. A commonly used exponentially decreasing function is given by

$$\alpha(t) = \alpha(0)\left(\frac{\alpha(f)}{\alpha(0)}\right)^{\frac{t}{t_{max}}} \tag{18.8}$$

where $\alpha(f)$ is the final learning rate. A function inversely proportional to time is given with the form

$$\alpha(t) = \frac{1}{m(t-1)+1}, \quad m = \frac{1 - N/t_{max}}{N - N/t_{max}} \tag{18.9}$$

where N is the total number of neurons. Using the learning rate function in Equation 18.9 ensures that earlier and later input samples have approximately equal effects on the training result.

The learning rate and the neighborhood function together determine which neurons and how much these neurons are allowed to learn. These two parameters are usually altered during training through two phases. In the first phase, namely the ordering phase, relatively large initial learning rate and neighborhood radius are used. The parameters keep decreasing with time. During this phase, a comparatively large number of weight vectors are to be updated and they move in big steps toward the input samples. In the second phase, the fine tuning phase, both parameters start with small vales from the beginning. They continue to decrease but very slowly. The number of iterations for the second phase should be much larger than that in the first phase, as the tuning usually takes much longer [7].

As a result of the learning rule, the neuron whose weight vector is the closest to the input vector is updated to be even closer. Consequently, the winning unit is more likely to win the competition the next time a similar input sample is presented, while less likely to win when a very different input sample is presented. As more inputs are presented to the network, the SOM gradually learns to recognize groups of similar input patterns in such a way that neurons physically close together on the map respond to similar input vectors.

18.2 Dynamic SOM Models

In spite of the widespread use of the SOM, some shortcomings have been noted, which are related to the static architecture of the basic SOM model. First of all, the number of neurons and the layout of the neurons (i.e., the topology) have to be determined before training. The need for predetermining a fixed network structure is a significant limitation on the final mapping [8–11]. To address the issue of static SOM architecture, several variations based on the classical SOM have been developed recently. These dynamic SOM models usually employ an incremental growing architecture to cope with the lack of prior knowledge about the number of map units. Some of the models are summarized in this section.

18.2.1 Growing Cell Structure

One of the first models of such kind is the growing cell structures (GCS) [8]. In the GCS, the basic 2D grid of the SOM is replaced by a network of nodes whose basic building blocks are triangles. Starting with a triangle structure of three nodes, the algorithm both adds new nodes to and removes existing nodes from the network during the training process. The connections between nodes are adjusted in order to maintain the triangular connectivity. A local error measure is used to decide the position to insert a new node, which is usually between the node with the highest accumulated error and its most distant neighbor. The algorithm results in a network graph structure consisting of a set of nodes and the connections between them.

18.2.2 Growing Neural Gas

In addition to the GCS, Fritzke has also proposed the growing neural gas (GNG) [9] and the growing grid (GG) [10]. The GNG algorithm combines the GCS and the Neural Gas algorithm [12]. It starts with two nodes at random positions, and as in GCS, new nodes are inserted successively to support the node with high accumulated errors. Unlike the GCS, the GNG structure is not constrained. The nodes are connected by edges with a certain age. Once the age of an edge exceeds a threshold, it will be deleted. After a fixed number of iterations, a new node is added between the node with the highest accumulated error and the one with maximum accumulated error among all its neighbors. As an alternative form of growing network, the GG starts with 2×2 nodes, taking advantages of a rectangular structured map. The model adds rows and columns of neurons during the training process, and therefore is able to automatically determine the height/width ratio suitable for the data structure. The heuristics used to add and remove nodes and connections are the same as those used in GCS.

18.2.3 Incremental Grid Growing

Another approach is the incremental grid growing (IGG) [13]. Starting from a small number of initial nodes, the IGG generates new nodes only at the boundary of the map. This guarantees that the IGG network will always maintain a 2D structure, which results in easy visualization. Another feature of IGG is that connections between neighboring map units may be added and removed according to a threshold value of the inter-unit weight differences. This may result in several disconnected sub-networks, which represent different clusters of input patterns. The growing self-organizing maps (GSOM) [11], in similar spirit as IGG, introduce a spread factor to control the growing process of the map.

18.2.4 Other Growing Structure Models

Other modified models have also been proposed, including the plastic self-organizing maps (PSOM) [14], the grow when required (GWR) [15], etc. Figure 18.3 shows the simulation results of the original SOM and some of the dynamic models discussed above, which are given in Ref. [16]. The simulation

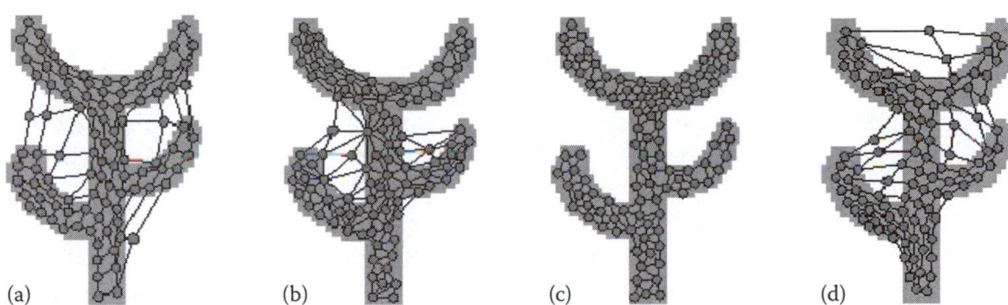

FIGURE 18.3 Simulation results of different models after 40,000 adaptation steps: (a) the SOM, (b) GCS, (c) GNG, and (d) GG. The distribution is uniform in the shaded area. Map units are denoted by circles.

results are generated using 40,000 input signals from a probability distribution that is uniform in the shaded area. The growing versions of the SOM aim at achieving an equal distribution of the input patterns across the map by adding new nodes near the nodes that represent an unproportionally high number of input data.

18.2.5 Hierarchical Models

Beside the limitation of the fixed structure, another deficiency of the classic SOM is the incapability of capturing the hierarchical structure commonly present in real-world data. The structural complexity of such data sets is usually lost during the mapping process by means of a single, low-dimensional map. In order to handle a data set with hierarchical relationships, hierarchical models should be used. These models try to organize data at different layers by displaying a representation of the entire data set at a top level and allowing the lower levels to reveal the internal structure of each cluster found in the higher-level representation, where such information might not be so apparent [17].

The hierarchical feature map [18] uses a hierarchical setup of multiple layers, where each layer is composed of a number of independent SOMs. Starting with one initial SOM at the top layer, a separate SOM is added to the next layer of the hierarchy for every unit in the current layer. Each map is trained with only a portion of the input data that is mapped onto the respective unit in the higher layer map. The amount of training data for a particular SOM is reduced as the hierarchy is traversed downward. As a result, the hierarchical feature map requires a substantially shorter training time than the basic SOM for the same data set. Moreover, it may be used to produce fairly isolated, or disjoint, clusters of the input data, while the basic SOM is incapable of performing the same [19].

Another hierarchical model, the hierarchical self-organizing map (HSOM) [20], focuses on speeding up the computation during winner selection by using a pyramidal organization of maps. However, like the hierarchical feature map, while representing the data in a hierarchical way this model does not provide a hierarchical decomposition of the input space.

As an extension to the Growing Grid and the hierarchical SOM models, the growing hierarchical self-organizing map (GHSOM) [21] builds a hierarchy of multiple layers, where each layer consists of several independent growing SOMs. Starting from a top-level SOM, each map grows incrementally to represent data at a certain level of detail in a manner similar to the GG. In GHSOM, the level of detail is measured in terms of the overall quantization error. For every map unit in a level, a new SOM might be added to a subsequent layer if this unit represents input data that are too diverse and thus more details are desirable for the respective data.

Once the training process is over, visual display of the map must be carried out in order for the underlying structure of data to be perceived. A variety of visualization techniques based on the SOM have been developed, which will be reviewed in the next section.

18.3 SOM Visualizations

Visualization potentiality is a key reason to apply the SOM for data analysis. Once the learning phase is over, visual display of the map can be carried out in order for the underlying structure of data to be observed. Extracting the visual information provided by the SOM is one of the primary motivations of this study.

The visualization of SOMs is motivated by the fact that a SOM achieves a nonlinear projection of the input distribution through a commonly 2D grid. This projection can be visualized in different ways by a variety of techniques. Some of them visualize the input vectors directly, whereas others take only the prototype vectors (or weight vectors) into account. Based on the object that is visualized, these techniques can be divided into several categories, which are reviewed in the remainder of this section.

18.3.1 Visualizing Map Topology

One category of the visualization techniques is to visualize the SOM topology through distance matrices. The most widely used method in this category is the unified distance matrix (U-matrix) [22], which enables visualization of the topological relations between the neurons in a trained SOM. The idea is to show the underlying data structure by graphically displaying the inter-neuron distances between neighboring units in the network. The distances of the prototype vector of each map unit to its immediate neighbors are calculated and form a matrix. The same metric is used to compute the distances between map units, as is used during the SOM training to find the BMU. By displaying the values in the matrix as a three-dimensional (3D) landscape or a gray-level image, the relative distances between adjacent units on the whole map becomes visible. The U-matrix is calculated in the prototype space and displayed using the map space.

A simplified approach is to calculate a single value for each map unit, such as the maximum or the sum of the distances to all immediate neighbors, and use it to control height or color [23]. High values in the U-matrix encode dissimilarity between neighboring units. Consequently, they correspond to cluster boundaries and are marked by mountains in a 3D landscape or dark shades of gray in a coloring scheme. Low values correspond to similarity between neighboring units, resulting in valleys or light shades of gray. A demonstration of the U-matrix is presented in Figure 18.4, which is based on a 10 × 10

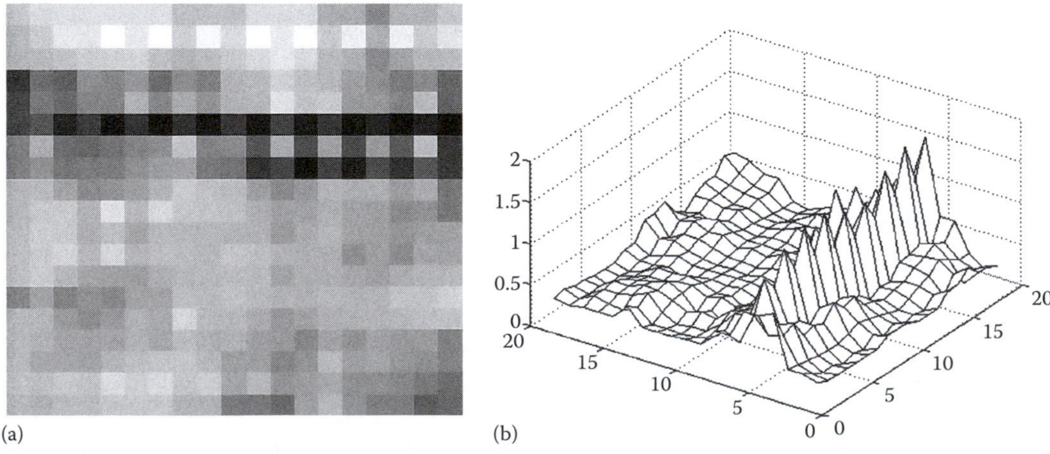

FIGURE 18.4 U-matrix presentations of a 10 × 10 rectangular SOM: (a) a gray-level image and (b) a 3D plot. The Iris data set is used to train the SOM.

rectangular SOM. The Iris data set [24] is used to train the SOM. Two essential clusters can be observed from both the gray-level presentation and the 3D landscape presentation.

18.3.2 Visualizing Data Density

Recently, a density-based visualization technique, the P-matrix [25,26], has been introduced, which estimates the data density in the input space sampled at the prototype vectors. The P-matrix is defined analogously to a U-matrix. Instead of local distances, this technique uses density values in data space measured at the position of each prototype vector as height values, called P-heights. The estimate of the data density is constructed using pareto density estimate (PDE) [25], which calculates the density as the number of input data points inside a hypersphere (Pareto sphere) within a certain radius (Pareto radius) around each prototype vector. In contrast to the U-matrix, neurons with large P-heights are situated in dense regions of the data space, while those with small P-heights are in sparse regions. Illustrations of different visualizations of a SOM are given in Figure 18.5, taken from Ref. [27]. Figure 18.5c shows the P-matrix of the Gaussian mixture data set in Figure 18.5a, where darker gray shades correspond to larger densities. Compared to the U-matrix presentation shown in Figure 18.5b, the P-matrix gives a complementary view of the same data set.

A combination of the U-matrix and the P-matrix has also been proposed by Ultsch, namely the U*-matrix [26]. Commonly viewed as an extension to the U-matrix, it takes both the prototype vectors and the data vectors into account. The values of the U-matrix are dampened in highly dense regions, unchanged in regions of average density, and emphasized in sparsely populated regions. It is designed for use with Emergent SOMs [27], which are SOMs trained with a high number of map units compared to the number of data samples. U*-matrix is advantageous over the U-matrix in data sets with clusters that are not clearly separated. The U*-matrix presentation of the Gaussian mixture data set in Figure 18.5d shows clearly two Gaussian distributions, which U-matrix fails to reveal.

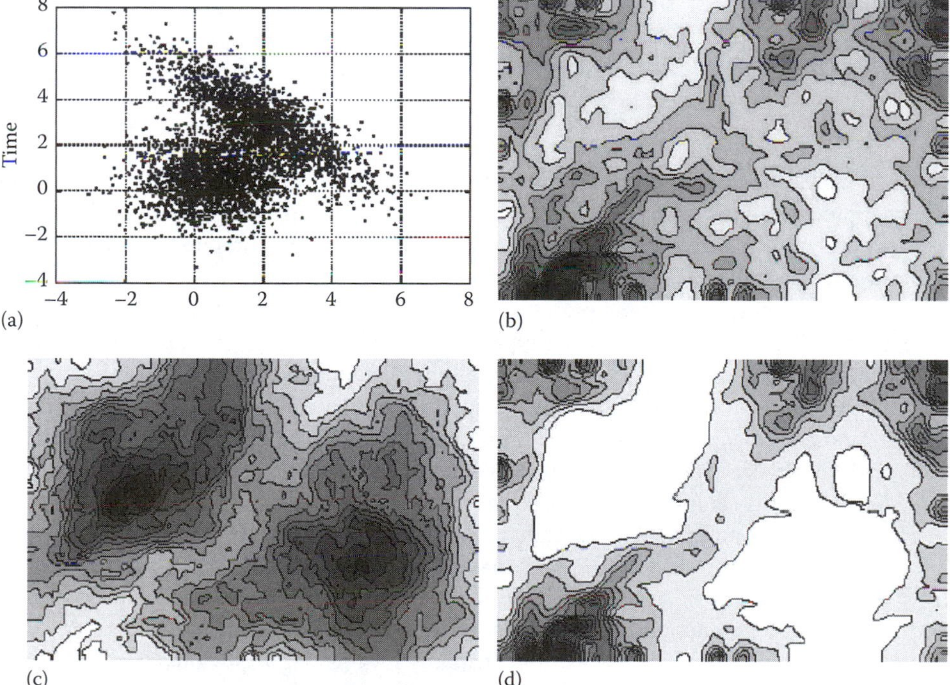

FIGURE 18.5 Different visualizations of the SOM: (a) the original data set of a mixture of two Gaussians, (b) U-matrix presentation, (c) P-matrix presentation, and (d) U*-matrix presentation.

18.3.3 Visualizing Prototype Vectors

An alternative way to visualize the SOM is to project the prototype vectors to a 2D output space using a generic projection method. Such methods include multidimensional scaling (MDS) [28] and Sammon's mapping [29]. MDS is a traditional technique for transforming a data set from a high-dimensional space to a space with lower dimensionality. It creates a mapping to a usually 2D coordinate space, where object can be represented as points. The inter-point distances in the original data space are approximated by the inter-point distances of the projected points in the projected space. Accordingly, more similar objects have representative points that are spatially nearer to each other. The error function to be minimized can be written as

$$E = \frac{\sum_{i \neq j} (d_{ij} - d_{ij}^{*})^2}{\sum_{i \neq j} (d_{ij}^{*})^2} \tag{18.10}$$

where

d_{ij} denotes the distance between vectors i and j in the original space

d_{ij}^{*} in the projected space

A gradient method is commonly used to optimize the above objective function. MDS methods are often computationally expensive.

Closely related MDS, Sammon's mapping also aims at minimizing an error measure that describes how well the pairwise distances in a data set are preserved [30]. The error function of Sammon's mapping is

$$E = \frac{1}{\sum_{i \neq j} d_{ij}} \sum_{i \neq j} \frac{(d_{ij} - d_{ij}^{*})^2}{d_{ij}} \tag{18.11}$$

Compared to MDS, the local distances in the original space are emphasized in Sammon's mapping. Since the mapping employs deepest descent procedure to minimize the error, it requires both first- and second-order derivatives of the objective function at each iteration. The computational complexity, as a result, is even higher than MDS [31].

Since the SOM provides a topology-preserving mapping of the input data, the MDS or Sammon's projection of the SOM can be used as a rough approximation of the shape of the input data. Both of these nonlinear projection approaches are iterative and computationally intensive. However, the computation load can be alleviated to an acceptable level when applied to the prototype vectors of a SOM instead of the original data set, provided a much smaller number of map units are used compared to the input vector number. The MDS projection and Sammon's mapping of a SOM are given in Figure 18.6. The map units are visualized as black dots, which are connected to their neighbors with lines. In this example, the Iris data set is used to train a 10×10 rectangular SOM. Roughly two clusters can be seen from both projections of the SOM. Apparently, the *setosa* class is distinct from the data set, while the other two linearly inseparable classes, *versicolor* and *virginica*, are still joined in the projection space.

In additional to the high computational cost, another drawback of MDS and Sammon's mapping is that they do not yield a mathematical or algorithmic mapping procedure for previously unseen data points [31]. That is, for any new input data point to be accounted for, the whole mapping procedure has to be repeated based on all available data. Mao and Jain have proposed a feed-forward neural network to solve this problem, which employs a specialized, unsupervised learning rule to learn Sammon's mapping [32].

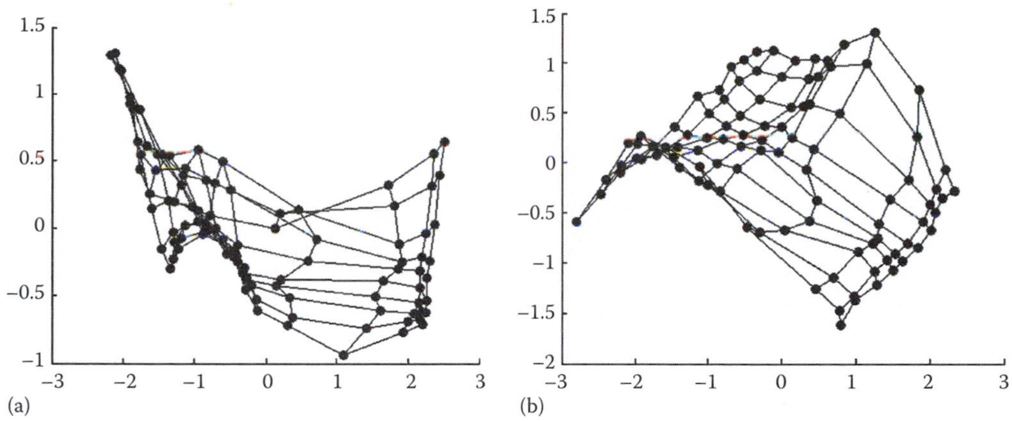

FIGURE 18.6 Different ways to visualize the prototype vectors: (a) MDS projection of a SOM and (b) Sammon's mapping of a SOM. Neighboring map units, depicted as black dots, are connected to each other.

18.3.4 Visualizing Component Planes

The prototype vectors can also be visualized using the component plane representation. Instead of a single plot, this technique provides a "sliced" version of the SOM, which shows the projection of each individual dimension of the prototype vectors on a separate plane [33]. The values of each component are taken from all prototype vectors and depicted by color coding. Each component plane shows the distribution of one prototype vector component. Similar patterns in different component planes indicate correlations between the corresponding vector components. This technique is hence useful when the correlation between different data features is of interest. However, one drawback of component planes is that cluster borders cannot be easily perceived. In addition, data with high dimensionality results in lots of plots.

The component planes of the 10 × 10 rectangular SOM trained with the Iris data set is presented in Figure 18.7. The color scheme of the map units has been set so that the lighter the color is, the smaller

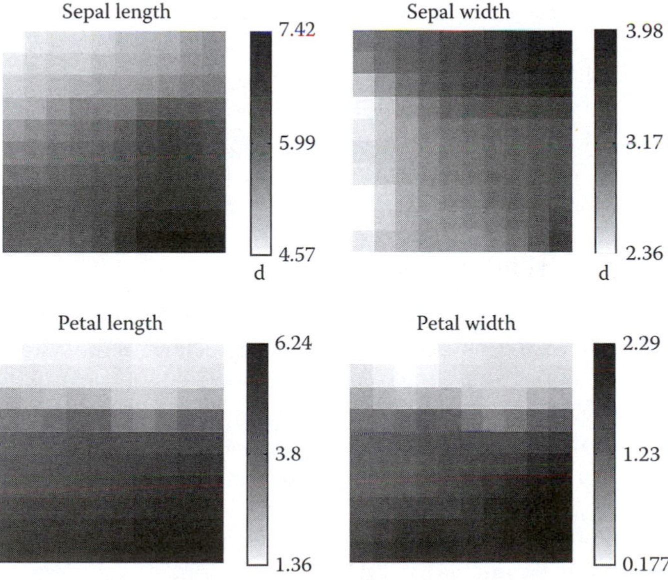

FIGURE 18.7 Component planes representation of a SOM trained with the Iris data set. The color bars beside each component plane show the maximum, mean, and minimum values and the corresponding colors.

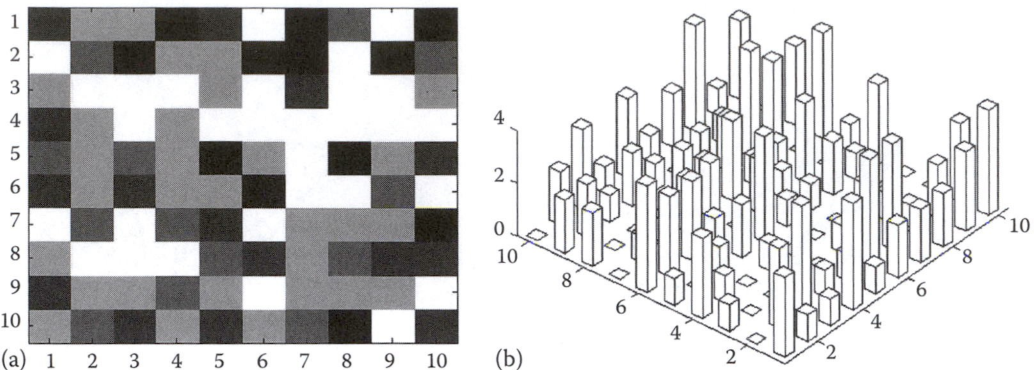

FIGURE 18.8 Different presentations of the hit histogram: (a) a gray level image and (b) a 3D plot. The Iris data set is used to train the SOM.

the component value of the corresponding prototype vector is. It can be seen, for instance, that the two components, petal length and petal width, are highly related.

18.3.5 Visualizing Best Matching Units

Another category of visualization is to display the BMUs of the input data set. Data vectors can be projected on the map by locating their BMUs. Because the prototype vectors are ordered on the map grid, nearby map units will have similar data projected to them. Projecting multiple input vectors will result in a histogram of the BMUs. For each data vector, the BMU is determined and the number of hits for that map unit is increased by one. The hit histogram shows the distribution of the data set on the map. Map units on cluster borders often have very few data samples, which imply very few hits in the histogram. Therefore, low-hit units can be used to indicate cluster borders. The values of a histogram can be depicted in different ways. Figure 18.8 illustrates the gray level presentation and the 3D presentation of a hit histogram. In Figure 18.8a, the darker the gray shade is, the higher the hit value of that unit is. In Figure 18.8b, the height directly corresponds to the value of the histogram.

However, hit histograms consider only the BMU for each data sample while real-world data is usually well represented by more than one unit. This inevitably causes distortions in the final map. A variation of the standard hit histogram, namely the smoothed data histogram (SDH) [34], has been developed counting the data sample's relativities to more than one map unit. The SDH allows a data sample to "vote" not only for the BMU but also for the next few good matches based on the ranks of distances between the data sample and the corresponding prototype vectors.

18.3.6 Other Visualizations

Aside from the above categories, other visualization techniques are also available for SOM. A rather different way to project the prototype vectors, the so-called adaptive coordinates [35], was proposed with a focus on cluster-boundary detection. This approach mirrors the movements of prototype vectors during the SOM training within a 2D "virtual" space, which is used for subsequent visualization of the clustering result. The initial positions of the prototype vectors are defined by the network structure, which are on top of the junctions of the map grid. The coordinates of the prototype vectors are adapted during the training. After convergence of the training process, the prototype vectors can be plotted in arbitrary positions in the projected space according to their coordinates. The algorithm offers an extension to both the basic training process and the fixed grid representation.

Another extended SOM model, called the visualization-induced SOM (ViSOM) [36], has been developed to directly preserve the distance information along with the topology on the map. The ViSOM updates the weight of the winning neuron using the same learning rule as the SOM. For the neighboring neurons, the weight adaptation is decomposed into two parts: a lateral movement toward the winner and an updating movement from the winner to the input vector. ViSOM places a constraint on the lateral contraction force between the neurons and hence regularizes the inter-neuron distances. As a result, the inter-neuron distances in the data space are in proportion to those in the map space. A scalable parameter λ is introduced in the constraint that controls the resolution of the map. If a high resolution is desirable, a small λ should be used, which will result in a large map.

18.4 SOM-Based Projection

Several challenges remain when using the SOM for visualizing document databases. First, the shape of the grid and the number of nodes have to be predetermined. This requires prior knowledge of the input data characteristics, which is usually unavailable before analysis. Second, the underlying hierarchical relations can hardly be detected by a single map. Such relations are commonly observed in document collections and, thus, their proper identification is highly desirable. A further limitation, which occurs when using the SOM projection, is that the map resolution depends solely on the size of the map. To have a high-resolution document map, which is desirable in most cases, it requires a considerably large number of neurons. To achieve a better visualization, a high-resolution SOM may even call for a higher number of neurons than that of input vectors [37]. As a result, the size of the SOM will become impractically huge when dealing with large data sets. The computational complexity grows quadratically with the number of neurons [38]. As a result, training huge maps may be exceedingly time-consuming.

To resolve the above limitations, a SOM-based visualization approach has been proposed [39]. Figure 18.9 shows the schematic diagram of the proposed approach. First, a similarity matrix is derived from the collection of documents of interest. The similarity matrix is then used to train a GHSOM, which clusters document items in a hierarchical manner and at the mean time allows for adaptation of the network architecture during training. Following the training of the GHSOM, a novel projection technique, the ranked centroid projection (RCP), is used to project the input vectors to a hierarchy of 2D output maps. Using the proposed approach, a hierarchy of multiple data projections can be achieved with comparatively low computational cost.

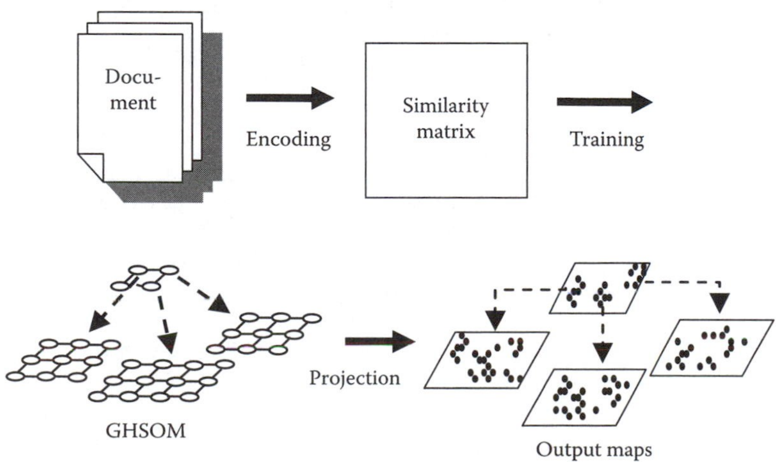

FIGURE 18.9 The schematic diagram of the proposed SOM-based approach.

18.4.1 User Architecture

The typical goal of document clustering and visualization is to discover subsets of large document collections that correspond to individual topics. Additionally, it can be applied hierarchically, yielding more refined groups within clusters. This leads to a large-to-small-scale presentation of the conceptual structure of the document collection, in which large-scale clusters correspond to more general topics and smaller scale ones correspond to more specific topics within the general topics. Cluster hierarchies thus serve as topic hierarchies [40]. To detect this hierarchical structure, the GHSOM [21] is employed in the proposed approach. The GHSOM combines the advantages of two principal extensions of the SOM, dynamic growth and hierarchical structure. As depicted in Figure 18.10, the GHSOM evolves to a multi-layered architecture composed of independent growing SOMs. At layer 0, a single-unit SOM serves as a representation of the complete data set. Only one map is used at the first layer of the hierarchy, which initially consists ofa grid of 2×2 units. For every unit in this map, a separate SOM can be added to the second layer. The model grows in two dimensions: in width (by increasing the size of each SOM) and in depth (by increasing the levels of the hierarchy). For growing in width, each SOM attempts to modify its layout and increase its total number of units systematically so that each unit is not representing too many input patterns. The basic steps are summarized in Table 18.1.

As for growing in depth, the general idea is to form a new map in the subsequent layer for the units representing a set of input vectors that are too diverse. The basic steps for the growth in depth are summarized in Table 18.2.

The growing process of the GHSOM is guided by two parameters, τ_1 and τ_2. τ_1 specifies the desired quality of input data representation at the end of the training process while τ_2 specifies the desired level

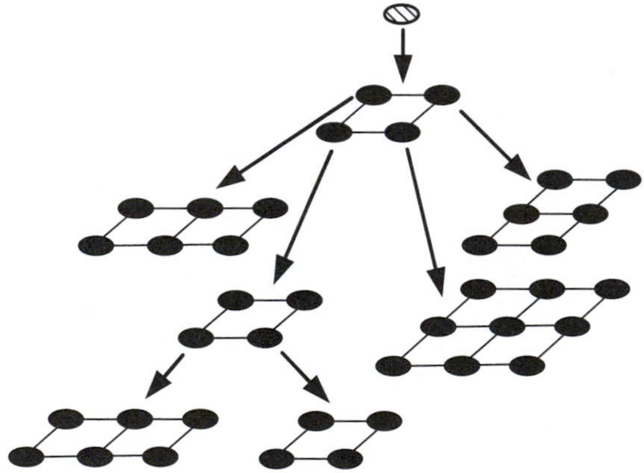

FIGURE 18.10 Graphic representation of a trained GHSOM.

TABLE 18.1 Steps of the Growth in Width

1. Initialize the weight of each unit with random values. Reset error variables E_i for every unit i
2. The standard SOM training algorithm is applied
3. For every input vector, the quantization error (*qe*) of the corresponding winner is measured in terms of the deviation between its weight vector and the input vector. Update the winner's error variable by adding the *qe* to E_i
4. After a fixed number λ of training iterations, identify the error unit q with the highest E_i
5. Insert a row or a column between the error unit q and its most dissimilar neighboring unit in terms of input space
6. Repeat steps 2–5 until the whole map's mean quantization error (MQE_m) reaches a given threshold so that $MQE_m < \tau_1 qe_u$ is satisfied, where qe_u is the quantization error of the corresponding unit u in the proceeding layer of the hierarchy and τ_1 is a fixed percentage

TABLE 18.2 Steps of the Growth in Depth

1. When the training of a map is finished, every unit is examined and those units fulfilling the criterion given as $qe_i > \tau_2 qe_0$ will be subject to a hierarchical expansion. qe_0 is the quantization error of the single unit in the layer 0
2. Train the newly added SOM with input vectors mapped to the unit map, which has just been expanded

of detail that is to be shown in a particular SOM. The smaller τ_1 is, the larger the emerging maps will be. Conversely, the larger τ_2 is, the deeper the hierarchy will be.

18.4.2 Rank Centroid Projection

The objective of the proposed RCP is to map input vectors onto the output space based on their similarities to the prototypes. Although this method is based on the standard SOM architecture [41,42], its application to a GHSOM is straightforward. Once the training process of the GHSOM is complete, a hierarchy of multiple layers consisting of several independent SOMs will be formed. The RCP can be applied to each individual map in the GHSOM afterward.

For each individual SOM in a GHSOM network, a set of prototype vectors is tuned and becomes topologically ordered during the training phase. The prototypes can be interpreted as cluster centers, while the coordinates of each map unit i indicate the position of the corresponding cluster center within the grid of the map. After convergence of the training process, for any input vector x_i, its similarity to each prototype vector can be calculated. A similarity measure can be defined as the inverse of the Euclidean distance between the respective vectors:

$$s_{ij} = d_{ij}^{-1} = \|x_i - w_j\|^{-1} \tag{18.12}$$

where

s_{ij} is the similarity value
d_{ij} is the distance between x_i and w_j

The map unit with the smallest distance to an input vector x_i is the BMU, as described in Equation 18.1. The BMU has the greatest similarity with x_i and corresponds to a cluster to which x_i is the most closely related. Hence, x_i should be projected to a position closer to the BMU than to other units. In a winner-takes-all case, in which only the BMU is considered, the data sample is to be mapped directly to its BMU. The coordinates of x_i and Cx_i can be represented as

$$Cx_i = Cw_1 \tag{18.13}$$

where Cw_1 represents the coordinates of the BMU. The resulting mapping is illustrated on the right side of Figure 18.11. Projecting multiple data samples will result in a hit histogram.

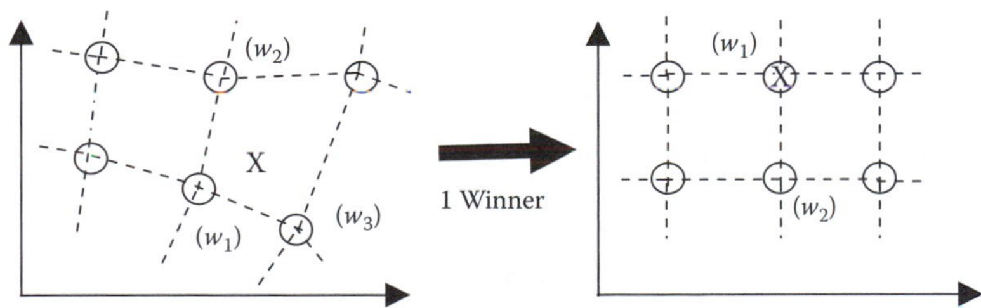

FIGURE 18.11 Illustration of mapping an input vector to its BMU.

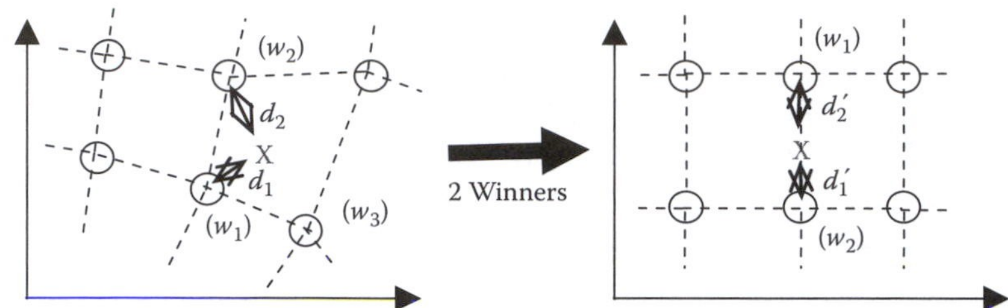

FIGURE 18.12 Illustration of mapping an input vector when two units are considered.

However, in most cases, there are usually several units that have almost as good matches as the BMU. As a result, pointing out only the BMU does not provide sufficient information of cluster membership, which is the problem with hit histograms. Intuitively, the data item should be projected to somewhere between the map units with a good match. Analogously, each map unit exerts an attractive force on the data item proportional to its response to that data item. The greater the force is, the closer the data item attracted to the map unit. The data item will end up in a position where these forces reach an equilibrium state.

In the projection process described in Figure 18.11, if W_1, the BMU and W_2, the second winning unit, are taken into account, intuitively the data sample should be projected to a position so that it is between these two units while closer to the BMU. This is illustrated in Figure 18.12, where d_1 and d_2 are the Euclidean distances between the data sample and two winners W_1 (BMU) and W_2. The responses of W_1 and W_2 to the data sample are, therefore, inversely proportional to d_1 and d_2. The projection of the data sample can be decided by the following weighted sum:

$$\mathbf{C}x_i = \frac{d_1^{-1}}{d_1^{-1} + d_2^{-1}}\mathbf{C}w_1 + \frac{d_2^{-1}}{d_1^{-1} + d_2^{-1}}\mathbf{C}w_2 \tag{18.14}$$

where $\mathbf{C}w_2$ is the coordinates of the second winner W_2.

Similarly, in the case of three winners, the coordinates of x_i are calculated as:

$$\mathbf{C}x_i = \frac{d_1^{-1}\mathbf{C}w_1 + d_2^{-1}\,\mathbf{C}w_2 + d_3^{-1}\,\mathbf{C}w_3}{d_1^{-1} + d_2^{-1} + d_3^{-1}} \tag{18.15}$$

where $\mathbf{C}w_3$ are the coordinates of the third winner. This can be extended to include all N units in the map grid in the computation of the projections. In general, the coordinates of x_i can be assigned by the following function:

$$\mathbf{C}x_i = \begin{cases} \dfrac{\sum_{j=1}^{N}(d_{ij})^{-1}\mathbf{C}w_j}{\sum_{j=1}^{N}(d_{ij})^{-1}}, & \text{if } d_{ij} \neq 0 \quad \text{for all } j \\ \mathbf{C}w_j, & \text{if } d_{ij} = 0 \end{cases} \tag{18.16}$$

where

d_{ij} is the distance between xi and prototype vector x_i

The inverse distance $(d_{ij})^{-1}$ is used as a measure of the similarity

As shown in Equation 18.16, the projected position of the data sample x_i is computed as a weighted average of the positions of nearby map units. The weighting is based on the distance d_{ij} between x_i and the map unit. As d_{ij} is inversely proportional to the similarity between the data sample x_i and

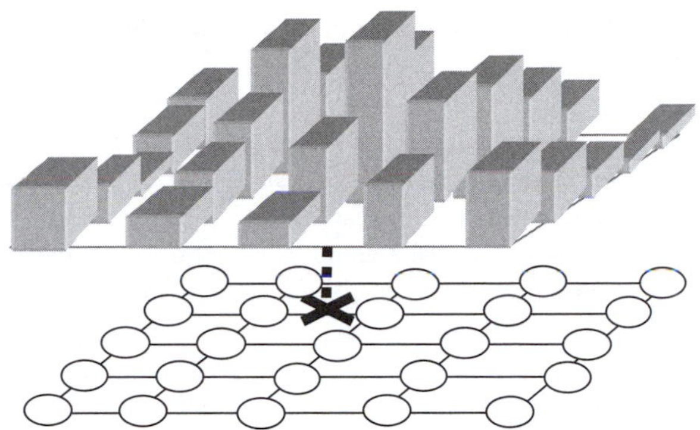

FIGURE 18.13 Illustration of mapping an input vector by finding the centroid of the spatial responses.

prototype vectors w_j, the weighting factors indicate the normalized responses of x_i to various prototypes. Since the map units are arranged in a rectangular grid, the set of weights may be characterized as a 2D histogram plotted across the map units as illustrated in Figure 18.13. The SOM projection procedure continues with finding the centroid of this spatial histogram, where the data sample is then mapped.

The basic centroid projection method [41,42] finds the projections in the output space by taking into account all N map units and calculating the weighted average, as shown in Equation 18.16. To enhance the performance of the projection method, the basic weighting function in Equation 18.16 is subject to modifications and correction terms. Instead of mapping the data sample directly onto the centroid of the spatial responses of all map units, a ranking scheme is applied to the weighting function. First, a constant R is set to select only a number of prototypes that are nearby the input vector in the input space. Only the positions of the associated R units will affect the calculation of the projection. R is in the range of one to the total number of neurons in the SOM. A membership degree of a data sample to a specific cluster is then defined based on the rank of closeness between the data vector and the unit associated with that cluster, which is given by

$$m_i = \begin{cases} \dfrac{R}{S}, & \text{for the closest unit} \\ \dfrac{R-1}{S}, & \text{for the 2nd closest unit} \\ \dots & \\ \dfrac{1}{S}, & \text{for the } R\text{th closest unit} \\ 0, & \text{for all other units} \end{cases} \tag{18.17}$$

where $S = \sum_{i=0}^{R-1} (R-i)$ ensures a normalized membership.

For a data sample x_i, the new weighting function is defined by applying the membership degree mi to Equation 18.16:

$$Cx_i = \begin{cases} \dfrac{\sum_{j=1}^{N} m_j d_{ij}^{-1} Cw_j}{\sum_{j=1}^{N} m_j d_{ij}^{-1}}, & \text{if } d_{ij} \neq 0 \quad \text{for all } j \\ Cw_j, & \text{if } d_{ij} = 0 \end{cases} \tag{18.18}$$

With the new weighting function, the spatial histogram shown in Figure 18.13 is first ranked by the corresponding membership degrees, whose centroid is then located. The projection method proceeds with mapping the data sample to this position.

The ranking scheme has a beneficial effect on the performance of the projection method. It introduces a membership degree factor into the weighting function in addition to the distance factor, which enables the proposed projection technique not only to reveal the inter-cluster relation of the input vectors but also to visualize information on cluster memberships. A positive side benefit of the ranking scheme is a considerable saving in computation as the result of selecting only the R closest units. This saving becomes more significant when the map size is large.

As discussed earlier, for each individual map in a trained GHSOM, the RCP is applied and the data samples are projected onto the map space. The data samples used for training in each layer are a fraction of the input data in the preceding layer. The projection process will result in multiple layers of rather small maps showing different degrees of details. Because the RCP algorithm allows the data points to be projected to any location across the SOM network, it can handle a large data set with a rather small map size and provide a high-resolution map in the mean time. Therefore, the presented procedure of mapping input vectors to the output grid using the RCP alleviates computational complexity considerably, making it possible to process a large data set.

18.4.3 Selecting the Ranking Parameter *R*

As shown in Figures 18.11 and 18.12, we can see that different R values result in different mapping positions of the input vector. The effect of the ranking parameter R can be further illustrated in the following example.

A 3D data set is shown in Figure 18.14, which consists of 300 data points randomly drawn from three Gaussian sources. The mean vectors of the three Gaussian sources are $[0,0,0]^T$, $[3,3,3]^T$, and $[9,0,0]^T$, respectively, while the variances are all 1. A SOM of 2×2 units is used to project the data points onto a 2D space. After training, the prototype vectors of the SOM are shown as plus signs in Figure 18.14, which span the input space with three of the map units representing the three cluster centers, respectively. The input vectors are then projected using the RCP method. The projection results produced with different R values are presented in Figure 18.15.

The effect of R can be seen in this figure. For the case of $R = 1$, where only the BMU is considered in the projection, the map is actually a hit histogram (a small random noise is added to the coordinates of each data point to show the volume of data points projected onto each map unit). Because it can only project input vectors to the map units on a rigid grid, this map does not provide much

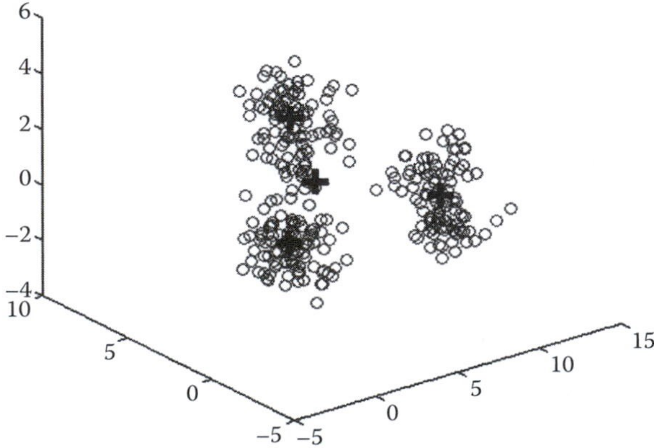

FIGURE 18.14 Data set I: samples in a 3D data space marked as small circles and prototype vectors as plus signs.

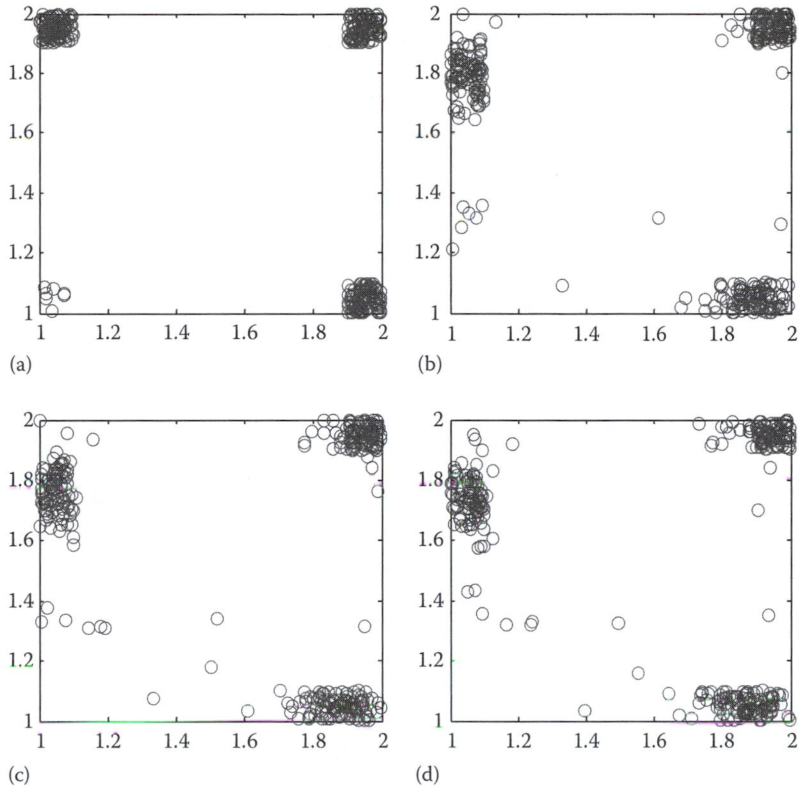

FIGURE 18.15 Projection results with different R: (a) $R = 1$, (b) $R = 2$, (c) $R = 3$, and (d) $R = 4$.

information about the global shape of the data. For all possible R values, three major clusters can be observed from the map. With R getting larger, the structure and shape of the data become more prominent. It is also noticeable that the cluster borders become obscure as R increases. The performance of the RCP depends heavily on the value of the ranking parameter R. It would be beneficial to determine the optimal R value automatically for each map based upon certain performance metrics.

If meaningful conclusions are to be drawn from the projection result, as much of the geometric relationships among the data patterns in the original space as possible should be preserved through the projection. At the mean time, it is desirable for the projection result to provide as much information about the shape and cluster structure of the data as possible. Members of each cluster should be close to one another while the clusters should be widely spaced from one another. Thus, a combination of two quantitative measures, Sammon's stress [29] and the Davies–Bouldin (DB) index [43], is used in this work to determine the optimal R. Sammon's stress measures the distortion between the pairwise distances in both the original and the projected spaces. To achieve good distance preservation, Sammon's stress should be minimized. The DB index attempts to maximize the inter-cluster distance while minimizing the intra-cluster distance at the same time. It is commonly used as a clustering validity index, low values indicating good clustering results.

For the projection results in Figure 18.15, both Sammon's stress and the DB index are calculated for each R value, which are shown in Figure 18.16a and b, respectively. It can be seen that the two quantitative measures have contradicting trends. As R grows larger, Sammon's mapping increases while the DB index decreases. It is, hence, impossible to optimize both of the objectives at the same time. We must identify the best compromise, which serves as the optimal R in this context. The task of selecting the

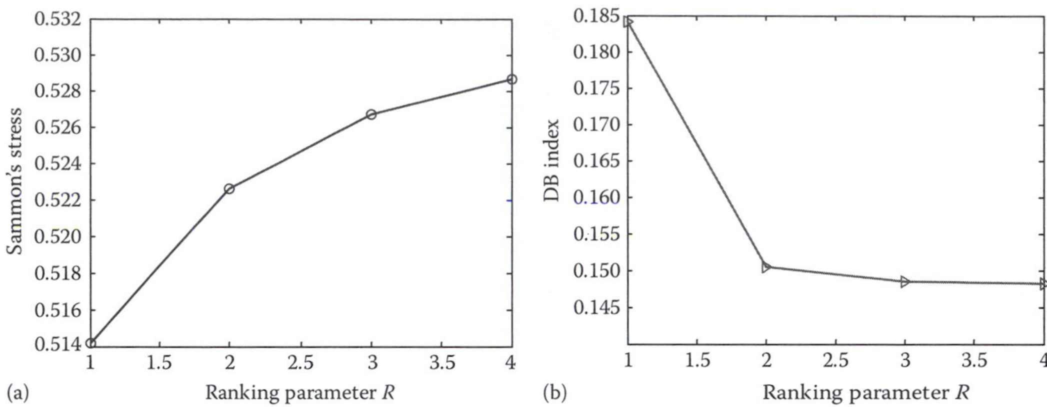

FIGURE 18.16 (a) Sammon's stress and (b) DB index for $R = 1, 2, 3, 4$.

optimal R now boils down to a bi-objective optimization problem. A typical way to solve the problem is to use the weighted sum method, which is stated in Ref. [44]:

$$\min\left[\alpha\frac{J_1(x)}{J_{1,0}(x)}+(1-\alpha)\frac{J_2(x)}{J_{2,0}(x)}\right] \tag{18.19}$$

where

J_1 and J_2 are two objective functions to be mutually minimized
$J_{1,0}$ and $J_{2,0}$ are normalization factors for J_1 and J_2, respectively
α is the weighting factor revealing the relative importance between J_1 and J_2

In the context of this work, J_1 and J_2 correspond to Sammon's stress and the DB index. Assuming these two objective functions have equal importance, α is set to be 0.5. By taking the weighted sum, the two objective functions are combined into a single cost function, which is shown in Figure 18.17. The optimization problem is, therefore, reduced to minimizing a scalar function. As shown in Figure 18.17, the objective function

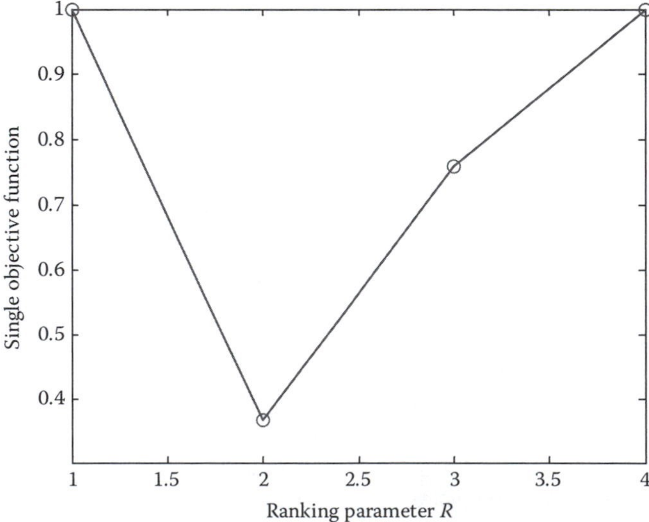

FIGURE 18.17 The optimal R is found at the minimal point of the single-cost function.

reaches its minimum when R equals 2. Consequently, for the 3D example, the condition of $R = 2$ leads to the best compromise between good distance preservation and good clustering quality.

18.5 Case Studies

To demonstrate the applicability of the proposed SOM visualization method, simulation results on two realistic document sets are presented in this section. The GHSOM Toolbox [45] was used in conducting this work.

18.5.1 Encoding of Documents Using Citation Patterns

To visualize a document collection, the inter-document relations are first encoded into a similarity matrix. Citation-based similarity matrices are used extensively when working with patents and journal articles from sources that provide citation data such as the SCI. The SCI provides access to current and retrospective citation information for scientific literature published in the physical, biological, and medical fields. It presents a great opportunity for citation-based document analysis.

Each dimension of the similarity matrix corresponds to one document in the set, and the value of each element is equal to the relative strength of the citation relationship between the corresponding document pair. These citation patterns provide explicit linkages among publications having particular points in common, and hence are considered as reliable indicators of intellectual connections among documents. Similarities calculated from citations generally produce meaningful document maps whose patterns expose clusters of documents and relations among those clusters [46].

In our approach, the similarity matrix is constructed based on one type of inter-document citations, in particular the bibliographic coupling [47]. Bibliographic coupling between a pair of papers is defined as the number of references both papers cite. In bibliometric studies [48], bibliographic coupling is used to cluster papers into research fronts, that is, groups of papers that cover the same topic. Using bibliographic coupling, inter-document similarities are calculated as

$$s_{ij} = \frac{bc_{ij}}{\sqrt{N_i N_j}} \tag{18.20}$$

where

bc_{ij} is the number of documents cited by both document i and j

N_i and N_j are the total number of document citations for document i and j, respectively

The similarity matrix is, therefore, a symmetric matrix that contains the bibliographic coupling counts between all pairs of documents in the database. The rows, or columns, of the similarity matrix are vectors corresponding to individual documents. Documents are subsequently clustered using the similarity matrix.

18.5.2 Collection of Journal Papers on Self-Organizing Maps

The first data set is constructed based on a collection of journal papers from the ISI Science Citation Index on the subject of SOMs. Using the term “self-organizing maps” in the general search function of ISI Web of Science, a set of 1349 documents was collected corresponding to journal articles published from 1990 to early 2005. In this simulation, document citation patterns are used to describe the inter-document relationships between pairs of documents, based on which a similarity matrix is built to store pair-wise similarity values among papers. After removing the poorly related papers, 638 documents remain in the data set. Following the document encoding process, we end up with a 638×638 similarity

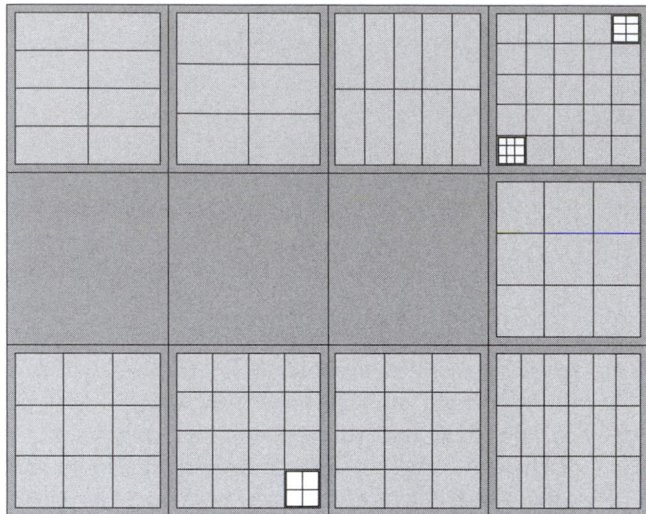

FIGURE 18.18 The resulting three-layer GHSOM for the collection of SOM papers.

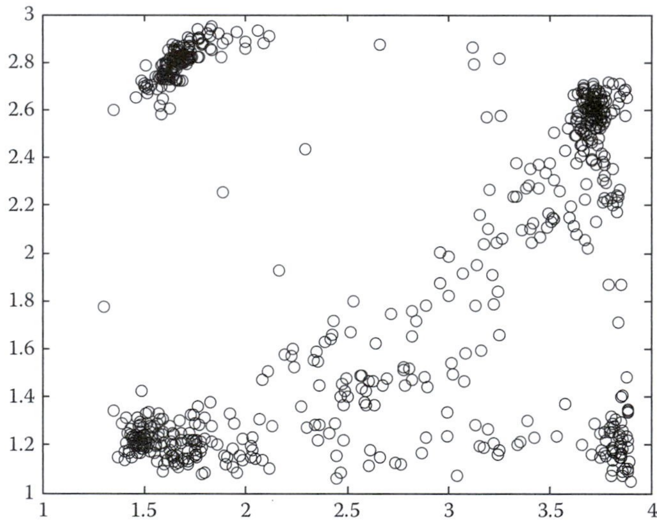

FIGURE 18.19 The projection result of the journal papers on the SOM, where documents are marked as circles.

matrix, each row (or column) vector describing the citation pattern of a document in a 638 dimensional space. These vectors are then used to train a GHSOM network.

In this simulation, a three-layer GHSOM was generated by setting the thresholds $\tau_1 = 0.8$ and $\tau_2 = 0.008$, which is illustrated in Figure 18.18. The first-layer map consists of 3×4 units. The projection of all the documents onto the first layer map is shown in Figure 18.19, which is obtained by using a ranking parameter $R = 5$. Five major topic groups can be observed from the map, four located around the four corners of the map and one in the lower center. The topics associated with each group are labeled. The labels are derived after examining manually the paper titles from each cluster for common subjects.

To enhance the visualization, the size of document markers can be made proportional to the number of times a document has been cited, as shown in Figure 18.20. Important papers, usually distinguished by large citation counts, are thus made to stand out on the document map. In this collection of SOM documents, three papers are extraordinarily heavily cited, as marked in Figure 18.20. The two large

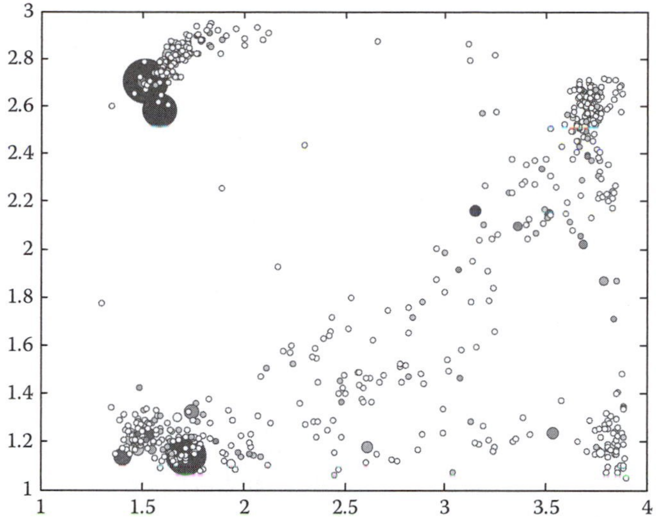

FIGURE 18.20 An enhanced visualization of the SOM papers, where the size of the document marker is proportional to the number of times a document has been cited.

circles in the top part of the map, "Toronen et al. 1999" and "Tamayo et al. 1999," which appear to be closely related, correspond to two important works on applying SOM for clustering gene expression data. Tamayo and colleagues used the SOM to cluster genes into various patterned time courses and also devised a gene expression clustering software, GeneCluster. Another implementation of SOM was developed by Toronen et al. in 1999 for clustering yeast genes. Another heavily cited paper is the Self-Organizing Map by Kohonen published in 1990, which is cited by a large portion of the documents in this data set. Note that the foundational papers Kohonen published in the 1980s, such as Ref. [1], are not available from the ISI Web service.

Based on the initial separation of the most dominant topical clusters, further maps were automatically trained to represent the various topics in more detail. Nine individual maps are developed in the second layer, each representing the documents of the respective higher layer unit in more detail. One example is illustrated in Figure 18.21. This figure shows the projection result of a submap, expanded

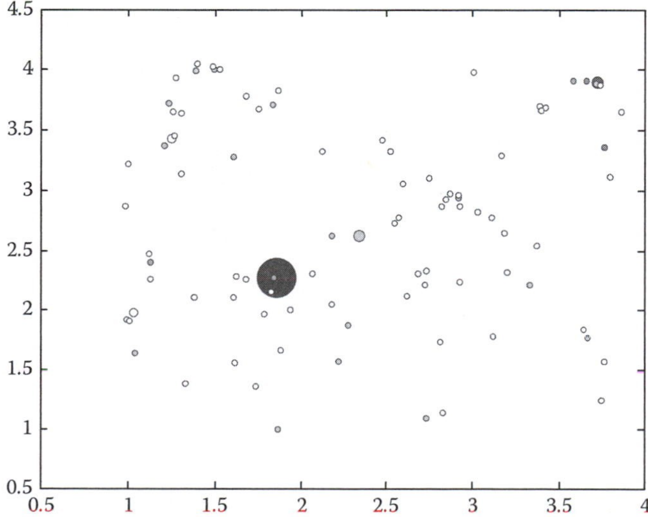

FIGURE 18.21 The projection result of a submap.

from the second left node in the bottom of the first-layer map. The submap consists of 3 × 3 units, representing a cluster of papers covering the theoretical aspect of the SOM. The Kohonen's seminal paper is located in this submap. Some of the nodes on these second-layer maps are further expanded as distinct SOMs in the third layer. Due to the incompleteness of this document collection, limited information about this subject domain is revealed from the map displays.

18.5.3 Collection of Papers on Anthrax Research

The second data set is a collection of journal papers on anthrax research, which is also obtained from ISI Web of Science. Anthrax research makes an excellent example for testing the performances of document clustering and visualization. The subject is well covered by the Science Citation Index. A great deal of the research has been performed in the past 20 years. A review paper [49] is available where the names of key papers in this field are identified and discussed. The anthrax paper set we collected for this simulation contains 987 documents corresponding to journal papers published from 1981 to the end of 2001.

A 987 × 987 similarity matrix was formed to train a GHSOM. A three-layer GHSOM was generated by setting the thresholds $\tau_1 = 0.78$ and $\tau_2 = 0.004$, as illustrated in Figure 18.22. The first-layer map consists of 3 × 4 units. The projection result of all documents onto this layer is shown in Figure 18.23, which was produced by setting the ranking parameter R to 3.

In Figure 18.23, several clusters can be seen on the map with their topics labels. Starting from the upper left corner of the map and going clockwise, we can see the topics of the papers change with different locations on the map. The cluster of documents in the upper left corner is focused on how anthrax moves, interacts with, and enters host cells. Note that several smaller groups are visible inside this cluster, which implies expansion of this cluster in a further layer would reveal several sub-topics. To the right, the documents in the upper center of the map are found to deal with anthrax genes. The documents located in the upper right corner cover biological effects of anthrax, while the cluster right below covers the effect of anthrax on the immune system. In the lower right corner of the map, another group of documents exist, which deals with the comparison of anthrax and other *Bacillus* strains. A tight cluster is formed in the lower left corner of the map, which discusses the use of anthrax as a bio-weapon. As a whole, several obvious groups of documents are formed on the map, which relate to different research focuses in the context of anthrax research. Fundamental research topics are located in the upper portion of the map, which are somewhat in vicinity of one another. There are no obvious borders between these groups as the topics are closely interrelated. On the contrary, other relevant topics on anthrax are mapped to the lower

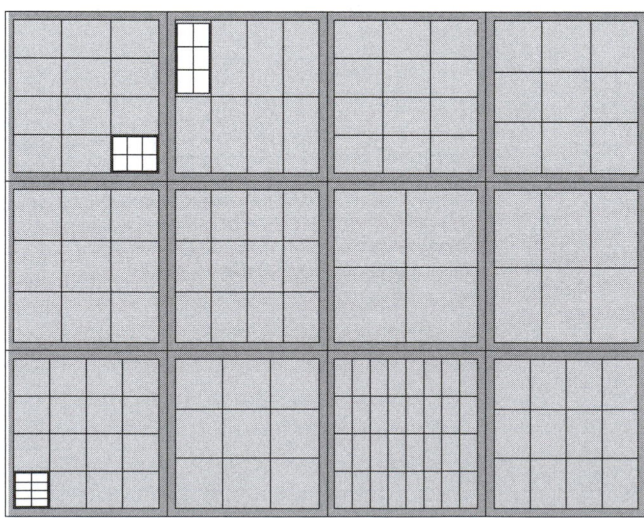

FIGURE 18.22 The resulting three-layer GHSOM for the collection of anthrax papers.

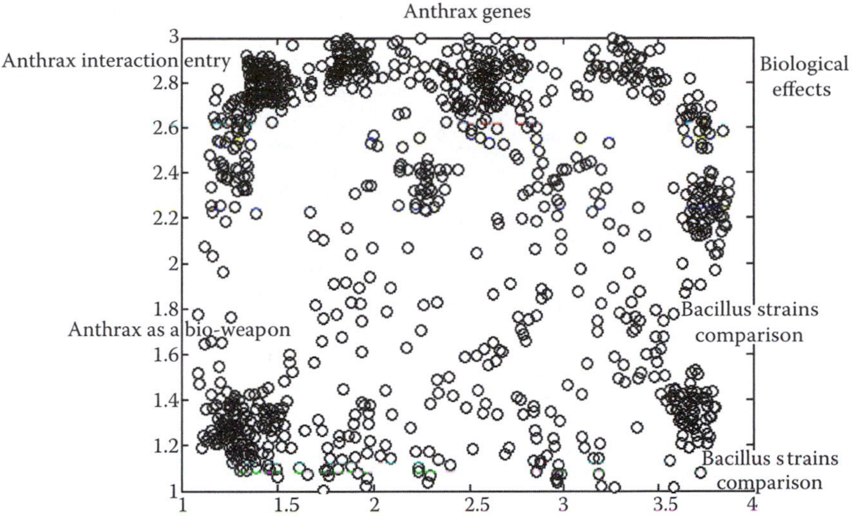

FIGURE 18.23 First-layer projection of the anthrax journal papers.

portion, which are rather far from the fundamental research topics and from one another. It can be seen that the geometric distance indicates the degree of relevance among documents. There are also many documents sitting between clusters, which are the result of the heavily overlapped research coverage.

More information about the document set can be gained by identifying the seminal papers in it. For this purpose, the document marker sizes are made proportional to the number of times they have been cited. The result is shown in Figure 18.24. Several seminal papers can be identified from Figure 18.24, five of which will be identified in the following as examples. The earliest seminal paper is the Gladstone paper published in 1946, in which Gladstone reported on the discovery of protective antigen. In the 1950s, Smith and Keppie showed in their paper that anthrax kills through a toxin. These papers are landmark papers forming the foundation for anthrax research and they fall into the cluster of anthrax effect on immunity. Later, another influential paper was published in 1962 when Beall showed

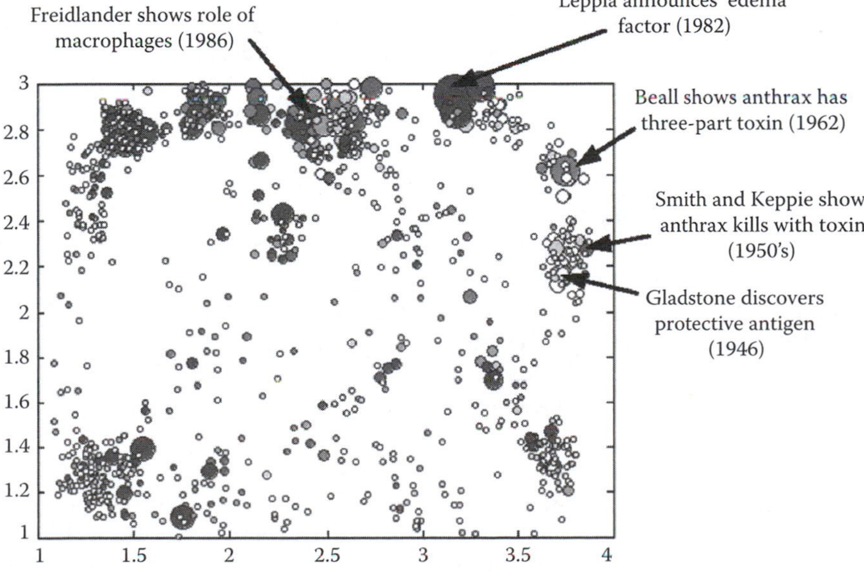

FIGURE 18.24 First-layer projection of the anthrax papers set.

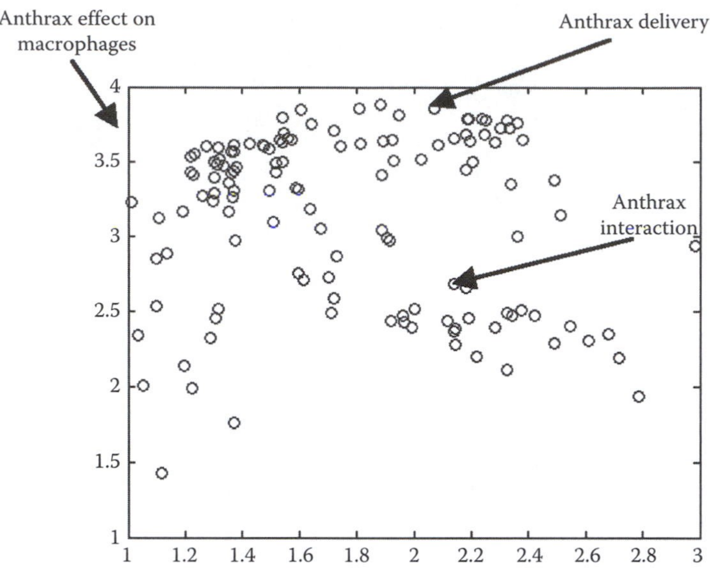

FIGURE 18.25 Second-layer document projection.

anthrax has a three-part toxin. Leppla announced the edema factor in his paper in 1982. These papers mainly deal with the effect of anthrax on host cells. Another heavily cited paper was on macrophages published by Freidlander in 1986, which became the key paper in this area.

All of the neurons of the first-layer SOM are expanded to the second layer to represent the respective topics in more detail. The resulting second-layer map of the upper-left node is shown in Figure 18.25, which has a total of 192 documents. In the second layer, the documents are further clustered into three groups: anthrax effect on macrophages, anthrax delivery, and anthrax interaction. This result is consistent with the first-layer representation. One unit on this second-layer map is further expanded in the third layer.

18.6 Conclusion

This chapter introduces an approach for clustering and visualizing high-dimensional data, especially textual data. The devised approach, which is an extension of the SOM, acts as an analysis tool as well as a direct interface to the data, making it a very useful tool for processing textual data. It clusters documents and presents them on a 2D display space. Documents with a similar concept are grouped into the same cluster, and clusters with similar concepts are located nearby on a map.

In the training phase, the proposed approach employs a GHSOM architecture, which grows both in depth according to the data distribution, allowing a hierarchical decomposition and navigation in portions of the data, and in width, implying that the size of each individual map adapts itself to the requirements of the input space. After convergence of the training process, a novel approach, the RCP, is used to project the input vectors to the hierarchy of 2D output maps of the GHSOM. The performance of the presented approach has been illustrated using two real-world document collections. The two document collections are scientific journal articles on two different subjects obtained from the Science Citation Index. The document representation relies on a citation-based model that depicts the inter-document similarities using the bibliographic coupling counts between all pairs of documents in the collection. For the given document collections, it is rather easy to judge the quality of the clustering result. Although the resulting SOM maps are graphical artifacts, the simulation results have demonstrated that the approach is highly effective in producing fairly detailed and objective visualizations that are easy to understand. These maps, therefore, have the potential of providing insights into the information hidden in a large collection of documents.

References

1. T. Kohonen, Self-organized formation of topologically correct feature maps, *Biological Cybernetics*, 43, 59–69, 1982.
2. D. J. Willshaw and C. von der Malsburg, How patterned neural connections can be set up by self-organization, *Proceedings of the Royal Society London*, B194, 431–445, 1976.
3. T. Kohonen, *Self-Organizing Maps*. Berlin, Germany: Springer-Verlag, 1995.
4. G. Pölzlbauer, M. Dittenbach, and A. Rauber, A visualization technique for self-organizing maps with vector fields to obtain the cluster structure at desired levels of detail, *Proceedings of the International Joint Conference on Neural Network*, Montreal, Canada, pp. 1558–1563, 2005.
5. A. Ultsch and H. P. Siemon, Kohonen's self-organizing feature maps for exploratory data analysis, *Proceedings of the International Neural Network Conference*, Dordrecht, the Netherlands, pp. 305–308, 1990.
6. J. Vesanto and E. Alhoniemi, Clustering of the self-organizing map, *IEEE Transactions on Neural Networks*, 11, 586–600, 2000.
7. H. Demuth and M. Beale, *Neural Network Toolbox User's Guide*. Natick, MA: The MathWorks, Inc., 1998.
8. B. Fritzke, Growing cell structures—A self-organizing network for unsupervised and supervised learning, *Neural Networks*, 7, 1441–1460, 1994.
9. B. Fritzke, A growing neural gas network learns topologies. In G. Tesauro, D.S. Touretzky, and T. K. Leen (eds.) *Advances in Neural Information Processing Systems 7*. Cambridge, MA: MIT Press, pp. 625–632, 1995.
10. B. Fritzke, Growing grid—a self-organizing network with constant neighborhood range and adaption strength, *Neural Processing Letters*, 2, 9–13, 1995.
11. D. Alahakoon, S. K. Halgarmuge, and B. Srinivasan, Dynamic self-organizing maps with controlled growth for knowledge discovery, *IEEE Transactions on Neural Networks*, 11, 601–614, 2000.
12. M. Martinetz and K. J. Schulten, A 'neural-gas' network learns topologies. In K. M. T. Kohonen, O. Simula, and J. Kangas (eds.) *Artificial Neural Networks*. Amsterdam, the Netherlands: North-Holland, pp. 397–402, 1991.
13. J. Blackmore and R. Miikkulainen, Incremental grid growing: Encoding high-dimensional structure into a two-dimensional feature map, *Proceedings of the IEEE International Conference on Neural Networks*, San Francisco, CA, pp. 450–455, 1993.
14. R. Lang and K. Warwick. The plastic self organising maps, *Proceedings of International Joint Conference on Neural Networks*, Honolulu, HI, pp. 727–732, 2002.
15. S. Marsland, J. Shapiro, and U. Nehmzow, A self-organizing network that grows when required, *IEEE Transactions on Neural Networks*, 15, 1041–1058, 2002.
16. B. Fritzke, Some competitive learning methods, Retrieved January 19, 2006, from the World Wide Web: http://www.ki.inf.tu-dresden.de/~fritzke/JavaPaper/t.html
17. D. Vicente and A. Vellido, Review of hierarchical models for data clustering and visualization. In R. Giráldez, J. C. Riquelme, J. S. Aguilar-Ruiz (eds.) *Tendencias de la Minería de Datos en España*. Red Española de Minería de Datos, 2004.
18. R. Miikkulainen, Script recognition with hierarchical feature maps, *Connection Science*, 2, 83–101, 1990.
19. D. Merkl and A. Rauber, CIA's view of the world and what neural networks learn from it: A comparison of geographical document space representation metaphors, *Proceedings of the 9th International Conference on Database and Expert Systems Applications*, Vienna, Austria, pp. 816–825, 1998.
20. J. Lampinen and E. Oja, Clustering properties of hierarchical self-organizing maps, *Journal of Mathematical Imaging and Vision*, 2, 261–272, 1992.
21. A. Rauber, D. Merkl, and M. Dittenbach, The growing hierarchical self-organizing map: Exploratory analysis of high-dimensional data, *IEEE Transactions on Neural Networks*, 13(6), 1331–1341, 2002.

22. A. Ultsch and H. P. Siemon, Kohonen's self-organizing feature maps for exploratory data analysis, *Proceedings of the International Neural Network Conference*, Dordrecht, Netherlands, pp. 305–308, 1990.
23. M. A. Kraaijveld, J. Mao, and A. K. Jain, A nonlinear projection method based on Kohonen's topology preserving maps, *IEEE Transactions on Neural Networks*, 6, 548–559, 1995.
24. R. A. Fisher, The use of multiple measurements in taxonomic problems, *Annual Eugenics*, 7, 178–188, 1936.
25. A. Ultsch, Maps for the visualization of high-dimensional data spaces, *Proceedings of Workshop on Self-Organizing Maps*, Kyushu, Japan, pp. 225–230, 2003.
26. A. Ultsch, U*-matrix: A tool to visualize clusters in high dimensional data, Technical Report No. 36, Department of Mathematics and Computer Science, University of Marburg, Marburg, Germany, 2003.
27. A. Ultsch, and F. Mörchen, ESOM-maps: Tools for clustering, visualization, and classification with Emergent SOM, Technical Report No. 46, Department of Mathematics and Computer Science, University of Marburg, Marburg, Germany, 2005.
28. M. L. Davison, *Multidimensional Scaling*. New York: John Wiley & Sons, 1983.
29. J. W. Sammon, A nonlinear mapping for data structure analysis, *IEEE Transactions on Computers*, 18, 401–409, 1969.
30. S. Kaski, Data exploration using self-organizing maps, PhD thesis, Helsinki University of Technology, Helsinki, Finland, 1997.
31. D. de Ridder and R. P. W. Duin, Sammon's mapping using neural networks: A comparison, *Pattern Recognition Letters*, 18, 1307–1316, 1997.
32. J. Mao and A. K. Jain, Artificial neural networks for feature extraction and multivariate data projection, *IEEE Transactions on Neural Networks*, 6, 296–317, 1995.
33. O. Simula, J. Vesanto, and P. Vasara, Analysis of industrial systems using the self-organizing map, *Proceedings of Second International Conference on Knowledge-Based Intelligent Electronic Systems*, Adelaide, SA, pp. 61–68, April 1998.
34. E. Pampalk, A. Rauber, and D. Merkl, Using smoothed data histograms for cluster visualization in self-organizing maps, *Proceedings of the International Conference on Artificial Neural Network*, Madrid, Spain, pp. 871–876, 2002.
35. D. Merkl and A. Rauber, Alternative ways for cluster visualization in self-organizing maps, *Proceedings of Workshop on Self-Organizing Maps*, Espoo, Finland, pp. 106–111, 1997.
36. H. Yin, ViSOM—A novel method for multivariate data projection and structure visualization, *IEEE Transactions on Neural Networks*, 13, 237–243, 2002.
37. B. Bienfait, Applications of high-resolution self-organizing maps to retrosynthetic and QSAR analysis, *Journal of Chemical Information and Computer Sciences*, 34, 890–898, 1994.
38. J. Vesanto and E. Alhoniemo, Clustering of the self-organizing map, *IEEE Transactions on Neural Networks*, 11, 586–600, 2000.
39. G. G. Yen and Z. Wu, Rank centroid projection: A data visualization approach for self-organizing maps, *IEEE Transactions on Neural Networks*, 19, 245–259, 2008.
40. S. Noel, V. Raghavan, and C.-H. H. Chu, Document clustering, visualization, and retrieval via link mining. In W. Wu, H. Xiong, and S. Shekhar (eds.) *Clustering and Information Retrieval*. Norwell, MA: Kluwer Academic Publisher, 2002.
41. S. A. Morris, Z. Wu, and G. Yen, A SOM mapping technique for visualizing documents in a database, *Proceedings of the IEEE International Conference on Neural Networks*, Washington, DC, pp. 1914–1919, 2001.
42. Z. Wu, and G. Yen, A SOM projection technique with the growing structure for visualizing high-dimensional data, *International Journal of Neural Systems*, 13, 353–365, 2003.
43. D. L. Davies and D. W. Bouldin, A cluster separation measure, *IEEE Transactions on Pattern Analysis and Machine Intelligence*, 1, 224–227, 1979.
44. I. Y. Kim and O. de Weck, Adaptive weighted sum method for i-objective optimization, *Structural and Multidisciplinary Optimization*, 29, 149–158, 2005.

45. A. Chan and E. Pampalk, Growing hierarchical self-organizing map (GHSOM) toolbox: Visualizations and enhancements, *Proceedings of the International Conference on Neural Information Processing*, Singapore, pp. 2537–2541, 2002.
46. S. Morris, C. Deyong, Z. Wu, S. Salman, and D. Yemenu, DIVA: A visualization system for exploring document databases for technology forecasting, *Computer and Industrial Engineering*, 43, 841–862, 2002.
47. M. M. Kessler, Bibliographic coupling between scientific papers, *American Documentation*, 14, 10–25, 1963.
48. E. Garfield, The concept of citation indexing: A unique and innovative tool for navigating the research literature, *Current Contents*, January 3, 1994.
49. R. Bhatnagar and S. Batra, Anthrax toxin, *Critical Reviews in Microbiology*, 27, 167–200, 2001.

III

Fuzzy Systems

19

Fuzzy Logic Controllers

Teresa Orlowska-Kowalska
Wroclaw University of Technology

Krzysztof Szabat
Wroclaw University of Technology

19.1 Introduction

Fuzzy models have become extremely popular in recent years [DHR96,P93,P01,CL00,YF94]. They are widely utilized as the model of systems or nonlinear controllers. Because of their nonlinear characteristics, they are especially suitable for modeling or control of complex ill-defined dynamic processes. The application of fuzzy systems has different advantages. For example, they allow designing the controller for the processes where mathematical models do not exist. Unlike the classical methodology, which requires the existence of the model, the fuzzy system can be designed using only the information on the process behavior.

In this chapter, the relations between classical linear PI/PID controllers and fuzzy-logic-based controllers as well as an overview of different fuzzy models are presented. Due to the limited length of the work, only the major properties of the considered systems are described and selected, and, in the authors' opinion, the most important, fuzzy systems are presented.

19.2 Fuzzy versus Classical Control

According to the paper [ÄH01], more than 90% of industrial processes are still controlled by means of the classical PI/PID. They are commonly applied in industry due to the following factors:

- The classical methods are commonly known and well understood by working engineers.
- There are a lot of tuning methods for the linear controller, which can be easily implemented for a variety of industrial plants.
- The stability analysis of the systems with linear controllers is much simpler than for the plants with nonlinear controllers.
- The area of fuzzy modeling and control in industry lacks specialists.

The output of the PID controller is a sum of the signal from three paths: proportional (P), integrating (I), and differentiating (D). The relationship between system output and input is described by the following equation:

$$u = K_P e_P + K_I e_I + K_D e_D \tag{19.1}$$

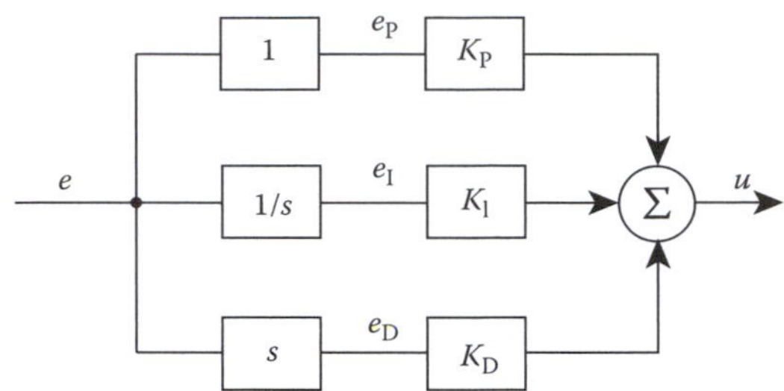

FIGURE 19.1 Block diagram of the classical PID controller.

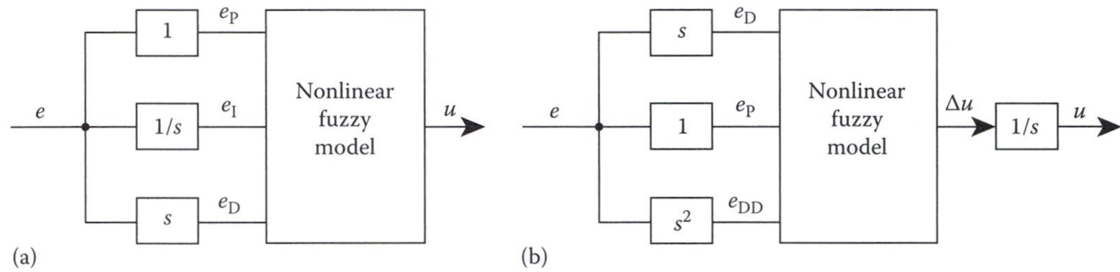

FIGURE 19.2 Direct (a) and the incremental form (b) of the fuzzy PID controller.

By setting the coefficient K_D to zero the PI controller is obtained, and by setting the K_I part to zero the PD controller is achieved. In Figure 19.1 the ideal PID controller is presented. In real-time application, the output saturation as well as one of the selected anti-windup strategies have to be implemented. Also, in the differentiating path the low-pass filter should be used in order to decrease the effects of the high-frequency noises.

Based on the structure presented in Figure 19.1, it is easy to go into fuzzy representation of PID controller. Two general structures of the PID fuzzy controllers are presented in Figure 19.2. The form directly related to Figure 19.1 is presented in Figure 19.2a, whereas the incremental form is shown in Figure 19.2b. The incremental form of the system is more commonly applied for the fuzzy PI controller due to the simplicity of limitation of the output signal.

The fuzzy PID controller connects the controller outputs and its inputs by the following relationship for the direct form:

$$u = f(e_P, e_I, e_D) \tag{19.2}$$

and for the incremental form

$$\Delta u = f\left(e_P, e_D, e_{DD}\right) \tag{19.3}$$

Unlike the classical PID controller, where the linear addition of the incoming signals is realized, the fuzzy controller transforms the incoming signals to the output using nonlinear fuzzy relationship. This transformation can be regarded as the nonlinear addition. In order to illustrate this feature, a hypothetical control surfaces for the classical PI and fuzzy PI controllers are shown in Figure 19.3.

As can be seen from Figure 19.3, the control surface of the classical system is linear, only the slope of this surface can be changed by modifying the controller coefficients. In contrast to the classical controller, the fuzzy system can realize any nonlinear control surface. This demonstrates the power of the fuzzy control because it is suitable for any nonlinear plant. It should also be highlighted that

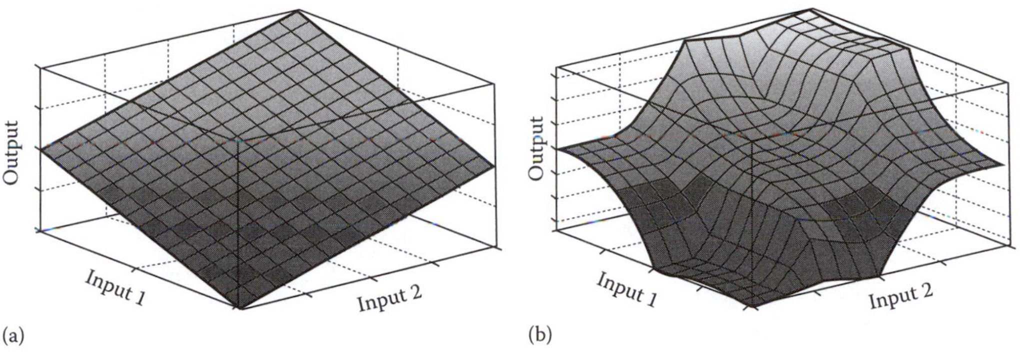

FIGURE 19.3 Hypothetical control surface of the classical (a) and fuzzy (b) PI controllers.

the control surface of the fuzzy controller can be linear, so the properties of the controlled object with fuzzy controller can never be worse (or at least the same) than with the classical controller.

It is widely established that the fuzzy controllers can be applied in the following situations:

- When there is no mathematical model of the plant in the form of differential or difference equations.
- Where due to the nonlinearities, the application of the classical methods is impossible (or not effective).
- When the aim of the control is given in a vague way (e.g., such term as *smoothly* is used).

The application of fuzzy control to the linear plants is sometimes questioned by scientists. It may be difficult to justify and everything depends on the form of the used control index, which can be represented in the following general form:

$$K = \min\left(\alpha\int_0^\infty e^2t^2dt + (1-\alpha)\int_0^\infty u^2dt\right) \tag{19.4}$$

where

α is an weight factor
t is time
e is the control error

Even if the controller and the object are linear, the presented control index is nonlinear. It means that due to the form of the control index, the control problem becomes nonlinear. The nonlinear fuzzy controller generates many possibilities to minimize the nonlinear control index. Therefore, the application of the fuzzy controller to linear objects is justified.

The nonlinear fuzzy model (Figure 19.2) relies on the nonlinear relationship between the inputs and outputs of the fuzzy controller. In the literature, different systems can be found, so the next section is devoted to the general structure of the fuzzy system. Next an overview of different fuzzy models is provided.

19.3 Fuzzy Models

19.3.1 General Structure of Fuzzy Models

The issues of system modeling have been widely investigated in a vast numbers of papers and books. Reliable models which can describe the input(s)–output(s) relationship(s) accurately enough are sought after. This problem is especially important in such fields as monitoring, fault detection, control, etc. [DHR96,P93,P01]. A number of classical systems have been proposed in the literature. However, due to the system nonlinearity and complexity the application of the classical methods is not always preferable.

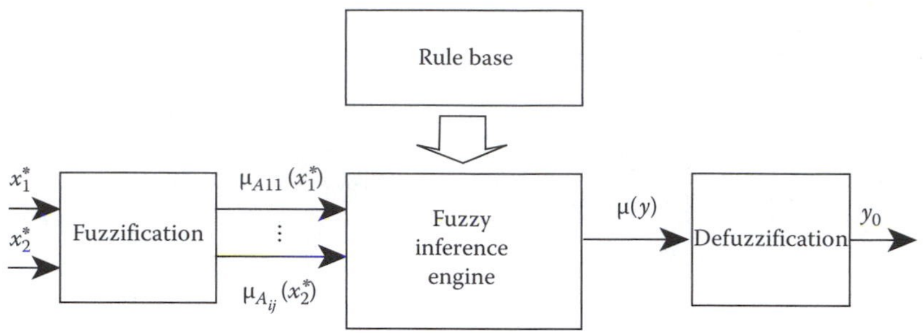

FIGURE 19.4 Basic structure of the fuzzy system.

Contrary to them, the fuzzy models possess the flexibility and intuition of the human reasoning; therefore, they are designated to the model nonlinear ill-defined processes.

In the literature, different types of fuzzy models can be found, varying in membership functions, inference methods, paths of the flowing signals, etc. At first glance, they can appear totally dissimilar, when comparing, for example, the Mamdani and wavelet neuro-fuzzy systems. Nonetheless, all of them have a common basic structure shown in Figure 19.4. This structure consists of four basic blocks: fuzzyfication, rule base with the fuzzy inference engine, and defuzzyfication [DHR96,P93,P01,CL00,YF94].

In the fuzzyfication block, the incoming sharp values are transferred to fuzzy values. In order to conduct this operation, input membership functions of the system have to be defined unambiguously. The shape of the membership functions exerts influence on the obtained fuzzy models. Due to their low computational effort, the triangular or trapezoidal functions are widely applied. In the ranges where the system should react very sensitively to the input values, the membership function should be narrow, which allows to distinguish different values more accurately. However, it should be pointed out that the number of rules grows rapidly with the number of membership functions. The output of the fuzzyfication block is a vector of the parameters $\mathbf{X}_f = [\mu_{A1}, \ldots, \mu_{B2}]$ describing the degree of membership of the input signals to the related membership functions. A hypothetical fuzzyfication process for one input value $x_1 = 3$ and two membership functions A_1 i A_2 is presented in Figure 19.5.

The next block of the system is called fuzzy inference engine. On the basis of the input vector $\mathbf{X}_f$, it calculates the resulting membership function which is passed to its output. The separate block entitled rule base (Figure 19.4) of *if–then* type includes several rules exerting the most important influence on the fuzzy model. Fuzzy inference engine connects the knowledge incorporated in *if–then* rules using fuzzy approximate reasoning.

In the first step, the fuzzy inference engine calculates the degree of fulfillment of the rule premises on the basis of $\mathbf{X}_f$. The bigger the degree, the more influence on the output. Depending on the form of the rule, a different operation is used.

For the AND premise, it is the *t-norm*, for the OR premise it is the *s-norm*. It should be pointed out that in the literature different types of the *t-norm* such as min, prod, Hamacher, Einstein, etc., and various types of *s-norms* such as max, sum, Hamacher, Einstein, etc., are cited. In Figure 19.6, the way of calculating the AND and OR premises is shown.

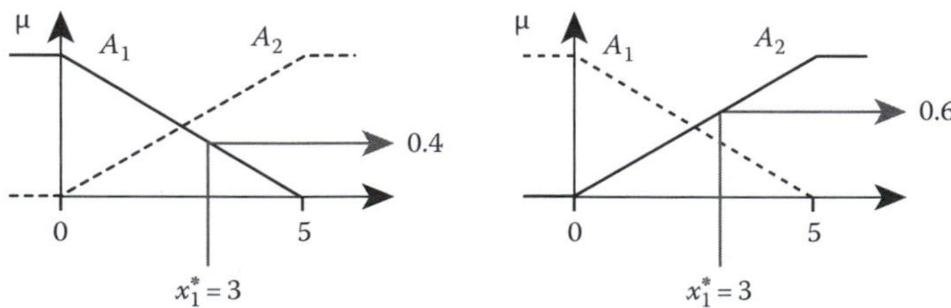

FIGURE 19.5 Hypothetical fuzzyfication process.

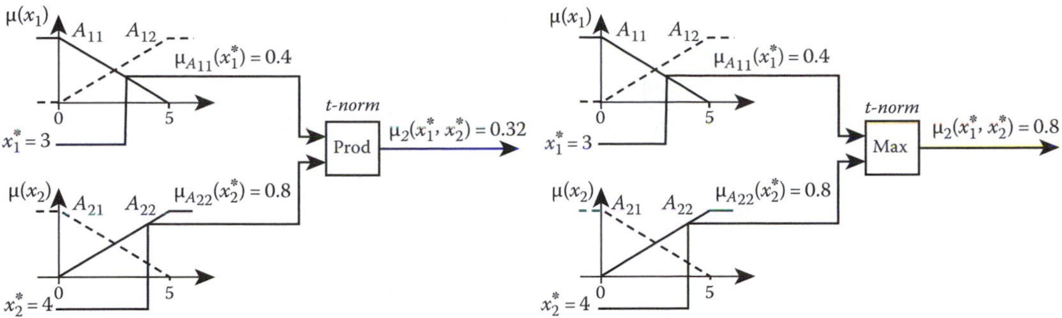

FIGURE 19.6 Hypothetical calculating of the premises using AND and OR operations.

Then the shape of output membership functions of each rule is determined. This operation is called implication and is conducted with the help of a fuzzy implication method. Two commonly used implication methods, *Mamdani* and *prod*, are shown in Figure 19.7.

As a result of the implication process, a specific number of conclusion membership functions is obtained. Thereafter, the aggregation process is carried out. From the output membership functions using the *s-norm* operation, one resulting fuzzy set is obtained. A hypothetical aggregation process with two output membership functions and *max s-norm* method is presented in Figure 19.8.

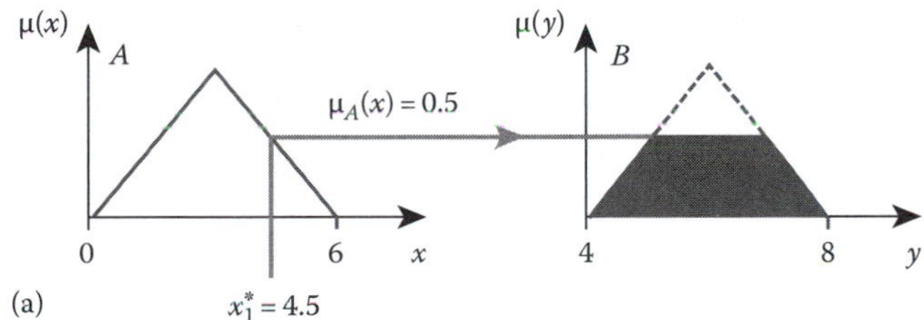

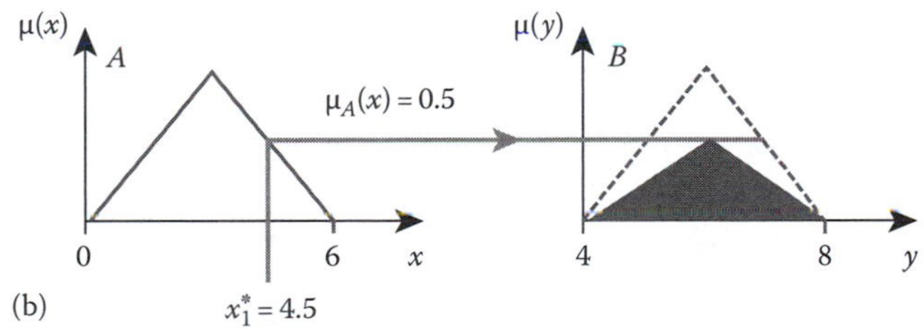

FIGURE 19.7 The idea of the *Mamdani* and *prod* implication methods.

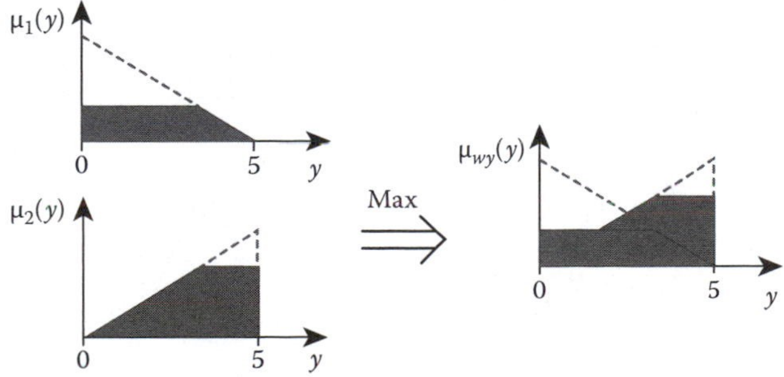

FIGURE 19.8 Hypothetical aggregation process using two fuzzy functions and *max s-norm* operation.

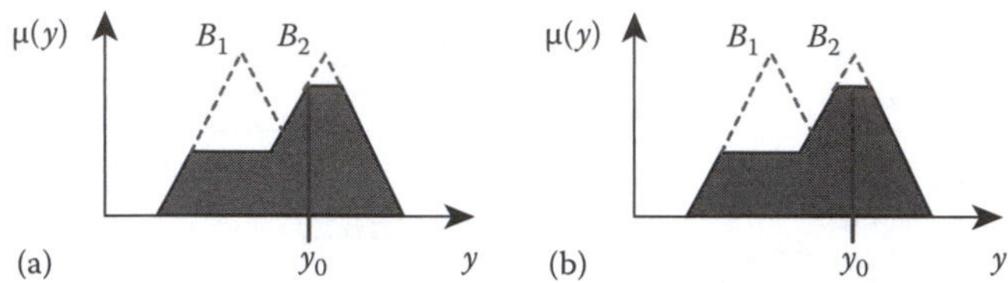

FIGURE 19.9 (a) The first maximum and (b) the mean of maximum defuzzyfication methods.

The last block of the fuzzy system includes the defuzzyfication procedure. Its goal is to compute the sharp value from the resulting fuzzy set. This value is submitted to the output of the whole fuzzy system. From different defuzzyfication methodologies, the following methods are worth mentioning:

- The first (last) maximum
- The mean of maximum
- The center of gravity
- The high (singleton)

As the output value of the first (last) maximum defuzzyfication method, the first (the last) maximum value of the resulting fuzzy set is taken. In the mean of maximum method, the output value is selected as the point between the first and the last maximum values. The graphic representation of the two described methods is presented in Figure 19.9.

The main advantage of the presented methods is their computational simplicity. However, there are also very serious drawbacks to them. The output sharp value is influenced only by the most activated fuzzy set which means that the fuzzy system is insensitive to changes of the other sets. Additionally, when these methods are used, the output of the system can change rapidly and due to this reason they are very rarely applied in real systems.

The center of gravity method is a commonly used defuzzyfication procedure. The output sharp value is related to the center of gravity of the output fuzzy set. All the activated fuzzy membership functions (related to specific rules) affect the system output, which is the advantage of this method. This method has also significant drawbacks though. One commonly referred disadvantage is its computational complexity resulting from the integration of the nonregular surface. Still, it should be mentioned that this problem can be solved by off-line calculation of the resulting output for all combinations of the inputs. Then those values are stored in processor memory and are quickly accessible. The other drawbacks are as follows. Firstly, when only one rule is activated, the system output does not depend on the level of degree of firing. This situation is presented in Figure 19.10.

The center of gravity method requires the application of adjacent fuzzy sets with a similar weight. Otherwise, the activation level of the narrow set cannot influence the system output significantly. It can be desirable in some applications, yet in the most cases is not preferable. A fuzzy system with membership functions of different weights is shown below (Figure 19.11).

As can be concluded from Figure 19.11, the change of the firing of the fuzzy sets B_2 influences the resulting output y_0 of the system in an insubstantial way.

This defuzzification method can have a narrow output range. Even if only the first (the last) fuzzy set is fully activated, the output of the system does not reach the minimum (maximum) value

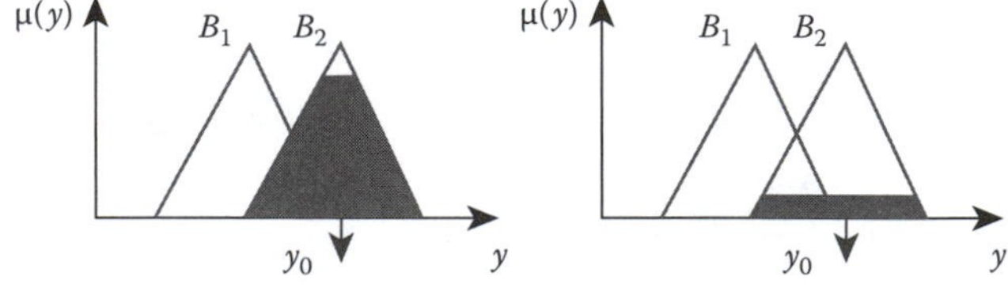

FIGURE 19.10 Insensitivity of the system output in the case of one activated set.

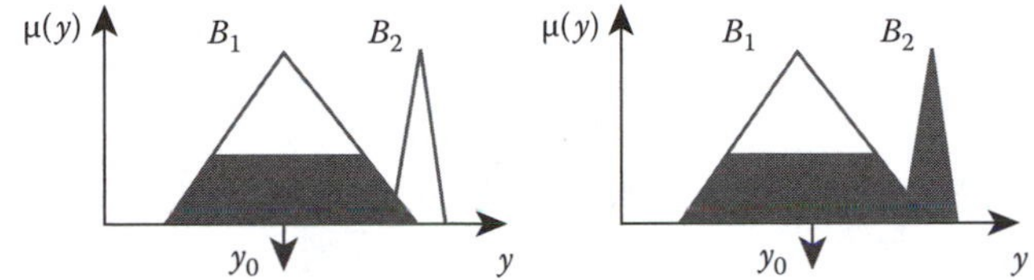

FIGURE 19.11 Output of the fuzzy system with membership functions with different weights.

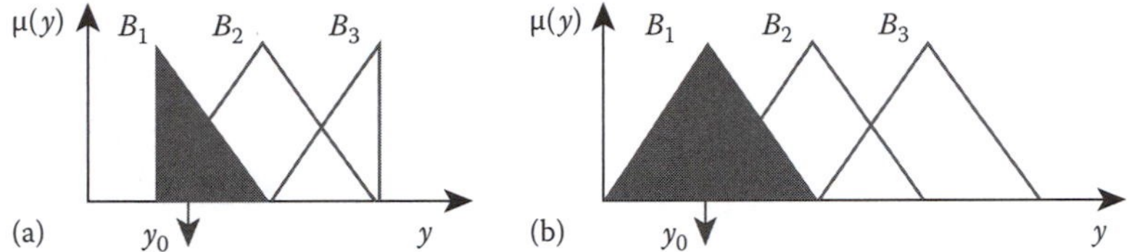

FIGURE 19.12 Narrow of (a) the output range and (b) the way of its elimination.

(Figure 19.12a). This problem is solved by extending the fuzzy sets to the regions outside the universe of discourse (Figure 19.12b). Still, it should be ensured that the system cannot yield the output outside the range of the permitted output values.

One of the most popular defuzzyfication methods applied in the real system, especially in time-consuming applications, is the height (singleton) defuzzyfication method, which relies on the replacement of the output fuzzy sets by singleton values. Then the resulting output is calculated according to the following equation:

$$y_0 = \frac{\sum_{j=1}^{m} y_j \mu_{Bj^*}}{\sum_{j=1}^{m} \mu_{Bj^*}} \tag{19.5}$$

where

y_j is the value of the suitable singleton
μ_{Bj} is the value of the antecedent part of the suitable rule

The major advantage of this method is its computational simplicity resulting from the replacement of the integration of nonregular shape of the set (in previous method) by sum and product operations. Sensitivity and continuity are its further strong points.

19.3.2 Mamdani (Mamdani–Assilian) Model

The first fuzzy system to be introduced is called Mamdani (in some papers Mamdani–Assilian) fuzzy model. It was applied to control a real steam engine system in 1975 [AM74].

This model consists of several *if–then* rules in the following form:

$$\begin{aligned} &\text{R1: IF } x_1 = A_{11} \text{ AND } x_2 = A_{12} \text{ AND} \ldots \text{AND } x_j = A_{1j} \text{ THEN } y = B_1 \\ &\ldots \\ &\text{Rn: IF } x_1 = A_{n1} \text{ AND } x_2 = A_{n2} \text{ AND} \ldots \text{AND } x_j = A_{nj} \text{ THEN } y = B_n \end{aligned} \tag{19.6}$$

In its original form, it used the following operators: *t-norm—min*, implication—*Mamdani*, aggregation—*max*, defuzzyfication—*center of gravity method*. Nowadays, the Mamdani system employing different operators is cited in the literature. The Mamdani system is very popular in various applications, from the simulation works to real-time systems [AM74].

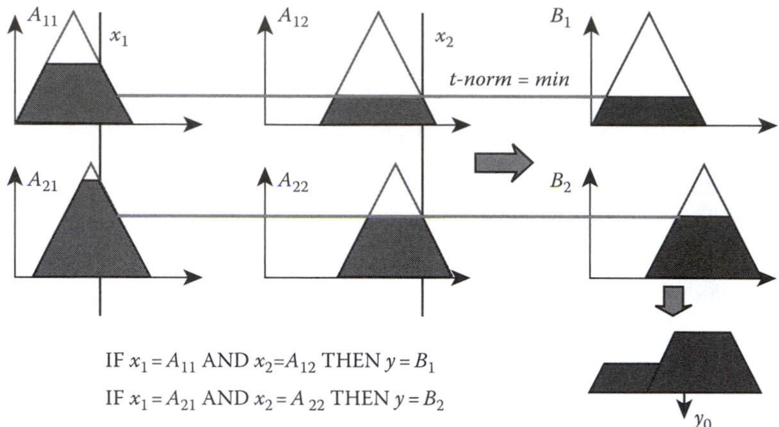

FIGURE 19.13 Illustration of the Mamdani system computation scheme.

A calculation of the output value of the hypothetical Mamdani system is presented in Figure 19.13. It consists of two rules in the following form:

$$\begin{aligned} &\text{R1: IF } x_1 = A_{11} \text{ AND } x_2 = A_{12} \text{ THEN } y = B_1 \\ &\text{R2: IF } x_1 = A_{21} \text{ AND } x_2 = A_{22} \text{ THEN } y = B_2 \end{aligned} \tag{19.7}$$

Initially, the fuzzyfication procedure of the two input signals is conducted. Then the activation degree of the premises using the *min* operation is determined and the shape of the conclusions of membership functions is obtained through the Mamdani implication method. Afterward, the resulting fuzzy set is determined using the max aggregation procedure. With the help of the center of gravity defuzzyfication method the output of the whole fuzzy system is worked out.

The Mamdani model consists of several fuzzy rules each of which determines one fuzzy point in the fuzzy surface. The set of the fuzzy points forms the fuzzy graph in which interpolation between

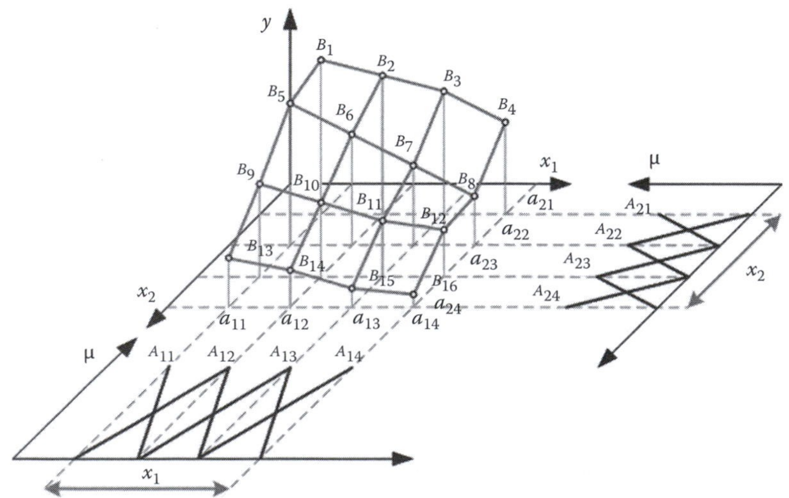

FIGURE 19.14 Division of the fuzzy surface to separate sectors.

points depends on the operators used in the fuzzy model. The fuzzy surface possesses specific properties, if the input triangular membership functions and replacement of conclusion fuzzy sets with singleton values are applied. This is illustrated in Figure 19.14, where the system with 16 rules and parameters presented in Figure 19.15 is considered.

Every fuzzy rule defines the fuzzy surface around the fuzzy point with suitable coordinates. For instance, the rule:

x_2 \ x_1	A_{11}	A_{12}	A_{13}	A_{14}
A_{21}	B_1	B_2	B_3	B_4
A_{22}	B_5	B_6	B_7	B_8
A_{23}	B_9	B_{10}	B_{11}	B_{12}
A_{24}	B_{13}	B_{14}	B_{15}	B_{16}

FIGURE 19.15 Rule base of the system presented in Figure 19.14.

$$\text{R7}: \text{IF } x_1 = A_{13} \text{ AND } x_2 = A_{24} \text{ THEN } y = B_{15}$$

defines the neighborhood of the point with the following coordinates (a_{13}, a_{24}). The control surface of the analyzed fuzzy system consists of nine sectors. The support points are determined by the singleton of each rule. A change of the specific singleton value brings about a slope of the four neighboring sectors. It should be noted that this change does not influence the rest of the sectors, that is, the modification of the singleton value works only locally. A modification of the selected input membership function affects each rule of this set. For instance, a change of the A_{12} set moves the following support points, a_{12}–a_{21}, a_{12}–a_{22}, a_{12}–a_{23}, a_{12}–a_{24}. Therefore, this modification has a global character because it changes the whole cross section.

19.3.3 Takagi–Sugeno Model

The next well-known system is called Takagi–Sugeno–Kang (TSK) model. It was proposed in 1985 by Takagi and Sugeno [TS85] and later in 1988 by Sugeno and Kang [SK88]. Nowadays, it is one of the most frequently applied fuzzy systems [EG03,ELNLN02]. It consists of several rules in the following form:

$$\begin{gathered} \text{R1}: \text{IF } x_1 = A_{11} \text{ AND } x_2 = A_{12} \text{ AND} \ldots \text{AND } x_j = A_{1j} \text{ THEN } y = f(x_1, x_2 \ldots x_j, x_0) \\ \ldots \\ \text{Rn}: \text{IF } x_1 = A_{n1} \text{ AND } x_2 = A_{n2} \text{ AND} \ldots \text{AND } x_j = A_{nj} \text{ THEN } y = f(x_1, x_2 \ldots x_j, x_0) \end{gathered} \tag{19.8}$$

where x_0 is a constant value.

The difference between the Mamdani and TSK model is evident in the conclusion part of the rule. In the Mamdani system, it is rendered by the fuzzy set, while in the TSK system it is a function of the input variables and constant value.

The output of the TSK model is calculated by means of the following equations:

$$y_0 = \frac{\sum_{j=1}^{n} \mu_{Aj} f\left(x_1, x_2 \ldots, x_j, x_0\right)}{\sum_{j=1}^{n} \mu_{Aj}} \tag{19.9}$$

An illustration of the TSK system computational scheme is presented in Figure 19.16. After the fuzzyfication procedure of the two input values x_1 and x_2, the degree of the premises part of each rule is computed using the *min* operator as the t-norm. Unlike in the Mamdani system, the consequent part of the rule is a function of the input variables. After its calculation, the implication and aggregation methods are applied to the system. Then the output of the system is conducted by means of the singleton defuzzyfication strategy.

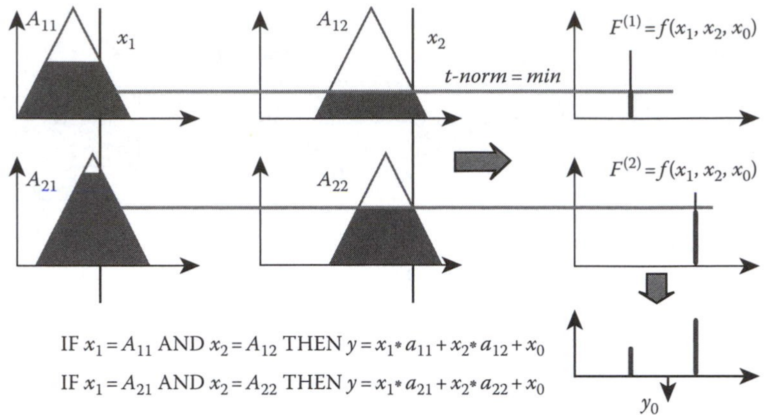

FIGURE 19.16 Illustration of the TSK system computation scheme.

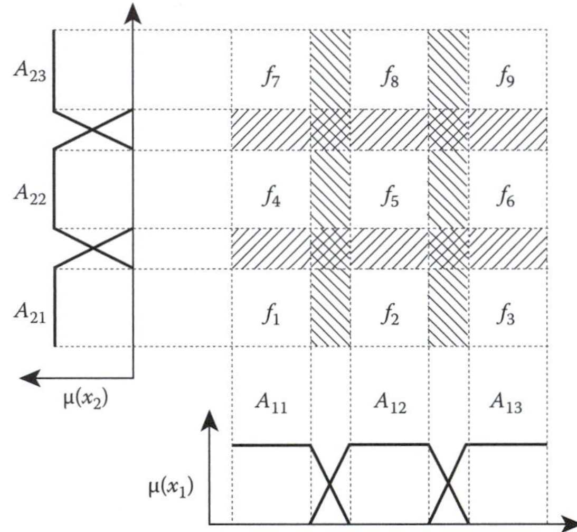

FIGURE 19.17 Hypothetical division of the model surface in TSK system.

As compared to the Mamdani system, the TSK model has the following advantages. First of all, it allows reducing the computational complexity of the whole system. This stems from the fact that the integration of the nonlinear surface is replaced by the *sum* and *prod* operations (in defuzzyfication methodology). Furthermore, by suitable selection of the input membership functions, it is possible to obtain the sectors in the control surface depending only on one rule, which simplifies optimization of the fuzzy system. Due to this reason, the TSK model is often called a quasi-linear fuzzy model. A schematic division of the model surface in a system with two inputs and with trapezoid input membership functions is presented in Figure 19.17.

As can be concluded from the figure, the use of the trapezoid membership functions allows obtaining the model surface with the sectors relying on one rule. Those sectors are marked with f_1–f_9 in Figure 19.17. In the shaded region, the system output is determined by two or four rules.

19.3.4 Tsukamoto Model

The Tsukamoto model was proposed in 1979 in [T79]. The main difference between the TSK and Tsukamoto models lies in the conclusion part of the rule. In the TSK system, the position of the singletons

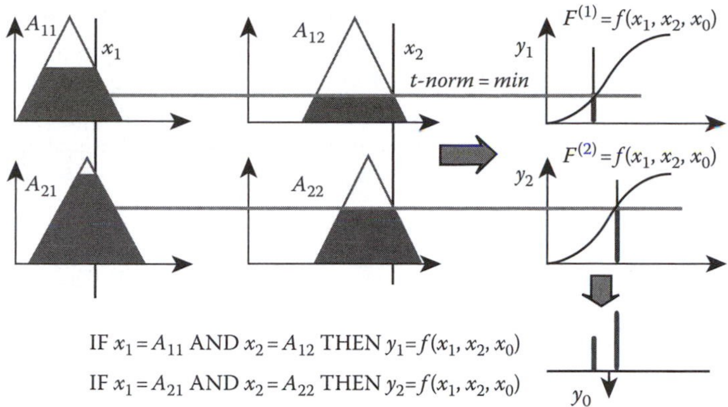

FIGURE 19.18 Illustration of the Tsukamoto system computation scheme.

is the function of the input signals and its amplitude depending on the firing degree of premise the rules. In the Tsukamoto system, on the other hand, the position and at the same time, the amplitude of the output singletons are functions of the degree of activation of each rule. The Tsukamoto system computational scheme is shown in Figure 19.18.

The Tsukamoto system has monotonic functions in the consequent part of the rule. The level of the firing of each rule defines the position and amplitude of the output singletons. Then the output of the Tsukamoto model is calculated similarly as in the TSK mode. Nevertheless, the Tsukamoto model is very rarely applied due to its complexity and difficulty of identifying the functions in the consequent part of the rules.

19.3.5 Models with Parametric Consequents

The system with parametric conclusions was proposed in 1997 by Leski and Czogala [LC99]. The parameters of the fuzzy sets in their conclusions are a function of the input variables in this system. The considered parameters are as follows: location, width or core, height of the input membership functions, and others. This system is also called model with the moving consequences. Figure 19.19 presents an example of fuzzy reasoning for the system with parametric conclusions with two rules.

The operation of computing the premises of individual rules is identical to this in Mamdani or TSK system. The firing degrees of the premises are using to calculate the shape of the conclusion membership

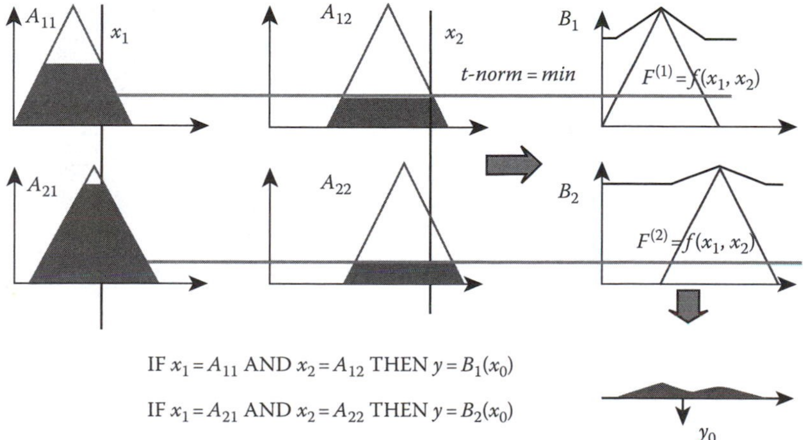

FIGURE 19.19 Process of reasoning in the system with parametric conclusions.

functions. In the discussed case, the Reichenbach implication is used. The fuzzy sets in conclusions are divided into the informative and non-informative part. Employing a selected operation of aggregation and defuzzyfication the output value of the fuzzy system is achieved. A vice of the presented system is the difficulty in obtaining rules with parametric conclusions from a human expert. The authors recommend utilizing this system for the procedure of automatic rule extraction on the basis of measurement data.

19.3.6 Models Based on Sets of the II-Type Fuzzy Sets

Fuzzy systems of the II type were put forward by Mendel in 1999 [KMQ99]. They were introduced as a result of discovered contradiction between uncertainty of information and exact defining of the values of classical fuzzy sets [KMQ99,LCCS09]. Data acquired from different operators can vary hence the information is not determined unambiguously. Similarly, in systems generating rules automatically on the basis of measurement data disturbances have to be accounted for. In such situations, classical sets (the so-called type 1 sets) do not prove useful due to their precisely defined shape. Type II Gauss sets with various types of uncertainty of information are displayed in Figure 19.20. The shape and width of the uncertainty area depends on the considered applications.

II-type fuzzy sets can appear both in the premises and conclusions of the fuzzy rules. In order to simplify the computational algorithm, most authors apply II-type fuzzy sets only in the premises of the rules and the conclusions can be the Mamdani, TSK, or a different system. The scheme of the output computation in the system with the fuzzy input II-type membership functions is presented in Figure 19.21.

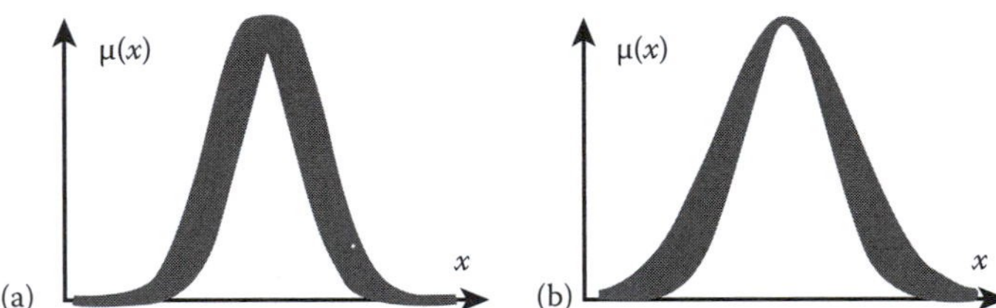

FIGURE 19.20 Gauss II-type fuzzy sets with uncertainty of (a) modal value and (b) weight value.

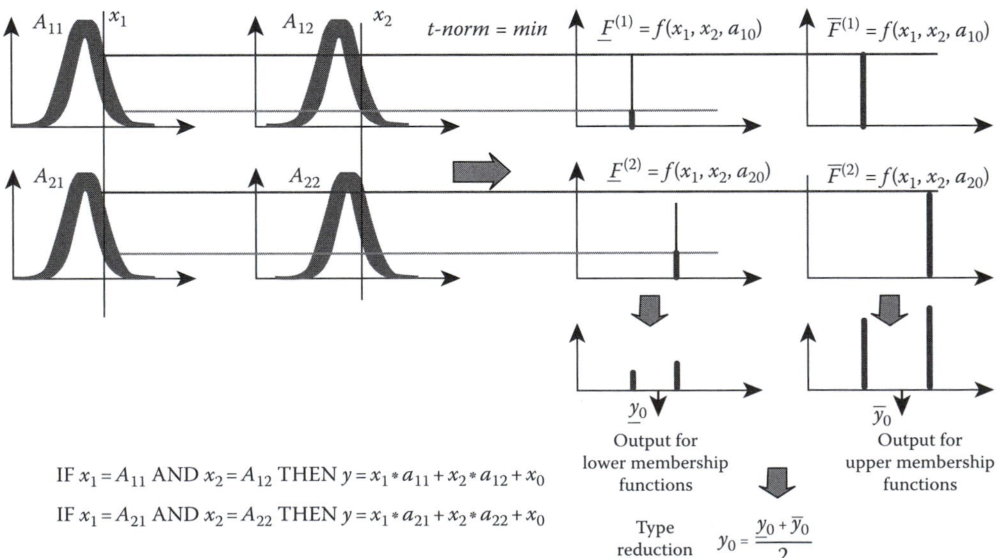

FIGURE 19.21 Illustration of the system computation scheme with the II-type input fuzzy sets.

The consequent parts of the rules have the TSK form. The fuzzyfication of the input variables is carried out for lower and upper ranges of the membership function. So the firing of the premises is represented by two separate values: the lower and the upper degree. Next, the implication and aggregation operations are carried out for the upper and lower values separately. The additional block in this fuzzy system is called the type reduction. The one sharp value y_0 is calculated on the basis of the two values through the following equation (among others):

$$y_0 = \frac{\underline{y}_0 + \overline{y}_0}{2} \quad (19.10)$$

where $\underline{y}_0 + \overline{y}_0$ denote the sharp values determined for the lower and upper values of the II-type membership function.

19.3.7 Neuro-Fuzzy Models

Optimization of the fuzzy system is a very important issue. A variety of methods, such as cluster, evolutionary strategy, etc., can be used for this purpose. One of the methods relies on the transformation of the classical system to a form of the neuro-fuzzy structure and the use of one of the classical neural network training methods (e.g., back propagation algorithm) [P01,CL00]. The types of neuro-fuzzy systems found in the literature are classical TSK, wavelet neuro-fuzzy systems and other. They all have the same general structure presented in Figure 19.22.

In the neuro-fuzzy system, the following layers can be distinguished:

Input layer. Each input node in this layer corresponds to a specific input variable (x_1, $x_2 \ldots x_n$). These nodes only pass input signals to the first layer.

Layer 1. Each node performs a membership function A_{ij} that can be referred to as the fuzzyfication procedure.

Layer 2. Each node in this layer represents the precondition part of fuzzy rule and is denoted by *T*, which conducts a *t-norm* operation and sends the results out.

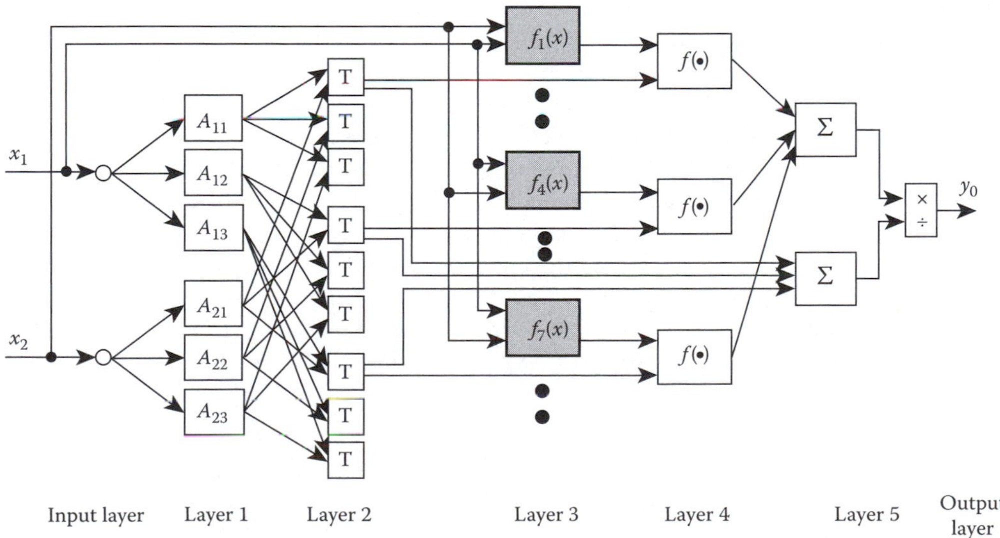

FIGURE 19.22 General structure of neuro-fuzzy with two inputs, six fuzzy input membership functions, and nine rules.

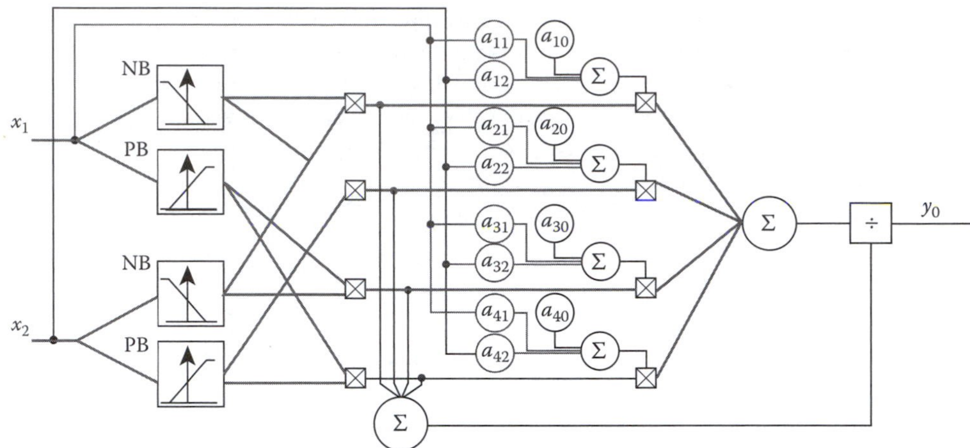

FIGURE 19.23 Hypothetical TSK neuro-fuzzy system with four rules.

Layer 3. In this layer, the function in the consequent part of the rule is calculated. It is a function of the input variables and/or a constant value.

Layer 4. In this layer, the output membership functions are computed. It combines, using a selected inference method, the activation level of each premise with the value of the consequent function.

Layer 5. This layer acts as a defuzzyfier. The single node is denoted by Σ and performs the summing up of all incoming signals. Then a selected defuzzyfication strategy is carried out.

Output layer. This is solely the output of the neuro-fuzzy system.

The wavelet neuro-fuzzy inference system is obtained by putting selected wavelet functions into Layer 3 (conclusion part of the rules). The rest of the structure remains the same. The II-type fuzzy inference system can be realized by putting the II-type membership functions into the first layer or/and into the third layer. In this case, a block which can realize the type reduction of the system (from the II-type to the single value) has to be added to the presented structure.

In Figure 19.23, hypothetical TSK neuro-fuzzy system for parameters shown in Figure 19.24 is presented. It is adequate to the classical TSK system. The name neuro-fuzzy refers to a different way of presenting the system structure (Figure 19.16).

From this figure, the relationship of the consequent parts of the rules form the input signals and the constant value can be clearly seen. In some adaptive systems, the sum of the premises is not calculated and the sum of singletons multiplied by suitable firing values is given as an output value.

As it was pointed previously, the most important advantage of the neuro-fuzzy system is the possibility to use the training methods developed for classical neural networks. Examples of the off-line or on-line training system can be found in the literature. One of the most popular adaptive neuro-fuzzy systems is the Adaptive Neuro-Fuzzy Inference System (ANFIS) as proposed by Jang in [J93]. Other types of adaptive neuro-fuzzy models are the Neural Fuzzy CONtroller (NEFCON) [NNK99] or Artificial Neural Network Based Fuzzy Inference System (ANNBFIS) [LC99], etc.

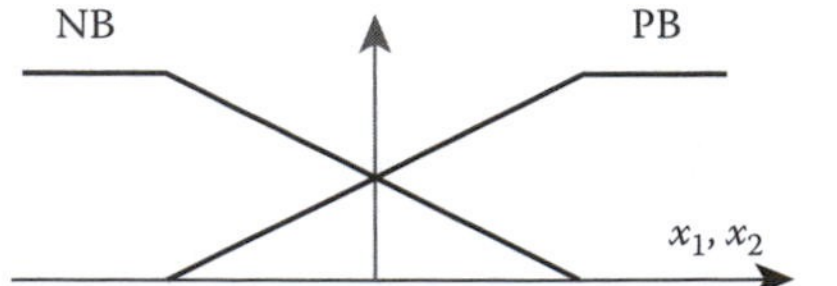

x_2 \ x_1	NB	PB
NB	f_1	f_2
PB	f_3	f_4

FIGURE 19.24 Input membership functions and rule base for the system presented in Figure 19.23.

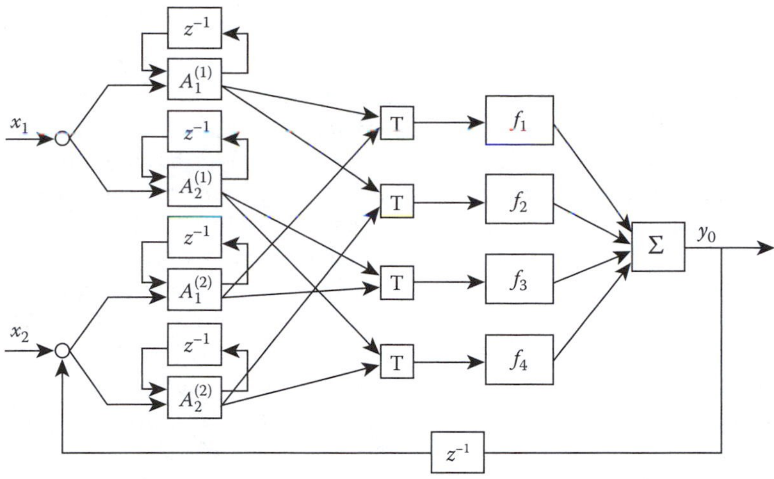

FIGURE 19.25 Recurrent neuro-fuzzy system with simple and hierarchical feedbacks.

In contrast to the pure feed-forward architecture of the classical neuro-fuzzy system, the recurrent models have the advantage of using the information from the past, which is especially useful to modeling and analyzing dynamic plants [ZM99,JC06,J02]. Recurrent neuro-fuzzy systems can be constructed in the same way as the standard feed-forward systems. However, due to their complexity optimization of the system parameters is significantly more difficult. A neuro-fuzzy system with two types of recurrent feedbacks is presented in Figure 19.25.

The first are simple feedback units put to the membership function in the antecedent part of the rules. The second type is hierarchical feedback units, which connect the output of the whole system with the second input.

Examples of significantly more complicated structures of recurrent neuro-fuzzy systems exist in the literature. An example of system consisting of two subsystems is presented in Figure 19.26.

At the beginning, the output of the first subsystem is calculated. Next, it is used as an additional feedback to the second subsystem. The delayed output of the second system is fed back to the first and alternatively also to the second subsystem. The recurrent feedbacks can be combined with every type of fuzzy system. For example, the II-type recurrent wavelet neuro-fuzzy system can be designed yet the optimization difficulty of such a complicated structure should be pointed out.

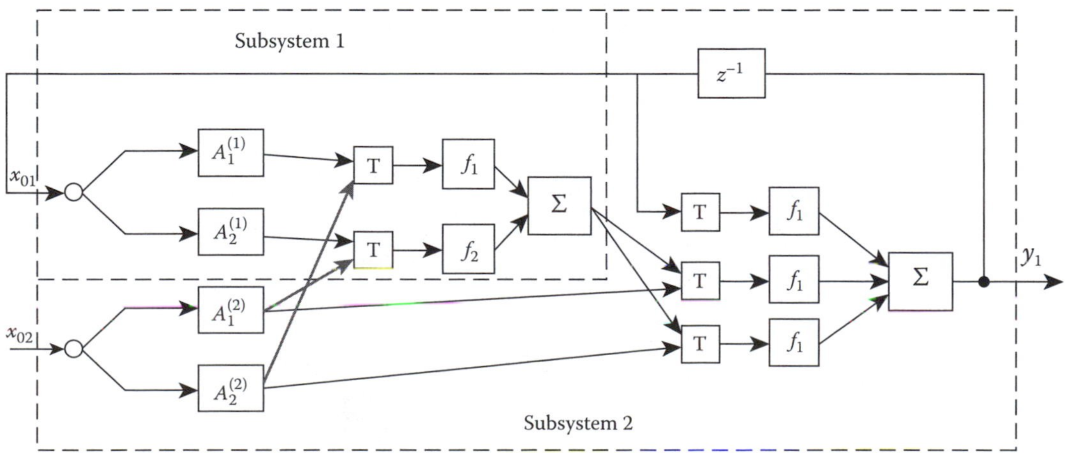

FIGURE 19.26 Recurrent neuro-fuzzy system consisting of two subsystems.

19.3.8 Local and Global Models

The first fuzzy models analyzed in the literature have the global character, that is, the whole universe of discourse has been divided evenly by the grid. In a real system, though, there exist flat and at the same time also very steep regions. In order to model the steep regions' accuracy, the grid lines have to be very dense. This increases the number of rules of the whole system and the optimization procedure of this type of system can be complicated [P01,ZS96]. The described situation is presented in Figure 19.27a. The displayed system has two "peeks." In order to model the surface of these peeks with sufficient accuracy, the grid is quite dense and the number of rules for the whole system is as high as 144. In order to reduce the number of rules, the whole system is divided into four separate regions (Figure 19.27b). Two flat regions are modeled by 8 rules, two steep regions by 72 rules. So when the local models are used, the reduction of the number of fuzzy rules from 144 to 72 in this particular case is achieved.

The desired feature of local fuzzy models is the continuity of the modeled surface on the points of contact of different models. Because the parameters of the local models are obtained separately, this condition is usually not fulfilled. Thus, the system presented in Figure 19.28 can be applied to ensure the continuity of the global model.

The total output of the global model is calculated on the basis on the responses of the local models. Then those signals go to the aggregation block, usually with trapezium functions with changeable value of the membership function in the areas of contact of the local models, which calculates the output of the system.

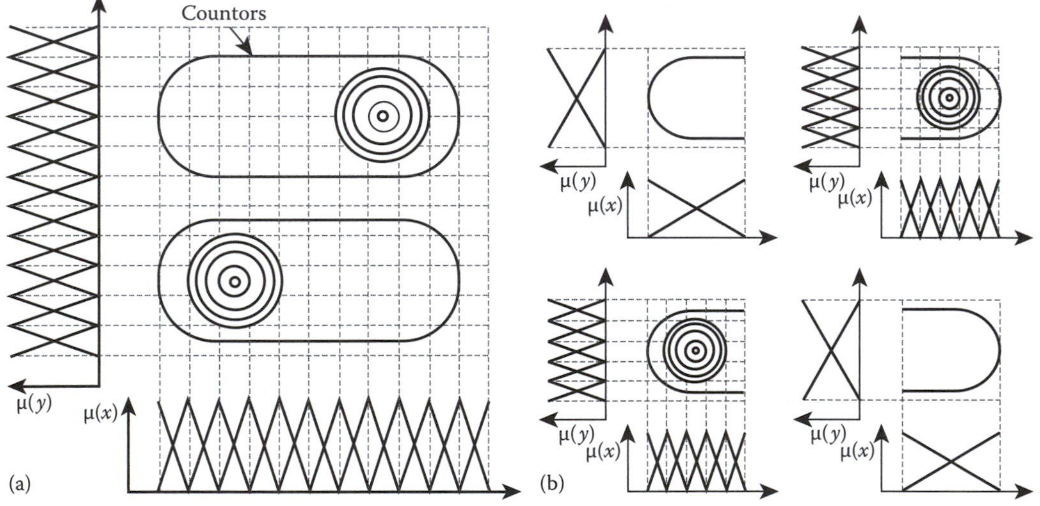

FIGURE 19.27 (a) Global model and (b) four corresponding local models.

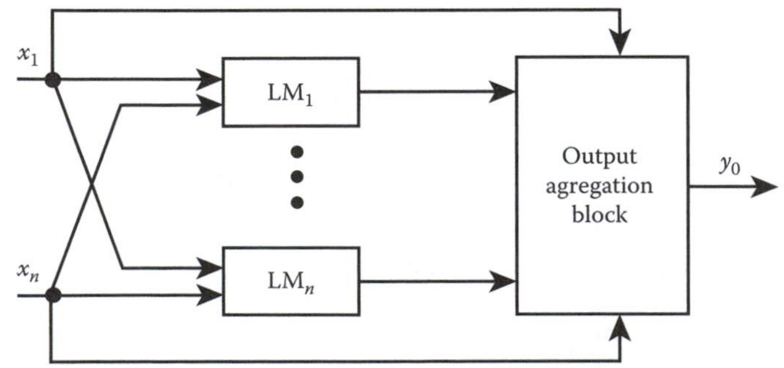

FIGURE 19.28 Aggregation block of the local fuzzy systems.

19.4 Summary

This chapter is devoted to the presentation of the properties of the fuzzy controller. Similarities and differences between the classical and the fuzzy PID controller are pointed out and the fields of application of fuzzy systems are briefly described. Then the general structure of the fuzzy system is presented. Basic operations of the fuzzy models are described and a survey emphasizing distinctions between different fuzzy models is put forward. Due to the limited length of the chapter, other very important issues such as stability analysis or optimization of the fuzzy controller etc. are not considered here. Readers are referred to the variety of books in the following topic, for example, [DHR96,P93,P01,CL00,YF94].

References

[ÄH01] K. J. Äström and T. Hägglund, The future of PID control, *Control Engineering Practice* 9, 1163–1175, 2001.

[AM74] S. Assilian and E. H. Mamdani, An experiment in linguistic synthesis with a fuzzy logic controller, *International Journal of Man-Machine Studies*, 7, 1–13, 1974.

[CL00] E. Czogala and J. Leski, *Fuzzy and Neuro-fuzzy Intelligent Systems*, Springer, Heidelberg, Germany, 2000.

[DHR96] D. Driankov, H. Hellendoorn, and M. Reinfrank, *An Introduction to Fuzzy Control*, Springer, Berlin, Germany, 1996.

[EG03] M. J. Er and Y. Gao, Robust adaptive control of robot manipulators using generalized fuzzy neural networks, *IEEE Transactions on Industrial Electronics*, 50(3), 620–628, 2003.

[ELNLN02] M. J. Er, Ch. B. Low, K. H. Nah, M. H. Lim, and S. Y. Ng, Real-time implementation of a dynamic fuzzy neural networks controller for a SCARA, *Microprocessors and Microsystems*, 26(9–10), 449–461, 2002.

[J93] J. S. R. Jang, ANFIS: Adaptive-network-based fuzzy inference system, *IEEE Transactions on Systems, Man and Cybernetics*, 23(3), 665–685, 1993.

[J02] C. F. Juang, A TSK-type recurrent fuzzy network for dynamic systems processing by neural network and genetic algorithms, *IEEE Transactions on Fuzzy Systems*, 10(2), 155–170, 2002.

[JC06] C. F. Juang and J. S. Chen, Water bath temperature control by a recurrent fuzzy controller and its FPGA implementation, *IEEE Transactions on Industrial Electronics*, 53(3), 941–949, 2006.

[LC99] J. Leski and E. Czogala E, A new artificial neural network based fuzzy inference system with moving consequents in if then rules, *Fuzzy Sets and Systems*, 108, 289–297, 1999.

[LCCS09] F. J. Lin, S. Y. Chen, P. H. Chou, and P. H. Shieh, Interval type-2 fuzzy neural network control for *X*–*Y*–theta motion control stage using linear ultrasonic motors, *Neurocomputing*, 72(4–6), 1138–1151, 2009.

[KMQ99] N. N. Karnik, J. M. Mendel, and L. Qilian, Type-2 fuzzy logic systems, *IEEE Transactions on Fuzzy Systems*, 7(6), 643–658, 1999.

[NNK99] A. Nurnberger, D. Nauck, and R. Kruse, Neuro-fuzzy control based on the NEFCON-model: Recent developments, *Soft Computing—A Fusion of Foundations, Methodologies and Applications*, 2(4), 168–182, 1999.

[P93] W. Pedrycz, *Fuzzy Control and Fuzzy Systems*, Research Studies Press Ltd., Taunton, U.K., 1993.

[P01] A. Piegat, *Fuzzy Modeling and Control*, Springer, Heidelberg, Germany, 2001.

[SK88] M. Sugeno and G. T. Kang, Structure identification of fuzzy model, *Fuzzy Sets and Systems*, 26, 15–33, 1988.

[T79] Y. Tsukamoto, An approach to fuzzy reasoning method, In: M. M. Gupta, R. K. Ragade, and R. R. Yager, Editors, *Advances in Fuzzy Set Theory and Applications*, North-Holland, Amsterdam, 1979, pp. 137–149.

[TS85] T. Takagi and M. Sugeno, Fuzzy identification of systems and its applications to modeling and control, *IEEE Transactions on System Man and Cybernetics*, 15, 116–132, 1985.

[YF94] R. R. Yager and D. P. Filev, *Essential of Fuzzy Modelling and Control*, John Wiley & Sons, Inc., New York, 1994.

[ZM99] J. Zhang and A. J. Morris, Recurrent neuro-fuzzy networks for nonlinear process modeling, *IEEE Transactions on Neural Networks*, 10(2), 313–326, 1999.

[ZS96] X. J. Zeng and M. G. Singh, Decomposition property of fuzzy systems and its applications, *IEEE Transactions on Fuzzy Systems*, 4(2), 149–165, 1996.

20

Neuro-Fuzzy System

Tiantian Xie
Auburn University

Hao Yu
Auburn University

Bogdan M. Wilamowski
Auburn University

20.1 Introduction

Conventional controllers, such as a PID controller, are broadly used for linear processes. In real life, most processes are nonlinear. Nonlinear control is considered as one of the most difficult challenges in modern control theory. While linear control system theory has been well developed, it is the nonlinear control problems that cause most headaches. Traditionally, a nonlinear process has to be linearized first before an automatic controller can be effectively applied [WB01]. This is typically achieved by adding a reverse nonlinear function to compensate for the nonlinear behavior so the overall process input–output relationship becomes somewhat linear.

The issue becomes more complicated if a nonlinear characteristic of the system changes with time and there is a need for an adaptive change of the nonlinear behavior. These adaptive systems are best handled with methods of computational intelligence such as neural networks and fuzzy systems [W02,W07].

In this chapter, the neuro-fuzzy system [WJK99,W09], as a combination of fuzzy system and neural networks, will be introduced, and compared with classic fuzzy systems, based on a simple case (Figure 20.1).

The studying case can be described as the nonlinear control surface, shown in Figure 20.1. All points (441 points in Figure 20.1a and 36 points in Figure 20.1b) in the surface are calculated by the equation

$$z = 1.1\,exp\,(-0.07(x-5)^2 - 0.07(y-5)^2) - 0.9 \tag{20.1}$$

20.2 Fuzzy System

The most commonly used architectures for fuzzy system development are the Mamdani fuzzy system [M74,MW01] and the TSK (Takagi, Sugeno, and Kang) fuzzy system [TS85,SK88,WB99], as shown in Figure 20.2. Both of them consist of three blocks: fuzzification, fuzzy rules, and defuzzification/normalization. Each of the blocks could be designed differently.

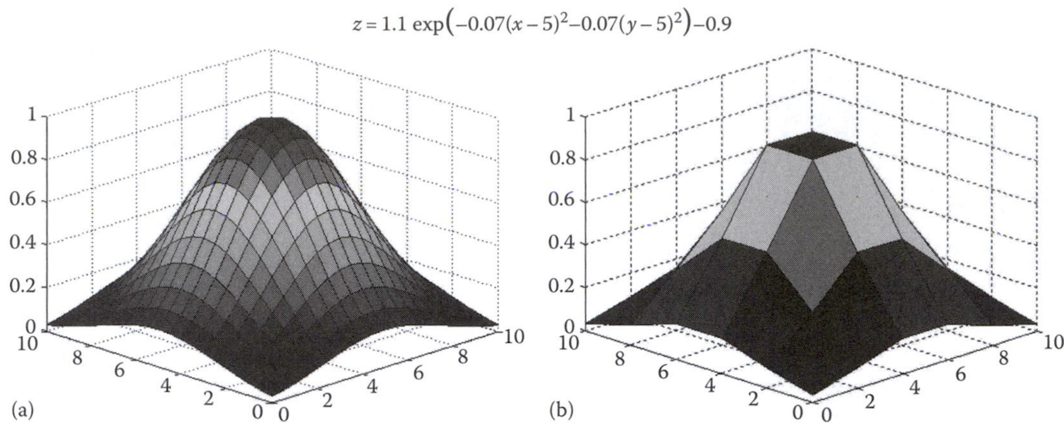

FIGURE 20.1 Required surface obtained from Equation 20.1: (a) $21 \times 21 = 441$ points and (b) $6 \times 6 = 36$ points.

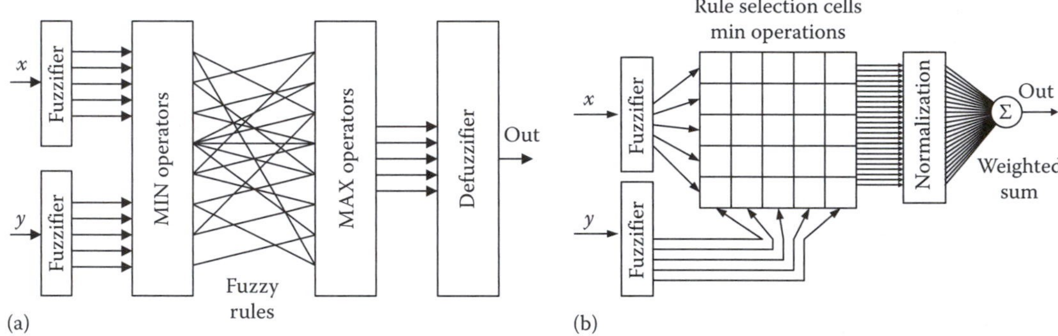

FIGURE 20.2 Block diagram of the two types of fuzzy systems: (a) Mamdani fuzzy system and (b) TSK fuzzy system.

20.2.1 Fuzzification

Fuzzification is supposed to convert the analog inputs into sets of fuzzy variables. For each analog input, several fuzzy variables are generated with values between 0 and 1. The number of fuzzy variables depends on the number of member functions in the fuzzification process. Various types of member functions can be used for conversion, such as triangular and trapezoidal. One may consider using a combination of them and different types of membership functions result in different accuracies. Figure 20.3 shows the surfaces and related accuracies obtained by using Mamdani fuzzy system with different membership functions, for solving the problem in Figure 20.1.

One may notice that using the triangular membership functions one can get better surface than from using the trapezoidal membership functions.

The more membership functions are used, the higher accuracy will be obtained. However, very dense functions may lead to frequent controller actions (known as "hunting"), and sometimes this may lead to system instability; on the other hand, more storage is required, because the size of the fuzzy table is increased exponentially to the number of membership functions.

20.2.2 Fuzzy Rules

Fuzzy variables are processed by fuzzy logic rules, with MIN and MAX operators. The fuzzy logic can be interpreted as the extended Boolean logic. For binary "0" and "1," the MIN and MAX operators

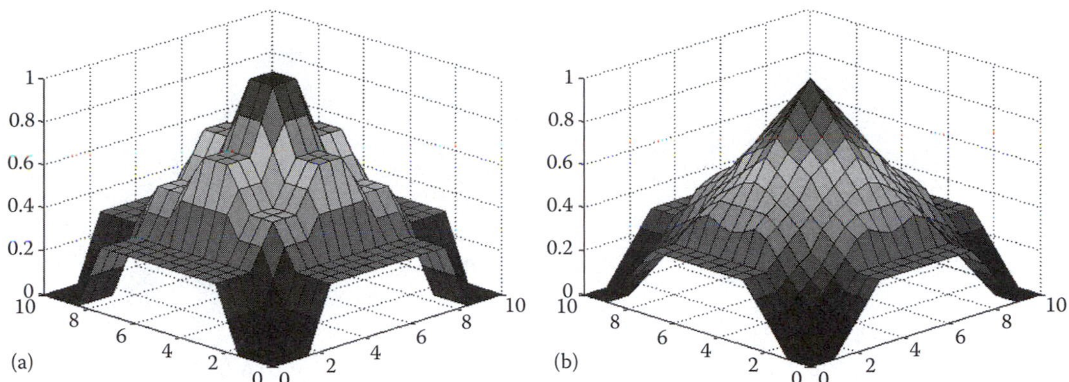

FIGURE 20.3 Control surface using Mamdani fuzzy systems and five membership functions per input: (a) trapezoidal membership function, error = 3.8723 and (b) triangular membership function, error = 2.4799.

TABLE 20.1 Binary Operation Using Boolean Logic and Fuzzy Logic

A	B	A AND B	MIN(A,B)	A OR B	MAX(A,B)
0	0	0	0	0	0
0	1	0	0	1	1
1	0	0	0	1	1
1	1	1	1	1	1

in the fuzzy logic perform the same calculations as the AND and OR operators in Boolean logic, respectively, see Table 20.1; for fuzzy variables, the MIN and MAX operators work as shown in Table 20.2.

TABLE 20.2 Fuzzy Variables Operation Using Fuzzy Logic

A	B	MIN(A,B)	MAX(A,B)
0.3	0.5	0.3	0.5
0.3	0.7	0.3	0.7
0.6	0.4	0.4	0.6
0.6	0.8	0.6	0.8

20.2.3 Defuzzification

As a result of "MAX of MIN" operations in the Mamdani fuzzy systems, a new set of fuzzy variables is generated, which later has to be converted to an analog output value by defuzzification blocks (Figure 20.1a). In the TSK fuzzy systems, the defuzzification block was replaced with normalization and weighted average; MAX operations are not required, instead, a weighted average is applied directly to regions selected by MIN operators.

Figure 20.4 shows the result surfaces using the TSK fuzzy architecture, with different membership functions.

20.3 Neuro-Fuzzy System

A lot of research is devoted to improve the ability of fuzzy systems [WJ96,DGKW02,GNW08,MW01,OW99], such as evolutionary strategy and neural networks [CW94]. The combination of fuzzy logic and neural networks is called neuro-fuzzy system, which is supposed to result in a hybrid intelligent system by combining the human-like reasoning style of neural networks.

20.3.1 Structure One

Figure 20.5 shows the neuro-fuzzy system which attempts to present a fuzzy system in a form of neural network [RH99].

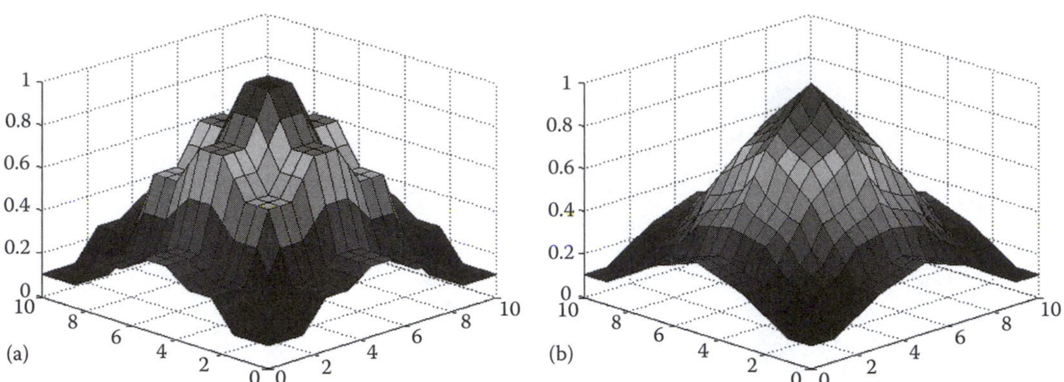

FIGURE 20.4 Control surface using TSK fuzzy systems and five membership functions per input: (a) trapezoidal membership function, error = 2.4423, and (b) triangular membership function, error = 1.5119.

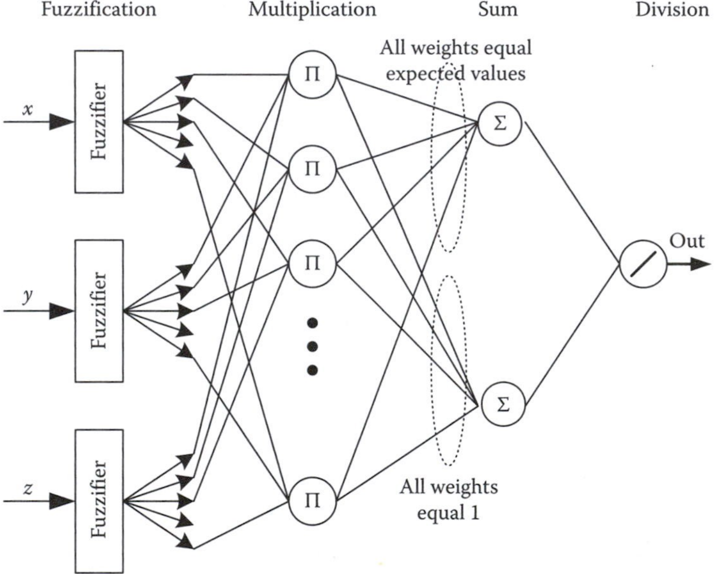

FIGURE 20.5 Neuro-fuzzy system.

The neuro-fuzzy system consists of four blocks: fuzzification, multiplication, summation, and division. The Fuzzification block translates the input analog signals into fuzzy variables by membership functions. Then, instead of MIN operations in classic fuzzy systems, product operations (signals are multiplied) are performed among fuzzy variables. This neuro-fuzzy system with product encoding is more difficult to implement [OW96], but it can generate a slightly smoother control surface (see Figures 20.6 and 20.7). The summation and division layers perform defuzzification translation. The weights on upper sum unit are designed as the expecting values (both the Mamdani and TSK rules can be used); while the weights on the lower sum unit are all "1."

Figures 20.6 and 20.7 show the surfaces obtained using the neuro-fuzzy system in Figure 20.5, which is smoother than the surfaces in Figures 20.3 and 20.4.

Note that, in this type of neuro-fuzzy systems, only architecture resembles neural networks because cells there perform different functions than neurons, such as signal multiplication or division.

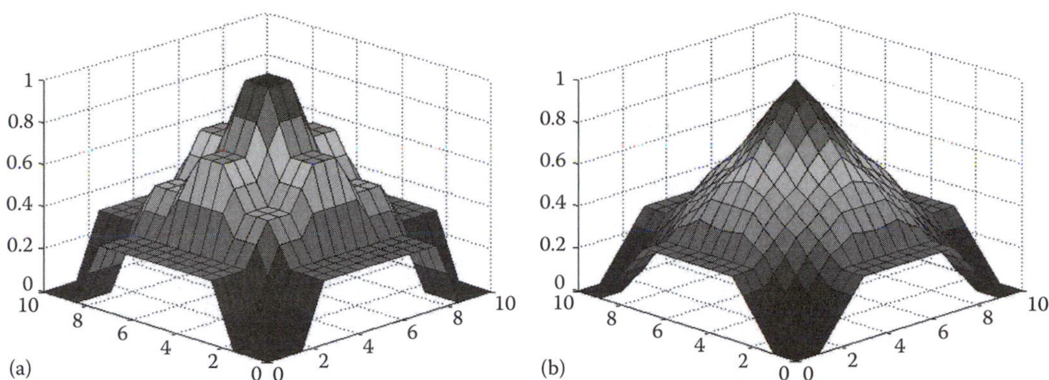

FIGURE 20.6 Control surface using neuro-fuzzy system in Figure 20.5, Mamdani rule for weight initialization of the upper sum unit and five membership functions per input: (a) trapezoidal membership function, error = 3.8723 and (b) triangular membership function, error = 2.4468.

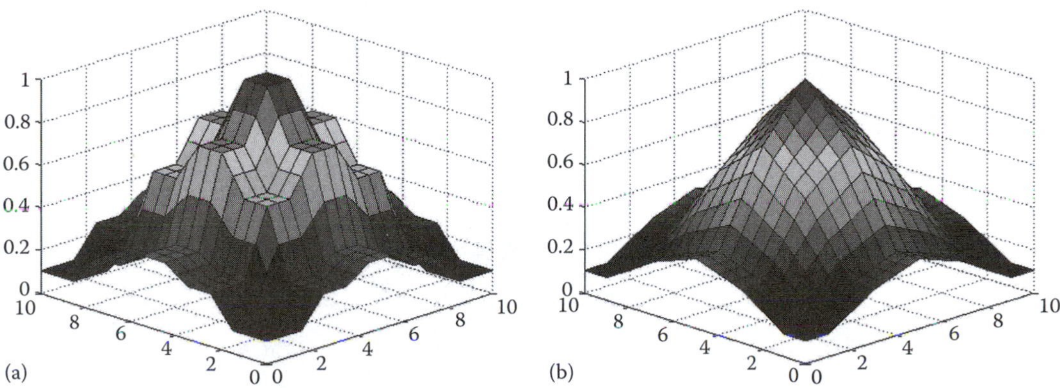

FIGURE 20.7 Control surface using neuro-fuzzy system in Figure 20.5, TSK rule for weight initialization of the upper sum unit and five membership functions per input: (a) trapezoidal membership function, error = 2.4423 and (b) triangular membership function, error = 1.3883.

20.3.2 Structure Two

A single neuron can divide input space by line, plane, or hyper plane, depending on the problem dimensionality. In order to select just one region in n-dimensional input space, more than $(n+1)$ neurons are required. For example, to separate a rectangular pattern, four neurons are required, as is shown in Figure 20.8. If more input clusters should be selected, then the number of neurons in the hidden layer should be properly multiplied. If the number of neurons in the hidden layer is not limited, then all classification problems can be solved using the three-layer network.

With the concept shown in Figure 20.8, fuzzifiers and MIN operators used for region selection can be replaced by simple neural network architecture [XYW10]. In this example, the two analog inputs, each with five membership functions, can be organized as a two-dimensional input space was divided by six neurons horizontally (from line a to line f) and by six neurons vertically (from line g to line l), as shown in Figure 20.9. The corresponding neural network is shown in Figure 20.10. Neurons in the first layer are corresponding to the lines indexed from a to l. Each neuron is connected only to one input. For each neuron input, weight is equal to +1 and the threshold is equal to the value of the crossing point on the x or y axis. The type of activation functions of neurons in the first layer decides the type of membership functions of the fuzzy system, as shown in Figure 20.11.

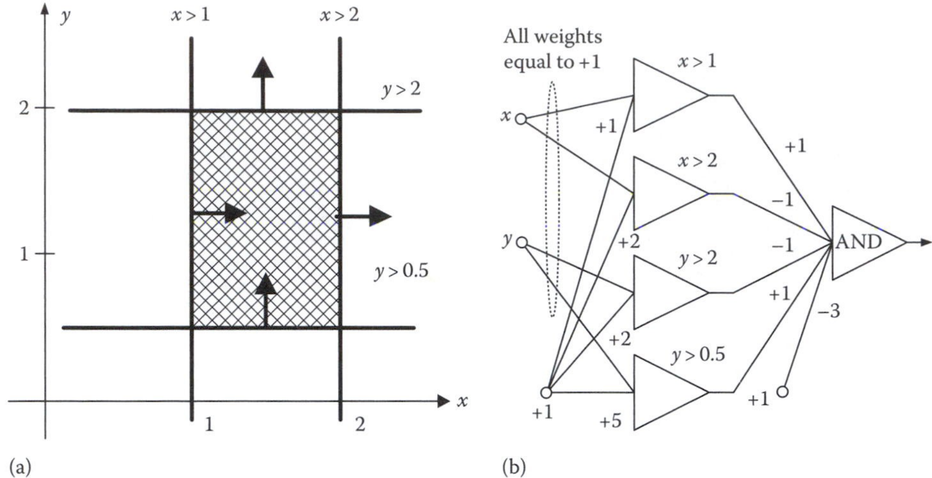

FIGURE 20.8 Separation of the rectangular area on a two dimensional input space and desired neural network to fulfill this task.

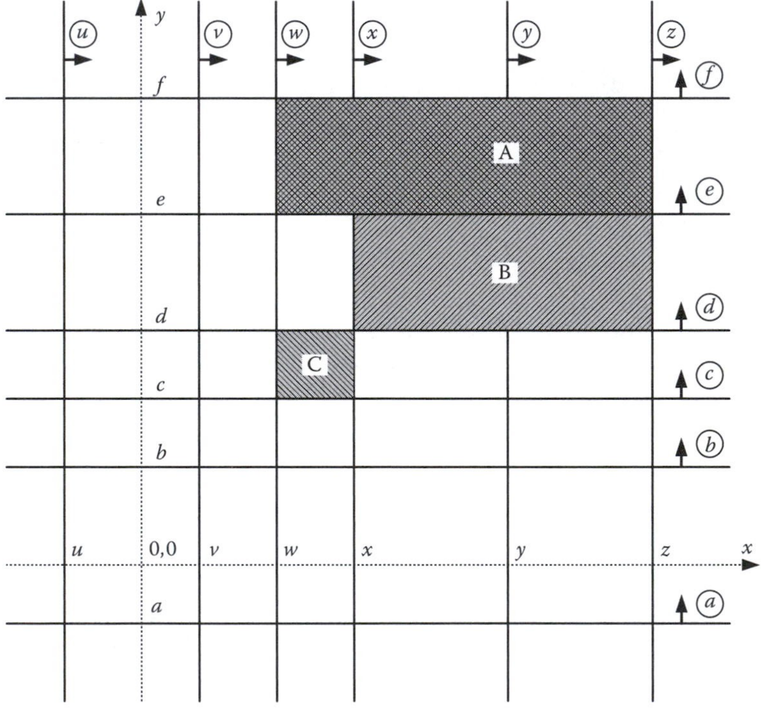

FIGURE 20.9 Two-dimensional input plane separated vertically and horizontally by six neurons in each direction.

Neurons in the second layers are corresponding to the sections indexed from 1 to 25. Each of them has two connections to lower boundary neurons with weights of +1 and two connections to upper boundary neurons with weights of −1. Thresholds for all these neurons in the second layer are set to 3.

Weights of the upper sum unit in the third layer have values corresponding to the specified values in the selected areas. The specified values can be obtained from either the fuzzy table (by Mamdani rule), or the expected function values (by TSK rule). Weights of the lower sum unit are equal to “1.”

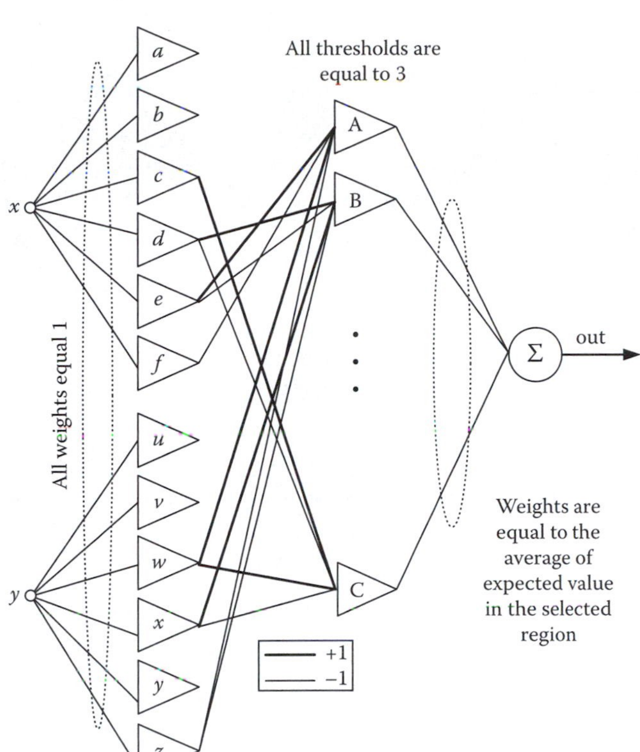

FIGURE 20.10 The neural network performing the function of fuzzy system.

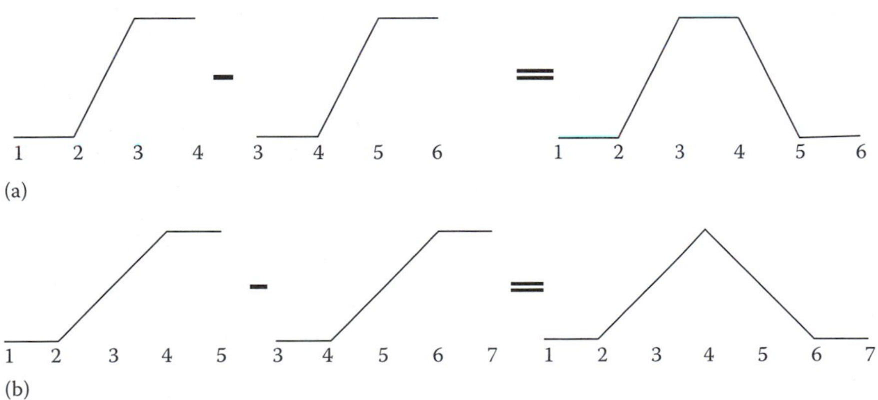

FIGURE 20.11 Construction of membership functions by neurons' activation functions: (a) trapezoidal membership function and (b) triangular membership function.

All neurons in Figure 20.8 have a unipolar activation function and if the system is properly designed, then for any input vector in certain areas only the neuron of this area produces +1 while all remaining neurons have zero values. In the case of when the input vector is close to a boundary between two or more regions, then all participating neurons are producing fractional values and the system output is generated as a weighted sum. The fourth layer performs such a calculation: the upper sum

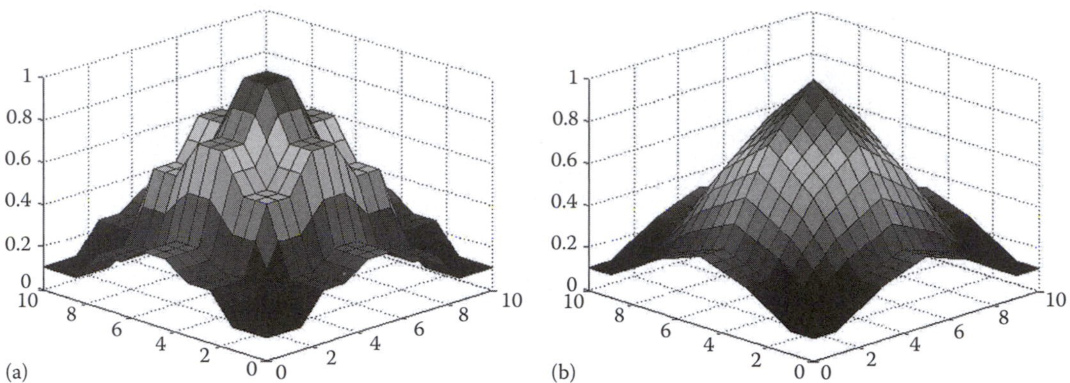

FIGURE 20.12 Control surface using neuro-fuzzy system in Figure 20.10: (a) using combination of activation functions in Figure 20.11a, error = 2.4423 and (b) using combination of activation functions in Figure 20.11b, error = 1.3883.

divided by the lower sum. Like the neuro-fuzzy system in Figure 20.5, the last two layers are used for defuzzification.

Using this concept of the neuro-fuzzy system, the result surfaces with different combination of activation functions, can be obtained as shown in Figure 20.12.

It was shown above that a simple neural network of Figure 20.10 can replace a TSK neuro-fuzzy system in Figure 20.5. All parameters of this network are directly derived from requirements specified for a fuzzy system and there is no need for a training process.

20.4 Conclusion

The chapter introduced two types of neuro-fuzzy architectures, in order to improve the performance of classic fuzzy systems. Based on a given example, the classic fuzzy systems and the neuro-fuzzy systems, with different settings, are compared. From the comparison results, the following conclusions can be drawn:

- In the same type of fuzzy system, using triangular membership functions can get better results than those from using the same number of trapezoidal membership functions.
- With the same membership function, the TSK (Takagi, Sugeno, and Kang) fuzzy systems perform more accurate calculation than the Mamdani fuzzy system.
- The neuro-fuzzy system in Figure 20.5 makes a slight improvement on the accuracy, with the cost of using signal multiplication units, which are difficult for hardware implementation.
- The neuro-fuzzy system in Figure 20.10 does the same job as the neuro-fuzzy system with the TSK rule in Figure 20.5.

References

[CW94] J. J. Cupal and B. M. Wilamowski, Selection of fuzzy rules using a genetic algorithm, *Proceedings of Word Congress on Neural Networks*, San Diego, CA, June 4–9, 1994, vol. 1, pp. 814–819.

[DGKW02] N. Dharia, J. Gownipalli, O. Kaynak, and B. M. Wilamowski, Fuzzy Controller with Second Order Defuzzification Algorithm, *Proceedings of the 2002 International Joint Conference on Neural Network s- IJCNN 2002*, Honolulu, USA, May 12–17, 2002, pp. 2327–2332.

[GNW08] K. Govindasamy, S. Neeli, and B. M. Wilamowski, Fuzzy System with Increased Accuracy Suitable for FPGA Implementation *12th INES 2008 -International Conference on Intelligent Engineering Systems*, Miami, FL, February 25–29, 2008, pp. 133–138.

[M74] E. H. Mamdani, Application of fuzzy algorithms for control of simple dynamic plant, *IEEE Proceedings*, 121(12), 1585–1588, 1974.

[MW01] M. McKenna and B. M. Wilamowski, Implementing a fuzzy system on a field programmable gate array, *International Joint Conference on Neural Networks (IJCNN'01)*, Washington DC, July 15–19, 2001, pp. 189–194.

[OW99] Y. Ota and B. M. Wilamowski, CMOS implementation of a voltage-mode fuzzy min-max controller, *Journal of Circuits, Systems and Computers*, 6(2), 171–184, April 1999.

[RH99] D. Rutkowska and Y. Hayashi, Neuro-fuzzy systems approaches, *International Journal of Advanced Computational Intelligence*, 3(3), 177–185, 1999.

[SK88] M. Sugeno and G. T. Kang, Structure identification of fuzzy model, *Fuzzy Sets and Systems*, 28(1), 15–33, 1988.

[TS85] T. Takagi and M. Sugeno, Fuzzy identification of systems and its application to modeling and control, *IEEE Transactions on System, Man, Cybernetics*, 15(1), 116–132, 1985.

[W02] B. M. Wilamowski, Neural networks and fuzzy systems, Chapter 32 in *Mechatronics Handbook*, (Ed.), R. R. Bishop CRC Press, Boca Raton, FL, 2002, pp. 33-1–32-26.

[W07] B. M. Wilamowski, Neural Networks and Fuzzy Systems for Nonlinear Applications, *11th INES 2007-11th INES 2007-International Conference on Intelligent Engineering Systems*, Budapest, Hungary, June 29–July 1 2007, pp. 13–19.

[W09] B. M. Wilamowski, Neural Networks or Fuzzy Systems, Workshop on Intelligent Systems, Budapest, Hungary, August 30, 2009, pp. 1–12.

[WB01] B. M. Wilamowski and J. Binfet, Microprocessor implementation of fuzzy systems and neural networks, *International Joint Conference on Neural Networks (IJCNN'01)*, Washington DC, July 15–19, 2001, pp. 234–239.

[WB99] B. M. Wilamowski and J. Binfet, Do fuzzy controllers have advantages over neural controllers in microprocessor implementation, *Proceedings of the Second International Conference on Recent Advances in Mechatronics—ICRAM'99*, Istanbul, Turkey, May 24–26, 1999, pp. 342–347.

[WJ96] B. M. Wilamowski and R. C. Jaeger, Implementation of RBF type networks by MLP networks, *IEEE International Conference on Neural Networks*, Washington, DC, June 3–6, 1996, pp. 1670–1675.

[WJK99] B. M. Wilamowski, R. C. Jaeger, and M. O. Kaynak, Neuro-fuzzy architecture for CMOS implementation, *IEEE Transaction on Industrial Electronics*, 46(6), 1132–1136, December 1999.

[XYW10] T. T. Xie, H. Yu, and B. M. Wilamowski, Replacing Fuzzy Systems with Neural Networks, in *Proc. 3rd IEEE Human System Interaction Conf. HSI 2010*, Rzeszow, Poland, May 13–15, 2010, pp. 189–193.

21

Introduction to Type-2 Fuzzy Logic Controllers

Hani Hagras
University of Essex

21.1 Introduction

Fuzzy control is regarded as the most widely used application of fuzzy logic [Mendel 2001]. A fuzzy logic controller (FLC) is credited with being an adequate methodology for designing robust controllers that are able to deliver a satisfactory performance in the face of uncertainty and imprecision. In addition, FLCs provide a way of constructing controller algorithms by means of linguistic labels and linguistically interpretable rules in a user-friendly way closer to human thinking and perception.

FLCs have successfully outperformed the traditional control systems (like PID controllers) and have given a satisfactory performance similar (or even better) to the human operators. According to Mamdani [Mamdani 1994]: "When tuned, the parameters of a PID controller affect the shape of the entire control surface. Because fuzzy logic control is a rule-based controller, the shape of the control surface can be individually manipulated for the different regions of the state space, thus limiting possible effects to neighboring regions only." FLCs have been applied with great success to many applications, where the first FLC was developed in 1974 by Mamdani and Assilian for controlling a steam generator [Mamdani 1975]. In 1976, Blue Circle Cement and SIRA in Denmark developed a cement kiln controller—which is the first industrial application of fuzzy logic. The system went to operation in 1982 [Holmblad 1982]. In the 1980s, several important industrial applications of fuzzy logic were launched successfully in Japan, where Hitachi put a fuzzy logic based automatic train control system into operation in Sendai city's subway system in 1987 [Yasunobu 1985]. Another early successful industrial application of fuzzy logic is a water-treatment system developed by Fuji Electric [Yen 1999]. These and other applications motivated many Japanese engineers to investigate a wide range of novel fuzzy logic applications. This led to the fuzzy boom in Japan which was a result of close collaboration and technology transfer between universities and industries where large-scale national research initiatives (like the Laboratory for International Fuzzy Engineering Research [LIFE]) were established by Japanese government agencies [Yen 1999]. In late January 1990, Matsushita Electric Industrial Co. introduced their newly developed fuzzy controlled automatic washing machine and launched a major commercial campaign for the "fuzzy" product. This campaign turns out to be a successful marketing effort not only for the product, but also for the fuzzy logic technology [Yen 1999]. Many other home electronics companies followed Matsushita's approach and introduced fuzzy vacuum cleaners, fuzzy rice cookers, fuzzy refrigerators, fuzzy camcorders, and others products. As a result, the consumers in Japan recognized the Japanese word "fuzzy," which won the gold prize for the new word in 1990 [Hirota 1995]. This fuzzy boom in Japan triggered a broad and serious interest in this technology in Korea, Europe, and the United States. Boeing, NASA, United Technologies, and other aerospace companies have developed FLCs for space and aviation applications [Munakata 1994]. Other control applications include control of alternating current induction motors, engine spark advance control, control of autonomous robots, and many other applications [Yen 1999]. Following this, the recent years have witnessed a wide-scale deployment of FLCs to numerous successful applications [Langari 1995, Yen 1999].

However, there are many sources of uncertainty facing the FLC in dynamic real-world unstructured environments and many real-world applications; some of the uncertainty sources are as follows:

- Uncertainties in inputs to the FLC, which translate into uncertainties in the antecedents' membership functions (MFs) as the sensors measurements are affected by high noise levels from various sources. In addition, the input sensors can be affected by the conditions of observation (i.e., their characteristics can be changed by the environmental conditions such as wind, sunshine, humidity, rain, etc.).
- Uncertainties in control outputs, which translate into uncertainties in the consequents' MFs of the FLC. Such uncertainties can result from the change of the actuators' characteristics, which can be due to wear, tear, environmental changes, etc.
- Linguistic uncertainties as the meaning of words that are used in the antecedents' and consequents' linguistic labels can be uncertain, as words mean different things to different people [Mendel 2001]. In addition, experts do not always agree and they often provide different consequents for the same antecedents. A survey of experts will usually lead to a histogram of possibilities for the consequent of a rule; this histogram represents the uncertainty about the consequent of a rule [Mendel 2001].
- Uncertainties associated with the change in the operation conditions of the controller. Such uncertainties can translate into uncertainties in the antecedents' and/or consequents' MFs.
- Uncertainties associated with the use of noisy training data that could be used to learn, tune, or optimize the FLC.

All of these uncertainties translate into uncertainties about fuzzy-set MFs [Mendel 2001]. The vast majority of the FLCs that have been used to date were based on the traditional type-1 FLCs. However, type-1 FLCs cannot fully handle or accommodate the linguistic and numerical uncertainties associated with dynamic unstructured environments, as they use type-1 fuzzy sets. Type-1 fuzzy sets handle the uncertainties associated with the FLC inputs and outputs by using *precise and crisp* MFs that the user believes capture the uncertainties. Once the type-1 MFs have been chosen, all the uncertainty disappears because type-1 MFs are precise [Mendel 2001]. The linguistic and numerical uncertainties associated with dynamic unstructured environments cause problems in determining the exact and precise antecedents' and consequents' MFs during the FLC design. Moreover, the designed type-1 fuzzy sets can be suboptimal under specific environment and operation conditions; however, because of the environment changes and the associated uncertainties, the chosen type-1 fuzzy sets might not be appropriate anymore. This can cause degradation in the FLC performance, which can result in poor control and inefficiency and we might end up wasting time in frequently redesigning or tuning the type-1 FLC so that it can deal with the various uncertainties.

A type-2 fuzzy set is characterized by a fuzzy MF, i.e., the membership value (or membership grade) for each element of this set is a fuzzy set in [0,1], unlike a type-1 fuzzy set where the membership grade is a crisp number in [0,1] [1]. The MFs of type-2 fuzzy sets are three dimensional (3D) and include a footprint of uncertainty. It is the new third dimension of type-2 fuzzy sets and the footprint of uncertainty that provide additional degrees of freedom that make it possible to directly model and handle uncertainties [Mendel 2001]. The type-2 fuzzy sets are useful where it is difficult to determine the exact and precise MFs. Type-2 FLCs that use type-2 fuzzy sets have been used to date with great success where the type-2 FLCs have outperformed their type-1 counterparts in several applications where there is high level of uncertainty [Hagras 2007b].

In the next section, we will introduce the type-2 fuzzy sets and their associated terminologies. Section 21.3 introduces briefly the interval type-2 FLC and its various components. Section 21.4 provides a practical example to clarify the various operations of the type-2 FLC. Finally, conclusions and future directions are presented in Section 21.4.

21.2 Type-2 Fuzzy Sets

Type-1 FLCs employ crisp and precise type-1 fuzzy sets. For example, consider a type-1 fuzzy set representing the linguistic label of "Low" temperature in Figure 21.1a: if the input temperature x is 15°C, then the membership of this input to the "Low" set will be the certain and crisp membership value of 0.4. However, the center and endpoints of this type-1 fuzzy set can vary due to uncertainties (which could arise, e.g., from noise) in the measurement of temperature (numerical uncertainty) and in the situations in which 15°C could be called low (linguistic uncertainty) (in the Arctic 15°C might be considered "High," while in the Caribbean it would be considered "Low"). If this linguistic label was employed with a fuzzy logic controller, then the type-1 FLC would need to be frequently tuned to handle such uncertainties. Alternatively, one would need to have a group of separate type-1 sets and type-1 FLCs where each FLC will handle a certain situation.

On the other hand, a type-2 fuzzy set is characterized by a fuzzy membership function (MF), i.e., the membership value (or membership grade) for each element of this set is itself a fuzzy set in [0,1]. For example, if the linguistic label of "Low" temperature is represented by a type-2 fuzzy set as shown in Figure 21.1b, then the input x of 15°C will no longer have a single value for the MF. Instead, the MF takes on values wherever the vertical line intersects the area shaded in gray. Hence, 15°C will have *primary membership* values that lie in the interval [0.2, 0.6]. Each point of this interval will have also a weight associated with it. Consequently, this will create an amplitude distribution in the third dimension to form what is called a *secondary MF*, which can be a triangle as shown in Figure 21.1c. In case the secondary MF is equal to 1 for all the points in the primary membership and if this is true for $\forall x \in X$, we have

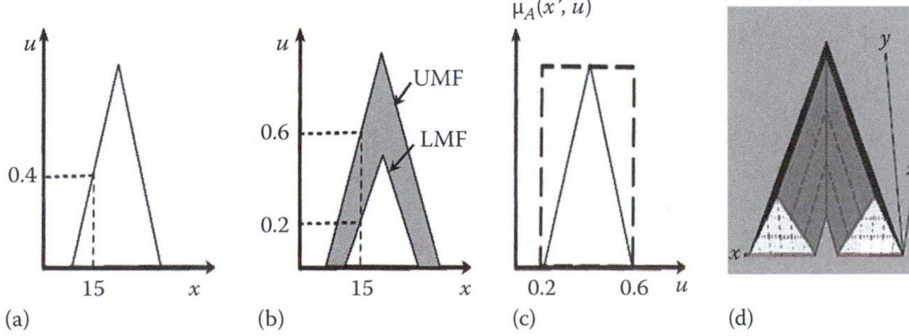

FIGURE 21.1 (a) A type-1 fuzzy set. (b) A type-2 fuzzy set—primary MF. (c) An interval type-2 fuzzy set secondary MF (drawn with dotted lines) and a general type-2 MF (solid line) at a specific point x'. (d) 3D view of a general type-2 fuzzy set.

the case of an interval type-2 fuzzy set. The input x of 15°C will now have a primary membership and an associated secondary MF. Repeating this for all $x \in X$, creates a 3D MF (as shown in Figure 21.1d)—a type-2 MF—that characterizes a type-2 fuzzy set. The MFs of type-2 fuzzy sets are 3D and include a footprint of uncertainty (FOU) (shaded in gray in Figure 21.1b). It is the new third-dimension of type-2 fuzzy sets and the FOU that provide additional degrees of freedom and that make it possible to directly model and handle the numerical uncertainties and linguistic uncertainties.

21.2.1 Type-2 Fuzzy Set Terminologies and Operations

A type-2 fuzzy set $\tilde{A}$ is characterized by a type-2 MF $\mu_{\tilde{A}}(x,u)$ [Mendel 2001] where $x \in X$ and $u \in J_x \subseteq [0,1]$, i.e.,

$$\tilde{A} = \left\{((x,u),\ \mu_{\tilde{A}}(x,u)) \mid \forall x \in X,\ \forall u \in J_x \subseteq [0,1]\right\} \tag{21.1}$$

in which $0 \le \mu_{\tilde{A}}(x,u) \le 1$. $\tilde{A}$ can also be expressed as follows [Mendel 2001]:

$$\tilde{A} = \int_{x\in X}\int_{u\in J_x} \mu_{\tilde{A}}(x,u)/(x,u) \quad J_x \subseteq [0,1] \tag{21.2}$$

where $\int\int$ denotes union over all admissible x and u. For discrete universes of discourse $\int$ is replaced by Σ [Mendel 2001].

At each value of x say $x = x'$, the two-dimensional plane whose axes are u and $\mu_{\tilde{A}}(x',u)$ is called a vertical slice of $\mu_{\tilde{A}}(x,u)$. A *secondary MF* is a vertical slice of $\mu_{\tilde{A}}(x,u)$. It is $\mu_{\tilde{A}}(x = x,u)$ for $x \in X$ and $\forall u \in J_{x'} \in [0,1]$ [Mendel 2001], i.e.,

$$\mu_{\tilde{A}}(x = x',u) \equiv \mu_{\tilde{A}}(x') = \int_{u\in J_{x'}} f_{x'}(u)/(u) \quad J_{x'} \subseteq [0,1] \tag{21.3}$$

in which $0 \le f_{x'}(u) \le 1$. Because $\forall x' \in X$, the prime notation on $\mu_{\tilde{A}}(x')$ is dropped and we refer to $\mu_{\tilde{A}}(x)$ as a secondary MF [Mendel 2002a]; it is a type-1 fuzzy set which is also referred to as a secondary set. Many choices are possible for the secondary MFs. According to Mendel [Mendel 2001], the name that we use to describe the entire type-2 MF is associated with the name of the secondary MFs; so, for example, if the secondary MF is triangular then we refer to $\mu_{\tilde{A}}(x,u)$ as a triangular type-2 MF. Figure 21.1c shows a triangular secondary MF at x' which is drawn using the thick line. Based on the concept of secondary sets, type-2 fuzzy sets can be written as the union of all secondary sets [Mendel 2001].

The domain of a secondary MF is called *primary membership* of x [Mendel 2001]. In Equation 21.1, J_x is the primary membership of x, where $J_x \subseteq [0,1]$ for $\forall x \in X$ [Mendel 2001].

When $f_x(u) = 1$, $\forall u \in J_x \subseteq [0,1]$, then the secondary MFs are interval sets, and, if this is true for $\forall x \in X$, we have the case of an *interval type-2 MF* [Mendel 2001]. Interval secondary MFs reflect a uniform uncertainty at the primary memberships of x. Figure 21.1c shows the secondary membership at x' (drawn in dotted lines in Figure 21.1c) in case of interval type-2 fuzzy sets.

21.2.1.1 Footprint of Uncertainty

The uncertainty in the primary memberships of a type-2 fuzzy set $\tilde{A}$, consists of a bounded region that is called the *footprint of uncertainty* (FOU) [Mendel 2002a]. It is the union of all primary memberships [Mendel 2002a], i.e.,

$$\text{FOU}(\tilde{A}) = \bigcup_{x\in X} J_x \tag{21.4}$$

The shaded region in Figure 21.1b is the FOU. It is very useful, because according to Mendel and John [Mendel 2002a], it not only focuses our attention on the uncertainties inherent in a specific type-2 MF, whose shape is a direct consequence of the nature of these uncertainties, but it also provides a very convenient verbal description of the entire domain of support for all the secondary grades of a type-2 MF. The shaded FOU implies that there is a distribution that sits on top of it—the new third dimension of type-2 fuzzy sets. What that distribution looks like depends on the specific choice made for the secondary grades. When they all equal one, the resulting type-2 fuzzy sets are called *interval type-2 fuzzy sets*. Establishing an appropriate FOU is analogous to establishing a probability density function (pdf) in a probabilistic uncertainty situation [Mendel 2001]. The larger the FOU, the more uncertainty there is. When the FOU collapses to a curve, then its associated type-2 fuzzy set collapses to a type-1 fuzzy set, in much the same way that a pdf collapses to a point when randomness disappears. Recently, it has been shown that regardless of the choice of the primary MF (triangle, Gaussian, trapezoid), the resulting FOU is about the same [Mendel 2002b]. According to Mendel and Wu [Mendel 2002b], the FOU of a type-2 MF also handles the rich variety of choices that can be made for a type-1 MF, i.e., by using type-2 fuzzy sets instead of type-1 fuzzy sets, the issue of which type-1 MF to choose diminishes in importance.

21.2.1.2 Embedded Fuzzy Sets

For continuous universes of discourse X and U, an embedded type-2 set $\tilde{A}_e$ is defined as follows [Mendel 2001]:

$$\tilde{A}_e = \int_{x\in X} [f_x(u)/u]/x \quad u \in J_x \subseteq U = [0,1] \tag{21.5}$$

Set $\tilde{A}_e$ is embedded in $\tilde{A}$ and there is an uncountable number of embedded type-2 sets is $\tilde{A}$ [Mendel 2002b]. For discrete universes of discourse X and U, *an embedded type-2 set* $\tilde{A}_e$ has N elements, where $\tilde{A}_e$ contains exactly one element from $J_{x_1}, J_{x_2}, \dots J_{x_N}$, namely $u_1, u_2, \dots u_N$, each with its associated secondary grade $f_{x_1}(u_1), f_{x_2}(u_2), \dots f_{x_N}(u_N)$ [Mendel 2001], i.e.,

$$\tilde{A}_e = \sum_{d=1}^{N} [f_{x_d}(u_d)/u_d]/x_d \quad u_d \in J_{x_d} \subseteq U = [0,1] \tag{21.6}$$

Set $\tilde{A}_e$ is embedded in $\tilde{A}$ and there is a total of $\prod_{d=1}^{N} M_d \tilde{A}_e$ [23], where M_d is the discretization levels of u_d^j at each x_d.

For continuous universes of discourse X and U, an embedded type-1 set A_e is defined as follows [Mendel 2002a]

$$A_e \int_{x\in X} u/x \quad u \in J_x \subseteq U = [0,1] \tag{21.7}$$

Set A_e is the union of all the primary memberships of set $\tilde{A}_e$ in Equation 21.5 and there is an uncountable number of A_e.

For discrete universes of discourse X and U *an embedded type-1 set* A_e has N elements, one each from $J_{x_1}, J_{x_2}, \dots J_{x_N}$, namely $u_1, u_2, \dots u_N$, [Mendel 2002b], i.e.,

$$A_e = \sum_{d=1}^{N} u_d/x_d \quad u_d \in J_{x_d} \subseteq U = [0,1] \tag{21.8}$$

There is a total of $\prod_{d=1}^{N} M_d A_e$ [Mendel 2002a].

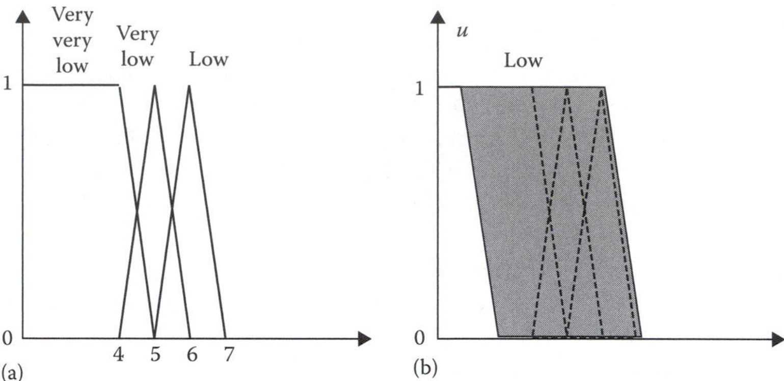

FIGURE 21.2 (a) Three type-1 fuzzy sets representing an input to the FLC. (b) The three type-1 fuzzy sets in Figure 2.2a are embedded in the Low type-2 fuzzy set.

It has proven by Mendel and John [Mendel 2002a] that a type-2 fuzzy set $\tilde{A}$ can be represented as the union of its type-2 embedded sets, i.e.,

$$\tilde{A} = \sum_{l=1}^{n''} \tilde{A}_e^l \quad \text{where } n'' \equiv \prod_{d=1}^{N} M_d \tag{21.9}$$

Figure 21.2a shows three type-1 fuzzy sets (*Very Very Low, Very Low,* and *Low*) used to express in detail the different fuzzy levels of Low for an input to the FLC. In Figure 21.2b notice that the type-1 fuzzy sets for Very Very Low, Very Low, and Low are embedded in the interval type-2 fuzzy set Low, not only this but there is a large number of other embedded type-1 fuzzy sets (uncountable for continuous universes of discourse).

21.2.1.3 Interval Type-2 Fuzzy Sets

In Equation 21.3 when $f_x(u) = 1$, $\forall u \in J_x \subseteq [0,1]$, then the secondary MFs are interval sets, and, if this is true for $\forall x \in X$, we have the case of an *interval type-2 MF*, which characterizes the interval type-2 fuzzy sets. Interval secondary MFs reflect a uniform uncertainty at the primary memberships of x. Interval type-2 sets are very useful when we have no other knowledge about secondary memberships [Liang 2000]. The membership grades of the interval type-2 fuzzy sets are called “interval type-1 fuzzy sets.” Since all the memberships in an interval type-1 set are unity, in the sequel, an interval type-1 set is represented just by its domain interval, which can be represented by its left and right end-points as $[l,r]$ [Liang 2000]. The two end-points are associated with two type-1 MFs that are referred to as *lower* and *upper* MFs $(\underline{\mu}_{\tilde{A}}(x), \bar{\mu}_{\tilde{A}}(x))$ [Liang 2000].

The upper and lower MFs are two type-1 MFs which are bounds for the footprint of uncertainty *FOU* $(\tilde{A})$ of a type-2 fuzzy set $\tilde{A}$.

According to Mendel [Mendel 2001], we can re-express Equation 21.3 as follows to represent the interval type-2 fuzzy set $\tilde{A}$ in terms of upper and lower MFs as follows:

$$\tilde{A} = \int_{x \in X} \left[\int_{u \in [\underline{\mu}_{\tilde{A}}(x), \bar{\mu}_{\tilde{A}}(x)]} 1/u \right] \Big/ x \tag{21.10}$$

For type-2 fuzzy sets, there are new operators named the meet and join to account for the intersection and union, respectively. Liang and Mendel [Liang 2000] had derived the expressions for meet and join in interval type-2 fuzzy sets in which we need to compute the join, meet of secondary MFs which are type-1 interval fuzzy sets.

Let $F = \int_{v \in F} 1/v$ be an interval type-1 set with domain $[l_f, r_f] \subseteq [0,1]$ and $G = \int_{w \in G} 1/w$ is another interval type-1 set with domain $[l_g, r_g] \subseteq [0,1]$.

The *meet* Q between F and G under the product t-norm which is used in our type-2 FLC is written as follows [Linag 2000]:

$$Q = F \sqcap G = \int_{q \in [l_f l_g, r_f r_g]} 1/q \tag{21.11}$$

From Equation 21.11 each term in $F \sqcap G$ is equal to the product of $v.w$ for some $v \in F$ and $w \in G$ in which the smallest term being $l_f l_g$ and the largest is $r_f r_g$. Since both F and G have continuous domains, $F \sqcap G$ has a continuous domain, therefore, $F \sqcap G$ is an interval type-1 set with domain $[l_f l_g, r_f r_g]$ [Liang 2000]. In a similar manner, the meet under product t-norm of n interval type-1 sets $F_1,\ldots, F_n$ having domains $[l_1, r_1], \ldots [l_n, r_n]$, respectively, is an interval set with domain with domain $\left[\prod_{o=1}^{n} l_o, \prod_{o=1}^{n} r_o\right]$. The meet under minimum t-norm is calculated in a similar manner [Liang 2000].

The join between F and G is given by

$$Q = F \sqcup G \int_{q \in [l_f \vee l_g, r_f \vee r_g]} 1/q \tag{21.12}$$

where $q = v \vee w$, where $\vee$ denotes the maximum operation used in our type-2 FLC. The join of n interval type-1 sets $F_1,\ldots, F_n$ having domains $[l_1, r_1], \ldots [l_n, r_n]$, respectively, is an interval set with domain $[(l_1 \vee l_2 \vee \ldots \vee l_n), (r_1 \vee r_2 \vee \ldots \vee r_n)]$, i.e., with domain equal $[\max(l_1, l_2,\ldots, l_n), \max(r_1, r_2, \ldots, r_n)]$ [18].

After reviewing the definition of the type-2 fuzzy sets and their associated terminologies, we can realize that using type-2 fuzzy sets to represent the inputs and outputs of a FLC has many advantages when compared to the type-1 fuzzy sets, we summarize some of these advantages as follows:

- As the type-2 fuzzy sets MFs are fuzzy and contain a FOU, they can model and handle the linguistic and numerical uncertainties associated with the inputs and outputs of the FLC. Therefore, FLCs that are based on type-2 fuzzy sets will have the potential to produce a better performance than the type-1 FLCs when dealing with uncertainties [Hagras 2004].
- Using type-2 fuzzy sets to represent the FLC inputs and outputs will result in the reduction of the FLC rule base when compared to using type-1 fuzzy sets, as the uncertainty represented in the FOU of the type-2 fuzzy sets lets us cover the same range as type-1 fuzzy sets with a smaller number of labels and the rule reduction will be greater when the number of the FLC inputs increases [Mendel 2001, Hagras 2004].
- Each input and output will be represented by a large number of type-1 fuzzy sets, which are embedded in the type-2 fuzzy sets [Mendel 2001, Hagras 2004]. The use of such a large number of type-1 fuzzy sets to describe the input and output variables allows for a detailed description of the analytical control surface as the addition of the extra levels of classification give a much smoother control surface and response. In addition, the type-2 FLC can be thought of as a collection of many different embedded type-1 FLCs [Mendel 2001].
- It has been shown in [Wu 2005] that the extra degrees of freedom provided by the FOU enables a type-2 FLC to produce outputs that cannot be achieved by type-1 FLCs with the same number of MFs. It has been shown that a type-2 fuzzy set may give rise to an equivalent type-1 membership grade that is negative or larger than unity. Thus, a type-2 FLC is able to model more complex input–output relationships than its type-1 counterpart and, thus, can give better control response.

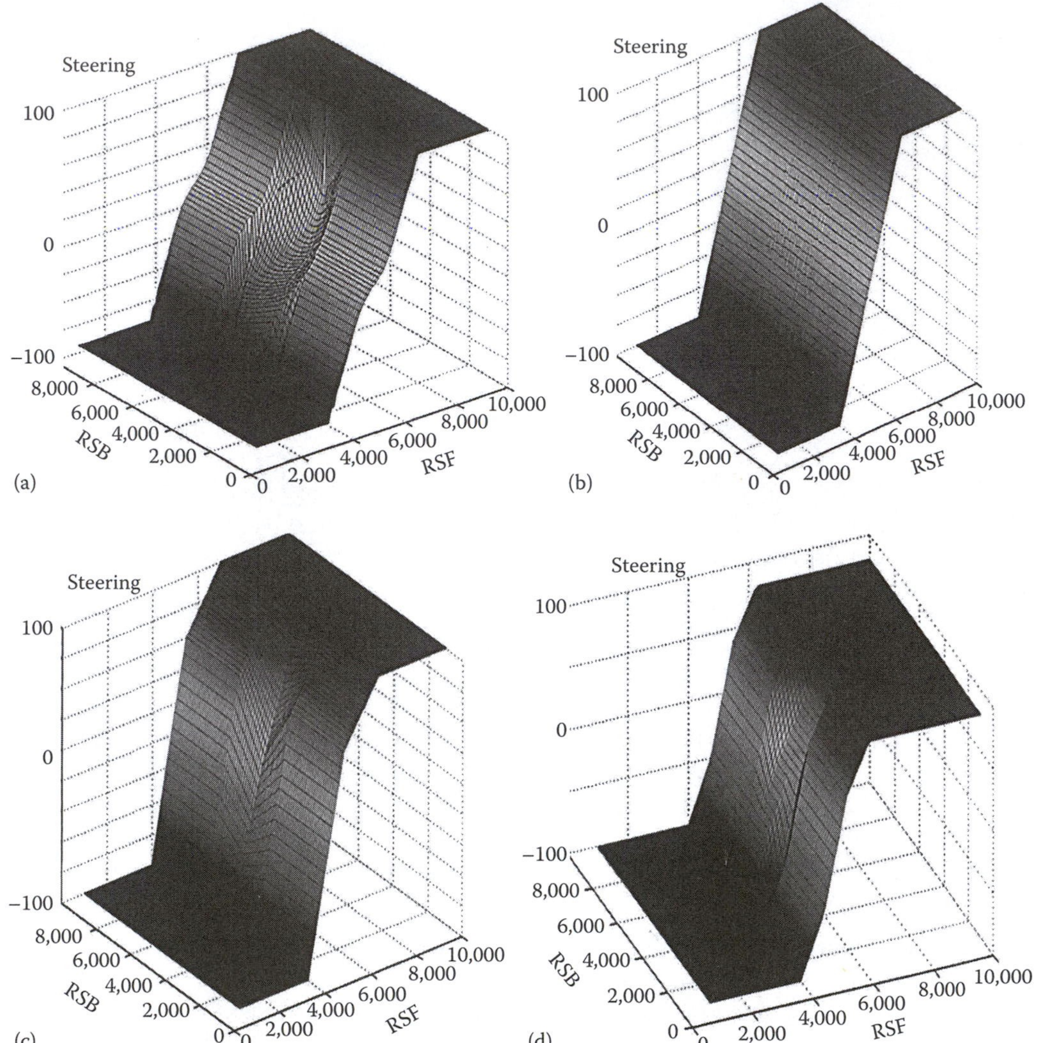

FIGURE 21.3 (a) Control surface of a robot type-2 FLC with 4 rules. (b) Control surface of a robot type-1 FLC with 4 rules. (c) Control surface of a robot type-1 FLC with 9 rules. (d) Control surface of a robot type-1 FLC with 25 rules.

The above points could be shown in Figure 21.3 which shows for an outdoor mobile robot how a type-2 FLC with a rule base of only four rules could produce a smoother control surface as shown in Figure 21.3a and hence better result than its type-1 counterpart that used a rule base of 4, 9, and 25 rules as shown in Figure 21.3b through d, respectively [Hagras 2004]. It is also shown that as the type-1 FLC rule base increases, its response approaches that of the type-2 FLC, which encompasses a huge number of embedded type-1 FLCs.

21.3 Interval Type-2 FLC

The interval type-2 FLC uses interval type-2 fuzzy sets to represent the inputs and/or outputs of the FLC. The interval type-2 FLC is a special case of the general type-2 FLC. The vast majority of type-2 FLC applications to date employ the interval type-2 FLC. This is because the general type-2 FLC is computationally intensive and the computation simplifies a lot when using the interval type-2 FLC, which will enable us to design a FLC that operates in real time.

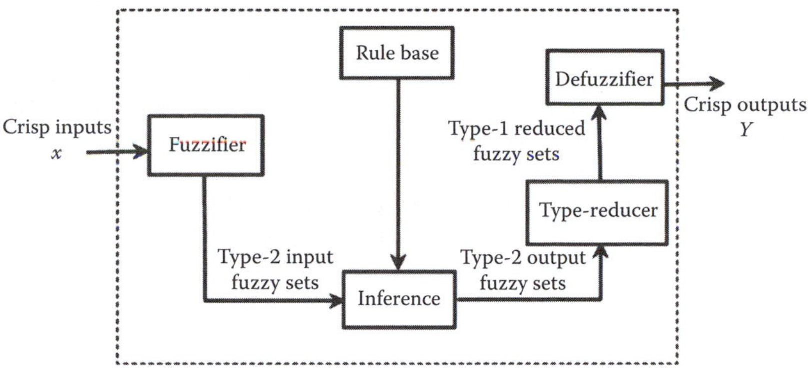

FIGURE 21.4 Type-2 FLC.

The interval type-2 FLC is depicted in Figure 21.4 and it consists of a Fuzzifier, Inference Engine, Rule Base, Type-reducer, and Defuzzifier. The interval type-2 FLC operate as follows: the crisp inputs from the input sensors are first fuzzified into input type-2 fuzzy sets; singleton fuzzification is usually used in interval type-2 FLC applications due to its simplicity and suitability for embedded processors and real-time applications. The input type-2 fuzzy sets then activate the inference engine and the rule base to produce output type-2 fuzzy sets. The type-2 FLC rules will remain the same as in the type-1 FLC, but the antecedents and/or the consequents will be represented by interval type-2 fuzzy sets. The inference engine combines the fired rules and gives a mapping from input type-2 fuzzy sets to output type-2 fuzzy sets. The type-2 fuzzy outputs of the inference engine are then processed by the type-reducer, which combines the output sets and performs a centroid calculation that leads to type-1 fuzzy sets called the type-reduced sets. The type-reduction process uses the iterative Karnik–Mendel (KM) procedure to calculate the type-reduced fuzzy sets [Mendel 2001]. The KM procedure convergence is proportional to the number of fired rules. After the type-reduction process, the type-reduced sets are then defuzzified (by taking the average of the type-reduced set) to obtain crisp outputs that are sent to the actuators. More information about the interval type-2 FLC can be found in [Hagras 2004]. The following sections will give a brief overview of the interval type-2 FLC.

21.3.1 Fuzzifier

The fuzzifier maps a crisp input vector with p inputs $\mathbf{x} = (x_1,\ldots,x_p)^T \in X_1 \times X_2 \ldots \times X_p \equiv \mathbf{X}$ into input fuzzy sets, these fuzzy sets can, in general, be type-2 fuzzy input sets $\tilde{A}_x$ [13], [17]. However, we will use singleton fuzzification as it is fast to compute and thus suitable for the robot real-time operation. In the singleton fuzzification, the input fuzzy set has only a single point of nonzero membership, i.e., $\tilde{A}_x$ is a type-2 fuzzy singleton if $\mu_{\tilde{A}x}(\mathbf{x}) = 1/1$ for $\mathbf{x} = \mathbf{x}'$ and $\mu_{\tilde{A}x}(\mathbf{x}) = 1/0$ for all other $\mathbf{x} \neq \mathbf{x}'$ [Mendel 2001].

21.3.2 Rule Base

The rules will remain the same as in type-1 FLC but the antecedents and the consequents will be represented by interval type-2 fuzzy sets. Consider an interval type-2 FLC having p inputs $x_1 \in X_1,\ldots, x_p \in X_p$ and c outputs $y_1 \in Y_1,\ldots, y_c \in Y_c$. The ith rule in this multi input multi output (MIMO) FLC can be written as follows:

$$R^i_{MIMO}: \text{IF } x_1 \text{ is } \tilde{F}^i_1 \text{ and } \ldots \text{ and } x_p \text{ is } \tilde{F}^i_p \quad \text{THEN } y_1 \text{ is } \tilde{G}^i_1 \ldots y_c \text{ is } \tilde{G}^i_1 \quad i = 1,\ldots M \tag{21.13}$$

where M is the number of rules in the rule base.

21.3.3 Fuzzy Inference Engine

The inference engine combines rules and gives a mapping from input type-2 sets to output type-2 sets. In the inference engine, multiple antecedents in the rules are connected using the *Meet* operation, the membership grades in the input sets are combined with those in the output sets using the extended sup-star composition, multiple rules are combined using the *Join* operation. In our interval type-2 FLC, we will use the *meet* under product *t*-norm so the result of the input and antecedent operations, which are contained in the firing set $\sqcap_{a=1}^{p}\mu_{\tilde{F}_a^i}(x'_a) \equiv F^i(\mathbf{x}')$, is an interval type-1 set, as follows [Mendel 2001]:

$$F^i(x') = \left[\underline{f}^i(\mathbf{x}'), \overline{f}^i(\mathbf{x}')\right] \equiv [\underline{f}^i, \overline{f}^i] \tag{21.14}$$

where $\underline{f}^i(\mathbf{x}')$ and $\overline{f}^i(\mathbf{x}')$ can be written as follows, where $*$ denotes the product or the minimum operations,

$$\underline{f}^i(\mathbf{x}') = \underline{\mu}_{\tilde{F}_1^i}(x'_1) * \ldots * \underline{\mu}_{\tilde{F}_p^i}(x'_p) \tag{21.15}$$

$$\overline{f}^i(\mathbf{x}') = \overline{\mu}_{\tilde{F}_1^i}(x'_1) * \ldots \overline{\mu}_{\tilde{F}_1^i}(x'_p) \tag{21.16}$$

21.3.4 Type Reduction

Type reduction is the operation that takes us from the type-2 output sets of the inference engine to a type-1 set that is called the "the type-reduced set." These type-reduced sets are then defuzzified to obtain crisp outputs that are sent to the outputs of the FLC.

The calculation of the type-reduced sets is divided into two stages; the first stage is the calculation of centroids of the type-2 interval consequent sets of each rule, which is conducted ahead of time and before starting the FLC operation. For each output, we determine the centroids of all the output type-2 interval fuzzy sets representing this output, then the centroid of the type-2 interval consequent set for the *i*th rule will be one of the pre-calculated centroids of type-2 output sets, which corresponds to the rule consequent. To calculate the centroids of the output interval type-2 fuzzy sets, we used the KM iterative procedure as explained in [Mendel 2001].

The second stage of type reduction happens each control cycle to calculate the type-reduced sets. For any output k in order to compute the type-reduced set, we need to compute its two end points y_{lk} and y_{rk} which can be found using the KM iterative procedures [Mendel 2001, Liang 2000].

21.3.5 Defuzzification

From the type-reduction stage, we have for each output a type-reduced set $Y_{\cos}(\mathbf{x})_k$ determined by its left-most point y_{lk} and right-most point y_{rk}. We defuzzify the interval set by using the average of y_{lk} and y_{rk} hence the defuzzified crisp output for each output k is

$$Y_k(\mathbf{x}) = \frac{y_{lk} + y_{rk}}{2} \tag{21.17}$$

21.4 Illustrative Example to Summarize the Operation of the Type-2 FLC

In this section, we summarize the operation of the type-2 FLC through an example of a type-2 FLC that realizes the right-edge following behavior for an outdoor robot. The objective of this behavior is to follow an edge to the right of the robot at a desired distance. This type-2 FLC will have two inputs

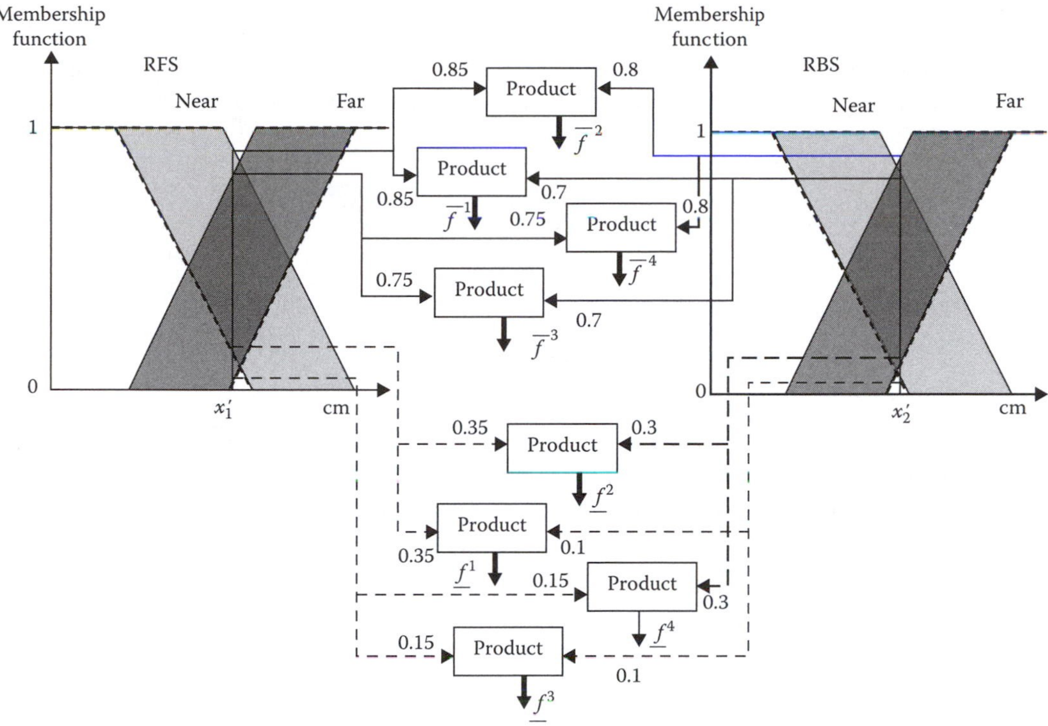

FIGURE 21.5 Pictorial description of the input and antecedent operations for a robot type-2 FLC.

from the two right-side sonar sensors, the first input is from the right-side front sensor (RSF) and the second input is from the right-side back sensor (RSB). The RSF crisp input will be denoted by x_1 and the RSB crisp input will be denoted by x_2. The type-2 FLC controls two outputs, which are the robot speed denoted by y_1 and the robot steering denoted by y_2. Each input will be represented only by two type-2 fuzzy sets, which are *Near* and *Far* as shown in Figure 21.5. The output robot speed will be represented by three type-2 fuzzy sets, which are *Slow*, *Medium*, and *Fast* while the robot steering will be represented by two type-2 fuzzy sets, which are *Left* and *Right*. In what follows, we will follow a crisp input vector through the various components of the type-2 FLC until we get crisp output signals to the robot actuators. The crisp input vector will consist of two crisp inputs, the first one represents the reading of RSF which we term x_1' and the second input represents the reading of RSB which we term x_2', the type-2 FLC crisp outputs corresponding to x_1' and x_2' are y_1' for the robot speed and y_2' for the robot steering.

21.4.1 Fuzzification

Figure 21.5 shows a pictorial description of the input and antecedent operations in Equations 21.14 through 21.16 in our type-2 FLC. In the fuzzification stage, as we are using singleton fuzzification, each input is matched against its MFs to calculate the upper and lower membership values for each fuzzy set. The input x_1' of the RSF is matched against its MF in Figure 21.5a and it was found that the lower membership value for the *Near* type-2 fuzzy set is 0.35 while the upper membership value is 0.85. For the *Far* fuzzy set, the lower membership value is 0.15 while the upper membership value is 0.75. The input x_2' of the RSB is matched against its MF in Figure 21.5b and it was found that the lower membership value for the *Near* fuzzy set is 0.1 and the upper membership value is 0.7. For the *Far* fuzzy set, the lower membership value is 0.3 while upper membership value is 0.8.

21.4.2 Rule Base

The rule base for this type-2 FLC is shown in Table 21.1, where p (the number of inputs) is 2 and c (the number of outputs) is 2. Any MIMO rule from Table 21.1 can be written according to Equation 21.13, for example rule (1) can be written as follows:

$$R^1_{MIMO}\text{: IF } x_1 \text{ is } \tilde{F}^1_1 \text{ and } x_2 \text{ is THEN } y_1 \text{ is } \tilde{G}^1_1, y_2 \text{ is } \tilde{G}^1_2$$

where
- $\tilde{F}^1_1$ is the *Near* type-2 fuzzy set for RSF
- $\tilde{F}^1_2$ is the *Near* type-2 fuzzy set for RSB
- $\tilde{G}^1_1$ is the *Slow* output type-2 fuzzy set for the robot speed
- $\tilde{G}^1_2$ is the *Left* output type-2 fuzzy set for the robot steering

In the fuzzy inference engine, we need to calculate the firing strength of each rule. According to Equation 21.14, the firing strength of each rule is an interval type-1 set $[\underline{f}^i, \overline{f}^i]$ where $\underline{f}^i$ is calculated according to Equation 21.15 and $\overline{f}^i$ is calculated according to Equation 21.16. Note that in our FLC, we use the meet under the product t-norm. So for rule (1), we can calculate $\underline{f}^1$ and $\overline{f}^1$ as follows:

$$\underline{f}^1 = \underline{\mu}_{\tilde{F}^1_1}(x'_1).\underline{\mu}_{\tilde{F}^1_2}(x'_2) == 0.35 * 0.1 = 0.035$$

where $\underline{\mu}_{\tilde{F}^1_1}(x'_1)$ is the lower membership value for x'_1 for the *Near* type-2 fuzzy set for RSF, which is 0.35 and $\underline{\mu}_{\tilde{F}^1_2}(x'_2)$ is the lower membership value for x'_2 for the *Near* type-2 fuzzy set for RSB which is 0.1, these membership values were calculated before in the fuzzification stage.

In a similar manner by using the upper membership values, we can calculate $\overline{f}^1$ as follows:

$$\overline{f}^1 = \overline{\mu}_{\tilde{F}^1_1}(x'_1).\overline{\mu}_{\tilde{F}^1_2}(x'_2) = 0.85 * 0.7 = 0.595.$$

In a similar manner, we can calculate the rest of the firing strengths for all the rules as follows:

$$\underline{f}^2 = \underline{\mu}_{\tilde{F}^2_1}(x'_1)\cdot\underline{\mu}_{\tilde{F}^2_2}(x'_2) = 0.35 * 0.3 = 0.105, \quad \overline{f}^2 = \overline{\mu}_{\tilde{F}^2_1}(x'_1)\cdot\overline{\mu}_{\tilde{F}^2_2}(x'_2) = 0.85 * 0.8 = 0.68$$

$$\underline{f}^3 = \underline{\mu}_{\tilde{F}^3_1}(x'_1)\cdot\underline{\mu}_{\tilde{F}^3_2}(x'_2) = 0.15 * 0.1 = 0.015, \quad \overline{f}^3 = \overline{\mu}_{\tilde{F}^3_1}(x'_1)\cdot\overline{\mu}_{\tilde{F}^3_2}(x'_2) = 0.75 * 0.7 = 0.525$$

$$\underline{f}^4 = \underline{\mu}_{\tilde{F}^4_1}(x'_1)\cdot\underline{\mu}_{\tilde{F}^4_2}(x'_2) = 0.15 * 0.3 = 0.045, \quad \overline{f}^4 = \overline{\mu}_{\tilde{F}^4_1}(x'_1)\cdot\overline{\mu}_{\tilde{F}^4_2}(x'_2) = 0.75 * 0.8 = 0.6$$

TABLE 21.1 Example Rule Base of a Right-Edge Following Behavior Implemented by a Type-2 FLC

Rule Number	RSF	RSB	Speed	Steering
1	Near	Near	Slow	Left
2	Near	Far	Slow	Left
3	Far	Near	Medium	Right
4	Far	Far	Fast	Right

21.4.3 Type Reduction

21.4.3.1 Calculating the Centroids of the Rule Consequents

In this stage, we need to calculate for each output the centroids of all the output type-2 fuzzy sets, so that we can calculate the centroid of the consequent of each rule, which will be one of the centroids of the output fuzzy sets that corresponds to the rule consequent. To calculate these centroids, we will use the iterative KM procedure explained in [Mendel 2001, Hagras 2004] using 100 sampling points. For each output, we calculate the centroids of all the type-2 fuzzy sets $y_k^t t = 1,\ldots T$. For the speed output, the number of output fuzzy sets $T = 3$. Assume for illustrative purposes that the centroid of the *Slow* output fuzzy set is [0.43, 0.55], the centroid of the *Medium* output fuzzy set is [0.63, 0.76] and the centroid of the *High* output fuzzy set is [1.03, 1.58]. Next, we can determine the centroids of the rule consequents of the output speed y_1^t as follows:

$$y_1^1 = [y_{l1}^1, y_{r1}^1] = y_1^2 = [y_{l1}^2, y_{r1}^2] = [0.43, 0.55]$$

$$y_1^3 = [y_{l1}^3, y_{r1}^3] = [0.63, 0.76], \quad y_1^4 = [y_{l1}^4, y_{r1}^4] = [1.03, 1.58]$$

For the output steering, the number of outputs fuzzy sets $T = 2$. The steering values are in percentage where right steering values are positive values and the left steering values are negative. Again for illustrative purposes assume that the centroid for the *Left* output fuzzy set is [56.8, 85.4] and the centroid for *Right* output fuzzy set is [−85.4, −56.8]. Next, we can determine the centroids of the rule consequents y_2^t as follows:

$$y_2^1 = [y_{l2}^1, y_{r2}^1] = y_2^2 = [y_{l2}^2, y_{r2}^2] = [-85.4, -56.8], y_2^3 = [y_{l2}^3, y_{r2}^3] = y_2^4 = [_{l2}^4, y_{r2}^4] = [56.8, 85.4]$$

21.4.3.2 Calculating the Type-Reduced Set

For each output k to compute the type-reduced $Y_{\cos}(\mathbf{x})_k$, we need to compute its two end points y_{lk} and y_{rk}. Using the iterative KM procedure for type reduction explained in [Mendel 2001, Liang 2000], we can determine switching point L and R needed to calculate the type-reduced sets. For the speed output, we do not need to reorder y_{l1}^i as they are already ordered in ascending order where $y_{l1}^1 \leq y_{l1}^2 \leq y_{l1}^3 \leq y_{l1}^4$ the same applies for y_{r1}^i. By using the iterative procedure in Figure 21.5, it was found that $L = 2$ so y_{l1} can be as follows:

$$y_{l1} = \frac{\sum_{u=1}^{2} \overline{f}^u y_{l1}^u + \sum_{v=3}^{4} \underline{f}^v y_{l1}^v}{\sum_{u=1}^{2} \underline{f}^u + \sum_{v31}^{4} \overline{f}^v} = \frac{\overline{f}^1 y_{l1}^1 + \overline{f}^2 y_{l1}^2 + \underline{f}^3 y_{l1}^3 + \underline{f}^4 y_{l1}^4}{\overline{f}^1 + \overline{f}^2 + \underline{f}^3 + \underline{f}^4}$$

$$= \frac{0.595 \times 0.43 + 0.68 \times 0.43 + 0.015 \times 0.63 + 0.045 \times 1.03}{0.595 + 0.68 + 0.015 + 0.045}$$

$$= 0.452$$

To calculate y_{r1}, we use the iterative KM procedure and it was found that $R = 3$, so y_{r1} can be found by substituting in the following equation:

$$y_{r1} = \frac{\sum_{u=1}^{3} \underline{f}^u y_{r1}^u + \sum_{v=4}^{4} \overline{f}^v y_{r1}^v}{\sum_{u=1}^{3} \underline{f}^u + \sum_{v=4}^{4} \overline{f}^v} = \frac{\underline{f}^1 y_{r1}^1 + \underline{f}^2 y_{r1}^2 + \underline{f}^3 y_{r1}^3 + \overline{f}^4 y_{r1}^4}{\underline{f}^1 + \underline{f}^2 + \underline{f}^3 + \overline{f}^4}$$

$$= \frac{0.035 \times 0.55 + 0.105 \times 0.55 + 0.015 \times 0.76 + 0.6 \times 1.58}{0.035 + 0.105 + 0.015 + 0.6}$$

$$= 1.37$$

For the steering output, we do not need to reorder y_{l2}^i as they are already ordered in ascending order where $y_{l2}^1 \le y_{l2}^2 \le y_{l2}^3 \le y_{l2}^4$, the same applies for y_{r2}^i. By using the iterative procedure, it was found that $L = 2$, so y_{l2} can be found as follows:

$$y_{l2} = \frac{\sum_{u=1}^{2} \overline{f}^u y_{l2}^u + \sum_{v=3}^{4} \underline{f}^v y_{l2}^v}{\sum_{u=1}^{2} \underline{f}^e + \sum_{v=3}^{4} \overline{f}^v} = \frac{\overline{f}^1 y_{l2}^1 + \overline{f}^2 y_{l2}^2 + \underline{f}^3 y_{l2}^3 + \underline{f}^4 y_{l2}^4}{\overline{f}^1 + \overline{f}^2 + \underline{f}^3 + \underline{f}^4}$$

$$= \frac{0.595 \times -85.4 + 0.68 \times -85.4 + 0.015 \times 56.8 + 0.045 \times 56.8}{0.595 + 0.68 + 0.015 + 0.045}$$

$$= -79.01$$

To calculate y_{r2}, we use the KM procedure and it was found that $R = 2$, so y_{r2} can be found by as follows:

$$y_{r2} = \frac{\sum_{u=1}^{2} \underline{f}^u y_{r2}^u + \sum_{v=3}^{4} \overline{f}^v y_{r2}^v}{\sum_{u=1}^{2} \underline{f}^u + \sum_{v=3}^{4} \overline{f}^v} = \frac{\underline{f}^1 y_{r2}^1 + \underline{f}^2 y_{r2}^2 + \overline{f}^3 y_{r2}^3 + \overline{f}^4 y_{r2}^4}{\underline{f}^1 + \underline{f}^2 + \overline{f}^3 + \overline{f}^4}$$

$$= \frac{0.035 \times -56.8 + 0.105 \times -56.8 + 0.525 \times 85.4 + 0.6 \times 85.4}{0.035 + 0.105 + 0.525 + 0.6}$$

$$= 69.66$$

21.5 Defuzzification

From the type-reduction stage, we have for each output k a type-reduced set; we defuzzify the interval set by calculating the average of y_{lk} *and* y_{rk} using Equation 21.17 for both outputs as follows:

$$\text{The speed crisp output } y_1 = \frac{y_{l1} + y_{r1}}{2} = \frac{0.452 + 1.37}{2} = 0.911 \text{ m/s}$$

$$\text{And the steering crisp output } y_2' = \frac{y_{l2} + y_{r2}}{2} = \frac{-79.01 + 69.66}{2} = -4.675\%$$

21.6 Conclusions and Future Directions

In this chapter, we presented an introduction and brief overview to the interval type-2 FLC and we highlighted its benefits, especially in highly uncertain environments. There have been recent work to avoid the computational overheads of the interval type-2 FLC and thus speeding its response to achieve satisfactory real-time performance. More information about these techniques could be found in [Hagras 2008].

It has been shown in various applications that as the level of imprecision and uncertainty increases, the type-2 FLC will provide a powerful paradigm to handle the high level of uncertainties present in real-world environments [Hagras 2004, 2007a, 2008; Lynch 2006; Melin 2003; Shu 2005; Wu 2004; Figueroa 2005]. It has been also shown in various applications that the type-2 FLCs have given very good and smooth responses that have always outperformed their type-1 counterparts. Thus, using a type-2 FLC in real-world applications can be a better choice than type-1 FLCs since the amount of uncertainty in real systems most of the time is difficult to estimate.

Current research has started to explore the general type-2 FLC. Recent research is looking at generating general type-2 FLCs that embed a group of interval type-2 FLCs. This will enable building on the existing theory of interval type-2 FLC while exploring the power of general type-2 FLCs.

Thus, with the latest developments in interval type-2 FLCs, we can see that type-2 FLC overcomes the limitations of type-1 FLCs and will present a way forward to fuzzy control and especially in highly uncertain environments, which includes most of the real-world applications. Hence, it is envisaged to see a wide spread of type-2 FLCs in many real-world application in the next decade.

References

[Figueroa 2005] J. Figueroa, J. Posada, J. Soriano, M. Melgarejo, and S. Roj, A type-2 fuzzy logic controller for tracking mobile objects in the context of robotic soccer games, *Proceedings of the 2005 IEEE International Conference on Fuzzy Systems*, Reno, NV, pp. 359–364, May 2005.

[Hagras 2004] H. Hagras, A hierarchical type-2 fuzzy logic control architecture for autonomous mobile robots. *IEEE Transactions on Fuzzy Systems*, 12, 524–539, April 2004.

[Hagras 2007a] H. Hagras, Type-2 FLCs: A new generation of fuzzy controllers, *IEEE Computational Intelligence Magazine*, 2(1), 30–43, February 2007.

[Hagras 2007b] H. Hagras, F. Doctor, A. Lopez, and V. Callaghan, An incremental adaptive life long learning approach for type-2 fuzzy embedded agents in ambient intelligent environments, *IEEE Transactions on Fuzzy Systems*, 15(1), 41–55, February 2007.

[Hagras 2008] H. Hagras, Type-2 fuzzy logic controllers: A way forward for fuzzy systems in real world environments, In *Computational Intelligence Research Frontiers*, J. Zurada, G. Yen, and J. Wang (eds.), Berlin/Heidelberg, Germany: Springer, pp. 181–200, June 2008.

[Hirota 1995] K. Hirota, History of industrial applications of fuzzy logic in Japan, In *Industrial Applications of Fuzzy Logic and Intelligent Systems*, J. Yen, R. Langari, and L.A. Zadeh (eds.), Piscataway, NJ: IEEE Press, pp. 43–54.

[Holmblad 1982] L. Holmblad and I. Ostergaard, Control of a cement Kiln by fuzzy logic. In *Fuzzy Information and Decision-Processes*, M. Gupta and E. Sanchez (eds.), Amsterdam, the Netherlands: North-Holland, pp. 389–399, 1982.

[Langari 1995] R. Langari and L.A. Zadeh, *Industrial Applications of Fuzzy Logic and Intelligent Systems*, Piscataway, NJ: IEEE Press, pp. 43–54, 1995.

[Liang 2000] Q. Liang and J. Mendel, Interval type-2 fuzzy logic systems: Theory and design, *IEEE Transactions on Fuzzy Systems*, 8, 535–550, October 2000.

[Lynch 2006] C. Lynch, H. Hagras, and V. Callaghan, Using uncertainty bounds in the design of an embedded real-time type-2 neuro-fuzzy speed Controller for marine diesel engines, *Proceeding of the 2006 IEEE International Conference on Fuzzy Systems*, Vancouver, Canada, pp. 7217–7224, July 2006.

[Mamdani 1975] E. Mamdani and S. Assilian, An experiment in linguistic synthesis with a fuzzy logic controller, *International Journal of Machine Studies*, 7(1), 1–13, 1975.

[Mamdani 1994] E. Mamdani, Fuzzy control—A misconception of theory and application, *IEEE Expert*, 9(4), 27–28, 1994.

[Melin 2003] P. Melin and O. Castillo, Fuzzy logic for plant monitoring and diagnostics, *Proceedings of the 2003 NAFIPS International Conference*, New York, pp. 20–25, July 2003.

[Mendel 2001] J. Mendel, *Uncertain Rule-Based Fuzzy Logic Systems: Introduction and New Directions*, Upper Saddle River, NJ: Prentice-Hall, 2001.

[Mendel 2002a] J. Mendel and R. John, Type-2 fuzzy sets made simple, *IEEE Transactions on Fuzzy Systems*, 10, 117–127, April 2002.

[Mendel 2002b] J. Mendel and H. Wu, Uncertainty versus choice in rule-based fuzzy logic systems, *Proceedings of IEEE International Conference on Fuzzy Systems*, Honolulu, HI, pp. 1336–1342, 2002.

[Munakata 1994] T. Munakata and Y. Jani, Fuzzy systems: An overview, *Communications of the ACM*, 37(3), 69–96, March 1994.

[Shu 2005] H. Shu and Q. Liang, Wireless sensor network lifetime analysis using interval type-2 fuzzy logic systems, *Proceedings of the 2005 IEEE International Conference on Fuzzy Systems*, Reno, NV, pp. 19–24, May 2005.

[Wu 2004] D. Wu and W. Tan, Type-2 fuzzy logic controller for the liquid-level process, *Proceedings of the 2004 IEEE Interntional Conference on Fuzzy Systems*, Budapest, Hungary, pp. 248–253, July 2004.

[Wu 2005] D. Wu and W. Tan, Type-2 FLS modeling capability analysis, *Proceedings of the 2005 IEEE International Conference on Fuzzy Systems*, Reno, NV, pp. 242–247, May 2005.

[Yasunobu 1985] S. Yasunobu and S. Miyamoto, Automatic train operation by fuzzy predictive control, In *Industrial Applications of Fuzzy Control*, M. Sugeno (ed.), Amsterdam, the Netherlands: North Holland, 1985.

[Yen 1999] J. Yen, Fuzzy logic—A modern perspective, *IEEE Transactions on Knowledge and Data Engineering*, 11(1), 153–165, January/February 1999.

22

Fuzzy Pattern Recognition

Witold Pedrycz
University of Alberta

and

Polish Academy of Sciences

22.1 Introduction

Fuzzy pattern recognition has emerged almost at the time fuzzy sets came into existence. One of the first papers authored by Bellman–Kalaba–Zadeh [BKZ66] has succinctly highlighted the key aspects of the technology of fuzzy sets being cast in the setting of pattern recognition. Since then we have witnessed a great deal of developments with a number of comprehensive review studies and books [P90,K82]. The recent years saw a plethora of studies in all areas of pattern recognition including the methodology, algorithms, and case studies [T73,R78,D82,LMT02].

Our key objective is to discuss the main aspects of the conceptual framework and algorithmic underpinnings of fuzzy pattern recognition. The key paradigm of pattern recognition becomes substantially augmented by the principles of fuzzy sets. There are new developments that go far beyond the traditional techniques of pattern recognition and bring forward novel concepts and architectures that have not been contemplated so far.

It is assumed that the reader is familiar with the basic ideas of pattern recognition and fuzzy sets; one may consult a number of authoritative references in these areas [DHS01,F90,PG07,H99,B81]. As a matter of fact, one can position the study in a more comprehensive setting of Granular Computing [BP03a,PB02] in which fuzzy sets are just instances of information granules. A number of results in fuzzy pattern recognition can be extended to the granular pattern recognition; throughout the text we will be making some pertinent observations with this regard.

Given the breadth of the area of fuzzy pattern recognition, it is impossible to cover all of its essentials. This study serves as an introduction to the area with several main objectives: We intend to demonstrate the main aspects of the technology of fuzzy sets (or Granular Computing) in the context of the paradigm of pattern recognition. With this regard, it is of interest to revisit the fundamentals and see how fuzzy sets enhance them at the conceptual, methodological, and algorithmic level and what area of applications could benefit the most from the incorporation of the technology of fuzzy sets.

The presentation follows a top-down approach. We start with some methodological aspects of fuzzy sets that are of paramount relevance in the context of pattern recognition. Section 22.3 is devoted to information granulation, information granules, and Granular Computing where we elaborate on the concept of abstraction and its role in information processing. In the sequel, Section 22.4 is focused on supervised learning with fuzzy sets by showing how several main categories of classifiers are constructed by taking into consideration granular information. Unsupervised learning (clustering) is discussed afterwards and here we show that fuzzy sets play a dominant role given the unsupervised character of the learning processes. Fuzzy sets offer an interesting option of quantifying available domain knowledge, giving rise to an idea of partial supervision or knowledge-based clustering. Selected ideas of data and feature reduction are presented in Section 22.6.

As far as the notation is concerned, we follow the symbols being in common usage. Patterns $\mathbf{x}_1, \mathbf{x}_2, \ldots, \mathbf{x}_N$ are treated as vectors in n-dimensional space $\mathbf{R}^n$, $\|.\|$ is used to describe a distance (Euclidean, Mahalanobis, Hamming, Tchebyshev, etc.). Fuzzy sets will be described by capital letters; the same notation is being used for their membership functions. Class labels will be denoted by $\omega_1, \omega_2, \ldots$, etc., while sets of integers will be described as $\boldsymbol{K} = \{1, 2, \ldots, K\}$, $\boldsymbol{N} = \{1, 2, \ldots, N\}$.

22.2 Methodology of Fuzzy Sets in Pattern Recognition

The concept of fuzzy sets augments the principles of pattern recognition in several ways. The well-established techniques are revisited and their conceptual and algorithmic aspects are extended. Let us briefly highlight the main arguments which also trigger some intensive research pursuits and exhibit several far reaching consequences from the applied perspective.

The leitmotiv is that fuzzy sets help realize user-centricity of pattern recognition schemes. Fuzzy sets are information granules of well-defined semantics which form a vocabulary of basic conceptual entities using which the problems are being formalized, models built, and decisions articulated. By expressing a certain most suitable point of view at the problem at hand and promoting a certain level of specificity, fuzzy sets form an effective conceptual framework for pattern recognition. There are two essential facets of the overall aspects of the nature of user centricity:

Class membership are membership grades. This quantification is of interest as there could be patterns whose allocation to classes might not be completely described in a Boolean (yes-no) manner. The user is more comfortable talking about levels of membership of particular patterns. It is also more instrumental to generate classification results where presented are intermediate values of membership values. There is an associated flagging effect: membership values in the vicinity of 0.5 are indicative of further needs to analyze the classification results or engage some other classifiers to either gather evidence in favor of belongingness to the given class (which is quantified through higher values of the membership functions) or collect evidence that justifies a reduction of such membership degrees.

Fuzzy sets contribute to the specialized, user-centric feature space. The original feature space is transformed via fuzzy sets and produces a new feature space that is easier to understand and enhances higher effectiveness of the classifiers formed at the next phase. Similarly, through the use of fuzzy sets, one could achieve a reduction of dimensionality of the original feature space. The nonlinearity effect introduced by fuzzy sets could be instrumental in reducing learning time of classifiers and enhancing their discriminative properties as well as improving their robustness. The tangible advantage results from the nonlinear character of membership functions. A properly adjusted nonlinearity could move apart patterns

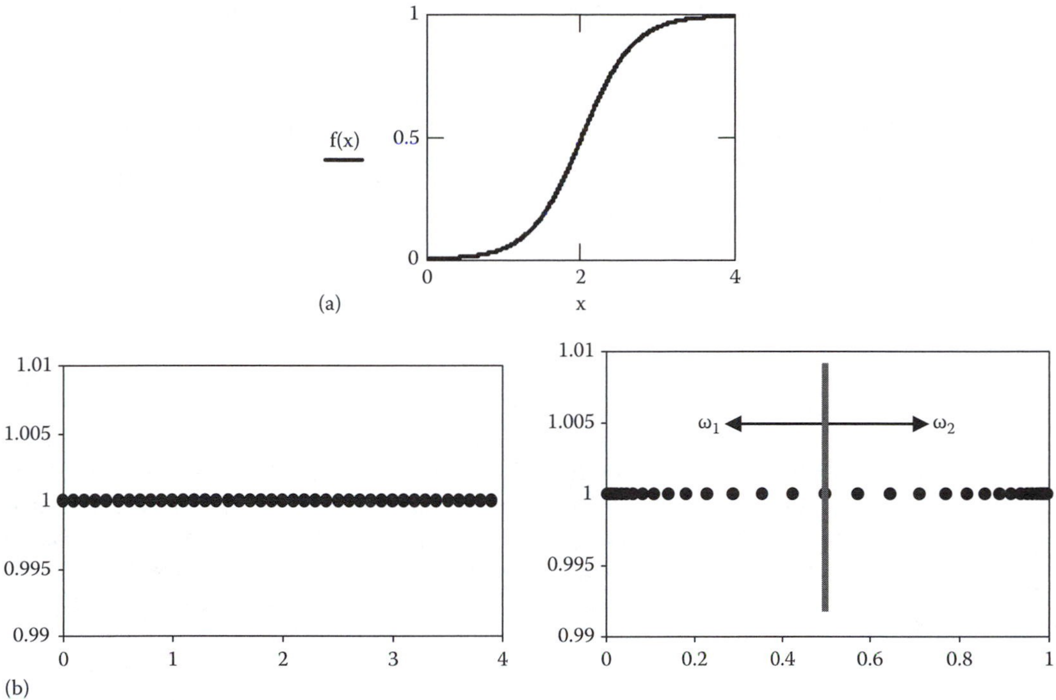

FIGURE 22.1 Nonlinear transformation realized by a sigmoid membership function: (a) sigmoid membership function and (b) original patterns distributed uniformly in the feature are grouped into quite distantly positioned groups.

belonging to different classes and bring closer those regions in which the patterns belong to the same category. For instance, patterns belonging to two classes and distributed uniformly in a one-dimensional space, see Figure 22.1, become well separated when transformed through a sigmoid membership function A and described in terms of the corresponding membership grades. In essence, fuzzy sets play a role of a nonlinear transformation of the feature space. We note that while the patterns are distributed uniformly, Figure 22.1b left, their distribution in the space of membership degrees [0,1] u = A(x) shows two groups of patterns that are located on the opposite ends of the unit interval with a large gap in between.

These two facets of fuzzy set-based user-centricity might be looked at together in a sense of an overall interface layer of the core computing faculties of pattern recognition, as illustrated in Figure 22.2.

Fuzzy pattern recognition dwells on the concepts of information granules and exploits their underlying formalism. Information granules giving rise to the general idea of Granular Computing help offer

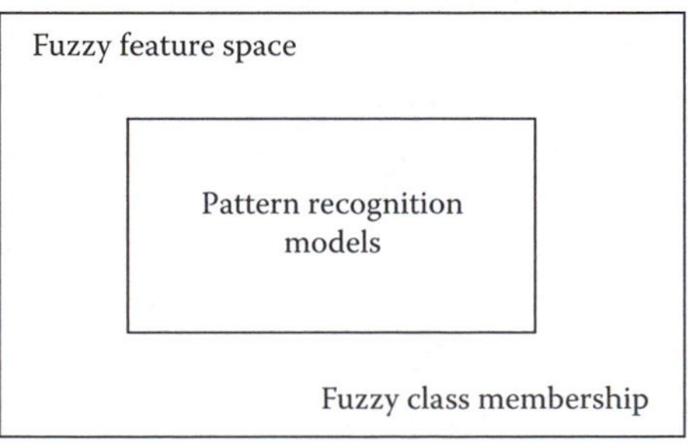

FIGURE 22.2 Fuzzy sets forming an interface layer (feature space and class membership) and wrapping the core computational faculties of pattern recognition.

a sound level of abstraction classification problems that need to be considered and address an issue of complexity when reducing the level of detail pattern recognition tasks are exposed to.

22.3 Information Granularity and Granular Computing

Information granules permeate numerous human endeavors [BP03a,BP03b,Z97]. No matter what problem is taken into consideration, we usually express it in a certain conceptual framework of basic entities, which we regard to be of relevance to the problem formulation and problem solving. This becomes a framework in which we formulate generic concepts adhering to some level of abstraction, carry out processing, and communicate the results to the external environment. Consider, for instance, image processing. In spite of the continuous progress in the area, a human being assumes a dominant and very much uncontested position when it comes to understanding and interpreting images. Surely, we do not focus our attention on individual pixels and process them as such but group them together into semantically meaningful constructs—familiar objects we deal with in everyday life. Such objects involve regions that consist of pixels or categories of pixels drawn together because of their proximity in the image, similar texture, color, etc. This remarkable and unchallenged ability of humans dwells on our effortless ability to construct information granules, manipulate them, and arrive at sound conclusions. As another example, consider a collection of time series. From our perspective we can describe them in a semi-qualitative manner by pointing at specific regions of such signals. Specialists can effortlessly interpret ECG signals. They distinguish some segments of such signals and interpret their combinations. Experts can interpret temporal readings of sensors and assess the status of the monitored system. Again, in all these situations, the individual samples of the signals are not the focal point of the analysis and the ensuing signal interpretation. We always granulate all phenomena (no matter if they are originally discrete or analog in their nature). Time is another important variable that is subjected to granulation. We use seconds, minutes, days, months, and years. Depending upon a specific problem we have in mind and who the user is, the size of information granules (time intervals) could vary quite dramatically. To the high level management time intervals of quarters of year or a few years could be meaningful temporal information granules on basis of which one develops any predictive model. For those in charge of everyday operation of a dispatching plant, minutes and hours could form a viable scale of time granulation. For the designer of high-speed integrated circuits and digital systems, the temporal information granules concern nanoseconds, microseconds, and perhaps microseconds. Even such commonly encountered and simple examples are convincing enough to lead us to ascertain that (a) information granules are the key components of knowledge representation and processing, (b) the level of granularity of information granules (their size, to be more descriptive) becomes crucial to the problem description and an overall strategy of problem solving, and (c) there is no universal level of granularity of information; the size of granules is problem-oriented and user dependent.

What has been said so far touched a qualitative aspect of the problem. The challenge is to develop a computing framework within which all these representation and processing endeavors could be formally realized. The common platform emerging within this context comes under the name of Granular Computing. In essence, it is an emerging paradigm of information processing. While we have already noticed a number of important conceptual and computational constructs built in the domain of system modeling, machine learning, image processing, pattern recognition, and data compression in which various abstractions (and ensuing information granules) came into existence, Granular Computing becomes innovative and intellectually proactive in several fundamental ways:

- It identifies the essential commonalities between the surprisingly diversified problems and technologies used there which could be cast into a unified framework we usually refer to as a granular world. This is a fully operational processing entity that interacts with the external world (which could be another granular or numeric world) by collecting necessary granular information and returning the outcomes of the granular computing.

- With the emergence of the unified framework of granular processing, we get a better grasp as to the role of interaction between various formalisms and visualize a way in which they communicate.
- It brings together the existing formalisms of set theory (interval analysis) [M66,Z65,Z05,P82,P91, PS07] under the same roof by clearly visualizing that in spite of their visibly distinct underpinnings (and ensuing processing), they exhibit some fundamental commonalities. In this sense, Granular Computing establishes a stimulating environment of synergy between the individual approaches.
- By building upon the commonalities of the existing formal approaches, Granular Computing helps build heterogeneous and multifaceted models of processing of information granules by clearly recognizing the orthogonal nature of some of the existing and well-established frameworks (say, probability theory coming with its probability density functions and fuzzy sets with their membership functions).
- Granular Computing fully acknowledges a notion of variable granularity whose range could cover detailed numeric entities and very abstract and general information granules. It looks at the aspects of compatibility of such information granules and ensuing communication mechanisms of the granular worlds.
- Interestingly, the inception of information granules is highly motivated. We do not form information granules without reason. Information granules arise as an evident realization of the fundamental paradigm of abstraction.

Granular Computing forms a unified conceptual and computing platform. Yet, it directly benefits from the already existing and well-established concepts of information granules formed in the setting of set theory, fuzzy sets, rough sets, and others. In the setting of this study it comes as a technology contributing to pattern recognition.

22.3.1 Algorithmic Aspects of Fuzzy Set Technology in Pattern Recognition: Pattern Classifiers

Each of these challenges comes with a suite of their own quite specific problems that do require a very careful attention both at the conceptual as well as algorithmic level. We have highlighted the list of challenges and in the remainder of this study present some of the possible formulations of the associated problems and look at their solutions. It is needless to say that our proposal points at some direction that deems to be of relevance however does not pretend to offer a complete solution to the problem. Some algorithmic pursuits are also presented as an illustration of some possibilities emerging there.

Indisputably, geometry of patterns belonging to different classes is a focal point implying an overall selection and design of pattern classifiers. Each classifier comes with its geometry and this predominantly determines its capabilities. While linear classifiers (built on a basis of some hyperplanes) and nonlinear classifiers (such as neural networks) are two popular alternatives, there is another point of view at the development of the classifiers that dwells on the concept of information granules. Patterns belonging to the same class form information granules in the feature space. A description of geometry of these information granules is our ultimate goal when designing effective classifiers.

22.4 Fuzzy Linear Classifiers and Fuzzy Nearest Neighbor Classifiers as Representatives of Supervised Fuzzy Classifiers

Linear classifiers [DHS01] are governed by the well-known linear relationship $y(\mathbf{x}) = \mathbf{w}^T\mathbf{x} + w_0$, where $\mathbf{w}$ and w_0 are the parameters (weights and bias) of the classifier. The classification rule in case of two classes (ω_1 and ω_2) reads as follows: classify $\mathbf{x}$ to ω_1 if $y(\mathbf{x}) > 0$ and assign to ω_2 otherwise. The design of such classifier (perceptron) has been intensively discussed in the literature and has resulted in a wealth of algorithms. The property of linear separability of patterns assures us that the learning method converges in a finite number of iterations. The classification rule does not quantify how close the pattern is to the linear boundary, which could be regarded as a certain drawback of this classifier. The analogue of the linear classifier expressed in the language of fuzzy sets brings about a collection of the parameters of the classifier which are represented as fuzzy numbers, and triangular fuzzy numbers, in particular. The underlying formula of the fuzzy classifier comes in the form

$$Y(\mathbf{x}) = W_1 \otimes x_1 \oplus W_2 \otimes x_2 \ldots W_n \otimes x_n \oplus W_0 \tag{22.1}$$

where W_i, i = 1, 2,…, n are triangular fuzzy numbers. As a result, the output of the classifier is a fuzzy number as well. Note that we used the symbols of addition of multiplication to underline the fact that the computing is concerned with fuzzy numbers rather than plain numeric entities. Triangular fuzzy number A can be represented as a triple $A = \langle a_-, a, a_+ \rangle$ with "a" being the modal value of A, and a_- and a_+ standing for the bounds of the membership function. The design criterion considered here stresses the separability of the two classes in the sense of the membership degrees produced for a given pattern **x**. More specifically we have the following requirements:

$$\text{Max } Y(\mathbf{x}) \quad \text{if } \mathbf{x} \in \omega_1 \quad \text{Min } Y(\mathbf{x}) \quad \text{if } \mathbf{x} \in \omega_2 \tag{22.2}$$

Similarly, as commonly encountered in fuzzy regression, one could consider a crux of the design based on Linear Programming. In contrast to linear classifiers, fuzzy linear classifiers produce classification results with class quantification; so rather than a binary decision is being generated, we come up with the degree of membership of pattern to class ω_1. An illustration of the concept is illustrated in Figure 22.3.

In virtue of the classification rule, the membership of Y(**x**) is highly asymmetric. The slope at one side of the classification line is reflective of the distribution of patterns belonging to class ω_1. The geometry of the classifier is still associated with a linear boundary. What fuzzy sets offer is a fuzzy set of membership associated with this boundary. Linear separability is an idealization of the classification problem. In reality there could be some patterns located in the boundary region which does not satisfy the linearity assumption. So even though the linear classifier comes as a viable alternative as a first attempt, further refinement is required. The concept of the nearest neighbor (NN) classifier could form a sound enhancement of the fuzzy linear classifier. The popularity of the NN classifiers stems from the fact that in their development we rely on lazy learning, so no optimization effort is required at all. Any new pattern is assigned to the same class as its closest neighbor. The underlying classification rule reads as follows: given **x**, determine $\mathbf{x}_{i0}$ in the training set such that $i0 = \arg_i \min\|\mathbf{x} - \mathbf{x}_i\|$ assuming that

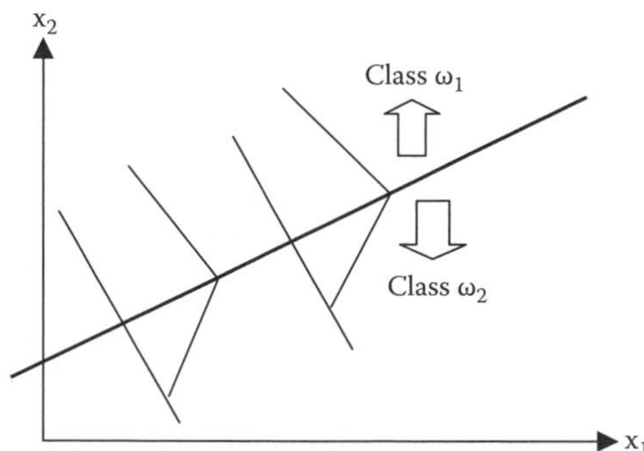

FIGURE 22.3 Fuzzy linear classifier; note asymmetric nature of membership degrees generated around the classification boundary.

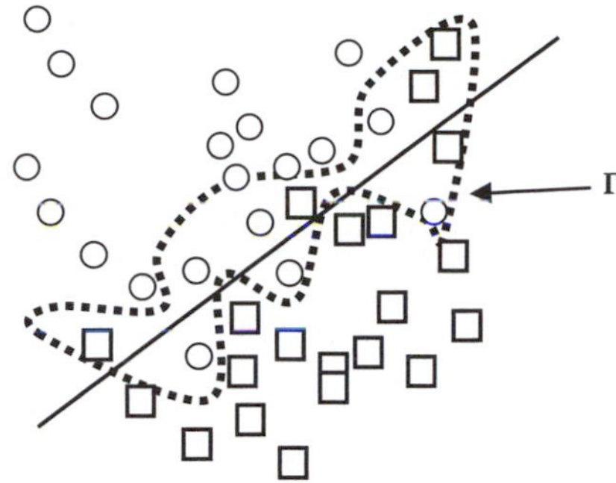

FIGURE 22.4 Linear classifier with a collection of patterns in the boundary region (**Γ**) whose treatment is handled by NN classifier.

the class membership of $\mathbf{x}_{i0}$ is ω_2, $\mathbf{x}$ is classified as ω_2 as well. The NN neighbor classifier could involve a single closest neighbor (in which case the classification rule is referred to as 1-NN classifier), 3 neighbors giving rise to 3-NNs, 5 neighbors resulting in 5-NNs, or k-NNs where "k" is an odd number (k-NN classifier). The majority vote implies the class membership of **x**. The extension of the k-NN classification rule can be realized in many ways. The intuitive one is to compute a degree of membership of **x** to a certain class by looking at the closest "k" neighbors, determining the membership degrees $u_i(\mathbf{x})$, i = 1, 2,…, L, and choosing the highest one as reflective of the allocation of **x** to given class. Here Card (**Γ**) = L (Figure 22.4).

More specifically, we have

$$u_i(\mathbf{x}) = \frac{1}{\sum_{x_j \in \Gamma}^{L} \left(\frac{\|\mathbf{x} - \mathbf{x}_i\|}{\|\mathbf{x} - \mathbf{x}_j\|} \right)^2} \tag{22.3}$$

(we will note a resemblance of this expression to the one describing membership degrees computed in the FCM algorithm). Given that pattern i0 where $i0 = \arg\max_{i=1,2,\ldots,L} u_i(\mathbf{x})$ belongs to class ω_1, we assign **x** to the same class with the corresponding membership degree. The patterns positioned close to the boundary of the linear classifier are engaged in the NN classification rule. The architecture illustrated in Figure 22.5 comes as an aggregate of the fuzzy linear classifier and the fuzzy NN classifier which modifies the original membership degrees coming from the linear classifier by adjusting them on a basis of some local characteristics of the data. The output of the NN classifier (generating the highest

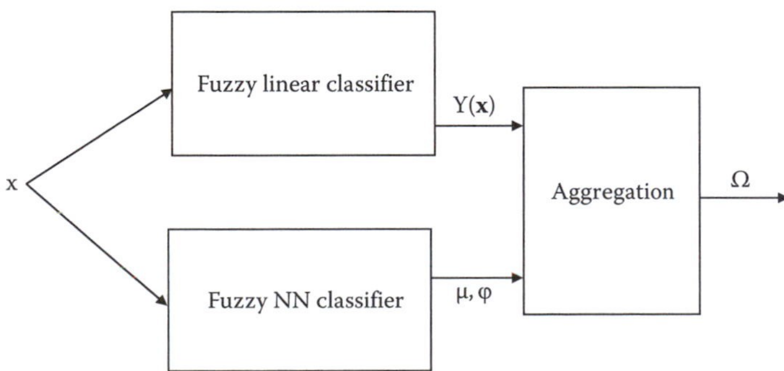

FIGURE 22.5 Combination of two fuzzy classifiers: fuzzy linear classifier focused on the global nature of classification is adjusted by the results formed by the fuzzy NN classifier acting on a local basis.

membership degree) is governed by the expression $\mu = \max_{i=1,2} u_i(\mathbf{x})$ and $i_0 = \arg\max_{i=1,2} u_i(\mathbf{x})$. Let us introduce the following indicator (characteristic) function:

$$\varphi(\mathbf{x}) = \begin{cases} 1 & \text{if } i_0 = 1 \text{ (class } \omega_1) \\ 0 & \text{if } i_0 = 2 \text{ (class } \omega_2) \end{cases} \tag{22.4}$$

The aggregation of the classification results produced by the two classifiers is completed in the following fashion with the result Ω being a degree of membership to class ω_1:

$$\Omega = \begin{cases} \max(1, Y(\mathbf{x}) + \mu\phi(\mathbf{x})) & \text{if } \varphi(\mathbf{x}) = 1 \\ \min(0, Y(\mathbf{x}) - \mu(1 - \phi(\mathbf{x})) & \text{if } \varphi(\mathbf{x}) = 0 \end{cases} \tag{22.5}$$

Note that if the NN classifier has assigned $\mathbf{x}$ to class ω_1, this class membership elevates the membership degree produced by the fuzzy linear classifier, hence we arrive at the clipped sum max $(1, Y(\mathbf{x}) + \mu)$. In the opposite case where the NN classification points at the assignment to the second class, the overall class membership to ω_1 becomes reduced by μ. As a result, given a subset of patterns $\mathbf{\Gamma}$ in the boundary region of the fuzzy linear classifier, the classification region is adjusted accordingly. Its overall geometry is more complicated and nonlinear adjusting the original linear form to the patterns located in this boundary region.

22.4.1 Fuzzy Logic–Oriented Classifiers

Fuzzy sets and information granules, in general, offer a structural backbone of fuzzy classifiers. The crux of the concept is displayed in Figure 22.6. Information granules are formed in the feature space. They are *logically* associated with classes in the sense that for each class its degree of class membership is a logic expression of the activation levels (matching degrees) of the individual information granules. The flexibility of the logic mapping is offered through the use of the collection of logic neurons (fuzzy neurons) whose connections are optimized during the design of the classifier.

22.4.2 Main Categories of Fuzzy Neurons

There are two main types of logic neurons: aggregative and referential neurons. Each of them comes with a clearly defined semantics of its underlying logic expression and is equipped with significant parametric flexibility necessary to facilitate substantial learning abilities.

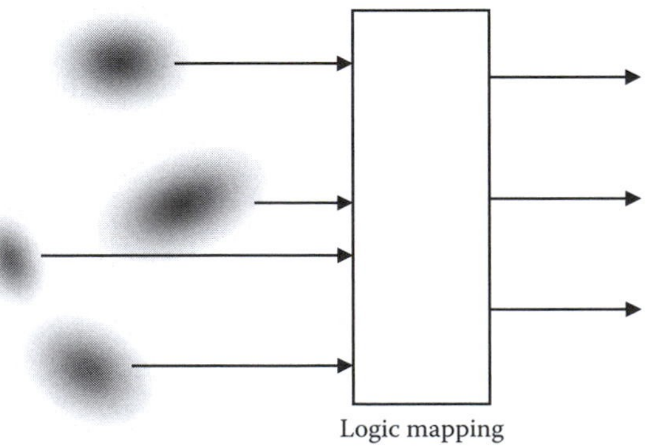

FIGURE 22.6 An overall scheme of logic mapping between information granules—fuzzy sets formed in the feature space and the class membership degrees.

22.4.2.1 Aggregative Neurons

Formally, these neurons realize a logic mapping from $[0, 1]^n$ to $[0, 1]$. Two main classes of the processing units exist in this category [P95,PR93,HP93].

OR neuron: This realizes an *and* logic aggregation of inputs $\mathbf{x} = [x_1\ x_2 \ldots x_n]$ with the corresponding connections (weights) $\mathbf{w} = [w_1\ w_2 \ldots w_n]$ and then summarizes the partial results in an *or*-wise manner (hence the name of the neuron). The concise notation underlines this flow of computing, y = OR($\mathbf{x}$; $\mathbf{w}$) while the realization of the logic operations gives rise to the expression (commonly referring to it as an s-t combination or s-t aggregation)

$$y = \mathop{\mathbf{S}}_{i=1}^{n} (x_i t w_i) \tag{22.6}$$

Bearing in mind the interpretation of the logic connectives (t-norms and t-conorms), the OR neuron realizes the following logic expression being viewed as an underlying logic description of the processing of the input signals:

$$(x_1 \textit{ and } w_1) \textit{ or } (x_2 \textit{ and } w_2) \textit{ or} \ldots \textit{or } (x_n \textit{ and } w_n) \tag{22.7}$$

Apparently, the inputs are logically "weighted" by the values of the connections before producing the final result. In other words we can treat "y" as a truth value of the above statement where the truth values of the inputs are affected by the corresponding weights. Noticeably, lower values of w_i discount the impact of the corresponding inputs; higher values of the connections (especially those being positioned close to 1) do not affect the original truth values of the inputs resulting in the logic formula. In limit, if all connections w_i, i =1, 2, ..., n are set to 1, then the neuron produces a plain *or*-combination of the inputs, $y = x_1$ *or* x_2 *or* ... *or* x_n. The values of the connections set to zero eliminate the corresponding inputs. Computationally, the OR neuron exhibits nonlinear characteristics (that is inherently implied by the use of the t- and t-conorms (that are evidently nonlinear mappings). The connections of the neuron contribute to its adaptive character; the changes in their values form the crux of the parametric learning.

AND neuron: the neurons in the category, described as y = AND($\mathbf{x}$; $\mathbf{w}$) with $\mathbf{x}$ and $\mathbf{w}$ being defined as in case of the OR neuron, are governed by the expression

$$y = \mathop{\mathbf{T}}_{i=1}^{n} (x_i s w_i) \tag{22.8}$$

Here the *or* and *and* connectives are used in a reversed order: first the inputs are combined with the use of the t-conorm and the partial results produced in this way are aggregated *and*-wise. Higher values of the connections reduce impact of the corresponding inputs. In limit w_i =1 eliminates the relevance of x_i. With all w_i set to 0, the output of the AND neuron is just an *and* aggregation of the inputs

$$y = x_1 \textit{ and } x_2 \textit{ and } \ldots \textit{ and } x_n \tag{22.9}$$

Let us conclude that the neurons are highly nonlinear processing units whose nonlinear mapping depends upon the specific realizations of the logic connectives. They also come with potential plasticity whose usage becomes critical when learning the networks including such neurons.

At this point, it is worth contrasting these two categories of logic neurons with "standard" neurons we encounter in neurocomputing. The typical construct there comes in the form of the weighted sum of the inputs $x_1, x_2, \ldots, x_n$ with the corresponding connections (weights) $w_1, w_2, \ldots, w_n$ being followed by a nonlinear (usually monotonically increasing) function that reads as follows:

$$y = g\left(\mathbf{w}^T\mathbf{x} + \tau\right) = g\left(\sum_{i=1}^{n} w_i x_i + \tau\right) \tag{22.10}$$

where
- $\mathbf{w}$ is a vector of connections
- τ is a constant term (bias)
- "g" denotes some monotonically non-decreasing nonlinear mapping

The other less commonly encountered neuron is a so-called π-neuron. While there could be some variations as to the parametric details of this construct, we can envision the following realization of the neuron:

$$y = g\left(\prod |x_i - t_i|^{w_i}\right) \tag{22.11}$$

where
- $\mathbf{t} = [t_1\ t_2 \ldots t_n]$ denotes a vector of translations
- $\mathbf{w}$ (>$\mathbf{0}$) denotes a vector of all connections

As before, the nonlinear function is denoted by "g." While some superficial and quite loose analogy between these processing units and logic neurons could be derived, one has to cognizant that these neurons do not come with any underlying logic fabric and hence cannot be easily and immediately interpreted.

Let us make two observations about the architectural and functional facets of the logic neurons we have introduced so far.

Incorporation of the bias term (bias) in the fuzzy logic neurons. In analogy to the standard constructs of a generic neuron as presented above, we could also consider a bias term, denoted by $w_0 \in [0, 1]$, which enters the processing formula of the fuzzy neuron in the following manner:

For the OR neuron

$$y = \mathop{\mathbf{S}}_{i=1}^{n} (x_i t w_i) s w_0 \tag{22.12}$$

For the AND neuron

$$y = \mathop{\mathbf{T}}_{i=1}^{n} (x_i s w_i) t w_0 \tag{22.13}$$

We can offer some useful interpretation of the bias by treating it as some nonzero initial truth value associated with the logic expression of the neuron. For the OR neuron it means that the output does not reach values lower than the assumed threshold. For the AND neuron equipped with some bias, we conclude that its output cannot exceed the value assumed by the bias. The question whether the bias is essential in the construct of the logic neurons cannot be fully answered in advance. Instead, we may include it into the structure of the neuron and carry out learning. Once its value has been obtained, its relevance could be established considering the specific value it has been produced during the learning. It may well be that the optimized value of the bias is close to zero for the OR neuron or close to one in the case of the AND neuron which indicates that it could be eliminated without exhibiting any substantial impact on the performance of the neuron.

Dealing with inhibitory character of input information. Owing to the monotonicity of the t-norms and t-conorms, the computing realized by the neurons exhibits an excitatory character. This means that higher values of the inputs (x_i) contribute to the increase in the values of the output of the neuron. The inhibitory nature of computing realized by "standard" neurons by using negative values of the connections or the inputs is not available here as the truth values (membership grades) in fuzzy sets are confined to the unit interval. The inhibitory nature of processing can be accomplished by considering the complement of the original input, $=1 - x_i$. Hence, when the values of x_i increase, the associated values of the complement decrease and subsequently in this configuration we could effectively treat such an input as having an inhibitory nature.

22.4.3 Architectures of Logic Networks

The logic neurons (aggregative and referential) can serve as building blocks of more comprehensive and functionally appealing architectures. The diversity of the topologies one can construct with the aid of the proposed neurons is surprisingly high. This architectural diversity is important from the application point of view as we can fully reflect the nature of the problem in a flexible manner. It becomes essential to capture the underlying nature of the problem and set up a logic skeleton of the network along with an optimization of its parameters. Throughout the entire development process we are positioned quite comfortably by monitoring the optimization of the network as well as interpreting its semantics.

22.4.3.1 Logic Processor in the Processing of Fuzzy Logic Functions: A Canonical Realization

The typical logic network that is at the center of logic processing originates from the two-valued logic and comes in the form of the famous Shannon theorem of decomposition of Boolean functions. Let us recall that any Boolean function $\{0, 1\}^n \rightarrow \{0, 1\}$ can be represented as a logic sum of its corresponding miniterms or a logic product of maxterms. By a minterm of "n" logic variables $x_1, x_2, \ldots, x_n$ we mean a logic product involving all these variables either in direct or complemented form. Having "n" variables we end up with 2^n minterms starting from the one involving all complemented variables and ending up at the logic product with all direct variables. Likewise by a maxterm we mean a logic sum of all variables or their complements. Now in virtue of the decomposition theorem, we note that the first representation scheme involves a two-layer network where the first layer consists of AND gates whose outputs are combined in a single OR gate. The converse topology occurs for the second decomposition mode: there is a single layer of OR gates followed by a single AND gate aggregating *or*-wise all partial results.

The proposed network (referred here as a logic processor) generalizes this concept, as shown in Figure 22.7. The OR-AND mode of the logic processor comes with the two types of aggregative neurons being

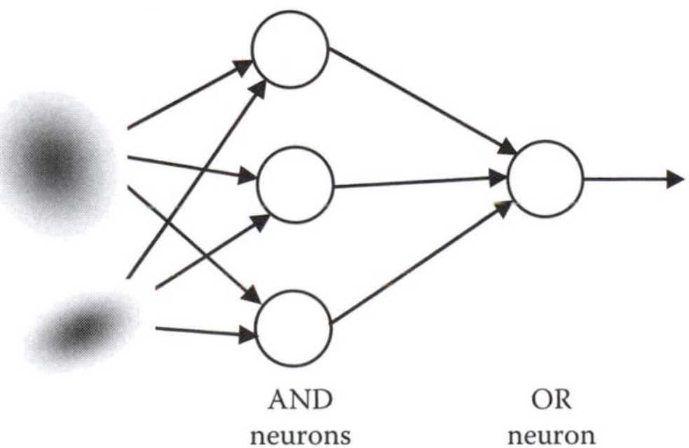

FIGURE 22.7 A topology of the logic processor in its AND-OR mode.

swapped between the layers. Here the first (hidden) layer is composed of the OR neuron and is followed by the output realized by means of the AND neuron.

The logic neurons generalize digital gates. The design of the network (viz. any fuzzy function) is realized through learning. If we confine ourselves to Boolean {0,1} values, the network's learning becomes an alternative to a standard digital design, especially a minimization of logic functions. The logic processor translates into a compound logic statement (we skip the connections of the neurons to underline the underlying logic content of the statement):

- if (input$_1$ *and* ... *and* input$_j$) *or* (input$_d$ *and* ... *and* input$_f$) then class membership

The logic processor's topology (and underlying interpretation) is standard. Two LPs can vary in terms of the number of AND neurons, their connections but the format of the resulting logic expression stays quite uniform (as a sum of generalized minterms) and this introduces a significant level of interpretability.

22.4.4 Granular Constructs of Classifiers

The fundamental development strategy pursued when dealing with this category of classifiers dwells upon the synergistic and highly orchestrated usage of two fundamental technologies of Granular Computing, namely intervals (hyperboxes) and fuzzy sets. Given this, the resulting constructs will be referred to as granular hyperbox-driven classifiers. The architecture of the hyperbox-driven classifier (HDC, for brief) comes with two well-delineated architectural components that directly imply its functionality. The core (primary) part of the classifier which captures the essence of the structure is realized in terms of interval analysis. Sets are the basic constructs that form the regions of the feature space where there is a high homogeneity of the patterns (which implies low classification error). We may refer to it as a *core* structure. Fuzzy sets are used to cope with the patterns outside the core structure and in this way contribute to a refinement of the already developed core structure. This type of the more detailed structure will be referred to as a *secondary* one. The two-level granular architecture of the classifier reflects a way in which classification processes are usually carried out: we start with a core structure where the classification error is practically absent and then consider the regions of high overlap between the classes where there is a high likelihood of the classification error. For the core structure, the use of sets as generic information granules is highly legitimate: there is no need to distinguish between these elements of the feature space. The areas of high overlap require more detailed treatment hence here arises a genuine need to consider fuzzy sets as the suitable granular constructs. The membership grades play an essential role in expressing levels of confidence associated with the classification result. In this way, we bring a detailed insight into the geometry of the classification problem and identify regions of very poor classification. One can view the granular classifier as a two-level hierarchical classification scheme whose development adheres to the principle of a stepwise refinement of the construct with sets forming the core of the architecture and fuzzy sets forming its specialized enhancement. Given this, a schematic view of the two-level construct of the granular classifier is included in Figure 22.8.

One of the first approaches to the construction of set-based classifiers (hyperboxes) were presented by Simpson [S92,S93] both in supervised and unsupervised mode. Abe et al. [ATK98] presented an efficient method for extracting rules directly from a series of activation hyperboxes, which capture the existence region of data for a given class and inhibition hyperboxes, which inhibit the existence of data of that class. Rizzi et al. [RMM00,RPM02] proposed an adaptive resolution classifier (ARC) and its pruned version (PARC) in order to enhance the constructs introduced by Simpson. ARC/PARC generates a regularized min-max network by a series of hyperbox cuts. Gabrys and Bargiela [GB00] described a general fuzzy min-max (GFMM) neural network which combines mechanisms of supervised and unsupervised learning into a single unified framework.

The design of the granular classifiers offers several advantages over some "standard" pattern classifiers. First, the interpretability is highly enhanced: both the structure and the conceptual organization appeals in a way in which an interpretation of the topology of patterns is carried out. Second, one can resort himself to the existing learning schemes developed both for set-theoretic classifiers and fuzzy classifiers [PP08,PS05].

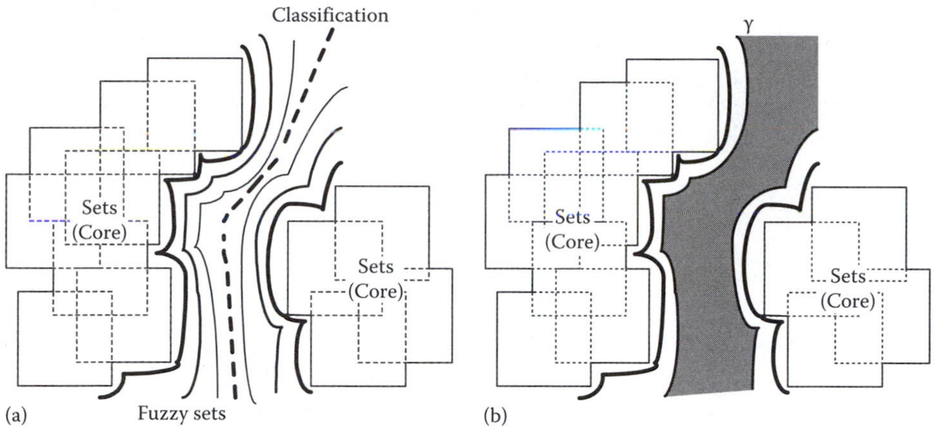

FIGURE 22.8 From sets to fuzzy sets: (a) a principle of a two-level granular classifier exploiting the successive usage of the formalisms of information granulation and (b) further refinements of the information granules realized on a basis of the membership degrees.

22.5 Unsupervised Learning with Fuzzy Sets

Unsupervised learning, quite commonly treated as an equivalent of clustering is the subdiscipline of pattern recognition which is aimed at the discovery of structure in data and its representation in the form of clusters—groups of data.

Clusters, in virtue of their nature, are inherently *fuzzy*. Fuzzy sets constitute a natural vehicle to quantify strength of membership of patterns to a certain group. An example shown in Figure 22.9 clearly demonstrates this need. The pattern positioned in between the two well-structured and compact groups exhibits some level of resemblance (membership) to each of the clusters. Surely enough, one could be hesitant to allocate it fully to either of the clusters. The membership values such as, e.g., 0.55 and 0.45 are not only reflective of the structure in the data but they flag the distinct nature of this data and maybe trigger some further inspection of this pattern. In this way we remark a user-centric character of fuzzy sets which make interaction with users more effective and transparent.

22.5.1 Fuzzy C-Means as an Algorithmic Vehicle of Data Reduction through Fuzzy Clusters

Fuzzy sets can be formed on a basis of numeric data through their clustering (groupings). The groups of data give rise to membership functions that convey a global more abstract and general view at the available data. With this regard Fuzzy C-Means (FCM, for brief) is one of the commonly used mechanisms of fuzzy clustering [B81,P05].

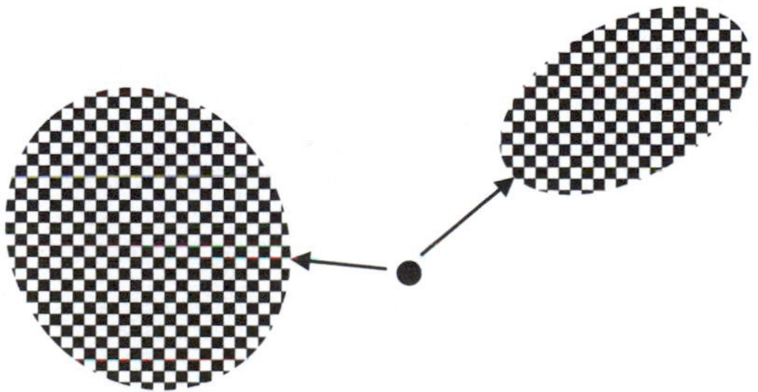

FIGURE 22.9 Example of two-dimensional data with patterns of varying membership degrees to the two highly visible and compact clusters.

Let us review its formulation, develop the algorithm and highlight the main properties of the fuzzy clusters. Given a collection of n-dimensional data set $\{\mathbf{x}_k\}$, k = 1,2,…,N, the task of determining its structure—a collection of "c" clusters—is expressed as a minimization of the following objective function (performance index) Q being regarded as a sum of the squared distances between data and their representatives (prototypes):

$$Q = \sum_{i=1}^{c} \sum_{k=1}^{N} u_{ik}^{m} \|\mathbf{x}_k - \mathbf{v}_i\|^2 \tag{22.14}$$

Here

$\mathbf{v}_i$s are n-dimensional prototypes of the clusters, i = 1, 2, ..., c

U = $[u_{ik}]$ stands for a partition matrix expressing a way of allocation of the data to the corresponding clusters

u_{ik} is the membership degree of data $\mathbf{x}_k$ in the ith cluster

The distance between the data $\mathbf{z}_k$ and prototype $\mathbf{v}_i$ is denoted by ||.||. The fuzzification coefficient m (>1.0) expresses the impact of the membership grades on the individual clusters. It implies as certain geometry of fuzzy sets which will be presented later in this study.

A partition matrix satisfies two important and intuitively appealing properties:

$$0 < \sum_{k=1}^{N} u_{ik} < N, \quad i = 1,2,\ldots,c \tag{22.15a}$$

$$\sum_{i=1}^{c} u_{ik} = 1, \quad k = 1,2,\ldots,N \tag{22.15b}$$

Let us denote by **U** a family of matrices satisfying (22.15a) through (22.15b). The first requirement states that each cluster has to be nonempty and different from the entire set. The second requirement states that the sum of the membership grades should be confined to 1.

The minimization of Q completed with respect to U ∈ **U** and the prototypes $\mathbf{v}_i$ of V = $\{\mathbf{v}_1, \mathbf{v}_2, \ldots \mathbf{v}_c\}$ of the clusters. More explicitly, we write it down as follows:

$$\min Q \text{ with respect to } U \in \mathbf{U}, \mathbf{v}_1, \mathbf{v}_2, \ldots, \mathbf{v}_c \in \mathbf{R}^n \tag{22.16}$$

From the optimization standpoint, there are two individual optimization tasks to be carried out separately for the partition matrix and the prototypes. The first one concerns the minimization with respect to the constraints given the requirement of the form (22.15b) which holds for each data point $\mathbf{x}_k$. The use of Lagrange multipliers converts the problem into its constraint-free version. The augmented objective function formulated for each data point, k = 1, 2,…, N, reads as

$$V = \sum_{i=1}^{c} u_{ik}^{m} d_{ik}^{2} + \lambda\left(\sum_{i=1}^{c} u_{ik} - 1\right) \tag{22.17}$$

where $d_{ik}^2 = \|\mathbf{x}_k - \mathbf{v}_i\|^2$.

It is instructive to go over the details of the optimization process. Starting with the necessary conditions for the minimum of V for k = 1,2,… N, one obtains

$$\frac{\partial V}{\partial u_{st}} = 0 \quad \frac{\partial V}{\partial \lambda} = 0 \tag{22.18}$$

s = 1, 2 … c, t = 1, 2 … N. Now we calculate the derivative of V with respect to the elements of the partition matrix in the following way:

$$\frac{\partial V}{\partial u_{st}} = m u_{st}^{m-1} d_{st}^2 + \lambda \tag{22.19}$$

Given this relationship, and using (22.15) we calculate u_{st}

$$u_{st} = -\left(\frac{\lambda}{m}\right)^{1/(m-1)} d_{st}^{2/(m-1)} \tag{22.20}$$

Taking into account the normalization condition $\sum_{j=1}^{c} u_{jt} = 1$ and plugging it into (22.20) one has

$$-\left(\frac{\lambda}{m}\right)^{1/(m-1)} \sum_{j=1}^{c} d_{jt}^{2/(m-1)} = 1 \tag{22.21}$$

We compute

$$-\left(\frac{\lambda}{m}\right)^{1/(m-1)} = \frac{1}{\sum_{j=1}^{c} d_{jt}^{2/(m-1)}} \tag{22.22}$$

Inserting this expression into (22.20), we obtain the successive entries of the partition matrix:

$$u_{st} = \frac{1}{\sum_{j=1}^{c} \left(\frac{d_{st}^2}{d_{jt}^2}\right)^{1/(m-1)}} \tag{22.23}$$

The optimization of the prototypes $\mathbf{v}_i$ is carried out assuming the Euclidean distance between the data and the prototypes that is $\|\mathbf{x}_k - \mathbf{v}_i\|^2 = \sum_{j=1}^{n} (x_{kj} - v_{ij})^2$. The objective function reads now as follows $Q = \sum_{i=1}^{c} \sum_{k=1}^{N} u_{ik}^m \sum_{j=1}^{n} (x_{kj} - v_{ij})^2$, and its gradient with respect to $\mathbf{v}_i$, $\nabla_{\mathbf{v}_i} Q$ made equal to zero yields the system of linear equations:

$$\sum_{k=1}^{N} u_{ik}^m (x_{kt} - v_{st}) = 0 \tag{22.24}$$

s = 1, 2,.., c; t = 1, 2, …, *n*.

Thus,

$$v_{st} = \frac{\sum_{k=1}^{N} u_{ik}^m x_{kt}}{\sum_{k=1}^{N} u_{ik}^m} \tag{22.25}$$

Overall, the FCM clustering is completed through a sequence of iterations where we start from some random allocation of data (a certain randomly initialized partition matrix) and carry out the following updates by adjusting the values of the partition matrix and the prototypes. The iterative process is continued until a certain termination criterion has been satisfied. Typically, the termination condition is quantified by looking at the changes in the membership values of the successive partition matrices.

TABLE 22.1 Main Features of the FCM Clustering Algorithm

Feature of the FCM Algorithm	Representation and Optimization Aspects
Number of clusters (c)	Structure in the data set and the number of fuzzy sets estimated by the method; the increase in the number of clusters produces lower values of the objective function however given the semantics of fuzzy sets one should maintain this number quite low (5–9 information granules)
Objective function Q	Develops the structure aimed at the minimization of Q; iterative process supports the determination of the local minimum of Q
Distance function $\|\|\cdot\|\|$	Reflects (or imposes) a geometry of the clusters one is looking for; essential design parameter affecting the shape of the membership functions
Fuzzification coefficient (m)	Implies a certain shape of membership functions present in the partition matrix; essential design parameter. Low values of "m" (being close to 1.0) induce characteristic function. The values higher than 2.0 yield spiky membership functions
Termination criterion	Distance between partition matrices in two successive iterations; the algorithm terminated once the distance below some assumed positive threshold (ε) that is $\|\|U(iter + 1) - U(iter)\|\| < \varepsilon$

Denote by U(t) and U(t+1) the two partition matrices produced in the two consecutive iterations of the algorithm. If the distance $\|U(t+1) - U(t)\|$ is less than a small predefined threshold ε (say, $\varepsilon = 10^{-5}$ or 10^{-6}), then we terminate the algorithm. Typically, one considers the Tchebyschev distance between the partition matrices meaning that the termination criterion reads as follows:

$$\max_{i,k}|u_{ik}(t+1) - u_{ik}(t)| \leq \varepsilon \tag{22.26}$$

The key components of the FCM and a quantification of their impact on the form of the produced results are summarized in Table 22.1.

With regard to the computing supported by the FCM algorithm, an essential point should be made. The calculations of the prototypes as given in (22.25) are feasible considering the nature of the distance. The use of the Euclidean distance invokes (22.24) while any other distance (which could be potentially quite appealing given the geometry of the induced regions in the feature space) does not lead to this closed-type of optimization scheme.

The fuzzification coefficient exhibits a direct impact on the geometry of fuzzy sets generated by the algorithm. Typically, the value of "m" is assumed to be equal to 2.0. Lower values of m (that are closer to 1) yield membership functions that start resembling characteristic functions of sets; most of the membership values become localized around 1 or 0. The increase of the fuzzification coefficient (m = 3, 4, etc.) produces "spiky" membership functions with the membership grades equal to 1 at the prototypes and a fast decline of the values when moving away from the prototypes. Furthermore the average values of the membership function are equal to 1/c. Several illustrative examples of the membership functions are included in Figure 22.10. In addition to the varying shape of the membership functions, observe that the requirement put on the sum of membership grades imposed on the fuzzy sets yields some rippling effect: the membership functions are not unimodal but may exhibit some ripples whose intensity depends upon the distribution of the prototypes and the values of the fuzzification coefficient.

The membership functions offer an interesting feature of evaluating an extent to which a certain data point is shared between different clusters and in this sense become difficult to allocate to a single cluster (fuzzy set). Let us introduce the following index which serves as a suitable separation measure between the clusters:

$$\varphi(u_1, u_2, \ldots, u_c) = 1 - c^c \prod_{i=1}^{c} u_i \tag{22.27}$$

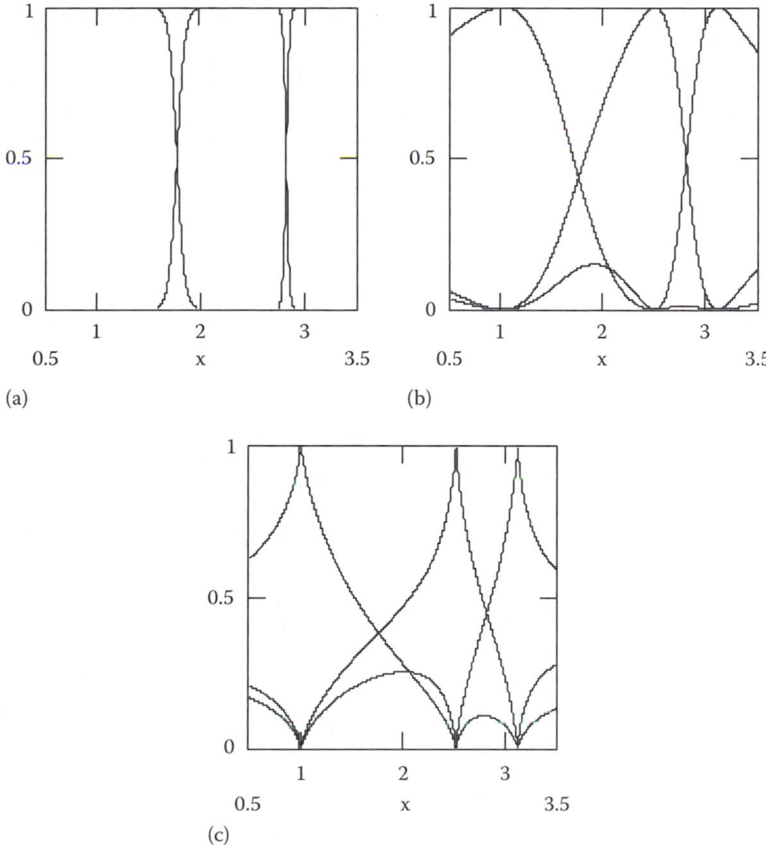

FIGURE 22.10 Examples of membership functions of fuzzy sets; the prototypes are equal to 1, 3.5, and 5 while the fuzzification coefficient assumes values of (a) 1.2, (b) 2.0, and (c) 3.5. The intensity of the rippling effect is affected by the values of "m" and increases with the higher values of "m."

where $u_1, u_2, \ldots, u_c$ are the membership degrees for some data point. If only one of membership degrees, say $u_i = 1$, and the remaining are equal to zero, then the separation index attains its maximum equal to 1. On the other extreme, when the data point is shared by all clusters to the same degree equal to 1/c, then the value of the index drops down to zero. This means that there is no separation between the clusters as reported for this specific point.

22.5.2 Knowledge-Based Clustering

As is well-known, clustering and supervised pattern recognition (classification) are the two opposite poles of the learning paradigm. In reality, there is no "pure" unsupervised learning. There is no fully supervised learning as some labels might not be completely reliable (as those encountered in case of learning with probabilistic teacher).

There is some domain knowledge and it has to be carefully incorporated into the generic clustering procedure. Knowledge hints can be conveniently captured and formalized in terms of fuzzy sets. Altogether with the underlying clustering algorithms, they give rise to the concept of knowledge-based clustering—a unified framework in which data and knowledge are processed together in a uniform fashion.

We discuss some of the typical design scenarios of knowledge-based clustering and show how the domain knowledge can be effectively incorporated into the fabric of the original data-driven only clustering techniques.

We can distinguish several interesting and practically viable ways in which domain knowledge is taken into consideration:

A subset of labeled patterns. The knowledge hints are provided in the form of a small subset of labeled patterns $\boldsymbol{K} \subset \boldsymbol{N}$ [PW97a,PW97b]. For each of them we have a vector of membership grades f_k, $k \in \boldsymbol{K}$ which consists of degrees of membership the pattern is assigned to the corresponding clusters. As usual, we have $f_{ik} \in [0, 1]$ and $\sum_{i=1}^{c} f_{ik} = 1$.

Proximity-based clustering. Here we are provided a collection of pairs of patterns [LPS03] with specified levels of closeness (resemblance) which are quantified in terms of proximity, prox (k, l) expressed for $\mathbf{x}_k$ and $\mathbf{x}_l$. The proximity offers a very general quantification scheme of resemblance: we require reflexivity and symmetry, that is, prox(k, k) = 1 and prox(k, l) = prox(l, k); however, no transitivity is needed.

"Belong" and "not-belong" Boolean relationships between patterns. These two Boolean relationships stress that two patterns should belong to the same clusters, $R(\mathbf{x}_k, \mathbf{x}_l) = 1$ or they should be placed apart in two different clusters, $R(\mathbf{x}_k, \mathbf{x}_l) = 0$. These two requirements could be relaxed by requiring that these two relationships return values close to one or zero.

Uncertainty of labeling/allocation of patterns. We may consider that some patterns are "easy" to assign to clusters while some others are inherently difficult to deal with meaning that their cluster allocation is associated with a significant level of uncertainty. Let $\boldsymbol{\Phi}(\mathbf{x}_k)$ stands for the uncertainty measure (e.g., entropy) for $\mathbf{x}_k$ (as a matter of fact, $\boldsymbol{\Phi}$ is computed for the membership degrees of $\mathbf{x}_k$ that is $\boldsymbol{\Phi}(\mathbf{u}_k)$ with $\mathbf{u}_k$ being the kth column of the partition matrix. The uncertainty hint is quantified by values close to 0 or 1 depending upon what uncertainty level a given pattern is coming from.

Depending on the character of the knowledge hints, the original clustering algorithm needs to be properly refined. In particular the underlying objective function has to be augmented to capture the knowledge-based requirements. Shown below are several examples of the extended objective functions dealing with the knowledge hints introduced above.

When dealing with some labeled patterns, we consider the following augmented objective function:

$$Q = \sum_{i=1}^{c} \sum_{k=1}^{N} u_{ik}^{m} ||\mathbf{x}_k - \mathbf{v}_i||^2 + \alpha \sum_{i=1}^{c} \sum_{k=1}^{N} (u_{ik} - f_{ik} b_k)^2 ||\mathbf{x}_k - \mathbf{v}_i||^2 \tag{22.28}$$

where the second term quantifies distances between the class membership of the labeled patterns and the values of the partition matrix. The positive weight factor (α) helps set up a suitable balance between the knowledge about classes already available and the structure revealed by the clustering algorithm. The Boolean variable b_k assumes values equal to 1 when the corresponding pattern has been labeled.

The proximity constraints are accommodated as a part of the optimization problem where we minimize the distances between proximity values being provided and those generated by the partition matrix $P(k_1, k_2)$:

$$Q = \sum_{i=1}^{c} \sum_{k=1}^{N} u_{ik}^{m} ||\mathbf{x}_k - \mathbf{v}_i||^2$$

$$||\text{prox}(k_1, k_2) - P(k_1, k_2)|| \;\rightarrow\; \text{Min } k_1, k_2 \in \boldsymbol{K} \tag{22.29}$$

with $\boldsymbol{K}$ being a pair of patterns for which the proximity level has been provided. It can be shown that given the partition matrix the expression $\sum_{i=1}^{c} \min(u_{ik1}, u_{ik2})$ generates the corresponding proximity value.

For the uncertainty constraints, the minimization problem can be expressed as follows:

$$Q = \sum_{i=1}^{c}\sum_{k=1}^{N} u_{ik}^{m} ||\mathbf{x}_k - \mathbf{v}_i||^2$$

$$||\Phi(\mathbf{u}_k) - \gamma_k|| \rightarrow \text{Min } k\ \boldsymbol{K} \tag{22.30}$$

where $\boldsymbol{K}$ stands for the set of patterns for which we are provided with the uncertainty values γ_k.

Undoubtedly the extended objective functions call for the optimization scheme that is more demanding as far as the calculations are concerned. In several cases we cannot modify the standard technique of Lagrange multipliers which leads to an iterative scheme of successive updates of the partition matrix and the prototypes. In general, though, the knowledge hints give rise to a more complex objective function in which the iterative scheme cannot be useful in the determination of the partition matrix and the prototypes. Alluding to the generic FCM scheme, we observe that the calculations of the prototypes in the iterative loop are doable in case of the Euclidean distance. Even the Hamming or Tchebyshev distance brings a great deal of complexity. Likewise, the knowledge hints lead to the increased complexity: the prototypes cannot be computed in a straightforward way and one has to resort himself to more advanced optimization techniques. Evolutionary computing arises here as an appealing alternative. We may consider any of the options available there including genetic algorithms, particle swarm optimization, ant colonies, to name some of them. The general scheme can be schematically structured as follows:

- repeat {EC (prototypes); compute partition matrix U;}

22.6 Data and Dimensionality Reduction

The problem of dimensionality reduction [DHS01] and complexity management in pattern recognition is by no means a new endeavor. This has led to a number of techniques which as of now are regarded classic and are used quite intensively. There have been a number of approaches deeply rooted in classic statistical analysis. The ideas of principal component analysis, Fisher analysis, and alike are the techniques of paramount relevance. What has changed quite profoundly over the decades is the magnitude of the problem itself which has forced us to the exploration of new ideas and optimization techniques involving advanced techniques of global search including tabu search and biologically inspired optimization mechanisms.

In a nutshell, we can distinguish between two fundamental reduction processes involving (a) data and (b) features (attributes). Data reduction is concerned with grouping patterns and revealing their structure in the form of clusters (groups). Clustering is regarded as one of the fundamental techniques within the domain of data reduction. Typically, we start with thousand of data points and arrive at 10–15 clusters. The nature of the clusters could vary depending upon the underlying formalisms. While in most cases, the representatives of the clusters are numeric entities such as prototypes or medoids, we can encounter granular constructs such as, e.g., hyperboxes.

Feature or attribute reduction [F90,M03,UT07,DWM02] deals with (a) transformation of the feature space into another feature space of a far lower dimensionality or (b) selection of a subset of features that are regarded to be the most essential (dominant) with respect to a certain predefined objective function. Considering the underlying techniques of feature transformation, we encounter a number of classic linear statistical techniques such as, e.g., principal component analysis or more advanced nonlinear mapping mechanisms realized by, e.g., neural networks.

The criteria used to assess the quality of the resulted (reduced) feature space give rise to the two general categories, namely *filters* and *wrappers*. Using filters we consider some criterion that pertains to the statistical *internal* characteristics of the selected attributes and evaluate them with this respect.

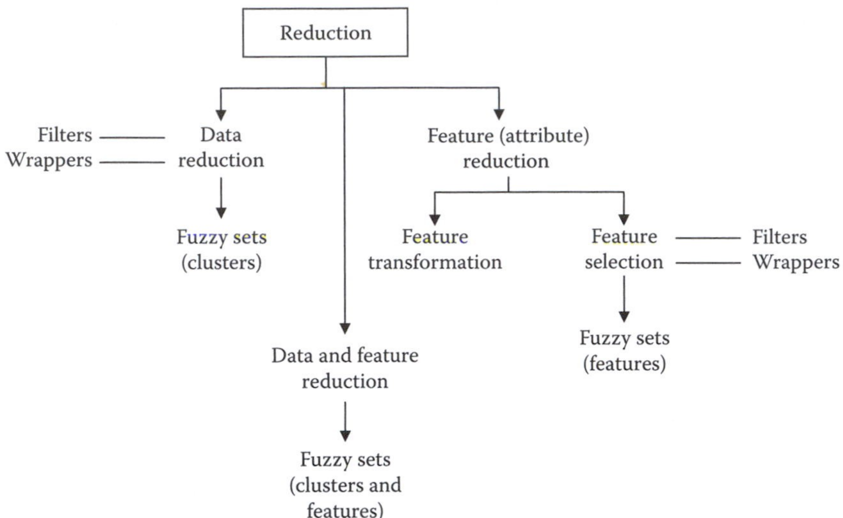

FIGURE 22.11 Categories of reduction problems in pattern recognition: feature and data reduction, use of filters, and wrappers criteria.

In contrast, when dealing with wrappers, we are concerned with the effectiveness of the features as a vehicle to carry out classification so in essence there is a mechanism (e.g., a certain classifier) which effectively evaluates the performance of the selected features with respect to their discriminatory (that is *external*) capabilities. In addition to feature and data reduction being regarded as two separate processes, we may consider their combinations in which both features and data are reduced. The general view at the reduction mechanisms is presented in Figure 22.11. Fuzzy sets augment the principles of the reduction processes. In case of data, fuzzy clusters reveal a structure in data and quantifying the assignment through membership grades. The same clustering mechanism could be applied to features and form collections of features which could be treated en block. The concept of biclustering engages the reduction processes realized together and leads to a collection of fuzzy sets described in the Cartesian product of data and features. In this sense, with each cluster one associates a collection of the features that describe it to the highest extent and make it different from the other clusters.

22.7 Conclusions

Fuzzy pattern recognition comes as a coherent and diversified setting of pattern recognition. The human-centricity is one of its dominant aspects, which is well supported by the essence of fuzzy sets. They are effectively used in the realization of more focused and specialized feature space being quite often of reduced dimensionality in comparison with the original one. The ability to quantify class membership by departing from the rigid 0–1 quantification and allowing for in-between membership grades is another important facet of pattern recognition, facilitating interaction with humans (designers and users of the classifiers). The logic fabric of the classifiers is yet another advantage of fuzzy classifiers enhancing their interpretability.

It is worth stressing that there are numerous ways in which the algorithmic aspects of fuzzy sets can be incorporated into fuzzy pattern recognition. The existing research, albeit quite diversified, comes with a number of conceptual avenues to investigate more thoroughly. This may involve (a) the use of other constructs of Granular Computing in a unified fashion, (b) synergy between fuzzy classification and probabilistic/statistical pattern classifiers, and (c) exploration of hierarchies of fuzzy pattern recognition constructs.

Acknowledgments

Support from the Natural Sciences and Engineering Research Council of Canada (NSERC) and Canada Research Chair (CRC) is gratefully acknowledged.

References

[ATK98] S. Abe, R. Thawonmas, and Y. Kobayashi, Feature selection by analyzing class regions approximated by ellipsoids, *IEEE Transactions on Systems, Man and Cybernetics-C*, 28, 2, 1998, 282–287.

[BP03a] A. Bargiela and W. Pedrycz, Recursive information granulation: Aggregation and interpretation issues, *IEEE Transactions on Systems, Man and Cybernetics-B*, 33, 1, 2003, 96–112.

[BP03b] A. Bargiela and W. Pedrycz, *Granular Computing: An Introduction*, Kluwer Academic Publishers, Dordrecht, the Netherlands, 2003.

[BKZ66] R. Bellman, R. Kalaba, and L. Zadeh, Abstraction and pattern classification, *Journal of Mathematical Analysis and Applications*, 13(1), 1966, 1–7.

[B81] J.C. Bezdek, *Pattern Recognition with Fuzzy Objective Function Algorithms*, Plenum Press, New York, 1981.

[DWM02] M. Daszykowski, B. Walczak, and D.L. Massart, Representative subset selection, *Analytica Chimica Acta*, 468, 2002, 91–103.

[DHS01] R.O. Duda, P.E. Hart, and D.G. Stork, *Pattern Classification*, 2nd edn, Wiley, New York, 2001.

[D82] R.P.W. Duin, The use of continuous variables for labeling objects, *Pattern Recognition Letters*, 1(1), 1982, 15–20.

[F90] K. Fukunaga, *Introduction to Statistical Pattern Recognition*, 2nd edn, Academic Press, San Diego, CA, 1990.

[GB00] B. Gabrys and A. Bargiela, General fuzzy min-max neural networks for clustering and classification, *IEEE Transactions on Neural Networks*, 11(3), 2000, 769–783.

[HP93] K. Hirota, W. Pedrycz, Logic based neural networks, *Information Sciences*, 71, 1993, 99–130.

[H99] F. Hoppner et al., *Fuzzy Cluster Analysis*, Wiley, Chichester, U.K., 1999.

[K82] A. Kandel, *Fuzzy Techniques in Pattern Recognition*, Wiley, Chichester, U.K., 1982.

[LMT02] R.P. Li, M. Mukaidono, and I.B. Turksen, A fuzzy neural network for pattern classification and feature selection, *Fuzzy Sets and Systems*, 130(1), 2002, 101–108.

[LPS03] V. Loia, W. Pedrycz, S. Senatore, P-FCM: A proximity-based fuzzy clustering for user-centered web applications, *International Journal of Approximate Reasoning*, 34, 2003, 121–144.

[M03] F. Marcelloni, Feature selection based on a modified fuzzy C-means algorithm with supervision, *Information Sciences*, 151, 2003, 201–226.

[M66] R. Moore, *Interval Analysis*, Prentice-Hall, Englewood Cliffs, NJ, 1966.

[P82] Z. Pawlak, Rough sets, *International Journal of Computing and Information Sciences*, 11, 1982, 341–356.

[P91] Z. Pawlak, *Rough Sets. Theoretical Aspects of Reasoning About Data*, Kluwer Academic Publishers, Dordrecht, the Netherlands, 1991.

[PS07] Z. Pawlak and A. Skowron, Rough sets and Boolean reasoning, *Information Sciences*, 177(1), 2007, 41–73.

[PB02] W. Pedrycz and A. Bargiela, Granular clustering: a granular signature of data, *IEEE Transactions on Systems, Man and Cybernetics*, 32(2), 2002, 212–224.

[P90] W. Pedrycz, Fuzzy sets in pattern recognition: Methodology and methods, *Pattern Recognition*, 23 (1–2), 1990, 121–146.

[PR93] W. Pedrycz and A. Rocha, Knowledge-based neural networks, *IEEE Transactions on Fuzzy Systems*, 1, 1993, 254–266.

[P95] W. Pedrycz, Distributed fuzzy systems modeling, *IEEE Transactions on Systems, Man, and Cybernetics*, 25, 1995, 769–780.

[PS05] W. Pedrycz and G. Succi, Genetic granular classifiers in modeling software quality, *Journal of Systems and Software*, 76(3), 2005, 277–285.

[PW97a] W. Pedrycz and J. Waletzky, Neural network front-ends in unsupervised learning, *IEEE Transactions on Neural Networks*, 8, 1997, 390–401.

[PW97b] W. Pedrycz and J. Waletzky, Fuzzy clustering with partial supervision, *IEEE Transactions on Systems, Man, and Cybernetics*, 5, 1997, 787–795.

[P05] W. Pedrycz, *Knowledge-Based Clustering: From Data to Information Granules*, Wiley, Hoboken, NJ, 2005.

[PG07] W. Pedrycz and F. Gomide, *Fuzzy Systems Engineering*, Wiley, Hoboken, NJ, 2007.

[PP08] W. Pedrycz, B.J. Park, and S.K. Oh, The design of granular classifiers: A study in the synergy of interval calculus and fuzzy sets in pattern recognition, *Pattern Recognition*, 41(12), 2008, 3720–3735.

[RMM00] A. Rizzi, F.M.F. Mascioli, and G. Martinelli, Generalized min-max classifiers, *Proceedings of the 9th IEEE International Conference on Fuzzy Systems, Fuzz-IEEE 2000*, San Antonio, TX, Vol. 1, pp. 36–41, 2000.

[RPM02] A. Rizzi, M. Panella, and F.M.F. Mascioli, Adaptive resolution min-max classifiers, *IEEE Transactions on Neural Networks*, 13(2), 2002, 402–414.

[R78] M. Roubens, Pattern classification problems and fuzzy sets, *Fuzzy Sets and Systems*, 1(4), 1978, 239–253.

[S92] P.K. Simpson, Fuzzy min-max neural networks-Part 1; Classification, *IEEE Transactions on Neural Networks*, 3(5), 1992, 776–786.

[S93] P.K. Simpson, Fuzzy min-max neural networks-Part 2; Clustering, *IEEE Transactions on Fuzzy Systems*, 1(1), 1993, 32–45.

[T73] M.G. Thomason, Finite fuzzy automata, regular fuzzy languages, and pattern recognition, *Pattern Recognition*, 5(4), 1973, 383–390.

[UT07] O. Uncu and I.B. Türksen, A novel feature selection approach: Combining feature wrappers and filters, *Information Sciences*, 177(2), 2007, 449–466.

[Z65] L.A. Zadeh, Fuzzy sets, *Information & Control*, 8, 1965, 338–353.

[Z97] L.A. Zadeh, Towards a theory of fuzzy information granulation and its centrality in human reasoning and fuzzy logic, *Fuzzy Sets and Systems*, 90, 1997, 111–117.

[Z05] L.A. Zadeh, Toward a generalized theory of uncertainty (GTU)—An outline, *Information Sciences*, 172, 2005, 1–40.

23

Fuzzy Modeling of Animal Behavior and Biomimcry: The Fuzzy Ant*

Valeri Rozin
Tel Aviv University

Michael Margaliot
Tel Aviv University

23.1 Introduction

Mathematical models are indispensable when we wish to rigorously analyze dynamic systems. Such a model summarizes and interprets the empirical data. It can also be used to simulate the system on a computer and to provide predictions for future behavior. Mathematical models of the atmosphere, which can be used to provide weather predictions, are a classic example.

In physics, and especially in classical mechanics, it is sometimes possible to derive mathematical models using *first principles* such as the Euler–Lagrange equations [1]. In other fields of science, like biology, economics, and psychology, no such *first principles* are known. In many cases, however, researchers have provided descriptions and explanations of various phenomena stated in *natural language*. Science can greatly benefit from transforming these verbal descriptions into mathematical models. This raises the following problem.

---

* This chapter is based on the paper "The fuzzy ant," by V. Rozin and M. Margaliot which appeared in the *IEEE Computational Intelligence Magazine*, 2, 18–28, 2007. © IEEE.

Problem 23.1 *Find an efficient way to transform verbal descriptions into a mathematical model or computer algorithm.*

This problem has already been addressed in the field of *artificial expert systems* (AESs) [2]. These are computer algorithms that emulate the functioning of a human expert, for example, a physician who can diagnose diseases or an operator who can successfully control a specific system. One approach to constructing AESs is based on questioning the human expert in order to extract information on his/her functioning. This leads to a verbal description, which must then be transformed into a computer algorithm.

Fuzzy logic theory has been associated with human linguistics ever since it was first suggested by Zadeh [3,4]. In particular, *fuzzy modeling* (FM) is routinely used to transform the knowledge of a human expert, stated in natural language, into a *fuzzy expert system* that imitates the expert's functioning [5,6]. The knowledge extracted from the human expert is stated as a collection of If–Then rules expressed using natural language. Defining the verbal terms in the rules using suitable membership functions, and inferring the rule base, yields a well-defined mathematical model. Thus, the verbal information is transformed into a form that can be programmed on a computer. This approach has been used to develop AESs that diagnose diseases, control various processes, and much more [2,6,7].

The overwhelming success of fuzzy expert systems suggests that FM may be a suitable approach for solving Problem 23.1. Indeed, decades of successful applications (see, e.g., [8–13]) suggest that the real power of fuzzy logic lies in its ability to handle and manipulate verbally-stated information based on perceptions rather than equations [14–18].

Recently, FM was applied in a different context, namely, in transforming verbal descriptions and explanations of *natural phenomena* into a mathematical model. The goal here is not to replace a human expert with a computer algorithm, but rather to assist a researcher in transforming his/her understanding of the phenomena, stated in words, into a well-defined mathematical model.

The applicability and usefulness of this approach was demonstrated using examples from the field of ethology. FM was applied to transform verbal descriptions of various animal behaviors into mathematical models. Examples include the following: (1) territorial behavior in the stickleback [19], as described by Nobel Laureate Konrad Lorenz in [20]; (2) the mechanisms governing the orientation to light in the planarian *Dendrocoleum lacteum* [21]; (3) flocking behavior in birds [22]; (4) the self-regulation of population size in blow–flies [23]; and (5) the switching behavior of an epigenetic switch in the lambda virus [24].

There are several reasons that FM seems particularly suitable for modeling animal behavior. First, many animal (and human) actions are "fuzzy." For example, the response to a (low intensity) stimulus might be what Heinroth called *intention movements*, that is, a slight indication of what the animal is tending to do. Tinbergen [25, Ch. IV] states: "As a rule, no sharp distinction is possible between intention movements and more complete responses; they form a continuum." In this respect, it is interesting to recall that Zadeh [3] defined a *fuzzy set* as "a class of objects with a continuum of grades of membership." Hence, FM seems an appropriate tool for studying such behaviors. The second reason is that studies of animal behavior often provide a *verbal* description of both field observations and interpretations. For example, Fraenkel and Gunn describe the behavior of a cockroach that becomes stationary when a large part of its body surface is in contact with a solid object as "A high degree of contact causes low activity ..." [26, p. 23]. Note that this can be immediately stated as the fuzzy rule: If degree of contact is *high*, then activity is *low*. In fact, it is customary to describe the behavior of simple organisms using simple rules of thumb [27].

23.1.1 Fuzzy Modeling and Biomimicry

Considerable research is currently devoted to the field of *biomimicry*—the development of artificial products or machines that mimic biological phenomena [28–31]. Over the course of evolution, living systems have developed efficient and robust solutions to various problems. Some of these problems are also encountered in engineering applications. For example, plants had to develop efficient mechanisms for absorbing and utilizing solar energy. Engineers who design solar cells face a similar challenge. More generally, many natural beings have developed the capabilities to reason, learn, evolve, adapt, and heal. Scientists and engineers are interested in implementing such capabilities in artificial systems.

An important component in the design of artificial systems based on natural phenomena is the ability to perform *reverse engineering* of the functioning of the natural phenomena. We believe that FM may be suitable for addressing this issue in a systematic manner [32]. Namely, start with a verbal description of the biological system's behavior (e.g., foraging in ants) and, using fuzzy logic theory, obtain a mathematical model of this behavior that can be immediately implemented by artificial systems (e.g., autonomous robots).

In this chapter, we describe the application of FM to develop a mathematical model for the foraging behavior of ants. The resulting model is simpler, more plausible, and more amenable to analysis than previously suggested models. Simulations and rigorous analysis of the resulting model show that it is congruent with the behavior actually observed in nature. Furthermore, the new model establishes an interesting link between the averaged behavior of a colony of foraging ants and mathematical models used in the theory of artificial neural networks (ANNs) (see Section 23.8).

The next section provides a highly simplified, yet hopefully intuitive, introduction to fuzzy modeling. Section 23.3 reviews the foraging behavior of social ants. Section 23.4 applies FM to transform the verbal description into a simple mathematical model describing the behavior of a single ant. In Section 23.5, this is used to develop a stochastic model for the behavior of a colony of identical ants. Section 23.6 reviews an averaged model of the colony. Sections 23.7 and 23.8 are devoted to studying this averaged model using simulations and rigorous analysis, respectively. The final section concludes and describes some possible directions for further research.

23.2 Fuzzy Modeling: A Simple Example

We begin by presenting the rudiments of FM using a very simple example. More information on FM can be found in many textbooks (e.g., [11,33]). Readers familiar with FM may skip to Section 23.3.

Consider the scalar control system:

$$\dot{x}(t) = u(t),$$

where

$x: \mathbb{R} \to \mathbb{R}$ is the state of the system
$u: \mathbb{R} \to \mathbb{R}$ is the control

Suppose that our goal is to design a state–feedback control (i.e., a control in the form $u(t) = u(x(t))$) guaranteeing that $\text{Lim}_{t\to\infty} x(t) = 0$ for any initial condition $x(0)$. It is clear that in order to achieve this, the control must be negative (positive) when $x(t)$ is positive (negative). This suggests the following two rules:

Rule 1: If (x is *positive*), then $u = -c$,

Rule 2: If (x is *negative*), then $u = c$,

where $c > 0$.

FM provides an efficient mechanism for transforming such rules into a well-defined mathematical formula: $u = u(x)$. The first step is to define the terms in the If part of the rules. To do this, we use two functions: $\mu_{positive}(x)$ and $\mu_{negative}(x)$. Roughly speaking, for a given x, $\mu_{positive}(x)$ measures how true the proposition (x *is positive*) is. For example, we may take

$$\mu_{positive}(x) = \begin{cases} 1, & \text{if } x > 0, \\ 0, & \text{if } x \le 0. \end{cases}$$

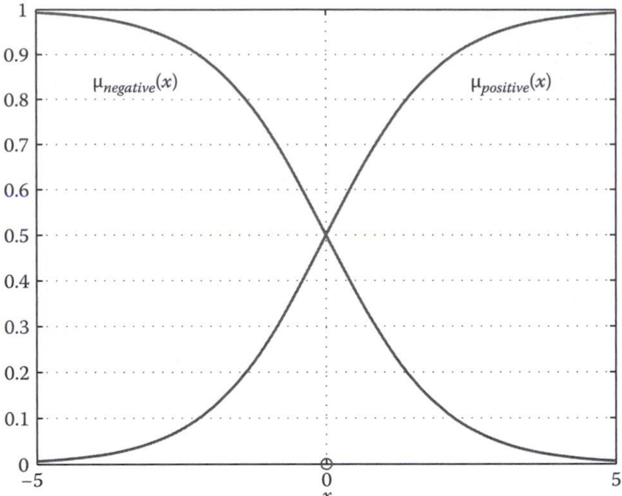

FIGURE 23.1 Membership functions $\mu_{positive}(x)$ and $\mu_{negative}(x)$.

However, using such a binary, 0/1, function will lead to a control that changes abruptly as x changes sign. It may thus be better to use a smoother function, say

$$\mu_{negative}(x) := 1 - (1 + \exp(-x))^{-1}.$$

This is a continuous function taking values in the interval [0,1] and satisfying (see Figure 23.1).

$$\lim_{x \to -\infty} \mu_{positive}(x) = 0, \lim_{x \to +\infty} \mu_{positive}(x) = 1.$$

We may also view $\mu_{positive}(x)$ as providing the *degree of membership* of x in the set of *positive numbers*. A smoother membership function seems more appropriate for sets that are defined using verbal terms [34]. To demonstrate this, consider the membership in the set of *tall people*. A small change in a person's height should not lead to an abrupt change in the degree of membership in this set.

The second membership function is defined by $\mu_{negative}(x) := 1 - (1 + \exp(-x))^{-1}$. Note that this implies that

$$\mu_{positive}(x) + \mu_{negative}(x) = 1, \quad \text{for all } x \in \mathbb{R},$$

i.e., the *total* degree of membership in the two sets is always one.

Once the membership functions are specified, we can define the *degree of firing* (DOF) of each rule, for a given input x, as $DOF_1(x) = \mu_{positive}(x)$ and $DOF_2(x) = \mu_{negaitive}(x)$. The output of the first (second) rule in our fuzzy rule base is then defined by $-cDOF_1(x)$ $(cDOF_2(x))$. In other words, the output of each rule is obtained by multiplying the DOF with the value in the Then part of the rule. Finally, the output of the entire fuzzy rule base is given by suitably combining the outputs of the different rules. This can be done in many ways. One standard choice is to use the so-called *center of gravity inferencing method* yielding:

$$u(x) = \frac{-cDOF_1(x) + cDOF_2(x)}{DOF_1(x) + DOF_2(x)}.$$

The numerator is the sum of the rules' outputs, and the denominator plays the role of a scaling factor. Note that we may also express this as

$$u(x) = -c\frac{DOF_1(x)}{DOF_1(x) + DOF_2(x)} + c\frac{DOF_2(x)}{DOF_1(x) + DOF_2(x)},$$

which implies that the output is always a convex combination of the rules' outputs.

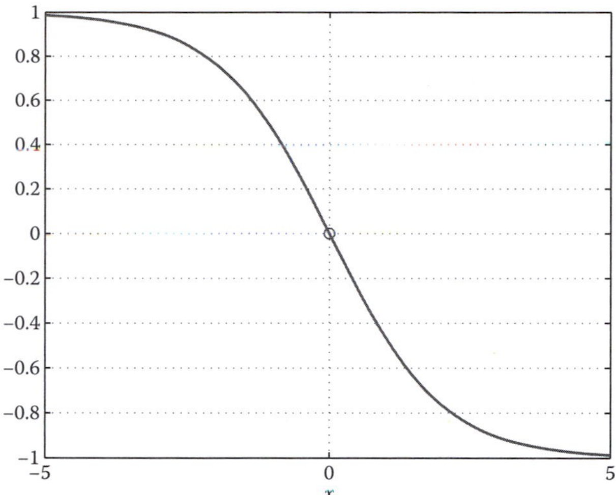

FIGURE 23.2 The function $u(x) = -c\tanh(x/2)$ for $c = 1$.

Substituting the membership functions yields the controller:

$$u(x) = -c(1+\exp(-x))^{-1} + c(1-(1+\exp(-x))^{-1})$$
$$= -c\tanh\left(\frac{x}{2}\right)$$

(see Figure 23.2). Note that this can be viewed as a smooth version of the controller:

$$u(x) = \begin{cases} -c, & \text{if } x > 0, \\ c, & \text{if } x < 0. \end{cases}$$

Summarizing, FM allows us to transform verbal information, stated in the form of If–Then rules, into a well-defined mathematical function. Note that the fuzziness here stems from the inherent vagueness of verbal terms. This vagueness naturally implies that any modeling process based on verbal information would include many degrees of freedom [32]. Yet, it is important to note that the final result of the FM process is a completely well-defined mathematical formulation.

23.3 Foraging Behavior of Ants

A foraging animal may have a variety of potential paths to a food item. Finding the *shortest* path minimizes time, effort, and exposure to hazards. For mass foragers, such as ants, it is also important that *all* foragers reach a consensus when faced with a choice of paths. This is not a trivial task, as ants have very limited capabilities of processing and sharing information. Furthermore, this consensus is not reached by means of an ordered chain of hierarchy.

Ants and other social insects have developed an efficient technique for solving these problems [35]. While walking from a food source to the nest, or vice versa, ants deposit a chemical substance called *pheromone*, thus forming a pheromone trail. Following ants are able to smell this trail. When faced by several alternative paths, they tend to choose those that have been marked by pheromones. This leads to a positive feedback mechanism: a marked trail will be chosen by more ants that, in turn, deposit more pheromone, thus stimulating even more ants to choose the same trail.

Goss et al. [36] designed an experiment in order to study the behavior of the Argentine ant *Iridomyrmex humilis* while constructing a trail around an obstacle. A laboratory nest was connected to a food source

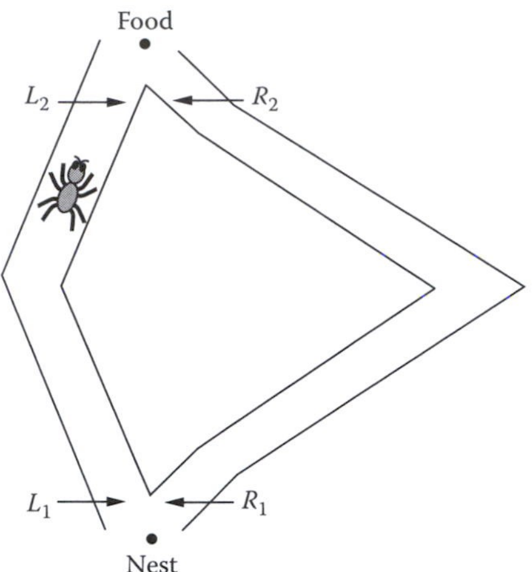

FIGURE 23.3 Experimental setup with two branches: the left branch is shorter.

by a double bridge (see Figure 23.3). Ants leaving the nest or returning from the food item to the nest must choose a branch. After making the choice, they mark the chosen branch. Ants that take the shorter of the two branches return sooner than those using the long branch. Thus, in a given time unit, the short branch receives more markings than the long branch. This small difference in the pheromone concentrations is amplified by the positive feedback process. The process generally continues until nearly all the foragers take the same branch, neglecting the other one. In this sense, it appears that the entire colony has decided to use the short branch.

The positive feedback process is counteracted by negative feedback due to pheromone evaporation. This plays an important role: the markings of obsolete paths, which lead to depleted food sources, disappear. This increases the chances of detecting new and more relevant paths.

Note that in this model, no single ant compares the length of the two branches directly. Furthermore, the ants only communicate indirectly by laying pheromones and thus locally modifying their environment. This form of communication is known as *stigmergy* [37]. The net result, however, is that the entire colony appears to have made a well informed choice of using the shorter branch.

The fact that simple individual behaviors can lead to a complex *emergent behavior* has been known for centuries. King Solomon marveled at the fact that "the locusts have no king, yet go they forth all of them by bands" (Proverbs 30:27). More recently, it was noted that this type of emergent collective behavior is a desirable property in many artificial systems. From an engineering point of view, the solution of a complex problem using simple agents is an appealing idea, which can save considerable time and effort. Furthermore, the specific problem of detecting the shortest path is important in many applications, including robot navigation and communication engineering (see, e.g., [38–40]).

23.4 Fuzzy Modeling of Foraging Behavior

In this section, we apply FM to transform a verbal description of the foraging behavior into a mathematical model. The approach consists of the following stages: (1) identification of the variables, (2) stating the verbal information as a set of fuzzy rules relating the variables, (3) defining the fuzzy terms using suitable membership functions, and (4) inferring the rule-base to obtain a mathematical model [19].

When creating a mathematical model from a verbal description there are always numerous degrees of freedom. In the FM approach, this is manifested in the freedom in choosing the components of the fuzzy model: the type of membership functions, logical operators, inference method, and the values of the different parameters. The following guidelines may be helpful in selecting the different components of the fuzzy model (see also [33] for details on how the various elements in the fuzzy model influence its behavior).

First, it is important that the resulting mathematical model has the simplest possible form, in order to be amenable to analysis. Thus, for example, a Takagi–Sugeno model with singleton consequents might be more suitable than a model based on Zadeh's compositional rule of inference [33].

Second, when modeling real-world systems, the variables are physical quantities with dimensions (e.g., length, time). *Dimensional analysis* [41,42], the process of introducing dimensionless variables, can often simplify the resulting equations and decrease the number of parameters.

Third, sometimes the verbal description of the system is accompanied by measurements of various quantities in the system. In this case, methods such as fuzzy clustering, neural learning, or least squares approximation (see, e.g., [43–45] and the references therein) can be used to fine-tune the model using the discrepancy between the measurements and the model's output.

For the foraging behavior in the simple experiment described above, we need to model the choice-making process of an ant facing a fork in a path. We use the following verbal description [46]: "If a mass forager arrives at a fork in a chemical recruitment trail, the probability that it takes the left branch is all the greater as there is more trail pheromone on it than on the right one."

An ant is a relatively simple creature, and any biologically feasible description of its behavior must also be simple, as is the description above. Naturally, transforming this description into a set of fuzzy rules will lead to a simple rule-base. Nevertheless, we will see below that the resulting fuzzy model, although simple, has several unique advantages.

23.4.1 Identification of the Variables

The variables in the model are the pheromone concentrations on the left and right branches denoted L and R, respectively. The output is $P = P(L, R)$, which is the probability of choosing the left branch.

23.4.2 Fuzzy Rules

According to the verbal description given above, the probability P of choosing the left branch at the fork is directly correlated with the difference in pheromone concentrations $D: = L - R$. We state this using two fuzzy rules:

Rule 1: If D is *positive* Then $P = 1$.

Rule 2: If D is *negative* Then $P = 0$.

23.4.3 Fuzzy Terms

A suitable membership function for the term *positive*, $\mu_{pos}(\cdot)$, must satisfy the following constraints: $\mu_{pos}(D)$ is a monotonically increasing function, $\lim_{D\to-\infty} \mu_{pos}(D) = 0$, and $\lim_{D\to\infty} \mu_{pos}(D) = 1$. There are good reasons for using the hyperbolic tangent function in both ANNs and fuzzy models [18,47], so we use the membership function $\mu_{pos}(D): = (1 + \tanh(qD))/2$. The parameter $q > 0$ determines the slope of $\mu_{pos}(D)$. The term *negative* is modeled using $\mu_{neg}(D) = 1 - \mu_{pos}(D)$.

As we will see below, this choice of membership functions also leads to a mathematical model for the behavior of a colony of ants that is more amenable to analysis than previously suggested models.

23.4.4 Fuzzy Inferencing

We use center of gravity inference. This yields

$$P(D) = \frac{\mu_{pos}(D)}{\mu_{pos}(D) + \mu_{neg}(D)}$$

$$= \frac{1 + \tanh(qD)}{2}. \tag{23.1}$$

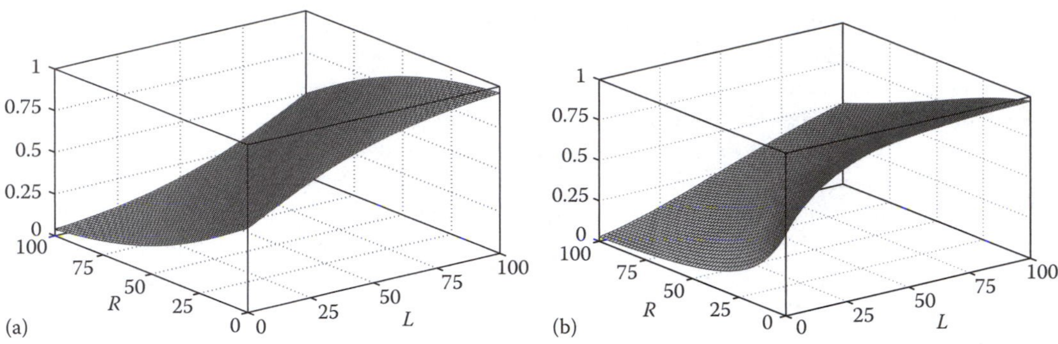

FIGURE 23.4 The functions: (a) $P(L - R)$ with $q = 0.016$ and (b) $P_{2,20}(L, R)$.

Note that $P(D) \in (0, 1)$ for all $D \in \mathbb{R}$. For $q = 0$, $P(D) = 1/2$ for all D, i.e., the selection between the two branches becomes completely random. For $q \to \infty$, $P(D) \approx 1$ ($P(D) \approx 0$) for $L > R$ ($L < R$). In other words, as q increases, $P(D)$ becomes more sensitive to the difference between L and R.

23.4.5 Parameter Estimation

Goss et al. [36] and Deneubourg et al. [48] suggested the following probability function:

$$P_{n,k}(L,R) = \frac{(k+L)^n}{(k+L)^n + (k+R)^n}. \tag{23.2}$$

As noted in [48], the parameter n determines the degree of nonlinearity of $P_{n,k}$. The parameter k corresponds to the degree of attraction attributed to an unmarked branch: as k increases, a greater marking is necessary for the choice to become significantly nonrandom.

Note that for $L \gg R$ ($L \ll R$), both (23.1) and (23.2) yield that the probability of choosing the left branch goes to one (zero). Our model is simpler and seems more plausible, as it depends only on the difference $D = L - R$, and it includes only a single parameter.

Deneubourg et al. [48] found that for $n = 2$ and $k = 20$, the function (23.2) agrees well with the actual behavior observed in an experiment involving *Iridomyrmex humilis*. Deneubourg et al. do not provide the exact biological data they used. In order to obtain (indirectly) a reasonable match with the real biological behavior, we tried to match $P(D)$ with the function $P_{2,20}(L, R)$. To do so, we (numerically) solved the problem:

$$\min_q \sum_{(L,R)\in A} |P(L-R) - P_{2,20}(L,R)|^2,$$

where $A = [0, 1, \ldots, 100] \times [0, 1, \ldots, 100]$. The best match is obtained for $q = 0.016$ (see Figure 23.4).

In the next sections, we simulate and rigorously analyze the behavior of a colony of "fuzzy" ants, that is, ants that choose between two alternative paths according to the function $P(D)$.

23.5 Stochastic Model

We model the scenario depicted in Figure 23.3 as a sequence of stochastic events. Initially, at time $t = 0$, all trails are unmarked: $L_1(0) = R_1(0) = L_2(0) = R_2(0) = 0$. Let τ denote the time needed to travel from the nest to the food item using the left branch. The corresponding time for the right branch is $r\tau$, with $r \geq 1$.

At every time step $t = 0, 1, \ldots, 1000$, a new ant heads out of the nest and chooses a branch at the fork near the nest. The choice is made according to the probability $P(L_1(t), R_1(t))$. If the choice is to follow the left [right] branch, then L_1 $[R_1]$ is increased by 1. This ant reaches the fork near the food source at time $t + \tau$ $[t + r\tau]$, adding 1 to $L_2(t + \tau)$ $[R_2(t + r\tau)]$, and then chooses which branch to use on its return according to $P(L_2(t + \tau), R_2(t + \tau))$ $[P(L_2(t + r\tau), R_2(t + r\tau))]$. Consequently, either L_2 or R_2 is increased, and after τ or $r\tau$ time steps, 1 is added to L_1 or R_1, respectively. The effect of pheromone evaporation, with rate $s \in (0, 1]$, is modeled by setting $L_I(t + 1) = (1 - s)\, L_i(t)$, $R_I(t + 1) = (1 - s)\, R_i(t)$ at every time step.

To estimate the traffic at steady state, we numbered the left/right decisions consecutively, and the results presented below are based on decisions 501–1000.

Figure 23.5 summarizes the results of 1000 simulations with $\tau = 20$, $s = 0.01$, and $r = 1$ (equal branches). Using (23.1), almost all simulations end up with the colony choosing one of the two branches. In 523 simulations, 80%–100% of the ants end up choosing the left branch. In almost all other simulations, 80%–100% of the ants end up choosing the right branch.

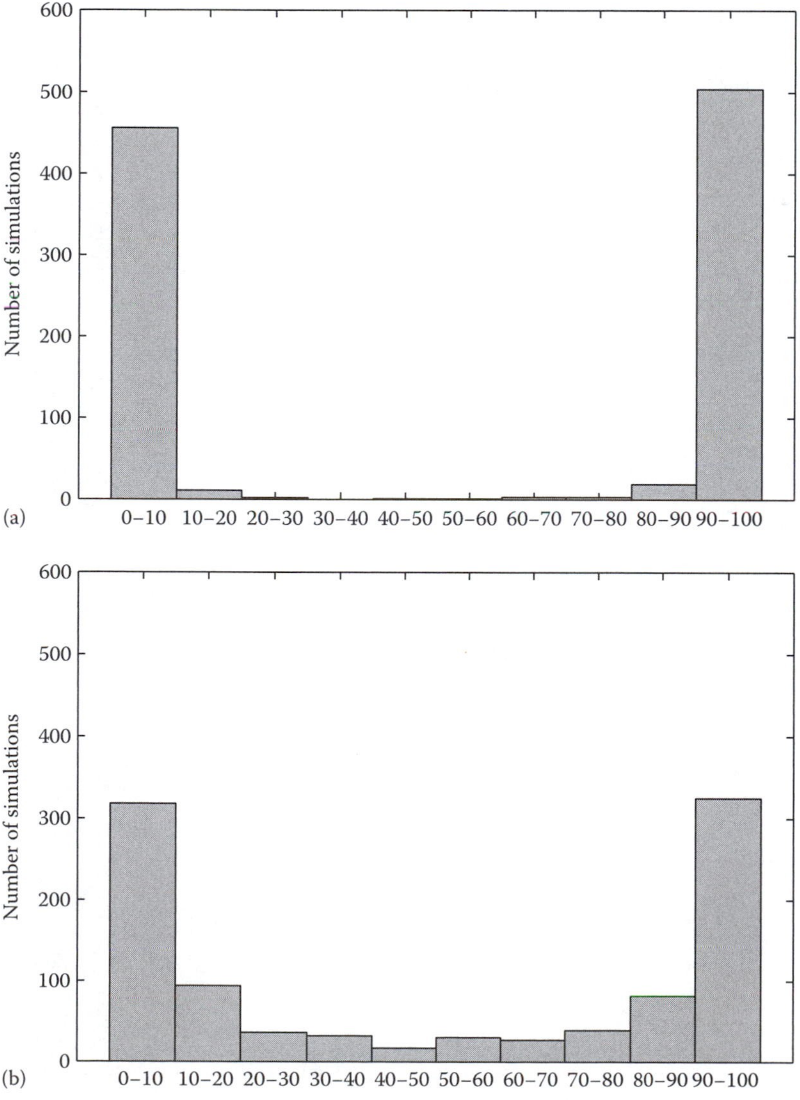

FIGURE 23.5 Percentage of ants that chose the left branch, for $r = 1$ and $s = 0.01$: (a) Using $P(L - R)$ with $q = 0.016$ and (b) using $P_{2,20}(L, R)$.

These results seem to agree with the behavior actually observed in experiments using a double bridge with equal length branches: "Should more ants use one of the branches at the beginning of the experiment, either by chance or for some other reason, then that branch will be most strongly marked and attract more ants, and so on until most of the ants use that branch" [46, p. 403].

Similar behavior is seen when using the probability function (23.2), but the distribution is more "blurry," as there are considerably more simulations ending with no clear-cut choice of one of the branches.

Figure 23.6 summarizes the results of 1000 simulations with $\tau = 20$, $s = 0.01$, and $r = 2$, that is, the time needed to follow the right branch is twice as long as that of the left branch. It may be seen that using the probability function (23.1) leads to a clear-cut distribution: in 807 simulations, 80%–100% of the ants choose the shorter branch. In 186 simulations, 0%–20% of the ants chose the shorter branch. Thus, in 993 of the 1000 simulations the colony converges to a favorable branch, and in 81% of the simulations this is indeed the shorter branch.

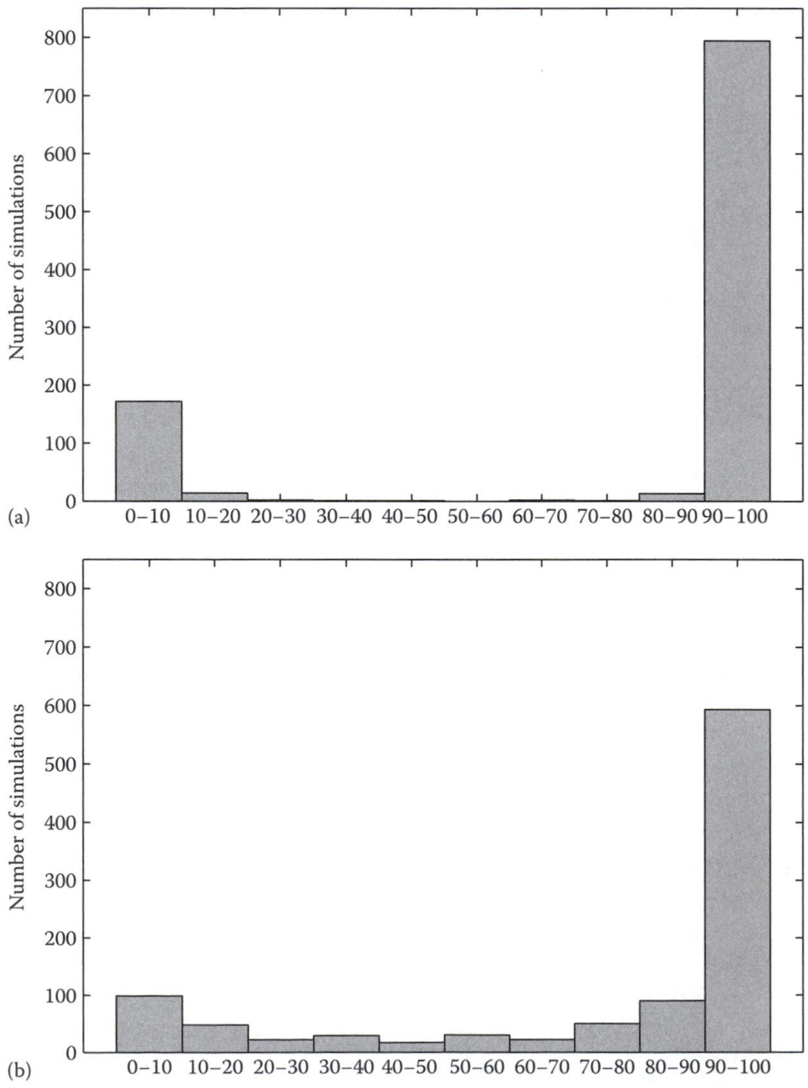

FIGURE 23.6 Percentage of ants that chose the left branch, for $r = 2$ and $s = 0.01$: (a) Using $P(L - R)$ with $q = 0.016$ and (b) using $P_{2,20}(L, R)$.

These results agree with the behavior actually observed in nature: "The experiments show that *L. niger* colonies nearly always select the shorter of two branches, and do so with a large majority of foragers" [46, p. 413].

Using the probability function (23.2) leads again to a more "blurry" distribution, where in more simulations there is no clear convergence toward a favorable branch.

23.6 Averaged Model

Following [36], we now consider a deterministic model that describes the "average" concentration of pheromones in the system. This averaged model is a set of four nonlinear delay differential equations (DDEs):

$$\dot{L}_1(t) = FP_1(t) + FP_2(t-\tau) - sL_1, \tag{23.3a}$$

$$\dot{L}_2(t) = FP_2(t) + FP_1(t-\tau) - sL_2, \tag{23.3b}$$

$$\dot{R}_1(t) = F(1-P_1(t)) + F(1-P_2(t-r\tau)) - sR_1, \tag{23.3c}$$

$$\dot{R}_2(t) = F(1-P_2(t)) + F(1-P_1(t-r\tau)) - sR_2, \tag{23.3d}$$

where

$P_1[P_2]$ is the probability of choosing the left branch at fork 1 [fork 2] as defined in (23.1)
F is the number of ants per second leaving the nest

The first equation can be explained as follows: the change in the pheromone concentration $L_1(t)$ is due to (1) the ants that choose to use the left branch at fork 1 and deposit pheromone as they start going; (2) the ants that choose the left branch at point 2 at time $t - \tau$. These reach point 1, and deposit pheromone on the left branch, after τ s; and (3) the reduction of pheromone due to evaporation. The other equations follow similarly.

Equation 23.3 is similar to the model used in [36,46], but we use P rather than $P_{n,k}$. It turns out that the fact that $P = P(L - R)$ allows us to transform the averaged model into a two-dimensional model, which is easier to analyze. Furthermore, using P leads to a novel and interesting link between the averaged behavior of the ant colony and mathematical models used in the theory of ANNs (see Section 23.8).

23.7 Simulations

We simulated (23.3) and compared the results to a Monte Carlo simulation of the stochastic model with a colony of 1000 foragers. Note that (23.3) describes pheromone concentrations, not ant numbers. Yet there is, of course, a correspondence between the two since the pheromones are laid by ants. We used the parameters $F = 1$, $\tau = 20$, $r = 2$, and the initial conditions $L_1(t) = 1$, $L_2(t) = R_1(t) = R_2(t) = 0$, for all $t \in [-r\tau, 0]$. To analyze the effect of evaporation, we considered two different values of s.

Figure 23.7 depicts the pheromone concentrations as a function of time for $s = 0.012$. The stochastic model and the averaged model behave similarly. Initially, the concentrations on both branches are equal. As time progresses, the left branch, which is the shorter one, receives more and more markings. The difference $L_1(t) - R_1(t)$ converges to a steady state value of 164.98.

Figure 23.8 depicts the results of the simulations for a higher value of evaporation rate, namely, $s = 0.2$. In this case, the traffic tends to distribute equally along the two branches. The reason is that the positive feedback process of pheromone laying is ineffective because the pheromones evaporate faster than the ants can lay them. This makes it impossible to detect the shorter branch. This is in agreement with the behavior actually observed in nature: "Only if the amount of ants arriving at the fork is insufficient

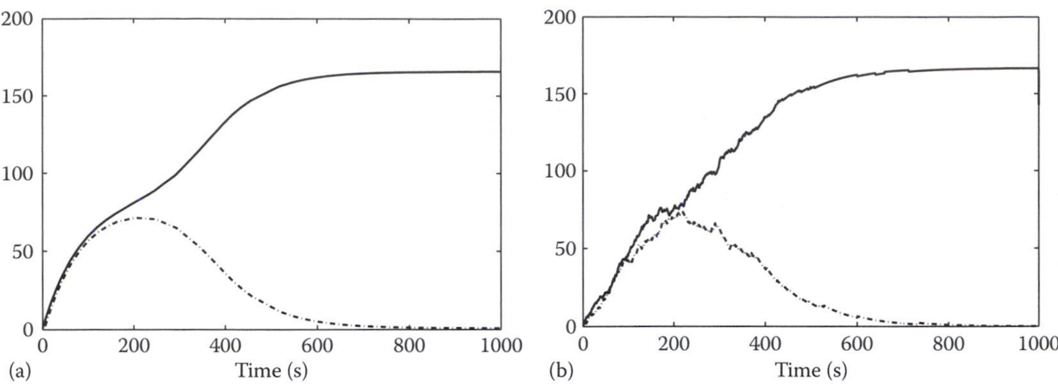

FIGURE 23.7 $L_1(t)$ (solid line) and $R_1(t)$ (dashed line) for $r = 2$, $s = 0.012$: (a) Solution of DDE (23.3) with $q = 0.016$ and (b) Monte Carlo simulation.

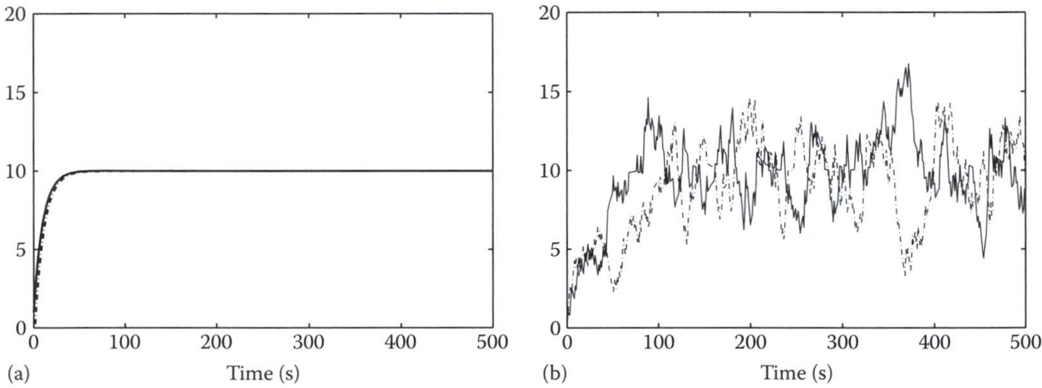

FIGURE 23.8 $L_1(t)$ (solid line) and $R_1(t)$ (dashed line) for $r = 2$, $s = 0.01$: (a) Solution of DDE (23.3) with $q = 0.016$ and (b) Monte Carlo simulation.

to maintain the pheromone trail in the face of evaporation will no choice be arrived at and the traffic distributed equally over the two branches" [46, p. 408].

23.8 Analysis of the Averaged Model

In this section, we provide a rigorous analysis of the averaged model. Let $v_j(t)$: = $(L_j(t) - R_j(t))/F$, $j = 1, 2$, that is, the scaled difference between the pheromone concentrations on the left-hand and right-hand sides of fork j. Using (23.3) and (23.1) yields

$$\begin{aligned}\dot{v}_1(t) &= -sv_1(t) + \tanh(pv_1(t)) + \frac{\tanh(pv_2(t-\tau)) + \tanh(pv_2(t-r\tau))}{2},\\ \dot{v}_2(t) &= -sv_2(t) + \tanh(pv_2(t)) + \frac{\tanh(pv_1(t-\tau)) + \tanh(pv_1(t-r\tau))}{2},\end{aligned} \tag{23.4}$$

where p: = $qF > 0$. Note that this simplification from a fourth-order to a second-order DDE is possible because our probability function, unlike (23.2), depends only on the difference $L - R$.

Models in the form (23.4) were used in the context of Hopfield-type ANNs with time delays (see [49] and the references therein). In this context, (23.4) represents a system of two dynamic neurons, each possessing nonlinear feedback, and coupled via nonlinear connections. The time delays represent propagation times along these connections. This yields an interesting and novel connection between the aggregated behavior of the colony and classical models used in the theory of ANNs. The set of ants choosing the left (right) path corresponds to the state of the first (second) neuron. The effect of the chemical communication between the ants corresponds to the time-delayed feedback connections between the neurons.

23.8.1 Equilibrium Solutions

The equilibrium solutions of (23.4) are $v(t) \equiv (v, v)^T$, where v satisfies

$$sv - 2\tanh(pv) = 0. \tag{23.5}$$

The properties of the hyperbolic tangent function yield the following result.

Proposition 23.1 If $s > 2p$, then the unique solution of (23.5) is $v = 0$, so (23.4) admits a unique equilibrium solution $v(t) \equiv 0$. If $s \in (0, 2p)$, then (23.5) admits three solutions $0, \underline{v}, -\underline{v}$, with $\underline{v} > 0$, and (23.4) admits three equilibrium solutions: $v(t) \equiv 0$, $v(t) \equiv v^1 := (v, v)^T$, and $v(t) \equiv -v^1$.

23.8.2 Stability

For the sake of completeness, we recall some stability definitions for DDEs (for more details, see [50–52]). Consider the DDE

$$x(t) = f(x(t), x(t-d)), \quad t \geq t_0, \tag{23.6}$$

with the initial condition $x(t) = \phi(t)$, $t \in [t_0 - d, t_0]$, and suppose that $x(t) \equiv 0$ is an equilibrium solution. For a continuous function $\phi: [t_0 - d, t_0] \to \mathbb{R}^n$, define the *continuous norm* by $\|\phi\|_c := \max\{\|\phi(\theta)\|: \theta \in [t_0 - d, t_0]\}$.

Definition 23.1 The solution 0 is said to be *uniformly stable* if for any $t_0 \in \mathbb{R}$ and any $\in > 0$ there exists $\delta = \delta(\in) > 0$ such that $\|\phi\|_c < \delta$ implies that $\|x(t)\| < \in$ for $t \geq t_0$. It is *uniformly asymptotically stable* if it is uniformly stable, and there exists $\delta_a > 0$ such that for any $\alpha > 0$, there exists $T = T(\delta_a, \alpha)$, such that $\|\phi\|_c < \delta_a$ implies that $\|x(t)\| < \alpha$ for $t \geq t_0 + T$. It is *globally uniformly asymptotically stable* (GUAS) if it is uniformly asymptotically stable and δ_a can be an arbitrarily large number.

Proposition 23.1 suggests that we need to consider the two cases $s > 2p$ and $s \in (0, 2p)$ separately.

23.8.2.1 High Evaporation

The next result shows that if $s > 2p$, then $L_i(t) \equiv R_i(t)$, $i = 1, 2$, is a GUAS solution of (23.3). In other words, when the evaporation rate is high and the pheromones cannot accumulate, the positive feedback process leading to a favorable trail cannot take place and eventually the traffic will be divided equally along the two possible branches.

Theorem 23.1 If $s > 2p > 0$, then 0 is a GUAS solution of (23.4) for any $\tau > 0$ and $r \geq 1$.
Proof: See [53].

23.8.2.2 Low Evaporation

For $s \in (0, 2p)$, the system admits three equilibrium solutions: 0, v^1, and $-v^1$.

Proposition 23.2 If $s \in (0, 2p)$, then 0 is an unstable solution of (23.4), and both v^1 and $-v^1$ are uniformly asymptotically stable solutions.
Proof: See [53].

Thus, $L_1(t) - R_1(t) \equiv L_2(t) - R_2(t) \equiv Fv^1$ and $L_1(t) - R_1(t) \equiv L_2(t) - R_2(t) \equiv -Fv^1$ are stable solutions of the averaged model. In other words, for low evaporation the system has a tendency toward a non-symmetric state, where one trail is more favorable than the other.

Summarizing, the analysis of the model implies that the system undergoes a bifurcation when $s = 2p$. Recall that s is the evaporation rate, and p is the product of the parameter q that determines the "sensitivity" of $P(D)$, and the rate of ants leaving the nest F. If $s < 2p$ (i.e., low evaporation), the pheromone laying process functions properly and leads to steady state solutions where one trail is more favorable than the other. If $s > 2p$ (high evaporation), then all solutions converge to the steady state where the traffic is divided equally along the two possible branches. These results hold for any delay $\tau > 0$, as both Theorem 23.1 and Proposition 23.2 are *delay-independent*.

Thus, the model predicts that if the evaporation rate s increases, the ants must respond by increasing either (1) the rate of ants leaving the nest or (2) their sensitivity to small differences in the pheromone concentrations. It might be interesting to try and verify this prediction in the real biological system.

23.9 Conclusions

In many fields of science, researchers provide verbal descriptions of various phenomena. FM is a simple and direct approach for transforming these verbal descriptions into well-defined mathematical models.

The development of such models can also be used to address various engineering problems. This is because many artificial systems must function in the real world and address problems similar to those encountered by biological agents such as plants or animals. The field of biomimicry is concerned with developing artificial systems inspired by the behavior of biological agents. An important component in this field is the ability to perform reverse engineering of an animal's functioning, and then implement this behavior in an artificial system. We believe that the FM approach may be suitable for addressing biomimicry in a systematic manner. Namely, start with a verbal description of an animal's behavior (e.g., foraging in ants) and, using fuzzy logic theory, obtain a mathematical model of this behavior which can be implemented by artificial systems (e.g., autonomous robots).

In this chapter, we described a first step in this direction by applying FM to transform a verbal description of the foraging behavior of ants into a well-defined mathematical model. Simulations and rigorous analysis of the resulting model demonstrate good fidelity with the behavior actually observed in nature. Furthermore, when the fuzzy model is substituted in a mathematical model for the colony of foragers, it leads to an interesting connection with models used in the theory of ANNs. Unlike previous models, the fuzzy model is also simple enough to allow a rather detailed analytical analysis.

The collective behavior of social insects has inspired many interesting engineering designs (see, e.g., [39,54]). Further research is necessary in order to study the application of the model studied here to various engineering problems.

References

1. H. Goldstein, *Classical Mechanics*. Addison-Wesley, Reading, MA, 1980.
2. J. Giarratano and G. Riley, *Expert Systems: Principles and Programming*, 3rd edn. PWS Publishing Company, Boston, MA, 1998.
3. L. A. Zadeh, Fuzzy sets, *Information and Control*, 8, 338–353, 1965.
4. L. A. Zadeh, Outline of a new approach to the analysis of complex systems and decision processes, *IEEE Transactions Systems, Man, Cybernetics*, 3, 28–44, 1973.

5. W. Siler and J. J. Buckley, *Fuzzy Expert Systems and Fuzzy Reasoning*. Wiley-Interscience, Hoboken, NJ, 2005.
6. A. Kandel, Ed., *Fuzzy Expert Systems*. CRC Press, Boca Raton, FL, 1992.
7. K. Hirota and M. Sugeno, Eds., *Industrial Applications of Fuzzy Technology in the World*. World Scientific, River Edge, NJ, 1995.
8. D. Dubois, H. Prade, and R. R. Yager, Eds., *Fuzzy Information Engineering*. Wiley, New York, 1997.
9. K. Tanaka and M. Sugeno, Introduction to fuzzy modeling, in *Fuzzy Systems: Modeling and Control*, H. T. Nguyen and M. Sugeno, Eds. Kluwer, Norwell, MA, 1998, pp. 63–89.
10. T. Terano, K. Asai, and M. Sugeno, *Applied Fuzzy Systems*. AP Professional, Cambridge, MA, 1994.
11. R. R. Yager and D. P. Filev, *Essentials of Fuzzy Modeling and Control*. John Wiley & Sons, Chichester, U.K., 1994.
12. J. Yen, R. Langari, and L. Zadeh, Eds., *Industrial Applications of Fuzzy Logic and Intelligent Systems*. IEEE Press, Piscataway, NJ, 1995.
13. W. Pedrycz, Ed., *Fuzzy Sets Engineering*. CRC Press, Boca Raton, FL, 1995.
14. D. Dubois, H. T. Nguyen, H. Prade, and M. Sugeno, Introduction: The real contribution of Fuzzy systems, in *Fuzzy Systems: Modeling and Control*, H. T. Nguyen and M. Sugeno, Eds. Kluwer, Norwell, MA, 1998, pp. 1–17.
15. M. Margaliot and G. Langholz, Fuzzy Lyapunov based approach to the design of fuzzy controllers, *Fuzzy Sets Systems*, 106, 49–59, 1999.
16. M. Margaliot and G. Langholz, *New Approaches to Fuzzy Modeling and Control—Design and Analysis*. World Scienticfic, Singapore, 2000.
17. L. A. Zadeh, Fuzzy logic = computing with words, *IEEE Transactions Fuzzy Systems*, 4, 103–111, 1996.
18. E. Kolman and M. Margaliot, *Knowledge-Based Neurocomputing: A Fuzzy Logic Approach*. Springer, Berlin, Germany, 2009.
19. E. Tron and M. Margaliot, Mathematical modeling of observed natural behavior: a fuzzy logic approach, *Fuzzy Sets Systems*, 146, 437–450, 2004.
20. K. Z. Lorenz, *King Solomon's Ring: New Light on Animal Ways*. Methuen & Co., London, U.K., 1957.
21. E. Tron and M. Margaliot, How does the Dendrocoleum lacteum orient to light? A fuzzy modeling approach, *Fuzzy Sets Systems*, 155, 236–251, 2005.
22. I. L. Bajec, N. Zimic, and M. Mraz, Simulating flocks on the wing: The fuzzy approach, *Journal of Theoretical Biology*, 233, 199–220, 2005.
23. I. Rashkovsky and M. Margaliot, Nicholson's blowies revisited: A fuzzy modeling approach, *Fuzzy Sets Systems*, 158, 1083–1096, 2007.
24. D. Laschov and M. Margaliot, Mathematical modeling of the lambda switch: A fuzzy logic approach, *Journal of Theoretical Biology*, 260, 475–489, 2009.
25. N. Tinbergen, *The Study of Instinct*. Oxford University Press, London, U.K., 1969.
26. G. S. Fraenkel and D. L. Gunn, *The Orientation of Animals: Kineses, Taxes, and Compass Reactions*. Dover Publications, New York, 1961.
27. S. Schockaert, M. De Cock, C. Cornelis, and E. E. Kerre, Fuzzy ant based clustering, in *Ant Colony, Optimization and Swarm Intelligence*, M. Dorigo, M. Birattari, C. Blum, L. M. Gambardella, F. Mondada, and T. Stutzle, Eds. Springer-Verlag, Berlin, Germany, 2004, pp. 342–349.
28. Y. Bar-Cohen and C. Breazeal, Eds., *Biologically Inspired Intelligent Robots*. SPIE Press, Bellingham, WA, 2003.
29. C. Mattheck, *Design in Nature: Learning from Trees*. Springer-Verlag, Berlin, Germany, 1998.
30. C. Chang and P. Gaudiano, Eds., Robotics and autonomous systems, Special Issue on *Biomimetic Robotics*, 30, 39–64, 2000.
31. K. M. Passino, *Biomimicry for Optimization, Control, and Automation*. Springer, London, U.K., 2004.
32. M. Margaliot, Biomimicry and fuzzy modeling: A match made in heaven, *IEEE Computational Intelligence Magazine*, 3, 38–48, 2008.

33. J. M. C. Sousa and U. Kaymak, *Fuzzy Decision Making in Modeling and Control.* World Scientific, Singapore, 2002.
34. V. Novak, Are fuzzy sets a reasonable tool for modeling vague phenomena? *Fuzzy Sets Systems*, 156, 341–348, 2005.
35. E. O. Wilson, *Sociobiology: The New Synthesis*. Harvard University Press, Cambridge, MA, 1975.
36. S. Goss, S. Aron, J. L. Deneubourg, and J. M. Pasteels, Self-organized shortcuts in the Argentine ant, *Naturwissenschaften*, 76, 579–581, 1989.
37. M. Dorigo, M. Birattari, and T. Stutzle, Ant colony optimization: Artificial ants as a computational intelligence technique, *IEEE Computational Intelligence Magazine*, 1, 28–39, 2006.
38. H. V. D. Parunak, Go to the ant: Engineering principles from natural multi agent systems, *Annals Operations Research*, 75, 69–101, 1997.
39. E. Bonabeau, M. Dorigo, and G. Theraulaz, *Swarm Intelligence: From Natural to Artificial Systems.* Oxford University Press, Oxford, U.K., 1999.
40. R. Schoonderwoerd, O. E. Holland, J. L. Bruten, and L. J. M. Rothkrantz, Ant-based load balancing in telecommunications networks, *Adaptive Behavior*, 5, 169–207, 1997.
41. G. W. Bluman and S. C. Anco, *Symmetry and Integration Methods for Differential Equations.* Springer-Verlag, New York, 2002.
42. L. A. Segel, Simplification and scaling, *SIAM Review*, 14, 547–571, 1972.
43. S. Guillaume, Designing fuzzy inference systems from data: An interpretability-oriented review, *IEEE Transactions Fuzzy Systems*, 9, 426–443, 2001.
44. G. Bontempi, H. Bersini, and M. Birattari, The local paradigm for modeling and control: From neuro-fuzzy to lazy learning, *Fuzzy Sets Systems*, 121, 59–72, 2001.
45. J. S. R. Jang, C. T. Sun, and E. Mizutani, *Neuro-Fuzzy and Soft Computing: A Computational Approach to Learning and Machine Intelligence*. Prentice-Hall, Englewood Cliffs, NJ, 1997.
46. R. Beckers, J. L. Deneubourg, and S. Goss, Trails and U-turns in the selection of a path by the ant Lasius niger, *Journal of Theoretical Biology*, 159, 397–415, 1992.
47. H. T. Nguyen, V. Kreinovich, M. Margaliot, and G. Langholz, Hyperbolic approach to fuzzy control is optimal, in *Proceedings of the 10th IEEE International Conference on Fuzzy Systems (FUZZ-IEEE'2001)*, Melbourne, Australia, 2001, pp. 888–891.
48. J. L. Deneubourg, S. Aron, S. Goss, and J. M. Pasteels, The self-organizing exploratory pattern of the Argentine ant, *Journal of Insect Behavior*, 3, 159–168, 1990.
49. L. P. Shayer and S. A. Campbell, Stability, bifurcation, and multistability in a system of two coupled neurons with multiple time delays, *SIAM Journal on Applied Mathematics*, 61, 673–700, 2000.
50. K. Gu, V. L. Kharitonov, and J. Chen, *Stability of Time-Delay Systems*. Birkhauser, Boston, MA, 2003.
51. S.-I. Niculescu, E. I. Verriest, L. Dugard, and J.-M. Dion, Stability and robust stability of time-delay systems: A guided tour, in *Stability and Control of Time-Delay Systems*, L. Dugard and E. I. Verriest, Eds. Springer, Berlin, Germany, 1998, pp. 1–71.
52. V. Lakshmikantham and S. Leela, *Differential and Integral Inequalities: Theory and Applications.* Academic Press, New York, 1969.
53. V. Rozin and M. Margaliot, The fuzzy ant, *IEEE Computational Intelligence Magazine*, 26(2), 18–28, 2007.
54. M. Dorigo and T. Stutzle, *Ant Colony Optimization*. MIT Press, Cambridge, MA, 2004.

IV

Optimizations

24

Multiobjective Optimization Methods

Tak Ming Chan
University of Minho

Kit Sang Tang
City University of Hong Kong

Sam Kwong
City University of Hong Kong

Kim Fung Man
City University of Hong Kong

24.1 Introduction

Evolutionary computation in multiobjective optimization (MO) has now become a popular technique for solving problems that are considered to be conflicting, constrained, and sometimes mathematically intangible [KD01,CVL02]. This chapter aims to bring out the main features of MO starting from the classical approaches to state-of-the-art methodologies, all of which can be used for practical designs.

24.1.1 Classical Methodologies

The popular optimization methods for solving MO problems are generally classified into three categories: (1) enumerative methods, (2) deterministic methods, and (3) stochastic methods [CVL02,MF00,RL98]. Enumerative methods are computationally expensive. Hence, these are now seldom used by researchers. On the other hand, a vast literature for either the deterministic or stochastic methods has become available. Some useful references for these methodologies are presented in Table 24.1.

24.1.1.1 Deterministic Methods

The deterministic methods do not suffer the enormous search space problem, which is usually confronted by enumerative methods. These methods incorporate domain knowledge (heuristics) so as to limit the search space for finding acceptable solutions in time and produce acceptable solutions in practice [MF00,BB88,NN96]. However, these are often ineffective to cope with high dimensional, multimodal, and non-deterministic polynomial time (NP)-complete MO problems. The deficiency is largely due to their heuristic search performance, which in turn limits the search space [DG89].

TABLE 24.1 Deterministic and Stochastic Optimization Methods

Types of Optimization Methods			
Deterministic		Stochastic	
Greedy	[CLR01,BB96]	Random search/walk	[ZZ03,FS76]
Hill-climbing	[HS95]	Monte Carlo	[JG03,RC87]
Branch & bound	[HT96,FKS02]	Simulated annealing	[VLA87,AZ01]
Depth-first	[JP84]	Tabu search	[GL97,FG89,GTW93]
Breadth-first	[JP84]	Evolutionary computation/algorithms	[DLD00,DF00,TMK08,CPL04]
Best-first	[JP84]	Mathematical programming	[MM86,HL95]
Calculus-based	[HA82]		
Mathematical programming	[MM86,HL95]		

The other disadvantage is the point-by-point approach [KD01], which only yields one optimized solution per single simulation run. Multiple runs are therefore required should a set of optimized solutions be sought, but then, at their best, would only yield suboptimal solutions.

24.1.1.2 Stochastic Methods

Rather than the use of enumerative and deterministic methods for solving irregular MO problems, stochastic methods are the other alternative [CVL02]. Stochastic methods can acquire multiple solutions in a single simulation run. Also, specific schemes can be devised to prevent the solutions from falling into the suboptimal domain [KD01].

These methods need an encode/decode mapping mechanism for coordinating between the problem and algorithm domains. Furthermore, a specific function is required for assigning the fitness values to possible solutions. This approach does not guarantee true Pareto-optimal solutions but can offer reasonably adequate solutions to a number of MO problems where the deterministic methods failed to deliver [DG89].

24.2 Multiobjective Evolutionary Algorithms

Evolutionary algorithms belong to a class of stochastic optimizing schemes. These algorithms emulate the process of the natural selection for which the noted philosopher, Herbert Spencer, coined the phrase "Survival of the fittest." These algorithms share a number of common properties and genetic operations.

The evolution of each individual is subject to the rules of genetics-inspired operations, i.e., selection, crossover, and mutation. Each individual in the population represents a potential solution to a certain problem. Also, a fitness value is assigned to each of individuals so as to evaluate them within the measurement-checking mechanism. The selection tends to allow the high-fitness individuals to reproduce more as compared to the low-fitness individuals. The genetic exploration allows the crossover to exchange the information amongst the individuals, while the mutation introduces a small perturbation within the structure of an individual.

Three common evolutionary algorithms are (1) evolutionary programming, (2) evolution strategies, and (3) genetic algorithms. The differences between these three evolutionary algorithms are listed in Table 24.2. But the same principle applies to each one of them. Further, details of evolutionary programming and evolution strategies can be obtained in [HS95,DF00]. Whereas for genetic algorithms, see [DG89,GC00,DC99,CL04].

The essential properties of multiobjective evolutionary algorithms (MOEAs) are outlined [TMC06]. These include (1) decision variables, (2) constraints, (3) pareto optimality, and (4) convergence and diversity.

TABLE 24.2 Differences between Three Evolutionary Algorithms

Evolutionary Algorithm	Representation	Genetic Operators
Evolutionary programming	Real values	Mutation and $(\mu + \lambda)$ selection
Evolution strategies	Real values and strategy parameters	Crossover, mutation, and $(\mu + \lambda)$ or (μ, λ) selection
Genetic algorithms	Binary or real values	Crossover, mutation, and selection

Note: μ is the number of parents and λ is the number of offsprings.

Currently, a number of MOEAs can be applied in practice:

1. Multiobjective genetic algorithm (MOGA) [FF98,FF93]
2. Niched Pareto genetic algorithm 2 (NPGA2) [EMH01]
3. Nondominated sorting genetic algorithm 2 (NSGA2) [DAM00,DAM02]
4. Strength pareto evolutionary algorithm 2 (SPEA2) [ZLT01]
5. Pareto archived evolution strategy (PAES) [KC09,KC00]
6. Micro-genetic algorithm (MICRO-GA) [CTP01,CP01]
7. Jumping genes (JG) [TMC06,TMK08]
8. Particle swarm optimization (PSO) [CPL04,VE06]

All of these MOEAs are genetic algorithm-based methods except PAES, which is an evolution strategy-based method. The Pareto-based fitness assignment is used to identify non-dominated Pareto-optimal solutions. The basic principles of these algorithms are briefly outlined in this chapter.

24.2.1 Multiobjective Genetic Algorithm

MOGA was proposed by Fonseca and Fleming [FF98,FF93]. It has three features: (1) a modified version of Goldberg's ranking scheme, (2) modified fitness assignment, and (3) niche count. The flowchart of the MOGA is shown in Figure 24.1.

A modified ranking scheme, which is slightly different from Goldberg's one [FF93], is used in the MOGA. The scheme can be represented graphically as depicted in Figure 24.2. In this scheme, the rank of an individual I is given by the number of individuals (q) that dominates I plus 1. That is,

$$\text{Rank}(I) = 1 + q,$$

if the individual I is dominated by other q individuals.

Modified fitness assignment

The procedure of modified rank-based fitness assignment is shown as follows:

1. Sort population according to rank.
2. Assign fitness to individuals by interpolation from the best (rank 1) individual to the worst (rank $n \leq M$ where M is the population size) on the basis of some function, usually linear or exponential, but possibly with some other types.
3. Average the fitness assigned to individuals with the same rank, so that all of them are sampled at the same rate while keeping the global population fitness constant.

Fitness sharing

To obtain a uniformly distributed and widespread set of Pareto-optimal solutions in multiobjective optimization, there exist at least two major difficulties: (1) the finite population size and stochastic selection errors and (2) convergence of the population to a small region of Pareto-optimal front (i.e., a phenomenon occurs in both natural and artificial evolution, also called genetic drift).

The technique of fitness sharing was suggested to avoid such a problem [FF98]. It utilizes individual competition for finite resources in a closed environment. Similar individuals reduce each other's fitness by competing for the same resource. The similarity of two individuals is measured either in the

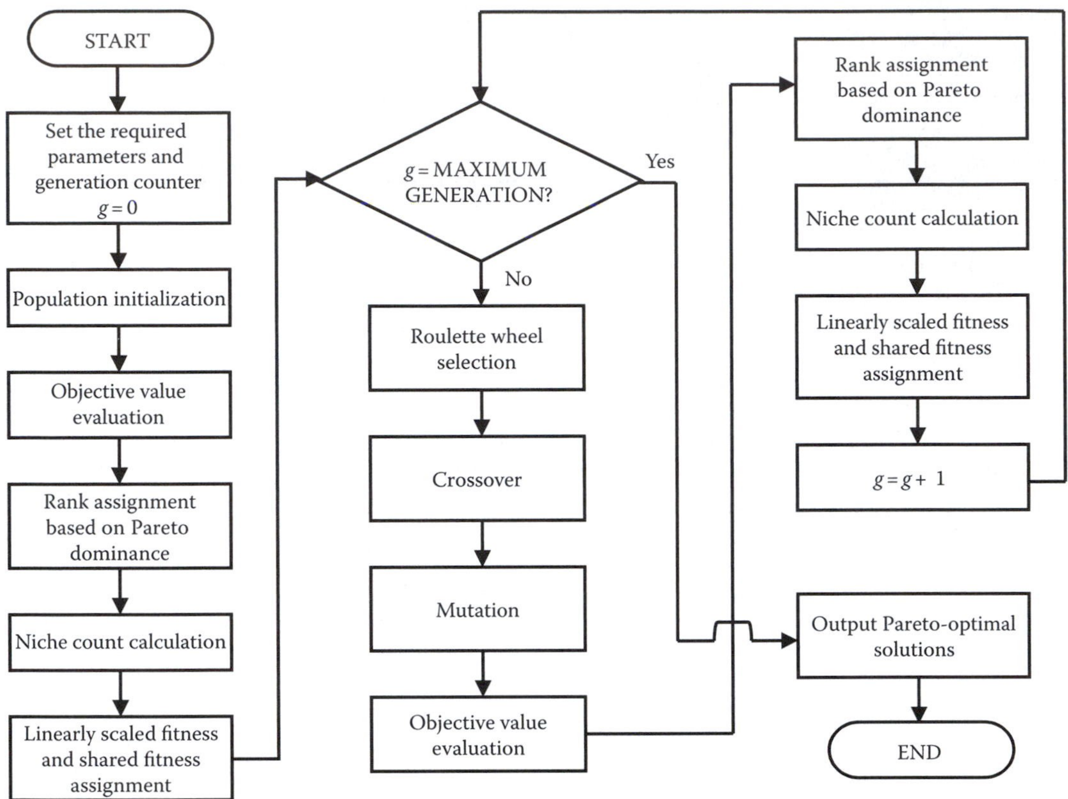

FIGURE 24.1 Flowchart of MOGA.

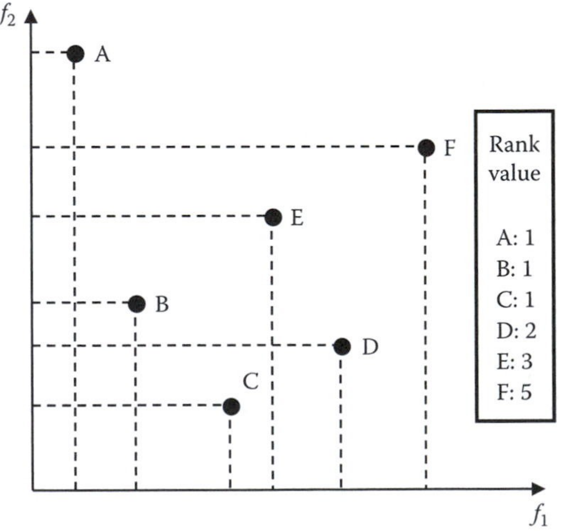

FIGURE 24.2 Fonseca and Fleming's ranking scheme (for minimizing f_1 and f_2).

genotypic space (the number of gene difference between two individuals) or in the phenotypic space (the distance in objective values of two individuals).

To perform fitness sharing, a sharing function is required to determine the fitness reduction of a chromosome on the basis of the crowding degree caused by its neighbors. The sharing function commonly used is

$$sf(d_{ij}) = \begin{cases} 1-\left(\dfrac{d_{ij}}{\sigma_{share}}\right)^{\alpha}, & \text{if } d_{ij} < \sigma_{share} \\ 0, & \text{otherwise} \end{cases} \tag{24.1}$$

where

α is a constant for adjusting the shape of the sharing function

σ_{share} is the niche radius chosen by the user for minimal separation desired

d_{ij} is the distance between two individuals used in encoding space (genotypic sharing) or decoding space (phenotypic sharing)

Then, the shared fitness of a chromosome can be acquired by dividing its fitness $f(j)$ by its niche count n_i,

$$f'(j) = \frac{f(j)}{n_i} \tag{24.2}$$

$$n_i = \sum_{j=1}^{population\_size} sf\left(d_{ij}\right) \tag{24.3}$$

where

$f(j)$ is the fitness of the chromosome considered

n_i is the niche count including the considered chromosome itself

24.2.2 Niched Pareto Genetic Algorithm 2

NPGA2 is an improved version of NPGA [HNG94] suggested by Erickson et al. [EMH01]. The flowchart of the NPGA2 is shown in Figure 24.3. The characteristic of this algorithm is the use of fitness sharing when the tournament selection ends in a tie. A tie means that both picked two individuals are dominated or nondominated. If happens, the niche count n_i in Equation 24.3 is computed for each selected individual. The individual with the lowest niche count will be the winner, and therefore be included in the mating pool.

Niche counts are calculated by using individuals in the partially filled next generation population rather than the current generation population. This method is called continuously updated fitness sharing [OGC94]. The combination of tournament selection and fitness sharing would lead to chaotic perturbations of the population composition. Note that the values of the objective functions should be scaled to equal ranges for estimating the niche count, i.e.,

$$f_i' = \frac{f_i - f_{i,\min}}{f_{i,\max} - f_{i,\min}} \tag{24.4}$$

where

f_i' is the scaled value of objective f_i

$f_{i,\min}$ is the minimum value of objective f_i

$f_{i,\max}$ is the maximum value of objective f_i

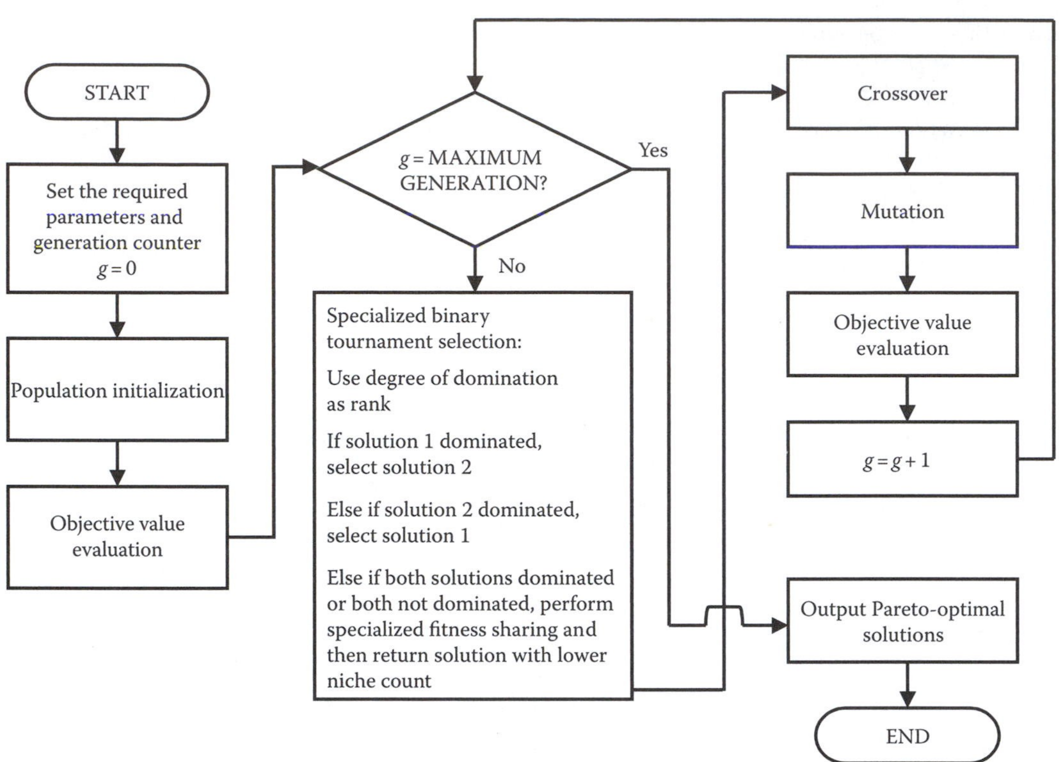

FIGURE 24.3 Flowchart of NPGA2.

24.2.3 Non-Dominated Sorting Genetic Algorithm 2

NSGA2 is the enhanced version of NSGA [SD94] proposed by Deb et al. [DAM00,DAM02]. The flowchart of the NSGA2 is shown in Figure 24.4. It has four peculiarities: (1) fast non-dominated sorting, (2) crowding distance assignment, (3) crowded-comparison operator, and (4) elitism strategy.

24.2.3.1 Fast Non-Dominated Sorting Approach

The invention of the fast non-dominated sorting approach in the NSGA2 is aimed at reducing the high computational complexity $O(MN^3)$, of traditional non-dominated sorting algorithm adopted in the NSGA.

In the traditional non-dominated sorting approach, to find solutions lying on the first non-dominated front for a population size N, each solution should be compared with every other solution in the population. Therefore, $O(MN)$ comparisons are needed for each solution; and the total complexity is $O(MN^2)$ for searching all solutions of the first non-dominated front where M is the total number of objectives.

To obtain the second non-dominated front, the solutions of the first front are temporarily discounted and the above procedure is to be repeated. In the worst possible circumstance, the second front $O(MN^2)$ computations are required. This applies to the subsequent non-dominated fronts (i.e., third, fourth, and so on). Hence, when there are N fronts and each front has only one solution as the worst case scenario, overall $O(MN^3)$ computations are necessary.

On the other hand, the fast non-dominated sorting approach, whose pseudo code and flowchart are shown in Figures 24.5 and 24.6, respectively, can firstly to initiate domination count n_p (the number of solutions which dominate the solution p) and calculate S_p (a set of solutions that the solution p dominates) to proceed $O(MN^2)$ comparisons. Then, all solutions with their $n_p = 0$ are said to be the members of the first non-dominated front.

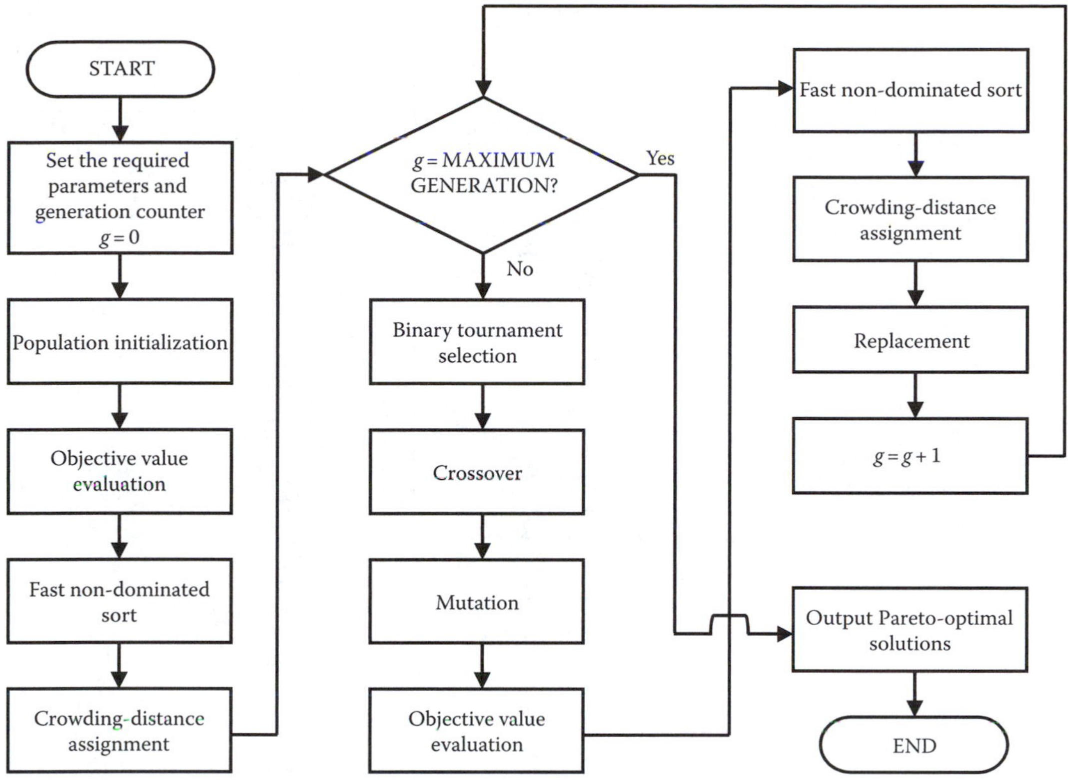

FIGURE 24.4 Flowchart of NSGA2.

```
for each p ∈ P
    S_p = ϕ
    n_p = 0
    for each q ∈ P
        if (p ≺ q)                // If p dominates q
            S_p = S_p ∪ {q}       // Add q to the set of solutions dominated by p
        else if (q ≺ p)
            n_p = n_p + 1         // Increase the domination counter of p
    if n_p = 0                    // p belongs to the first front
        p_rank = 1
        F_1 = F_1 ∪ {p}
i = 1                             // Initialize the front counter
while F_i ≠ ϕ
    Q = ϕ                         // Used to store the members of the next front
    for each p ∈ F_i
        for each q ∈ S_p
            n_q = n_q − 1
            if n_q = 0            // q belongs to the next front
                q_rank = i + 1
                Q = Q ∪ {q}
    i = i + 1
    F_i = Q
```

FIGURE 24.5 Pseudo code of the fast non-dominated sort.

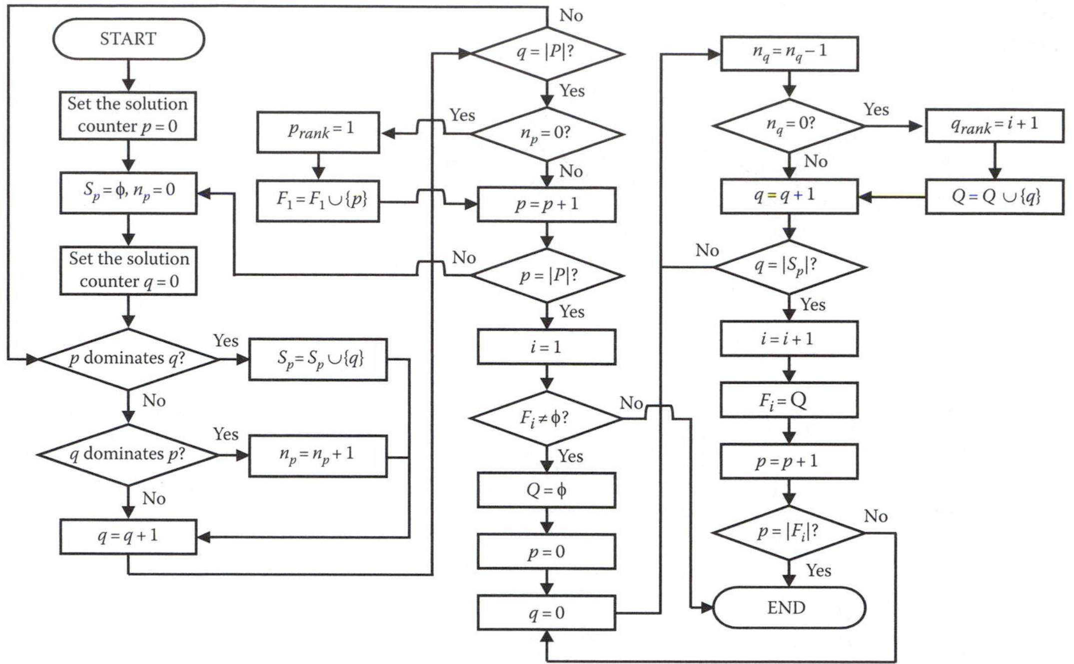

FIGURE 24.6 Flowchart of the fast non-dominated sort.

For each solution p belonging to the first front, n_q of each solution q belonging to S_p is decreased by one. If $n_q = 0$ for any solution q, the solution q is placed to a separate list Q. All solutions in Q are the members of the second non-dominated front. The above procedure is repeated until all non-dominated fronts are completed.

To calculate the computational complexity as referred to Figure 24.5, the first inner loop (for each $p \in F_i$) is executed N times since each individual can be the member of utmost non-dominated front. Also, the second inner loop (for each $q \in S_p$) can be executed at maximum $(N - 1)$ times for each individual (i.e., each individual dominates $(N - 1)$ and each domination check requires at most M comparisons). Therefore, this results overall $O(MN^2)$ computations.

24.2.3.2 Crowded-Comparison Approach

The goal of using the crowded-comparison approach in the NSGA2 is to eliminate the difficulties arises from the well-known sharing function in the NSGA. The merit of this new approach is the preservation of population diversity without using any user-defined parameter. This approach consists of the crowding-distance assignment and crowded-comparison operator.

Crowding-Distance Assignment

In order to obtain diversified Pareto-optimal solutions, the crowding-distance assignment can be devised. Its purpose is to estimate the density of solutions surrounding a particular solution within the population. The preferable solutions in the less crowded area may be chosen on the basis of their assigned crowding distances.

The crowding distance measures the perimeter of the cuboid, which is formed by using the nearest neighbors as the vertices. As indicated in Figure 24.7, the crowding distance of the ith non-dominated solution marked with the solid circle is the average side length of the cuboid represented by a dashed box assuming that the non-dominated solutions i, $i - 1$, and $i + 1$ in Figure 24.7 have the function values (f_1, f_2), (f_1', f_2') and (f_1'', f_2''), respectively. The crowding distance $I[i]_{distance}$ of the solution i in the non-dominated set I is given by

$$I[i]_{distance} = |f_1' - f_1| + |f_1'' - f_1| + |f_2'' - f_2'| \tag{24.5}$$

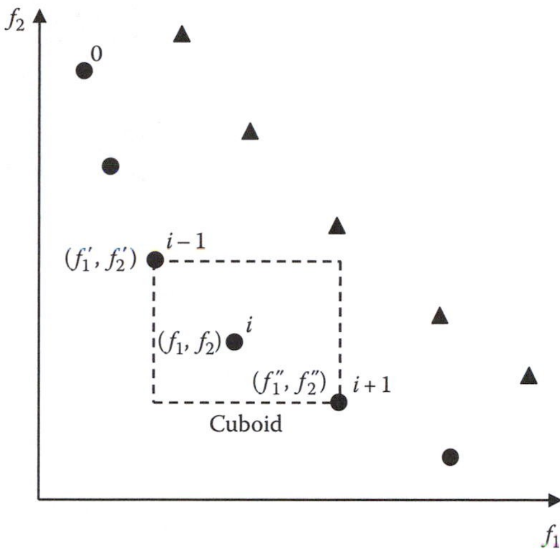

FIGURE 24.7 Crowding-distance computation.

```
l = |I|                                   // Number of solutions in I
for each i
    set I[i]distance = 0                  // Initialize distance
for each objective m
    I = sort (I, m)                       // Sort using each objective value
    I[1]distance = I[I]distance = ∞       // Boundary points must be selected
    for i = 2 to (l-1)
```

$$I[i]_{distance} = I[i]_{distance} + \frac{I[i+1].m - I[i-1].m}{f_m^{max} - f_m^{min}}$$

FIGURE 24.8 Pseudo code of the crowding-distance assignment.

The pseudo code and flowchart of the crowding-distance assignment are shown in Figures 24.8 and 24.9, respectively. The procedures are given as follows:

Step 1: Set the crowding distance $I[i]_{distance}$ of each non-dominated solution i as zero.

Step 2: Set the counter of objective function $m = 1$.

Step 3: Sort the population according to each objective value of the mth objective function in the ascending order.

Step 4: For the mth objective function, assign a nominated relatively large $I[i]_{distance}$ for the extreme solutions of the non-dominated front, i.e., the smallest and largest function values. This selected $I[i]_{distance}$ must be greater than the crowding distance values of other solutions within the same non-dominated front.

Step 5: For the mth objective function, calculate $I[i]_{distance}$ for each of remaining non-dominated solutions $i = 2, 3, \ldots, l-1$ by using the following equation

$$I[i]_{distance} = I[i]_{distance} + \frac{I[i+1].m - I[i-1].m}{f_m^{max} - f_m^{min}} \quad (24.6)$$

where

l is the total number of solutions in a non-dominated set I

$I[i].m$ is the mth objective function value of the ith solution in I

f_m^{max} and f_m^{min} are the maximum and minimum values of the mth objective function

Step 6: $m = m + 1$. If $m \leq M$ where M is the total number of objective functions, go to Step 3.

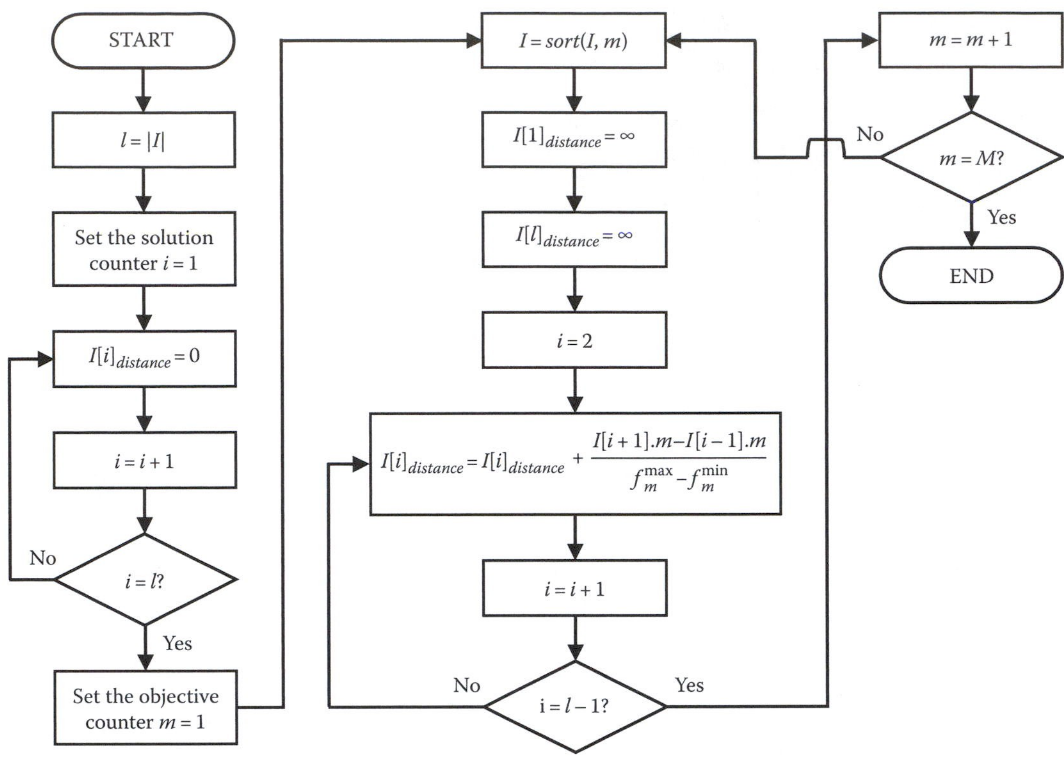

FIGURE 24.9 Flowchart of the crowding-distance assignment.

A solution with a smaller crowding distance value means that it is more crowded. In other words, this solution and its other surrounding solutions are close at distance.

24.2.3.3 Crowded-Comparison Operator

The crowded-comparison operator, which is implemented in the selection process, is to obtain uniformly distributed Pareto-optimal solutions. Should any two picked solutions have different ranks, the solution with the lower rank will be chosen. However, if their ranks are the same, the solution with the larger crowding distance (i.e., located in a less crowded region) is preferred.

24.2.3.4 Elitism Strategy

The procedure of executing the elitism is outlined as follows:

Step 1: Assuming that the tth generation is considered, combine a parent population P_t and an offspring population Q_t to form a population R_t with size $2N$.

Step 2: Sort R_t using the fast non-dominated sorting technique and then identify the different non-dominated fronts $F_1, F_2, \ldots$.

Step 3: Include these fronts into the new parent population P_{t+1} one by one until P_{t+1} with size N is full.

However, not all the members of a particular front can be fully added to P_{t+1}. For example, the fronts F_1, F_2 are completely added to P_{t+1} and therefore P_{t+1} has only j vacancies. If F_3 has k members where $k > j$, they will be sorted by using the crowded-comparison operator in descending order of the crowding distance. Then, the first j members with larger crowding distance will be selected to fill P_{t+1}. The pseudo code and flowchart of the elitism strategy are shown in Figures 24.10 and 24.11, respectively.

```
R_t = P_t ∪ Q_t                          // Combine parent and offspring population
F = fast non-dominated sort (R_t)        // F = (F_1, F_2, ...) all non-dominated fronts of R_t
P_t+1 = ϕ and i = 1
while |P_t+1| + |F_i| ≤ N                // until the parent population is filled
      crowding distance                  // Calculate crowding distance in F_i
        assignment (F_i)
      P_t+1 = P_t+1 ∪ F_i                // Include ith non-dominated front in the parent
                                            population
      i = i + 1                          // Check the next front for inclusion
sort (F_i, ≺_n)                          // Sort in descending order by using ≺_n
P_t+1 = P_t+1 ∪ F_i[1: (N- |P_t+1|)]     // Choose the first (N-|P_t+1|) elements of F_i
```

FIGURE 24.10 Pseudo code of the elitism strategy.

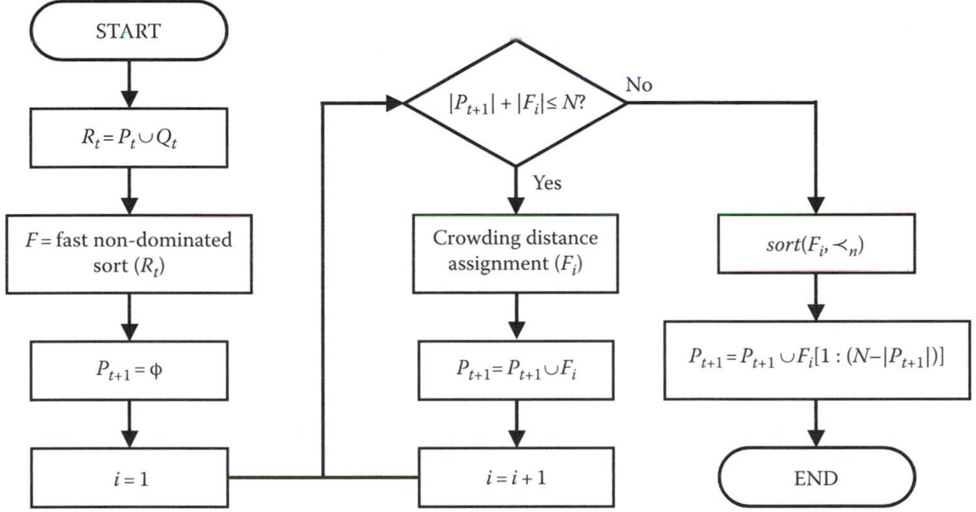

FIGURE 24.11 Flowchart of the elitism strategy.

24.2.4 Strength Pareto Evolutionary Algorithm 2

SPEA2 [ZLT01] is an improved version of SPEA [ZT99]. It consists of three main features: (1) strength value and raw fitness, (2) density estimation, and (3) archive truncation method. The flowchart of the SPEA2 is shown in Figure 24.12.

24.2.4.1 Strength Value and Raw Fitness

The strength value of each individual i in the population with size $|P|$ and archive with size $|A|$ is equal to the number of solutions that it dominates:

$$S(i) = \left|\{j \mid j \in P + A \wedge i \succ j\}\right| \tag{24.7}$$

where

$|\bullet|$ is the cardinality of a set
"+" is the multiset union
"$\succ$" is the Pareto dominance relation

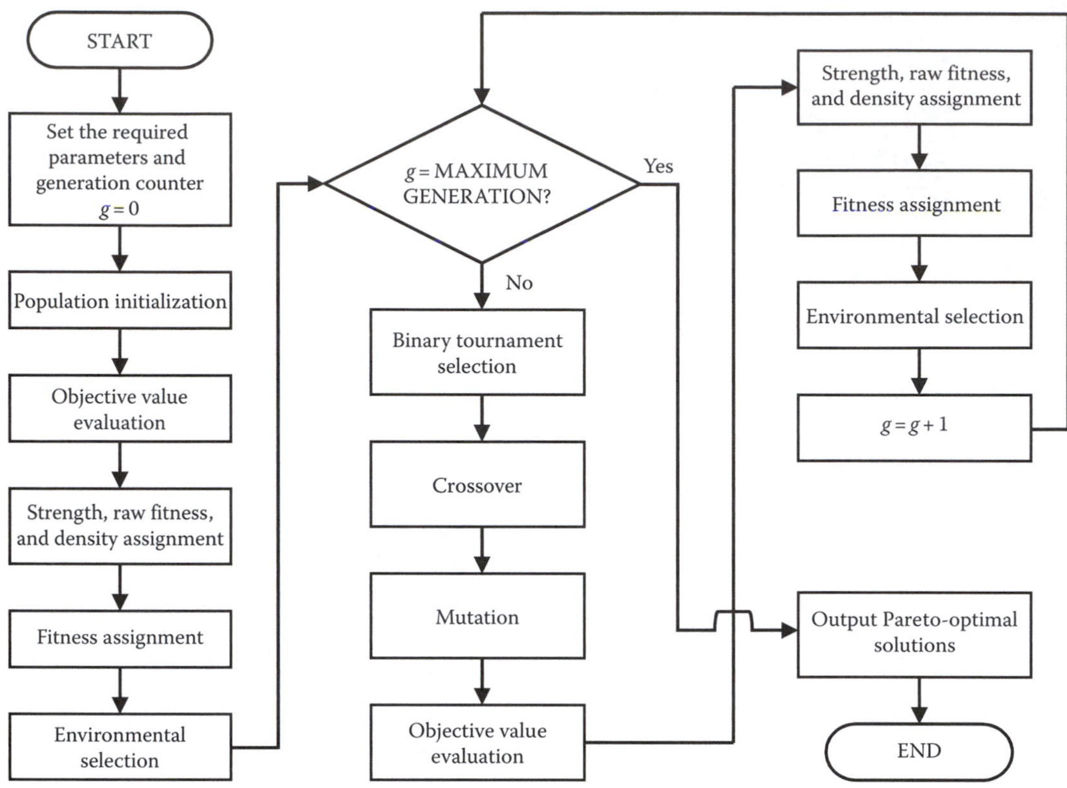

FIGURE 24.12 Flowchart of SPEA2.

The raw fitness value of each individual i is acquired by summing the strength values of its dominators in both population and archive, and is calculated as

$$R(i) = \sum_{j \in P+A, j \succ i} S(j) \tag{24.8}$$

As for a minimization problem, an individual with zero raw fitness value is a non-dominated individual, while an individual with high raw fitness value means that it is dominated by many individuals.

24.2.4.2 Density Estimation

Even though the raw fitness assignment offers a sorting of niching mechanism on the basis of Pareto dominance, this may fail when most individuals do not dominate each other. As a result, the density estimation is employed to discriminate between each individual that is having identical raw fitness values.

The density estimation technique used in the SPEA2 is called kth nearest neighbor method [BS86], where the density at any point is a decreasing function of the distance to the kth nearest data point. The density value of each individual i is given by

$$D(i) = \frac{1}{d_i^k + 2} \tag{24.9}$$

where d_i^k is the desired distance value.

'2' is biased value added in the denominator to ensure that its value is indeed larger than zero. Thus, it is guaranteed that the density value is smaller than 1. The steps of finding the value of d_i^k are as follows:

1. For each individual *i*, the distances in objective space to all individuals *j* in archive and population are calculated and stored in a list.
2. Sort the list in increasing order.
3. The *k*th element gives the value of d_i^k where $k = \sqrt{|P| + |A|}$.

24.2.4.3 Archive Truncation Method

In each generation, it is necessary to update the archive by copying all non-dominated individuals (i.e., those having a fitness lower than one) from archive and population to the archive of the next generation. If the number of all those non-dominated individuals exceeds the archive size $|A|$, the archive truncation method is used to iteratively remove some non-dominated individuals until the number reaches $|A|$. The merit of this archive truncation method is that the removal of boundary non-dominated solutions can be avoided.

In each generation, an individual *i* is selected for the removal for which $i \leq_d j$ for all $j \in A_{t+1}$ with

$$i \leq_d j :\Leftrightarrow \forall 0 < k < |A_{t+1}| : \sigma_i^k = \sigma_j^k \vee \exists 0 < k < |A_{t+1}| : \left[\left(\forall 0 < l < k : \sigma_i^l = \sigma_j^l \right) \wedge \sigma_i^k < \sigma_j^k \right]$$

where
σ_i^k is the distance of *i* to its *k*th nearest neighbor in A_{t+1}
t is the counter of generation

That is, the individual having the minimum distance to another individual is chosen at each stage. If there are several individuals with the minimum distance, the tie is broken by considering the second smallest distances and so forth.

24.2.5 Pareto Archived Evolution Strategy

PAES is a local-search-based algorithm [KC09,KC00]. It imitates an algorithm called Evolution Strategy and its flowchart is shown in Figure 24.13. In the PAES, the unique genetic operation, mutation, is the main hill-climbing strategy, and an archive with a limited size is comprised to store the previously found non-dominated solutions.

The PAES has three versions which are [(1 + 1) – PAES], [(1 + λ) – PAES], and [(μ + λ) – PAES]. The first one means that a single parent generates a single offspring. The second one represents a single parent for producing λ offsprings. The last one means that a population of μ parents for generating λ offsprings. By comparison, [(1 + 1) – PAES] has the lower computational overhead (i.e., it is a faster algorithm) and also the simplest but the most reliable performer.

The unique characteristic of the PAES is its adaptive grid scheme. Its notion is the use of a new crowding procedure on the basis of recursively dividing up *d*-dimensional objective space for tracing the crowding degrees of various regions of the archive of non-dominated solutions. This way ensures diversified non-dominated solutions. Also, it helps remove excessive non-dominated solutions located in the crowded grids (i.e., the grids with high crowding degree) if the number of those solutions exceeds the archive size.

An adaptive grid scheme is also possible. When each solution is generated, it is necessary to determine its grid location in the objective space. Suppose that the range of the space is defined in each objective, the required grid location can be acquired by repetitively bisecting the range in each objective in which half the solution can be found. The location of the solution is marked with a binary string with a length $2^{b \times n}$ where *b* is the number of bisections of the space for each objective and *n* is the number of objectives.

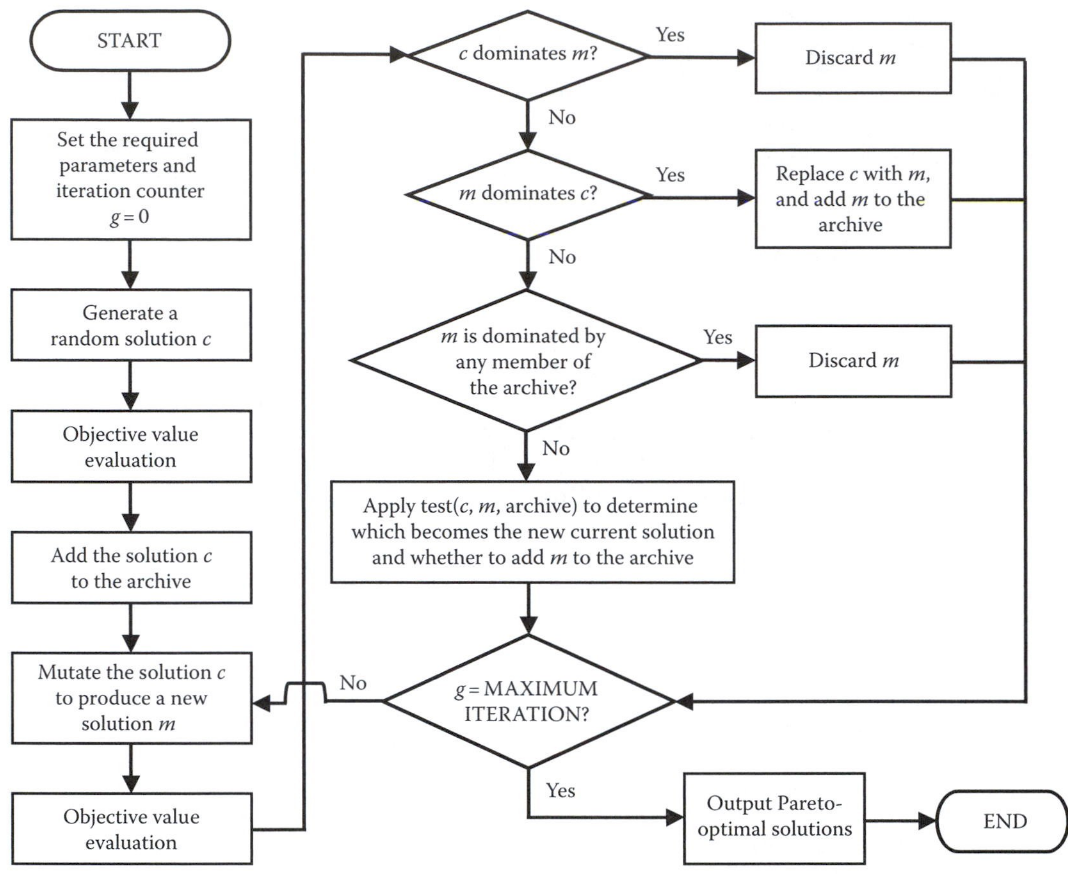

FIGURE 24.13 Flowchart of PAES.

If the solution is located at the larger half of the bisection of the space, the corresponding bit in the binary string is set. In order to record the number and which is the non-dominated solutions residing in each grid, a map of the grid is therefore maintained throughout the run. Besides, grid locations are recalculated when the range of the objective space of archived solutions changes by a threshold amount in order to avoid recalculating the ranges too frequently. Only one parameter represents the number of divisions of the space required.

24.2.6 Micro Genetic Algorithm

Micro genetic algorithm (MICROGA) was suggested by Coello Coello and Toscano Pulido [CTP01,CP01]. The flowchart of the MICROGA is shown in Figure 24.14 and it has three peculiarities: (1) population memory, (2) adaptive grid algorithm, and (3) three types of elitism.

24.2.6.1 Population Memory

The population memory is divided into two parts, which can be replaceable and non-replaceable. The replaceable part has some changes after each cycle of the MICROGA. In contrast, the non-replaceable part is never changed during the run and aims to provide the required diversity for the algorithm.

At the beginning of each cycle, the population is taken from both portions of the population memory so as to have a mixture of randomly generated individuals (non-replaceable) and evolved individuals (replaceable).

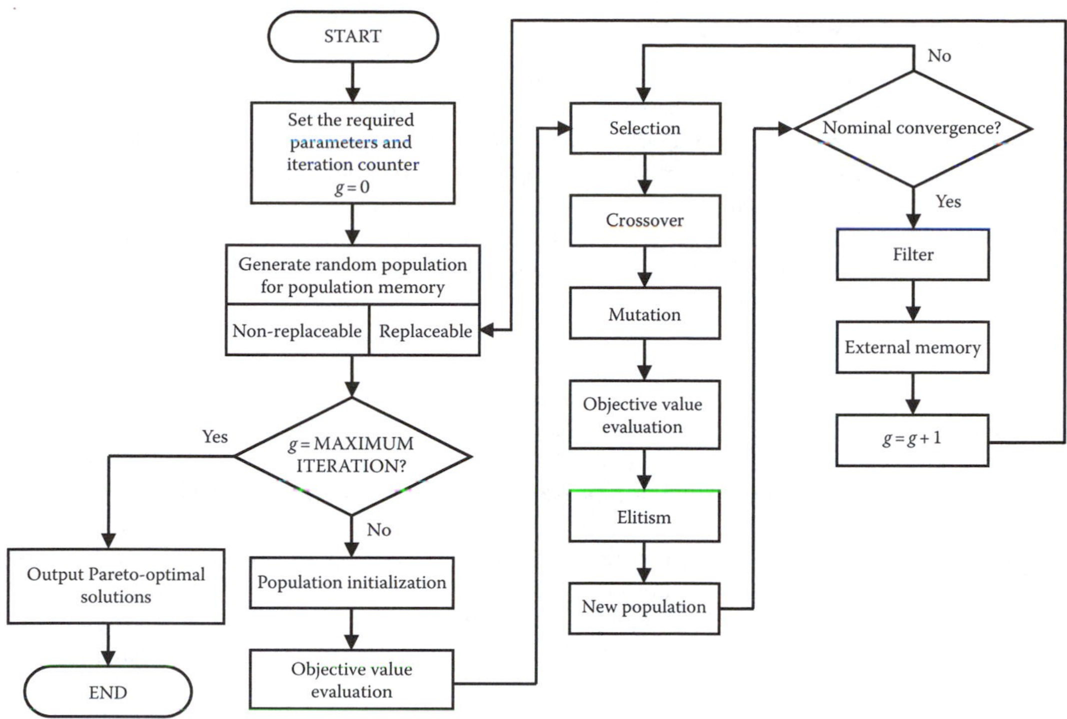

FIGURE 24.14 Flowchart of MICROGA.

24.2.6.2 Adaptive Grid Algorithm

The adaptive grid algorithm is similar to PAES offering the diversity to non-dominated solutions. Once the archive storing the non-dominated solutions reaches its limit, the objective space covered by the archive is divided into a number of grids. Then, each solution in the archive is assigned a set of coordinates.

When a new non-dominated solution is generated, it will be accepted if it is located at a grid where the number of stored non-dominated individuals is smaller than that of the most crowded grid, or located outside the previously specified boundaries. Note that this adaptive grid algorithm requires two parameters: (1) the expected size of the Pareto front, and (2) the number of positions in which the solution space will be divided for each objective.

24.2.6.3 Types of Elitisms

In the first type of elitism, the non-dominated solutions found within the internal cycle of the MICROGA are stored in case the valuable information obtained from the evolutionary process is not lost.

The second type of elitism is the nominal solutions (i.e., the best solutions found when the nominal convergence is reached) added to the replaceable part of the population memory. This enhances speedily converged solutions. It is because of the crossover and mutation that have a higher probability of reaching the true Pareto front of the problem over time.

The last type is a certain number of solutions from all the regions of the Pareto front being picked uniformly including those in the replaceable part of the population memory. The purpose of this type is to utilize the best solutions generated so far right from the starting point as improvement goes (either by getting closer to the true Pareto front or by getting a better distribution).

24.2.7 Jumping Genes

This is the latest evolutionary algorithms for solving MO problems [TMC06,TMK08]. JG merely comprises of simple operations in the evolutionary process. But the usual suffocation in terms of convergence and diversity for reaching the appropriately solutions can then be greatly alleviated. The mimicking of JG phenomenon can fully explore and exploit the solutions space that are considered to be as accurate and wide spread along the Pareto-optimal solutions front.

The very first discovery of JG was reported by the Nobel laureate, Barbara McClintock, based on her work on corn plant [FB92]. To emulate the analogy for JG computation, a transposition of gene(s) into the same chromosome or even to other chromosomes was devised. As a result, this operation can further enhance the genetic operations like crossover, mutation, and selection for improving the fitness quality of chromosomes from generation to generation.

JG comprises of autonomous and nonautonomous transposable elements, called Activator (Ac) and Dissociation (Ds) elements. These mobile genetic elements transpose (jump) from one position to another position within the same chromosome or even to another chromosome. The difference of Ac and Ds is that the transposition occurs itself for Ac, while Ds can only transpose when Ac is activated.

Further, experimental observation also revealed that these jumping genes could move around the genome in two different ways. The first one is called *cut-and-paste*, which means a piece of DNA is cut and pasted somewhere else. The second one is known as *copy-and-paste*, meaning that the genes remain at the same location, while the message in the DNA is copied into RNA and then copied back into DNA at another place in the genome.

The classical gene transmission from generation to generation is termed as "vertical transmission" (VT), i.e., from parent to children. Then, the genes that can "jump" are considered as horizontal transmission (HT). The genes in HT can jump within the same individual (chromosome) or even from one individual to another at the same generation. Then, through genes manipulation, the HT can benefit the VT to gain various natural building blocks. However, this process of placing a foreign set of genes into the new location is not streamlined, nor can it be planned in advance as natural selection tends to be opportunistic, not foresighted. Therefore, the most of genetic takeover, acquisition, mergers, and fusions are usually ensued under the conditions of environmental hardship, i.e., *stress* [MS02].

In MO, the purpose of fitness functions is to examine the quality of the chromosome through the evolutionary process. This will serve as a means of inducing '*stress*' to the chromosome. Then, the movement of JG comes into an exploring effect to the chromosome as its genes have been horizontally disturbed to create the necessary building blocks to form a new individual (solution).

Motivated by these biogenetic observations, computational JG operations have been recently proposed for the enhancement of the searching ability in genetic algorithms, in particular for MO problems. In order to emulate the jumping behavior (transposition process), the following issues are assumed in the design and implementation of the JG:

1. Each chromosome has some consecutive genes, which are randomly selected as the transposon. There may be more than one transposon with different lengths (e.g., 1 or 2 bit for binary code).
2. The jumped locations of the transposons are randomly assigned, and the operations can be carried on the same chromosome or to another chromosome in the population pool.
3. Two JG operations, namely cut-and-paste and copy-and-paste, as depicted in Figures 24.15 and 24.16, respectively, were devised. The actual manipulation of the former operation is that element is cut from an original position and pasted into a new position of a chromosome. As for the later operation, the element replicates itself; and the copy is inserted into a new location of the chromosome while the original one remains unchanged.
4. The JG operations are not limited to binary encoded chromosomes, but can also be extended for other coding methods, such as integer or real number [RKM07].

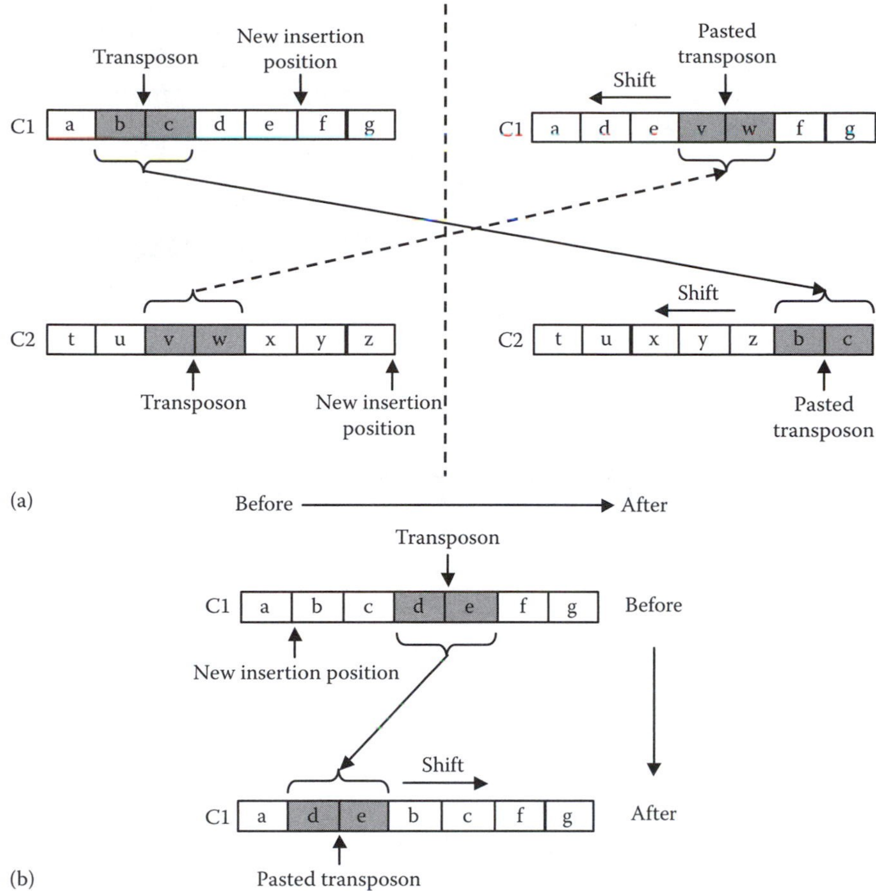

FIGURE 24.15 Cut-and-paste operation on (a) two different chromosomes and (b) the same chromosome.

Similar to the crossover and mutation operations, these two jumping gene operations can be integrated into any general framework of evolutionary algorithms. However, it has been found that the combination of JG with the commonly used sorting scheme [DAM00,DAM02] would create a better performance. A flowchart of JG transposition is shown in Figure 24.17, whereas the flowcharts of the detailed operations of cut-and-paste and copy-and-paste are given in Figure 24.18.

To enhance the search, these two operations are to be inserted after the parent selection process. The entire flow diagram is given in Figure 24.19, in which the shaded part is added to the normal flow of a classical GA (the design of GA can be referred to [MTK96,TMH96] and the references cited in). This common computational flow is now generally known as JGEA.

24.2.8 Particle Swarm Optimization

PSO is an evolutionary computation belongs to model-based search technique. It was inspired by the social action of searching for food of a flock of bird. Each bird (particle) adjusts its search direction for food in accordance to three factors: (1) its own velocity $v_{ij}(k)$, (2) its own best previous experience ($pBest_{ij}(k)$), and (3) the best experience of all the folks ($gBest_{ij}(k)$).

Coello et al. [CPL04] firstly extended PSO to the multiobjective problems. The historical records of best solutions found by particle(s) (*pBest and gBest*) are used to store non-dominated solutions generated in the past. The particle flies through the problem space with a velocity, which is constantly updated by

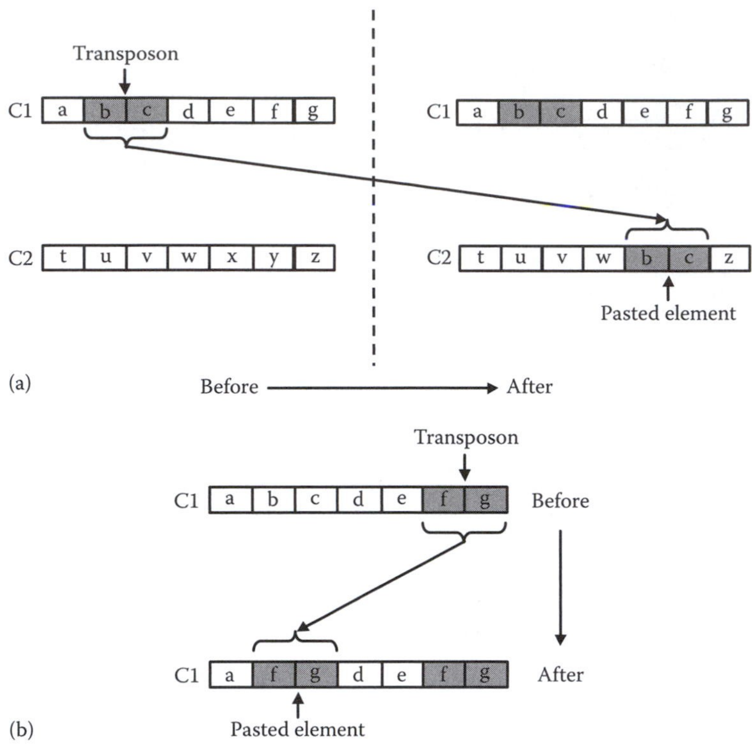

FIGURE 24.16 Copy-and-paste operation on (a) two different chromosomes and (b) the same chromosome.

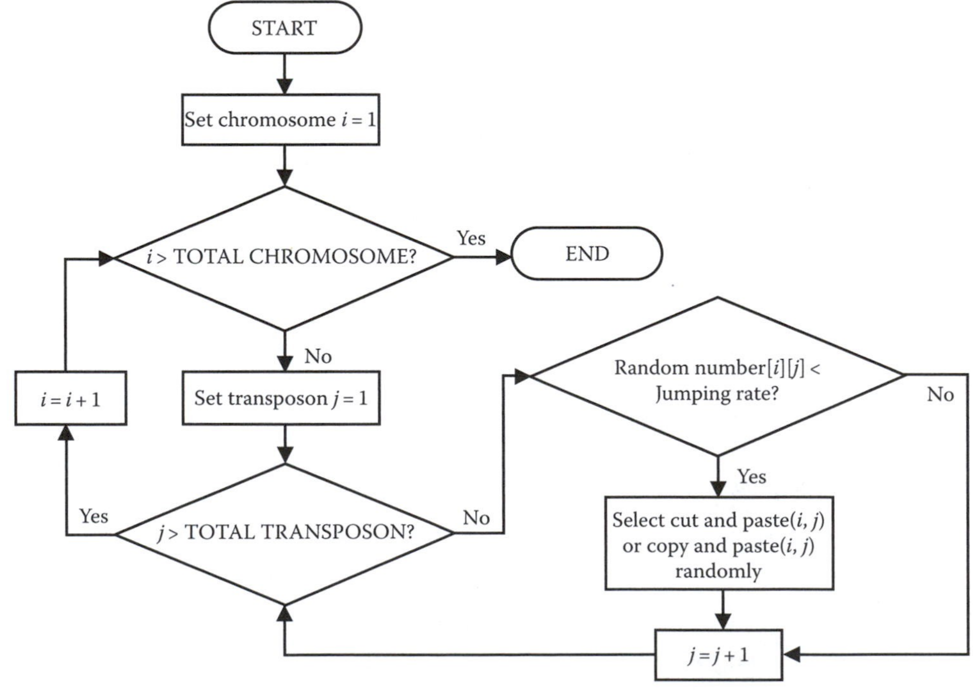

FIGURE 24.17 Flowchart of JG transposition.

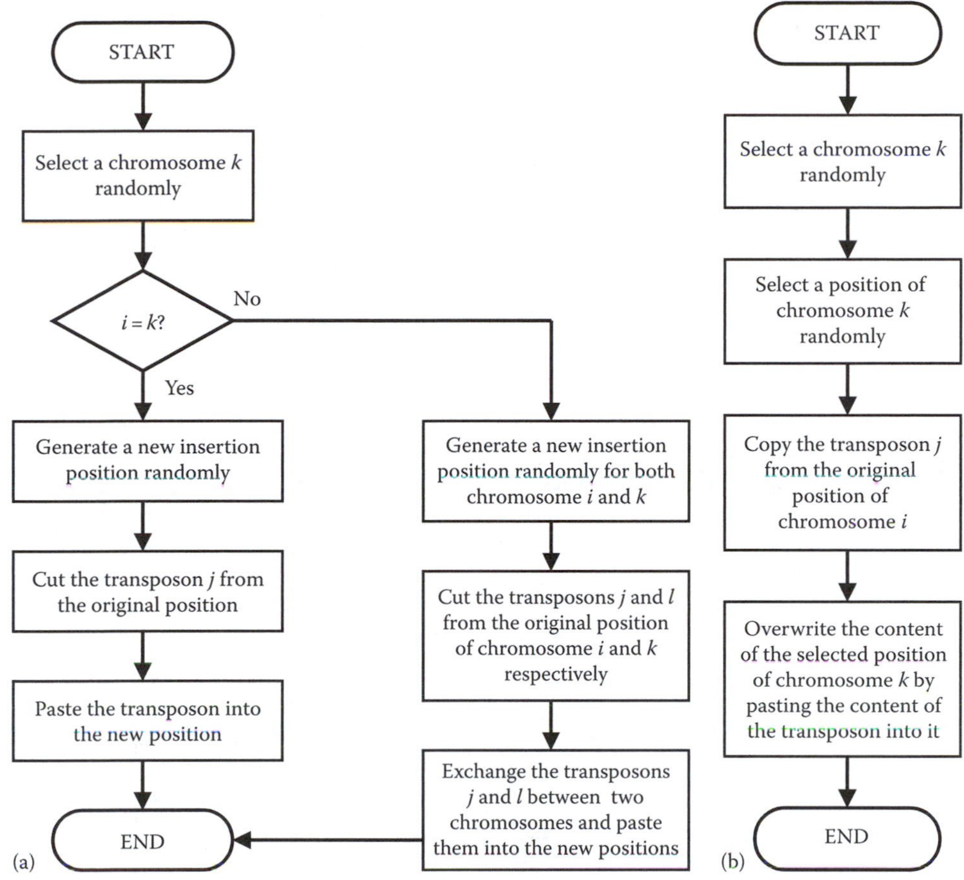

FIGURE 24.18 Flowchart of JG (a) cut-and-paste and (b) copy-and-paste operations.

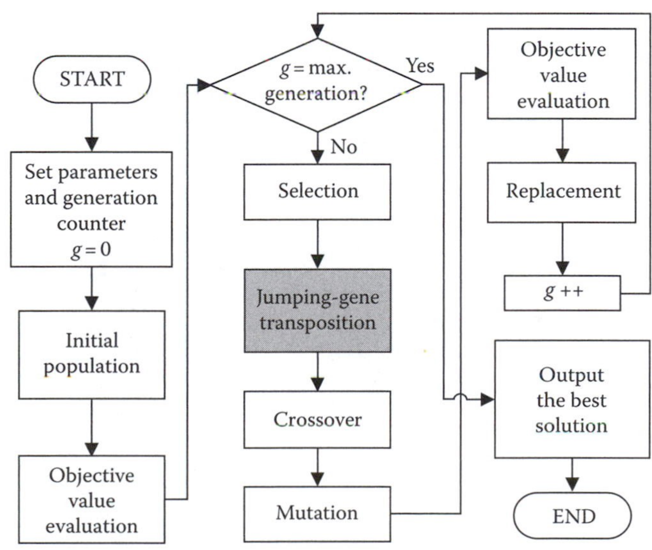

FIGURE 24.19 Genetic cycle of JGEA.

the particle's own experience and the best experience of its neighbors in order to locate the optimum point iteration by iteration. In each iteration, the velocity and position of each particle are updated as follows:

Velocities:

$$v_{ij}(k+1) = \omega v_{ij}(k) + c_1 r_{1j}(k)\left[pBest_{ij}(k) - pos_{ij}(k)\right] \\ + c_2 r_{2j}(k)\left[gBest_{ij}(k) - pos_{ij}(k)\right], \quad i = 1, 2, \ldots, pop; \quad j = 1, \ldots, n \tag{24.10}$$

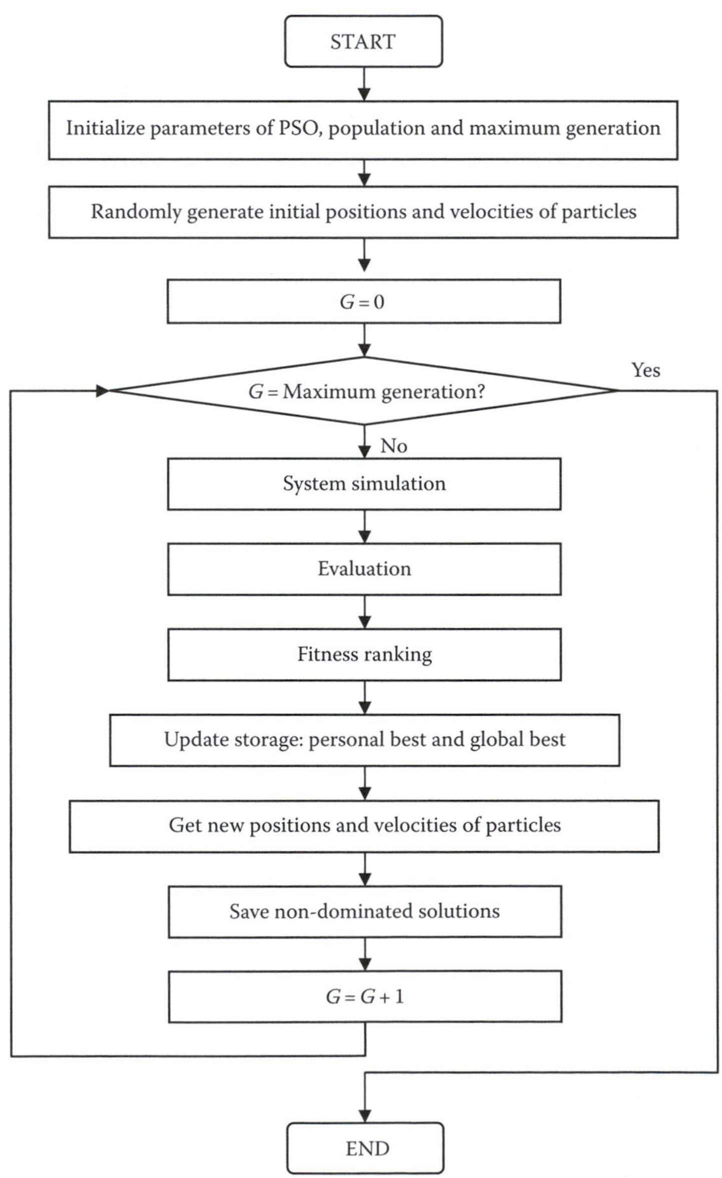

FIGURE 24.20 Flowchart of PSO.

Positions:

$$pos_{ij}(k+1) = pos_{ij}(k) + v_{ij}(k+1), \quad i = 1, 2, \ldots, pop; \quad j = 1, \ldots, n \tag{24.11}$$

where

- w is inertia weight
- c_1 is the cognition factor
- c_2 is social-learning factor
- $r_{1j}(k) = r_{2j}(k) \sim U(0, 1)$
- pop is the population size
- $v_i(k)$ and $pos_i(k)$ are velocity and position vectors
- k is time step

The appropriate values of ω, c_1 and c_2 were studied in [VE06]. The nominal values are $\omega = 0.4$ and $c_1 = c_2 = 1.5$, for most general applications. The flowchart of multiobjective PSO (MOPSO) is shown in Figure 24.20; whereas the PSO pseudo-code is listed in Figure 24.21. In general, PSO tends to suffer in diversity but performs faster in convergence as the obtained solutions are seemingly clustered to a small region.

```
Particle Swarm Optimization
[popParticle, MAXGEN, w, c1, c2] =         // set initial parameters of PSO
  initialpara();
Pareto=[ ]; movePareto=[ ];Fit=[ ];
                                           // Note: n is dimension of solution
                                              space
position = initialPop(popParticle, n)      // generate initial positions of
                                              particles
velocity = initialV(n);                    // initialize velocities of particles
for a=1:MAXGEN
    for b=1: popParticle
      ObjVal = sim(position(b));           // perform system simulation and
                                              compute
                                           // the objectives of a particle.
      Fit=[Fit; ObjVal];                   // save the fitness
end
popRank=Rank(Fit, move);                   // ranking solutions
[pBest, gBest]=                            // find and save personal best &
   findBest(Fit, position);                // global best
Update (position, velocity)                // update the velocities and the
                                              positions
                                           // of the population
Pareto =                                   // get the non-dominated solutions
   pareto(Pareto, popRank, Fit, position, popParticle);
end
```

FIGURE 24.21 Pseudocode for PSO.

24.3 Concluding Remarks

A historical account for MO has been introduced in this chapter. Although, due to the constraint of page limit, there is no specific engineering design in the example given here, nonetheless, the pros and cons of each recommended MO scheme were outlined. The computational procedures, organization flow charts, and the necessary equations of each scheme were vividly described and stated. The material should be adequate enough for both experienced scientists and engineers to follow, and for beginners in the field to absorb the knowledge for their practical uses.

It has been generally recognized that the conundrum of applying MO for engineering designs is largely hindered to the conflicting requirements of rate of convergence and diversity of obtained non-dominated solutions that can lead to or close to the ultimate Pareto-optimal front. A thorough study [TMK08] has provided an accountable investigation in this respect amongst the popular schemes. On the other hand, the PSO method [CPL04] has gained considerable attention in the recent years claiming that a quicker rate of convergence could be achieved.

Nonetheless, the incorporated long list of relevant references has provided a decent platform for further enhancing the transparence of MO techniques and serves the purpose of offering a quick and easy access to this intriguing technology that were once considered to be a far cry for computational optimization community.

References

[AZ01] A. Y. Zomaya, Natural and simulated annealing, *IEEE Computing in Science & Engineering*, 3(6), 97–99, November–December 2001.

[BB88] G. Brassard and P. Bratley, *Algorithms: Theory and Practice*, Englewood Cliffs, NJ: Prentice Hall, 1988.

[BB96] G. Brassard and P. Bratley, *Fundamentals of Algorithmics*, Englewood Cliffs, NJ: Prentice Hall, 1996.

[BS86] B. W. Silverman, *Density Estimation for Statistics and Data Analysis*, London, U.K.: Chapman & Hall, 1986.

[CL04] C. A. Coello Coello and G. B. Lamont (Eds.), *Applications of Multi-Objective Evolutionary Algorithms*, Singapore: World Scientific, 2004.

[CLR01] T. H. Cormen, C. E. Leiserson, R. E. Rivest, and C. Stein, *Introduction to Algorithms*, Cambridge, MA: MIT Press, 2001.

[CP01] C. A. Coello Coello and G. Toscano Pulido, Multiobjective optimization using a micro-genetic algorithm, in *Proceedings of the Genetic and Evolutionary Computation Conference (GECCO-2001)*, L. Spector et al., Eds. San Francisco, CA: Morgan Kaufmann Publishers, July 2001, pp. 274–282.

[CPL04] C. A. Coello Coello, G. T. Pulido, and M. S. Lechuga, Handling multiple objectives with particle swarm optimization, *IEEE Transactions on Evolutionary Computation*, 8(3), 256–279, 2004.

[CTP01] C. A. Coello Coello and G. Toscano Pulido, A micro-genetic algorithm for multiobjective optimization, in *Proceedings of the First International Conference on Evolutionary Multi-Criterion Optimization*, E. Zitzler et al., Eds. Berlin, Germany: Springer, 2001, pp. 126–140.

[CVL02] C. A. Coello Coello, D. A. Van Veldhuizen, and G. B. Lamont, *Evolutionary Algorithms for Solving Multi-objective Problems,* New York: Kluwer, 2002.

[DAM00] K. Deb, S. Agrawal, A. Pratap, and T. Meyarivan, A fast elitist non-dominated sorting genetic algorithm for multi-objective optimization: NSGA-II, in *Proceedings of the Sixth International Conference on Parallel Problem Solving from Nature (PPSN VI)*, M. Schoenauer et al., Eds. Berlin, Germany: Springer, 2000, pp. 849–858.

[DAM02] K. Deb, A. Pratap, S. Agrawal, and T. Meyarivan, A fast and elitist multiobjective genetic algorithm: NSGA-II, *IEEE Transactions on Evolutionary Computation*, 6(2), 182–197, April 2002.

[DC99] D. A. Coley, *An Introduction to Genetic Algorithms for Scientists and Engineers*, Singapore: World Scientific, 1999.

[DF00] D. B. Fogel, *Evolutionary Computation: Toward a New Philosophy of Machine Intelligence,* New York: IEEE Press, 2000.

[DG89] D. E. Goldberg, *Genetic Algorithms in Search, Optimization and Machine Learning,* Reading, MA: Addison-Wesley, 1989.

[DLD00] D. Dumitrescu, B. Lazzerini, L. C. Jain, and A. Dumitrescu, *Evolutionary Computation*, Boca Raton, FL: CRC Press, 2000.

[EMH01] M. Erickson, A. Mayer, and J. Horn, The niched Pareto genetic algorithm 2 applied to the design of groundwater remediation systems, in *Proceedings of the First International Conference on Evolutionary Multi-criterion Optimization*, E. Zitzler et al., Eds. Berlin, Germany: Springer, 2001, pp. 681–695.

[FB92] N. Fedoroff and D. Botstein, Eds, *The Dynamic Genome: Barbara McClintock's Ideas in the Century of Genetics*, Cold Spring Harbor, NY: Cold Spring Harbor Laboratory Press, 1992.

[FF93] C. M. Fonseca and P. J. Fleming, Genetic algorithms for multiobjective optimization: Formulation, discussion and generalization, in *Proceedings of the Fifth International Conference on Genetic Algorithms*, S. Forrest, Ed. San Mateo, CA: Morgan Kaufmann, 1993, pp. 416–423.

[FF98] C. M. Fonseca and P. J. Fleming, Multiobjective optimization and multiple constraint handling with evolutionary algorithms—Part I: A unified formulation, *IEEE Transactions on System, Man and Cybernetic Part A: Systems and Humans*, 28(1), 26–37, January 1998.

[FG89] F. Glover, Tabu search: Part I, *ORSA Journal on Computing*, 1(3), 190–206, 1989.

[FKS02] U. Faigle, W. Kern, and G. Still, *Algorithmic Principles of Mathematical Programming*, Dordrecht, the Netherlands: Kluwer Academic Publishers, 2002.

[FS76] F. Spitzer, *Principles of Random Walk*, New York: Springer, 1976.

[GC00] M. Gen and R. Cheng, *Genetic Algorithms and Engineering Optimization*, New York: Wiley, 2000.

[GL97] F. Glover and M. Laguna, *Tabu Search*, Boston, MA: Kluwer Academic Publishers, 1997.

[GTW93] F. Glover, E. Taillard, and D. De Werra, A user's guide to Tabu search, *Annuals of Operations Research*, 41, 3–28, 1993.

[HA82] H. Anton, *Calculus with Analytic Geometry,* New York: Wiley, 1992.

[HL95] F. S. Hillier and G. J. Lieberman, *Introduction to Mathematical Programming*, New York: McGraw-Hill, 1995.

[HNG94] J. Horn, N. Nafpliotis, and D. E. Goldberg, A niched Pareto genetic algorithm for multiobjective optimization, in *Proceedings of the First IEEE Conference on Evolutionary Computation*, Orlando, FL, June 27–29, 1994, vol. 1, pp. 82–87.

[HS95] H.-P. Schwefel, *Evolution and Optimum Seeking*, New York: Wiley, 1995.

[HT96] R. Horst and H. Tuy, *Global Optimization: Deterministic Approaches*, Berlin, Germany: Springer, 1996.

[JG03] J. E. Gentle, *Random Number Generation and Monte Carlo Methods*, New York: Springer, 2003.

[JP84] J. Pearl, *Heuristics: Intelligent Search Strategies for Computer Problem Solving,* Reading, MA: Addison-Wesley, 1984.

[KC00] J. D. Knowles and D. W. Corne, Approximating the nondominated front using the Pareto archived evolution strategy, *Evolutionary Computation*, 8(2), 149–172, Summer 2000.

[KC09] J. D. Knowles and D. W. Corne, The Pareto archived evolution strategy: A new baseline algorithm for Pareto multiobjective optimization, in *Proceedings of the 1999 Congress on Evolutionary Computation*, Washington, DC, July 6–9, 1999, vol. 1, pp. 98–105.

[KD01] K. Deb, *Multi-Objective Optimization Using Evolutionary Algorithms*, Chichester, U.K.: Wiley, 2001.

[MF00] Z. Michalewicz and D. B. Fogel, *How to Solve It: Modern Heuristics*, Berlin, Germany: Springer, 2000.

[MM86] M. Minoux, *Mathematical Programming: Theory and Algorithms*, Chichester, U.K.: Wiley, 1986.

[MS02] L. Margulis and D. Sagan, *Acquiring Genomes: A Theory of the Origins of Species*, New York: Basic Books, 2002.

[MTK96] K. F. Man, K. S. Tang, and S. Kwong, Genetic algorithms: Concepts and applications, *IEEE Transactions on Industrial Electronics*, 43(5), 519–534, October 1996.

[NN96] R. E. Neapolitan and K. Naimipour, *Foundations of Algorithms*, Lexington, MA: DC Health & Company, 1996.

[OGC94] C. K. Oei, D. E. Goldberg, and S.-J. Chang, Tournament selection, niching, and the preservation of diversity, Technical Report (94002), Illinois Genetic Algorithms Laboratory, University of Illinois, Urbana, IL, 1994.

[RC87] C. P. Robert and G. Casella, *Monte Carlo Statistical Methods*, New York: Springer, 1999.

[RKM07] K. S. N. Ripon, S. Kwong, and K. F. Man, A real-coding jumping gene genetic algorithm (RJGGA) for multiobjective optimization, *Journal of Information Sciences*, 177(2), 632–654, January 2007.

[RL98] R. L. Rardin, *Optimization in Operations Research*, Upper Saddle River, NJ: Prentice Hall, 1998.

[SD94] N. Srinivas and K. Deb, Multiobjective function optimization using nondominated sorting genetic algorithms, *Evolutionary Computation*, 2(3), 221–248, 1994.

[TMC06] T. M. Chan, A generic jumping gene paradigm: Concept, verification and application, PhD thesis, City University of Hong Kong, Hong Kong, 2006.

[TMH96] K. S. Tang, K. F. Man, S. Kwong, and Q. He, Genetic algorithms and their application, *IEEE Signal Processing Magazine*, pp. 22–37, November 1996.

[TMK08] T. M. Chan, K. F. Man, K. S. Tang, and S. Kwong, A jumping gene paradigm for evolutionary multiobjective optimization, *IEEE Transactions on Evolutionary Computation*, 12(2), 143–159, April 2008.

[VE06] F. van den Bergh and A.P. Engelbrecht, A study of particle swarm optimization, *Journal of Information Science*, 176, 937–971, 2006.

[VLA87] P. J. M. van Laarhoven and E. H. L. Aarts, *Simulated Annealing: Theory and Applications*, Dordrecht, the Netherlands: D. Reidel, 1987.

[ZLT01] E. Zitzler, M. Laumanns, and L. Thiele, SPEA2: Improving the strength Pareto evolutionary algorithm, Technical Report (TIK-Report 103), Swiss Federal Institute of Technology, Lausanne, Switzerland, May 2001.

[ZT99] E. Zitzler and L. Thiele, Multiobjective evolutionary algorithms: A comparative case study and the strength Pareto approach, *IEEE Transactions on Evolutionary Computation*, 3(4), 257–271, November 1999.

[ZZ03] Z. B. Zabinsky, *Stochastic Adaptive Search for Global Optimization*, Boston, MA: Kluwer Academic Publishers, 2003.

25

Fundamentals of Evolutionary Multiobjective Optimization

Carlos A. Coello Coello*
Centro de Investigación y de Estudios Avanzados del Instituto Politécnico Nacional

The solution of optimization problems having two or more (often conflicting) criteria has become relatively common in a wide variety of application areas. Such problems are called "multiobjective," and their solution has raised an important amount of research within operations research, particularly in the last 35 years [52]. In spite of the large number of mathematical programming methods available for solving multiobjective optimization problems, such methods tend to have a rather limited applicability (e.g., when dealing with differentiable objective functions, or with convex Pareto fronts). This has motivated the use of alternative solution approaches such as evolutionary algorithms.

The use of evolutionary algorithms for solving multiobjective optimization problems was originally hinted at in the late 1960s [65], but the first actual implementation of a multiobjective evolutionary algorithm (MOEA) was not produced until 1985 [68]. However, this area, which is now called "evolutionary multiobjective optimization," or EMO, has experienced a very important growth, mainly in the last 15 years [8,15].

This chapter presents a basic introduction to EMO, focusing on its main concepts, the most popular algorithms in current use, and some of its applications. The remainder of this chapter is organized as follows. In Section 25.1, we provide some basic concepts from multiobjective optimization. The use of evolutionary algorithms in multiobjective optimization is motivated in Section 25.2. Some of the main topics of research that are currently attracting a lot of attention in the EMO field are briefly discussed in Section 25.3. A set of sample applications of MOEAs is provided in Section 25.4. Some of the main topics of research in the EMO field that currently attract a lot of attention are briefly discussed in Section 25.5. Finally, some conclusions are provided in Section 25.6.

---

* The author is also associated to the UMI-LAFMIA 3175 CNRS.

25.1 Basic Concepts

We are interested in the solution of multiobjective optimization problems (MOPs) of the form

$$\text{minimize}\left[f_1(\vec{x}), f_2(\vec{x}), \ldots, f_k(\vec{x})\right] \tag{25.1}$$

subject to the m inequality constraints:

$$g_i(\vec{x}) \leq 0 \quad i = 1,2,\ldots,m \tag{25.2}$$

and the p equality constraints:

$$h_i(\vec{x}) = 0 \quad i = 1,2,\ldots,p \tag{25.3}$$

where k is the number of objective functions f_i: $\mathbb{R}^n \to \mathbb{R}$. We call $\vec{x} = [x_1, x_2, \ldots, x_n]^T$ the vector of decision variables. We wish to determine from the set $\mathcal{F}$ of all vectors which satisfy (25.2) and (25.3) the particular set of values $x_1^*, x_2^*, \ldots, x_n^*$ which yield the optimum values of all the objective functions.

25.1.1 Pareto Optimality

It is rarely the case that there is a single point that simultaneously optimizes all the objective functions.* Therefore, we normally look for "trade-offs," rather than single solutions when dealing with multiobjective optimization problems. The notion of "optimality" normally adopted in this case is the one originally proposed by Francis Ysidro Edgeworth [20] and later generalized by Vilfredo Pareto [57]. Although some authors call this notion *Edgeworth–Pareto optimality*, we use the most commonly adopted term *Pareto optimality*.

We say that a vector of decision variables $\vec{x}^* \in \mathcal{F}$ is *Pareto optimal* if there does not exist another $\vec{x} \in \mathcal{F}$ such that $f_i(\vec{x}) \leq f_i(\vec{x}^*)$ for all $i = 1, \ldots, k$ and $f_j(\vec{x}) < f_j(\vec{x}^*)$ for at least one j (assuming minimization).

In words, this definition says that $\vec{x}^*$ is Pareto optimal if there exists no feasible vector of decision variables $\vec{x} \in \mathcal{F}$ which would decrease some criterion without causing a simultaneous increase in at least one other criterion. It is worth noting that the use of this concept normally produces a set of solutions called the *Pareto optimal set*. The vectors $\vec{x}^*$ corresponding to the solutions included in the Pareto optimal set are called *nondominated*. The image of the Pareto optimal set under the objective functions is called *Pareto front*.

25.2 Use of Evolutionary Algorithms

The idea of using techniques based on the emulation of the mechanism of natural selection (described in Darwin's evolutionary theory) to solve problems can be traced back to the early 1930s [25]. However, it was not until the 1960s that the three main techniques based on this notion were developed: genetic algorithms [35], evolution strategies [70], and evolutionary programming [26].

* In fact, this situation only arises when there is no conflict among the objectives, which would make unnecessary the development of special solution methods, since this single solution could be reached after the sequential optimization of all the objectives, considered separately.

These approaches, which are now collectively denominated "evolutionary algorithms," have been very effective for single-objective optimization [27,30,71].

The basic operation of an evolutionary algorithm (EA) is the following. First, they generate a set of possible solutions (called a "population") to the problem at hand. Such a population is normally generated in a random manner. Each solution in the population (called an "individual") encodes all the decision variables of the problem. In order to assess their suitability, a fitness function must be defined. Such a fitness function is a variation of the objective function of the problem that we wish to solve. Then, a selection mechanism must be applied in order to decide which individuals will "mate." This selection process is normally based on the fitness contribution of each individual (i.e., the fittest individuals have a higher probability of being selected). Upon mating, a set of "offspring" are generated. Such offspring are "mutated" (this operator produces a small random change, with a low probability, on the contents of an individual), and constitute the population to be evaluated at the following iteration (called a "generation"). This process is repeated until reaching a stopping condition (normally, a maximum number of generations).

EAs are considered a good choice for solving multiobjective optimization problems because they adopt a population of solutions, which allows them (if properly manipulated) to find several elements of the Pareto optimal set in a single run. This contrasts with mathematical programming methods, which normally generate a single nondominated solution per execution. Additionally, EAs tend to be less susceptible to the discontinuity and the shape of the Pareto front, which is an important advantage over traditional mathematical programming methods [21].

Multiobjective evolutionary algorithms (MOEAs) extend a traditional evolutionary algorithm in two main aspects:

- The selection mechanism. In this case, the aim is to select nondominated solutions, and to consider all the nondominated solutions in a population as equally good.
- A diversity maintenance mechanism. This is necessary to avoid convergence to a single solution, which is something that will eventually happen with an EA (because of stochastic noise) if run for a sufficiently long time.

Regarding selection, although in their early days, several MOEAs relied on aggregating functions [34] and relatively simple population-based approaches [68], today, most of them adopt some form of *Pareto ranking*. This approach was originally proposed by David E. Goldberg [30], and it sorts the population of an EA based on Pareto dominance, such that all nondominated individuals are assigned the same rank (or importance). The aim is that all nondominated individuals get the same probability of being selected, and that such probability is higher than the one corresponding to individuals which are dominated. Although conceptually simple, this sort of selection mechanism allows for a wide variety of possible implementations [8,15].

A number of methods have been proposed in the literature to maintain diversity in an EA. Such approaches include fitness sharing and niching [16,32], clustering [78,84], geographically based schemes [42], and the use of entropy [12,39], among others. Additionally, some researchers have proposed the use of mating restriction schemes [72,84]. Furthermore, the use of relaxed forms of Pareto dominance has also become relatively popular in recent years, mainly as an archiving technique which encourages diversity, while allowing the archive to regulate convergence (see for example, ε-dominance [45]).

A third component of modern MOEAs is elitism, which normally consists of using an external archive (called a "secondary population") that can (or cannot) interact in different ways with the main (or "primary") population of the MOEA. The main purpose of this archive is to store all the nondominated solutions generated throughout the search process, while removing those that become dominated later in the search (called local nondominated solutions). The approximation of the Pareto optimal set produced by a MOEA is thus the final contents of this archive.

25.3 Multiobjective Evolutionary Algorithms

Despite the considerable volume of literature on MOEAs that is currently available,* very few algorithms are used by a significant number of researchers around the world. The following are, from the author's perspective, the most representative MOEAs in current use:

1. Strength Pareto Evolutionary Algorithm (SPEA): This MOEA was conceived as the merge of several algorithms developed during the 1990s [84]. It adopts an external archive (called the external nondominated set), which stores the nondominated solutions previously generated, and participates in the selection process (together with the main population). For each individual in this archive, a *strength* value is computed. This strength is proportional to the number of solutions which a certain individual dominates. In SPEA, the fitness of each member of the current population is computed according to the strengths of all external nondominated solutions that dominate it. Since the external nondominated set can grow too much, this could reduce the selection process and could slow down the search. In order to avoid this, SPEA adopts a technique that prunes the contents of the external nondominated set so that its size remains below a certain (predefined) threshold. For that sake, the authors use a clustering technique.
2. Strength Pareto Evolutionary Algorithm 2 (SPEA2): This approach has three main differences with respect to its predecessor [83]: (1) it incorporates a fine-grained fitness assignment strategy which takes into account for each individual the number of individuals that dominate it and the number of individuals by which it is dominated; (2) it uses a nearest-neighbor density estimation technique which guides the search more efficiently, and (3) it has an enhanced archive truncation method that guarantees the preservation of boundary solutions.
3. Pareto Archived Evolution Strategy (PAES): This is perhaps the most simple MOEA than one can conceive, and was introduced by Knowles and Corne [44]. It consists of a (1 + 1) evolution strategy (i.e., a single parent that generates a single offspring) in combination with a historical archive that stores the nondominated solutions previously found. This archive is used as a reference set against which each mutated individual is being compared. Such (external) archive adopts a crowding procedure that divides objective function space in a recursive manner. Then, each solution is placed in a certain grid location based on the values of its objectives (which are used as its "coordinates" or "geographical location"). A map of such a grid is maintained, indicating the number of solutions that reside in each grid location. When a new nondominated solution is ready to be stored in the archive, but there is no room for them (the size of the external archive is bounded), a check is made on the grid location to which the solution would belong. If this grid location is less densely populated than the most densely populated grid location, then a solution (randomly chosen) from this heavily populated grid location is deleted to allow the storage of the newcomer. This aims to redistribute solutions, favoring the less densely populated regions of the Pareto front. Since the procedure is adaptive, no extra parameters are required (except for the number of divisions of the objective space).
4. Nondominated Sorting Genetic Algorithm II (NSGA-II): This is a heavily revised version of the nondominated sorting genetic algorithm (NSGA), which was introduced in the mid-1990s [74]. The NSGA-II adopts a more efficient ranking procedure than its predecessor. Additionally, it estimates the density of solutions surrounding a particular solution in the population by computing the average distance of two points on either side of this point along each of the objectives of the problem. This value is the so-called *crowding distance*. During selection, the NSGA-II uses a crowded-comparison operator which takes into consideration both the nondomination rank of an individual in the population and its crowding distance (i.e., nondominated solutions are

* The author maintains the EMOO repository, which, as of December 2008, contains over 3600 bibliographic references on evolutionary multiobjective optimization. The EMOO repository is available at: http://delta.cs.cinvestav.mx/~ccoello/EMOO/

preferred over dominated solutions, but between two solutions with the same nondomination rank, the one that resides in the less crowded region is preferred). The NSGA-II does not use an external archive as most of the modern MOEAs in current use. Instead, the elitist mechanism of the NSGA-II consists of combining the best parents with the best offspring obtained (i.e., a $(\mu + \lambda)$-selection). Due to its clever mechanisms, the NSGA-II is much more efficient (computationally speaking) than its predecessor, and its performance is so good that it has become very popular in the last few years, triggering a significant number of applications, and becoming some sort of landmark against which new MOEAs have to be compared in order to merit publication.

5. Pareto Envelope-Based Selection Algorithm (PESA): This algorithm was proposed by Corne et al. [11], and uses a small internal population and a larger external (or secondary) population. PESA adopts the same adaptive grid from PAES to maintain diversity. However, its selection mechanism is based on the crowding measure used by the aformentioned grid. This same crowding measure is used to decide what solutions to introduce into the external population (i.e., the archive of nondominated vectors found along the evolutionary process). Therefore, in PESA, the external memory plays a crucial role in the algorithm since it determines not only the diversity scheme, but also the selection performed by the method. There is also a revised version of this algorithm, called PESA-II [10], which is identical to PESA, except for the fact that region-based selection is used in this case. In region-based selection, the unit of selection is a hyperbox rather than an individual. The procedure consists of selecting (using any of the traditional selection techniques [31]) a hyperbox and then randomly selecting an individual within such hyperbox. The main motivation of this approach is to reduce the computational costs associated with traditional MOEAs (i.e., those based on Pareto ranking).

Many other MOEAs have been proposed in the specialized literature (see for example [9,17,81]), but they will not be discussed here due to obvious space limitations. A more interesting issue, however, is to devise which sort of MOEA will become predominant in the next few years. Efficiency is, for example, a concern nowadays, and several approaches have been developed in order to improve the efficiency of MOEAs (see for example [37,41]). There is also an interesting trend consisting on designing MOEAs based on a performance measure (see for example [3,82]). However, no clear trend exists today, from the author's perspective, that seems to attract the interest of a significant portion of the EMO community, regarding algorithmic design.

25.4 Applications

Today, there exists a very important volume of applications of MOEAs in a wide variety of domains. Next, we provide a brief list of sample applications classified in three large groups: engineering, industrial, and scientific. Specific areas within each of these large groups are also identified.

By far, engineering applications are the most popular in the current EMO literature. This is not surprising if we consider that engineering disciplines normally have problems with better understood mathematical models, which facilitates the use of MOEAs. A representative sample of engineering applications is the following:

- Electrical engineering [1,63]
- Hydraulic engineering [51,62]
- Structural cngineering [56,58]
- Aeronautical engineering [38,47]
- Robotics [2,77]
- Control [5,6]
- Telecommunications [59,75]
- Civil engineering [22,36]
- Transport engineering [50,73]

Industrial applications are the second most popular in the EMO literature. A representative sample of industrial applications of MOEAs is the following:

- Design and manufacture [19,33]
- Scheduling [29,40]
- Management [53,64]

Finally, there are several EMO papers devoted to scientific applications. For obvious reasons, computer science applications are the most popular in the EMO literature. A representative sample of scientific applications is the following:

- Chemistry [60,76]
- Physics [61,67]
- Medicine [24,28]
- Computer science [14,48]

This sample of applications should give at least a rough idea of the increasing interest of researchers for adopting MOEAs in practically all types of disciplines.

25.5 Current Challenges

The existence of challenging, but solvable problems, is a key issue to preserve the interest in a research discipline. Although EMO is a discipline in which a very important amount of research has been conducted, mainly within the last 10 years, several interesting problems still remain open. Additionally, the research conducted so far has also led to new, and intriguing topics. The following is a small sample of open problems that currently attract a significant amount of research within EMO:

- Scalability: In spite of the popularity of MOEAs in a plethora of applications, it is known that Pareto ranking is doomed to fail as we increase the number of objectives, and it is also known that with about 10 objectives, it behaves like random sampling [43]. The reason is that most of the individuals in a population will become nondominated, as the number of objectives increases. In order to deal with this problem, researchers have proposed selection schemes different from Pareto ranking [18,23], as well as mechanisms that allow to reduce the number of objectives of a problem [4,49]. However, there is still a lot of work to be done in this regard, and this is currently a very active research area.
- Incorporation of user's preferences: It is normally the case, that the user does not need the entire Pareto front of a problem, but only a certain portion of it. For example, solutions lying at the extreme parts of the Pareto front are unlikely necessary since they represent the best value for one objective, but the worst for the others. Thus, if the user has at least a rough idea of the sort of trade-offs that aims to find it is desirable to be able to explore in more detail only the nondominated solutions within the neighborhood of such trade-offs. This is possible, if we use, for example, biased versions of Pareto ranking [13] or some multi-criteria decision making technique, from the many developed in Operations Research [7]. Nevertheless, this area has not been very actively pursued by EMO researchers, in spite of its usefulness.
- Parallelism: Although the use of parallel MOEAs is relatively common in certain disciplines such as aeronautical engineering [55], the lack of serious research in this area is remarkable [8,79]. Thus, it is expected to see much more research around this topic in the next few years, for example, related to algorithmic design, the role of local search in parallel MOEAs and convergence analysis, among others.
- Theoretical Foundations: Although an important effort has been made in recent years to develop theoretical work related to MOEAs, in areas such as convergence [66,80], archiving [69], algorithm complexity [54], and run-time analysis [46], a lot of work still remains to be done in this regard.

25.6 Conclusions

In this chapter, we have provided some basic concepts related to evolutionary multiobjective optimization, as well as a short description of the main multiobjective evolutionary algorithms in current use. The main application areas of such algorithms have also been included, in order to provide a better idea of their wide applicability and of the increasing interest to use them.

In the last part of the chapter, we provided a short discussion of some challenging topics that are currently very active within this research area. The main objective of this chapter is to serve as a general (although brief) overview of the EMO field. Its main aim is to motivate researchers and newcomers from different areas, who have to deal with multiobjective problems, to consider MOEAs as a viable choice.

Acknowledgment

The author acknowledges support from CONACYT project no. 103570.

References

1. M.A. Abido. Multiobjective evolutionary algorithms for electric power dispatch problem. *IEEE Transactions on Evolutionary Computation*, 10(3):315–329, June 2006.
2. G. Avigad and K. Deb. The sequential optimization-constraint multi-objective problem and its applications for robust planning of robot paths. In *IEEE Congress on Evolutionary Computation (CEC'2007)*, pp. 2101–2108, Singapore, September 2007. IEEE Press.
3. N. Beume, B. Naujoks, and M. Emmerich. SMS-EMOA: Multiobjective selection based on dominated hypervolume. *European Journal of Operational Research*, 181(3):1653–1669, September 16, 2007.
4. D. Brockhoff, T. Friedrich, N. Hebbinghaus, C. Klein, F. Neumann, and E. Zitzler. Do additional objectives make a problem harder? In D. Thierens, ed., *2007 Genetic and Evolutionary Computation Conference (GECCO'2007)*, Vol. 1, pp. 765–772, London, U.K., July 2007. ACM Press.
5. M. Brown and N. Hutauruk. Multi-objective optimisation for process design and control. *Measurement & Control*, 40(6):182–187, July 2007.
6. G. Capi. Multiobjective evolution of neural controllers and task complexity. *IEEE Transactions on Robotics*, 23(6):1225–1234, 2007.
7. C.A.C. Coello. Handling preferences in evolutionary multiobjective optimization: A survey. In *2000 Congress on Evolutionary Computation* Vol. 1, pp. 30–37, Piscataway, NJ, July 2000. IEEE Service Center.
8. C.A.C. Coello, G.B. Lamont, and D.A. Van Veldhuizen. *Evolutionary Algorithms for Solving Multi-Objective Problems*. Springer, New York, 2nd edn., September 2007. ISBN 978-0-387-33254-3.
9. C.A.C. Coello and G.T. Pulido. Multiobjective optimization using a micro-genetic algorithm. In L. Spector, E.D. Goodman, A. Wu, W.B. Langdon, H.-M. Voigt, M. Gen, S. Sen, M. Dorigo, S. Pezeshk, M.H. Garzon, and E. Burke, eds., *Proceedings of the Genetic and Evolutionary Computation Conference (GECCO'2001)*, pp. 274–282, San Francisco, CA, 2001. Morgan Kaufmann Publishers.
10. D.W. Corne, N.R. Jerram, J.D. Knowles, and M.J. Oates. PESA-II: Region-based selection in evolutionary multiobjective optimization. In L. Spector, E.D. Goodman, A. Wu, W.B. Langdon, H.-M. Voigt, M. Gen, S. Sen, M. Dorigo, S. Pezeshk, M.H. Garzon, and E. Burke, eds., *Proceedings of the Genetic and Evolutionary Computation Conference (GECCO'2001)*, pp. 283–290, San Francisco, CA, 2001. Morgan Kaufmann Publishers.
11. D.W. Corne, J.D. Knowles, and M.J. Oates. The Pareto envelope-based selection algorithm for multiobjective optimization. In M. Schoenauer, K. Deb, G. Rudolph, X. Yao, E. Lutton, J.J. Merelo, and H.-P. Schwefel, eds., *Proceedings of the Parallel Problem Solving from Nature VI Conference*, pp. 839–848, Paris, France, 2000. Springer, Berlin. Lecture Notes in Computer Science No. 1917.

12. X. Cui, M. Li, and T. Fang. Study of population diversity of multiobjective evolutionary algorithm based on immune and entropy principles. In *Proceedings of the Congress on Evolutionary Computation 2001 (CEC'2001)* Vol. 2, pp. 1316–1321, Piscataway, NJ, May 2001. IEEE Service Center.
13. D. Cvetković and I.C. Parmee. Preferences and their application in evolutionary multiobjective optimisation. *IEEE Transactions on Evolutionary Computation*, 6(1):42–57, February 2002.
14. B. de la Iglesia, G. Richards, M.S. Philpott, and V.J. Rayward-Smith. The application and effectiveness of a multi-objective metaheuristic algorithm for partial classification. *European Journal of Operational Research*, 169:898–917, 2006.
15. K. Deb. *Multi-Objective Optimization using Evolutionary Algorithms*. John Wiley & Sons, Chichester, U.K., 2001. ISBN 0-471-87339-X.
16. K. Deb and D.E. Goldberg. An investigation of niche and species formation in genetic function optimization. In J.D. Schaffer, ed., *Proceedings of the Third International Conference on Genetic Algorithms*, pp. 42–50, San Mateo, CA, June 1989. George Mason University, Morgan Kaufmann Publishers.
17. K. Deb, M. Mohan, and S. Mishra. Evaluating the ε-domination based multi-objective evolutionary algorithm for a quick computation of pareto-optimal solutions. *Evolutionary Computation*, 13(4):501–525, Winter 2005.
18. F. di Pierro, S.-T. Khu, and D.A. Savić. An investigation on preference order ranking scheme for multiobjective evolutionary optimization. *IEEE Transactions on Evolutionary Computation*, 11(1):17–45, February 2007.
19. C. Dimopoulos. Explicit consideration of multiple objective in cellular manufacturing. *Engineering Optimization*, 39(5):551–565, July 2007.
20. F.Y. Edgeworth. *Mathematical Psychics*. P. Keagan, London, England, 1881.
21. A.E. Eiben and J.E. Smith. *Introduction to Evolutionary Computing*. Springer, Berlin, 2003. ISBN 3-540-40184-9.
22. K. El-Rayes and K. Hyari. Optimal lighting arrangements for nighttime highway construction projects. *Journal of Construction Engineering and Management-ASCE*, 131(12):1292–1300, December 2005.
23. M. Farina and P. Amato. A fuzzy definition of "optimality" for many-criteria optimization problems. *IEEE Transactions on Systems, Man, and Cybernetics Part A—Systems and Humans*, 34(3):315–326, May 2004.
24. P. Fazendeiro, J.V. de Oliveira, and W. Pedrycz. A multiobjective design of a patient and anaesthetist-friendly neuromuscular blockade controller. *IEEE Transactions on Biomedical Engineering*, 54(9):1667–1678, September 2007.
25. D.B. Fogel. *Evolutionary Computation. Toward a New Philosophy of Machine Intelligence*. The Institute of Electrical and Electronic Engineers, New York, 1995.
26. L.J. Fogel. *Artificial Intelligence through Simulated Evolution*. John Wiley, New York, 1966.
27. L.J. Fogel. *Artificial Intelligence through Simulated Evolution. Forty Years of Evolutionary Programming*. John Wiley & Sons, Inc., New York, 1999.
28. B. Gaal, I. Vassanyi, and G. Kozmann. A novel artificial intelligence method for weekly dietary menu planning. *Methods of Information in Medicine*, 44(5):655–664, 2005.
29. M.J. Geiger. On operators and search space topology in multi-objective flow shop scheduling. *European Journal of Operational Research*, 181(1):195–206, August 16, 2007.
30. D.E. Goldberg. *Genetic Algorithms in Search, Optimization and Machine Learning*. Addison-Wesley Publishing Company, Reading, MA, 1989.
31. D.E. Goldberg and K. Deb. A comparison of selection schemes used in genetic algorithms. In G.J.E. Rawlins, ed., *Foundations of Genetic Algorithms*, pp. 69–93. Morgan Kaufmann, San Mateo, CA, 1991.
32. D.E. Goldberg and J. Richardson. Genetic algorithm with sharing for multimodal function optimization. In J.J. Grefenstette, ed., *Genetic Algorithms and Their Applications: Proceedings of the Second International Conference on Genetic Algorithms*, pp. 41–49. Lawrence Erlbaum, Hillsdale, NJ, 1987.

33. A. Gurnani, S. Ferguson, K. Lewis, and J. Donndelinger. A constraint-based approach to feasibility assessment in preliminary design. *AI EDAM-Artificial Intelligence for Engineering Design Analysis and Manufacturing*, 20(4):351–367, Fall 2006.
34. P. Hajela and C.Y. Lin. Genetic search strategies in multicriterion optimal design. *Structural Optimization*, 4:99–107, 1992.
35. J.H. Holland. Concerning efficient adaptive systems. In M.C. Yovits, G.T. Jacobi, and G.D. Goldstein, eds., *Self-Organizing Systems—1962*, pp. 215–230. Spartan Books, Washington, DC, 1962.
36. K. Hyari and K. El-Rayes. Optimal planning and scheduling for repetitive construction projects. *Journal of Management in Engineering*, 22(1):11–19, 2006.
37. M.T. Jensen. Reducing the run-time complexity of multiobjective EAs: The NSGA-II and other algorithms. *IEEE Transactions on Evolutionary Computation*, 7(5):503–515, October 2003.
38. S. Jeong, K. Chiba, and S. Obayashi. Data mining for aerodynamic design space. *Journal of Aerospace Computing, Information, and Communication*, 2:452–469, November 2005.
39. H. Kita, Y. Yabumoto, N. Mori, and Y. Nishikawa. Multi-objective optimization by means of the thermodynamical genetic algorithm. In H.-M. Voigt, W. Ebeling, I. Rechenberg, and H.-P. Schwefel, eds., *Parallel Problem Solving from Nature—PPSN IV*, Lecture Notes in Computer Science, pp. 504–512, Berlin, Germany, September 1996. Springer-Verlag.
40. M.P. Kleeman and G.B. Lamont. Scheduling of flow-shop, job-shop, and combined scheduling problems using MOEAs with fixed and variable length chromosomes. In K.P. Dahal, K.C. Tan, and P.I. Cowling, eds., *Evolutionary Scheduling, Studies in Computational Intelligence (SCI)*, pp. 49–99. Springer, Berlin, 2007. ISBN 3-540-48582-1.
41. J. Knowles. ParEGO: A hybrid algorithm with on-line landscape approximation for expensive multiobjective optimization problems. *IEEE Transactions on Evolutionary Computation*, 10(1):50–66, February 2006.
42. J. Knowles and D. Corne. Properties of an adaptive archiving algorithm for storing nondominated vectors. *IEEE Transactions on Evolutionary Computation*, 7(2):100–116, April 2003.
43. J. Knowles and D. Corne. Quantifying the effects of objective space dimension in evolutionary multiobjective optimization. In S. Obayashi, K. Deb, C. Poloni, T. Hiroyasu, and T. Murata, eds., *4th International Conference Evolutionary Multi-Criterion Optimization (EMO 2007)*, pp. 757–771, Matshushima, Japan, March 2007. Springer, Berlin. Lecture Notes in Computer Science Vol. 4403.
44. J.D. Knowles and D.W. Corne. Approximating the nondominated front using the Pareto archived evolution strategy. *Evolutionary Computation*, 8(2):149–172, 2000.
45. M. Laumanns, L. Thiele, K. Deb, and E. Zitzler. Combining convergence and diversity in evolutionary multi-objective optimization. *Evolutionary Computation*, 10(3):263–282, Fall 2002.
46. M. Laumanns, L. Thiele, and E. Zitzler. Running time analysis of multiobjective evolutionary algorithms on pseudo-Boolean functions. *IEEE Transactions on Evolutionary Computation*, 8(2):170–182, April 2004.
47. M.R. Lavagna. Multi-objective PSO for interplanetary trajectory design. In D. Thierens, ed., *2007 Genetic and Evolutionary Computation Conference (GECCO'2007)*, vol. 1, p. 175, London, U.K., July 2007. ACM Press.
48. Z.-J. Lee, S.-W. Lin, S.-F. Su, and C.-Y. Lin. A hybrid watermarking technique applied to digital images. *Applied Soft Computing*, 8(1):798–808, January 2008.
49. A.L. Jaimes, C.A.C. Coello, and D. Chakraborty. Objective reduction using a feature selection technique. In *2008 Genetic and Evolutionary Computation Conference (GECCO'2008)*, pp. 674–680, Atlanta, GA, July 2008. ACM Press. ISBN 978-1-60558-131-6.
50. H.R. Lourenço, J.P. Paix ao, and R. Portugal. Multiobjective metaheuristics for the bus-driver scheduling problem. *Transportation Science*, 35(3):331–343, August 2001.
51. C.E. Mariano-Romero, V.H. Alcocer-Yamanaka, and E.F. Morales. Multi-objective optimization of water-using systems. *European Journal of Operational Research*, 181(3):1691–1707, September 16, 2007.
52. K.M. Miettinen. *Nonlinear Multiobjective Optimization*. Kluwer Academic Publishers, Boston, MA, 1999.

53. M.K. Muleta. *A Decision Support System for the Management of Non-Point Source Pollution from Watersheds*. PhD thesis, College of Engineering, Southern Illinois University Carbondale, February 2003.
54. F. Neumann. Expected runtimes of a simple evolutionary algorithm for the multi-objective minimum spanning tree problem. *European Journal of Operational Research*, 181(3):1620–1629, September 16, 2007.
55. S. Obayashi and D. Sasaki. Multiobjective aerodynamic design and visualization of supersonic wings by using adaptive range multiobjective genetic algorithms. In C.A.C. Coello and G.B. Lamont, eds., *Applications of Multi-Objective Evolutionary Algorithms*, pp. 295–315. World Scientific, Singapore, 2004.
56. M. Ohsaki, T. Kinoshita, and P. Pan. Multiobjective heuristic approaches to seismic design of steel frames with standard sections. *Earthquake Engineering & Structural Dynamics*, 36(11):1481–1495, September 2007.
57. V. Pareto. *Cours D'Economie Politique*, Vols. I and II. F. Rouge, Lausanne, 1896.
58. R. Perera and A. Ruiz. A multistage FE updating procedure for damage identification in large-scale structures based on multiobjective evolutionary optimization. *Mechanical Systems and Signal Processing*, 22(4):970–991, May 2008.
59. L. Raisanen and R.M. Whitaker. Comparison and evaluation of multiple objective genetic algorithms for the antenna placement problem. *Mobile Networks & Applications*, 10(1–2):79–88, February–April 2005.
60. N. Ramzan and W. Witt. Multi-objective optimization in distillation unit: A case study. *Canadian Journal of Chemical Engineering*, 84(5):604–613, October 2006.
61. W. Raza and K.-Y. Kim. Multiobjective optimization of a wire-wrapped LMR fuel assembly. *Nuclear Technology*, 162(1):45–52, April 2008.
62. P. Reed, J.B. Kollat, and V.K. Devireddy. Using interactive archives in evolutionary multiobjective optimization: A case study for long-term groundwater monitoring design. *Environmental Modelling & Software*, 22(5):683–692, May 2007.
63. F. Rivas-Dávalos and M.R. Irving. An approach based on the strength Pareto evolutionary algorithm 2 for power distribution system planning. In Carlos A. Coello Coello, Arturo Hernández Aguirre, and Eckart Zitzler, eds., *Evolutionary Multi-Criterion Optimization. Third International Conference, EMO 2005*, pp. 707–720, Guanajuato, México, March 2005. Springer. Lecture Notes in Computer Science Vol. 3410.
64. S. Rochet and C. Baron. An evolutionary algorithm for decisional assistance to project management. In Jean-Philippe Rennard, ed., *Handbook of Research on Nature Inspired Computing for Economy and Management*, Vol. 2, pp. 444–464. Idea Group Reference, Hershey, U.K., 2006. ISBN 1-59140-984-5.
65. R.S. Rosenberg. Simulation of genetic populations with biochemical properties. PhD thesis, University of Michigan, Ann Harbor, MI, 1967.
66. G. Rudolph and A. Agapie. Convergence properties of some multi-objective evolutionary algorithms. In *Proceedings of the 2000 Conference on Evolutionary Computation*, Vol. 2, pp. 1010–1016, Piscataway, NJ, July 2000. IEEE Press.
67. D.E. Salazar, C.M. Rocco, and E. Zio. Robust reliability design of a nuclear system by multiple objective evolutionary optimisation. *International Journal of Nuclear Knowledge Management*, 2(3):333–345, 2007.
68. J.D. Schaffer. Multiple objective optimization with vector evaluated genetic algorithms. In *Genetic Algorithms and their Applications: Proceedings of the First International Conference on Genetic Algorithms*, pp. 93–100. Lawrence Erlbaum, 1985.
69. O. Schuetze, M. Laumanns, E. Tantar, C.A.C. Coello, and E.G. Talbi. Convergence of stochastic search algorithms to gap-free Pareto front approximations. In D. Thierens, ed., *2007 Genetic and Evolutionary Computation Conference (GECCO'2007)*, Vol. 1, pp. 892–899, London, U.K., July 2007. ACM Press.
70. H.-P. Schwefel. Kybernetische evolution als strategie der experimentellen forschung in der strömungstechnik. Dipl.-Ing. thesis, 1965. (in German).

71. H.-P. Schwefel. *Numerical Optimization of Computer Models*. Wiley, Chichester, U.K., 1981.
72. K.J. Shaw and P.J. Fleming. Initial study of practical multi-objective genetic algorithms for scheduling the production of chilled ready meals. In *Proceedings of Mendel'96, the 2nd International Mendel Conference on Genetic Algorithms*, Brno, Czech Republic, September 1996.
73. M.M. Ould Sidi, S. Hayat, S. Hammadi, and P. Borne. A novel approach to developing and evaluating regulation strategies for urban transport disrupted networks. *International Journal of Computer Integrated Manufacturing*, 21(4):480–493, 2008.
74. N. Srinivas and K. Deb. Multiobjective optimization using nondominated sorting in genetic algorithms. *Evolutionary Computation*, 2(3):221–248, Fall 1994.
75. E.G. Talbi and H. Meunier. Hierarchical parallel approach for GSM mobile network design. *Journal of Parallel and Distributed Computing*, 66(2):274–290, February 2006.
76. A. Tarafder, G.P. Rangaiah, and Ajay K. Ray. A study of finding many desirable solutions in multiobjective optimization of chemical processes. *Computers & Chemical Engineering*, 31(10):1257–1271, October 2007.
77. J. Teo, L.D. Neri, M.H. Nguyen, and H.A. Abbass. Walking with EMO: Multi-objective robotics for evolving two, four, and six–legged locomotion. In Lam Thu Bui and Sameer Alam, eds., *Multi-Objective Optimization in Computational Intelligence: Theory and Practice*, pp. 300–332. Information Science Reference, Hershey, PA, 2008. ISBN 978-1-59904-498-9.
78. G.T. Pulido and C.A.C. Coello. Using clustering techniques to improve the performance of a particle swarm pptimizer. In Kalyanmoy Deb et al., eds., *Genetic and Evolutionary Computation–GECCO 2004. Proceedings of the Genetic and Evolutionary Computation Conference. Part I*, pp. 225–237, Seattle, WA, June 2004. Springer-Verlag, Lecture Notes in Computer Science Vol. 3102.
79. D.A. Van Veldhuizen, J.B. Zydallis, and G.B. Lamont. Considerations in engineering parallel multiobjective evolutionary algorithms. *IEEE Transactions on Evolutionary Computation*, 7(2):144–173, April 2003.
80. M. Villalobos-Arias, C.A.C. Coello, and O. Hernández-Lerma. Asymptotic convergence of metaheuristics for multiobjective optimization problems. *Soft Computing*, 10(11):1001–1005, September 2006.
81. S.Y. Zeng, L.S. Kang, and L.X. Ding. An orthogonal multi-objective evolutionary algorithm for multi-objective optimization problems with constraints. *Evolutionary Computation*, 12(1):77–98, Spring 2004.
82. E. Zitzler and S. Künzli. Indicator-based selection in multiobjective search. In Xin Yao et al., eds., *Parallel Problem Solving from Nature—PPSN VIII*, pp. 832–842, Birmingham, U.K., September 2004. Springer-Verlag. Lecture Notes in Computer Science Vol. 3242.
83. E. Zitzler, M. Laumanns, and L. Thiele. SPEA2: Improving the strength Pareto evolutionary algorithm. In K. Giannakoglou, D. Tsahalis, J. Periaux, P. Papailou, and T. Fogarty, eds., *EUROGEN 2001. Evolutionary Methods for Design, Optimization and Control with Applications to Industrial Problems*, pp. 95–100, Athens, Greece, 2002.
84. E. Zitzler and L. Thiele. Multiobjective evolutionary algorithms: A comparative case study and the strength Pareto approach. *IEEE Transactions on Evolutionary Computation*, 3(4):257–271, November 1999.

26

Ant Colony Optimization

Christian Blum
Universitat Politècnica de Catalunya

Manuel López-Ibáñez
Université Libre de Bruxelles

26.1 Introduction

Many practical problems in logistics, planning, design, engineering, biology, and other fields can be modeled as optimization problems, where the goal is the minimization (respectively, maximization) of a particular objective function. The objective function assigns an objective cost value to each possible candidate solution. The domain of the objective function is called the search space, which may be either discrete or continuous. Optimization problems with discrete search space are also called combinatorial optimization problems (COPs). In principle, the purpose of an optimization algorithm is to find a solution that minimizes the objective function, that is, an optimal solution.

When dealing with optimization problems, algorithms that guarantee to find an optimal solution within bounded time are called *complete* algorithms. Nonetheless, some optimization problems may be too large or complex for complete algorithms to solve. In particular, there exists a class of problems, known as NP-hard, for which it is generally assumed that no complete algorithm with polynomial running time will ever be found. However, many practical situations only require to find a solution that, despite not being the optimal, is "good enough," specially if such solution is found within a reasonable computation time. This compromise explains the interest in the development of *approximate* (incomplete) algorithms, which aim to find a solution with objective cost close to the optimal within a computation time much shorter than any complete algorithm.

Ant colony optimization (ACO) [26] was one of the first techniques for approximate optimization inspired by the collective behavior of social insects. From the perspective of operations research, ACO algorithms belong to the class of metaheuristics [11,33,38]. At the same time, ACO algorithms are part of a research field known as swarm intelligence [10]. ACO takes its inspiration from the foraging behavior of ant colonies. At the core of this behavior is the indirect communication between the ants by means of chemical pheromone trails, which enables them to find short paths between their nest and food sources. This characteristic of real ant colonies is exploited in ACO algorithms for solving combinatorial and continuous optimization problems. In this introductory chapter, we focus, for space reasons, on combinatorial optimization, which is the traditional field of application of ACO. The interested reader may find comprehensive information on ACO algorithms for continuous optimization in the work of Socha and Dorigo [66].

The motivation of this chapter is to introduce the basic concepts of ACO without presuming any elaborate knowledge of computer science. Toward this goal, Section 26.2 briefly defines what a COP is. Section 26.3 introduces the concepts of complete versus approximate optimization algorithms and the definition of metaheuristic. Section 26.4 covers the origin and fundamental concepts of the ACO metaheuristic. Section 26.5 provides examples of modern ACO algorithms and their characteristics. Section 26.6 describes how ACO algorithms have been combined with other optimization techniques to further improve performance. Finally, Section 26.7 offers a representative sample of the wide range of applications of ACO algorithms.

26.2 Combinatorial Optimization Problems

In computer science, an *optimization problem* is composed of a search space S defined over a finite set of decision variables x_i ($i = 1,\ldots,n$), a set of constraints Ω among these variables, and an objective function $f: S \rightarrow \mathbb{R}^+$ that assigns a cost value to each element of S. Elements of the search space S are called *candidate solutions* to the problem. Candidate solutions that do not satisfy the constraints in Ω are called *infeasible*. COPs are those where the domains of the decision variables are discrete, hence the search space is finite. Nonetheless, the search space of a COP may be too large to be enumerated.

The definition of an optimization problem usually contains several parameters that are not fully specified, such as the number n of decision variables. An *instance* of a particular optimization problem is one possible specification of all parameters of an optimization problem, excluding the decision variables themselves. Given an instance of an optimization problem, the goal is to find a feasible solution, which minimizes the objective function.*

A notable example of a COP is the travelling salesman problem (TSP). In the classical definition of the TSP, the goal is to find the shortest route that traverses a given number of cities just once and returns to the origin. This problem is generalized as the problem of finding the minimal Hamiltonian cycle on a completely connected graph with weighted arcs. In this case, an instance of the TSP problem is defined by a particular graph, typically given as a distance matrix between nodes. The difficulty of a TSP instance greatly depends on the number of nodes in the graph. Despite its simplicity, the TSP represents various real-world problems.

Given an instance of an optimization problem, the goal of an optimization algorithm is to find an optimal solution, that is, a feasible solution that minimizes the objective function.

26.3 Optimization Algorithms

Ideally, an optimization algorithm for a given problem should be able to find an optimal solution for any problem instance. When this is the case, such algorithm is called *complete* (or *exact*). A trivial example of a complete algorithm is exhaustive search, where all possible solutions are iteratively enumerated and evaluated.

In some cases, for example, when the problem is trivial enough or the instance size is small, complete algorithms are able to find the optimal solution in a reasonable amount of time. For many interesting problems, however, the computation time required by complete algorithms is excessively large. In particular, there is a class of COPs—called NP-hard—for which it is generally assumed that no complete algorithm that requires a polynomial computation time will ever be found. The TSP is an example of an NP-hard problem and, hence, any complete algorithm will require exponential time with respect to the instance size.

* The rest of the chapter will concern only minimization problems, without loss of generality, because maximizing over an objective function f is the same as minimizing over $-f$.

Approximate (or heuristic) algorithms, on the other hand, do not guarantee returning the optimal solution. Instead, approximate algorithms aim to find a good approximation to the optimal solution in a short time. Simple heuristics use some kind of rule of thumb specific to the problem at hand. In the case of the TSP, an often used rule of thumb concerns the selection of short arcs. By comparison, more sophisticated *metaheuristics* are approximate algorithms that can be adapted to different problems [11,33,38]. Examples of metaheuristics are simulated annealing, iterated local search, tabu search, evolutionary algorithms, and ACO.

26.4 Ant Colony Optimization

The first ACO algorithm, Ant System (AS) [22,25], was inspired by the observation that ant colonies are able to find short paths between food sources and their nests. Ants searching for food explore the environment around their nest in a seemingly random wandering. As they move around, ants deposit a chemical substance, called *pheromone*. At the same time, ants are influenced by the pheromone present in their surroundings, having a probabilistic tendency to follow the direction where the concentration of pheromone is stronger. An ant that finds a food source returns to the nest, laying on the ground on its way back an amount of pheromone that depends on the quality of the food source. In this manner, an ant is able to indicate to other ants from the same colony the location and quality of the food source without any direct communication. Experiments on real ants [20] have shown that this indirect coordination between ants via pheromone trails—a mechanism generally known as *stigmergy*—produces a self-organizing collective behavior, where shorter paths between their nest and food sources are progressively followed by more ants, reinforcing pheromone trails of successful routes and eventually finding the shortest path.

ACO is a term that designates the number of algorithms based on this collective behavior of real ant colonies. When moving from real ants to artificial ants, we must bear in mind that the goal of ACO is not to faithfully model the natural world, but rather to find a good approximation to the unknown optimal solution of a given COP. In the context of COPs, pheromone values are probabilities associated with solution components. Solution components are elementary units that are assembled to form complete solutions to the particular problem. The task of each artificial ant consists in selecting solution components to construct a feasible candidate solution to the particular problem. Other notions from the natural world, such nest and food source, do not need an equivalent in ACO.

The basic elements of an ACO algorithm are graphically described by Figure 26.1. When applying an ACO algorithm to a particular COP, the first step is to define a finite set $C = \{c_1, c_2, \ldots\}$ of *solution components*, such that complete solutions may be constructed by iteratively selecting elements from this set. Second, one has to define a set $\mathcal{T}$ of pheromone values, which are numerical values associated with solution components: $\exists c_i \in C, \forall \tau_i \in \mathcal{T}$. In the example of the TSP, each edge of the given graph may be considered a solution component, where τ_{ij} represents the pheromone information associated with the

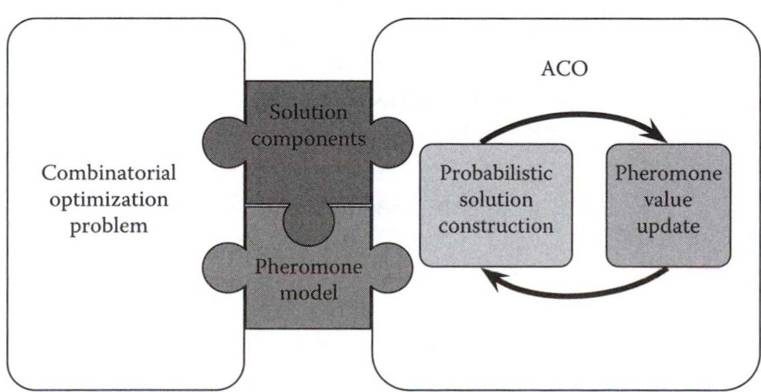

FIGURE 26.1 The fundamental components of the ACO framework.

Algorithm 26.1 Ant Colony Optimization (ACO)

```
Initialization()
while termination criteria not met do
    for each ant a ∈ {1, ..., N^a} do
        AntBasedSolutionConstruction()          /* see Algorithm 2 */
    end for
    PheromoneUpdate()
    DaemonActions()          /* optional */
end while
```

edge connecting nodes i and j. An alternative definition considers solution components as assignments of nodes to absolute positions in the tour, and hence, pheromone information τ_{ij} would be associated with visiting node j as the ith node of the tour. The former approach has repeatedly shown to produce better results for the TSP, but the opposite is true for other problems. In fact, the study of different pheromone representations is an ongoing research subject [57].

Pheromone values (implicitly) define a probability distribution over the search space. Artificial ants assemble complete candidate solutions by probabilistically choosing a sequence of solution components. In addition, pheromone values are modified taking into account the quality of the candidate solutions in order to bias future construction of solutions toward high-quality solutions. This cycle of probabilistic solution construction and pheromone update conforms the two fundamental steps of algorithms based on the ACO metaheuristic. Hence, most ACO algorithms follow the algorithmic schema shown in Algorithm 26.1, where three main procedures, AntBasedSolutionConstruction(), PheromoneUpdate(), and DaemonActions(), are performed at each iteration of the algorithm. Hereby, the order of execution of these procedures is up to the algorithm designer. From the algorithmic point of view, AntBasedSolutionConstruction() and PheromoneUpdate() are the two basic operations that must be defined by any ACO algorithm and we describe them in more detail in the following sections. The optional procedure DaemonActions() denotes any additional operation that is not related to solution construction or pheromone update, such as applying local search.

26.4.1 Solution Construction

At every iteration of ACO, a number of artificial ants construct complete solutions following procedure AntBasedSolutionConstruction() (Algorithm 26.2). Each artificial ant assembles a complete solution by probabilistically choosing a sequence of elements from the set C of solution components. Each ant starts with an empty sequence $s = <>$. At each construction step, the current sequence s is extended by adding a solution component from the set $\mathcal{N}(s)$ C. The set $\mathcal{N}(s)$ is defined such that the extension of the partial solution s may still result in a valid solution for the problem under consideration. For example, in the case of the TSP, $\mathcal{N}(s)$ is the set of nodes not visited yet in partial tour s. The choice of a solution component $c_i \in \mathcal{N}(s)$ is performed probabilistically with respect to the pheromone information $\tau_i \in \mathcal{T}$. In most ACO algorithms, the *transition probabilities* are defined as follows:

Algorithm 26.2 Procedure AntBasedSolutionConstruction()

```
s = <>                      /* start with an empty solution */
repeat
    determine N(s)
    c ← ChooseFrom(N(s))
    s ← s ∪ {c} /* extend s with solution component c */
until s is a complete solution
```

$$\mathbf{p}(c_i \mid s) = \frac{[\tau_i]^\alpha \cdot [\eta_i]^\beta}{\sum_{c_j \in \mathcal{N}(s)} [\tau_j]^\alpha \cdot [\eta_j]^\beta} \quad \forall\, c_i \in \mathcal{N}(s), \tag{26.1}$$

where η is the optional *heuristic information*, which assigns a heuristic value η_j to each solution component $c_j \in \mathcal{N}(s)$. The specification of the heuristic information depends on the problem. In the most simple case, the heuristic information depends only on the solution components. More sophisticated versions also consider the current partial solution. In general, its calculation should be fast, typically a constant value or a computationally inexpensive function. Finally, the exponents α and β determine the relative influence of pheromone and heuristic information on the resulting probabilities.

26.4.2 Pheromone Update

The pheromone information is modified in order to increase the probability of constructing better solutions in future iterations. Normally, this is achieved by performing both reinforcement—increasing the pheromone values associated with some solution components—and *evaporation*. Pheromone evaporation uniformly decreases all pheromone values, effectively reducing the influence of previous pheromone updates in future decisions. This has the effect of slowing down convergence. An algorithm has converged when subsequent iterations repeatedly construct the same solutions, and hence, further improvement is unlikely. Premature convergence prevents the adequate exploration of the search space and leads to poor solution quality. On the other hand, excessive exploration prevents the algorithm from quickly obtaining good solutions.

Pheromone reinforcement is performed by, first, selecting one or more solutions from those already constructed by the algorithm, and second, by increasing the values of the pheromone information associated with solution components that are part of these solutions. A general form of the pheromone update would be

$$\tau_i \leftarrow (1-\rho)\cdot\tau_i + \rho\cdot \sum_{\{s \in \mathcal{S}_{\text{upd}} \mid c_i \in s\}} \Delta\tau(s) \quad \forall \tau_i \in \mathcal{T}, \tag{26.2}$$

where

$\rho \in (0, 1)$ is a parameter called evaporation rate

$\mathcal{S}_{\text{upd}}$ denotes the set of solutions that are selected for updating the pheromone information

$\Delta\tau(s)$ is the pheromone amount deposited by each ant, which may be a function of the objective value of s

The only requirement is that $\Delta\tau(s)$ is nonincreasing with respect to the value of the objective function, that is, $f(s) < f(s') \Rightarrow \Delta\tau(s) \geq \Delta\tau(s'), \forall s \neq s' \in \mathcal{S}$. In the simplest case, $\Delta\tau(s)$ may be a constant value.

ACO algorithms often differ in the specification of $\mathcal{S}_{\text{upd}}$. In most cases, $\mathcal{S}_{\text{upd}}$ is composed of some of the solutions generated in the respective iteration (denoted by $\mathcal{S}_{\text{iter}}$) and the best solution found since the start of the algorithm, the best-so-far solution, henceforth denoted by s_{bf}. For example, AS selects for update all solutions constructed in the latest iteration, $\mathcal{S}_{\text{upd}} = \mathcal{S}_{\text{iter}}$. Other successful alternatives are the *iteration-best* and *best-so-far* update strategies, respectively ib-update and bf-update. The ib-update strategy utilizes only the best solution from the current iteration, that is, $\mathcal{S}_{\text{upd}} = \{s_{\text{ib}}\}$, where $s_{\text{ib}} = \arg\min \{f(s) \mid s \in \mathcal{S}_{\text{iter}}\}$. The ib-update rule focuses the search on the best solutions found in recent iterations. Similarly, the bf-update strategy utilizes only the *best-so-far* solution (s_{bf}), which produces an even faster convergence. In practice, the most successful ACO variants use variations of ib-update or bf-update rules and additionally include mechanisms to avoid premature convergence.

TABLE 26.1 Selection of Successful ACO Variants

ACO Variant	Authors	Main Reference
Elitist AS	Dorigo	[22]
	Dorigo, Maniezzo, and Colorni	[25]
Rank-based AS	Bullnheimer, Hartl, and Strauss	[16]
$\mathcal{MAX}$–$\mathcal{MIN}$ Ant System ($\mathcal{MMAS}$)	Stützle and Hoos	[70]
Ant Colony System	Dorigo and Gambardella	[23]
Hyper-Cube Framework	Blum and Dorigo	[9]

26.5 Modern ACO Algorithms

Despite proving the feasibility of a discrete optimization algorithm based on ants' foraging behavior, the results of AS—the first ACO algorithm—were inferior to state-of-the-art algorithms for the TSP. Hence, several improved variants have been proposed over time. Table 26.1 enumerates the most prevalent ACO variants, and we give a brief summary of their characteristics.

The variants elitist ant system (EAS) [25] and rank-based ant system (RAS) [16] mainly differ in the update method. EAS updates pheromone values using all solutions constructed in the current iteration plus the best-so-far solution ($\mathcal{S}_{\text{upd}} = \mathcal{S}_{\text{iter}} \cup \{s_{\text{bf}}\}$) hereby assigning a larger pheromone amount to solution s_{bf}. RAS considers for update a limited number $m - 1$ of solutions from the current iteration plus s_{bf}. These solutions are ranked according to their objective value, and the highest ranked solution s contributes an amount of pheromone of $m \cdot \Delta\tau(s)$, the one with the second-best rank s' contributes $(m - 1) \cdot \Delta\tau(s')$, and so on for all m solutions.

$\mathcal{MAX}$–$\mathcal{MIN}$ Ant System ($\mathcal{MMAS}$) [70] and Ant Colony System (ACS) [23]—perhaps the most successful ACO variants nowadays—introduced more sophisticated characteristics. In $\mathcal{MMAS}$, bounds of the pheromone values are dynamically calculated to avoid premature convergence and favor exploration of new solutions. In addition, $\mathcal{MMAS}$ uses a combination of ib-update and bf-update rules in order to focus the search on the (seemingly) most promising solution components.

ACS adds a greedy alternative to the probabilistic solution construction described by Equation 26.1. When choosing the next solution component, an ant has a certain probability of q_0 of choosing the solution component that maximizes $[\tau_i]^\alpha \cdot [\eta_i]^\beta$, otherwise the ant chooses probabilistically following Equation 26.1. A higher value of q_0 focuses the search around the best-solution components, accelerating convergence. ACS follows a bf-update rule; however, evaporation is applied only to the pheromone values associated with the best-so-far solution s_{bf}. Finally, ACS includes a local pheromone update that is performed after each solution construction step and decreases the pheromone values of solution components already visited.

A different proposal is the hyper-cube framework (HCF) [9], which is a framework for implementing ACO algorithms, including the ones mentioned above. The HCF has allowed to obtain theoretical results about the convergence of ACO. The practical benefits of the HCF include limiting pheromone values to the interval [0, 1], and the pheromone update is not affected by the scale of the objective function values.

26.6 Extensions of the ACO Metaheuristic

The ACO metaheuristic described above has been frequently complemented with additional algorithmic features. In particular, the combination with local search has been a standard approach, where local search can be considered a step to refine the solutions constructed by ACO before using them for pheromone update. Another notable example is the *candidate list* strategy [29], where the number of available choices at each solution construction step is restricted to a set of best choices. The set of best choices is usually selected with respect to their transition probabilities (Equation 26.1). The rationale for this approach is that for the construction of high-quality solutions, it is often enough to consider only the *promising* choices at each construction step. Moreover, for large instances with many solution components, limiting the number of choices greatly speeds up the search.

The ideas behind ACO have also been integrated into hybrid techniques. Hybrid metaheuristics [8] combine different metaheuristics with other optimization ideas in order to complement the strengths of different approaches. In the following, we enumerate a few hybrid metaheuristics based on ACO, namely, hybridization with beam search (Beam-ACO) and with constraint programming (CP), and the application of ACO in multilevel frameworks.

26.6.1 Hybridization with Beam Search

Beam search (Bs) is a classical tree search method for combinatorial optimization [59]. In particular, Bs is a deterministic approximate algorithm that relies heavily on bounding information for selecting among partial solutions. In Bs, a number of solutions—the beam—are iteratively constructed interdependently and in parallel. Each construction step consists of a first step where a number of candidate extensions of the partial solutions in the beam are selected based on heuristic information. In a second step, the candidate extensions form new partial solutions, and the Bs algorithm selects a limited number of these partial solutions by means of bounding information. The combination of ACO and Bs—labeled Beam-ACO—replaces the ant-based solution construction procedure of ACO with a probabilistic beam search procedure [6,7]. The probabilistic beam search utilizes pheromone information for the selection of candidate extensions of the beam. A further improvement, in cases where the bounding information is computationally expensive or unreliable, is replacing the bounding information with stochastic sampling of partial solutions [44]. Stochastic sampling in Beam-ACO executes the ant-based solution construction procedure of ACO to obtain a number of complete solutions (samples) from each partial solution in the beam. The best sample of each partial solution is considered an estimation of its quality.

26.6.2 ACO and Constraint Programming

Highly constrained problems, such as scheduling or timetabling, pose a particular challenge to metaheuristics, since the difficulty lies not simply in finding a good solution among many feasible solutions, but in finding feasible solutions among many infeasible ones. Despite the fact that ACO algorithms generally obtain competitive results for many problems, the performance of classical ACO algorithms has not been entirely satisfactory in the case of overly constrained problems. These problems have been targeted by means of CP techniques [52]. Hence, the application of CP techniques for restricting the search performed by an ACO algorithm to promising regions of the search space [54] is not too far-fetched.

26.6.3 Multilevel Frameworks Based on ACO

Multilevel techniques [14,71] start from the original problem instance and generate smaller and smaller instances by successive coarsening until some stopping criteria are satisfied. This creates a hierarchy of problem instances in which the problem instance of a given level is always smaller than the problem instance of the next lower level. Then, a solution is generated for the smallest problem instance, and successively transformed into a solution of the next higher level until a solution for the original problem instance is obtained. In a multilevel framework based on ACO, an ACO algorithm is applied at each level of the hierarchy to improve the solutions obtained at lower levels [13,42,43].

26.7 Applications of ACO Algorithms

Since the first application of AS to the travelling salesman problem in the early 1990s, the scope of ACO algorithms has widened considerably. Researchers have applied ACO to classical optimization problems such as assignment problems, scheduling problems, graph coloring, the maximum clique problem, and vehicle routing problems. Recent real-world applications of ACO include cell placement problems arising in circuit design, the design of communication networks, bioinformatics problems, and the optimal

TABLE 26.2 ACO Applications

Problem	References
Traveling salesman problem	[22–25,70]
Quadratic assignment problem	[49,51,70]
Scheduling problems	[6,12,19,28,53,69]
Vehicle routing problems	[31,61]
Timetabling	[67]
Set packing	[32]
Graph coloring	[18]
Shortest supersequence problem	[55]
Sequential ordering	[30]
Constraint satisfaction problems	[68]
Data mining	[60]
Maximum clique problem	[15]
Edge-disjoint paths problem	[5]
Cell placement in circuit design	[1]
Communication network design	[50]
Bioinformatics problems	[13,39,40,58,62,63]
Industrial problems	[2,7,17,34,64]
Water distribution networks	[47,48]
Continuous optimization	[4,27,41,56,65,66]
Non-static problems	[3,36]
Multi-objective problems	[21,37,45,46]
Music	[35]

design and operation of water distribution networks. Furthermore, there exists ongoing work on the application of ACO to non-static problems. Finally, a recent research trend is the extension of ACO to deal with problems with multiple objectives. Table 26.2 provides a list of representative applications of ACO algorithms. Dorigo and Stützle [26] have compiled a comprehensive list of references.

26.8 Concluding Remarks

Finally, we shortly want to elaborate on the question when ACO should be used. First of all, ACO is, in general, not superior to any other general purpose optimizer. This results from work that is known as *no-free-lunch* [72]. However, for specific problems (see Table 26.2), ACO might, of course, work better than other techniques. In general, ACO can be expected to work well for optimization problems for which well-working constructive heuristics are known. Moreover, ACO can only work if the search space is such that good solutions are concentrated in certain areas of the search space. On the contrary, if good solutions are scattered all over the search space, there is nothing that can be learned from already visited solutions. Unfortunately, it is currently impossible to make more specific claims about the general suitability of ACO for different classes of problems. However, this is not limited to ACO but happens for all general purpose optimizers.

References

1. S. Alupoaei and S. Katkoori. Ant colony system application to macrocell overlap removal. *IEEE Transactions on Very Large Scale Integration (VLSI) Systems*, 12(10):1118–1122, 2004.
2. J. Bautista and J. Pereira. Ant algorithms for a time and space constrained assembly line balancing problem. *European Journal of Operational Research*, 177(3):2016–2032, 2007.

3. L. Bianchi, L. M. Gambardella, and M. Dorigo. An ant colony optimization approach to the probabilistic traveling salesman problem. In J. J. Merelo, P. Adamidis, H.-G. Beyer, J.-L. Fernández-Villacanas, and H.-P. Schwefel, eds, *Proceedings of PPSN-VII, Seventh International Conference on Parallel Problem Solving from Nature*, volume 2439 of Lecture Notes in Computer Science, pp. 883–892. Springer-Verlag, Berlin, Germany, 2002.
4. B. Bilchev and I. C. Parmee. The ant colony metaphor for searching continuous design spaces. In T. C. Fogarty, ed, *Proceedings of the AISB Workshop on Evolutionary Computation*, volume 993 of Lecture Notes in Computer Science, pp. 25–39. Springer-Verlag, Berlin, Germany, 1995.
5. M. J. Blesa and C. Blum. Finding edge-disjoint paths in networks by means of artificial ant colonies. *Journal of Mathematical Modelling and Algorithms*, 6(3):361–391, 2007.
6. C. Blum. Beam-ACO—Hybridizing ant colony optimization with beam search: An application to open shop scheduling. *Computers & Operations Research*, 32(6):1565–1591, 2005.
7. C. Blum. Beam-ACO for simple assembly line balancing. *INFORMS Journal on Computing*, 20(4):617–628, 2008.
8. C. Blum, M. J. Blesa, A. Roli, and M. Sampels, eds. *Hybrid Metaheuristics—An Emerging Approach to Optimization*. Number 114 in Studies in Computational Intelligence. Springer-Verlag, Berlin, Germany, 2008.
9. C. Blum and M. Dorigo. The hyper-cube framework for ant colony optimization. *IEEE Transactions on Systems, Man, and Cybernetics Part B*, 34(2):1161–1172, 2004.
10. C. Blum and D. Merkle, eds. *Swarm Intelligence–Introduction and Applications. Natural Computing*. Springer-Verlag, Berlin, Germany, 2008.
11. C. Blum and A. Roli. Metaheuristics in combinatorial optimization: Overview and conceptual comparison. *ACM Computing Surveys*, 35(3):268–308, 2003.
12. C. Blum and M. Sampels. An ant colony optimization algorithm for shop scheduling problems. *Journal of Mathematical Modelling and Algorithms*, 3(3):285–308, 2004.
13. C. Blum and M. Yábar Vallès. Ant colony optimization for DNA sequencing by hybridization. *Computers & Operations Research*, 11:3620–3635, 2008.
14. A. Brandt. Multilevel computations: Review and recent developments. In S. F. Mc-Cormick, ed, *Multigrid Methods: Theory, Applications, and Supercomputing, Proceedings of the 3rd Copper Mountain Conference on Multigrid Methods*, volume 110 of Lecture Notes in Pure and Applied Mathematics, pp. 35–62. Marcel Dekker, New York, 1988.
15. T. N. Bui and J. R. Rizzo Jr. Finding maximum cliques with distributed ants. In K. Deb et al., ed, *Proceedings of the Genetic and Evolutionary Computation Conference (GECCO 2004)*, volume 3102 of Lecture Notes in Computer Science, pp. 24–35. Springer-Verlag, Berlin, Germany, 2004.
16. B. Bullnheimer, R. Hartl, and C. Strauss. A new rank-based version of the Ant System: A computational study. *Central European Journal for Operations Research and Economics*, 7(1):25–38, 1999.
17. P. Corry and E. Kozan. Ant colony optimisation for machine layout problems. *Computational Optimization and Applications*, 28(3):287–310, 2004.
18. D. Costa and A. Hertz. Ants can color graphs. *Journal of the Operational Research Society*, 48:295–305, 1997.
19. M. L. den Besten, T. Stützle, and M. Dorigo. Ant colony optimization for the total weighted tardiness problem. In M. Schoenauer, K. Deb, G. Rudolph, X. Yao, E. Lutton, J. J. Merelo, and H.-P. Schwefel, eds, *Proceedings of PPSN-VI, Sixth International Conference on Parallel Problem Solving from Nature*, volume 1917 of Lecture Notes in Computer Science, pp. 611–620. Springer-Verlag, Berlin, Germany, 2000.
20. J.-L. Deneubourg, S. Aron, S. Goss, and J. M. Pasteels. The self-organizing exploratory pattern of the Argentine ant. *Journal of Insect Behavior*, 3(2):159–168, 1990.
21. K. Doerner, W. J. Gutjahr, R. F. Hartl, C. Strauss, and C. Stummer. Pareto ant colony optimization: A metaheuristic approach to multiobjective portfolio selection. *Annals of Operations Research*, 131:79–99, 2004.

22. M. Dorigo. Optimization, learning and natural algorithms. Phd thesis, Dipartimento di Elettronica, Politecnico di Milano, Italy, 1992 (in Italian).
23. M. Dorigo and L. Gambardella. Ant Colony System: A cooperative learning approach to the traveling salesman problem. *IEEE Transactions on Evolutionary Computation*, 1(1):53–66, 1997.
24. M. Dorigo, V. Maniezzo, and A. Colorni. Positive feedback as a search strategy. Technical Report 91-016, Dipartimento di Elettronica, Politecnico di Milano, Italy, 1991.
25. M. Dorigo, V. Maniezzo, and A. Colorni. Ant System: Optimization by a colony of cooperating agents. *IEEE Transactions on Systems, Man, and Cybernetics Part B*, 26(1):29–41, 1996.
26. M. Dorigo and T. Stützle. *Ant Colony Optimization*. MIT Press, Cambridge, MA, 2004.
27. J. Dréo and P. Siarry. A new ant colony algorithm using the heterarchical concept aimed at optimization of multiminima continuous functions. In M. Dorigo, G. Di Caro, and M. Sampels, eds, *Ant Algorithms—Proceedings of ANTS 2002—Third International Workshop*, volume 2463 of Lecture Notes in Computer Science, pp. 216–221. Springer-Verlag, Berlin, Germany, 2002.
28. C. Gagné, W. L. Price, and M. Gravel. Comparing an ACO algorithm with other heuristics for the single machine scheduling problem with sequence-dependent setup times. *Journal of the Operational Research Society*, 53:895–906, 2002.
29. L. Gambardella and M. Dorigo. Solving symmetric and asymmetric TSPs by ant colonies. In T. Baeck, T. Fukuda, and Z. Michalewicz, eds, *Proceedings of the 1996 IEEE International Conference on Evolutionary Computation (ICEC'96)*, pp. 622–627. IEEE Press, Piscataway, NJ, 1996.
30. L. M. Gambardella and M. Dorigo. Ant Colony System hybridized with a new local search for the sequential ordering problem. *INFORMS Journal on Computing*, 12(3):237–255, 2000.
31. L. M. Gambardella, É. D. Taillard, and G. Agazzi. MACS-VRPTW: A multiple ant colony system for vehicle routing problems with time windows. In D. Corne, M. Dorigo, and F. Glover, eds, *New Ideas in Optimization*, pp. 63–76. McGraw Hill, London, U.K., 1999.
32. X. Gandibleux, X. Delorme, and V. T'Kindt. An ant colony optimisation algorithm for the set packing problem. In M. Dorigo, M. Birattari, C. Blum, L. M. Gambardella, F. Mondada, and T. Stützle, eds, *Proceedings of ANTS 2004—Fourth International Workshop on Ant Colony Optimization and Swarm Intelligence*, volume 3172 of Lecture Notes in Computer Science, pp. 49–60. Springer-Verlag, Berlin, Germany, 2004.
33. F. Glover and G. Kochenberger, eds. *Handbook of Metaheuristics*. Kluwer Academic Publishers, Norwell, MA, 2002.
34. J. Gottlieb, M. Puchta, and C. Solnon. A study of greedy, local search, and ant colony optimization approaches for car sequencing problems. In S. Cagnoni, J. J. Romero Cardalda, D. W. Corne, J. Gottlieb, A. Guillot, E. Hart, C. G. Johnson, E. Marchiori, J.-A. Meyer, M. Middendorf, and G. R. Raidl, eds, *Applications of Evolutionary Computing, Proceedings of EvoWorkshops 2003*, volume 2611 of Lecture Notes in Computer Science, pp. 246–257. Springer-Verlag, Berlin, Germany, 2003.
35. C. Guéret, N. Monmarché, and M. Slimane. Ants can play music. In M. Dorigo, M. Birattari, C. Blum, L. M. Gambardella, F. Mondada, and T. Stützle, eds, *Proceedings of ANTS 2004—Fourth International Workshop on Ant Colony Optimization and Swarm Intelligence*, volume 3172 of Lecture Notes in Computer Science, pp. 310–317. Springer-Verlag, Berlin, Germany, 2004.
36. M. Guntsch and M. Middendorf. Pheromone modification strategies for ant algorithms applied to dynamic TSP. In E. J. W. Boers, J. Gottlieb, P. L. Lanzi, R. E. Smith, S. Cagnoni, E. Hart, G. R. Raidl, and H. Tijink, eds, *Applications of Evolutionary Computing: Proceedings of EvoWorkshops 2001*, volume 2037 of Lecture Notes in Computer Science, pp. 213–222. Springer-Verlag, Berlin, Germany, 2001.
37. M. Guntsch and M. Middendorf. Solving multi-objective permutation problems with population based ACO. In C. M. Fonseca, P. J. Fleming, E. Zitzler, K. Deb, and L. Thiele, eds, *Proceedings of the Second International Conference on Evolutionary Multi-Criterion Optimization (EMO 2003)*, volume 2636 of Lecture Notes in Computer Science, pp. 464–478. Springer-Verlag, Berlin, Germany, 2003.
38. H. H. Hoos and T. Stützle. *Stochastic Local Search: Foundations and Applications*. Elsevier, Amsterdam, the Netherlands, 2004.

39. O. Karpenko, J. Shi, and Y. Dai. Prediction of MHC class II binders using the ant colony search strategy. *Artificial Intelligence in Medicine*, 35(1–2):147–156, 2005.
40. O. Korb, T. Stützle, and T. E. Exner. PLANTS: Application of ant colony optimization to structure-based drug design. In M. Dorigo, L. M. Gambardella, A. Martinoli, R. Poli, and T. Stützle, eds, *Ant Colony Optimization and Swarm Intelligence—Proceedings of ANTS 2006—Fifth International Workshop*, volume 4150 of Lecture Notes in Computer Science, pp. 247–258. Springer-Verlag, Berlin, Germany, 2006.
41. P. Korosec, J. Silc, K. Oblak, and F. Kosel. The differential ant-stigmergy algorithm: An experimental evaluation and a real-world application. In *Proceedings of CEC 2007—IEEE Congress on Evolutionary Computation*, pp. 157–164. IEEE Press, Piscataway, NJ, 2007.
42. P. Korošec, J. Šilc, and B. Robič. Mesh-partitioning with the multiple ant-colony algorithm. In M. Dorigo, M. Birattari, C. Blum, L. M. Gambardella, F. Mondada, and T. Stützle, eds, *Proceedings of ANTS 2004—Fourth International Workshop on Ant Colony Optimization and Swarm Intelligence*, volume 3172 of Lecture Notes in Computer Science, pp. 430–431. Springer-Verlag, Berlin, Germany, 2004.
43. P. Korošec, J. Šilc, and B. Robič. Solving the mesh-partitioning problem with an ant-colony algorithm. *Parallel Computing*, 30:785–801, 2004.
44. M. López-Ibáñez and C. Blum. Beam-ACO based on stochastic sampling: A case study on the TSP with time windows. In *Proceedings of Learning and Intelligent Optimization (LION 3)*, Lecture Notes in Computer Science. Springer-Verlag, Berlin, Germany, 2009.
45. M. López-Ibáñez, L. Paquete, and Thomas Stützle. On the design of ACO for the biobjective quadratic assignment problem. In M. Dorigo, L. Gambardella, F. Mondada, Thomas Stützle, M. Birratari, and C. Blum, eds, *ANTS'2004, Fourth International Workshop on Ant Algorithms and Swarm Intelligence*, volume 3172 of Lecture Notes in Computer Science, pp. 214–225. Springer-Verlag, Berlin, Germany, 2004.
46. M. López-Ibáñez, L. Paquete, and T. Stützle. Hybrid population-based algorithms for the bi-objective quadratic assignment problem. *Journal of Mathematical Modelling and Algorithms*, 5(1):111–137, April 2006.
47. M. López-Ibáñez, T. Devi Prasad, and B. Paechter. Ant colony optimisation for the optimal control of pumps in water distribution networks. *Journal of Water Resources Planning and Management, ASCE*, 134(4):337–346, 2008.
48. H. R. Maier, A. R. Simpson, A. C. Zecchin, W. K. Foong, K. Y. Phang, H. Y. Seah, and C. L. Tan. Ant colony optimization for design of water distribution systems. *Journal of Water Resources Planning and Management, ASCE*, 129(3):200–209, May/June 2003.
49. V. Maniezzo. Exact and approximate nondeterministic tree-search procedures for the quadratic assignment problem. *INFORMS Journal on Computing*, 11(4):358–369, 1999.
50. V. Maniezzo, M. Boschetti, and M. Jelasity. An ant approach to membership overlay design. In M. Dorigo, M. Birattari, C. Blum, L. M. Gambardella, F. Mondada, and T. Stützle, eds, *Proceedings of ANTS 2004—Fourth International Workshop on Ant Colony Optimization and Swarm Intelligence*, volume 3172 of Lecture Notes in Computer Science, pp. 37–48. Springer-Verlag, Berlin, Germany, 2004.
51. V. Maniezzo and A. Colorni. The Ant System applied to the quadratic assignment problem. *IEEE Transactions on Data and Knowledge Engineering*, 11(5):769–778, 1999.
52. K. Marriott and P. Stuckey. *Programming with Constraints*. MIT Press, Cambridge, MA, 1998.
53. D. Merkle, M. Middendorf, and H. Schmeck. Ant colony optimization for resource-constrained project scheduling. *IEEE Transactions on Evolutionary Computation*, 6(4):333–346, 2002.
54. B. Meyer and A. Ernst. Integrating ACO and constraint propagation. In M. Dorigo, M. Birattari, C. Blum, L. M. Gambardella, F. Mondada, and T. Stützle, eds, *Proceedings of ANTS 2004—Fourth International Workshop on Ant Colony Optimization and Swarm Intelligence*, volume 3172 of Lecture Notes in Computer Science, pp. 166–177. Springer-Verlag, Berlin, Germany, 2004.
55. R. Michel and M. Middendorf. An island model based ant system with lookahead for the shortest supersequence problem. In A. E. Eiben, T. Bäck, M. Schoenauer, and H.-P. Schwefel, eds, *Proceedings of PPSN-V, Fifth International Conference on Parallel Problem Solving from Nature*, volume 1498 of Lecture Notes in Computer Science, pp. 692–701. Springer-Verlag, Berlin, Germany, 1998.

56. N. Monmarché, G. Venturini, and M. Slimane. On how *pachycondyla apicalis* ants suggest a new search algorithm. *Future Generation Computer Systems*, 16:937–946, 2000.
57. J. Montgomery, M. Randall, and T. Hendtlass. Solution bias in ant colony optimisation: Lessons for selecting pheromone models. *Computers & Operations Research*, 35(9):2728–2749, 2008.
58. J. D. Moss and C. G. Johnson. An ant colony algorithm for multiple sequence alignment in bioinformatics. In D. W. Pearson, N. C. Steele, and R. F. Albrecht, eds, *Artificial Neural Networks and Genetic Algorithms*, pp. 182–186. Springer-Verlag, Berlin, Germany, 2003.
59. P. S. Ow and T. E. Morton. Filtered beam search in scheduling. *International Journal of Production Research*, 26:297–307, 1988.
60. R. S. Parpinelli, H. S. Lopes, and A. A. Freitas. Data mining with an ant colony optimization algorithm. *IEEE Transactions on Evolutionary Computation*, 6(4):321–332, 2002.
61. M. Reimann, K. Doerner, and R. F. Hartl. D-ants: Savings based ants divide and conquer the vehicle routing problems. *Computers & Operations Research*, 31(4):563–591, 2004.
62. A. Shmygelska, R. Aguirre-Hernández, and H. H. Hoos. An ant colony optimization algorithm for the 2D HP protein folding problem. In M. Dorigo, G. Di Caro, and M. Sampels, eds, *Ant Algorithms—Proceedings of ANTS 2002—Third International Workshop*, volume 2463 of Lecture Notes in Computer Science, pp. 40–52. Springer-Verlag, Berlin, Germany, 2002.
63. A. Shmygelska and H. H. Hoos. An ant colony optimisation algorithm for the 2D and 3D hydrophobic polar protein folding problem. *BMC Bioinformatics*, 6(30):1–22, 2005.
64. C. A. Silva, T. A. Runkler, J. M. Sousa, and R. Palm. Ant colonies as logistic processes optimizers. In M. Dorigo, G. Di Caro, and M. Sampels, eds, *Ant Algorithms—Proceedings of ANTS 2002—Third International Workshop*, volume 2463 of Lecture Notes in Computer Science, pp. 76–87. Springer-Verlag, Berlin, Germany, 2002.
65. K. Socha and C. Blum. An ant colony optimization algorithm for continuous optimization: An application to feed-forward neural network training. *Neural Computing & Applications*, 16(3):235–247, 2007.
66. K. Socha and M. Dorigo. Ant colony optimization for continuous domains. *European Journal of Operational Research*, 185(3):1155–1173, 2008.
67. K. Socha, M. Sampels, and M. Manfrin. Ant algorithms for the university course timetabling problem with regard to the state-of-the-art. In S. Cagnoni, J. J. Romero Cardalda, D. W. Corne, J. Gottlieb, A. Guillot, E. Hart, C. G. Johnson, E. Marchiori, J.-A. Meyer, M. Middendorf, and G. R. Raidl, eds, *Applications of Evolutionary Computing, Proceedings of EvoWorkshops 2003*, volume 2611 of Lecture Notes in Computer Science, pp. 334–345. Springer-Verlag, Berlin, Germany, 2003.
68. C. Solnon. Ant can solve constraint satisfaction problems. *IEEE Transactions on Evolutionary Computation*, 6(4):347–357, 2002.
69. T. Stützle. An ant approach to the flow shop problem. In *Proceedings of the 6th European Congress on Intelligent Techniques & Soft Computing (EUFIT'98)*, pp. 1560–1564. Verlag Mainz, Aachen, Germany, 1998.
70. Thomas Stützle and H. H. Hoos. $\mathcal{MAX}$–$\mathcal{MIN}$ Ant System. *Future Generation Computer Systems*, 16(8):889–914, 2000.
71. C. Walshaw and M. Cross. Mesh partitioning: A multilevel balancing and refinement algorithm. *SIAM Journal on Scientific Computing*, 22(1):63–80, 2000.
72. D. H. Wolpert and W. G. Macready. No free lunch theorems for optimization. *IEEE Transactions on Evolutionary Computation*, 1(1):67–82, 1997.

27

Heuristics for Two-Dimensional Bin-Packing Problems

Tak Ming Chan
University of Minho

Filipe Alvelos
University of Minho

Elsa Silva
University of Minho

J.M. Valério de Carvalho
University of Minho

27.1 Introduction

In the wood cutting industry, the central step of the manufacturing process is cutting large two-dimensional boards of raw material to obtain the desired items. The operation of the cutting machines is often driven by intelligent software systems with graphical user interfaces (GUIs), which help the operator planning the cutting operations, and may include other features, such as establishing the sequence in which the boards should be cut.

The characteristics of the set of items that are grouped to be cut in a set of large boards may vary largely. There are companies that produce large quantities of items of a small number of different sizes, as happens for instance in furniture companies that produce semifinished articles with standard sizes. On the other hand, there are make-to-order furniture companies that produce finished goods, which typically solve different problems every day, with a large variety of item sizes, in small quantities.

Intelligent software systems for this industry may include different heuristic algorithms tailored to provide good quality solutions to instances with a wide variety of characteristics, because optimal solutions may be beyond reach, due to the difficulty of the problem. In this chapter, we address heuristic procedures that are more suitable to make-to-order companies.

In Section 27.2, the concept of bin packing related to those cutting operations and its characteristics such as cutting methods and packing strategies are presented. In Section 27.3, we review several heuristics for solving bin packing problems as the level-oriented (shelf-oriented) one-phase and two-phase heuristics, and introduce new local search heuristics.

In Section 27.4, computational results of benchmark and real-world problems obtained by using a recently proposed heuristic, a greedy heuristic (one-phase heuristic) together with stochastic neighborhood structures (local search heuristic), and by considering various problem scenarios, i.e., various combinations of the following three parameters: (1) with or without item rotation, (2) horizontal or vertical cut(s) applied in the first-stage cutting, (3) two or three-stage exact/non-exact case, are summarized. Finally, in Section 27.5, the main conclusions are drawn.

27.2 Bin-Packing Problems

Bin-packing problems are well-known combinatorial optimization problems. They are closely related to cutting stock problems and these two problem types are conceptually equal (Wäscher et al. 2007). Thus, some terminologies borrowed from cutting stock problems will be used in the following discussion. The common objective of two-dimensional bin-packing problems is to pack a given set of rectangular items to an unlimited number of identical rectangular bins such that the total number of used bins is minimized and subject to three limitations: (1) all items must be packed to bins, (2) all items cannot overlap, and (3) the edges of items are parallel to those of the bins. Nowadays, these problems are always faced by wood, glass, paper, steel, and cloth industries.

Since determining the packing location of each item relies on the cutting approaches used, those approaches are firstly presented. Guillotine cutting and free cutting are the two possible types of cutting methods. In particular, the former approach is frequently needed because of technological characteristics of automated cutting machines. A guillotine cutting means the one from an edge of the rectangle to the opposite edge. Guillotine cuttings that are applied n times are referred to n-stage guillotine cutting. Examples of the two-stage and three-stage guillotine cutting are portrayed in Figures 27.1 and 27.2, respectively. In these two figures, the bold lines represent horizontal (or vertical) cuts in the corresponding stages. Also, the gray rectangles stand for the waste materials. For both the two-stage and three-stage guillotine cutting, processes start with horizontal cut(s) in the first stage and then vertical cut(s) in the second stage. However, further horizontal cut(s) are applied in the third stage of the three-stage guillotine cutting. Although the process of guillotine cutting described above starts with horizontal cut(s) in the first stage, the direction of cutting(s) can be chosen as vertical in the first stage, as an alternative. As long as this happens, horizontal cut(s) and vertical cut(s) will be performed in the second and third stage (for three-stage cutting only), respectively. Furthermore, additional cut(s) can be applied to separate the waste from items after the final stage and this is known as trimming. However, these cuts are not considered as an additional stage. A free cutting means that the cutting does not have any restriction, i.e., non-guillotine. An instance of a free cutting is depicted in Figure 27.3. This chapter focuses on the guillotine cutting.

In order to easily fit the guillotine-cutting approach, the most common way of packing items to bins is the level-oriented (shelf-oriented) method. The idea of this method is that shelves are created by packing items from left to right. The height of each shelf is determined by the tallest item resided in the leftmost of the shelf. Also, whether an item is packed on the top or to the right of another packed item or

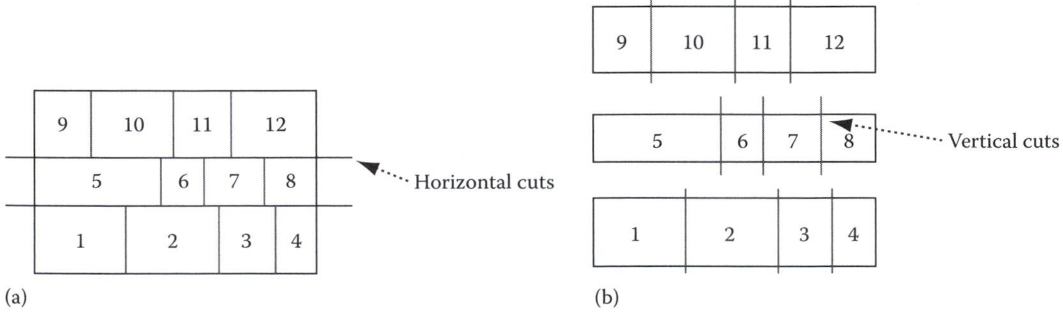

FIGURE 27.1 An example of two-stage guillotine cutting.

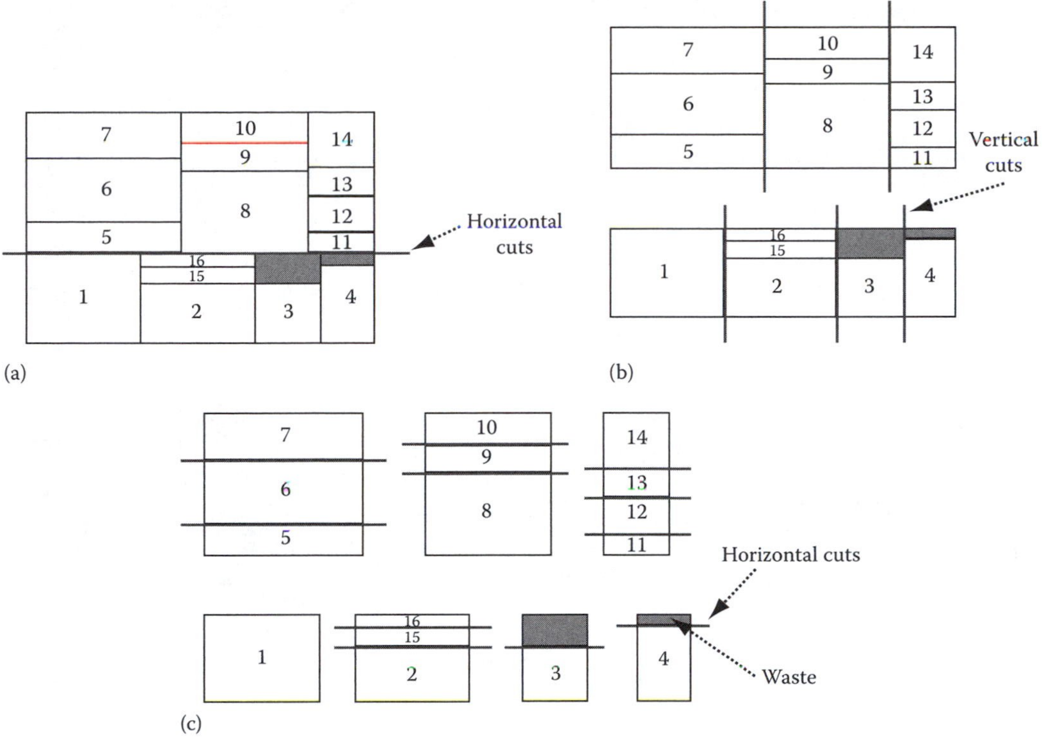

FIGURE 27.2 An example of three-stage guillotine cutting.

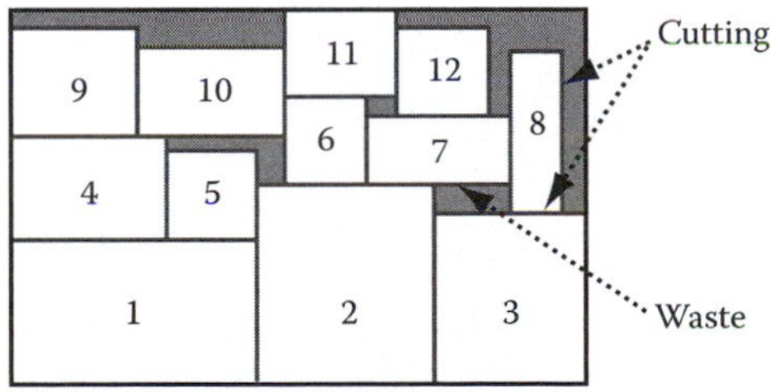

FIGURE 27.3 An example of free cutting.

in a new shelf (i.e., the packing location of the item) depends on the following factors: the size of the free space available, the number of cutting stages required and whether trimming is adopted. In this chapter, the numbers of stages considered are two and three. Accordingly, four different level-oriented scenarios can be formed: (1) two-stage without trimming, (2) two-stage with trimming, (3) three-stage without trimming, and (4) three-stage with trimming. Figure 27.4a through d illustrates examples of these four scenarios, respectively. Assume that there are 13 and 23 items packed in a bin for the two-stage and three-stage problems, respectively, and the packing order of items is based on the ascending order of the item index. For both scenarios (a) and (b), since they are two-stage problems, no item can be packed on the top of any packed item in each shelf. In scenario (a), trimming is not allowed after the final stage and therefore the height of each item in the same shelf must be identical. Nevertheless, it is permitted in scenario (b) and thus the free space can exist on the top of any packed item except the leftmost item in each shelf. For both scenarios (c) and (d), they are three-stage problems and therefore an item can be packed on the top of any packed item except the leftmost item in each shelf. As trimming is not allowed

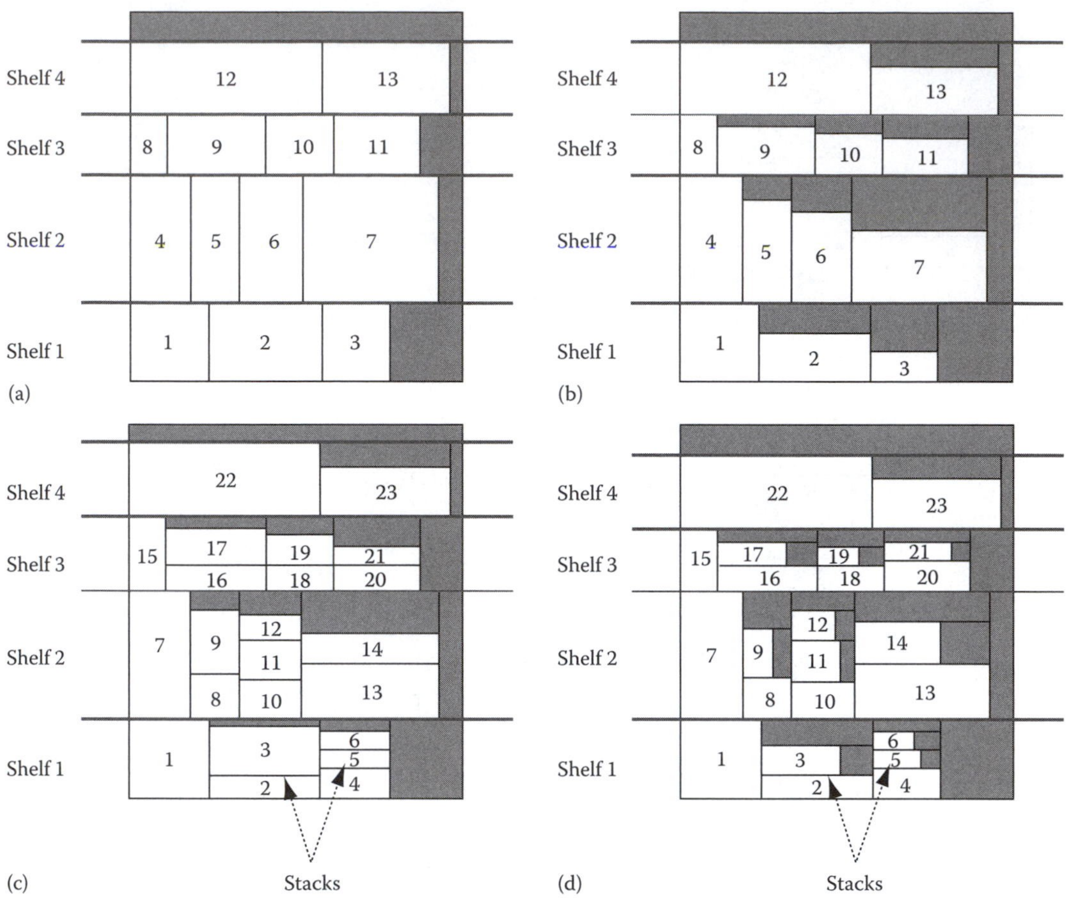

FIGURE 27.4 Examples of four different level-oriented scenarios.

in scenario (c), the width of each item in the same stack must be equivalent. However, this restriction is not applied in scenario (d) because trimming is permitted. Note that the leftmost item in each shelf is also a stack, which determines the height of the shelf.

There are two possible ways to deal with items when packing them into bins: (1) rotation of items is not permitted, i.e., fixed orientation, and (2) rotation of items is permitted by 90°. Items in the form of some raw materials cannot be rotated. To cite an example, some wood and cloth have decorative patterns, which are required to be fixed in a particular direction when pieces are assembled. Nevertheless, the items may be rotated provided that they are plain materials.

Based on different combinations of feasibility of item rotation and cutting methods, Lodi et al. (1999) proposed a classification of two-dimensional bin-packing problems (2BP) for four cases, which are (1) 2BP|O|G, (2) 2BP|R|G, (3) 2BP|O|F, and (4) 2BP|R|F, where O denotes that items are oriented, i.e., they cannot be rotated; R denotes that items may be rotated by 90°; G signifies that guillotine cutting is used; and F signifies that free cutting is adopted. This kind of classification can help unify definitions and notations, facilitate communication between researchers in the field, and offer a faster access to the relevant literature (Wäscher et al. 2007).

Cutting and packing problems have been extensively studied in the last decade. Some websites such as EURO Special Interest Group on Cutting and Packing (ESICUP), PackLibï[2], and OR-Library were established in order to facilitate the researchers to collect benchmark problems proposed in the past to test the efficiency and effectiveness of their suggested algorithms.

The possible optimization methods to solve bin-packing problems are exact methods, heuristics, and meta-heuristics (Lodi et al. 1999; Carter and Price 2001; Sait and Youssef 1999). Even though it is guaranteed that exact methods can find an optimal solution, the difficulty of obtaining an optimal solution increases drastically if the problem size increases, due to the fact that it is an NP-hard problem (Garey and Johnson 1979). For large instances, alternatively, heuristics or meta-heuristics approaches are able to search a good-quality solution in a reasonable amount of time. In particular, this chapter will focus on presenting the implementation of different level-oriented heuristics and local search–based heuristics proposed in the past and recently for solving bin-packing problems.

27.3 Heuristics

This section is devoted to discussing various heuristics proposed for tackling bin-packing problems. There are two scenarios for bin-packing problems: (1) The information of all items is known prior to solving the problem. It is called an offline approach. (2) When an item is being packed, the information of the next item is unknown. It is named an online approach. In this chapter, we will focus on heuristics for only the former one. Heuristics for the offline approach can be classified into two types (Lodi et al. 2002): (1) one-phase heuristics and (2) two-phase heuristics. The idea of the first one is to pack items into bins directly while the second one is to aggregate items to form strips (shelves) in the first stage and then pack strips into bins in the second stage. In the following, the descriptions of one-phase heuristics, two-phase heuristics, and local search heuristics will be provided in order.

27.3.1 One-Phase Heuristics

27.3.1.1 Finite First Fit

This algorithm was proposed by Berkey and Wang (1987). First, the items are sorted by nonincreasing height. Starting from the lowest to the highest level of the first used bin, the current item is packed into the level currently considered if it fits. If no level can fit it but the residual height is sufficient, it is packed into a new level created in the same bin. Otherwise, the same steps are applied to the subsequently used bins. If no bin can accommodate it, it is loaded into a new level of a new bin.

Figure 27.5a depicts an instance of this method. Sorted items 1 and 2 are first loaded into the first level of the first bin. As the empty space at the right of item 2 is not large enough to accommodate item 3, it is packed into the newly created second level. Then, items 4 and 5 are packed into the "first-fit" free spaces of the first and second levels, respectively. It is needed to load item 6 into the newly created third level because the empty spaces of the first or second level cannot fit it. Similarly, item 7 is loaded into a new bin since no level of the first bin can fit it.

27.3.1.2 Finite Next Fit

This heuristic was proposed by Berkey and Wang (1987). First, the items are sorted by nonincreasing height. The current item is packed into the current level of the current bin if it fits. Otherwise, it is packed into a new level created in the current bin if the residual height is sufficient. Otherwise, it is loaded into a new level of a new bin.

Figure 27.5b shows an example of this approach. After sorting the items, items 1 and 2 are first loaded into the first level of the first bin. Since the empty space at the right of item 2 is not large enough to fit item 3, it is packed into the newly created second level. After packing item 4, it is required to load item 5 into the newly created third level because the empty space at the right of item 4 is not sufficiently large. After packing item 6, item 7 is loaded into the first level of a new bin since the third level of the first bin cannot accommodate it.

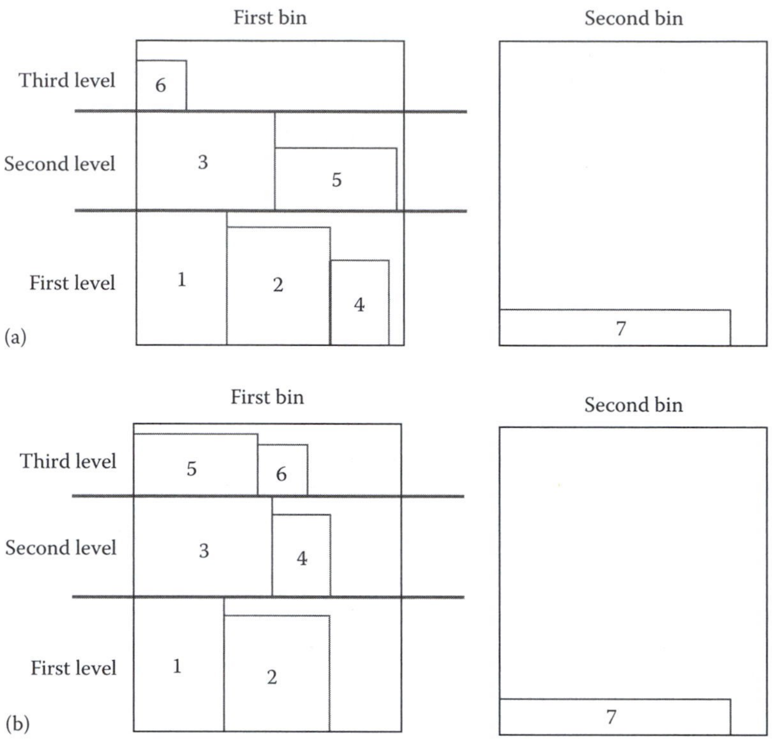

FIGURE 27.5 Examples of finite first fit and finite next fit.

27.3.1.3 Greedy Heuristic

A greedy heuristic was proposed by Alvelos et al. (2009). The two elements involved in the heuristic are described as follows:

- A stack is a set of items, each of which, except the bottom one, is placed on the top of another. For the two-stage problems, since no further horizontal cut(s) is/are permitted after the second-stage guillotine cutting, except trimming(s), each stack must possess only one item but such a constraint is not required in the three-stage problems.
- A shelf, which is a row of a bin, contains at least one stack or item.

The idea of the heuristic is based on the sorting of item types for defining an initial packing sequence and iterative trials of packing items into existing stacks, shelves, or bins according to the following criteria. For the former one, the three criteria considered to sort the item types in the descending order are (1) by width, (2) by height, (3) by area. For the latter one, it is possible that more than one existing stack, shelf, or bin can accommodate each item. Therefore, criteria should be established to determine which existing stack, shelf, and bin should be selected for packing the items. The three criteria used to achieve this purpose are: (1) minimize the residual width after packing the item, (2) minimize the residual height after packing the item, (3) minimize the residual area after packing the item. Note that different criteria can be adopted for stack, shelf, and bin selection.

The first step of the heuristic is to define a packing sequence for items by sorting the item types based on a specified criterion. Then, iteratively, each item, which can or cannot be rotated, is packed into an existing stack that minimizes a specified criterion. If this is not possible, we try to pack it into an existing shelf, which minimizes a specified criterion. If this is not possible, we try to pack it into a used bin, which minimizes a specified criterion. If this is not possible again, it is placed into a new bin.

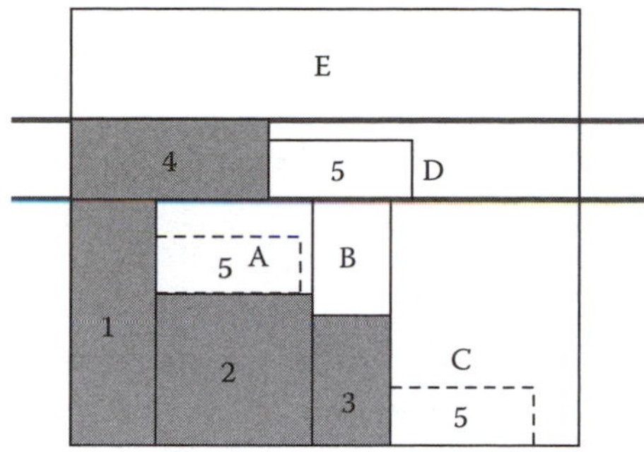

FIGURE 27.6 An example of constructive heuristic.

Finally, a solution is obtained by running the heuristic with several sets of criteria and then selecting the best one among the obtained solutions.

Figure 27.6 illustrates an example of the greedy heuristic. Assume that the items are sorted by height and four items have already been loaded into the bin. The item 5 currently considered can be loaded into an existing stack (free space A) (for three-stage problems only), an existing shelf (free space C or D), on the top of shelves of the bin (free space E), or a new empty bin. However, the free space B cannot accommodate it. The place where the item is packed depends on the preselected criterion and the type of the problem tackled.

Suppose that a two-stage problem with trimming is now being solved. In this case, only three possible choices of empty spaces, C, D, and E, are available for packing item 5. If the criterion (1) is set, it will be packed into C to fulfill the criterion; if criterion (2) is selected, it will be packed into D; if criterion (3) is chosen, it will be placed into D. Note that the same applies to a three-stage problem without trimming. Nevertheless, if it is a three-stage problem with trimming, and the criterion (1) or (3) is set, it will be placed into A instead of C or D.

27.3.2 Two-Phase Heuristics

27.3.2.1 Hybrid First Fit

The hybrid first fit algorithm was proposed by Chung et al. (1982). In the first phase, a strip packing is performed by using the first-fit decreasing height strategy whose principle is given as follows. An item is packed in the left-justified way in the "first-fit" level. If no level can accommodate the item, a new level is created and the item is packed in the left-justified way in this level. Now, the problem will become a one-dimensional bin-packing problem. In the second phase, this problem is solved by means of the first-fit decreasing algorithm whose procedures are provided as follows. The first bin is created for packing the first strip. For subsequent strips, the current strip is packed into the "first-fit" bin. If no bin can accommodate the strip, a new bin is initialized.

Figure 27.7a illustrates an example. First, the strip packing is conducted in a bin with the same width as an actual bin and infinite height. Its packing procedures are the same as the example given in the finite first fit described in Section 27.3.1.1. Then, after packing the first strip into the first bin, the remaining empty space is not large enough to fit the second strip. Thus, it is packed into the second bin. The third strip is packed into the first bin because it can first fit the remaining empty space. The same applies to the last strip in the second bin.

27.3.2.2 Hybrid Next Fit

This algorithm was suggested by Frenk and Galambos (1987). The principle of this approach is similar to that of hybrid first fit. In the first stage, the next-fit decreasing height strategy is adopted to carry out a strip packing. This strategy is described as follows. If the current level can accommodate an item, it

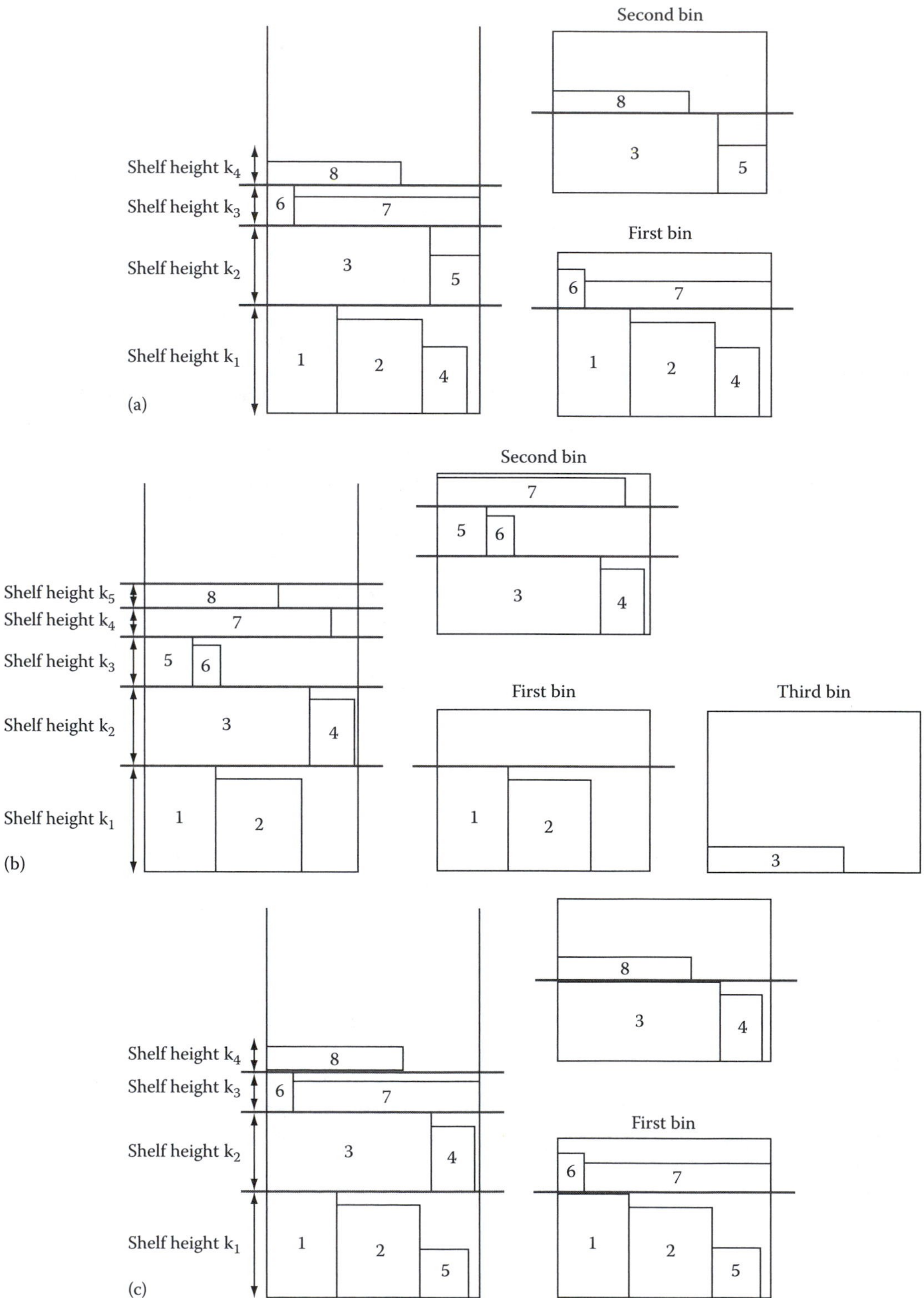

FIGURE 27.7 Examples of hybrid first fit, hybrid next fit, and hybrid best fit (finite best strip).

is packed in the left-justified way in this level. Otherwise, a new level is created and the item is packed in the left-justified way in the new level. Then, in the second stage, the one-dimensional bin-packing problem is solved by using the next-fit decreasing algorithm, the description of which is given as follows. The current strip is packed into the current bin where it fits. If the current bin cannot accommodate the strip, a new bin is initialized.

Figure 27.7b shows an example. First, the strip packing is performed in a bin with the same width as an actual bin and infinite height. Its packing procedures are the same as the example given in the finite next fit described in Section 27.3.1.2. Then, after loading the first strip into the first bin, the remaining empty space is not large enough to fit the second strip. Thus, it is packed into the second bin. Since the remaining empty space of the second bin can accommodate both the third and fourth strips, they are placed into that bin. The last strip is loaded into the third bin as the remaining empty space of the second bin cannot accommodate it.

27.3.2.3 Hybrid Best Fit (Finite Best Strip)

The hybrid best fit approach was suggested by Berkey and Wang (1987). The implementation of this approach is similar to that of hybrid first fit. The best-fit decreasing height strategy is adopted to conduct a strip packing in the first stage. The idea of this strategy is that an item is packed in the left-justified way in the level satisfying two criteria: (1) it fits the item and (2) the residual width is minimized. If no level can accommodate the item, a new level is created and the item is loaded in the left-justified way in this level. In the second stage, the one-dimensional bin-packing problem is attacked by the best-fit decreasing algorithm whose procedures are given as follows. The current strip is packed into the bin fulfilling two criteria: (1) it fits the strip and (2) the residual height is minimized. If no bin can accommodate the strip, a new bin is created.

Figure 27.7c portrays an instance. First, the strip packing is implemented in a bin with the same width as an actual bin and infinite height. After packing the sorted items 1 and 2, item 3 does not fit the empty space at the right of item 2. Therefore, it is packed into the second strip. Item 4 is then loaded into the second strip rather than the first strip because the residual width is minimized in the former strip. The same applies to item 5 loaded into the first strip. Finally, items 6, 7, and 8 are placed into the third and fourth strips, respectively. Now, it is needed to pack strips into bins. After packing the first strip into the first bin, the remaining empty space is not large enough to fit the second strip and, therefore, it is packed into the second bin. The third strip is loaded into the first bin because this can minimize the residual height. The last strip does not fit the remaining empty space of the first bin and thus it is loaded into the second bin.

27.3.2.4 Floor Ceiling

This approach was proposed by Lodi et al. (1999). In this algorithm, ceiling of a level is defined as the horizontal line touching the upper edge of the tallest item packed in the level. Floor ceiling packs the items not only from left to right on the floor of a level, but also from right to left on the ceiling of the level. However, the condition of packing the first item on the ceiling is that this item cannot be packed on the floor.

In the first phase, the levels are created by packing items to them in the following order: (1) on a ceiling by using a best-fit algorithm if the condition aforementioned is satisfied, (2) on a floor by means of a best-fit algorithm, and (3) on the floor of a new level. In the second phase, the levels are packed into bins by using either the best-fit decreasing algorithm or an exact algorithm for the one-dimensional bin-packing problem, halted after a prefixed number of iterations.

Floor ceiling was initially designed for non-guillotine bin packing. However, it can be amended to support guillotine bin packing. Also, the modified variant can be used to solve only three-stage problems because of its intrinsic characteristic. Figure 27.8a and b illustrates the difference between packing without and with guillotine constraint, respectively, in the first phase of floor ceiling. In the former case shown in Figure 27.8a, after packing the first five sorted items on the floor of the first level, the empty

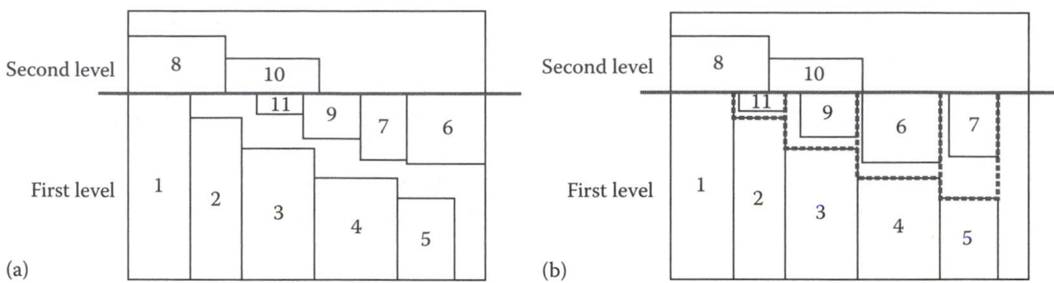

FIGURE 27.8 Difference between packing without and with guillotine constraint in the first phase of floor ceiling.

space at the right of item 5 is not large enough to fit item 6 and, thus, it is the first one packed on the ceiling of the same level. The same applies to item 7. Since item 8 cannot fit the spaces both on the floor and ceiling of the first level, it is loaded onto the floor of the second level. Similarly, items 9, 11, and 10 are placed into the first and second level, respectively.

In the latter case shown in Figure 27.8b, the first five sorted items are placed in the same way as the former case in Figure 27.8a. Now, it is needed to delimit the empty spaces as four smaller areas bounded by dotted lines, where subsequent items can be placed to meet the guillotine constraint in the first level and this is also applied to each of the other levels. Since the empty space at the right of item 5 is not sufficiently large to accommodate item 6, it is required to search an empty space bounded by dotted lines, whose width is equal to or larger than that of item 6 from right to left on the ceiling. The empty space above item 4 satisfies this condition and thus item 6 is loaded there with either one of its vertical sides touching a vertical dotted line. In this example, the right side is used. The same rule applies to item 7. Like the example in Figure 27.8a, as item 8 cannot fit the spaces both on the floor and ceiling of the first level, it is packed on the floor of the second level. Items 9, 10, and 11 are loaded by using the same rules described above. Note that the scenario being tackled is a three-stage problem with trimming. For the non-trimming case, items 11, 9, 6, and 7 placed at those empty spaces must have the same widths as items 2, 3, 4, and 5, respectively.

27.3.3 Local Search Heuristics

A local search algorithm aims at hopefully and efficiently finding a good solution by conducting a sequence of tiny perturbation on an initial solution and deals with one solution (current solution) at a time. Before its implementation, the neighborhoods, i.e., the operation of how to obtain a different solution from the current solution, have to be defined. The procedures of a local search algorithm are given as follows. First, an initial solution is generated and the current solution is set as the initial solution. The value of the current solution is calculated. The iteration loop starts with acquiring a neighborhood of the current solution through the defined operation and its objective value is computed. In the first improvement scheme, if the value of the neighborhood solution outperforms that of the current solution, the current solution is replaced by the neighborhood. Otherwise, the current solution remains unchanged. In the best improvement (steepest descent) scheme, all the neighbor solutions are evaluated and the best one is selected for comparing with the current solution.

The iteration loop ends here and repeats again until no neighbor solution improves the current solution. The two local search heuristics named variable neighborhood descent (VND) and stochastic neighborhood structures were built by means of the concept of the local search algorithm and will be introduced as follows.

27.3.3.1 Variable Neighborhood Descent

VND is a meta-heuristic proposed by Mladenovic and Hansen (Hansen and Mladenovic 1999, 2001; Mladenovic and Hansen 1997). The concept of VND is to systematically utilize different neighborhood structures. The brief description of VND is given as follows. First, a set of neighborhood structures, which will be adopted in the descent, is chosen and an initial solution is found. Now, an iteration loop starts. The first neighborhood structure is used to search the best neighbor of the current solution. If the best neighborhood solution outperforms the current solution, the latter one is replaced by the former one. Otherwise, the next (second, third, and so on) neighborhood structure is considered. Finally, the iteration loop repeats until all neighborhood structures are utilized.

This chapter will present three neighborhood structures devised in a particular order for the sequential VND that was proposed by Alvelos et al. (2009) and different from the common one described above. In their study, the three neighborhood structures arranged in the fixed order are implemented one by one as a loop until the time limit is reached or the current solution cannot be improved.

It is vital to realize the solution representation and evaluation function before devising neighborhood structures. A solution is represented by a sequence of items satisfying the demand of each item. Let us consider an example as follows. Assume that items 1, 2, 3 have demands 3, 5, 2, respectively. A feasible solution is represented by 2, 2, 2, 2, 2, 1, 1, 1, 3, 3. The first five items of type 2 will be packed first. Next, three items of type 1 will be loaded and finally the last two items of type 3. Note that the solution representation before implementing any neighborhood structure is the same as that after. Moreover, in order to examine the quality of a solution, an evaluation function f is required for evaluating the solution, which is given by

$$f = n_b \cdot a_b + o_b - m \tag{27.1}$$

where

n_b is the number of used bins
a_b is the area of a bin
o_b is the occupied area of the bin in solution with the smallest occupied area
m is the number of items packed in the same bin

The rationale behind this function is that a solution with fewer used bins always outperforms other solutions with more used bins and that if two solutions have the same number of used bins, the solution where it is easier to empty one used bin is better than the other one. In the following, these neighborhood structures are introduced in order.

27.3.3.1.1 First Neighborhood Structure: Swap Adjacent Item Types

"Swap adjacent item types" is aimed at swapping all items in two adjacent item types. Figure 27.9 illustrates an instance of this neighborhood structure. Assume that two highlighted adjacent item types, 2 and 4, are chosen. One item in the first type and three items in the second type are exchanged to produce the new solution.

27.3.3.1.2 Second Neighborhood Structure: Swap Adjacent Item Subsequences

The mission of "swap adjacent item subsequences" is to swap two adjacent item subsequences, both of which have the same size. A size parameter is used to define the size of the neighborhood. Figure 27.10a and b

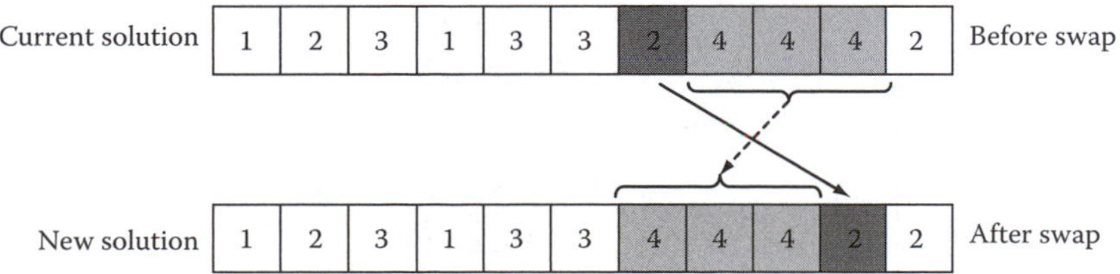

FIGURE 27.9 An example of "swap adjacent item types."

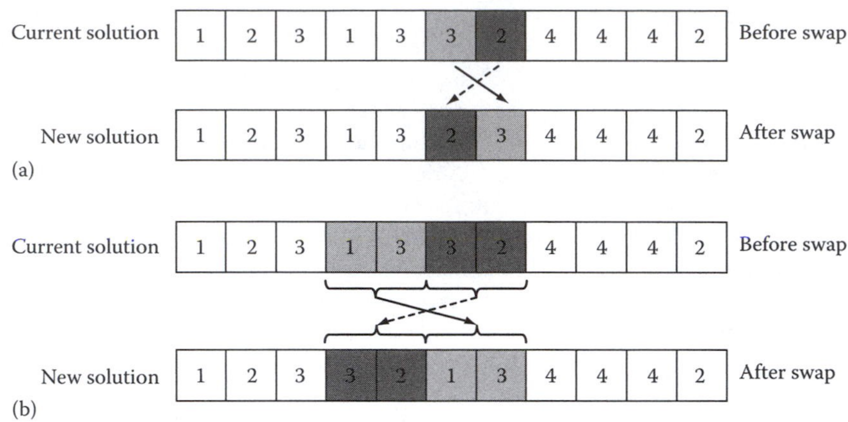

FIGURE 27.10 Examples of "swap adjacent item subsequences."

portrays instances of this neighborhood structure for two different sizes of the neighborhood. For the former one whose size of the neighborhood is one, suppose that two highlighted adjacent item subsequences, 3 and 2, are chosen in the whole item packing sequence (solution). Then, these two item subsequences are exchanged to complete the operation. For the latter one whose size of the neighborhood is two, two shaded adjacent item subsequences, 1, 3 and 3, 2, are selected in the current solution and then swapped to produce the new solution.

27.3.3.1.3 Third Neighborhood Structure: Reverse Item Subsequences

The objective of "reverse item subsequences" is to reverse the order of an item subsequence with a given size. A size parameter is utilized to define the size of the item subsequence. Figure 27.11a and b shows examples of this neighborhood structure for two various sizes of the item subsequences. The item subsequences 3, 2, and 3, 2, 4 are selected and then their packing orders are reversed to generate new solutions, respectively.

27.3.3.2 Stochastic Neighborhood Structures

Stochastic neighborhood structures (SNS) proposed by Chan et al. (2009) are adopted in a way similar to VND. Since a local optimal solution corresponding to one neighborhood structure is not necessarily the same as that corresponding to another neighborhood structure, the use of several different neighborhood structures as the basic concept of SNS/VND can further improve the current local optimal one. The differences between SNS and VND are that SNS (1) impose the restriction of using all stochastic

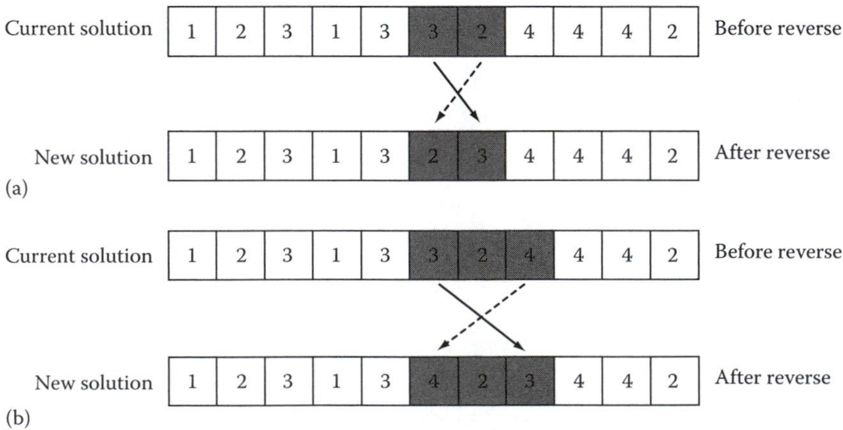

FIGURE 27.11 Examples of "reverse item subsequences."

neighborhood structures rather than deterministic or mixed ones, (2) use the fixed number of iterations to explore better neighborhood solutions for each or all neighborhood structures, and (3) do not use the iteration loop for neighborhood structures.

In their study, SNS is employed to improve the quality of the solution given by the greedy heuristic in the first phase. In each neighborhood structure of the proposed approach, instead of finding the best neighbor of the initial solution by complete enumeration, only one neighbor is randomly generated in each time and then compared with the initial solution. The advantage of this modification is that a large computational load is not required to search all neighbors of the initial solution, especially when the problem size is huge. Three neighborhood structures are proposed to be implemented in a fixed order. In the following, these neighborhood structures will be introduced in order.

27.3.3.2.1 First Neighborhood Structure: Cut-and-Paste

"Cut-and-paste" is a genetic operation, which was applied in the jumping-gene paradigm to solve multi-objective optimization problems (Chan et al. 2008). Its implementation is that the "jumping" element is cut from an original position and pasted into a new position of a chromosome. In this study, "cut-and-paste" is applied to the solution in a way that there is only one "jumping" segment in the solution, and the length, original position, and new position of the "jumping" segment are randomly chosen. Figure 27.12 depicts an example of this neighborhood structure. Given that the randomly generated length is 4, the original position is 6 and the new position is 2. In other words, the highlighted segment (4, 2, 1, 4) is randomly selected. The segment is cut from the original position and then pasted into the new position to complete the operation.

27.3.3.2.2 Second Neighborhood Structure: Split-and-Redistribute

The objective of "split-and-redistribute" is to split various blocks with a given length from the solution and redistribute them to the solution. The total number, length, original positions, and new positions of the blocks are randomly selected and all blocks have the same length. Figure 27.13 illustrates an instance of this neighborhood structure. Suppose that the randomly selected total number is 3, the length is 2, the original positions are 1, 5, and 10, and the new positions are 10, 1, and 5 for the three blocks, respectively. That is, the three highlighted blocks (1, 3), (3, 4), and (4, 4) are randomly chosen. These three blocks are split and then redistributed to their new positions to acquire the new solution.

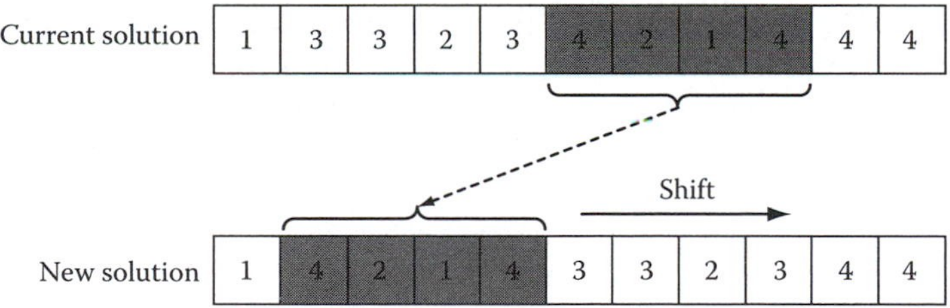

FIGURE 27.12 An example of "cut-and-paste."

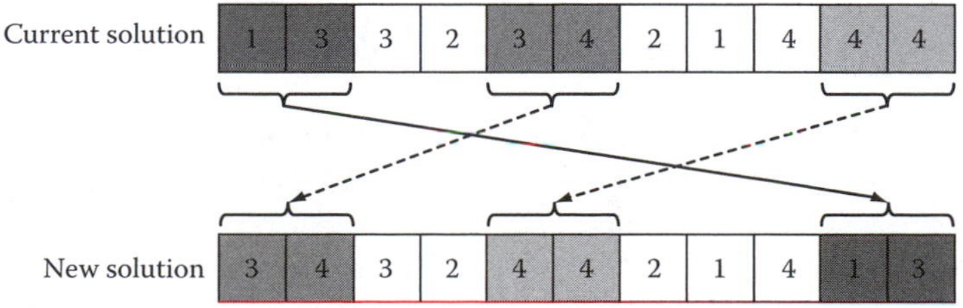

FIGURE 27.13 An example of "split-and-redistribute."

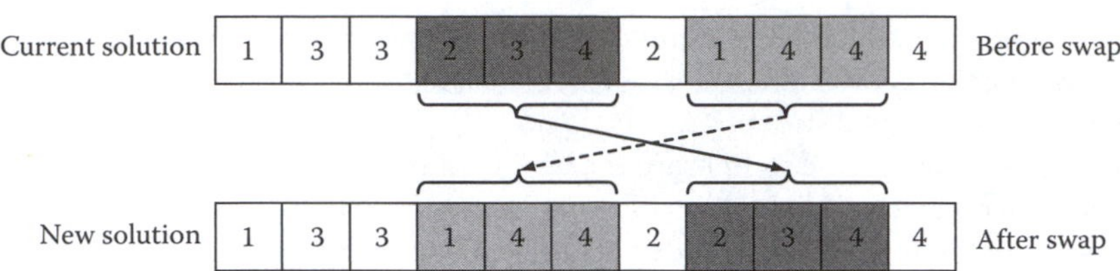

FIGURE 27.14 An example of "swap block."

27.3.3.2.3 Third Neighborhood Structure: Swap Block

"Swap block" is aimed at exchanging two different blocks with a given length in the solution. The length and the positions of the two blocks are randomly selected and the lengths of these two blocks are equivalent. Figure 27.14 portrays an example of this neighborhood structure. Assume that the randomly selected length is 3 and the randomly chosen positions are 4 and 8. That means the two highlighted blocks (2, 3, 4) and (1, 4, 4) are randomly selected. Then, these two blocks are swapped to produce the new solution.

27.4 Computational Results

In this section, the computational results and times obtained by using the recently proposed heuristic, i.e., a greedy heuristic (one-phase heuristic) together with stochastic neighborhood structures (local search heuristic), are summarized. Four different sets of benchmark problems were used to verify the effectiveness of the heuristic: (1) instances cgcut1—cgcut3 from the (Christofides and Whitlock 1977) study, (2) instances gcut1—gcut13 and ngcut1—ngcut12 from the two Beasley studies, respectively (Beasley 1985a, b), (3) 300 instances (they are named B&W1—B&W300 in this chapter) from the (Berkey and Wang 1987) study, and (4) 200 instances (they are named M&V1—M&V200 in this chapter) from the (Martello and Vigo 1998) study. Two sets of 47 and 121 real-world instances offered by furniture companies were also adopted. Different scenarios with all the possible combinations of the following three parameters were considered: (1) whether rotation by 90° is allowed and prohibited for each item, (2) horizontal or vertical cut(s) in the first-stage cutting, and (c) four cases: 2-stage without trimming, 2-stage with trimming, 3-stage without trimming, or 3-stage with trimming. In the following, the relative gaps between the heuristic solutions and the optimal solutions found by an exact method based on the pseudo-polynomial integer programming model (Silva et al. 2010) and averages of percentage of waste are given in order to reflect the quality of the proposed heuristic. A relative gap is calculated by means of the following equation:

$$GAP\ (\%) = \frac{Z_H - Z_{Opt}}{Z_{Opt}} \times 100 \tag{27.2}$$

where

Z_H is the heuristic solution
Z_{Opt} is the optimal solution obtained by the exact method

However, since optimal solutions for only the case of horizontal cut(s) in the first-stage cutting are known, only gaps belonging to this case will be reported. For the details about the implementation aforementioned, please refer to Chan et al. (2009).

Tables 27.1 and 27.2 show the ranges of the gaps and averages of percentage of waste with all possible combinations of parameters for all benchmark and real-world problems, respectively. From Table 27.1, it can be observed that all maximum percentages of the ranges of the gaps are small for all scenarios

TABLE 27.1 Ranges of Relative Gaps between the Heuristic Solutions and the Optimal Solutions Found with All Possible Combinations of Parameters for All Benchmark and Real-World Problems

		Horizontal Cut(s) at the First Stage			
Instances		2-Stage without Trimming	2-Stage with Trimming	3-Stage without Trimming	3-Stage with Trimming
cgcut1 – 3, gcut1 – 13, ngcut1 – 12	NR	[0%, 0.54%]	[0%, 0%]	[0%, 0%]	[0%, 0.92%]
	R	[0%, 8.70%]	[3.13%, 4.35%]	[0%, 3.23%]	[0%, 3.23%]
B&W1 – 300, M&V1 – 200	NR	[0%, 0.22%]	[0%, 1.67%]	[0%, 2.10%]	[0%, 2.82%]
	R	[0%, 4.83%]	[0%, 3.10%]	[0%, 0.81%]	[0%, 0.81%]
First set of 47 real-world instances	NR	[0%, 2.81%]	[0%, 4.38%]	[0%, 4.65%]	[0%, 4.65%]
	R	—	[0.93%, 1.85%]	[0.94%, 0.94%]	[0.94%, 0.94%]
Second set of 121 real-world instances	NR	[0.78%, 0.95%]	[1.66%, 3.03%]	[1.91%, 3.06%]	[1.98%, 3.09%]
	R	[1.09%, 4.91%]	[2.02%, 3.42%]	[2.79%, 3.92%]	[2.59%, 3.78%]

Note: NR: no rotation is allowed; R: rotation is allowed; $[x, y]$: x is the minimum percentage and y is the maximum percentage.

TABLE 27.2 Ranges of Averages of Percentage of Waste with All Possible Combinations of Parameters for All Benchmark and Real-World Problems

		Horizontal Cut(s) at the First Stage			
Instances		2-Stage without Trimming	2-Stage with Trimming	3-Stage without Trimming	3-Stage with Trimming
cgcut1 – 3, gcut1 – 13, ngcut1 – 12	NR	[44.02%, 59.04%]	[28.67%, 37.89%]	[26.11%, 32.28%]	[26.11%, 30.51%]
	R	[31.27%, 44.38%]	[22.65%, 30.07%]	[21.26%, 28.43%]	[21.26%, 28.43%]
B&W1 – 300, M&V1 – 200	NR	[23.69%, 79.12%]	[13.53%, 36.75%]	[12.48%, 36.75%]	[12.68%, 36.75%]
	R	[10.35%, 65.20%]	[8.88%, 36.43%]	[8.62%, 36.43%]	[8.62%, 36.43%]
First set of 47 real-world instances	NR	[19.54%, 37.90%]	[12.75%, 32.76%]	[12.75%, 30.40%]	[12.75%, 30.40%]
	R	[15.49%, 34.58%]	[10.86%, 28.59%]	[10.29%, 26.69%]	[9.02%, 26.23%]
Second set of 121 real-world instances	NR	[13.64%, 23.24%]	[11.48%, 20.45%]	[11.51%, 20.03%]	[11.44%, 20.28%]
	R	[12.55%, 19.36%]	[8.75%, 17.49%]	[8.68%, 17.12%]	[8.68%, 17.01%]
		Vertical Cut(s) at the First Stage			
cgcut1 – 3, gcut1 – 13, ngcut1 – 12	NR	[55.90%, 57.73%]	[30.25%, 41.78%]	[27.07%, 34.06%]	[27.95%, 34.06%]
	R	[34.73%, 44.38%]	[22.65%, 28.43%]	[22.65%, 28.43%]	[22.65%, 28.43%]
B&W1 – 300, M&V1 – 200	NR	[22.52%, 79.64%]	[12.87%, 38.11%]	[12.47%, 36.75%]	[12.67%, 36.75%]
	R	[10.41%, 65.20%]	[8.82%, 36.43%]	[8.62%, 36.43%]	[8.68%, 36.43%]
First set of 47 real-world instances	NR	[19.36%, 52.30%]	[14.98%, 36.06%]	[12.63%, 35.76%]	[13.22%, 36.06%]
	R	[14.43%, 39.93%]	[10.67%, 25.35%]	[9.87%, 25.35%]	[9.55%, 25.35%]
Second set of 121 real-world instances	NR	[13.44%, 32.08%]	[13.84%, 31.39%]	[13.56%, 30.83%]	[13.66%, 31.30%]
	R	[11.02%, 17.09%]	[9.20%, 16.30%]	[8.64%, 16.30%]	[8.56%, 16.30%]

Note: NR: no rotation is allowed; R: rotation is allowed; $[x, y]$: x is the minimum percentage and y is the maximum percentage.

(i.e., all of them are less than 5% except the case of 2-stage without trimming and rotation is allowed in instances cgcut1—3, gcut1—13, ngcut1—12). This implies that the heuristic solutions obtained are quite close to the optimal solutions. Some heuristic solutions found are even optimal. Also, the minimum and maximum percentages found in the two sets of the real-world instances are generally larger than those of the two sets of the benchmark instances because the former ones are harder problems. Note that since the optimal solutions cannot be attained for the case of 2-stage without trimming and rotation is allowed in the first set of 47 real-world instances, the range of the relative gaps is not given in this case.

Moreover, it can be seen from Table 27.2 that both the minimum and maximum percentages of the ranges of averages of the percentage of waste obtained in the case of 2-stage without trimming are larger than those of the other cases. The reason is that it restricts the items' heights to be the same as the height of the shelf in which they reside, and free spaces are wasted if no appropriate item that satisfies the constraint can be packed into these spaces. In addition, both the minimum and maximum percentages of the ranges of averages of the percentage of waste found in the rotation case are smaller than those of the nonrotation case because rotation is more flexible for various items to fit empty spaces with different dimensions, and it turns out that fewer empty spaces are wasted. It is noteworthy that allowing trimming in the three stage problem reduces the waste only marginally.

Table 27.3 gives the ranges of the sum of average and the average of standard deviation of computational times with all possible combinations of parameters for all benchmark and real-world problems.

TABLE 27.3 Ranges of the Sum of Average and the Average of Standard Deviation of Computational Times with All Possible Combinations of Parameters for All Benchmark and Real-World Problems (Unit: s)

Instances		2-Stage without Trimming	2-Stage with Trimming	3-Stage without Trimming	3-Stage with Trimming
		Horizontal Cut(s) at the First Stage			
cgcut1 – 3, gcut1 – 13, ngcut1 – 12	NR	avg: [0.4, 1.7] sd: [0.005, 0.020]	avg: [0.3, 1.0] sd: [0.006, 0.007]	avg: [0.3, 1.1] sd: [0.006, 0.007]	avg: [0.3, 1.0] sd: [0.005, 0.006]
	R	avg: [0.3, 1.5] sd: [0.006, 0.006]	avg: [0.3, 1.0] sd: [0.006, 0.007]	avg: [0.3, 1.1] sd: [0.006, 0.020]	avg: [0.3, 1.1] sd: [0.006, 0.017]
B&W1 – 300, M&V1 – 200	NR	avg: [7.3, 30.0] sd: [0.006, 0.139]	avg: [5.8, 20.6] sd: [0.007, 0.022]	avg: [7.0, 21.3] sd: [0.007, 0.036]	avg: [6.6, 21.1] sd: [0.007, 0.025]
	R	avg: [7.2, 29.1] sd: [0.006, 0.118]	avg: [6.2, 21.5] sd: [0.007, 0.024]	avg: [6.9, 23.0] sd: [0.006, 0.034]	avg: [6.5, 22.4] sd: [0.007, 0.030]
First set of 47 real-world instances	NR	avg: [0.9, 32.2] sd: [0.006, 0.056]	avg: [0.7, 22.8] sd: [0.007, 0.068]	avg: [0.7, 27.1] sd: [0.006, 0.065]	avg: [0.7, 25.7] sd: [0.006, 0.080]
	R	avg: [0.9, 35.1] sd: [0.006, 0.075]	avg: [0.7, 25.5] sd: [0.005, 0.063]	avg: [0.8, 30.5] sd: [0.007, 0.073]	avg: [0.7, 28.7] sd: [0.006, 0.081]
Second set of 121 real-world instances	NR	avg: [9.6, 1099.4] sd: [0.012, 3.053]	avg: [8.2, 1110.8] sd: [0.012, 3.596]	avg: [8.4, 1207.1] sd: [0.013, 4.838]	avg: [8.3, 1218.1] sd: [0.010, 4.025]
	R	avg: [10.3, 1316.6] sd: [0.011, 3.173]	avg: [8.7, 1160.5] sd: [0.011, 3.769]	avg: [9.1, 1237.9] sd: [0.012, 3.617]	avg: [8.9, 1228.9] sd: [0.012, 3.863]
		Vertical Cut(s) at the First Stage			
cgcut1 – 3, gcut1 – 13, ngcut1 – 12	NR	avg: [0.4, 1.5] sd: [0.005, 0.006]	avg: [0.3, 1.0] sd: [0.004, 0.006]	avg: [0.3, 1.0] sd: [0.005, 0.006]	avg: [0.3, 1.1] sd: [0.005, 0.006]
	R	avg: [0.3, 1.5] sd: [0.006, 0.006]	avg: [0.2, 1.0] sd: [0.005, 0.006]	avg: [0.3, 1.1] sd: [0.005, 0.020]	avg: [0.3, 1.1] sd: [0.006, 0.006]
B&W1 – 300, M&V1 – 200	NR	avg: [7.4, 29.3] sd:[0.008, 0.145]	avg: [6.2, 20.2] sd: [0.007, 0.028]	avg: [7.0, 21.8] sd: [0.006, 0.031]	avg: [6.7, 20.9] sd: [0.007, 0.024]
	R	avg: [7.2, 28.7] sd: [0.007, 0.123]	avg: [6.2, 21.5] sd: [0.007, 0.026]	avg: [7.0, 23.0] sd: [0.007, 0.032]	avg: [6.5, 22.4] sd: [0.006, 0.028]
First set of 47 real-world instances	NR	avg: [0.9, 25.4] sd: [0.007, 0.043]	avg: [0.7, 19.3] sd: [0.006, 0.037]	avg: [0.7, 23.2] sd: [0.007, 0.048]	avg: [0.7, 23.3] sd: [0.006, 0.057]
	R	avg: [1.0, 31.7] sd: [0.007, 0.056]	avg: [0.7, 24.5] sd: [0.006, 0.049]	avg: [0.7, 28.0] sd: [0.006, 0.058]	avg: [0.7, 26.7] sd: [0.006, 0.062]
Second set of 121 real-world instances	NR	avg: [9.3, 1024.7] sd: [0.009, 2.376]	avg: [8.4, 1004.7] sd: [0.013, 2.633]	avg: [8.6, 1184.8] sd: [0.010, 2.777]	avg: [8.7, 1091.7] sd: [0.014, 2.521]
	R	avg: [10.1, 1123.1] sd: [0.010, 2.378]	avg: [8.7, 1736.6] sd: [0.009, 2.878]	avg: [9.1, 1841.4] sd: [0.010, 3.220]	avg: [9.2, 1862.0] sd: [0.013, 3.158]

Note: NR: no rotation is allowed; R: rotation is allowed; avg: average; sd: standard deviation; $[x, y]$: x is the minimum time and y is the maximum time.

The sum of average is obtained by summing the averages of computational times of 30 simulation runs for all instances. The average of standard deviation is obtained by averaging the standard deviations of computational times of 30 simulation runs for all instances. Considering all scenarios of two- and three-stage problems with and without trimming, and with and without rotation in the case of horizontal cut(s) at the first stage, the sums of averages for the set of cgcut, gcut, and ngcut instances, the set of B&W and M&V instances, the first set of 47 real-world instances and the second set of 121 real-world instances are within 1.7, 30, 35.1, and 1316.6 s, respectively. The former three is short because they are easier problems (i.e., the number of items ranges from 7 to 809 in the instances). The last one is opposite since they are hard problems (i.e., the number of items ranges from 32 to 10,710 in the instances), but it is still acceptable. Also, the averages of standard deviations are within 0.020, 0.139, 0.081, and 4.838 s, respectively. This shows that the computational times of those instances are quite stable and consistent. The magnitude of the sum of average and the average of standard deviation achieved in the case of vertical cut(s) at the first stage is similar to that in the case of horizontal cut(s). This is expected because only bin width and bin height, and item widths and item heights are exchanged in the former case, and the numbers of algorithmic operations implemented in both two versions are the same.

27.5 Conclusions

Several variants of two-dimensional guillotine bin-packing problems were addressed. For two-stage problems with trimming, one-phase and two-phase heuristics were reviewed. A greedy heuristic based on the definition of a packing sequence of the items and on a set of criteria to pack one item was described. Recently proposed deterministic and stochastic neighborhood structures, based on modifications on the current sequence of items, were presented. Since a solution is coded as a sequence and decoded by the constructive heuristic, the neighborhood structures are independent of the particular variant being addressed, which gives the approach the flexibility to deal with different variants of the problem (in particular, variants with two and three stages with and without trimming, with and without rotation).

The computational tests revealed that the proposed heuristic approaches are able to find good-quality solutions within reasonable amounts of time for instances from the literature and for "real-world" instances.

Using heuristics to address bin-packing problems in the wood-cutting industry is a robust strategy, because companies often face problems with different characteristics and constraints, which can be addressed with relatively minor amendments on the core version of the heuristic.

References

Alvelos, F., T. M. Chan, P. Vilaça, T. Gomes, and E. Silva. 2009. Sequence based heuristics for two-dimensional bin packing problems. *Engineering Optimization* 41 (8):773–791.

Beasley, J. E. 1985a. Algorithms for unconstrained two-dimensional guillotine cutting. *Journal of the Operationla Research Society* 36 (4):297–306.

Beasley, J. E. 1985b. An exact two-dimensional non-guillotine cutting tree search procedure. *Operations Research* 33 (1):49–64.

Berkey, J. O. and P. Y. Wang. 1987. Two-dimensional finite bin packing problems. *The Journal of Operational Research Society* 38 (5):423–429.

Carter, M. W. and C. C. Price. 2001. *Operations Research: A Practical Introduction*. Boca Raton, FL: CRC Press.

Chan, T. M., K. F. Man, S. Kwong, and K. S. Tang. 2008. A jumping gene paradigm for evolutionary multiobjective optimization. *IEE Transactions on Evolutionary Computation* 12 (2):143–159.

Chan, T. M., F. Alvelos, E. Silva, and J. M. Valério de Carvalho. 2010. Heuristic with stochastic neighborhood structures for 2-dimensional bin packing problems. *Asia-Pacific Journal of Operations Research* (in press).

Christofides, N. and C. Whitlock. 1977. An algorithm for two-dimensional cutting problems. *Operations Research* 25 (1):30–44.

Chung, F. K. R., M. R. Garey, and D. S. Johnson. 1982. On packing two-dimensional bins. *SIAM Journal on Algebraic and Discrete Methods* 3 (1):66–76.

Frenk, J. B. and G. G. Galambos. 1987. Hybrid next-fit algorithm for the two-dimensional rectangle bin-packing problem. *Computing* 39 (3):201–217.

Garey, M. R. and D. S. Johnson. 1979. *Computers Intractability: A Guide to the Theory of NP-Completeness*. New York: W. H. Freeman.

Hansen, P. and N. Mladenovic. 1999. An introduction to variable neighborhood search. In *Meta-Heuristics, Advances and Trends in Local Search Paradigms for Optimization*, S. Voss et al. (Eds.) Boston, MA: Kluwer Academic Publishers.

Hansen, P. and N. Mladenovic. 2001. Variable neighborhood search: Principles and applications. *European Journal of Operational Research* 130 (3):449–467.

Lodi, A., S. Martello, and D. Vigo. 1999. Heuristic and metaheuristics approaches for a class of two-dimensional bin packing problems. *INFORMS Journal on Computing* 11 (4):345–357.

Lodi, A., S. Martello, and D. Vigo. 2002. Recent advances on two-dimensional bin packing problems. *Discrete Applied Mathematics* 123 (5):423–429.

Martello, S. and D. Vigo. 1998. Exact solution of the two-dimensional finite bin packing problem. *Management Science* 44 (3):388–399.

Mladenovic, N. and P. Hansen. 1997. Variable neighborhood search. *Computers & Operations Research* 24 (11):1097–1100.

Sait, S. M. and H. Youssef. 1999. *Iterative Computer Algorithms with Applications in Engineering: Solving Combinatorial Optimization Problems*. Los Alamitos, CA: IEEE Computer Society.

Silva, E., F. Alvelos, and J. M. Valério de Carvalho. 2010. An integer programming model for the two-staged and three-staged two dimensional cutting stock problems. *European Journal of Operational Research* 205(3): 699–708.

Wäscher, G., H. Haussner, and H. Schumann. 2007. An improved typology of cutting and packing problems. *European Journal of Operational Research* 183 (3):1109–1130.

28

Particle Swarm Optimization

Adam Slowik
Koszalin University of Technology

28.1 Introduction

Recently, optimization techniques based on behaviors of some animal species in natural environment are strongly developed. Algorithms simulating behaviors of bees colony [1], ants colony [2,3], and birds flock [4] (fish school) have appeared. The last algorithm from those techniques is named in the literature as a particle swarm algorithm (in short *PSO*—particle swarm optimization), and is a new technique dedicated to optimization problems having continuous domain. However, its modifications to optimize discreet problems [5] have been developed lately. The *PSO* algorithm has many common features with evolutionary computation techniques. This algorithm is operating on randomly created population of potential solutions, and is searching optimal solution through the creation of successive populations of solutions. Genetic operators like cross-over and mutation, which exist in evolutionary algorithms, are not used in the *PSO* algorithm. In this algorithm, potential solutions (also called the particles) are moving to the actual (dynamically changing) optimum in the solution space.

There exist two versions of *PSO* algorithm: local, *LPSO*, and global, *GPSO*, algorithm. In the *LPSO*, the process of optimization is based on velocity V changes of each particle P_i moving toward position P_{best}, which corresponds to the best position of a given particle, and L_{best}, which corresponds to the best position of another particle chosen from *Ne* nearest neighbors of the particle P_i, found up to the present step of the algorithm. In the *GPSO* algorithm, the process of optimization is based on velocity V changes (acceleration) of each particle P_i moving toward position P_{best} and position G_{best}, which represents the best position having been obtained in previous iterations of the algorithm.

The values of velocity in each D direction (the value of D is equal to the number of variables in optimized task) of solution space are computed for each particle P_i using the formula

- For *LPSO* algorithm

$$V[j] = V[j] + c_1 \cdot r_{1,j} \cdot (P_{best}[j] - X[j]) + c_2 \cdot r_{2,j} \cdot (L_{best}[j] - X[j]) \quad (28.1)$$

- For *GPSO* algorithm

$$V[j] = V[j] + c_1 \cdot r_{1,j} \cdot (P_{best}[j] - X[j]) + c_2 \cdot r_{2,j} \cdot (G_{best}[j] - X[j]) \quad (28.2)$$

where $j = 1, 2,\ldots, D$; c_1 and c_2 are coefficients of particle acceleration, usually positive values are chosen experimentally from the range [0, 2]; $r_{1,j}$, $r_{2,j}$ are random real numbers with uniform distribution from the range [0, 1], these values introduce random character to the algorithm.

The main advantage of *LPSO* algorithm is lower susceptibility of solution to be "trapped" in local minimum, than in the case of *GPSO* algorithm. This advantage is a consequence of higher spread of solution values in the population in *LPSO* algorithm (larger part of the solution space is considered). However, the main advantage of *GPSO* algorithm is faster convergence than in the case of *LPSO* algorithm; it is caused by smaller spread of solution values in the population (smaller part of the solution space is covered). These two versions of *PSO* algorithm have been successfully applied to different disciplines of science during latest years [6,7]. This chapter is only an introduction to the problems connected with *PSO*.

28.2 Particle Swarm Optimization Algorithm

Generally, the *PSO* algorithm can be described using six following parts.

In the first part, the objective function and values of all algorithm parameters such as M (number of particles in population), c_1, c_2, and D are determined. Additionally, for the *LPSO* algorithm, an *Ne* value, which determines the number of the nearest neighbors for each particle, is chosen. After parameter selection, the population P, which consists M particles (potential solutions), is randomly created. In *GPSO* algorithm, also the vector G_{best} is prepared. In this vector, the data representing the best position of particle (solution) found during algorithm operation are written down (at the start of the algorithm, vector $[G_{best}] = [0]$). Each particle P_i (i.e., [1, M]) from population P is composed of the following D-element vectors. Vector X represents current position of the particle P_i (solution of a given problem); vector P_{best} represents the best position of the particle P_i in the solution space found for this particle at a given step of the algorithm; vector L_{best} (only for *LPSO*) represents the best position of other particle from among *Ne* nearest neighbors of the particle P_i, which has been obtained during previous iterations of the algorithm; and vector V represents the values of velocity of particle P_i in each direction D of the solution space. At the start of the algorithm, the vector X is assigned to the vector P_{best}, and values of vectors V and L_{best} are cleared during creation of initial population.

In the second part, each particle P_i is evaluated using objective function. In the case of minimization tasks, when computed value of objective function for data written down in vector X of particle P_i is lower than the best value found for this particle at a given step of the algorithm (the value written down in vector P_{best}), then the values from vector X are written down in the vector P_{best}. In the case of maximization tasks, the values stored in vector X are written down in vector P_{best}, if and only if the computed value of the objective function for data from vector X of particle P_i is higher than the best value found for this particle at a given step of the algorithm. In the *GPSO* algorithm, updating of the vector G_{best} is performed after evaluation of all particles P_i in the population P. If there exists a particle P_i in population having lower value of the objective function (for minimization tasks) or having higher value of objective function (for maximization tasks) than the value of the objective function stored in the solution written down in the vector G_{best}, then the position of particle P_i is written down in the vector G_{best}.

```
Procedure Particle Swarm Optimization Algorithm
Begin
Determine objective function and algorithm parameters
Randomly create swarm (population) P composed of M particles
Repeat
For each particle i=1, ..., M do
Begin
Evaluate the quality of particle position X (solution) using objective function
Determine the best position X from obtained until current step of the algorithm, and write down this position in vector Pbest
//--- for local version - algorithm LPSO
Update the position written down in vector Lbest, if the better position has been found among Ne nearest neighbours of current particle
//--- for global version - algorithm GPSO
If position of current particle is better
Update position written down in vector Gbest
End
For each particle i=1, ..., M do
Begin
Randomly determine the values of ri,j coefficients
// --- for local version - algorithm LPSO
Update particle velocity V using formula (1)
// --- for global version - algorithm GPSO
Update particle velocity V using formula (2)
// --- for both versions - algorithm LPSO, and GPSO
Update particle position X using formula (3)
End
Until termination condition is fulfilled
End
```

FIGURE 28.1 Particle swarm optimization algorithm in pseudo-code form. (Adapted from Engelbrecht, A. P., *Computational Intelligence—An Introduction*, 2nd edn., Wiley, Chichester, U.K., 2007.)

In the third part (only for *LPSO*), the best particle among *Ne* nearest neighbors of the particle P_i is determined for each particle P_i. The neighborhood is determined based on positions of particular particles in the solution space (e.g., using euclidean distance). The position of the best neighboring particle with respect to particle P_i is written down in the vector L_{best} for particle P_i, if and only if this position is better than the position (solution) actually stored in vector L_{best}.

In the fourth part, the values of velocity in each *D* direction of solution space are computed for each particle P_i using formula (28.1) for *LPSO*, and formula (28.2) for *GPSO*.

In the fifth part, the vector *X* consisting position of particle P_i, in *D*-dimensional solution space, is updated for each particle P_i using velocity vector *V* according to the formula

$$X[j] = X[j] + V[j] \tag{28.3}$$

In the sixth part, a fulfilling of termination condition of the algorithm is checked. The termination condition can be the algorithm convergence (invariability of the best solution during the prescribed number of iterations) or assumed number of generations. If termination condition is fulfilled, then the solution written down in vector *X* of the particle having the lowest value of the objective function (in the case of minimization tasks) or particle having the highest value of the objective function (in the case of maximization tasks) is returned as a result of the *LPSO* algorithm operation. The solution written down in vector G_{best} is a result of algorithm operation in the case of *GPSO* algorithm. When the termination condition is not fulfilled, then again the second step of the algorithm is executed.

In Figure 28.1, the algorithm *PSO* in pseudo-code form is presented.

28.3 Modifications of *PSO* Algorithm

There exist many modifications of *PSO* algorithm, improving its convergence. Among these modifications, we can mention velocity clamping, inertia weight, and constriction coefficient.

28.3.1 Velocity Clamping

During research on *PSO* algorithm, it was noticed that the update of the velocity vector V using formula (28.1) or formula (28.2) causes fast increase of values stored in this vector. As a consequence, the particle positions are changed with increasingly higher values during algorithm operation. Due to these changes, new positions of particles can be located outside of the acceptable space for a given search space. In order to limit this drawback, constraint values of particle velocity $V_{max,j}$ are introduced in each jth dimension D of the search space. When computed value of the velocity $V_{i,j}$ for ith particle, in jth dimension is higher than value $V_{max,j}$, then the following formula is used:

$$V_{i,j} = V_{max,j} \tag{28.4}$$

Of course, it is important to choose suitable values of $V_{max,j}$. For small values of $V_{max,j}$, the algorithm convergence time will be higher, and swarm can stick in local extreme without any chances for escape from this region of the search space. However, if the values of $V_{max,j}$ are too high, then the particles can "jump" over good solutions, and continue the search in worse areas of the search space [8]. Usually, following formula (28.5) is used in order to guarantee a suitable selection of $V_{max,j}$ values for jth variable:

$$V_{max,j} = \delta \cdot (x_{max,j} - x_{min,j}) \tag{28.5}$$

where $x_{max,j}$ and $x_{min,j}$ are maximum and minimum values, respectively, which determine the range of jth variable ($j \in [x_{min,j}; x_{max,j}]$), δ—is a value from the range (0; 1), which is determined experimentally for solved problem.

28.3.2 Inertia Weight

The inertia weight ω is introduced to the *PSO* algorithm in order to better control particle swarm ability of exploration (searching over the whole solution space) and exploitation (searching in the neighborhood of "good" solutions). Inertia weight coefficient ω determines how a large part of the particle velocity from the previous "fly" will be used to create its new velocity vector. The introduction of ω coefficient causes the following modifications of formulas (28.1) and (28.2), respectively:

$$V[j] = \varpi \cdot V[j] + c_1 \cdot r_{1,j} \cdot (P_{best}[j] - X[j]) + c_2 \cdot r_{2,j} \cdot (L_{best}[j] - X[j]) \tag{28.6}$$

$$V[j] = \varpi \cdot V[j] + c_1 \cdot r_{1,j} \cdot (P_{best}[j] - X[j]) + c_2 \cdot r_{2,j} \cdot (G_{best}[j] - X[j]) \tag{28.7}$$

Many different techniques of determination of ω values exist in literature [8]. One of them is described by the following formula:

$$\varpi \geq \frac{1}{2} \cdot (c_1 + c_2) - 1 \tag{28.8}$$

In this case, the choice of ω value depends on the selection of c_1 and c_2 values. In paper [9], it is shown that if formula (28.8) is fulfilled, then the algorithm will converge; in other case, the oscillations of obtained results can occur. Other methods of determination of ω value can be found, for example, in paper [8].

28.3.3 Constriction Coefficient

In its operation, the constriction coefficient χ is similar to the inertia weight described in Section 28.3.2. The main task of constriction coefficient is balancing of the *PSO* algorithm properties between global and local searching of the solution space in the neighborhood of "good" solutions. Modified formulas (28.1) and (28.2) using the constriction coefficient χ are as follows:

$$V[j] = \chi \cdot \left[V[j] + c_1 \cdot r_{1,j} \cdot \left(P_{best}[j] - X[j]\right) + c_2 \cdot r_{2,j} \cdot \left(G_{best}[j] - X[j]\right)\right] \tag{28.9}$$

$$V[j] = \chi \cdot \left[V[j] + c_1 \cdot r_{1,j} \cdot \left(P_{best}[j] - X[j]\right) + c_2 \cdot r_{2,j} \cdot \left(L_{best}[j] - X[j]\right)\right] \tag{28.10}$$

and

$$\chi = \frac{2 \cdot \kappa}{\left|2 - \Phi - \sqrt{\Phi \cdot (\Phi - 4)}\right|} \tag{28.11}$$

where $\Phi = c_1 \cdot r_{1,j} + c_2 \cdot r_{2,j}$; usually, it is assumed that $\Phi \geq 4$ and $\kappa \in [0, 1]$.

Due to the application of constriction coefficient χ, the *PSO* algorithm convergence is assured without necessity of using of the velocity clamping model. The χ coefficient values are chosen from the range [0, 1] and, therefore the particle velocities are decreasing during each iteration of the algorithm. The value of the parameter κ influences swarm abilities for global or local searching of the solution space. If $\kappa \approx 0$, then faster algorithm convergence occurs together with local searching (this behavior is similar to the hill-climbing algorithm). In the case when $\kappa \approx 1$, the algorithm convergence is slower, and at the same time, searching of the solution space [8] is more exact.

28.4 Example

Minimize the following function:

$$FC = f(x_1, x_2) = \sum_{i=1}^{2} x_i^2, \quad -5.12 \leq x_i \leq 5.12, \quad \text{Global minimum} = 0 \text{ in } (x_1, x_2) = (0,0)$$

It is assumed that the PSO algorithm has following parameters: number of particles $M = 5$; $c_1 = c_2 = 0.3$. The dimension of the solution space (identical as a dimension of optimized function) is equal to $D = 2$. Two versions of the *PSO* algorithm are considered: local and global. Additionally, in *LPSO*, the number of nearest neighbors $Ne = 2$ is assumed.

28.4.1 Random Creation of the Population *P* Consisting *M* Particles

28.4.1.1 For *LPSO*

It is assumed that the particles have following parameters:

Particle P_1: $X = \{3.12; 4.01\}$; $P_{best} = X = \{3.12; 4.01\}$; $V = \{0; 0\}$; $L_{best} = \{0; 0\}$
Particle P_2: $X = \{-2.89; -1.98\}$; $P_{best} = X = \{-2.89; -1.98\}$; $V = \{0; 0\}$; $L_{best} = \{0; 0\}$
Particle P_3: $X = \{4.32; 2.11\}$; $P_{best} = X = \{4.32; 2.11\}$; $V = \{0; 0\}$; $L_{best} = \{0; 0\}$
Particle P_4: $X = \{2.11; -2.12\}$; $P_{best} = X = \{2.11; -2.12\}$; $V = \{0; 0\}$; $L_{best} = \{0; 0\}$
Particle P_5: $X = \{0.11; -2.71\}$; $P_{best} = X = \{0.11; -2.71\}$; $V = \{0; 0\}$; $L_{best} = \{0; 0\}$

28.4.1.2 For *GPSO*

It is assumed that the created population is composed of particles similar to those in *LPSO* algorithm. The difference is that, the vector L_{best} does not exist in created particles P_i, but instead, there exists the vector G_{best}.

28.4.2 Evaluation of Particle Positions Using Objective Function *FC*

28.4.2.1 For *LPSO*

The quality of positions is computed using objective function *FC* and data stored in the vector *X* for particular particles, the FC_i values are

$$FC_1 = 25.8145; \quad FC_2 = 12.2725; \quad FC_3 = 23.1145; \quad FC_4 = 8.9465; \quad FC_5 = 7.3562$$

Since it is the first iteration of the algorithm, there is no position *X* of any particle P_i that has lower value of the objective function than the position written down in its vector P_{best}. As a result, none of the vectors P_{best} will be updated. Of course, this update will occur in next steps of the algorithm.

28.4.2.2 For *GPSO*

The position *X* having the lowest value of the objective function *FC* (minimization task) is chosen and written down in the vector G_{best}. After this operation, considered vector is equal to $G_{best} = X_5 = \{0.11; -2.71\}$. In next generations the vector G_{best} will be updated, only if a particle in the population will have a better position *X* than the position *X* stored in the G_{best}.

28.4.3 Calculation of the Best Neighbors (Only for *LPSO* Algorithm)

At the beginning of the operation of the algorithm, the vectors L_{best} of all particles are updated, because at the start, all vectors $[L_{best}] = [0]$. In next generations, the vector L_{best} of the particle P_i will be updated only if the position *X* of the best particle from *Ne* nearest neighbors of particle P_i has lower value of the objective function than position *X* actually stored in the vector L_{best} in particle P_i.

In order to compute $Ne = 2$ nearest neighbors for each particle P_i, the distance between all particles in the population must be computed. The Euclidean distance $d(A, B)$ between two points $A = \{a_1, a_2, \ldots, a_n\}$ and $B = \{b_1, b_2, \ldots, b_n\}$ in *n*-dimensional space is determined as follows:

$$d(A,B) = \sqrt{\sum_{i=1}^{n} (a_i - b_i)^2} \tag{28.12}$$

For example, the distance between *X* position of particles P_1 and P_2 is equal to

$$d(P_1,P_2) = \sqrt{\left(3.12-(-2.89)\right)^2 + \left(4.01-(-1.98)\right)^2} = \sqrt{36.1201+35.8801} = 8.4853$$

In a similar way, the remaining distances between particles are as follows:

$d(P_1, P_2) = 8.4853$; $d(P_1, P_3) = 2.2472$; $d(P_1, P_4) = 6.2126$; $d(P_1, P_5) = 7.3633$; $d(P_2, P_3) = 8.2893$; $d(P_2, P_4) = 5.0020$; $d(P_2, P_5) = 3.0875$; $d(P_3, P_4) = 4.7725$; $d(P_3, P_5) = 6.3997$; $d(P_4, P_5) = 2.0852$.

After the analysis of distances between particles, the $Ne = 2$ nearest neighbors are determined for each particle. For example, the particle P_1 is a neighbor of particles P_3 and P_4, what is written as $P_1 = >\{P_3, P_4\}$. The two neighbor particles for the remaining particle P_i are as follows: $P_2 = >\{P_4, P_5\}$; $P_3 = >\{P_1, P_4\}$; $P_4 = >\{P_3, P_5\}$; $P_5 = >\{P_2, P_4\}$.

Next, only one particle of the two having the lowest value of the objective function is chosen from each pair of neighbor particles with respect to the particle P_i; then, the position X of chosen particle is written down in the vector L_{best} of particle P_i. For example, the vector L_{best} for particle P_1 is updated using position vector X of particle P_4. If we perform the same computations for other cases, we obtained the following values of the vector L_{best} for the remaining particles:

$$P_1: L_{best} = \{2.11; -2.12\}; \quad P_2: L_{best} = \{0.11; -2.71\}; \quad P_3: L_{best} = \{2.11; -2.12\};$$

$$P_4: L_{best} = \{0.11; -2.71\}; \quad P_5: L_{best} = \{2.11; -2.12\}$$

28.4.4 Calculation of New Values of Particle Velocity

28.4.4.1 For *LPSO*

The velocity vectors are computed for each particle using formula (28.1). For example, for particle P_1, the computations are as follows:

$$V[1] = V[1] + 0.3 \cdot r_{1,1} \cdot \left(P_{best}[1] - X[1]\right) + 0.3 \cdot r_{2,1} \cdot \left(L_{best}[1] - X[1]\right)$$

$$V[2] = V[2] + 0.3 \cdot r_{1,2} \cdot \left(P_{best}[2] - X[2]\right) + 0.3 \cdot r_{2,2} \cdot \left(L_{best}[2] - X[2]\right)$$

If it is assumed that for coefficients $r_{i,j}$ the followning real numbers are randomly chosen: $r_{1,1} = 0.22$, $r_{2,1} = 0.76$, $r_{1,2} = 0.55$, $r_{2,2} = 0.21$, we can obtain

$$V[1] = 0 + 0.3 \cdot 0.22 \cdot (3.12 - 3.12) + 0.3 \cdot 0.76 \cdot (2.11 - 3.12) = -0.2303$$

$$V[2] = 0 + 0.3 \cdot 0.55 \cdot (4.01 - 4.01) + 0.3 \cdot 0.21 \cdot (-2.12 - 4.01) = -0.3862$$

If the same computations are performed for the remaining particles and the same set of random numbers are assumed for coefficients $r_{i,j}$ (for simplification) $r_{1,1} = 0.22$, $r_{2,1} = 0.76$, $r_{1,2} = 0.55$, $r_{2,2} = 0.21$, the following vectors are obtained:

$$P_1: V = \{-0.2303; -0.3862\}; \quad P_2: V = \{0.6840; -0.0460\}; \quad P_3: V = \{-0.5039; -0.2665\}$$

$$P_4: V = \{-0.4560; -0.0372\}; \quad P_5: V = \{0.4560, 0.0372\}$$

28.4.4.2 For *GPSO*

The velocity vector is updated using formula (28.2) for each particle P_i. For example, for particle P_1, the new velocity vector is computed as follows:

$$V[1] = V[1] + 0.3 \cdot r_{1,1} \cdot \left(P_{best}[1] - X[1]\right) + 0.3 \cdot r_{2,1} \cdot \left(G_{best}[1] - X[1]\right)$$

$$V[2] = V[2] + 0.3 \cdot r_{1,2} \cdot \left(P_{best}[2] - X[2]\right) + 0.3 \cdot r_{2,2} \cdot \left(G_{best}[2] - X[2]\right)$$

if it is assumed, that for coefficients $r_{i,j}$ following real numbers have been randomly chosen: $r_{1,1} = 0.85$, $r_{2,1} = 0.23$, $r_{1,2} = 0.45$, $r_{2,2} = 0.63$ it is obtained:

$$V[1] = 0 + 0.3 \cdot 0.85 \cdot (3.12 - 3.12) + 0.3 \cdot 0.23 \cdot (0.11 - 3.12) = -0.2077$$

$$V[2] = 0 + 0.3 \cdot 0.45 \cdot (4.01 - 4.01) + 0.3 \cdot 0.63 \cdot (-2.71 - 4.01) = -1.2701$$

For other particles the velocity vectors are as follows:

$$P_1\colon V = \{-0.2077; -1.2701\}; \quad P_2\colon V = \{0.2070; -0.1380\}; \quad P_3\colon V = \{-0.2905; -0.9110\}$$

$$P_4\colon V = \{-0.1380; -0.1115\}; \quad P_5\colon V = \{0, 0\}$$

The identical values of $r_{1,1}$, $r_{2,1}$, $r_{1,2}$, $r_{2,2}$ are assumed in order to simplify this example; in real algorithm, the new values of parameters $r_{1,1}$, $r_{2,1}$, $r_{1,2}$, $r_{2,2}$ must be randomly chosen for each particle in each iteration of the algorithm.

28.4.5 Calculation of New Values of Particle Position Vectors

New values of position vectors X of particle P_i are computed using formula (28.3) for both versions of the *PSO* algorithm: *LPSO* and *GPSO*.

28.4.5.1 For *LPSO*

For *LPSO*, new values of position vectors X for particular particles P_i are computed using computed earlier velocity vectors V. For example, the new position X for particle P_1 is equal to

$$X[1] = X[1] + V[1]; \qquad X[2] = X[2] + V[2];$$

therefore new values are

$$X[1] = 3.12 + (-0.2303) = 2.8897; \quad X[2] = 4.01 + (-0.3862) = 3.6238.$$

Thus, the new position vector X for particle P_1 is equal to $X = \{2.8897; 3.6238\}$.

For other particles, new position vectors X are as follows:

$$P_1\colon X = \{2.8897; 3.6238\}; \quad P_2\colon X = \{-2.2060; -2.0260\}; \quad P_3\colon X = \{3.8161; 1.8435\};$$

$$P_4\colon X = \{1.6540; -2.1572\}; \quad P_5\colon X = \{0.5660; -2.6728\}$$

28.4.5.2 For *GPSO*

For *GPSO* of the algorithm, new values of position vectors X for particular particle P_i are as follows:

$$P_1\colon X = \{2.9123; 2.7399\}; \quad P_2\colon X = \{-2.6830; -2.1180\}; \quad P_3\colon X = \{4.0295; 1.1990\};$$

$$P_4\colon X = \{1.9720; -2.2315\}; \quad P_5\colon X = \{0.1100; -2.7100\}$$

In summary, the population of particles for *LPSO* of the algorithm after first iteration is as follows:

Particle P_1: $X = \{2.89; 3.63\}$; $P_{best} = \{3.12; 4.01\}$; $V = \{-0.23; -0.39\}$; $L_{best} = \{2.11; -2.12\}$
Particle P_2: $X = \{-2.21; -2.03\}$; $P_{best} = \{-2.89; -1.98\}$; $V = \{0.68; -0.05\}$; $L_{best} = \{0.11; -2.71\}$
Particle P_3: $X = \{3.82; 1.84\}$; $P_{best} = \{4.32; 2.11\}$; $V = \{-0.50; -0.27\}$; $L_{best} = \{2.11; -2.12\}$
Particle P_4: $X = \{1.65; -2.16\}$; $P_{best} = \{2.11; -2.12\}$; $V = \{-0.46; -0.04\}$; $L_{best} = \{0.11; -2.71\}$
Particle P_5: $X = \{0.57; -2.67\}$; $P_{best} = \{0.11; -2.71\}$; $V = \{0.46; 0.04\}$; $L_{best} = \{2.11; -2.12\}$

If we compute values of objective function *FC* for data stored in *X* vectors for all particles, it can be noticed that the best result is written down in vector *X* in particle P_4 for which the value of objective function *FC* is lowest, and equal to $FC_4 = 7.3881$.

Similarly, the population of particles for *GPSO* of the algorithm after first iteration is as follows:

Particle P_1: $X = \{2.91; 2.74\}$; $P_{best} = \{3.12; 4.01\}$; $V = \{-0.21; -1.27\}$;
Particle P_2: $X = \{-2.68; -2.12\}$; $P_{best} = \{-2.89; -1.98\}$; $V = \{0.21; -0.14\}$;
Particle P_3: $X = \{4.03; 1.20\}$; $P_{best} = \{4.32; 2.11\}$; $V = \{-0.29; -0.91\}$;
Particle P_4: $X = \{1.97; -2.23\}$; $P_{best} = \{2.11; -2.12\}$; $V = \{-0.14; -0.11\}$;
Particle P_5: $X = \{0.11; -2.71\}$; $P_{best} = \{0.11; -2.71\}$; $V = \{0; 0\}$;
Vector $G_{best} = \{0.11; -2.71\}$;

In the last step, the termination condition of the algorithm is checked. In the case when termination condition is fulfilled, for *LPSO* algorithm, the result (solution) having the lowest value of the objective function in current population of particles is returned, or for *GPSO* algorithm, the result stored in the vector G_{best} is returned. If termination condition of the algorithm is not fulfilled, then the algorithm jumps to the evaluation of particle positions in the population (Section 28.4.2), and whole process is repeated.

28.5 Summary

In this chapter, a basic version of the *PSO* algorithm is presented. Two versions of this algorithm: *LPSO* and *GPSO* are described. Also, some modifications of *PSO* algorithm, which improve its convergence, are shown. The *PSO* algorithm operation is illustrated in detail by examples of minimization of two variable functions.

References

1. D. T. Pham, A. Ghanbarzadeh, E. Koç, S. Otri, S. Rahim, and M. Zaidi, The bees algorithm—A novel tool for complex optimisation problems, *IPROMS 2006, Intelligent Production, Machines and Systems*, Elsevier, Oxford, U.K., 2006.
2. M. Dorigo, V. Maniezzo, and A. Colorni, Ant system: Optimization by a colony of cooperating agents, *IEEE Transactions on Systems, Man, and Cybernetics—Part B*, 26(1), 29–41, 1996.
3. A. Colorni, M. Dorigo, and V. Maniezzo, Distributed optimization by ant colonies, *Proceedings of the First European Conference on Artificial Life*, F. J. Varela and P. Bourgine (Eds.), MIT Press, Cambridge, MA, pp. 134–142, 1992.
4. J. Kennedy, R. C. Eberhart, and Y. Shi, *Swarm Intelligence*, Morgan Kaufmann Publishers, San Francisco, CA, 2001.
5. Q.-K. Pana, M. Fatih Tasgetirenc, and Y.-C. Liang, A discrete particle swarm optimization algorithm for the no-wait flowshop scheduling problem, *Computers and Operations Research*, 35(9), 2807–2839, 2008.
6. D. J. Krusienski and W. K. Jenkins, Particle swarm optimization for adaptive IIR filter structures, *Congress on Evolutionary Computation, 2004, CEC 2004*, Vol. 1, Honolulu, HI, pp. 965–970, June 19–23, 2004.
7. A. Slowik and M. Bialko, Design and optimization of IIR digital filters with non-standard characteristics using particle swarm optimization, *Proceedings of the 14th IEEE International Conference on Electronics, Circuits and Systems, ICECS 2007*, Marrakech, Morocco, pp. 162–165, December 11–14, 2007.
8. A. P. Engelbrecht, *Computational Intelligence—An Introduction*, 2nd edn., Wiley, Chichester, U.K., 2007.
9. F. van den Bergh and A. P. Engelbrecht, A study of particle swarm optimization particle trajectories, *Information Sciences*, 176(8), 937–971, 2006.

V

Applications

29

Evolutionary Computation

Adam Slowik
Koszalin University of Technology

29.1 Introduction

In the real-world, engineers often need to solve difficult optimization problems, such as digital filters designed with non-standard characteristics, floor planning of 2D elements, decomposition of digital circuit on subcircuits, etc. The majority of these problems are classified as NP-hard problems in which the solution space is very huge, and algorithms that can find an optimal solution in acceptable computational time do not exist. In these cases, the engineers must use heuristic methods, which do not guarantee that the solution found will be optimal. However, these heuristic methods can find acceptable suboptimal solutions in all required computational time. Among these methods, a special place is reserved for evolutionary algorithms [1,2]; their main advantages are as follows: the computation starts from a population (many points) of potential solutions (not from a single potential solution), and only a proper objective function, which describes a given problem is required (any other information and derivative of objective functions are not required). The existence of different genetic operators (like mutation, cross-over) in these algorithms allows easy "escape" from actual local minimum that can be found during computations, and prevents premature convergence of the algorithm. Evolutionary computation is widely used in real-world applications, such as digital filters design [3], design and optimization of digital circuits [4,5], partitioning of VLSI circuits on subcircuits with minimal number of external connections between them [6], optimization of placement of 2D elements [7], training of artificial neural networks, and optimization of parameters in grinding process.

29.2 Description of Evolutionary Algorithms

Algorithms in which the way of finding a solution, i.e., searching for potential solutions space (the way of information processing), is based on natural evolution process and Darwin's theory of natural selection (only individuals with the best fitness to the environment will survive in next generation) are named as evolutionary algorithms. The notions, which are used for description of parameters and processes in evolutionary algorithms have close union with genetics, and natural evolution. In general, the evolutionary algorithm processes the population of P individuals, each of which is also named a chromosome and represents one potential solution of the given problem. Evolutionary algorithm is operating in an artificial environment, which can be defined based on the problem

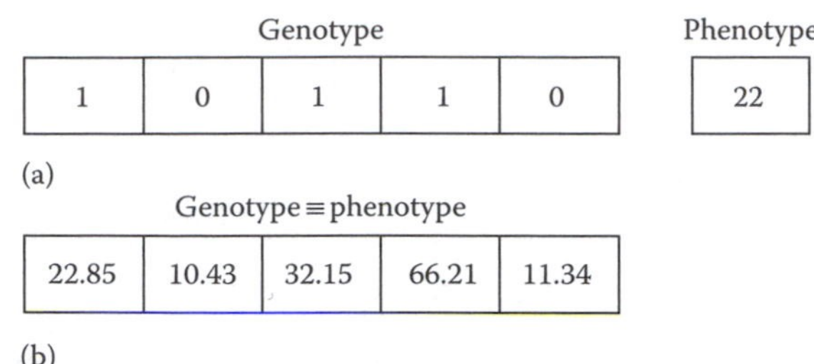

FIGURE 29.1 Example of individual with representation: binary (a) and real-number (b).

solved by this algorithm. During the operation of the algorithm, individuals are evaluated and those, better fitting to the environment, obtain better so-called "fitness value." This value is the main factor of evaluation. Particular individual consists of coded information named as a genotype. The phenotypes, which are the decoded form of potential solutions of a given problem, are created from genotypes, and are evaluated using fitness function. In some kinds of evolutionary algorithms, the phenotypes are identical to genotypes. A genotype is a point in the space of codes, while the phenotype is a point in the space of problem solutions. Each chromosome consists of elementary units named genes (in Figure 29.1, both individuals have five genes). Additionally, the values of a particular gene are called allels (e.g., in the case of binary representation, the allowed allele values are: 0 and 1). In Figure 29.1a, an example of individual with binary representation (genotype and phenotype) is presented, and in Figure 29.1b, an individual with real-number representation is shown (the phenotype is identical as the genotype).

The environment can be represented using fitness function, which is related to the objective function. The structure of evolutionary algorithm is shown in Figure 29.2.

It can be seen, that evolutionary algorithm is a probabilistic one in which new population of individuals $P(t) = \{x_1^t, \dots, x_n^t\}$ is generated in each iteration t. Each individual x_i^t represents possible solution of considered task, and is most often represented by data structure in the form of a single-layer chromosome [2] (however in paper [6], and especially in [8], a concept of multilayer chromosome is introduced together with its possible potential applications). Each solution x_i^t is evaluated using certain measure of fitness of chromosome. Therefore, the new population $P(t + 1)$ (in iteration $t + 1$) is created by selection of the best fitted individuals (selection phase). Additionally, in the new population, some individuals are transformed (exchange phase) using "genetics" operators, leading to the creation of a new solution. The transformations can be represented by mutation operator, in which new individuals are generated

```
Procedure of evolutionary algorithm
begin
      determine fitness function
      t ← 0
      randomly create individuals in initial population P(t)
      evaluate individuals in population P(t) using fitness function
      while (not terminate criterion) do
      begin
            t ← t + 1
            perform selection of individuals to population P(t) from  P(t - 1)
            change individuals of P(t) using cross-over and mutation operators
            evaluate individuals in population P(t) using fitness function
      end
end
```

FIGURE 29.2 Pseudo code of evolutionary algorithm. (Adapted from Michalewicz, Z., *Genetic Algorithms + Data Structures = Evolution Programs*, Springer-Verlag, Berlin/Heidelberg, Germany, 1992.)

by a small modification of a single individual, and cross-over operator, where new individuals in the form of single-layer chromosome are created by linking the fragments of chromosome from several (two or more) individuals [2]. These operators are described in Section 29.2.5. After several generations, the computation converges, and we can expect that the best individuals representing acceptable solution are located near the optimal solution.

29.2.1 Fitness Function

The fitness function in evolutionary algorithm is an element linking the considered problem and the population of individuals. The main task of this function is a determination of qualities of particular individuals (solutions) in the aspect of problem to be solved. In the case of problems related to minimization of objective function, the solutions having the smallest value of fitness function will be better solutions, while in the case of maximization problem, these solutions will be the worst ones. In evolutionary algorithms, typically, the maximization of objective function is considered. Therefore, in the case of minimization problem, we must convert the problem to maximization task. The simplest way to do this is a change of the sign of the objective function (multiply by −1), and assure positiveness of this function for all values of input arguments. This is necessary because selection based on roulette method, typically used in classical genetic algorithms, requires nonnegative fitness values for each individual. The solution of this problem is a suitable definition of the fitness function (*FF*) based on the objective function (*OF*). The following formula can be used for minimization tasks [9]:

$$FF\left(x_i^t\right) = f_{max} - OF\left(x_i^t\right) \tag{29.1}$$

Of course, the value f_{max} (the highest value of objective function) is usually not known a priori, therefore the highest value of objective function obtained during all previous generations of algorithm is assigned as a f_{max} value. Additionally, if we want to obtain fitness function values only in the range (0; 1], then the definition of fitness function for minimization problems is as follows [9]:

$$FF\left(x_i^t\right) = \frac{1}{1 + OF\left(x_i^t\right) - f_{min}} \tag{29.2}$$

where f_{min} is the lowest value of objective function observed during previous iterations of the algorithm. In the case of maximization tasks, the value of fitness of individual is scaled as follows [9]:

$$FF\left(x_i^t\right) = \frac{1}{1 + f_{max} - OF\left(x_i^t\right)} \tag{29.3}$$

29.2.2 Representation of Individuals—Creation of Population

Depending on the problem, typically used representations of individuals are binary, real-number, and integer-number. In the case of binary representation, the determination of the number of genes (*NG*) required for the coding of a given variable $V \in [x_{min}; x_{max}]$ with assumed precision β is very important. In this case, the following inequality must be fulfilled:

$$2^{NG} \geq \frac{x_{max} - x_{min}}{\beta} + 1 \tag{29.4}$$

In the case of coding in the form of real-number or integer-number, the total number of genes in an individual is equal to the number of variables in an optimized task. During the creation of a population consisting of M individuals having NG genes in each chromosome, the value of each gene is randomly selected from an assumed range. In the case of binary coding, each gene has value "0" or "1," and in the case of real-number coding or integer-number coding, the value represented by particular gene is randomly chosen from the range $[x_{min}; x_{max}]$, determined individually for each variable.

29.2.3 Evaluation of Individuals

In the case of binary representation of individuals, the fitness value can be evaluated for particular individual using defined fitness function and values for each individual. The phenotype can be computed from genotype using the following formula:

$$\text{phenotype} = x_{min} + \frac{x_{max} - x_{min}}{2^{NG} - 1} \cdot dec(\text{genotype}) \tag{29.5}$$

where $dec(\cdot)$ represents the decimal value corresponding to the chosen genotype.

In the case when genotype consists information related to several variables (located in several parts of the chromosome—see Figure 29.3a), the value of phenotype is computed for each variable separately using that part of genotype where this variable is written down (see Figure 29.3).

For individual representation in the form of real-number or integer-number, the genotype is identical to phenotype, and phenotype computing is not required (see Figure 29.1b). If we have the phenotype of a particular individual, then it is possible to compute a value of fitness function for each individual, which determines the quality of a given individual.

29.2.4 Selection

The selection, also named as reproduction, is a procedure of choosing given individuals from the population in order to create a new population in the next generation of evolutionary algorithm. The probability of selection of a given individual depends on its fitness value. When the given individual has a higher fitness value, then it possess higher chance to be selected to the new generation. The reproduction process is strictly connected with the two most important factors in evolutionary algorithms: preservation of the diversity of population and selection pressure. These factors are dependent on each other because increase of selection pressure causes decrease of population diversity (and inversely) [1]. Too high value of selection pressure (concentration of the search only on best individuals) leads to premature convergence, which is an undesirable effect in evolutionary algorithms, because the algorithm can stick in local extreme. However, too small value of selection pressure causes that search of solution space has almost random character. The main goal of the selection operators is the preservation of balance between these factors [2]. There exist many selection methods. The oldest one (most popular) is a proportional selection also named as a roulette selection. In this method, the probability of individual selection is proportional to the value of its fitness function [1]. For each individual, the sector size on roulette wheel is equal to the individual relative fitness (*rfitness*) value, that is, the fitness value divided by the sum of all fitness values *GF* (*global fitness*) in the population (see formula (29.6) and (29.7)). In Figure 29.4, an example of roulette wheel with scaled sectors for $M = 5$ individuals is presented.

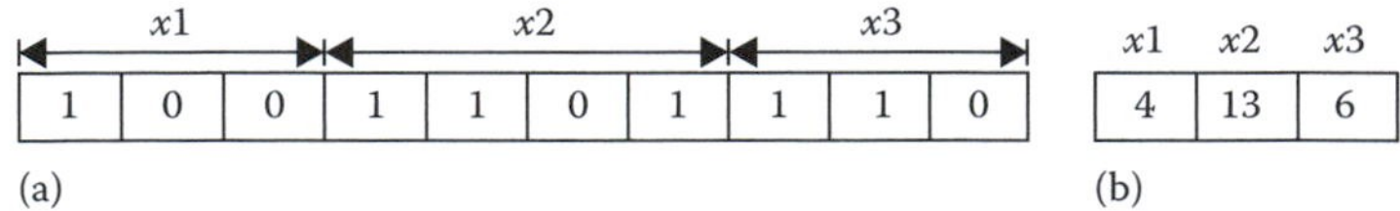

FIGURE 29.3 Genotype with coded 3 variables: $x1$, $x2$, and $x3$ (a), phenotype corresponding to it (b).

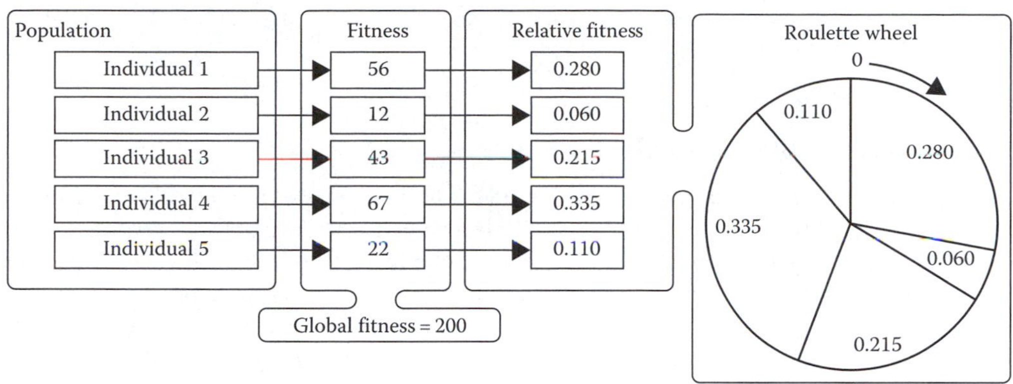

FIGURE 29.4 Example of roulette wheel with scaled sectors for $M = 5$ individuals.

To use the roulette selection, we must compute global fitness (GF) for the whole population:

$$GF = \sum_{i=1}^{M} fitness_i \tag{29.6}$$

where

$fitness_i$ represents fitness value for ith individual in population
M is the number of individuals in population

Then, the value of relative fitness ($rfitness$) is computed for each ith individual:

$$rfitness_i = \frac{fitness_i}{GF} \tag{29.7}$$

The value of relative fitness $rfitness_i$ represents the probability of the selection of ith individual to the new population (probability of the selection of the individual is higher for those having higher roulette sector). Next, the sector ranges on roulette wheel must be determined for particular individuals. Roulette sector is equal to $[min_i; max_i)$ for ith individual. The border values are computed as follows:

$$min_i = max_{i-1} \tag{29.8}$$

$$max_i = min_i + rfitness_i \tag{29.9}$$

For the first individual the value $min_i = 0$ (see Figure 29.4).

In the next step, a random value from the range $[0; 1)$ is chosen M – times in order to select M individuals to the new population. If randomly chosen value is inside the range $[min_i; max_i)$, then ith individual is selected to the new population.

Besides roulette selection, there exist many other selection methods. Among those, we can mention rank selection [2], tournament selection [2], and fan selection [10].

The roulette selection method described above could be equipped with mechanisms putting attention on the survival of better individuals. The most known of them is the elitist model, in which the best individual is introduced to the next generation with the omission of the standard selection procedure [2]. It is performed in the case when the best individual (with highest value of fitness function) does not survive in the next population. In such a case, the worst individual in the population is replaced by the best one from earlier population.

29.2.5 Mutation and Cross-Over

The selection procedure is the first step in the creation of new generations in evolutionary algorithms. However, the new created individuals are only duplicates of individuals from previous generation; therefore, the application of genetic operators is necessary in order to modify chosen individuals. In general, the operators used in evolutionary algorithms can be divided into two groups. First of these groups is one-argument operators, that is operating on a single individual. One-argument operators are called mutation, and are executed on the population of genes with probability $PM \in [0; 1]$; this value is one of the parameters of the algorithm. Mutation depends on the random selection of real-number $rand_m$ from the range [0; 1) for each gene in population. If randomly chosen real-number $rand_m < PM$ for jth gene in ith individual, then this gene is mutated. The scheme of simple mutation (for binary representation of individuals) depends on value exchange in chosen gene from "0" to "1," or inversely. In Figure 29.5a, the scheme of simple mutation is shown. In the real-number representation of individuals, the procedure of mutation is analogical as in binary representation, but new value of gene is randomly chosen from assumed range for each variable (see Figure 29.5b).

The second group of genetic operators is multi-argument operators named recombination or cross-over. In evolutionary algorithm, the cross-over operation is operating on the population of individuals with probability $PC \in [0; 1]$; it depends on the random choice of real-number $rand_k$ from the range [0; 1) for each individual. In the case when $rand_k < PC$ for ith individual, this individual is chosen for cross-over operation. In evolutionary algorithms, the simplest model of cross-over is a simple cross-over operator, also named as a one-point cross-over. In Figure 29.6a, the scheme of the one-point cross-over with crossing point equal to "*K*1" is graphically shown for binary representation of individuals (in Figure 29.6b), the scheme of one-point cross-over is presented for real-number representation of individuals). In general, two child individuals are created from two parent individuals using cross-over operator. However, in evolutionary algorithms, many other types of cross-over operators are used. These operators have been created in order to provide an effective exchange of information between two chromosomes. Usually, the types of the cross-over operators are suitably chosen to the solved problem. The examples of typical recombination (cross-over) operators that are dependent on the problem are *PMX* (partially mapped cross-over), *CX* (cycle cross-over), *OX* (order cross-over) [4], which are used to solve the traveling salesman problem.

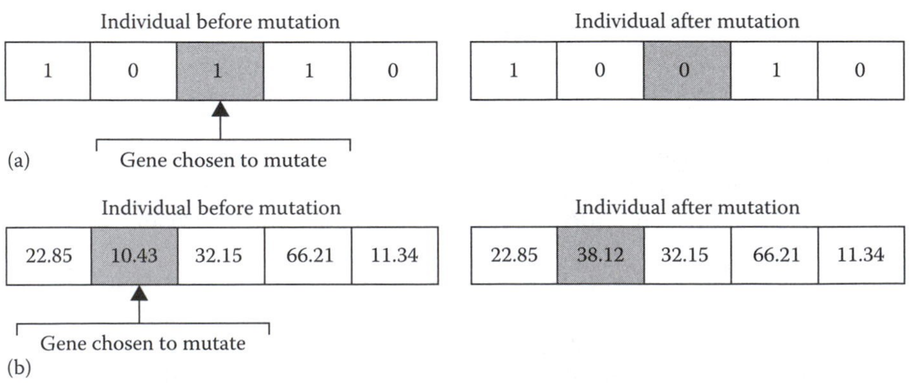

FIGURE 29.5 Scheme of simple mutation operator for individual with binary representation (a) and real-number representation (b).

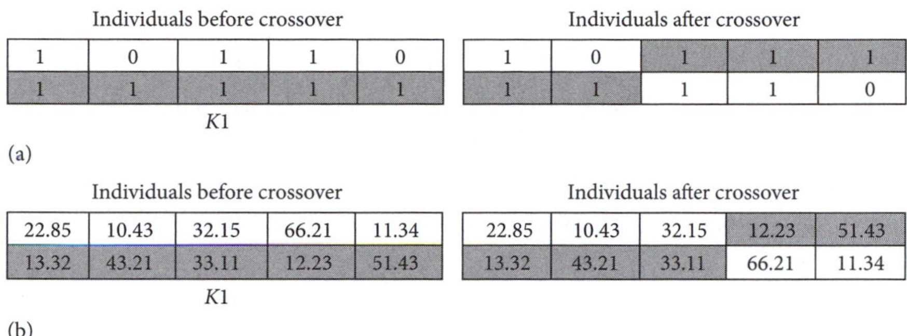

FIGURE 29.6 Scheme of one-point cross-over operator for individuals with binary representation (a) and real-number representation (b).

29.2.6 Terminate Conditions of the Algorithm

The terminate conditions, which are mostly used in evolutionary algorithms are: algorithm convergence, that is invariability of the best solution after assumed number of generations, or reaching the assumed number of generations by the algorithm.

29.2.7 Example

Minimize objective function *OF* having three variables:

$$OF = f(x_1, x_2, x_3) = \sum_{i=1}^{3} x_i^2, \quad -5.12 \le x_i \le 5.12, \quad \text{Global minimum} = 0 \text{ in}(x_1, x_2, x_3) = (0,0,0)$$

The parameters of evolutionary algorithm are $M = 5$, $PM = 0.05$, $PC = 0.5$. For better problem presentation two kinds of representation of individuals: binary and real-number are considered. It is assumed, that each variable must be coded with precision $\beta \ge 0.2$ for binary representation. Therefore, the lowest value of *NG* (number of genes representing one variable), which fulfills inequality (29.4)

$$2^{NG} \ge \frac{x_{max} - x_{min}}{\beta} + 1 \Rightarrow 2^{NG} \ge \frac{5.12 + 5.12}{0.2} + 1 \Rightarrow 2^{NG} \ge 52.2$$

is equal to 6; therefore, each individual will be composed of 18 genes (6 genes for each variable).

29.2.7.1 Determination of Fitness Function

In order to guarantee nonnegative values of fitness function for all values of input arguments, the fitness function *FF* is scaled according to the formula (29.2)—minimization task:

$$FF\left(x_i^t\right) = \frac{1}{1 + OF\left(x_i^t\right) - f_{min}}$$

This scaling will be performed during the process of selection of individuals. Of course, the process of scaling is connected with the application of roulette selection method in the evolutionary algorithm. In the case of other selection method, the above transformation is not required.

29.2.7.2 Random Creation of Population of Individuals O_i

For binary representation (each variable is represented using six genes)
O_1 = {001011000111110011}; O_2 = {000111111110001100}; O_3 = {001111100110010010};
O_4 = {010101100110101110}; O_5 = {001111001110000101}

For real-number representation (each variable is represented using one gene)
O_1 = {2.13; −0.56; −3.56}; O_2 = {4.11; 2.01; −1.96}; O_3 = {0.95; 0.43; −0.87};
O_4 = {1.12; −3.54; 1.67}; O_5 = {−0.87; 3.44; 2.55}.

29.2.7.3 Evaluation of Individuals

For binary representation
The phenotypes of individuals are computed using formula (29.5); for example, for the first variable from individual O_1, the computational process is as follows:

$$\text{phenotype} = x_{min} + \frac{x_{max} - x_{min}}{2^{NG} - 1} \cdot dec(\text{genotype})$$

$$\text{phenotype} = -5.12 + \frac{5.12 + 5.12}{2^6 - 1} \cdot dec(001011) = -3.33$$

For other individuals, phenotype values are as follows:

O_1 = {−3.33; −3.98; 3.17}; O_2 = {−3.98; 4.96; −3.17}; O_3 = {−2.68; 1.06; −2.19};
O_4 ={−1.71; 1.06; 2.36}; O_5 = {−2.68; −2.84; −4.31};

The values of objective function OF_i for ith individuals are

$$OF_1 = (-3.33)^2 + (-3.98)^2 + (3.17)^2 = 36.98; \quad OF_2 = 50.49; \quad OF_3 = 13.10; \quad OF_4 = 9.62; \quad OF_5 = 33.82$$

For real-number representation
The phenotype values are equal to genotype ones and, therefore, additional computations are not required; the values OF_i for ith individuals are as follows:

$$OF_1 = 17.52; \quad OF_2 = 24.77; \quad OF_3 = 1.84; \quad OF_4 = 16.57; \quad OF_5 = 19.09$$

29.2.7.4 Selection

The performance of selection process is identical for binary and real-number representation of individuals. Therefore, the selection of individuals only in real-number representation is considered.

For real-number representation
The values of fitness function FF_i are computed for ith individuals according to formula (29.2). The lowest value of objective function OF obtained in previous generations is assumed as a value of f_{min}; thus $f_{min} = OF_3 = 1.84$. As a result, the following values of the fitness function are obtained: $fitness_1 = FF_1 = 0.0599$, $fitness_2 = FF_2 = 0.0418$ $fitness_3 = FF_3 = 1$, $fitness_4 = FF_4 = 0.0636$ and $fitness_5 = FF_5 = 0.0548$.

The value of total (global) fitness of individuals computed according to formula (29.6) is equal to GF = 1.2201. The values of relative fitness (*rfitness*) computed for particular individuals, using formula (29.7), are

$$rfitness_1 = 0.049; \quad rfitness_2 = 0.034; \quad rfitness_3 = 0.820; \quad rfitness_4 = 0.052; \quad rfitness_5 = 0.045.$$

We can see that individual 3 has the highest chance to survive, and individual 2 has the lowest chance to survive. In the next step, it is possible to construct a roulette sector for each individual using relative fitness values (see formula (29.8) and (29.9)). These roulette sectors are as follows:

$$O_1 \Rightarrow [0; 0.049); \quad O_2 \Rightarrow [0.049; 0.083); \quad O_3 \Rightarrow [0.083; 0.903); \quad O_4 \Rightarrow [0.903; 0.955); \quad O_5 \Rightarrow [0.955; 1)$$

After roulette wheel is scaled, then five real-numbers from the range [0; 1) are randomly chosen as, for example, 0.3, 0.45, 0.8, 0.96, 0.04, therefore, individuals $\{O_3; O_3; O_3; O_5; O_1\}$ are selected to the new population.

In the next step, the new selected individuals are mutated and crossed-over (see Section 29.2.5, and Figures 29.5 and 29.6). After these operations, the solutions (individuals) are evaluated as in Section 29.2.7.3, and the whole process is repeated until the termination condition of the algorithm is reached.

29.3 Conclusions

In this chapter, fundamental information concerning evolutionary algorithms is presented. The evolutionary algorithms are widely used in many optimization tasks. At the present time, many different kinds of these algorithms, which are used in multimodal optimization, multi-objective optimization, optimization with constraints have been developed. In this chapter, only a few characteristics of evolutionary algorithm together with examples presenting the successive steps of evolutionary computation are described.

References

1. D. E. Goldberg, *Genetic Algorithms in Search, Optimization & Machine Learning*, Addison Wesley, Reading, MA, 1989.
2. Z. Michalewicz, *Genetic Algorithms + Data Structures = Evolution Programs*, Springer-Verlag, Berlin/Heidelberg, Germany, 1992.
3. M. Erba, R. Rossi, V. Liberali, and A. G. B. Tettamanzi, Digital filter design through simulated evolution, in *Proceedings of ECCTD '01*, Espoo, Finland, vol. 2, pp. 137–140, August 2001.
4. C. A. Coello Coello, A. D. Christiansen, and A. Hernández Aguirre, Towards automated evolutionary design of combinational circuits, *Computers and Electrical Engineering*, 27(1), 1–28, January 2001.
5. A. Slowik and M. Bialko, Design and optimization of combinational digital circuits using modified evolutionary algorithm, *Lecture Notes in Artificial Intelligence*, Vol. 3070/2004, pp. 468–473, Springer-Verlag, New York, 2004.
6. A. Slowik and M. Bialko, Partitioning of VLSI circuits on subcircuits with minimal number of connections using evolutionary algorithm, *Lecture Notes in Artificial Intelligence*, Vol. 4029/2006, pp. 470–478, Springer-Verlag, New York, 2006.
7. E. Hopper, B. C. H. Turton, An empirical investigation of meta-heuristic and heuristic algorithms for a 2D packing problem, *European Journal of Operational Research* 128(1), 34–57, 2000.
8. A. Slowik, Design and optimization of digital electronic circuits using evolutionary algorithms, Doctoral dissertation, Koszalin University of Technology, Department of Electronics and Computer Science, 2007, (in Polish).
9. A. P. Engelbrecht, *Computational Intelligence—An Introduction*, 2nd edn., Wiley, Chichester, U.K., 2007.
10. A. Slowik and M. Bialko, Modified version of Roulette selection for evolution algorithm—The fan selection, *Lecture Notes in Artificial Intelligence*, Vol. 3070/2004, pp. 474–479, Springer-Verlag, New York, 2004.

30

Data Mining

Milos Manic
University of Idaho, Idaho Falls

30.1 Introduction

Data mining has been attracting increasing attention in recent years. Automated data collection tools and major sources of abundant data ranging from remote sensing, bioinformatics, scientific simulations, via web, e-commerce, transaction and stock data, to social networking, YouTube, and other means of data recording have resulted in an explosion of data and the paradox known as "drowning in data, but starving for knowledge."

The Library of Congress had collected about 70 terabytes (TBs) of data through May 2007; by February 2009 this had increased to more than 95 TBs of data [LOC 98TB]. On July 4, 2007, the National Archives of Britain issued a press release announcing a memorandum of understanding (MoU) with Microsoft to preserve the UK's digital heritage, an estimated archive content of "580 TBs of data, the equivalent of 580 thousand encyclopaedias" [NAB 580TB]. Today's ubiquitous consumer electronics and mobile devices reveal astonishing jumps in storage capacity. In late 2008, TiVos, iMACs, Time Capsule, and various external hard disk drives regularly offered 1 TB storage capacity. Musical players such as iPods, personal digital assistants (PDAs), and pocket PCs offered storage options for hundreds of megabytes of multimedia material.

Hard disk drives technology have made a giant leap since their commercial inception in the mid-1950s [Hoagland 03], going from 5.25 in. drives to 3.5 in. drives, reducing the number of platters and heads while at the same time increasing the areal density (amount of data per square inch of media). In September 2006, the major innovator in disk drive technology, Hitachi Global Storage Technologies (Hitachi GST) based at the San Jose Research Center in San Jose, CA (previously IBM Storage Technology Division), demonstrated the 345 Gbits per square inch recording density, based on known but abandoned perpendicular recording technology (replacing the conventional longitudinal recording) [Hitachi PR,Hoagland 03]. By 2009, Hitachi has predicted that new technology would result in a two-terabyte (TB) 3.5-in. desktop drive, a 400-gigabyte (GB) 2.5-in. notebook drive or a 200-GB 1.8-in. drive (such is the one used in Apple iPods) [Hitachi PR,Hitachi SC]. These quickly growing and vastly available massive data sets have imposed a next logical necessity—the need for automated analysis of these massive data sets.

30.2 What Is Data Mining?

Data mining occurred as a natural evolution of database (DB) technology [Han 06]. Primitive file processing systems and database creation in the 1960s quickly evolved into powerful database systems in the 1970s, including hierarchical, relational databases, SQL query languages, and high speed transaction types called on-line transaction processing (OLTP) methods. The mid-1980s saw the introduction of advanced databases, leading to data warehouses (repositories of multiple heterogeneous data at a single site facilitating decision making) and on-line analytical processing (OLAP) techniques capable of functional techniques such as summarization, consolidation, and aggregation. Spatial, temporal, multimedia, web, and text mining became available with more in-depth analysis and sophisticated techniques of machine learning, pattern, time-series, and social data mining (Figure 30.1). As the growth of data continues, commercial tools are becoming more updated and aimed at solving diverse problems ranging from marketing, sales, and business intelligence, to counterterrorism and social networking [Thuraisingham 2003,Berry 04,Linoff 2002,Matignon 07].

Although different definitions of data mining exist in literature, the simplest designation that puts all of the above together is that data mining is an *intelligent process of mining knowledge from large amounts of data*. Despite referring to mining knowledge from massive amounts of data, "data mining" carries

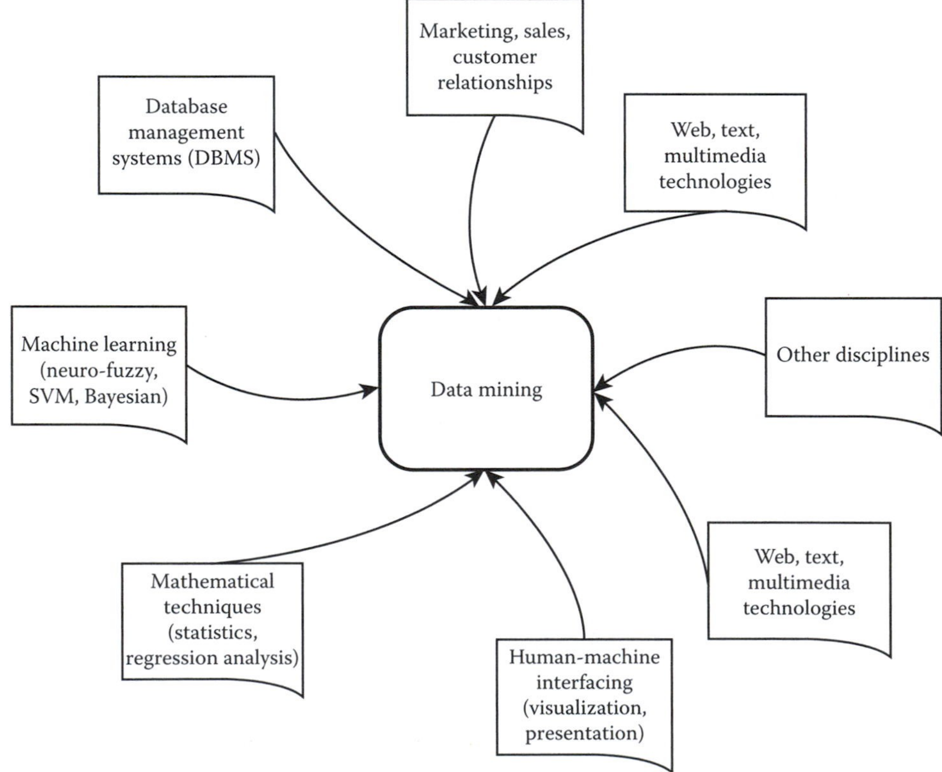

FIGURE 30.1 Data mining as fusion of disciplines.

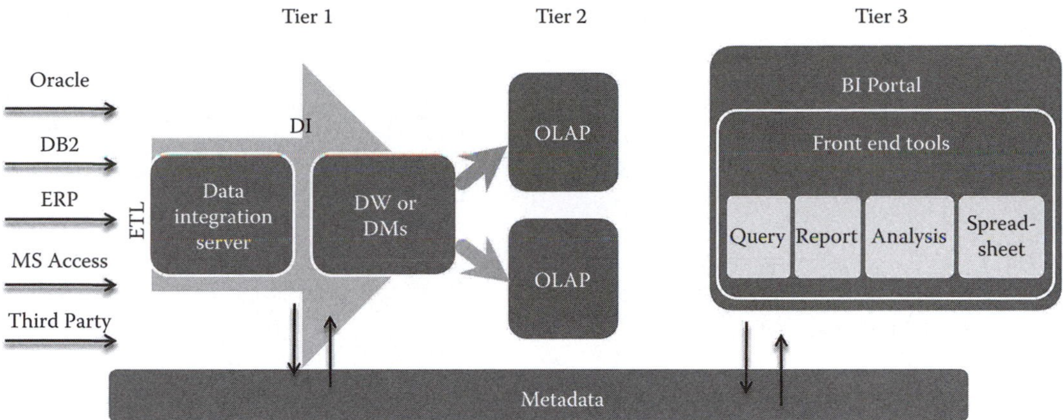

FIGURE 30.2 Steps of knowledge discovery process.

an incorrect designation—we are really *mining knowledge from data* (such as mining gold from rocks, which is referred to as gold mining, not rock mining). Probably the most correct naming is "knowledge mining from data." Because of its length, it never gained popularity, and applies similarly to expressions such as knowledge extraction, data archeology, and data dredging. Perhaps, the only other term that has gained popularity similar to data mining is knowledge discovery from data (KDD), data to knowledge (D2K), knowledge and data engineering, or the combination of data mining and knowledge discovery [FD2K fi,IBM kdd,UCI D2K,ELSE DKE,IEEE TKDE,Springer DMKD,Babovic 01,Babovic 99].

The phases of the typical knowledge discovery process (Figure 30.2) can be described by three main phases. These phases are data integration (DI), OLAP, and front end, knowledge presentation tools. The first phase, data integration, entails data preprocessing such as data cleaning, integration, selection, and transformation. While data cleaning pertains to removal of noisy, inconsistent data, data integration refers to disparate data sources merging (from Oracle to flat Ascii). Data selection relates to task-applicable data retrieval, and transformation involves data conversion into a format adequate for the next step, data mining. This phase is popularly known as ETL (extraction, transformation, loading), and results in the creation of data marts (DM) and data warehouses (DW). Data warehouses are large data repositories composed of data marts. The final phase is front-end tools often referred to as business intelligence (BI) portal type tools.

Data mining represents tasks ranging from association rules and regression analysis, to various intelligent and machine learning techniques such are neuro-fuzzy systems, support vector machines, and Bayesian techniques. Through further pattern evaluation, the selection of knowledge-representative mined data instances is further visualized in the final, knowledge presentation phase. While data mining is clearly only a phase in the entire process of knowledge discovery, the data mining designation has been widely accepted as a synonym for the whole data to knowledge process [Han 06].

Throughout all of the phases of the knowledge discovery process, the constant reference to metadata is maintained. Metadata contains the data about the data such as data names, definitions, time stamping, missing fields, and structure of the data warehouses (schemas, dimensions, hierarchies, data definitions, mart locations).

30.3 OLAP versus OLTP

As the operational databases are being targeted at different use and applications, performance reasons imply that they are kept separately from data warehouses. Transaction-oriented OLTP systems are typically responsible for known operations, such as daily searching for particular records, higher

OLTP	OLAP
Customer-oriented Data: Current, detailed data Model: Entity-Relationship (ER) Design: Appl.-oriented View: no historical or various organization Access: atomic trans. with concurrency control and recovery mechanisms	Market-oriented Data: Historical, summarized/aggreg. Model: Star/Snowflake model Design: Subject-oriented View: evolutionary, integrated diff. organizations, volumes Access: complex, read-mostly over historical WH

FIGURE 30.3 OLTP versus OLAP systems.

performance, and availability of flat relational, or entity-relationship (ER) types of transactions. These are online transaction and query processing operations, traditionally used for inventory or payroll (accounting type of processing).

Unlike the operational, OLTP database systems, data warehouses provide OLAP for an interactive, various granularity analysis of multidimensional data stored in n-dimensional data cubes. OLAP systems provide higher flexibility in data analysis and decision support capabilities, more complex than the operational processing offered by OLTP systems. The comparison between the two systems is given in Figure 30.3.

30.3.1 Data Cubes

With data warehouses and OLAP data analysis, the emphasis is put on multidimensional data cubes (DCs). DCs are multidimensional entities that offer a means of data modeling and processing. Data cubes are defined by dimension tables and fact tables. Dimension tables contain attributes that are entities used to organize and group data. Fact tables are numerical measures (quantities) of dimensions in dimension tables. For example, *Company1* would like to create a sales data warehouse with the following dimensions: *time, item, and state.* The fact table contains the measures (facts) of the sales data warehouse, as well as keys to each of the related dimension tables. For example, measures (facts) of the sales data warehouse could be *$SaleAmount,* and *SoldUnits.* A 3D table and 3D DC representation is given in Figures 30.4 and 30.5.

	Trading web site = Idaho				Trading web site = Wyoming				Trading web site = Montana			
	Item				Item				Item			
Time	Auto parts	Motorcycle parts	Farm equipment parts	Racing equipment parts	Auto parts	Motorcycle parts	Farm equipment parts	Racing equipment parts	Auto parts	Motorcycle parts	Farm equipment parts	Racing equipment parts
Q1	230	32	79	20	268	41	32	18	240	34	44	29
Q2	220	119	420	30	198	115	445	28	183	109	462	37
Q3	314	146	410	28	325	152	429	32	301	148	458	35
Q4	297	15	50	15	276	22	48	14	251	19	68	16

FIGURE 30.4 3D view of web based sales.

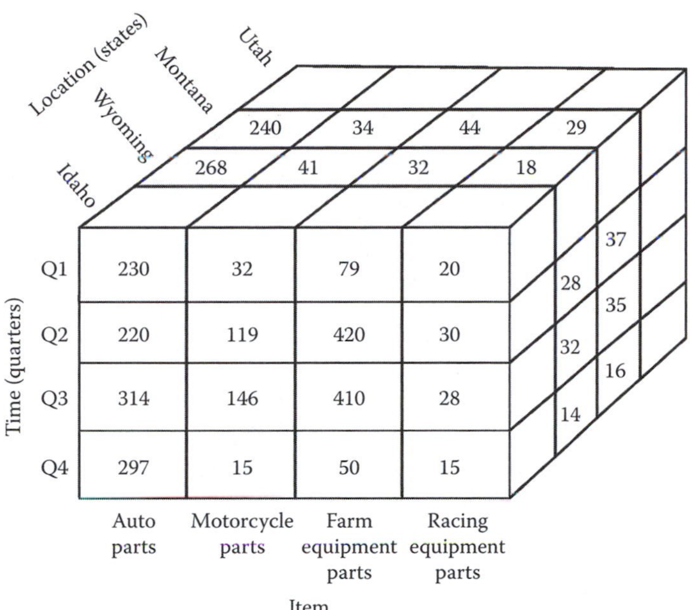

FIGURE 30.5 3D data cube web based sales.

Naturally, an n-dimensional DC can be composed of a series of $(n - 1)$-dimensional DCs:

$$DC^n := \text{series of } DC^{n-1} \tag{30.1}$$

or, as illustrated by Figure 30.6:

In a 4D cube, each possible subset of four dimensions will create a DC called a cuboid. All possible cuboids form a lattice of cuboids, each representing different levels of summarization, or "*group by*" SQL command (Figure 30.7).

The cuboid on top (apex, 0D cuboid, all), represents the apex cuboid, i.e., summarization over all four dimensions (highest summarization level). The cuboid at the bottom of the lattice (Figure 30.7) represents the base cuboid, i.e., a 4D cuboid for the four given dimensions (lower summarization level). Figure 30.8 illustrates various aggregations of quarters, parts, states, and finally by all.

30.3.2 OLAP Techniques on Data Cubes

Data warehouses (DWs) represent repositories of heterogeneous, disparate data sources (for instance MS Access, Excel, SQL, mySQL), stored at a single site under unified schema. Data warehouses result from data preprocessing (data cleaning, integration, selection, and transformation) tasks with the addition of data loading and refreshing. The physical structure of DWs is typically comprised of either relational DBs or multidimensional DCs provide a multidimensional view of stored data and enable various OLAP techniques for data analysis, such as drill-down, roll-up (drill-up), slice & dice, pivot (rotate), and others (drill-across, drill-through, ranking the top/bottom N items, computing moving averages and growths rates and depreciation, etc.).

Consider the example of web sales of different items (auto, motorcycle, farm equipment, and racing equipment parts). Imagine the web sale sites are located in the following states: Idaho, Wyoming, Montana, and Utah. Figure 30.9 illustrates this scenario, where web sales of auto parts by quarters recorded by the Idaho server were 230, 220, 314, and 297. This scenario will be used later to clarify basic OLAP techniques.

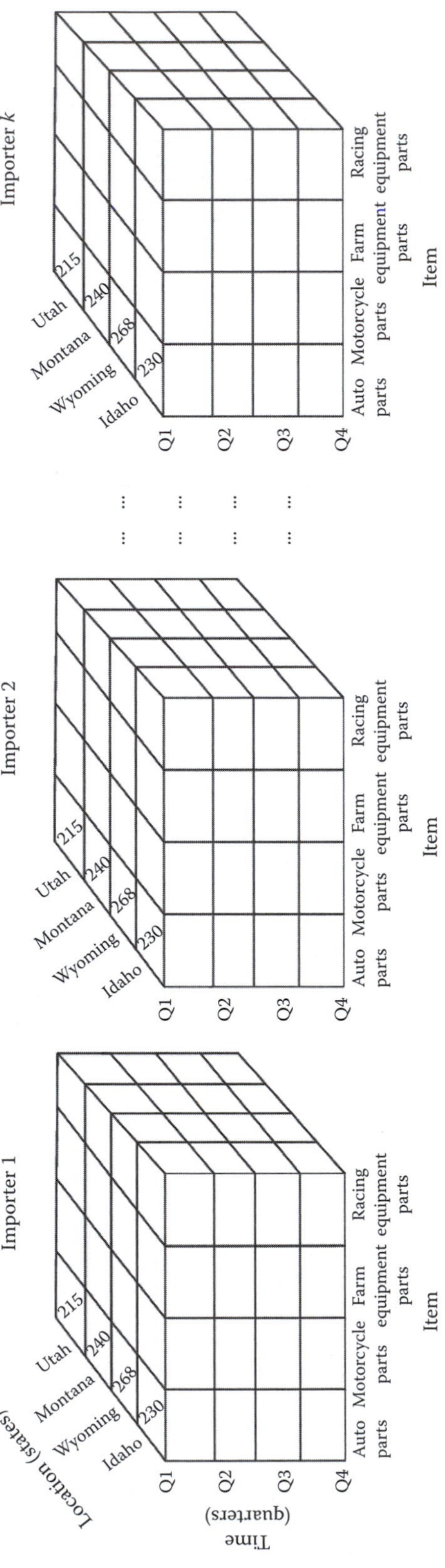

FIGURE 30.6 4D data cube composed of 3D data cubes.

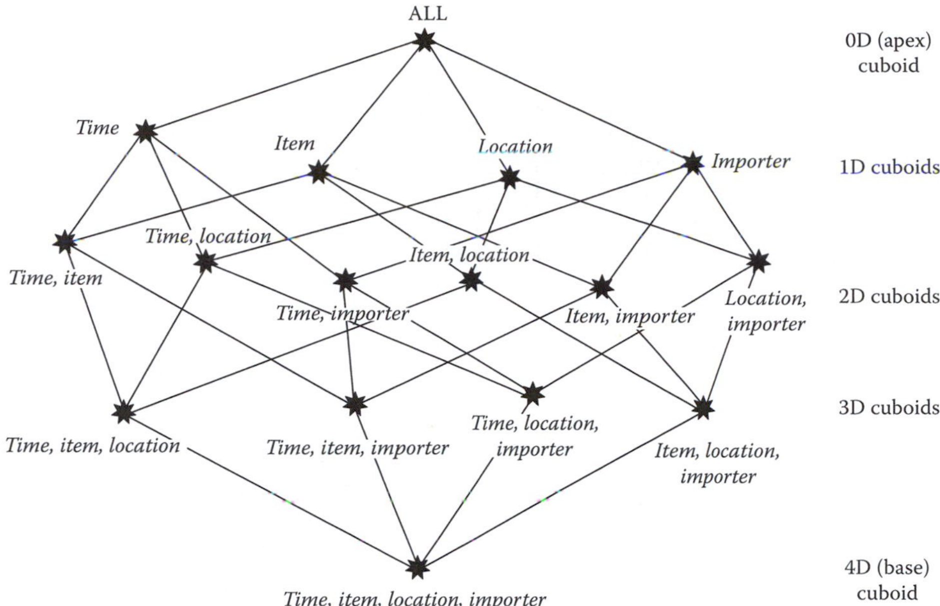

FIGURE 30.7 Lattice of cuboids, based on the 4D data in previous figure.

The roll-up (or drill-up) technique assumes mapping from low to high-level concepts (for example, aggregation or reduction in dimensions from quarters to years). The opposite technique, roll-down (or drill-down) assumes dimension expansion (for example, from quarters to months). Dicing assumes the technique of extracting sub-cubes of certain dimensions from original data cube. Dicing is illustrated in Figure 30.9, upper right corner (time = Q2 or Q3, location = Idaho or Wyoming and item = Auto Parts or Racing Parts). The technique of slicing represents a selection of one dimension, while pivoting simply rotates the existing dimensions states and items, also illustrated in Figure 30.9 (lower right corner).

30.4 Data Repositories, Data Mining Tasks, and Data Mining Patterns

30.4.1 Data Repositories

Data mining in a general sense can be applied to various kinds of data. In the temporal sense, the data can be static or transient (data streams). Thus, data repositories use varies from flat files, transactional databases, to relational databases, data marts and data warehouses.

Transactional databases typically consist of flat files in a table format (one record, one transaction). Relational databases (database management systems or DBMSs), are comprised of tables with unique names with columns (attributes, or fields), and rows (records, or tuples). A unique key described by a set of attributes then identifies each tuple. The semantic, entity-relationship (ER) data model then describes the database via a set of entities with their relationships.

30.4.2 Data Mining Tasks

In principle, data mining tasks can be categorized as *descriptive* (characterizing the general properties of DB data) and *predictive* (specific inference on data to enable predictions). Data on the other hand can be associated with classes (concepts), i.e., class/concept descriptions. These descriptions can be derived via data characterization (summarizing the general characteristics of the

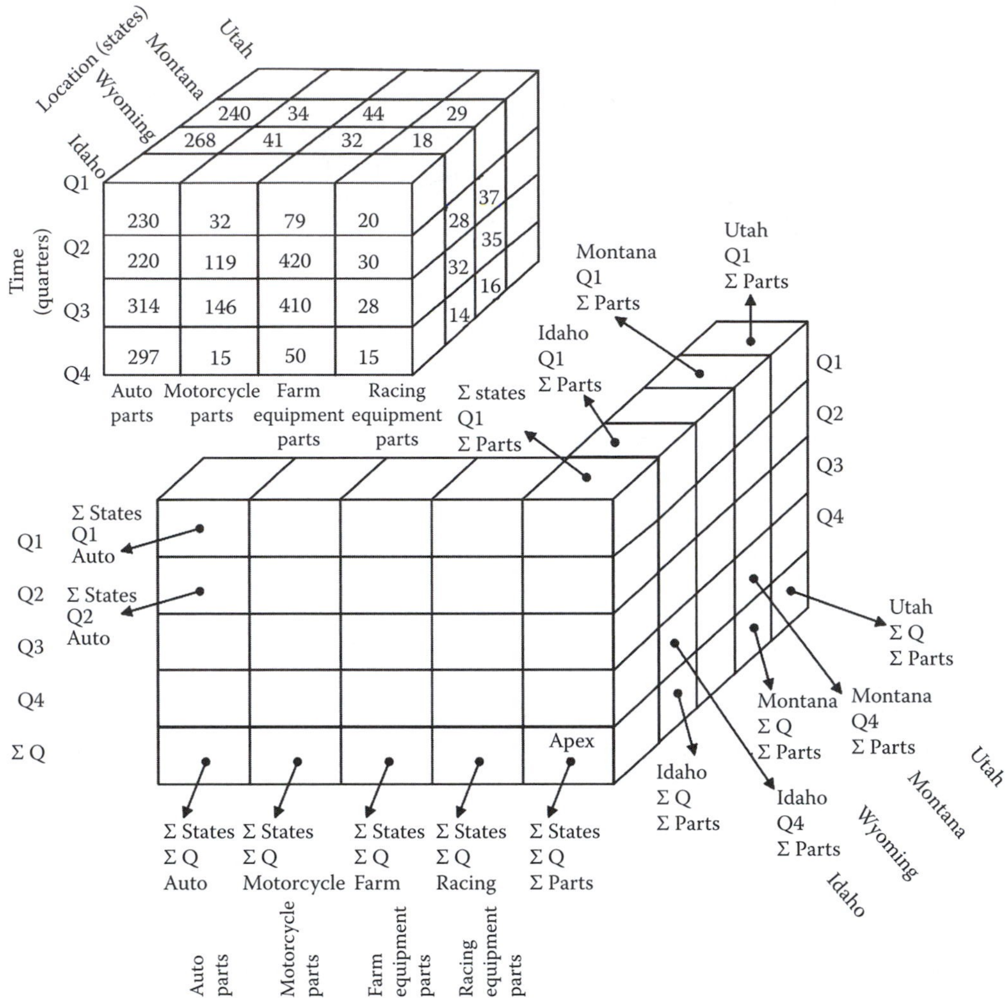

FIGURE 30.8 Sample data cube (above) and various aggregations it.

class of data), or data discrimination (comparison of the one class with other, contrasting classes). Summarization and characterization can be done using various techniques. Some of the known summarization techniques are OLAP roll-up operation to achieve data summarization along a specific dimension (Figure 30.9), statistical measures and plots, attribute-oriented induction. Data characterization techniques on the other hand can entail pie, bar, charts, *n*-dimensional DCs, multidimensional tables, and crosstabs.

An example of data characterization can be the task of summarizing the demographics of customers that spend about $25,000 on a new car every 3 years. This characterization may result in a profile of employed individuals who are 35–45 years old and who have credit ratings in preferable range. The system should allow a drill-down operation to obtain, for example, different sex, education level, or occupation type.

An example of data discrimination could be the comparison of a specific profile of convertible car customers in one state, say Florida, against a contrasting class of customers of the same type of vehicles from another state say Idaho. A further example can be discriminating the descriptions of the same profile of car customers today and 10 years ago.

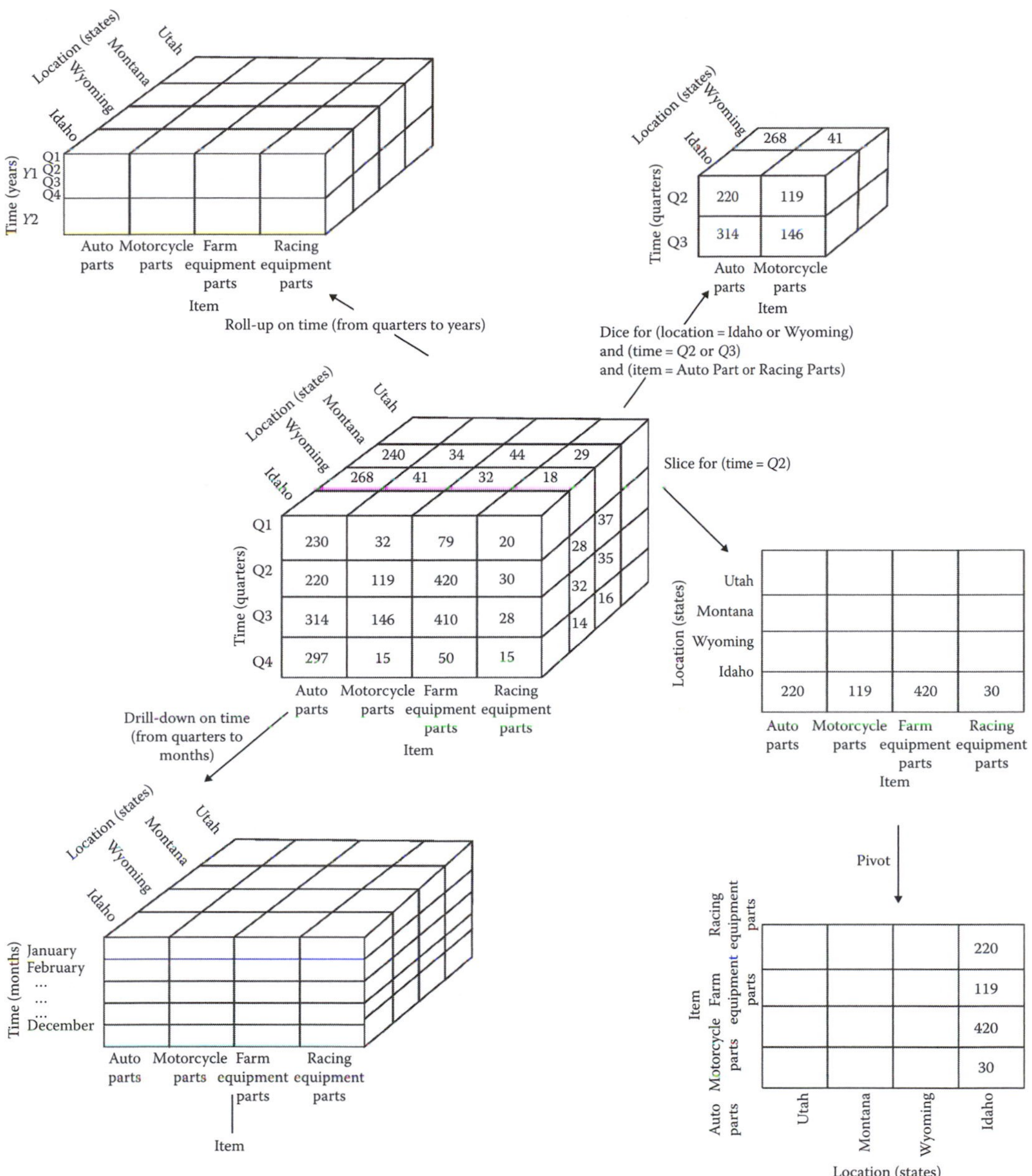

FIGURE 30.9 Basic OLAP techniques: roll-up, roll down, dice, slice, and pivot.

30.4.3 Data Mining Patterns

The frequent pattern recognition of data mining makes assumptions about identification of associations, correlations, classification, leading to description and prediction. For example, people who often deal with graphics or video processing applications may opt for certain type of Macintosh computers. Frequently, this purchase would be followed by an additional backup device such as Time Capsule. Younger gaming-oriented clientele will frequently follow a computer purchase with various gaming devices (joystick or gaming keyboard). These events are known as frequent sequential patterns. They can result in rules such as "80% of web designer professionals within a certain age and income group will buy Macintosh computers." In multidimensional databases, the previous statement would be recognized as a multidimensional association rule (dimensions here refer to "age," "income," and "buy," for example, with web designer professionals).

30.5 Data Mining Techniques

Numerous data mining techniques exist, ranging from rule-based, Bayesian belief networks, support vector machines; to artificial neural networks, *k*-nearest neighbor classifiers; to mining time series, graph mining, social networking, and multimedia (text, audio, video) data mining. A few select techniques will be addressed here. (For more details, please refer to [Han 06,Berry 04,Thuraisingham 2003,Matignon 07,Witten 2005,Witten 2000].)

30.5.1 Regression Analysis

The term "regression" was introduced in the eighteenth century by Sir Francis Galton, cousin of C. Darwin, and had a pure biological connotation. It was known by regression toward the mean, where offspring of exceptional individuals tend on average to be less exceptional than their parents and closer to their more distant ancestors.

Regression analysis models the predictor–response relationship between independent variables (known values, predictor), and dependent variables (responses, values to predict). By the use of regression, curve fitting can be done as generalized linear, Poisson regression, log-linear, regression trees, least square, spline, or fractal.

Regression can be linear (curve fitting to a line), or nonlinear (in closed-form or iterative such as steepest descent, Newton method, or Levenberg-Marquardt). Also, regression can be parametric where the regression function is defined by unknown parameters (LSM, least square method), or by nonparametric (functions such as polynomial regression). Fuzzy regression (fuzzy linear least square regression) can be used to address the phenomenon of data uncertainty driving the solution uncertainty (nonparametric regression) [Roychowdhury 98]. For example, trying to perform linear regression over the function y_i:

$$f(x_i) = \hat{y}_i; \quad \hat{y}_i \approx y_i, \quad \hat{y}_i = y_i + err_i, \quad y_i = w_1 x_i + w_0 \tag{30.2}$$

can be reduced to finding w_1 and w_0 by minimizing the sum of the squared residuals (SSR), or setting the derivatives of SSR with regard to both variables w_1 and w_0 to zero:

$$\begin{gathered} \text{SSR} = \sum_{i=1}^{n} \left(\hat{y}_i - (w_1 x_i + w_0) \right)^2 \\ \frac{\partial \text{SSR}}{\partial w_0} = 0; \quad \frac{\partial \text{SSR}}{\partial w_1} = 0 \end{gathered} \tag{30.3}$$

30.5.2 Decision Trees

Classification rules can easily substitute decision trees (*if buyer = WebDesigner => computer = Macintosh*) [Han 06,Witten 2005,Witten 2000]. These rules may have exceptions (*if buyer = WebDesigner except if buyerPreference = Windows => computer = Macintosh*). Association rules, in addition, can demonstrate how strong an association exists between frequently occurring attribute–value pairs, or items based on frequent itemset mining. For example:

if buyer = WebDesigner and buyer - GraphicsUser => computer = Macintosh [*support* 70%, *confidence* = 95%]

Here, support or coverage of 70% means that out of the customers under study, 70% of the buyers that are web designers and graphic users have purchased Macintosh computers (correct prediction vs. proportion of instances to which the rule applies). Accuracy or confidence in the 95% represents the probability that a customer of this type will purchase a Macintosh computer.

30.5.3 Neural Networks

Artificial neural networks (ANNs) can be used as universal approximators and classifiers. Artificial neurons represent summation, threshold elements. When weighted (w_i), the sum of the neuron inputs (x_i) exceeds the defined threshold value, the neuron then produces an *output*, i.e. "fires":

$$\text{net} = \sum_{i=1}^{n} w_i x_i + w_{n+1}, \quad \text{out} = \begin{cases} 1 & \text{if net} \geq 0 \\ 0 & \text{if net} < 0 \end{cases} \tag{30.4}$$

The weight (w_{i+1}) with default input +1 is called the bias, and can be understood as the threshold (T), but with the opposite sign (Figure 30.10).

Typically used threshold functions are bipolar sinusoidal threshold functions as described in the following equation:

$$o_{\text{bip}} = f_{\text{bip}}(k \cdot \text{net}) = \tanh(k \cdot \text{net}) = \frac{2}{1 + \exp(-2 \cdot k \cdot \text{net})} - 1 \tag{30.5}$$

The graphical representation of a single neuron operation can be described easily via analytic geometry. Thus, a single, two input neuron as illustrated by Figure 30.11, represents a linear classifier where the neuron definition is $1x + 3y - 3 > 0$. The weights used for inputs x and y are 1 and 3, respectively. The neuron divides the xOy space into two areas by selecting the upper one. This neuron correctly classifies the rectangular pattern producing the output +1 identical to the desired output, +1 (point 2,1). Further, by deselecting the lower part of xOy space, the neuron produces −1 on the output, again matching the desired output for the pattern (0.5, 0.5).

One of the most used ANN algorithms is error back propagation (EBP), proposed by Werbos in 1994 and Rumelhart in 1986 [Werbos 94,Rumelhart 86]. Other popular algorithms include modification of EBP (Quickprop, RPROP, Delta-Bar-Delta, Back Percolation), Levenberg-Marquardt, adaptive resonance theory (ART), counterpropagation networks (CPN), and cascade correlation networks.

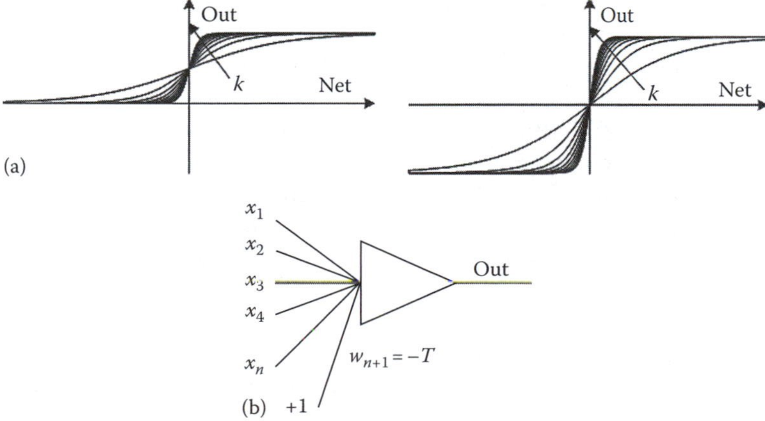

FIGURE 30.10 (a) Typically used threshold functions: unipolar and bipolar sigmoidal and (b) artificial neuron as weighted threshold element.

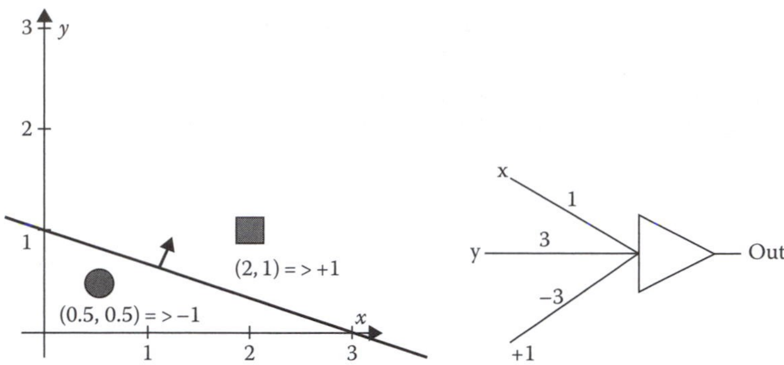

FIGURE 30.11 Graphical representation of single neuron operation.

30.6 Multidimensional Database Schemas

Operational databases are based on entity-relationship (E-R) data models or schemas, which describe the set of entities and relationships among them. Data warehouse schemas reflect the subject-oriented schemas, more suitable for OLAP. Three typical OLAP schemas are star, snowflake, and fact constellation.

A star schema is composed of the fact (central) table with keys to (four) dimension tables (redundancies are possible). A snowflake schema is composed of centralized fact tables connected to dimensions, hence resembling a snowflake (E-R relationship). The snowflake schema is easier to maintain, less effective in browsing (some dimension tables are normalized, no redundancies). A collection of stars produces fact constellation (two fact tables) schema (dimension tables are now shared among fact tables).

30.7 Mining Multimedia Data

Data mining of text, web, and other multimedia type data has experienced constant growth in the recent decade. The applications vary from research, scientific, social networking, to governmental, business, and marketing.

One of the famous algorithms developed specifically for the World Wide Web, but also generally applicable to search and structural analysis is the hyperlink-induced topic search (HITS) algorithm. The HITS algorithm was introduced by Jon Kleinberg from Cornell University while a visiting scientist in the CLEVER project at IBM's Almaden Research Laboratory [Kleinberg 98].

The HITS algorithm is also known as *hubs* and *authorities* and represents the distillation of broad search topics via authorities and hubs joined in the link structure. User queries present various obstacles. For example, specific queries are often associated with scarcity problem—answers are hard to find, while broad topic queries are often burdened by an abundance problem, resulting in too many hits for humans to digest. And, a "filter" idea based on the notion of authoritative pages. However, one may quickly realize that purely endogenous measures can be hard to establish. Kleinberg mentions several examples such as the term "Harvard" is not necessarily a word often used at www.harvard.edu; the term "search engines" may not be necessarily used on many of the natural authorities (Yahoo, AltaVista); or Honda or Toyota may not be using terms "auto manufacturer" on their pages [Kleinberg 98].

The solution Kleinberg pointed out lies in the analysis of the link structure where the creator of page p links to page q and confers the authority q. Collective endorsements in this way solve the problem of non self-descriptive pages. Hubs, set of pages that link to authorities, represent the mutual reinforcement relationship that facilitate automated discovery. The mutual reinforcement means that a good hub is a page that points to many good authorities, but also that a good authority is a page that is pointed to by many good hubs. This approach is not without risks (paid endorsements, commercial competitors deliberately omitting words).

The HITS algorithm defines an eigenvector of adjacency matrices associated with the link graph (weighted links among authorities and hubs). In this way, the algorithm produces a list of hubs and authorities with large weights, regardless of the weights initialization.

30.8 Accuracy Estimation and Improvement Techniques

Regardless of the specific data mining task, it is beneficial to estimate the accuracy of the process executed and also be able to improve the accuracy achieved. Although these statistical analysis techniques are presented here in light of data mining tasks, they can also be applied for general accuracy estimation and improvement of training-testing methods such as classification and prediction.

30.8.1 Accuracy Measures

The holdout method assumes partitioning of a data set into two sets, training 2/3, and testing 1/3, where testing is executed after the training process of a classifier has been completed. Random sampling assumes k repetitions of holdout method, with the average being the overall accuracy. *K-fold* cross-validation assumes splitting of initial data set into nearly equal size k subsets (folds), with the training and testing being executed k times (i.e., each fold is used for testing of a system trained on remaining $k - 1$ folds). Typically, the 10-fold approach is used. "Bootstrap," introduced by Efron in 1979, is a concept similar to pulling yourself up by your own bootstrap and is a resampling technique that assumes uniform selection of training patterns with repetition (one pattern has the same probability of being trained upon, comprising a virtual training population) [Efron 79]. Every resample consists of the same number of observations; bootstrap then can model the impact of the actual sample size [Fan 96]. One of the most popular approaches is the 0.632 bootstrap method in which, as it turns out, 63.2% of the original data set of d tuples will end up in bootstrap with 36.8%, forming the test data set [Efron 97]. This idea comes from the $1/d$ probability of a sample tuple being selected from a data set of d tuples:

$$\lim_{d\to\infty}\left(1-\frac{1}{d}\right)^{d} = e^{-1} \approx 0.368 \tag{30.6}$$

with the accuracy estimated on both training and test sets:

$$\text{Accuracy}(M) = \sum_{i=1}^{k}\left(0.632\cdot\text{Accuracy}(M_i)_{\text{test\_set}} + 0.368\cdot\text{Accuracy}(M_i)_{\text{train\_set}}\right) \tag{30.7}$$

30.8.2 Accuracy Improvement

Once the accuracy of the data mining task at hand is estimated, techniques for accuracy improvement can be applied. Commonly used techniques are known as bagging and boosting.

Bagging, known as bootstrap aggregation, is based on the number of bootstrap samples with patterns sampled with replacements. The algorithm returns its prediction (vote) when presented with a previously unseen pattern. The bagged algorithm (classifier for example) is based on majority of votes. The bagged classifier is therefore comprised of n classifiers, each trained on one bootstrap sample. The advantages of such classifiers are increased accuracy over the single model based on the initial set of patterns and robustness to noisy data.

Boosting assumes assigning weights relative to the "pattern's difficulty." The higher the weight, the more difficult it is to train on it. The weights are assigned after each of the n models is trained. The final model combines the votes of each of the n models, with the weight being the function of its accuracy.

One of the most often used boosting algorithms is the AdaBoost, introduced by Feund and Schapire in 1995 [Freund 97]. AdaBoost solved many of the practical difficulties of the earlier boosting algorithms [Freund 99].

30.9 Summary

Although already often recognized as the umbrella of vast array of data analysis techniques, data mining techniques will undoubtedly experience even further popularity with the perpetual growth of data, coupled with advances in data storage technology. While data mining underwent tremendous changes over the years, from early file processing systems, via hierarchical, relation databases, from customer (transaction-oriented) OLTP, via marketing (analytical oriented) OLAP, to multidimensional/hybrid MOLAP/HOLAP processing techniques, the future trends in data mining seem to be dependent on the development of sophisticated computational intelligence techniques for cluster analysis, trending, and prediction of various types of data (multimedia, web, stream, sequence, time-series, social networking, and others).

References

[Babovic 01] V. Babovic, M. Keijzer, D. R. Aguilera, and J. Harrington, Analysis of settling processes using genetic programming, D2K Technical Report 0501-1, http://www.d2k.dk, 2001.

[Babovic 99] V. Babovic and M. Keijzer, Data to knowledge—The new scientific paradigm, in *Water Industry Systems: Modelling and Optimisation Applications*, D. Savic and G. Walters (eds.), Research Studies Press: Exeter, U.K., 1999, pp. 3–14.

[Berry 04] J. A. Berry and G. S. Linoff, *Data Mining Techniques: For Marketing, Sales, and Customer Relationship Management*, 2nd edn., John Wiley & Sons Inc., New York, April 2004.

[Efron 79] B. Efron, Bootstrap methods: Another look at the jackknife. *The Annals of Statistics*, 7(1):1–26, 1979.

[Efron 97] B. Efron and R. Tibshirani, Improvements on cross-validation: The 632+ bootstrap method, *Journal of American Statistics Association*, 92(438):548–560, 1997.

[ELSE DKE] Elsevier, Data & Knowledge Engineering Jrnl, URL from Feb.09, http://www.elsevier.com/wps/find/journaldescription.cws_home/505608/description#description

[Fan 96] Fan, X. and Wang, L. Comparability of jackknife and bootstrap results: An investigation for a case of canonical correlation analysis. *Journal of Experimental Education*, 64:173–189, 1996.

[FD2K fi] From Data To Knowledge, Dept. of CS of Univ. of Helsinki, URL from Feb.09, http://www.cs.helsinki.fi/research/fdk/

[Freund 97] Y. Freund and R. E. Schapire, A decision-theoretic generalization of on-line learning and an application to boosting, *Journal of Computer and System Sciences*, 55(1):119–139, August 1997.

[Freund 99] Y. Freund and R. E. Schapire, A short introduction to boosting, Shannon Laboratory, *Journal of Japanese Society for Artificial Intelligence*, 14(5):771–780, September 1999.

[Han 06] J. Han and M. Kamber, *Data Mining: Concepts and Techniques*, 2nd edn., Elsevier Science Ltd. (The Morgan Kaufmann Series in Data Management Systems), San Francisco, CA, April 2006.

[Hitachi PR] Hitachi GST, URL from Feb.09, http://www.hitachigst.com/hdd/research/recording_head/pr/index.html

[Hitachi SC] Hitachi GST, URL from Feb.09, http://www.hitachigst.com/hdd/technolo/overview/storagetechchart.html

[Hoagland 03] A. S. Hoagland, History of magnetic disk storage based on perpendicular mage tic recording, *IEEE Transactions on Magnetics*, 39(4):1871–1875, July 2003.

[IBM kdd] IBM Knowledge Discovery & Data Mining, URL from Feb.09, http://domino.research.ibm.com/comm/research.nsf/pages/r.kdd.html

[IEEE TKDE] *IEEE Transactions on Knowledge & Data Engineering*, URL from Feb.09, http://www.computer.org/portal/site/transactions/menuitem.a66ec5ba52117764cfe79d108bcd45f3/index.jsp?&pName=tkde_home&

[Kleinberg 98] J. Kleinberg, Authoritative sources in a hyperlinked environment. In *Proceedings of the Ninth Annual ACM-SIAM Symposium Discrete Algorithms*, pp. 668–677, ACM Press, New York, 1998.

[Linoff 2002] G. Linoff and M. J. A. Berry, *Mining the Web: Transforming Customer Data into Customer Value*, John Wiley & Sons Inc., New York, February 2002.

[LOC 98TB] Library of Congress, URL from Feb.09, http://www.loc.gov/webcapture/faq.html

[Matignon 07] R. Matignon, *Data Mining Using SAS Enterprise Miner*, (*Wiley Series in Computational Statistics*), Wiley-Interscience, New York, August 2007.

[NAB 580TB] The National Archives of Britain, On July 4th, 2007 press release, URL from Feb.09, http://www.nationalarchives.gov.uk/news/stories/164.htm

[Roychowdhury 98] S. Roychowdhury, Fuzzy curve fitting using least square principles, *Computational Cybernetics, 29 Soft Computing*, San Diego, CA, 1998.

[Rumelhart 86] D. E. Rumelhart and J. L. McClelland, *Parallel Distributed Processing: Explorations in the Microstructure of Cognition*, Vol. 1, Cambridge, MA, MIT Press, 1986.

[Springer DMKD] Springer, Jrnl. On Data Mining and Knowledge Discovery, URL from Feb.09, http://www.springer.com/computer/database+management+&+information+retrieval/journal/10618

[Thuraisingham 2003] B. Thuraisingham, *Web Data Mining and Applications in Business Intelligence and Counter-Terrorism*, CRC Press, Boca Raton, FL, June 2003.

[UCI D2K] Office of Technology Management at Urbana-Champaign, D2K—Data to Knowledge tool, URL from Feb.09, http://www.otm.illinois.edu/node/229

[Werbos 94] Paul John, W., *The Roots of Backpropagation: From Ordered Derivatives to Neural Networks and Political Forecasting*, New York, Wiley-Interscience, January 1994.

[Witten 2000] I. H. Witten and E. Frank, *Data Mining: Practical Machine Learning Tools and Techniques with Java Implementations*, Morgan Kaufmann Publishers, San Francisco, CA, ISBN: 1-55860-552-5, 2000.

[Witten 2005] I. H. Witten and E. Frank, *Data Mining: Practical Machine Learning Tools and Techniques*, 2nd edn., Morgan Kaufmann Series in Data Management Systems, San Francisco, CA, 2005.

31

Autonomous Mental Development

Juyang Weng
Michigan State University

Artificial neural networks perform signal processing and they learn. However, they cannot autonomously learn and develop like a brain. Autonomous mental development models all or part of the brain and how a system develops autonomously through interactions with the environments. The most fundamental difference between traditional machine leaning and autonomous mental development is that a developmental program is task nonspecific so that it can autonomously generate internal representations for a wide variety of simple to complex tasks. This chapter first discusses why autonomous development is necessary based on a concept called task muddiness. No traditional methods can perform muddy tasks. If the electronic system that you design is meant to perform a muddy task, you need to enable it to develop its own mind. Then some basic concepts of autonomous development are explained, including the paradigm for autonomous development, mental architectures, developmental algorithm, a refined classification of types of machine learning, spatial complexity, and time complexity. Finally, the architecture of spatiotemporal machine that is capable of autonomous development is described.

31.1 Biological Development

A human being starts to develop from the time of conception. At that time, a single cell called a zygote is formed. In biology, the term *genotype* refers to all or part of the genetic constitution of an organism. The term *phenotype* refers to all or part of the visible properties of an organism that are produced through the interaction between the genotype and the environment. In the zygote, all the genetic constitution is called genome, which mostly resides in the nucleus of a cell. At the conception of a new human life, a biological program called the *developmental program* starts to run. The code of this program is the genome, but this program needs the entire cell as well as the cell's environment to run properly.

The biological developmental program handles two types of development, body development and mental development. The former is the development of everything in the body excluding the brain.

The latter is the development of the brain (or the central nervous system, CNS). Through the body development, a normal child grows in size and weight, along with many other physical changes. Through the mental development, a normal child develops a series of mental capabilities through interactions with the environment. Mental capabilities refer to all known brain capabilities, which include, but not limited to, perceptual, cognitive, behavioral, and motivational capabilities. In this chapter, the term development refers to mental development unless stated otherwise. The biological mental development takes place in concurrence with the body development and they are closely related. For example, if the eyes are not normally developed, the development of the visual capabilities is greatly affected. In the development of an artificial agent, the body can be designed and fixed (not autonomously developed), which helps to reduce the complexity of the autonomous mental development.

The genomic equivalence principle [36] is a very important biological concept for us to understand how biological development is regulated. This principle states that the set of genes in the nucleus of every cell (not only that in the zygote) is functionally complete—sufficient to regulate the development from a single cell into an entire adult life. This principle is dramatically demonstrated by cloning. This means that there are no genes that are devoted to more than one cell as a whole. Therefore, development guided by the genome is cell-centered. Carrying a complete set of genes and acting as an autonomous machine, every cell must handle its own learning while interacting with its external environment (e.g., other cells). Inside the brain, every neuron develops and learns in place. It does not need any dedicated learner outside the neuron. For example, it does not need an extracellular learner to compute the covariance matrix (or any other moment matrix or partial derivatives) of its input lines and store extracelullarly. If an artificial developmental program develops every artificial neuron based on only information that is available to the neuron itself (e.g., the cellular environment such as presynaptic activities, the developmental program inside the cell, and other information that can be biologically stored intracellularly), we call this type of learning in-place learning.

This in-place concept is more restrictive than a common concept called "local learning." For example, a local learning algorithm may require the computation of the covariance matrix of the presynaptic vector, which must store extracellularly. In electronics, the in-place learning principle can greatly reduce the required electronics and storage space, in addition to the biological plausibility. For example, suppose that every biological neuron requires the partial derivative matrix of its presynaptic vector. As the average number of synapses of a neuron in the brain is on the order of $n = 1000$. Each neuron requires about $n^2 = 1{,}000{,}000$ storage units outside every neuron. This corresponds to about 1,000,000 of the total number of synapses (10^{14}) in the brain!

Conceptually, the fate and function of a neuron is not determined by a "hand-designed" (i.e., genome specified) meaning of the external environment. This is another consequence of the genomic equivalence principle. The genome in each cell regulates the cell's mitosis, differentiation, migration, branching, and connections, but it does not regulate the meaning of what the cell does when it receives signals from other connected cells. For example, we can find a V1 cell (neuron) that responds to an edge of a particular orientation. This is just a facet of many emergent properties of the cell that are consequences of the cell's own biological properties and the activities of its environment. A developmental program does not need to, and should not, specify which neuron detects a pre-specified feature type (such as an edge or motion).

31.2 Why Autonomous Mental Development?

One can see that biological development is very "low level," regulating only individual neurons. Then, why is it necessary to enable our complex electronic machines to develop autonomously? Why do we not design high-level concepts into the machines and enable them to carry out our high-level directives? In fact, this is exactly what many symbolic methods have been doing for many years. Unfortunately, the resulting machines are brittle—they fail miserably in real world when the environment fall out of the domains that have been modeled by the programmer.

To appreciate what are faced by a machine to carry out a complex task, Weng [48] introduced a concept called *task muddiness*. The composite muddiness of a task is a multiplicative product of many individual muddiness measures. There are many possible individual muddiness measures. Those individual muddiness measures are not necessarily mutually independent or at the same level of abstraction, since such a requirement is not practical nor necessary for describing the muddiness of a task. They fall into five categories: (1) external environment, (2) input, (3) internal environment, (4) output, and (5) goal, as shown in Table 31.1. The term "external" means external with respect to the brain and "internal" means internal to the brain.

The composite muddiness of a task can be considered as a product of all individual muddiness measures. In other words, a task is extremely muddy when all the five categories have a high measure. A chess playing task with symbolic input and output is a clean problem because it is low in categories (1) through (5). A symbolic language translation problem is low in (1), (2), and (4), moderate in (3) but high in (5). A vision-guided navigation task for natural human environment is high in (1), (2), (3), and (5), but moderate in (4). A human adult handles extremely muddy tasks that are high in all the five categories.

From the muddiness table (Table 31.1), we have a more detailed appreciation what a human adult deals with even in a daily task, e.g., navigating or driving in a city environment. The composite muddiness of many tasks that a human or a machine can execute is proposed by Weng [48] as a metric for measuring required intelligence.

TABLE 31.1 List of Muddiness Factors for a Task

Category	Factor	Clean	Muddy
External environment	Awareness	Known	Unknown
	Complexity	Simple	Complex
	Controlledness	Controlled	Uncontrolled
	Variation	Fixed	Changing
	Foreseeability	Foreseeable	Nonforeseeable
Input	Rawness	Symbolic	Real sensor
	Size	Small	Large
	Background	None	Complex
	Variation	Simple	Complex
	Occlusion	None	Severe
	Activeness	Passive	Active
	Modality	Simple	Complex
	Multi-modality	Single	Multiple
Internal environment	Size	Small	Large
	Representation	Given	Not given
	Observability	Observable	Unobservable
	Imposability	Imposable	Nonimposable
	Time coverage	Simple	Complex
Output	Terminalness	Low	High
	Size	Small	Large
	Modality	Simple	Complex
	Multimodality	Single	Multiple
Goal	Richness	Low	High
	Variability	Fixed	Variable
	Availability	Given	Unknown
	Telling-mode	Text	Multimodal
	Conveying-mode	Simple	Complex

A human infant is not able to perform those muddy tasks that a human adult performs everyday. The process of mental development is necessary to develop such a wide array of mental skills. Much evidence in developmental psychology has demonstrated that not only a process of development is necessary for human intelligence, the environment of the development is also critical for normal development.

Likewise, it is not practical for a human programmer to program a machine to successfully execute a muddy task. Computers have done very well for clean tasks, such as playing chess games. But they have done poorly in performing muddy tasks, such as visual and language understanding. Enabling a machine to autonomously develop task skills in its real task environments is the only approach that has been proved successful for muddy tasks—no existing higher intelligence for muddy tasks is not developed autonomously.

31.3 Paradigm of Autonomous Development

By definition, an agent is something that senses and acts. A robot is an agent, so is a human. In the early days of artificial intelligence, smart systems that caught the general public's imagination were programmed by a set of task-specific rules. The field of artificial intelligence moved beyond that early stage when it started the trend of studying general agent methodology [40], although the agent is still a task-specific machine.

As far as we know, Cresceptron 1993 [50,51] was the first developmental model for visual learning from complex natural backgrounds. By developmental, we mean that the internal representation is fully emergent from interactions with environment, without allowing a human to manually instantiate a task-specific representation. By the mid-1990s, connectionists had started the exploration of the challenging domain of development [8,29,37].

Due to a lack of the breadth and depth of the multidisciplinary knowledge in the single mind of a researcher or a reviewer, there have been various doubts from domain experts, mainly due to the widespread lack of sufficient cross-disciplinary knowledge discussed above. Examples include the following assumptions: (1) Artificial intelligence does not need to follow the brain's way. (2) Modeling the human mind does not need to follow the brain's way. (3) Your commitment to understanding the brain is laudable but naive.

There is a lack of bylaws, guidelines, and due process that contain the negative effects of human nature that are well documented by Thomas Kuhn [25]. Such negative effects eroded "revolutionary advances" required by some programs. Serious overhauls and investments for the infrastructure for converging research on intelligence are urgently needed. Such infrastructure is necessary for the healthy development of science and technology in the modern time.

Not until the birth of the new AMD field marked by the NSF- and DARPA-funded Workshop on Development and Learning 2000 [54,55] has the concept of the task-nonspecific developmental program caught the attention of researchers. A hallmark difference between traditional artificial intelligence approaches and autonomous mental development [54] is the task specificity. All the existing approaches to artificial intelligence are task specific, except the developmental approach. Table 31.2 lists the major differences among existing approaches to artificial intelligence. An entry marked as "avoid modeling" means that the representation is emergent from experience.

TABLE 31.2 Comparison of Approaches to Artificial Intelligence

Approach	Species Architecture	World Knowledge	Agent Behaviors	Task Specific
Knowledge-based	Model	Model	Model	Yes
Learning-based	Model	Parametrically model	Model	Yes
Behavior-based	Model	Avoid modeling	Model	Yes
Genetic	Genetic search	Parametrically model	Model	Yes
Developmental	Parametrically model	Avoid modeling	Minimize modeling	No

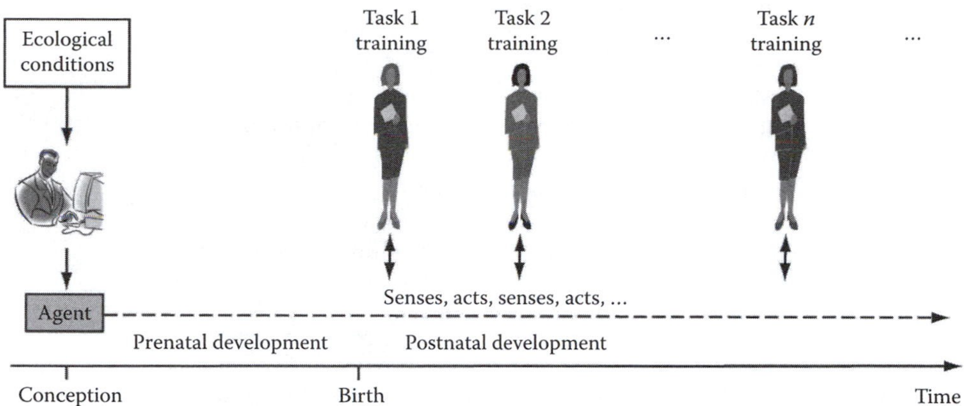

FIGURE 31.1 Illustration of the paradigm of developmental agents, inspired by human mental development. No task is given during the programming (i.e., conception) time, during which a general-purpose task-nonspecific developmental program is loaded onto the agent. Prenatal development is used for developing some initial processing pathways in the brain using spontaneous (internally generated) signals from sensors. After the birth, the agent starts to learn an open series of tasks through interactions with the physical world. The tasks that the agent learns are determined after the birth.

Traditionally, given a task to be executed by the machine, it is the human programmer who understands the task and, based on his understanding, designs a task-specific representation. Depending on different approaches, different techniques are used to produce the mapping from sensory inputs to effector outputs. The techniques used range from direct programming (knowledge-based approach), to learning the parameters (in the parametric model), to genetic search (genetic approach). Although genetic search is a powerful method, the chromosome representations used in artificial genetic search algorithms are task specific.

Using the developmental approach, the tasks that the robot (or human) ends up doing are unknown during the programming time (or conception time), as illustrated in Figure 31.1. The ecological conditions that the robot will operate under must be known, so that the programmer can design the body of the robot, including sensors and effectors, suited for the ecological conditions. The programmer may guess some typical tasks that the robot will learn to perform. However, world knowledge is not modeled and only a set of simple reflexes is allowed for the developmental program. During "prenatal" development, internally generated synthetic data can be used to develop the system before birth. For example, the retina may generate spontaneous signals to be used for the prenatal development of the visual pathway. At the "birth" time, the robot's power is turned on. The robot starts to interact with its environment, including its teachers, in real time. The tasks the robot learns are in the mind of its teachers. In order for the later learning to use the skills learned in early learning, a well-designed sequence of educational experience is an important practical issue.

31.4 Learning Types

In the machine learning literature, there have been widely accepted definitions of learning types, such as supervised, unsupervised, and reinforcement learning. However, these conventional definitions are too coarse to describe computational learning through autonomous development. For example, it is difficult to identify any type of learning that is completely unsupervised. Further, the traditional classification of animal learning models, such as classical conditioning and instrumental conditioning, is not sufficient to address computational considerations of every time instant of learning. A definition of a refined classification of learning types is necessary.

TABLE 31.3 Eight Types of Learning

Type (Binary)	Internal State	Effector	Biased Sensor
0 (000)	Autonomous	Autonomous	Communicative
1 (001)	Autonomous	Autonomous	Reinforcement
2 (010)	Autonomous	Imposed	Communicative
3 (011)	Autonomous	Imposed	Reinforcement
4 (100)	Imposable	Autonomous	Communicative
5 (101)	Imposable	Autonomous	Reinforcement
6 (110)	Imposable	Imposed	Communicative
7 (111)	Imposable	Imposed	Reinforcement

We use a variable i to indicate internal task-specific representation imposed by human programmer (called *internal-state imposed* $i = 1$) or not (called *internal-state autonomous* $i = 0$).

We use e to denote autonomy of effector. If the concerned effector is directly guided by the human teacher or other teaching mechanisms for the desired action, we call the situation *action imposed* ($e = 1$). Otherwise, the learning is effector autonomous ($e = 0$).

We need to distinguish the channels of reward (e.g., sweet and pain sensors) that are available at the birth time, and other channels of reward that are not ready to be used as reward at the birth time (e.g., auditory input "good" or "bad") but implies a value after a certain amount of development. We define (inborn) biased sensors:

If the machine has a predefined preference pattern to the signals from a sensor at the birth time, this sensor is an (inborn) biased sensor. Otherwise, it is an (inborn) unbiased sensor.

In fact, all the sensors become biased gradually through postnatal experience—the development of the value system. For example, the image of a flower does not give a newborn baby much reward, but the same image becomes pleasant to look at (high value) after the baby has gown up.

We use the third variable b to denote whether a biased sensor is used. If any biased sensor is activated (sensed) during the learning, we called the situation reinforcement ($b = 1$). Otherwise, the learning is called communicative ($b = 0$).

Using these three key factors, any type of learning can be represented by a 3-tuple (i, e, b), which contains three components i, e, and b, each of which can be either represented by 0 or 1. Thus, there are a total of eight different 3-tuples, representing a total of eight different learning types. If we consider *ieb* as three binary bits of the type index number of learning type, we have eight types of learning defined in Table 31.3. We can also name each type. For example, Type 0 is state-autonomous, effector-autonomous, communicative learning. Type 7 is state-imposable, effector-imposed, reinforcement learning, but it has not been included in the traditional definition of either supervised learning or reinforcement learning. However, this learning is useful when teaching a positive or negative lesson through supervision.

Using three key features, state-imposed, effector-imposed and reinforcement, eight learning types are defined. This refined definition is necessary to understanding various modes of developmental and nondevelopmental learning.

All learning types using a non-developmental learning method corresponding to Types 7 to 4, this is because the task-specific representation is at least partially handcrafted after the task is given. Autonomous mental development uses Types 0 to 3.

31.5 Developmental Mental Architectures

Weng [47] proposed a SASE model through which the agent can autonomously learn to think, while the thinking behavior is manifested as internal attention. Attention is a key to emergent intelligence.

31.5.1 Top-Down Attention Is Hard

Consider a car in a complex urban street environment. Attention and recognition are a pair of dual-feedback problems. Without attention, recognition cannot do well; recognition requires attended areas (e.g., the car area) for the further processing (e.g., to recognize the car). Without recognition, attention cannot do well; attention requires recognition for guidance of the next fixation (e.g., a possible car area).

31.5.1.1 Bottom-Up Attention

Studies in psychology, physiology, and neuroscience provided qualitative models for bottom-up attention, i.e., attention uses different properties of sensory inputs, e.g., color, shape, and illuminance to extract saliency. Several models of bottom-up attention have been published. The first explicit computational model of bottom-up attention was proposed by Koch and Ullman in 1985 [24], in which a "saliency map" is computed to encode stimuli saliency at every lactation in the visual scene. More recently, Itti and Koch et al. [17] integrated color, intensity, and orientation as basic features in multiple scales for attention control. An active-vision system, called NAVIS (neural active vision) by Backer et al., was proposed to conduct the visual attention selection in a dynamic visual scene [1]. Our SASE model to be discussed next indicates that saliency is not necessarily independent of learning: The top-down process in the previous time instant may affect the current bottom-up saliency.

31.5.1.2 Top-Down Attention

Volitional shifts of attention are also thought to be performed top-down, through spacial defined and feature-dependant controls. Olshausen et al. [33] proposed a model of how visual attention can be directed to address the position and scale invariance in object recognition, assuming that the position and size information is available from the top control. Tsotsos et al. [45] implemented a version of attention selection using a combination of a bottom-up feature extraction scheme and a top-down position selective tuning scheme. Rao and Ballard [39] described a pair of cooperating neural networks, to estimate object identity and object transformations, respectively. Schill et al. [42] presented a top-down, knowledge-based reasoning system with a low-level preprocessing where eye movement is to maximize the information about the scene. Deco and Rolls [5] wrote a model of object recognition that incorporates top-down attentional mechanisms on a hierarchically organized set of visual cortical areas. In the above studies, the model of Deco and Rolls [5] was probably the most biologically plausible, as it incorporates bottom-up and top-down flows into individual neuronal computation, but unfortunately the top-down connections were disabled during learning and no recognition performance data were reported.

In the Where-What Network 2 (WWN-2) experiment [18] discussed later, we found that the corresponding network that drops the L4-L2/3 laminar structure gave a recognition rate lower than 50%. In other words, a network that treats top-down connection similar to bottom-up connection (like a uniform liquid state machine [38]) is not likely to achieve an acceptable performance.

31.5.2 Motor Shapes Cortical Areas

On one hand, high-order (i.e., later) visual cortex of the adult brain includes functionally specific regions that preferentially respond to objects, faces, or places. For example, the fusiform face area (FFA) responds to face stimuli [11,21], and the parahippocampal place area (PPA) responds to place identity [2,7,32]. How does the brain accomplish this feat of localizing internal representation based on meaning? Why is such a representation necessary?

In the cerebral cortex, there is a dense web of anatomically prominent feedback (i.e., top-down) connections [9,20,22,23,35,41]. It has been reported that cortical feedback improves discrimination between figure and background and plays a role in attention and memory [12,14,44]. Do feedback connections perform attention? Furthermore, do feedback connections play a role in developing abstractive internal representation?

The computational roles of feedback connections in developing meaning-based internal representations have not been clarified in existing studies reviewed above. The self-abstractive architecture next indicates that in the cerebral cortex, each function layer (L4 and L2/3) is a state at this layer. We will show that, unlike the states in POMDP, HMM, Hopfield network and many others, the states in the self-abstractive architecture integrate information from bottom-up inputs (feature inputs), lateral inputs (collaborative context), and top-down inputs (abstract contexts) into a concise continuous vector representation, without the artificial boundaries of a symbolic representation.

31.5.3 Brain Scale: "Where" and "What" Pathways

Since the work of Ungerleider and Mishkin 1982 [30,46], a widely accepted description of visual cortical areas is illustrated in Figure 31.2 [9,33]. A ventral or "what" stream that runs from V1, to V2, V4, and IT areas TEO and TE computes properties of object identity such as shape and color. A dorsal or "where" stream that runs from V1, to V2, V3, MT, and the medial superior temporal areas MST, and on to the posterior parietal cortex (PP) computes properties of the location of the stimulus on the retina or with respect to the animal's head. Neurons in early visual areas have small spatial receptive fields (RFs) and code basic image features; neurons in later areas have large RFs and code abstract features such as behavioral relevance. Selective attention coordinates the activity of neurons to affect their competition and link distributed object representations to behaviors (e.g., see the review by Serences and Yantius [43]).

With the above rich, suggestive information from neuroscience, I propose that the development of the functions of the "where" and "what" pathways is largely due to the following:

1. Downstream motors. The motor ends of the dorsal pathway that perform position tasks (e.g., stretching an arm to reaching for an apple or a tool), and the motor ends of the ventral pathway that perform type classification and conceptual tasks (e.g., different limbic needs between a food and an enemy).
2. Top-down connections. The top-down connections from motor areas that shape the corresponding pathway representations.

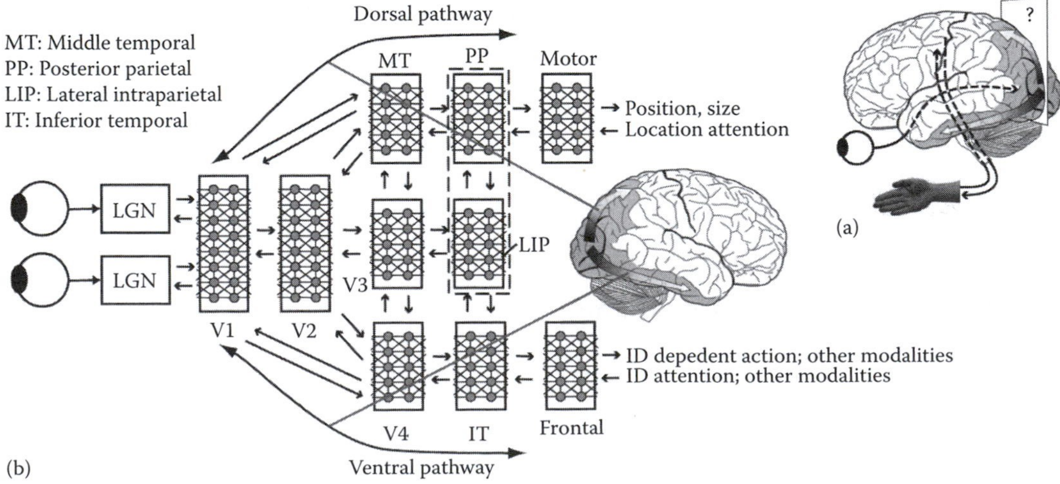

FIGURE 31.2 (a) How does the brain generate internal representation? The only external sources are sensors and effectors. The imaginary page slices the brain to "peek" into its internal representation. (b) The dorsal "where" pathway and the ventral "what" pathways. The nature of the processing along each pathway is shaped by not only sensory inputs but also the motor outputs.

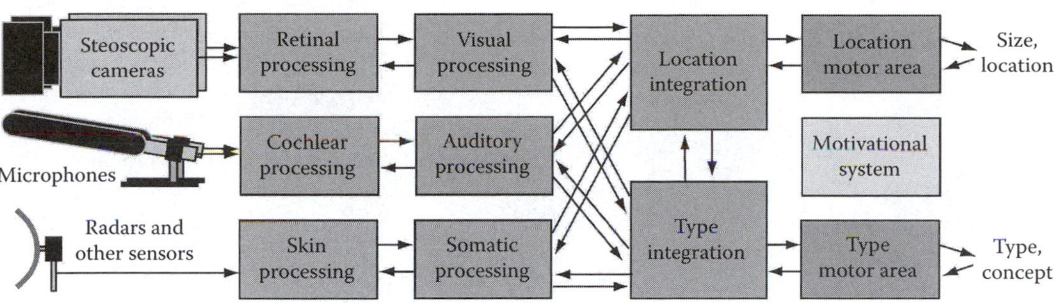

FIGURE 31.3 The system diagram: multi-sensory and multi-effector integration through learning.

Put in a short way, motor is abstract. Any meaning that can be communicated between humans is motorized: spoken, written, hand-signed, etc. Of course, "motor is abstract" does not mean that every stage of every motor action sequence is abstract. However, the sequences of motor actions provide statistically crucial information for the development of internal abstractive representation.

31.5.4 System

The system level architecture is illustrated in Figure 31.3.

An agent, either biological or artificial, can perform regression and classification.

Regression: The agent takes a vector as input (a set of receptors). For vision, the input vector corresponds to a retinal image. The output of the network corresponds to motor signals, with multiple components to be active (firing). The brain is a very complex spatiotemporal regressor.

Classification: The agent can perform classification before it has developed sophisticated human language capability to verbally tell us the name of a class. For example, each neuron in the output layer corresponds to a different class.

31.5.4.1 Two Signal Sources: Sensor and Motor

The brain faces a major challenge as shown in Figure 31.2a. It does not have the luxury of having a human teacher to implant symbols into it, as the brain is not accessible directly to the external human teacher. Thus, it must generate internal representations from the two signal sources: the sensors and the effectors (motors). This challenging goal is accomplished by the brain's where-what networks schematically illustrated in Figure 31.4. The system has two motor areas, the where motor that indicates where the attended object is and the what motor that tells what the attended object is. This specialization of each pathway makes computation of internal representation more effective.

31.5.5 Pathway Scale: Bottom-Up and Top-Down

It is known that cortical regions are typically interconnected in both directions [4,9,57]. However, computational models that incorporate both bottom-up and top-down connections have resisted full analysis [3,5,10,16,19,26,31]. The computational model, illustrated in Figure 31.5, provides further details about how each functional level in cortex takes inputs from the bottom-up signal representation space X and top-down signal representation space Z to generate and update self-organized cortical *bridge representation* space Y. This model further computationally predicts that a primary reason for the dorsal and ventral pathways to be able to deal with "where" and "what" (or achieving identity and positional invariances [19]), respectively, is that they receive top-down signals that drive their motors.

From where does the forebrain receive teaching signals that supervise its motors? Such supervised-motor signals can be generated either externally (e.g., a child passively learns writing while his teacher

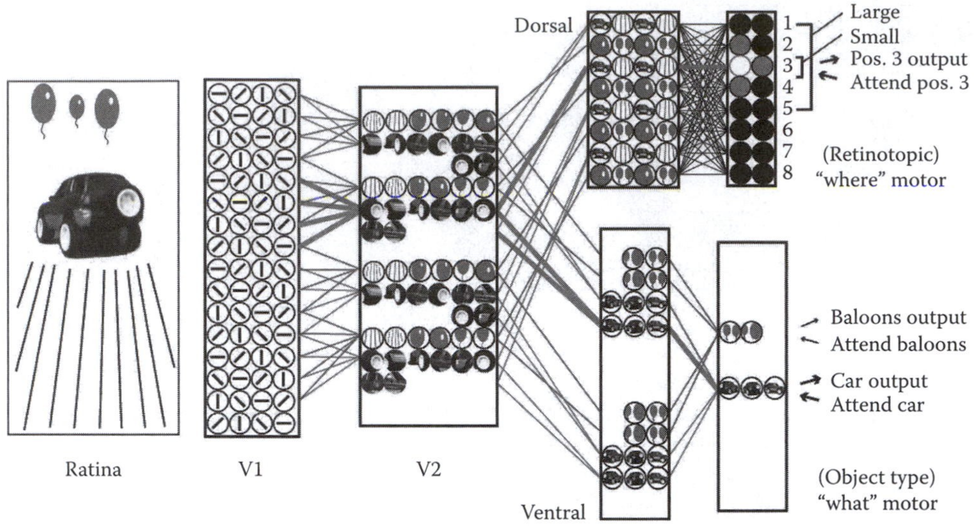

FIGURE 31.4 A schematic illustration of the visual where-what networks (WWNs). Features in the drawing are for intuition only since they are developed by the dually optimal LCA instead. V1 neurons have small receptive fields, which represent local features. Top-down connections enable V1 to recruit more neurons for action related features. V2 is similar to V1 but its neurons have larger receptive fields. They are both type specific and location specific, because they receive top-down information from both pathways. The pre-where-motor area in the dorsal pathway receives two sources of inputs, bottom-up from V2 and top-down from the "where" motor. Learning enables neurons to group according to position, to become less sensitive to type (or type quasi-invariant). In our WWN-2 experiments, the pre-where-motor neurons became almost positionally pure—each neuron only responds to a single retinal location. In the where-motor area, learning enables each neuron to link to premotor neurons that fired with it (when it was supervised), serving a logic-OR type of function. Thus, the where-motor becomes totally type-invariant but positional specific. The pre-what-motor area in the ventral pathway receives very different top-down signals from the what motor (in contrast with the pre-where-motor) to become less sensitive to position (or position quasi-invariant). The what motor becomes totally positionally invariant but type specific. There are multiple firing neurons in V2 at any time, some responding to the foreground and some responding to the background. Each pre-motor area enables global competition, causing only foreground neurons to win (without top-down supervision) or attended neurons to win (with top-down supervision). Therefore, although background pixels cause V1 neurons and V2 neurons to fire, their signals cannot pass the two premotor areas. The experimental results are available at Ji and Weng [18] for WWN-2 and Luciw and Weng [27] for WWN-3.

manually guides his hand) or internally (e.g., from the trials generated by the spinal cord or the mid brain). As illustrated in Figure 31.4, the model indicates that from early to later cortical areas, the neurons gradually increase their receptive fields and gradually reduce their effective fields as the processing of the corresponding *bridge representations* becomes less sensory and more motoric.

31.5.6 Cortex Scale: Feature Layers and Assistant Layers

The cerebral cortex contains six layers: layer L1 is the superficial layer and layer L6 is the deep layer. Weng et al. [53] reasoned that L4 and L2/3 are two feature detection layers as shown in Figure 31.5 with L5 assisting L2/3 and L6 assisting L4, in the sense of enabling long range lateral inhibition. Such long-range inhibitions encourage different neurons to detect different features. The model illustrated in Figure 31.5 was informed by the work of Felleman and Van Essen [9], Callaway and coworkers [4,57], and others (e.g., [12]). There are no top-down connections from L2/3 to L4, indicating that L4 uses unsupervised learning (U) while L2/3 uses supervised (S) learning. Weng et al. [53] reported that such a *paired* hierarchy USUS led to better recognition rates than the unpaired SSSS alternative.

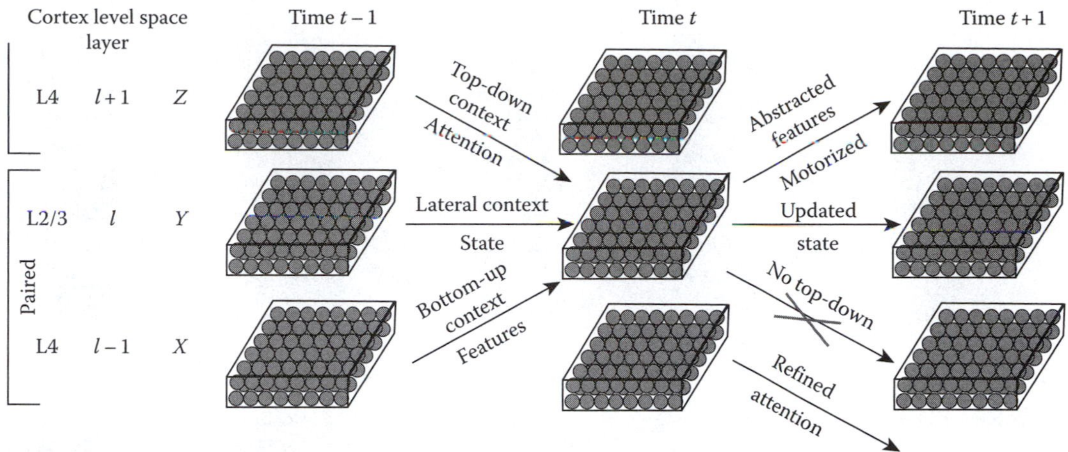

FIGURE 31.5 Cortex scale: The spatial SASE network for both spatial processing and temporal processing without dedicated temporal components. At each temporal unit shown above (two time frames), three basic operations are possible: link, drop prefix, and drop postfix. After proper training, the TCM is able to attend any possible temporal context up to the temporal sampling resolution.

31.5.6.1 Three-Source Attentive Spatiotemporal (TAS) Model

Sequential attentions (i.e., a cortical mechanism of thinking [47]) must deal with temporal contexts. How the brain treats time has been largely a mystery [6,15,28]. The following cortex-inspired spatiotemporal model is a new theory. It enables adaptive, arbitrary temporal lengths without using a fixed temporal window length as iconic memory (e.g., the brain has only one retina!). Bottom-up and top-down two-way connections mean mutual dependency, which requires iterations to reach an "attractor valley" [13] if the model is for one-shot solution. However, the brain is for sequential decisions instead as illustrated in Figure 31.5. Thus, the convergence is not guaranteed in general while the brain thinks. Specifically, the level L2/3 based on its current own content $L(t-1)$ takes three signal sources: bottom-up input $\mathbf{x}(t-1)$ as lower features, lateral input $\mathbf{y}(t-1)$ as the last temporal context, and top-down input $\mathbf{z}(t-1)$ as attention, all at time $t-1$, through the cortical function modeled as the lobe component analysis (LCA) [52,56] which generates its response $\mathbf{y}(t)$ at time t as the attention-selected context and to update its level to $L(t)$:

$$(\mathbf{y}(t), L(t)) = \text{Cortex}_{\text{LCA}}(\mathbf{x}(t-1), \mathbf{y}(t-1), \mathbf{z}(t-1) \mid L(t-1)) \tag{31.1}$$

We call this process attentive context folding. The response vector $\mathbf{y}(t)$ from L2/3 is used as more abstract features (more motoric) for the next higher level but not as the top-down input for the lower L4. The L4 context folding is analogous to L2/3, except that it does not take top-down input and its response is also used as top-down input to L2/3 in the lower cortex, as illustrated in Figure 31.5. The absence of top-down flow to L4 reduces the undesirable top-down hallucination that has shown unacceptable experimental results (lower than 50% recognition rate).

31.5.6.2 Sequential Decisions

In sequential attentions, an outcome depends on multiple attention decisions in sequence, as each decision depends on the outcomes of all the related previous decisions. For example, the motor output at time t is affected by the top-down attention signals from the previous motor output at time $t-2$. Figure 31.6 describes an example about how the attentive "machine" recursively makes sequential decisions—generates different top-down attentions and bottom-up attention-abstracted features at different times. Each top-down context directs the cortical region (L2/3-L4 combination) to attend to a different part

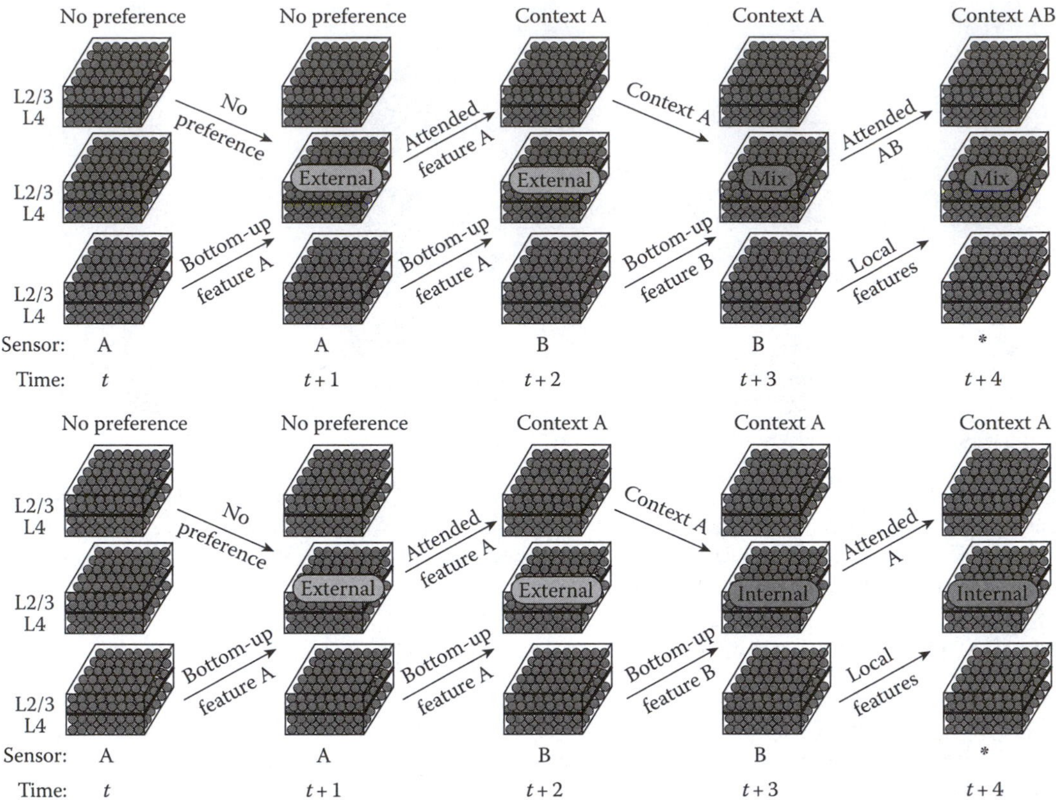

FIGURE 31.6 A schematic illustration of sequential decision making using top-down context (i.e., attention), while the sensory inputs (A, B, etc.) flow in one at a time. Two examples are presented here: Linking two sensory inputs (upper) and dropping a sensory input (lower). It has been proved [49] that any subset of the past context can be formed at the motor cortex, depending on the effectiveness of learning.

(or feature set) of the bottom-up input. This leads to a sequence of different top-down attentions and a sequence of different cortical responses, regardless whether the bottom-up input (or the environment) changes or not.

31.5.7 Level Scale: Dually Optimal CCI LCA

As shown in Figure 31.6, given parallel input space consisting of the bottom-up space X and the top-down input space Z, represented as $X \times Z$, the major developmental goal of each cortical level (L4 or L2/3 in Figure 31.5) is to have different neurons in the level to detect different features, but nearby neurons should detect similar features.

Each feature level faces two pairs of conflicting criteria which are probably implicit during biological evolution: (1) The spatial pair: with its limited number of neurons, the level must learn the best internal representation from the environment while keeping a stable long-term memory. (2) The spatiotemporal pair: with its limited child time for learning, the level must not only learn the best representation but also learn quickly without forgetting important mental skills acquired long time ago. The sparse coding principle [34] is useful to address the first pair: Allowing only a few neurons (best matched) to fire and update. Other neurons in the level are long-term memory because they are not affected. In other words, in each cortical region, only closely related mental skills are replaced each time. Therefore, the role of each neuron as working memory or long-term memory is dynamic, depending on the feature match (i.e., binding) with the input, as shown in Figure 31.7. However, this rough idea is not sufficient for optimality.

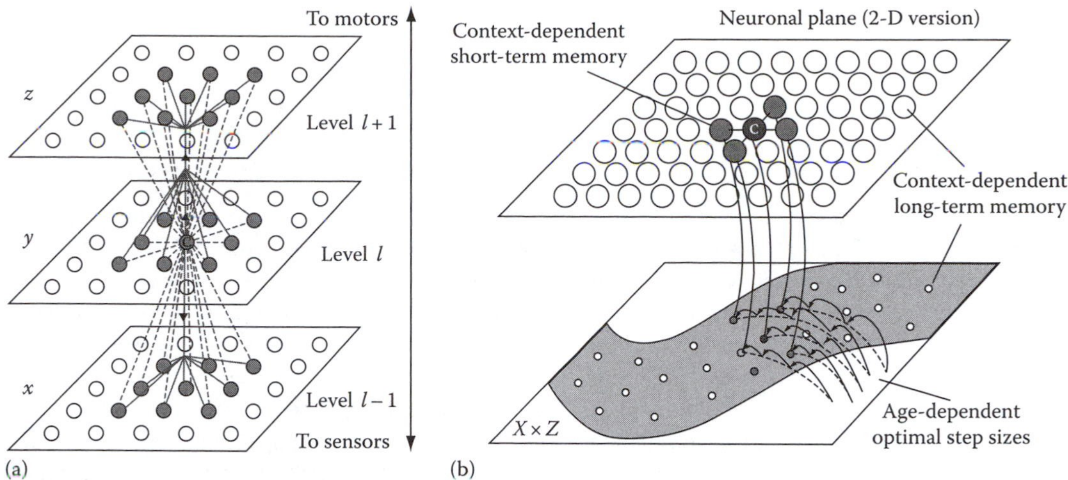

FIGURE 31.7 (a) The default connection pattern of every neuron in cortical layer L2/3 and L4 (no Z input for L4). The connections are local but two-way. (b) For each neuron in a layer, near neurons are connected to the neuron by excitatory connections (for layer smoothness) and far neurons are connected to the center neuron by inhibitory connections (competition resulting in detection of different features by different neurons). The upper layer indicates the positions for the neurons in the same layer: firing neurons are (context-dependent) working memory and those do not fire are (context dependent) long-term memory. The lower layer indicates the very high dimensional input space of the cortical layer ($X \times Z$) but illustrated in 2-D. The shaded area indicates the manifold of the input distribution. The connection curve from the upper neuron and lower small circle indicates the correspondence between the upper neuron and the feature that it detects. The neuronal weight vectors must quickly move to this manifold as the inputs are received and further the density of the neurons in the purple area should reflect the density of the input distribution. The challenge of fast adaptation at various maturation stages of development: The updating trajectory of every neuron is a highly nonlinear trajectory. The statistical efficiency theory for neuronal weight update (amnesic average) results in the nearly minimum error in each age-dependent update, meaning not only the direction of each update is nearly optimal, but also every step length.

The cortex inspired candid incremental covariance-free (CCI) LCA [53,56] has the desired dual optimality: spatial and spatiotemporal, as illustrated in Figure 31.7. CCI LCA models optimal self-organization by a cortical level with a limited resource: c neurons. The cortical level takes two parallel input spaces: the bottom-up space X and top-down space Z denoted as $P = X \times Z$ as illustrated by Figure 31.5. Each input vector is then denoted as $\mathbf{p} = (\mathbf{x}, \mathbf{z})$ where $\mathbf{x} \in X$ and $\mathbf{z} \in Z$. CCI LCA computes c feature vectors $\mathbf{v}_1, \mathbf{v}_2, \ldots, \mathbf{v}_c$. Associated with these c feature vectors is a partition of the input space P into c disjoint regions $R_1, R_2, \ldots, R_c$, so that the input space P is the union of all these regions. For the optimal distribution of neuronal resource, we consider that each input vector $\mathbf{p}$ is represented by the winner feature $\mathbf{v}_j$ which has the highest response r_j:

$$j = \arg \max_{1 \le i \le c} r_i$$

where r_i is the projection of input $\mathbf{p}$ onto the normalized feature vector $\mathbf{v}_i$: $\mathrm{r}_i = \mathrm{p} \cdot (\mathrm{vi}/\|\mathrm{vi}\|)$ $\mathbf{v}_i$: $r_i = \mathbf{p} \cdot (\mathbf{v}_i/\|\mathbf{v}_i\|)$. The form of approximation of $\mathbf{p}$ is represented by $\hat{\mathbf{p}} = r_i \mathbf{v}_i/\|\mathbf{v}_i\|$ and the error of this representation for $\mathbf{p}$ is $e(\mathbf{p}) = \|\hat{\mathbf{p}} - \mathbf{p}\|$.

31.5.7.1 Spatial Optimality

The spatial optimality requires that the spatial resource distribution in the cortical level is optimal in minimizing the representational error. For this optimality, the cortical-level developmental program modeled by CCI LCA computes the best feature vectors $V = (\mathbf{v}_1, \mathbf{v}_2, \ldots, \mathbf{v}_c)$ so that the expected square approximation error $\|\hat{\mathbf{p}}(V) - \mathbf{p}\|^2$ is statistically minimized:

$$V^* = (\mathbf{v}_1^*, \mathbf{v}_2^*, \ldots, \mathbf{v}_c^*) = \arg\min_V E \| \hat{\mathbf{p}}(V) - \mathbf{p} \|^2 . \quad (31.2)$$

where E denotes statistical expectation. The minimum error means the optimal allocation of limited neuronal resource: frequent experience is assigned with more neurons (e.g., human face recognition) but rare experience is assigned with fewer neurons (e.g., flower recognition for a nonexpert). This optimization problem must be computed incrementally, because the brain receives sensorimotor experience incrementally. As the feature vectors are incrementally updated from experience, the winner neurons for the past inputs are not necessarily the same if past inputs are fed into the brain again (e.g., parents' speech when their baby was little is heard again by the grown-up baby). However, while the feature vectors are stabilized through extensive experience, the partition of the input space becomes also stable. Given a fixed partition, it has been proved that the best feature set V^* consists of the c local first principal component vectors, one for each region R_i. The term "local" means that the principal component vector for region R_i only considers the samples that fall into region R_i. As the partition is tracking a slowly changing environment (e.g., while the child grows up), the optimal feature set V^* tracks the slowly changing input distribution (called nonstationary random process).

Intuitively speaking, the spatial optimality means that with the same cortical size, all the children will eventually perform at the best level allowed by the cortical size. However, to reach the same mental skill level one child may require more teaching than another. The spatiotemporal optimality is deeper. It requires the best performance for every time t. That is, the child learns quickest allowed by the cortical size at every stage of his age.

31.5.7.2 Temporal Optimality

The spatiotemporal optimality gives optimal step sizes of learning. Each neuron takes response weighted input $\mathbf{u}(t) = r(t)\mathbf{x}(t)$ at time t (i.e., Hebbian increment). From the mathematical theory of statistical efficiency, CCI LCA determines the optimal feature vectors $V^*(t) = (\mathbf{v}_1^*(t), \mathbf{v}_2^*(t), \ldots, \mathbf{v}_c^*(t))$ for every time instant t starting from the conception time $t = 0$, so that the distance from $V^*(t)$ to its target V^* is minimized:

$$V^*(t) = \arg\min_{V(t)} E \| V(t) - V^* \|^2 . \quad (31.3)$$

CCI LCA aims at this deeper optimality—the smallest average error from the starting time (birth of the network) up to the current time t, among all the possible estimators, under some regularity conditions. A closed form solution was found that automatically gives the optimal retention rate and the optimal learning rate (i.e., step size) at each synaptic update [56].

In summary, the spatial optimality leads to Hebbian incremental direction: response weighted presynamptic activity ($r\mathbf{p}$). The deeper spatiotemporal optimality leads to the best learning rates, automatically determined by the update age of each neuron. This is like different racers racing on a rough terrain along a self-determined trajectory toward an unknown target. The spatially optimal racers, guided by Hebbian directions, do not know step sizes. Thus, they cover other trajectories that require more steps. The spatiotemporally optimal racer, CCI LCA, correctly estimates not only the optimal direction at every step as illustrated in Figure 31.7, but also the optimal step size at every step. In our experiments, CCI LCA out performed the self-organization map (SOM) algorithm by an order (over 10 times) in terms of percentage distance covered from the initial estimate to the target. This work also predicts cell-age dependent plasticity schedule which needs to be verified biologically.

31.6 Summary

The material in this chapter outlines a series of tightly intertwined breakthroughs recently made in understanding and modeling how the brain develops and works. The grand picture of the human brain is getting increasingly clear. Developmental robots and machines urgently need industrial electronics for real-time, brain scale computation, and learning. This need is here and now. This has created a great challenge for the field of industrial electronics, but an exciting future as well.

References

1. G. Backer, B. Mertsching, and M. Bollmann. Data- and model-driven gaze control for an active-vision system. *IEEE Transactions on Pattern Analysis and Machine Intelligence*, 23(12):1415–1429, December 2001.
2. V. D. Bohbot and S. Corkin. Posterior parahippocampal place learning in h.m. *Hippocampus*, 17:863–872, 2007.
3. T. J. Buschman and E. K. Miller. Top-down versus bottom-up control of attention in the prefrontal and posterior parietal cortices. *Science*, 315:1860–1862, 2007.
4. E. M. Callaway. Local circuits in primary visual cortex of the macaque monkey. *Annual Review of Neuroscience*, 21:47–74, 1998.
5. G. Deco and E. T. Rolls. A neurodynamical cortical model of visual attention and invariant object recognition. *Vision Research*, 40:2845–2859, 2004.
6. P. J. Drew and L. F. Abbott. Extending the effects of spike-timing-dependent plasticity to behavioral timescales. *Proceedings of the National Academy of Sciences of the United States of America*, 103(23):8876–8881, 2006.
7. A. Ekstrom, M. Kahana, J. Caplan, T. Fields, E. Isham, E. Newman, and I. Fried. Cellular networks underlying human spatial navigation. *Nature*, 425:184–188, 2003.
8. J. L. Elman, E. A. Bates, M. H. Johnson, A. Karmiloff-Smith, D. Parisi, and K. Plunkett. *Rethinking Innateness: A Connectionist Perspective on Development*. MIT Press, Cambridge, MA, 1997.
9. D. J. Felleman and D. C. Van Essen. Distributed hierarchical processing in the primate cerebral cortex. *Cerebral Cortex*, 1:1–47, 1991.
10. M. D. Fox, M. Corbetta, A. Z. Snyder, J. L. Vincent, and M. E. Raichle. Spontaneous neuronal activity distinguishes human dorsal and ventral attention systems. *Proceedings of the National Academy of Sciences of the United States of America*, 103(26):10046–10051, 2006.
11. K. Grill-Spector, N. Knouf, and N. Kanwisher. The fusiform face area subserves face perception, not generic within-category identification. *Nature Neuroscience*, 7(5):555–562, 2004.
12. S. Grossberg and R. Raizada. Contrast-sensitive perceptual grouping and object-based attention in the laminar circuits of primary visual cortex. *Vision Research*, 40:1413–1432, 2000.
13. G. E. Hinton. Learning multiple layers of representation. *Trends in Cognitive Science*, 11(10):428–434, 2007.
14. J. M. Hupe, A. C. James, B. R. Payne, S. G. Lomber, P. Girard, and J. Bullier. Cortical feedback improves discrimination between figure and background by v1, v2 and v3 neurons. *Nature*, 394:784–787, August 20, 1998.
15. I. Ito, R. C. Ong, B. Raman, and M. Stopfer. Sparse odor representation and olfactory learning. *Nature Neuroscience*, 11(10):1177–1184, 2008.
16. L. Itti and C. Koch. Computational modelling of visual attention. *Nature Reviews Neuroscience*, 2:194–203, 2001.
17. L. Itti, C. Koch, and E. Niebur. A model of saliency-based visual attention for rapid scene analysis. *IEEE Transactions on Pattern Analysis and Machine Intelligence*, 20(11):1254–1259, November 1998.
18. Z. Ji and J. Weng. WWN-2: A biologically inspired neural network for concurrent visual attention and recognition. In *Proceedings of the IEEE International Joint Conference on Neural Networks*, Barcelona, Spain, July 18–23, 2010.

19. Z. Ji, J. Weng, and D. Prokhorov. Where-what network 1: "Where" and "What" assist each other through top-down connections. In *Proceedings of the IEEE International Conference on Development and Learning*, Monterey, CA, Aug. 9–12, 2008, pp. 61–66.
20. R. R. Johnson and A. Burkhalter. Microcircuitry of forward and feedback connections within rat visual cortex. *Journal of Comparative Neurology*, 368(3):383–398, 1996.
21. N. Kanwisher, D. Stanley, and A. Harris. The fusiform face area is selective for faces not animals. *NeuroReport*, 10(1):183–187, 1999.
22. L. C. Katz and E. M. Callaway. Development of local circuits in mammalian visual cortex. *Annual Review of Neuroscience*, 15:31–56, 1992.
23. H. Kennedy and J. Bullier. A double-labelling investigation of the afferent connectivity to cortical areas v1 and v2 of the macaque monkey. *Journal of Neuroscience*, 5(10):2815–2830, 1985.
24. C. Koch and S. Ullman. Shifts in selective visual attention: Towards the underlying neural circuitry. *Human Neurobiology*, 4:219–227, 1985.
25. T. S. Kuhn. *The Structure of Scientific Revolutions*, 2nd edn. University of Chicago Press, Chicago, IL, 1970.
26. T. S. Lee and D. Mumford. Hierarchical bayesian inference in the visual cortex. *Journal of the Optical Society of America A*, 20(7):1434–1448, 2003.
27. M. Luciw and J. Weng. Where What Network 3: Developmental top-down attention with multiple meaningful foregrounds. In *Proceedings of the IEEE International Joint Conference on Neural Networks*, Barcelona, Spain, July 18–23, 2010.
28. M. D. Mauk and D. V. Buonomano. The neural basis of temporal processing. *Annual Review of Neuroscience*, 27:307–340, 2004.
29. J. L. McClelland. The interaction of nature and nurture in development: A parallel distributed processing perspective. In P. Bertelson, P. Eelen, and G. d'Ydewalle (eds.), *International Perspectives on Psychological Science. Leading Themes* Vol. 1, Erlbaum, Hillsdale, NJ, 1994, pp. 57–88.
30. M. Mishkin, L. G. Unterleider, and K. A. Macko. Object vision and space vision: Two cortical pathways. *Trends in Neuroscicence*, 6:414–417, 1983.
31. J. Moran and R. Desimone. Selective attention gates visual processing in the extrastrate cortex. *Science*, 229(4715):782–784, 1985.
32. J. O'Keefe and J. Dostrovsky. The hippocampus as a spatial map: Preliminary evidence from unit activity in the freely-moving rat. *Brain Research*, 34(1):171–175, 1971.
33. B. A. Olshausen, C. H. Anderson, and D. C. Van Essen. A neurobiological model of visual attention and invariant pattern recognition based on dynamic routing of information. *Journal of Neuroscience*, 13(11):4700–4719, 1993.
34. B. A. Olshaushen and D. J. Field. Emergence of simple-cell receptive field properties by learning a sparse code for natural images. *Nature*, 381:607–609, June 13, 1996.
35. D. J. Perkel, J. Bullier, and H. Kennedy. Topography of the afferent connectivity of area 17 of the macaque monkey. *Journal of Computational Neuroscience*, 253(3):374–402, 1986.
36. W. K. Purves, D. Sadava, G. H. Orians, and H. C. Heller. *Life: The Science of Biology*, 7th edn. Sinauer, Sunderland, MA, 2004.
37. S. Quartz and T. J. Sejnowski. The neural basis of cognitive development: A constructivist manifesto. *Behavioral and Brain Sciences*, 20(4):537–596, 1997.
38. M. Rabinovich, R. Huerta, and G. Laurent. Transient dynamics for neural processing. *Science*, 321:48–50, 2008.
39. R. P. N. Rao and D. H Ballard. Probabilistic models of attention based on iconic representations and predictive coding. In L. Itti, G. Rees, and J. Tsotsos (eds.), *Neurobiology of Attention*. Academic Press, New York, 2004.
40. S. Russell and P. Norvig. *Artificial Intelligence: A Modern Approach*. Prentice-Hall, Upper Saddle River, NJ, 1995.

41. P. A. Salin and J. Bullier. Corticocortical connections in the visual system: structure and function. *Physiological Review*, 75(1):107–154, 1995.
42. K. Schill, E. Umkehrer, S. Beinlich, G. Krieger, and C. Zetzsche. Scene analysis with saccadic eye movements: Top-down and bottom-up modeling. *Journal of Electronic Imaging*, 10(1):152–160, 2001.
43. J. T. Serences and S. Yantis. Selective visual attention and perceptual coherence. *Trends in Cognitive Sciences*, 10(1):38–45, 2006.
44. T. J. Sullivan and V. R. de Sa. A model of surround suppression through cortical feedback. *Neural Networks*, 19:564–572, 2006.
45. J. K. Tsotsos, S. M. Culhane, W. Y. K. Wai, Y. Lai, N. Davis, and F. Nuflo. Modeling visual attention via selective tuning. *Artificial Intelligence*, 78:507–545, 1995.
46. L. G. Ungerleider and M. Mishkin. Two cortical visual systems. In D. J. Ingel (ed.), *Analysis of Visual Behavior*. MIT Press, Cambridge, MA, 1982, pp. 549–586.
47. J. Weng. On developmental mental architectures. *Neurocomputing*, 70(13–15):2303–2323, 2007.
48. J. Weng. Task muddiness, intelligence metrics, and the necessity of autonomous mental development. *Minds and Machines*, 19(1):93–115, 2009.
49. J. Weng. The SWWN connectionist models for the cortical architecture, spatiotemporal representations and abstraction. In *Proceedings of the Workshop on Bio-Inspired Self-Organizing Robotic Systems, IEEE International Conference on Robotics and Automation*, Anchorage, AK, May 3–8, 2010.
50. J. Weng, N. Ahuja, and T. S. Huang. Learning recognition and segmentation of 3-D objects from 2-D images. In *Proceedings of the IEEE Fourth International Conference on Computer Vision*, Berlin, Germany, May 1993, pp. 121–128.
51. J. Weng, N. Ahuja, and T. S. Huang. Learning recognition and segmentation using the Cresceptron. *International Journal of Computer Vision*, 25(2):109–143, November 1997.
52. J. Weng and M. Luciw. Dually optimal neuronal layers: Lobe component analysis. *IEEE Transactions on Autonomous Mental Development*, 1(1):68–85, 2009.
53. J. Weng, T. Luwang, H. Lu, and X. Xue. Multilayer in-place learning networks for modeling functional layers in the laminar cortex. *Neural Networks*, 21:150–159, 2008.
54. J. Weng, J. McClelland, A. Pentland, O. Sporns, I. Stockman, M. Sur, and E. Thelen. Autonomous mental development by robots and animals. *Science*, 291(5504):599–600, 2001.
55. J. Weng and I. Stockman. Autonomous mental development: Workshop on development and learning. *AI Magazine*, 23(2):95–98, 2002.
56. J. Weng and N. Zhang. Optimal in-place learning and the lobe component analysis. In *Proceedings of the IEEE World Congress on Computational Intelligence*, Vancouver, Canada, July 16–21, 2006.
57. A. K. Wiser and E. M. Callaway. Contributions of individual layer 6 pyramidal neurons to local circuitry in macaque primary visual cortex. *Journal of Neuroscience*, 16:2724–2739, 1996.

32

Synthetic Biometrics for Testing Biometric Systems and User Training

Svetlana N. Yanushkevich
University of Calgary

Adrian Stoica
Jet Propulsion Laboratory

Ronald R. Yager
Iona College

Oleg Boulanov
University of Calgary

Vlad P. Shmerko
University of Calgary

32.1 Introduction

Synthetic biometrics are understood as generated biometric data that are biologically meaningful for existing biometric systems. These synthetic data replicate possible instances of otherwise unavailable data, in particular corrupted or distorted data. For example, facial images, acquired by video cameras, can be corrupted due to their position and angle of observation (appearance variation), as well as lighting (environmental conditions), camera resolution, and other parameters (measurement conditions). The other reason for the use of synthetic data is the difficulty in collecting a statistically meaningful amount of biometric samples due to privacy issues and the unavailability of large databases, etc. In order to avoid these difficulties, synthetic biometric data can be used as samples, or tests, generated using the controllability of various parameters. This renders them capable of being used to test biometric tools and devices [17,28].

Synthetic biometric data can also be thought in terms of a forgery of biometric data. Properly created artificial biometric data provides an opportunity for the detailed and controlled modeling of a wide range of training skills, strategies, and tactics, thus enabling better approaches to enhancing system performance.

Contemporary techniques and achievements in biometrics are being developed in two directions: toward the analysis of biometric information (direct problems) and toward the synthesis of biometric information (inverse problems) [1,6,11,33,34] (Figure 32.1).

The crucial point of modeling in biometrics (inverse problems) is the analysis-by-synthesis paradigm. This paradigm states that synthesis of biometric data can verify the perceptual equivalence between

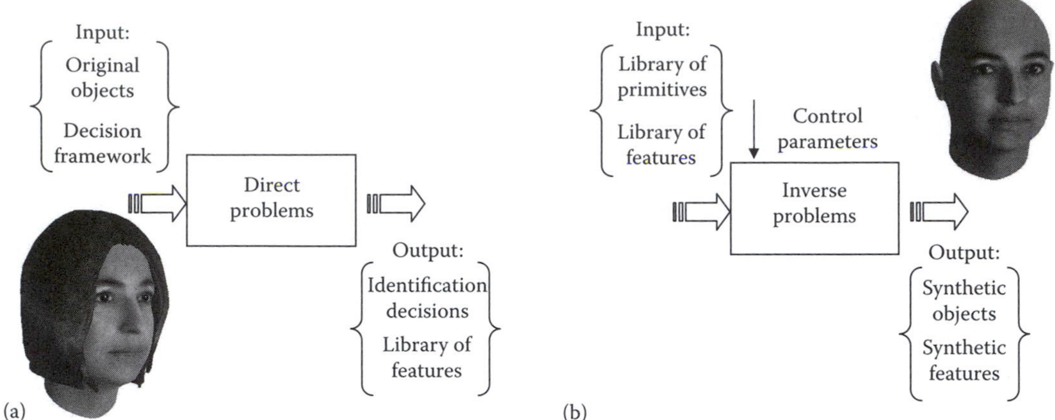

FIGURE 32.1 Direct (a) and inverse (b) problems of biometrics.

original and synthetic biometric data, i.e., synthesis-based feedback control. For example, facial analysis can be formulated as deriving a symbolic description of a real facial image. The aim of facial synthesis is to produce a realistic facial image from a symbolic facial expression model.

32.2 Synthetic Biometrics

In this section, examples of synthetic biometrics, such as synthetic fingerprints, iris, retina, and signatures, are introduced.

32.2.1 Synthetic Fingerprints

Today's interest in automatic fingerprint synthesis addresses the urgent problems of testing fingerprint identification systems, training security personnel, enhancing biometric database security, and protecting intellectual property [8,18,33].

Traditionally, two methods for fingerprint imitation are discussed with respect to obtaining unauthorized access to an information system: (1) the authorized user provides his/her fingerprint for making a copy and (2) a fingerprint is taken without the authorized user's consent, for example, from a glass surface (a classic example of spy work) in a routine forensic procedure.

Cappelli et al. [6,8] developed a commercially available synthetic fingerprint generator called SFinGe. In SFinGe, various models of fingerprint topologies are used: shape, directional map, density map, and skin deformation models. In Figure 32.2, two topological primitives are composed in various ways. These are examples of acceptable (valid) and unacceptable (invalid) synthesized fingerprints.

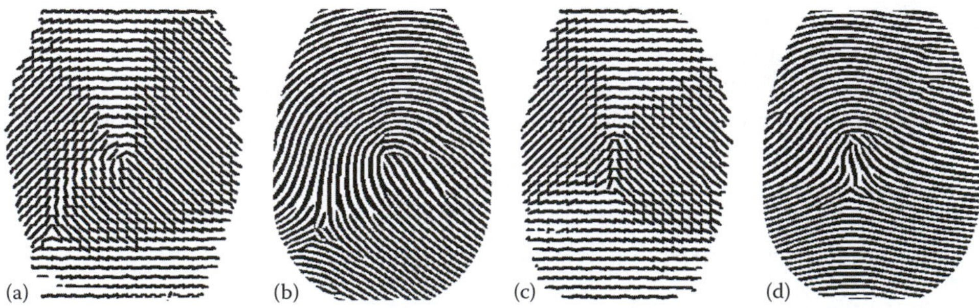

FIGURE 32.2 Synthetic fingerprints generated by the SFinGe system: Invalid (a,c) and valid (b,d) topological compositions of fingerprint primitives.

Kuecken [16] proposed an alternative method for synthetic fingerprint generation based on natural fingerprint formation, that is, embryological process. Kuecken's modeling approach started with an idea originating from Kollman (1883) that was promoted by Bonnevie in the 1920. In Kuecken's generator of synthetic fingerprints, the Karmen equations are used to describe the mechanical behavior of a thin curved sheet of elastic material.

32.2.2 Synthetic Iris and Retina Images

Iris recognition systems scan the surface of the iris to analyze patterns. Retina recognition systems scan the surface of the retina and analyze nerve patterns, blood vessels, and such features. Automated methods of iris and retina image synthesis have not been developed yet, except for an approach based on generation of iris layer patterns [33].

A synthetic image can be created by combining segments of real images from a database. Various operators can be applied to deform or warp the original iris image: translation, rotation, rendering, etc. Various models of the iris, retina, and eye used to improve recognition can be found in [3,4,7,33].

An example of the generating of posterior pigment epithelia of the iris using a Fourier transform on a random signal is considered below. A fragment of the FFT signal is interpreted as a gray-scaled vector: the peaks in the FFT signal represent lighter shades and valleys represent darker shades. This procedure is repeated for other fragments as well. The data plotted in 3D, a 2D slice of the data, and a round image generated from the slice using a polar transform. The superposition of the patterns of various iris layers forms a synthetic iris pattern. Synthetic collarette topology modeled by a randomly generated curve is shown in Figure 32.3. Figure 32.3b illustrates three different patterns obtained by this method. Other layer patterns can be generated based on wavelet, Fourier, polar, and distance transforms, as well as Voronoi diagrams [33].

32.2.3 Synthetic Signatures

The current interest in signature analysis and synthesis is motivated by the development of improved devices for human–computer interaction, which enable input of handwriting and signatures. The focus of this study is the formal modeling of this interaction [5,15,23].

To generate signatures with any automated technique, it is necessary to consider (a) the formal description of curve segments and their kinematical characteristics, (b) the set of requirements that

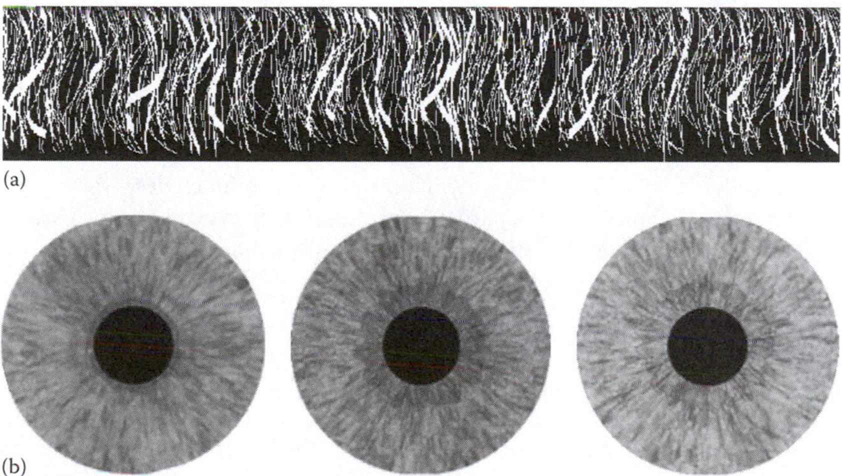

FIGURE 32.3 Synthetic collarette topology modelled by a randomly generated curve: spectral representation (a) and three different synthetic patterns (b).

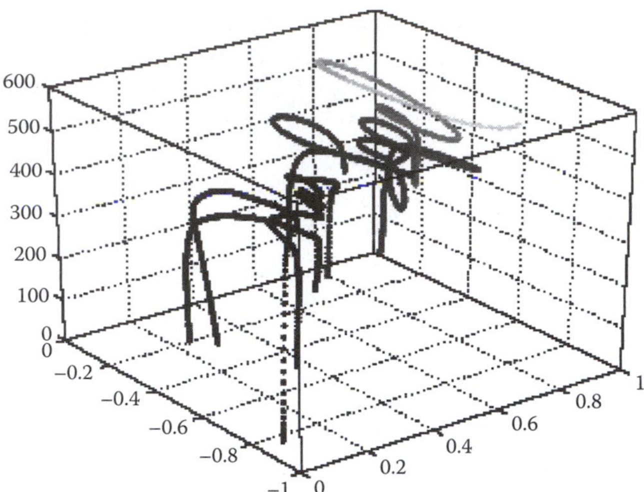

FIGURE 32.4 3D view of an on-line signature: the plain curve is given by the two-tuple (X, Y), the pressure is associated with the Z axis, the speed of writing (that is, an additional dimension) is depicted by the shade of the curve, where darker is slower speed. (Courtesy of Prof. D. Popel, Baker University, USA.)

should be met by any signature generation system, and (c) the possible scenarios for signature generation. The simplest method of generating synthetic signatures is based on formal 2D geometrical description of the curve segments. Spline methods and Bezier curves are used for curve approximation, given some control points; manipulations to the control points give variations on a single curve in these methods [33]. A 3D on-line representation of a signature is given in Figure 32.4.

32.3 Example of the Application of Synthetic Biometric Data

In this section, an application of synthetic biometrics for training users of a physical access control system (PASS) is introduced. The main purpose of the PASS is the efficient support of security personnel enhanced with the situational awareness paradigm and intelligent tools. A registration procedure is common in the process of personal identification (checkpoints in homeland and airport security applications, hospitals, and other places where secure physical admission is practiced). We refer to the Defense Advanced Research Projects Agency (DARPA) research program HumanID, which is aimed at the detection, recognition, and identification of humans at a distance in an early warning support system for force protection and homeland defense [29]. The access authorization process is characterized by *insufficiency* of information. The result of a customer's identification is a decision under uncertainty made by the user. Uncertainty (incompleteness, imprecision, contradiction, vagueness, unreliability) is understood in the sense that available information allows for several possible interpretations, and it is not entirely certain which is the correct one. The user must make a decision under uncertainty, that is, select an alternative before any complete knowledge is obtained [31,32].

The architecture of the PASS is shown in Figure 32.5. The system consists of sensors such as cameras in the visible and infrared bands, decision-support assistants, and a dialogue support device to support conversation based on the preliminary information obtained, and a personal file generating module. Three-level surveillance is used in the system: surveillance of the line (prescreening); surveillance during the walk between pre-screened and screened points, and surveillance during the authorization process at the officer's desk (screening).

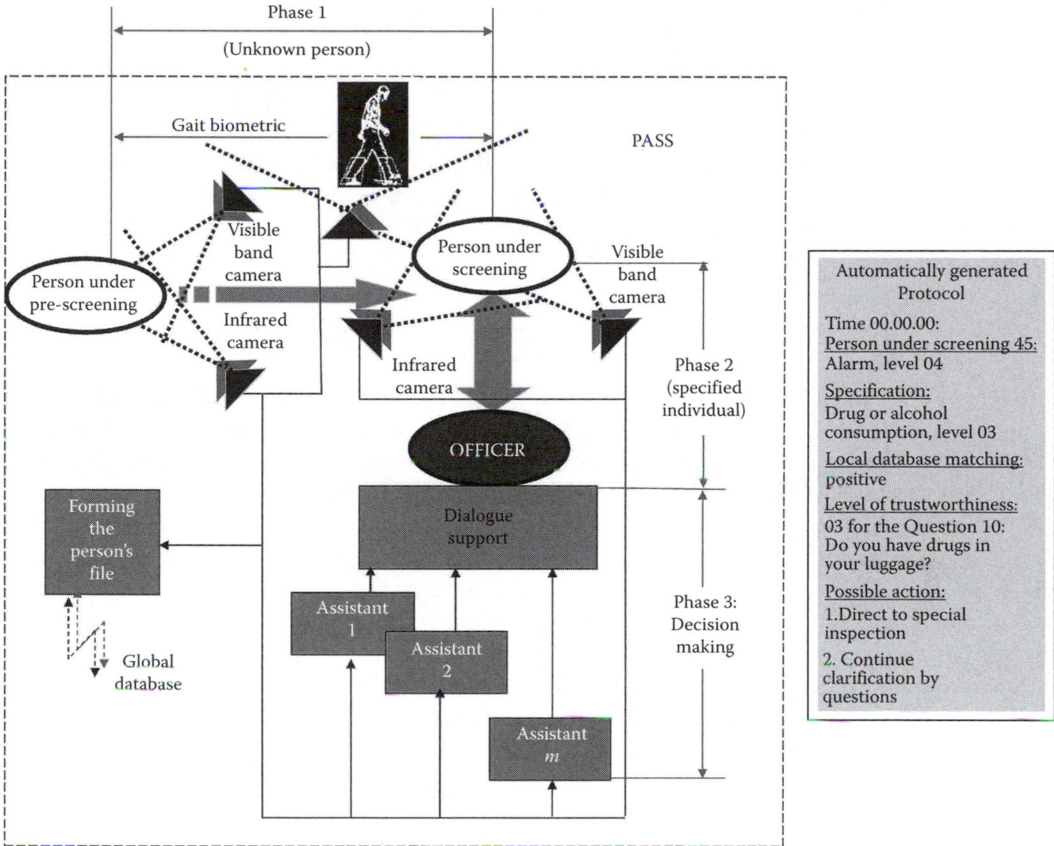

FIGURE 32.5 The PASS is a semi-automatic, application-specific distributed computer system that aims to support the officer's job in access authorization (left), and typical protocol of the authorization (right).

Decision-support assistants. The device gathered from the sensors and intelligent data processing for the situational awareness is called decision-support assistant. The PASS is a distributed computing network of biometric-based assistants. For example, an assistant can be based on noninvasive metrics such as temperature measurement, artificial accessory detection, estimation of drug and alcohol intoxication, and estimation of blood pressure and pulse [9,25,35,36]. The decision support is built upon the discriminative biometric analysis that means detecting features used for evaluation the physiological and psychoemotional states of a person. Devices for various biometrics can be used as the kernels of decision support assistants. The most assistants in PASS are multipurpose devices, that is, they can be used for the person authorization and user training.

The role of biometric device. In the PASS, the role of each biometric device is twofold: their primary function is to extract biometric data from an individual, and their secondary function is to support a dialog of a user and a customer. For example, if high temperature is detected, the question to this customer should be formulated as follows: "Do you need any medical assistance?" The key of the concept of the dialog support in PASS is the process of generating questions initiated by information from biometric devices. In this way, the system assists the user in access authorization.

The time of service, T, can be divided into three phases: T_1 (the prescreening phase of service or waiting), T_2 (individual's movement from the prescreened position to the officer's desk), and T_3 (the time of identification (document-check) and authorization).

32.3.1 Hyperspectral Facial Analysis and Synthesis in Decision-Support Assistant

The concept of a multipurpose platform is applied to the decision-support assistants, including an assistant for hyperspectral face analysis.

Hyperspectral face analysis. The main goal of face analysis in infrared band is detecting physical parameters such as temperature, blood flow rate and pressure, as well as physiological changes, caused by alcohol and substances. Another useful application of the infrared image analysis is detection of artificial accessories such as artificial hair and plastic surgery. This data can provide valuable information to support interviewing of the customers in the systems like PASS.

There are several models, which are implemented in decision-support assistant for hyperspectral facial analysis and synthesis, namely, a skin color model and a 3D hyperspectral face model.

Skin color models. In [30], a skin color model was developed and shown to be useful for the detection of changes due to alcohol intoxication. The fluctuation of temperature in various facial regions is primary due to the changing blood flow rate. In [14], the heat-conduction formulas at the skin surface are introduced. In [19], mass blind screening of potential SARS or bird flu patients was studied.

The human skin has a layered structure and the skin color is determined by how incident light is absorbed and scattered by the melanin and hemoglobin pigments in two upper skin layers, epidermis, and dermis.

The color of human skin can reveal distinct characteristics valuable for diagnostics. The dominant pigments in skin color formation are melanin and hemoglobin. Melanin and hemoglobin determine the color of the skin by selectively absorbing certain wavelengths of the incident light. The melanin has a dark brown color and predominates in the epidermal layer while the hemoglobin has a reddish hue or purplish color, depending on the oxygenation, and is found mainly in the dermal layer. It is possible to obtain quantitative information about hemoglobin and melanin by fitting the parameters of an analytical model with reflectance spectra. In [20], a method for visualizing local blood regions in the skin tissue using diffuse reflectance images was proposed.

A quantitative analysis of human skin color and temperature distribution can reveal a wide range of physiological phenomena. For example, skin color and temperature can change due to drug or alcohol consumption, as well as physical exercises [2].

32.3.2 Hyperspectral Analysis-to-Synthesis 3D Face Model

A decision-support assistant performs the hyperspectral face analysis based on a model that includes two constituents: a face shape model (represented by a 3D geometric mesh) and a hyperspectral skin texture model (generated from images in visible and infrared bands). The main advantage of a 3D face modeling is that the effect of variations in illumination, surface reflection, and shading from directional light can be significantly decreased. For example, a 3D model can provide controlled variations in appearance while the pose or illumination is changed. Also, the estimations of facial expressions can be made more accurately in 3D models compared with 2D models.

A face shape is modeled by a polygonal mesh, while the skin is represented by texture map images in visible and infrared bands (Figure 32.6). Any individual face shape can be generated from the generic face model by specifying 3D displacements for each vertex. Synthetic face images are rendered by mapping the texture image on the mesh model.

Face images in visible and infrared bands, acquired by the sensors, constitute the input of the module for hyperspectral face analysis and synthesis. The corresponding 3D models, one for video and one for infrared images, are generated by fitting the generic model to images (Figure 32.6).

The texture maps represent the hemoglobin and melanin content and the temperature distribution of the facial skin. These maps are the output of the face analysis and modeling module. This information is used for evaluating the physical and psychoemotional state of a person.

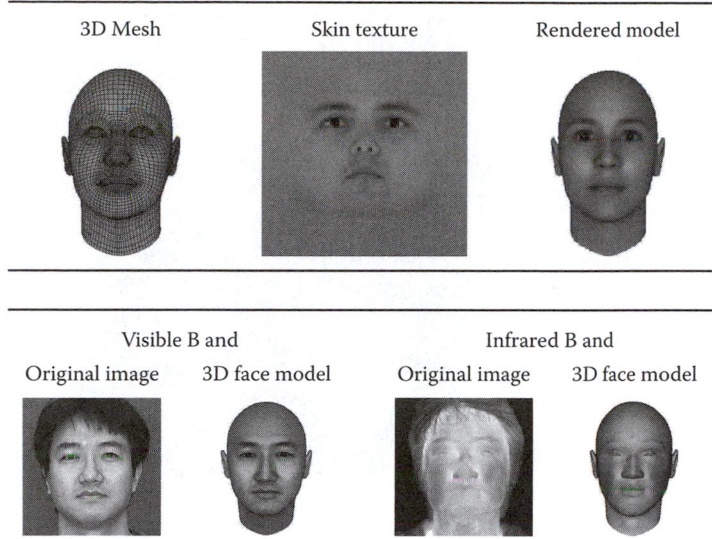

FIGURE 32.6 The generic 3D polygonal mesh, skin texture, and the resulting 3D rendered model (top). Face images in visible and infrared bands and their 3D models (down).

Facial action analysis. Face models are considered to convey emotions. In the systems like PASS, emotions play the role of "indicators" used for decision-making about the emotional and physiological state of the customer and for generating questions for further dialog. Visual band images along with thermal (infrared) images can be used in this task [21,22,27]. Facial expressions are formed by about 50 facial muscles [12] and are controlled by dozens of parameters in the model (Figure 32.7). The facial expression can be identified once the facial action units are recognized. This task involves facial feature extraction (eyes, eyebrow, nose, lips, chin lines), measuring geometric distances between the extracted points/lines, and then facial action units recognition based on these measurements. Decision-making is based on the analysis of changes in facial expression while a person listens and responds to questions.

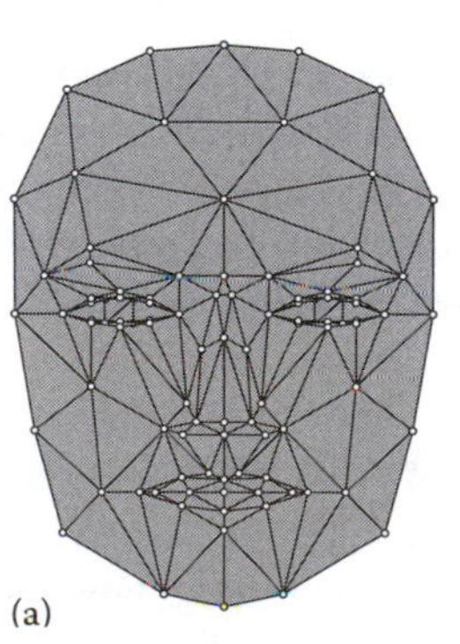
(a)

(b)

Action Unit	Muscular basis
Inner brow raiser	Frontalis, pars medialis
Outer brow raiser	Frontalis, pars lateralis
Upper lid raiser	Levator palpebrae, superioris
Cheek raiser	Orbicularis oculi, pars palebralis
Lip corner puller	Zygomatic major
Cheek puffer	Caninus
Chin raiser	Mentalis
Lip stretcher	Risorius
Lip funneler	Orbicularis oris
Lip tightner	Orbicularis oris
Mouth stretch	Pterygoid, digastric
Lip suck	Orbicularis oris
Nostril dilator	Nasalis, pars alaris
Slit	Orbicularis oculi

FIGURE 32.7 A 3D facial mesh model (a) and fragment of corresponding facial action units (b).

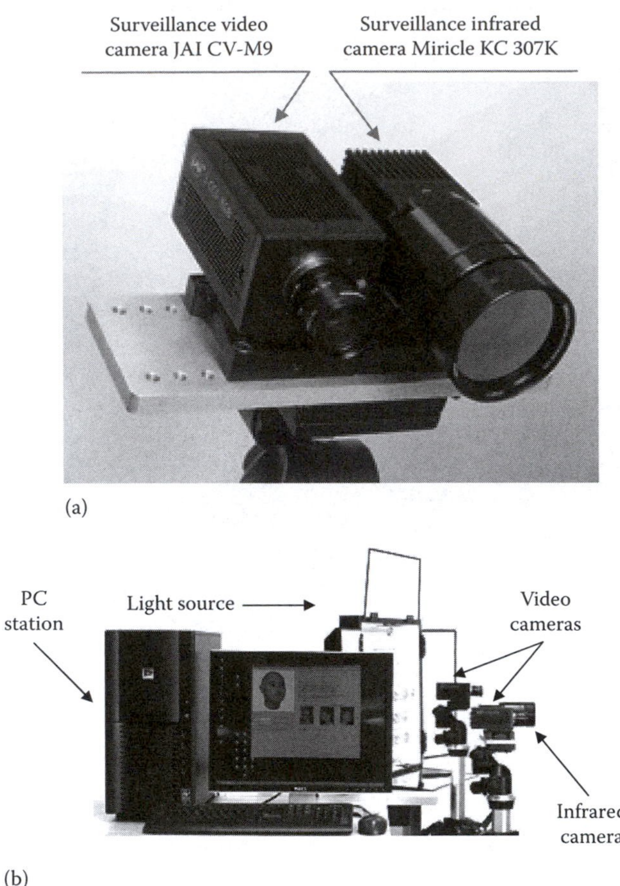

FIGURE 32.8 A setup of a pair of video and infrared cameras for surveillance (a) and experimental equipment for a 3D hyperspectral face modeling (b).

Hyperspectral data acquisition. A setup of paired video and thermal cameras for acquisition of facial images in both visible and infrared bands is shown in Figure 32.8. Two cameras can acquire full resolution images. Infrared facial images are provided by an uncooled microbolometer infrared camera. The network of various assistants is based on a PC station with acquisition boards.

32.4 Synthetic Data for User Training in Biometric Systems

The basic concept of the PASS is the collaboration of the user, the customer, and the machine. This is a dialogue-based interactions. Based on the premise that the user has priority in the decision making at the highest level of the system hierarchy, the role of the machine is defined as assistance, or support of the user.

The training methodology should be short-term, periodically repeated, and intensive. The PASS can be used as a training system (with minimal extension of tools) without changing of the place of deployment. In this way, we fulfill the criterion of cost efficiency and satisfy the above requirements.

Simulation of extreme scenarios is aimed at developing the particular skills of the personnel. The modeling of extreme situations requires developing specific training methodologies and techniques, including virtual environments.

Scenarios of decision-making support. The possible scenarios are divided into three groups: regular, nonstandard, and extreme. Let us consider an example of a scenario, in which the system generates the following data about the screened person.

```
Protocol for person #45 under
pre-screening
Time:  12.00.00:
Warning:  level 04
Specification:  Drug or alcohol
intoxication, level [03]
Possible action:
1.  Database inquiry
2.  Clarify in the dialogue
```

```
Protocol for person #45 under
screening
Time:  12.10.20:
Warning:  level 04
Specification:  Drug or alcohol
intoxication, level [03]
Local database matching: positive
Possible action:
1.  Further inquiry using dialogue
2.  Direct to the special inspection
```

FIGURE 32.9 Scenarios for user training: protocol of pre-screening (left) and screening (right).

```
Protocol of the person #45 under
screening
Time 00.00.00:
Warning, level 04
Specification: Drug or alcohol
consumption, level [03]
Local database matching: positive
Proposed dialogue questions:
Question 1: Do you need any medical
assistance?
Question 2: Any service problems
during the flight?
Question 3: Do you plan to rent a
car?
Question 4: Did you meet friends on
board?
Question 5: Did you consume wine or
whisky aboard?
Question 6: Do you have drugs in
your luggage?
```

```
Protocol of the person #45 under
screening (continuation)
Level of trustworthiness of Question 1
is 02:
Level of trustworthiness of Question 2
is 02:
Level of trustworthiness of Question 3
is 03:
Level of trustworthiness of Question 4
is 00:
Level of trustworthiness of Question 5
is 03:
Level of trustworthiness of Question 6
is 03:

Possible action:
1. Direct to special inspection
2. Further inquiry using dialogue
```

FIGURE 32.10 Protocol of the person during screening: the question generation (left) and their analysis (right) with corresponding level of trustworthiness.

According to the protocol shown in Figure 32.9, left, that the system estimates the third level of warning using automatically measured drug or alcohol intoxication for the screened customer. A knowledge-based subsystem evaluates the risks and generates two possible solutions. The user can, in addition to the automated analysis, evaluate the images acquired in the visible and infrared spectra.

The example in Figure 32.10 (left) introduces a scenario based on the analysis of behavioral biometric data. The results of the automated analysis of behavioral information are presented to the user (Figure 32.10, right). Let us assume that there are three classes of samples assigned to "Disability," "Alcohol intoxication," and "Normal." The following linguistic constructions can be generated by the system: Not enough data, but abnormality is detected, or Possible alcohol intoxication, or An individual with a disability.

The user must be able to communicate effectively with the customer in order to minimize uncertainty. Limited information will be obtained if the customer does not respond to inquiries or if his/her answers are not understood. We distinguish two types of uncertainty about the customer: the uncertainty that can be minimized by using customer responses, his/her documents, and information from databases; and the uncertainty of appearance (physiological and behavior) information such as specific features in the infrared facial image, gait, and voice. In particular, facial appearance alternating the document photos can be modeled using a software that models aging. The uncertainty of appearance can be minimized by specifically oriented questionnaire techniques. These techniques have been used in criminology, in particular, for interviewing and interrogation. The output of each personal assistant is represented in semantic form. The objective of each semantic construction is the minimization of uncertainly, that is, (a) choosing an appropriate set of questions (expert support) from the database, (b) alleviating the errors and temporal faults of biometric devices, and (c) maximizing the correlation between various biometrics.

Deception can be defined as a semantic attack that is directed against the decision-making process. Technologies for preventing, detecting, and prosecuting semantic attacks are still in their infancy. Some techniques of forensic interviewing and interrogation formalism with elements of detecting the semantic attack are useful in dialogue development. In particular, in training system, modeling is replaced by real-world conditions, and long-term training is replaced by periodically repeated short-term intensive computer-aided training.

The PASS extension for user training. In PASS, an expensive training system is replaced by an inexpensive extension of the PASS, already deployed at the place of application. In this way, an important effect is achieved: complicated and expensive modeling is replaced with real-world conditions, except some particular cases considered in this chapter. Furthermore, long-term training is replaced by periodically repeated short-time intensive computer-aided training. The PASS and T-PASS implement the concept of multi-target platforms, that is, the PASS can be easy reconfigured into the T-PASS and vice versa.

32.5 Other Applications

Simulators of biometric data are emerging technologies for educational and training purposes (immigration control, banking service, police, justice, etc.). They emphasize decision-making skills in non-standard and extreme situations.

Data bases for synthetic biometric data. Imitation of biometric data allows the creation of databases with tailored biometric data without expensive studies involving human subjects. An example of tool used to create databases for fingerprints is SFinGe system [6]. The generated databases were included in the Fingerprint Verification Competition FVC2004 and perform just as well as real fingerprints.

Synthetic speech and singing voices. A synthetic voice should carry information about age, gender, emotion, personality, physical fitness, and social upbringing [10]. A closely related but more complicated problem is generating a synthetic singing voice for the training of singers, by studying famous singers' styles and designing synthetic user-defined styles combining voice with synthetic music. An example of a direct biometric problem is identifying speech, given a video fragment without recorded voice. The inverse problem is mimicry synthesis (animation) given a text to be spoken (synthetic narrator).

Cancelable biometrics. The issue of protecting privacy in biometric systems has inspired the direction research referred to as cancelable biometrics [4]. Cancelable biometrics is aimed at enhancing the security and privacy of biometric authentication through the generation of "deformed" biometric data, that is, synthetic biometrics. Instead of using a true object (finger, face), the fingerprint or face image is intentionally distorted in a repeatable manner, and this new print or image is used.

Caricature is the art of making a drawing of a face, which makes part of its appearance more noticeable than it really is, and which can make a person look ridiculous. Specifically, a caricature is a synthetic facial expression, in which the distances of some feature points from the corresponding positions in the normal face have been exaggerated. The reason why the art-style of the caricaturist is of interest for image analysis, synthesis, and especially for facial expression recognition and synthesis is as follows [13]. Facial caricatures incorporate the most important facial features and a significant set of distorted features.

Lie detectors. Synthetic biometric data can be used in the development of a new generation of lie detectors [12,24,33]. For example, behavioral biometric information is useful in evaluation of truth in answers to questions, or evaluating the honesty of a person in the process of speaking [12]. Emotions contribute additionally to temperature distribution in the infrared facial image.

Humanoid robots are artificial intelligence machines whose design demands the resolution of certain direct and inverse biometric problems, such as, language technologies, recognition by means of facial expressions and gestures of the "mood" of instructor, following of cues; dialogue and logical reasoning; vision, hearing, olfaction, tactile, and other senses [26].

Ethical and social aspects of synthetic biometrics. Particular examples of the negative impact of synthetic biometrics are as follows: (a) Synthetic biometric information can be used not only for improving the characteristics of biometric devices and systems, but also by forgers to discover new strategies of attack. (b) Synthetic biometric information can be used for generating multiple copies of original biometric information.

References

1. J. Ahlberg, CANDIDE-3—An updated parameterised face, Technical Report LiTH-ISY-R-2326, Department of Electrical Engineering, Linköping University, Sweden, 2001.
2. M. Anbar, Clinical thermal imaging today, *IEEE Engineering in Medicine and Biology*, 17(4):25–33, July/August 1998.
3. W. Boles and B. Boashash, A human identification technique using images of the iris and wavelet transform, *IEEE Transactions on Signal Processing*, 46(4):1185–1188, 1998.
4. R. Bolle, J. Connell, S. Pankanti, N. Ratha, and A. Senior, *Guide to Biometrics*, Springer, New York, 2004.
5. J. J. Brault and R. Plamondon, A complexity measure of handwritten curves: Modelling of dynamic signature forgery, *IEEE Transactions on Systems, Man and Cybernetics*, 23:400–413, 1993.
6. R. Cappelli, Synthetic fingerprint generation, In D. Maltoni, D. Maio, A. K. Jain, and S. Prabhakar (eds.), *Handbook of Fingerprint Recognition*, Springer, New York, pp. 203–232, 2003.
7. A. Can, C. V. Steward, B. Roysam, and H. L. Tanenbaum, A feature-based, robust, hierarchical algorithm for registering pairs of images of the curved human retina, *IEEE Transactions on Analysis and Machine Intelligence*, 24(3):347–364, 2002.
8. R. Cappelli, SFinGe: Synthetic fingerprint generator, In *Proceedings of the International Workshop on Modeling and Simulation in Biometric Technology*, Calgary, Canada, pp. 147–154, June 2004.
9. S. Chague, B. Droit, O. Boulanov, S. N. Yanushkevich, V. P. Shmerko, and A. Stoica, Biometric-based decision support assistance in physical access control systems, In *Proceedings of the Bio-Iinspired, Learning and Intelligent Systems for Security* (*BLISS*) *Conference*, Edinburgh, U.K., pp. 11–16, 2008.
10. P. R. Cook, *Real Sound Synthesis for Interactive Applications*, A K Peters, Natick, MA, 2002.
11. Y. Du and X. Lin, Realistic mouth synthesis based on shape appearance dependence mapping, *Pattern Recognition Letters*, 23:1875–1885, 2002.
12. P. Ekman and E. L. Rosenberg (eds.), *What the Face Reveals: Basic and Applied Studues of Spontaneouse Expression Using the Facial Action Coding System* (*FACS*), Oxford University Press, New York, 1997.
13. T. Fujiwara, H. Koshimizu, K. Fujimura, H. Kihara, Y. Noguchi, and N. Ishikawa, 3D modeling system of human face and full 3D facial caricaturing. In *Proceedings of the Third IEEE International Conference on 3D Digital Imaging and Modeling*, Quebec, Canada, pp. 385–392, 2001.
14. I. Fujimasa, T. Chinzei, and I. Saito, Converting far infrared image information to other physiological data, *IEEE Engineering in Medicine and Biology Magazine*, (3):71–76, 2000.
15. J. M. Hollerbach, An oscilation theory of handwriting, *Biological Cybernetics*, 39:139–156, 1981.
16. M. U. Kuecken and A. C. Newell, A model for fingerprint formation, *Europhysics Letters*, 68(1):141–146, 2004.
17. A. Mansfield and J. Wayman, Best practice standards for testing and reporting on biometric device performance, National Physical Laboratory of UK, Teddington, U.K., 2002.
18. H. Matsumoto, K. Yamada, and S. Hoshino, Impact of artificial gummy fingers on fingerprint systems, In *Proceedings of the SPIE, Optical Security and Counterfeit Deterrence Techniques IV*, 4677:275–289, 2002.

19. E. Y. K. Ng, G. J. L. Kaw, and W. M. Chang, Analysis of IR thermal imager for mass blind fever screening, *Microvascular Research*, 68:104–109, 2004.
20. I. Nishidate, Y. Aizu, and H. Mishina, Depth visualization of a local blood region in skin tissue by use of diffuse reflectance images, *Optics Letters*, 30:2128–2130, 2005.
21. N. Oliver, A. P. Pentland, and F. Berard, LAFTER: A real-time face and lips tracker with facial expression recognition, *Pattern Recognition*, 33(8):1369–1382, 2000.
22. M. Pantic and L. J. M. Rothkrantz, Automatic analysis of facial expressions: The state-of-the-art, *IEEE Transactions on Pattern Analysis and Machine Intelligence*, 22(12):1424–1445, 2000.
23. R. Plamondon and W. Guerfali, The generation of handwriting with delta-lognormal synergies, *Bilogical Cybernetics*, 78:119–132, 1998.
24. *The Polygraph and Lie Detection*, The National Academies Press, Washington, DC, 2003.
25. V. P. Shmerko, S. N. Yanushkevich, H. Moon, A. Stoica, and R. R. Yager, Accelerating decision making support in biometric assistant for remote temperature measures, In *Proceedings of the ECSIS Symposium on Bio-Inspired, Learning, and Intelligent Systems for Security* (*BLISS 2007*), Edinburgh, U.K., August, pp. 11–14, 2007.
26. A. Stoica, Humanoids for urban operations, White Paper, Humanoid Project, NASA, Jet Propulsion Laboratory, Pasadena, CA, 2005, http://ehw.jpl.nasa.gov/humanoid
27. Y. Sugimoto, Y. Yoshitomi, and S. Tomita, A method for detecting transitions of emotional states using a thermal facial image based on a synthesis of facial expressions, *Robotics and Autonomous Systems*, 31:147–160, 2000.
28. C. J. Tilton, An emerging biometric standards, *IEEE Computer Magazine*, *Special Issue on Biometrics*, 1:130–135, 2001.
29. Total information awareness DAPRA's research program, *Information and Security*, 10:105–109, 2003.
30. N. Tsumura, N. Ojima, K. Sato, M. Shiraishi, H. Shimizu, H. Nabeshima, S. Akazaki, K. Hori, and Y. Miyake, Image-based skin color and texture analysis/synthesis by extracting hemoglobin and melanin information in the skin, *ACM Transactions on Graphics*, 22(3):770–779, 2003.
31. R. R. Yager, Uncertainty representation using fuzzy measures, *IEEE Transactions on System*, *Man*, *and Cybernetics—Part B: Cybernetics*, 32(1):13–20, 2002.
32. R. R. Yager, On the valuation of alternatives for decision-making under uncertainty, *International Journal of Intelligent Systems*, 17:687–707, 2002.
33. S. N. Yanushkevich, A. Stoica, V. P. Shmerko, and D. V. Popel, *Biometric Inverse Problems*, Taylor & Francis/CRC Press, Boca Raton, FL, 2005.
34. S. Yanushkevich, A. Stoica, V. Shmerko, Synthetic biometrics, *IEEE Computational Intelligence Magazine*, 2(2):60–69, 2007.
35. S. N. Yanushkevich, A. Stoica, and V. P. Shmerko, Fundamentals of biometric-based training system design, In S. N. Yanushkevich, P. Wang, S. Srihari, M. Gavrilova (eds.), and M. S. Nixon (Consulting ed.), *Image Pattern Recognition: Synthesis and Analysis in Biometrics*, World Scientific, Singapore, 2007.
36. S. N. Yanushkevich, A. Stoica, and V. P. Shmerko, Experience of design and prototyping of a multi-biometric early warning physical access control security system (PASS) and a training system (T-PASS), In *Proceedings of the 32nd Annual IEEE Industrial Electronics Society Conference*, Paris, France, pp. 2347–2352, 2006.

Index

A

B

C

D

G

H

I

J

K

L

N

Q

R

S

T

U

V

W